Acta Numerica 1997

Acta Numerica

Volume 6 1997

CAMBRIDGE
UNIVERSITY PRESS

CAMBRIDGE UNIVERSITY PRESS
Cambridge, New York, Melbourne, Madrid, Cape Town, Singapore,
São Paulo, Delhi, Dubai, Tokyo, Mexico City

Cambridge University Press
The Edinburgh Building, Cambridge CB2 8RU, UK

Published in the United States of America by Cambridge University Press, New York

www.cambridge.org
Information on this title: www.cambridge.org/9780521157643

First published 1997
First paperback edition 2010

A catalogue record for this publication is available from the British Library

ISBN 978-0-521-59106-5 Hardback
ISBN 978-0-521-15764-3 Paperback

Contents

Acta Numerica (1997), *pp.* 1–54

Constructing cubature formulae: the science behind the art

Ronald Cools
Katholieke Universiteit Leuven
Dept. of Computer Science
Celestijnenlaan 200A
B-3001 Heverlee, Belgium
E-mail: Ronald.Cools@cs.kuleuven.ac.be

In this paper we present a general, theoretical foundation for the construction of cubature formulae to approximate multivariate integrals. The focus is on cubature formulae that are exact for certain vector spaces of polynomials. Our main quality criteria are the algebraic and trigonometric degrees. The constructions using ideal theory and invariant theory are outlined. The known lower bounds for the number of points are surveyed and characterizations of minimal cubature formulae are given. We include references to all known minimal cubature formulae. Finally, some methods to construct cubature formulae illustrate the previously introduced concepts and theorems.

CONTENTS

1. What to expect

This is a paper for patient readers. The reader has to digest several pages before being enlightened on the direction taken by this paper. Following a short section with historical notes, Section 3 describes the problem this paper concentrates upon, approximating multivariate integrals, and presents my favourite quality criteria. Section 4 sketches several ways to construct such approximations, one of which is this paper's real subject. After introducing concrete integrals and a tool to deal with symmetries in Section 5, we are ready for the real work.

In Section 6, interpolatory cubature formulae are characterized and the connection with orthogonal polynomials and ideal theory is in the spotlight. Sections 7 and 8 are devoted to the determination of lower bounds and the characterization of minimal cubature formulae. Finally, Sections 9 and 10 concentrate on the art of constructing cubature formulae.

Readers familiar with the construction of quadrature formulae may find it helpful to spell out the meaning in one dimension of the definitions and theorems given for arbitrary dimensions.

2. On the origin of cubature formulae

2.1. The prehistory

According to the Oxford English Dictionary, cubature is the determination of the cubic contents of a solid, that is, the computation of a volume. We are interested in the construction of cubature formulae, that is, formulae to estimate volumes. The problem of measuring areas and volumes has always been present in everyday life. The ancient Babylonians and Egyptians already had precise and correct rules for finding the areas of triangles, trapezoids and circles (for the Babylonians π equalled 3, for the Egyptians 235/81) and the volumes of parallelepipeds, pyramids and cylinders. They thought of these figures in concrete terms, mainly as storage containers for grain. They discovered these rules empirically.

The first abstract proofs of rules for finding some areas and volumes are said to have been developed by Eudoxus of Cnidus in about 367 BC. About a century later, his method was further developed by Archimedes. In the middle of the 16th century Archimedes' work became available in Greek and Latin and in the 17th century his method became known as the method of exhaustion. It culminated in the 19th century in the isolation of the concept of Riemann integration, defined by approximating Riemann sums.

In southern Germany, due to increased commerce, measuring the contents of wine barrels became important in the 15th century, and therefore approximations were introduced. In 1613, Johannes Kepler witnessed a salesman using *one* gauging-rod to measure the contents of *all* Austrian wine barrels *without* further calculations. This was the motivation for what became his book *Nova Stereometria Doliorum Vinariorum*[1] (Kepler 1615). It turned out that for the type of barrels used in Austria, the approximation used by the salesman was quite good. At the end of the book Kepler wrote that his book was longer than he had expected and people could just as well continue to use the approximation. In his final sentence he philosophizes on the eternal compromise between approximations and exact calculations:

Et cum pocula mille mensi erimus,
Conturbabimus illa, ne sciamus.[2]

The start of the modern study of volume computation is usually linked with Kepler.

The word 'cubature' appeared in the written English language around the same time. The oldest known reference, according to the Oxford English Dictionary, is a letter from Collins in 1679 containing the sentence: 'In order to the quadrature of these figures and the cubature of their solids.' From 1877 we cite Williamson: 'The cube ... is ... the measure of all solids, as the square is the measure of all areas. Hence the finding the volume of a solid is called its cubature.'

The formulation of the problem of measuring in terms of integrals and functions is much more recent. The first cubature formula in the form we are now familiar with was constructed by Maxwell (1877). And that is when our story starts. For us, a cubature formula is a weighted sum of function evaluations used to approximate a multivariate integral. (The function is not necessarily the integrand, nor is the same function used for each evaluation.) The prehistory of our field of interest thus ends in 1877. In the following section we briefly sketch different approaches and specify the approach we follow in the rest of this paper.

2.2. In search of a pedigree

There are several criteria to specify and classify cubature formulae based on their behaviour for specific classes of functions. A classical way to present a survey is to sketch the pedigree of different approaches.

The oldest criterion is the *algebraic degree* of a cubature formula, used by James Clerk Maxwell in 1877[3]. This criterion is obviously inherited from the work on quadrature formulae. We have no idea what a cubature formula of algebraic degree d will give us when applied to a function that is not a polynomial of degree smaller than or equal to d.

[1] *Solid Geometry of Wine Barrels*

[2] After measuring a thousand cups, we will be so confused that we lose our head.

[3] When we write that something happened in a particular year, we in fact refer to the year the results were published.

The second oldest approach to approximate multivariate integrals does not have this problem. One evaluates the integrand function in a number of randomly selected points and uses the average function value. This is the classical *Monte Carlo method.* The idea came to Stanislaw Ulam, Nick Metropolis and John von Neumann while working on the Manhattan Project in 1945. From the Strong Law of Large Numbers it follows that the expected value this method delivers is the integral. If one restricts the integrands to the class of square integrable functions, the Central Limit Theorem gives rise to a probabilistic error bound known as the '$N^{1/2}$ law': for a fixed level of confidence, the error bound varies inversely as $N^{1/2}$.

Because truly random samples are not available and the error estimate of the Monte Carlo method is only probabilistic, researchers in the early 1950s became interested in *quasi-Monte Carlo methods.* The method received its name from R. D. Richtmyer (1952). In these methods one uses, as in the classical Monte Carlo method, an equal-weights cubature formula but chooses the points to be 'better than random'. One obtains rigorous error bounds that behave better than the $N^{1/2}$ law. The first quasi-Monte Carlo methods were based on low discrepancy sequences. Another type of quasi-Monte Carlo method is the *method of good lattice points* introduced by Nikolai M. Korobov (1959). The more general notion of a lattice rule was introduced by Konstantin K. Frolov (1977) and rediscovered by Ian H. Sloan and Philip Kachoyan (1987).

It should be noted that Frolov did not see his rules as quasi-Monte Carlo methods. He constructed cubature formulae that are exact for a set of trigonometric polynomials, that is, his criterion is the *trigonometric degree.* It is strange that his paper is not cited in the Russian literature on cubature formulae of trigonometric degree. We will see that there are many similarities between the construction of cubature formulae of algebraic degree and the construction of formulae of trigonometric degree. In addition, Frolov made the link with lattice rules. Hence the pedigree approach breaks down here and we will use another thread for this story.

We will focus on cubature formulae that are exact for a certain class of functions: polynomials, both algebraic and trigonometric. Cubature formulae of algebraic degree and lattice rules fit in this single framework.

Cubature formulae of algebraic degree play an important role for low dimensions and are essential building blocks for adaptive routines to compute integrals. Practical experience with lattice rules is still limited. Most people expect them to be important for high dimensions. However, there already exist two-dimensional applications that benefit from their properties.

3. Problem setting and criteria

An integral I is a linear continuous functional

$$I[f] := \int_\Omega w(\mathbf{x}) f(\mathbf{x})\,\mathrm{d}\mathbf{x} \tag{3.1}$$

where the region $\Omega \subset \mathbb{R}^n$. We use $\mathbf{x}$ as a shorthand for the variables $x_1, x_2, \ldots, x_n$. We will always assume that $w(\mathbf{x}) \geq 0$, for all $\mathbf{x} \in \Omega$, that is, I is a positive functional.

It is often desirable to approximate I by a weighted sum of (easier) functionals such that

$$I[f] \simeq Q[f] = \sum_{j=1}^{N} w_j L_j[f] \tag{3.2}$$

where $w_j \in \mathbb{R}$. We will only consider approximations that are exact for a given vector space of functions and we start with a very general result on the existence of such approximations, due to Sobolev (1962); see also Mysovskikh (1981).

We will need the following lemma.

Lemma 3.1 The system of linear equations

$$A\mathbf{x} = \mathbf{b} \quad \text{with} \quad A \in \mathbb{C}^{\mu\times\nu}, \quad \mathbf{b} \in \mathbb{C}^{\mu}, \quad \mathbf{x} \in \mathbb{C}^{\nu}$$

has a solution if and only if $\sum_{j=1}^{\mu} b_j \bar{\mathbf{y}}_j = 0$ for all solutions $\mathbf{y}$ of $A^\star \mathbf{y} = 0$ ($A^\star = \bar{A}^T$; $\bar{y}$ is the complex conjugate of y).

Proof. Let L be the vector space generated by the columns $a^{(1)}, \ldots, a^{(\nu)}$ of A, and $L^\perp \subset \mathbb{C}^n$ the orthogonal complement of L. Thus $\mathbf{y} \in L^\perp$ if and only if $\mathbf{y}$ is orthogonal to all columns of A. Hence $L^\perp$ is the subspace of all solutions of $A^\star\mathbf{y} = 0$.

$A\mathbf{x} = \mathbf{b}$ has a solution if and only if $\mathbf{b} \in L$. But $\mathbf{b} \in L$ if and only if $\mathbf{b}$ is orthogonal to $\mathbf{y}$ and $A^\star\mathbf{y} = 0$. □

Let F be a vector space of functions defined on $\Omega \subset \mathbb{R}^n$ and $F_1 \subset F$ a subspace. Let I be a linear, continuous functional defined on F, that is approximated by a linear combination of other functionals (3.2) with constant coefficients. Let

$$F_0 = \{f \in F_1 : L_j[f] = 0, j = 1, \ldots, N\} \subset F_1.$$

Theorem 3.1 A necessary and sufficient condition for the existence of an approximation (3.2) that is exact for all $f \in F_1$ is

$$f \in F_0 \Rightarrow I[f] = 0. \tag{3.3}$$

Proof. It is trivial that the condition is necessary. It remains to be proven that it is sufficient.

Let f_i, $i = 1, \ldots, \mu$ be a basis of F_1. Then the approximation (3.2) is exact for all $f \in F_1$ if and only if it is exact for f_i, $i = 1, \ldots, \mu$:

$$\sum_{j=1}^{N} w_j L_j[f_i] = I[f_i]. \tag{3.4}$$

(3.4) is a system of linear equations for the weights w_j, $j = 1, \ldots, N$.

Let $(a_1, \ldots a_\mu)^T$ be the solution of the adjoint homogeneous system:

$$\sum_{j=1}^{\mu} a_j L_i[f_j] = 0, \quad i = 1, \ldots, N. \tag{3.5}$$

The lemma implies that (3.4) has a solution if and only if $\sum_{j=1}^{\mu} a_j I[f_j] = 0$ which is equivalent to

$$I\left[\sum_{j=1}^{\mu} a_j f_j\right] = 0. \tag{3.6}$$

But (3.5) is equivalent to

$$L_i\left[\sum_{j=1}^{\mu} a_j f_j\right] = 0, \quad i = 1, \ldots, N,$$

which means that $f = \sum_{j=1}^{\mu} a_j f_j \in F_0$. Hence the necessary and sufficient condition (3.6) for the solvability of (3.4) can be written as $I[f] = 0$. From the solvability of (3.4) follows the sufficientness of (3.3). □

We will only consider functionals L_j that are point evaluations. Most often $L_j[f] = f(\mathbf{y}^{(j)})$ for a $\mathbf{y}^{(j)} \in \mathbb{R}^n$ but occasionally one encounters approximations that use partial derivatives of f, that is,

$$L_j[f] = \frac{\partial^{j_1+\cdots+j_n} f}{\partial x_1^{j_1} \ldots x_n^{j_n}}(\mathbf{y}^{(j)}).$$

Unless stated otherwise, we shall concentrate on approximations that use function values only. If $n = 1$, then Q is called a *quadrature formula*. If $n \geq 2$, then Q is called a *cubature formula*. If partial derivatives are used, Q is called a *generalized cubature formula*. So, for our purposes, a cubature formula Q has the form

$$I[f] \approx Q[f] := \sum_{j=1}^{N} w_j f(\mathbf{y}^{(j)}). \tag{3.7}$$

The choice of the points $\mathbf{y}^{(j)}$ and weights w_j is independent of the function f. They are chosen so that the formula gives a good approximation for some class of functions.

According to Rabinowitz and Richter (1969), a *good quadrature or cubature formula* has all points $\mathbf{y}^{(j)}$ inside the region Ω and all weights w_j positive. Positive weights imply that Q is also a positive functional. As Maxwell (1877) noted when he obtained a cubature formula for the cube with 27 points, some outside the cube, it might be difficult to apply a cubature formula when points are outside the region Ω:

This, of course, renders the method useless in determining the integral from the *measured values* of the quantity u, as when we wish to determine the weight of a brick from the specific gravities of samples taken from 27 selected places in the brick, for we are directed by the method to take some of the samples from places outside the brick.

In the remainder of the paper, we will only consider cubature formulae that are exact for algebraic or trigonometric polynomials.

Let $\alpha = (\alpha_1, \alpha_2, \ldots, \alpha_n) \in \mathbb{Z}^n$ and $|\alpha| = \sum_{j=1}^n |\alpha_j|$. An (algebraic) monomial in the variables $x_1, x_2, \ldots, x_n$ is a function of the form $\prod_{j=1}^n x_j^{\alpha_j}$, also denoted by $\mathbf{x}^\alpha$, for $\alpha \in \mathbb{N}^n$. A trigonometric monomial is a function of the form

$$\prod_{j=1}^{n} e^{2\pi\, \mathrm{i}\alpha_j x_j}, \quad \text{where} \quad \mathrm{i}^2 = -1,$$

also denoted by $e^{2\pi\, \mathrm{i}\alpha\mathbf{x}}$. An algebraic, respectively trigonometric, polynomial in n variables is a finite linear combination of monomials, that is,

$$p(\mathbf{x}) = \sum_{\alpha_1,\ldots,\alpha_n=0}^{\infty} a_\alpha \mathbf{x}^\alpha, \qquad \text{respectively} \quad t(\mathbf{x}) = \sum_{\alpha_1,\ldots,\alpha_n=-\infty}^{\infty} a_\alpha e^{2\pi\, \mathrm{i}\alpha\mathbf{x}}.$$

For trigonometric polynomials, some authors add the restriction that a_α and $a_{-\alpha}$ are complex conjugates. One can of course also use sine and cosine functions to describe real trigonometric polynomials. This restriction is unnecessary here.

The *degree* of a multivariate algebraic or trigonometric polynomial v is defined as

$$\deg(v) = \begin{cases} \max\{|\alpha| : a_\alpha \neq 0\} & \text{if } v \neq 0, \\ -\infty & \text{if } v = 0. \end{cases}$$

The vector space of all algebraic polynomials in n variables of degree at most d is denoted by $\mathcal{P}_d^n$. The vector space of all trigonometric polynomials in n variables of degree at most d is denoted by $\mathcal{T}_d^n$. The dimensions of these vector spaces are:

$$\begin{aligned} \dim \mathcal{P}_d^n &= \binom{n+d}{d} \quad \text{and} \\ \dim \mathcal{T}_d^n &= \sum_{i=0}^{n} \binom{n}{i}\binom{d}{i} 2^i. \end{aligned}$$

We can now formulate the criterion we will most often use.

Definition 3.1 A cubature formula Q for an integral I has algebraic, respectively trigonometric, degree d if it is exact for all polynomials of algebraic, respectively trigonometric, degree at most d and it is not exact for at least one polynomial of degree $d+1$.

Cubature formulae of algebraic degree are available for a large variety of regions and weight functions. For a survey, we refer to Stroud (1971) and Cools and Rabinowitz (1993). Cubature formulae of trigonometric degree are only published for $\Omega = [0,1]^n$ and $w(\mathbf{x}) \equiv 1$, and in this paper we will only consider this region. For a survey, we refer to Cools and Sloan (1996).

The *overall degree* of a multivariate polynomial v is defined as

$$\overline{\deg}(v) = \begin{cases} \max\{\max\{|\alpha_j| : j = 1, \ldots, n\} : a_\alpha \neq 0\} & \text{if } v \neq 0, \\ -\infty & \text{if } v = 0. \end{cases}$$

The vector space of all algebraic polynomials in n variables of overall degree at most d is denoted by $\overline{\mathcal{P}}_d^n$. The vector space of all trigonometric polynomials in n variables of overall degree at most d is denoted by $\overline{\mathcal{T}}_d^n$. The dimension of these vector spaces are:

$$\dim \overline{\mathcal{P}}_d^n = (d+1)^n \quad \text{and} \quad \dim \overline{\mathcal{T}}_d^n = (2d+1)^n.$$

We can now define another criterion for cubature formulae.

Definition 3.2 A cubature formula Q for an integral I has algebraic, respectively trigonometric, overall degree d if it is exact for all algebraic, respectively trigonometric polynomials of overall degree at most d, and it is not exact for at least one polynomial of overall degree $d+1$.

A notable example of cubature formulae with overall algebraic degree d is the family of Gauss-product rules, obtained from quadrature formulae of degree d.

Most known cubature formulae for integrals of periodic functions on the unit cube $[0,1)^n$ are so-called *lattice rules* and for them a criterion used much more often than the trigonometric degree is the *Zaremba index*. The Zaremba index is related to the dominant terms in the error of the lattice rule for a worst possible function in a particular class of functions.

Definition 3.3 A *multiple integration lattice* L in $\mathbb{R}^n$ is a subset of $\mathbb{R}^n$ which is discrete and closed under addition and subtraction and which contains $\mathbb{Z}^n$ as a subset. A *lattice rule* is a cubature formula for approximating integrals over $[0,1)^n$ where the N points are the points of a multiple integration lattice L that lie in $[0,1)^n$ and the weights are all equal to $1/N$.

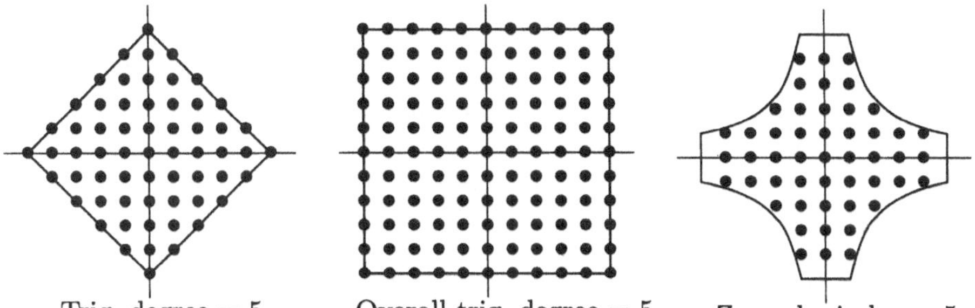

Trig. degree = 5 Overall trig. degree = 5 Zaremba index = 5

Fig. 1. Monomials for which a two-dimensional cubature formula is exact

Definition 3.4 A cubature formula Q for an integral I has Zaremba index d if it is exact for all trigonometric monomials $e^{2\pi\,\mathrm{i}\alpha\mathbf{x}}$ with

$$\prod_{i=1}^{n} r_i < d \quad \text{with} \quad r_i := \begin{cases} 1 & \text{if} \quad \alpha_i = 0, \\ |\alpha_i| & \text{if} \quad \alpha_i \neq 0, \end{cases}$$

and it is not exact for at least one monomial with $d = \prod_{i=1}^{n} r_i$.

Why is the algebraic degree of a cubature formula a measure of its quality? The main argument is that a well-behaved function is expected to be well approximated by a polynomial (for instance a Taylor series) and consequently its integral is expected to be well approximated by a cubature formula of a suffienciently high algebraic degree. Another argument, which applies only to some regions, is that the rate of convergence of a compound cubature formula as the mesh size shrinks is directly related to the algebraic degree of the basic cubature formula. This follows from the asymptotic error expansion for compound cubature formulae, which we will encounter in Section 4.

Why is the trigonometric degree, as well as other criteria based on trigonometric polynomials, a measure of the quality of a cubature formula? The main argument is that a well-behaved function is expected to be well approximated by a trigonometric polynomial (for instance its Fourier series) and consequently its integral is expected to be well approximated by a cubature formula of a sufficiently high trigonometric degree.

One uses other criteria, such as the overall algebraic or trigonometric degree or the Zaremba index, if one has reasons to believe that the corresponding set of monomials is more relevant. This is obviously connected to one's favourite way to study the error of a cubature formula when applied to a function for which it does not give the exact value of the integral.

We will use the symbol $\mathcal{V}_d^n$ to refer to one of the vector spaces $\mathcal{P}_d^n$, $\overline{\mathcal{P}}_d^n$, $\mathcal{T}_d^n$ or $\overline{\mathcal{T}}_d^n$. The results we present in this paper are also valid for other vector spaces, but one has to be cautious. A property that is needed to generalize several proofs is that the convex hull of the powers of α of the monomials in $\mathcal{V}_n^d$ contains only these monomials. In Figure 1 we illustrate this for the trigonometric degree, overall trigonometric degree and Zaremba index. It is obvious that the Zaremba index no longer has a role to play in this paper. The role of this criterion is important in the context of quasi-Monte Carlo methods. For readers who want to know more about this, we recommend Niederreiter (1992) and Sloan and Joe (1994).

4. Different ways to construct cubature formulae

In Section 2.2 we mentioned that there is more than one way to obtain a cubature formula. Much depends on the quality criterion used. As stated earlier, in this paper we restrict our attention to cubature formulae that are designed to be exact for a vector space of algebraic or trigonometric polynomials. Even then there are several ways to reach this goal and in this section we will briefly outline some of these. The examples in this section will only be two-dimensional but the ideas behind them are perfectly general.

4.1. Repeated quadrature

No doubt the field of quadrature is more threaded and explored than its multivariate counterpart. It is hence not surprising that even today many people use the product of two quadrature formulae to integrate over a square. Let

$$\int_0^1 g(x)\,\mathrm{d}x \simeq \sum_{j=1}^{N} w_j g(x^{(j)}) \tag{4.1}$$

be a quadrature formula of degree d_x, then

$$\int_0^1 \int_0^1 f(x,y)\,\mathrm{d}x\,\mathrm{d}y \simeq \sum_{i=1}^{N}\sum_{j=1}^{N} w_i w_j f(x^{(i)}, x^{(j)}). \tag{4.2}$$

If the quadrature formula (4.1) has algebraic degree d, then the cubature formula (4.2) has overall algebraic degree d.

One can use different quadrature formulae for each of the one-dimensional integrals. Even the one-dimensional integrals may have different limits or weight functions. If the quadrature formula in x has degree d_x with N_x points and the formula in y has degree d_y with N_y points, the resulting cubature formula will be exact for a space of polynomial 'between' $\bar{\mathcal{P}}_d^n$ and $\bar{\mathcal{P}}_D^n$ with $d := \min\{d_x, d_y\}$ and $D := \max\{d_x, d_y\}$, and has $N = N_x N_y$ points.

4.2. Change of variables

If one encounters a new problem, it is tempting to transform it into a problem for which a solution is familiar. For instance, an integral over a circle or triangle can be transformed into an integral over a square:

$$\begin{aligned} \int_0^1 \int_0^x f(x,y)\,\mathrm{d}y\,\mathrm{d}x &= \int_0^1 x \int_0^1 f(x,xt)\,\mathrm{d}t\,\mathrm{d}x, \\ \int_{-1}^1 \int_{-\sqrt{1-x^2}}^{\sqrt{1-x^2}} f(x,y)\,\mathrm{d}y\,\mathrm{d}x &= \int_0^1 \int_0^{2\pi} r f(r\cos\theta, r\sin\theta)\,\mathrm{d}\theta\,\mathrm{d}r. \end{aligned}$$

One can then use repeated quadrature, preferably using quadrature formulae that take the Jacobian of the transformation into account. For the above examples a possible choice is a combination of a Gauss–Legendre and an appropriate Gauss–Jacobi quadrature formula. This results in a so-called Conical Product rule for the triangle and a Spherical Product rule for the circle (Stroud 1971).

Transformations can have surprising advantages and disadvantages. For example, the above transformation of a triangle into a square, mentioned by Stroud (1971), but now usually referred to as the Duffy transformation (Duffy 1982), removes some types of singularity from the integrand (Lyness 1992, Lyness and Cools 1994), but the resulting cubature formula lacks symmetry. In fact there are three distinct Conical Products rules for each degree, depending on which vertex of the triangle is the preferred one.

4.3. Compound rules and copy rules

It can happen that the given integration region has an unusual shape for which no cubature formula is available, but that it can be subdivided into standard regions for which cubature formulae are available. The sum of all cubature formulae on all subregions is a so-called *compound rule*. If a cubature formula on a standard region does not give a result that is accurate enough, because it is applied to a function for which it was not designed to give the exact result, one can also subdivide and apply a properly scaled version of the given cubature formula on each subregion. And so on until one obtains the desired accuracy.

If the given region can be subdivided in congruent regions, a special kind of compound rule, the *copy rule*, becomes interesting. If, for example, the integration region is a square, one can divide this into m^2 identical squares, each of side $1/m$th the original side, and apply a properly scaled version of the given cubature formula to each. This approach looks expensive, especially if the dimension goes up, but is appealing because an error expansion is readily available.

So far, we have considered cubature formulae that are exact for a certain vector space. Almost all users will apply them to functions for which they do not give the exact result. So, we have arrived at a point where we need to say something about the error, that is, $Q[f] - I[f]$.

For regular $f(x,y) \in C^p$, $p \in \mathbb{N}$, the almost self-evident extension of the one-dimensional Euler–Maclaurin expansion may be expressed as

$$Q^{(m)}[f] - I[f] = \sum_{i=1}^{p-1} \frac{B_i(Q,f)}{m^i} + \mathcal{O}(m^{-p}), \tag{4.3}$$

where $Q^{(m)}$ is the m^2-copy of Q and the coefficients B_i depend on the cubature formula Q, the integral I and the integrand f.

Once it is known that an error expansion such as (4.3) exists, Richardson extrapolation (Richardson 1927) can be used to speed up convergence (by eliminating terms of the error expansion). In order to apply extrapolation one need not know all details: the value of the B_i need not be known.

The m-copy rules for cubes and simplices have received considerable attention, because for some classes of non-regular functions, error expansions are also available and Richardson extrapolation can be used to speed up convergence. It is beyond the scope of this text to pursue this further. The situation seems to be that for many algebraic or logarithmic singularities that occur at a vertex or along a side, an appropriate expansion exists. For a brief survey of what is available for a triangle we refer to Lyness and Cools (1994). Readers who want to know more about this topic will find it in Lyness and McHugh (1970), Lyness and Puri (1973), Lyness (1976), de Doncker (1979), Lyness and Monegato (1980), Lyness and de Doncker-Kapenga (1987), Lyness and de Doncker (1993), Verlinden and Haegemans (1993).

4.4. Direct construction of cubature formulae

In the previous subsections we described indirect approaches to constructing cubature formulae. These are not the main subject of the article. We are especially interested in the direct approach.

Suppose one wants a cubature formula that is exact for all functions of a vector space of functions. Because an integral and a cubature formula are linear operators, it is sufficient and necessary that the cubature formula is exact for all functions of a basis of the vector space. Hence, if one desires a cubature formula that is exact for a vector space $\mathcal{V}_d^n$ and if the functions f_i form a basis for $\mathcal{V}_d^n$, then it is necessary and sufficient that

$$Q[f_i] = I[f_i], \quad i = 1, \dots, \dim \mathcal{V}_d^n. \tag{4.4}$$

If the f_i are monomials, then the right-hand sides of (4.4), the so-called *moments*, are known in closed form or can be evaluated. When the left-hand sides of (4.4) are replaced by the weighted sum of function values (3.7) and the number of points N is fixed, then one obtains a system of nonlinear equations in the unknown points $\mathbf{y}^{(j)}$ and weights w_j:

$$\sum_{j=1}^{N} w_j f_i(\mathbf{y}^{(j)}) = I[f_i], \quad i = 1, \ldots, \dim \mathcal{V}_d^n. \tag{4.5}$$

We are interested in cubature formulae with a 'low' number of points. In Section 7.1 we will search for a lower bound for the number of points depending on $\mathcal{V}_d^n$.

At this point we want to mention that one can distinguish between two approaches to construct cubature formulae the direct way:

- one may proceed directly to solve the system of nonlinear equations, or
- one can search for polynomials that vanish at the points of the formula.

The foundation for successful application of the first approach is laid in Section 5. The building blocks for the second approach are presented in Section 6. The second approach has been very successful in (one-dimensional) quadrature. Most published cubature formulae were, however, obtained using the first approach.

5. On regions and symmetry

We will always try to be as general as possible but we will soon discover that, for instance, lower bounds for the number of points depend on the specific region Ω and weight function $w(\mathbf{x})$. In this section we will define some standard regions and describe their most important property, namely symmetry.

5.1. Standard regions

In this paper we will encounter the following regions and weight functions for the algebraic-degree case:

C_n: the n-dimensional cube

$$\Omega := \{(x_1, \ldots, x_n) : -1 \leq x_i \leq 1, i = 1, \ldots, n\}$$

with weight function $w(\mathbf{x}) := 1$,

C_2^α: the square

$$\Omega := \{(x_1, x_2) : -1 \leq x_i \leq 1, i = 1, 2\}$$

with weight function

$$w(x_1, x_2) := (1 - x_1^2)^\alpha (1 - x_2^2)^\alpha, \quad \alpha > -1,$$

S_n: the n-dimensional ball

$$\Omega := \left\{ (x_1, \ldots, x_n) : \sum_{j=1}^{n} x_i^2 \leq 1 \right\}$$

with weight function $w(\mathbf{x}) := 1$,

U_n: the n-dimensional sphere, that is, the surface of the ball

$$\Omega := \left\{ (x_1, \ldots, x_n) : \sum_{j=1}^{n} x_i^2 = 1 \right\}$$

with weight function $w(\mathbf{x}) := 1$,

T_n: the n-dimensional simplex

$$\Omega := \left\{ (x_1, \ldots, x_n) : \sum_{j=1}^{n} x_i \leq 1 \text{ and } x_i \geq 0, i = 1, \ldots, n \right\}$$

with weight function $w(\mathbf{x}) := 1$,

$E_n^{r^2}$: the entire n-dimensional space $\Omega := \mathbb{R}^n$ with weight function

$$w(\mathbf{x}) := e^{-r^2} \quad \text{with} \quad r^2 := \sum_{j=1}^{n} x_j^2,$$

E_n^r: the entire n-dimensional space $\Omega := \mathbb{R}^n$ with weight function

$$w(\mathbf{x}) := e^{-r}.$$

The trigonometric-degree case deals usually with the following region:

$C_n^\star$: the n-dimensional cube

$$\Omega := \{(x_1, \ldots, x_n) : 0 \leq x_i < 1, i = 1, \ldots, n\}$$

with weight function $w(\mathbf{x}) := 1$.

We will use the above notation to refer to both the region and weight function and to the integral over this region with this weight function.

5.2. *Symmetry groups*

The symmetry of an integral is described by its symmetry group. Let G be any group of orthogonal transformations that have a fixed point at the origin, and let $|G|$ denote the order of the group.

Definition 5.1 A set $\Omega \subset \mathbb{R}^n$ is said to be invariant with respect to (w.r.t.) a group G if Ω is left unchanged by each transformation of the group, that is, $g(\Omega) = \Omega$, for all $g \in G$. A function f is said to be invariant w.r.t. G

if it is left unchanged by each transformation of the group, that is, $f(\mathbf{x}) = f(g(\mathbf{x}))$ for all $g \in G$. An integral is invariant w.r.t. G if both its region and weight function are invariant w.r.t. G.

Note that S_n, U_n, $E_n^{r^2}$ and E_n^r are invariant w.r.t. each group of orthogonal transformations.

Definition 5.2 The *G-orbit* of a point $\mathbf{y} \in \mathbb{R}^n$ is the set $\{g(\mathbf{y}) : g \in G\}$.

A G-orbit of a given point is obviously an invariant set w.r.t. G. Observe that the number of points in an orbit depends on the given point.

Example 5.1 Let $n = 2$, $\Omega = \{(x, y) : -1 \le x, y \le 1\}$ and G the group of linear transformations for which Ω is G-invariant. The group can be represented by the following set of matrices:

$$\left\{\begin{pmatrix} 1 & 0 \\ 0 & 1 \end{pmatrix}, \begin{pmatrix} 0 & 1 \\ -1 & 0 \end{pmatrix}, \begin{pmatrix} -1 & 0 \\ 0 & -1 \end{pmatrix}, \begin{pmatrix} 0 & -1 \\ 1 & 0 \end{pmatrix},\right.$$

$$\left.\begin{pmatrix} 1 & 0 \\ 0 & -1 \end{pmatrix}, \begin{pmatrix} 0 & 1 \\ 1 & 0 \end{pmatrix}, \begin{pmatrix} -1 & 0 \\ 0 & 1 \end{pmatrix}, \begin{pmatrix} 0 & -1 \\ -1 & 0 \end{pmatrix}\right\}.$$

The orbit of an arbitrary point (a, b) is

$$\{(a, b), (b, -a), (-a, -b), (-b, a), (a, -b), (b, a), (-a, b), (-b, -a)\}.$$

Orbits can have less than 8 points:

- the orbit of (a, a), $a \ne 0$, is $\{(a, a), (-a, a), (a, -a), (-a, -a)\}$
- the orbit of $(a, 0)$, $a \ne 0$, is $\{(a, 0), (-a, 0), (0, a), (0, -a)\}$
- the orbit of $(0, 0)$ is $\{(0, 0)\}$.

The most important symmetries for our purposes are *central symmetry* and *shift symmetry.*

Definition 5.3 A set, integral or, respectively, cubature formula is called *centrally symmetric* if it remains unchanged under reflection through the origin, that is, it is invariant w.r.t. the group of transformations

$$G_{cs} := \{\mathbf{x} \mapsto \mathbf{x}, \mathbf{x} \mapsto -\mathbf{x}\}.$$

Given $\mathbf{a} \in \mathbb{R}^n$, let $\{\mathbf{a}\} \in [0, 1)^n$ denote the vector each of whose components is the fractional part of the corresponding component of $\mathbf{a}$.

Definition 5.4 A set, integral or cubature formula is called *shift symmetric* if it is invariant w.r.t. the group of transformations

$$G_{ss} := \left\{\mathbf{x} \to \mathbf{x}, \mathbf{x} \to \left\{\mathbf{x} + \left(\tfrac{1}{2}, \dots, \tfrac{1}{2}\right)\right\}\right\}.$$

Shift symmetry is for the trigonometric-degree case what central symmetry is for the algebraic case. C_n is centrally symmetric and $C_n^\star$ is shift symmetric.

The other important groups are the symmetry groups of regular polytopes and their subgroups. The most common are:

$A_n, n \geq 2$	:	symmetry group of a regular simplex
$B_n, n \geq 2$	:	symmetry group of a cube
$H_2^m, n = 2$	:	dihedral group, that is, symmetry group of regular m-gon
$I_3, n = 3$	:	symmetry group of a regular icosahedron.

(The origin is the barycentre of the regular polytopes.)

In addition, the associated group $A_n^\star$ is obtained from A_n by adding the reflection through the origin as generator to the group.

Regions and cubature formulae are often called *fully symmetric* when they are B_n-invariant, that is, when they are invariant w.r.t. the following group of transformations:

$$\begin{aligned} G_{B_n} \;:=\; & G_{FS} := \{(x_1, \ldots, x_n) \mapsto (s_1 x_{p_1}, \ldots, s_n x_{p_n}) : \\ & s_i \in \{-1, +1\}, i \in \{1, \ldots, n\}, \{p_1, \ldots, p_n\} = \{1, \ldots, n\}\}. \end{aligned}$$

Example 5.1 dealt with this group. Observe that fully symmetric regions are also centrally symmetric.

Regions and cubature formulae are often called *symmetric* when they are invariant w.r.t. the following subgroup of G_{FS}:

$$G_S := \{(x_1, \ldots, x_n) \mapsto (s_1 x_1, \ldots, s_n x_n) : s_i \in \{-1, +1\}, i \in \{1, \ldots, n\}\}.$$

Definition 5.5 A cubature formula is said to be invariant w.r.t. a group G if the region Ω and the weight function $w(\mathbf{x})$ are G-invariant and if the set of points is a union of G-orbits. All points of one and the same orbit have the same weight.

A G-invariant cubature formula can be written as

$$Q[f] := \sum_{j=1}^{K} w_j Q_G(\mathbf{y}^{(j)})[f], \tag{5.1}$$

where the functional $Q_G(\mathbf{y}^{(j)})$ is the average of the function values of f in the points of the G-orbit of $\mathbf{y}^{(j)}$. $Q_G(\mathbf{y})$ is called a *basic G cubature rule operator*.

5.3. Usefulness for cubature formula construction

The usefulness of symmetry groups in the context of constructing cubature formulae is highlighted by the following result, due to Sobolev (1962). Let F be a vector space of functions defined on $\Omega \subset \mathbb{R}^n$ that is G-invariant, so that $g(f) \in F$ for all $f \in F$ and $g \in G$.

The G-invariant functions of F

$$F(G) := \{f \in F : g(f) = f \text{ for all } g \in G\}.$$

form a subspace

Theorem 5.1 Let G be a finite group of linear transformations acting on F. Then, every G-invariant linear functional on F is determined by its restriction to $F(G)$.

Proof. For every $h \in G$ we have

$$h\left(\sum_{g\in G} g(f)\right) = \sum_{g\in G} h(g(f)) = \sum_{hg\in G} h(g(f)) = \sum_{g\in G} g(f),$$

hence $\sum_{g\in G} g(f) \in F(G)$.

Let I be a G-invariant linear functional on F, so that $I[g(f)] = I[f]$ for all $f \in F$ and $g \in G$. Hence we have

$$I[f] = \frac{1}{|G|}\sum_{g\in G} I[g(f)] = I\left[\frac{1}{|G|}\sum_{g\in G} g(f)\right].$$

This proves the theorem, since we showed that for each $f \in F$, a function in $F(G)$ exists such that the functional gives the same result for both. □

The usual formulation of this theorem is an obvious corollary and is generally known as Sobolev's theorem.

Corollary 5.1 (Sobolev's theorem) Let the cubature formula Q be G-invariant. The cubature formula has degree d if it is exact for all invariant polynomials of degree at most d and if it is not exact for at least one polynomial of degree $d+1$.

The exploitation of the symmetry of the region by imposing a structure to the cubature formula has a simplifying effect. If one wants a G-invariant cubature formula (5.1), the necessary and sufficient conditions (4.4) can be replaced by the reduced system of nonlinear equations

$$Q[\phi_i] = I[\phi_i], \quad i = 1, \ldots \dim \mathcal{V}_d^n(G), \tag{5.2}$$

where the ϕ_i form a basis for $\mathcal{V}_d^n(G)$. The larger the symmetry group G, the lower the dimension of the space of all G-invariant functions and, consequently, the easier it will be to determine a cubature formula.

Example 5.2 If $p(\mathbf{x})$ is an algebraic monomial, deg(p) is odd, and Q is a centrally symmetric cubature formula, then $I[f] = Q[f] = 0$. If $t(\mathbf{x})$ is a trigonometric monomial, deg(t) is odd and Q is a shift symmetric cubature formula, then $I[f] = Q[f] = 0$. So the symmetry of the cubature formula suffices to integrate odd-degree monomials exactly. This is in agreement with Sobolev's theorem because all invariant polynomials for both groups have even degree, and thus odd-degree monomials need not be taken into account.

5.4. Invariant theory

We will now mention some results from invariant theory, a tool for working with vector spaces of invariant polynomials. This will help us to set up the system of nonlinear equations (5.2).

Definition 5.6 The G-invariant polynomials $\phi_1, \ldots, \phi_l$ form an *integrity basis* for the invariant polynomials of G if and only if every invariant polynomial of G is a polynomial in $\phi_1, \ldots, \phi_l$. Each polynomial ϕ_i is called a *basic invariant polynomial* of G.

Because the degree of a polynomial is left unchanged by a linear transformation of the variables, one can restrict the search of basic invariant polynomials to homogeneous polynomials. If the number of basic invariant polynomials $l > n$, then there exist polynomials equations, called *syzygies*, relating $\phi_1, \ldots, \phi_l$. Syzygies come into play when calculating the dimension of a vector space of invariant polynomials.

Some properties are summarized by the following theorems.

Theorem 5.2 There always exists a finite integrity basis for the invariant polynomials of a finite group G.

Theorem 5.3 Let G be a finite group acting on the n-dimensional vector space $\mathbb{R}^n$. G is a finite reflection group if and only if the invariant polynomials of G have an integrity basis consisting of n homogeneous polynomials which are algebraically independent.

Example 5.3 For the symmetry group of a regular m-gon, H_2^m, it is very convenient to use basic invariant polynomials in the variables x and y, or in polar coordinates r and θ:

$$\begin{aligned}\sigma_2 &:= r^2 = x^2 + y^2, \\ \sigma_m &:= r^m \cos(m\theta) = \sum_{i=0}^{\lfloor \frac{m}{2} \rfloor} (-1)^i \binom{m}{2i} x^{m-2i} y^{2i}.\end{aligned}$$

In H_2^m one can distinguish two types of element: there are orientation-reversing transformations (reflections) and orientation-preserving transformations (rotations). The rotations of H_2^m form a subgroup R_2^m of order m. R_2^m is not a reflection group and thus an integrity basis consists of more than two polynomials. In addition to σ_2 and σ_m one can use as basic invariant polynomial

$$\sigma'_m := r^m \sin(m\theta).$$

The syzygy relating σ_2, σ_m and σ'_m is

$$\sigma_2^m - \sigma_m^2 - \sigma_m'^2 = 0.$$

For proofs of the theorems, basic invariant polynomials and other information we refer to Fisher (1967) and Flatto (1978).

6. Characterization of cubature formulae

6.1. Interpolatory cubature formulae

Because we are interested in cubature formulae with a 'low' number of points, we can restrict our attention to interpolatory cubature formulae. Indeed, when a non-interpolatory cubature formula is given, by applying Steinitz's *Austauschsatz* (Davis 1967) an interpolatory cubature formula that uses a subset of the given points can be constructed.

Definition 6.1 If the weights of a cubature formula of degree d are uniquely determined by the points, the cubature formula is called an *interpolatory cubature formula.*

A cubature formula that is exact for all elements of $\mathcal{V}_d^n$ is determined by a system of nonlinear equations (4.4) or (5.2):

$$Q[f_i] = I[f_i], \quad i = 1, \ldots, \dim \mathcal{V}_d^n, \tag{6.1}$$

where the f_i form a basis for $\mathcal{V}_d^n$. If the points of a cubature formula are given, then (6.1) is a system of $\dim \mathcal{V}_d^n$ linear equations in the N unknown weights. Hence an interpolatory cubature formula has $N \leq \dim \mathcal{V}_d^n$ and there exist N linearly independent polynomials $U_1, \ldots, U_N \in \mathcal{V}_d^n$ such that

$$\det \begin{pmatrix} U_1(\mathbf{y}^{(1)}) & \ldots & U_N(\mathbf{y}^{(1)}) \\ \vdots & & \vdots \\ U_1(\mathbf{y}^{(N)}) & \ldots & U_N(\mathbf{y}^{(N)}) \end{pmatrix} \neq 0.$$

These polynomials generate a maximal, not uniquely determined, vector space of polynomials that do not vanish at all given points.

One can always find $t := \dim \mathcal{V}_d^n - N$ polynomials $p_1, \ldots, p_t$ such that the polynomials

$$U_1, \ldots, U_N, p_1, \ldots, p_t$$

form a basis for $\mathcal{V}_d^n$. Then one can solve

$$\begin{pmatrix} U_1(\mathbf{y}^{(1)}) & \ldots & U_N(\mathbf{y}^{(1)}) \\ \vdots & & \vdots \\ U_1(\mathbf{y}^{(N)}) & \ldots & U_N(\mathbf{y}^{(N)}) \end{pmatrix} \begin{pmatrix} a_{i1} \\ \vdots \\ a_{iN} \end{pmatrix} = \begin{pmatrix} p_i(\mathbf{y}^{(1)}) \\ \vdots \\ p_i(\mathbf{y}^{(N)}) \end{pmatrix}, \qquad i = 1, \ldots, t,$$

and so obtain $t = \dim \mathcal{V}_d^n - N$ linearly independent polynomials

$$R_i = p_i - \sum_{j=1}^{N} a_{ij} U_j, \quad i = 1, \ldots, t \tag{6.2}$$

that vanish at the given points of the cubature formula. We can replace the polynomials p_i in the basis of $\mathcal{V}_d^n$ by the polynomials R_i.

With every cubature formula of degree d one can associate a basis of $\mathcal{V}_d^n$ that consists of $\dim \mathcal{V}_d^n - N$ polynomials R_i that vanish at all the points of the cubature formula and N polynomials U_i that do not vanish at all points. A cubature formula is thus fully characterized by the polynomials R_i. The polynomials U_i give rise to a linear system that determines the weights.

These characterizing polynomials provide the links between cubature formulae on one hand, and orthogonal polynomials and ideal theory on the other hand.

6.2. Orthogonal polynomials

Because each R_i (6.2) vanishes at all points of the cubature formula,

$$Q[R_i P] = 0, \quad \text{for all } P \in \mathcal{V}^n.$$

Because the cubature formula has degree d,

$$I[R_i P] = Q[R_i P] = 0 \quad \text{whenever} \quad R_i P \in \mathcal{V}_d^n.$$

And that brings us to orthogonality.

Definition 6.2 A polynomial $f \in \mathcal{V}^n$ is called d-orthogonal (w.r.t. a given integral I), if $I[fg] = 0$ whenever $fg \in \mathcal{V}_d^n$.

Definition 6.3 A polynomial $f \in \mathcal{V}^n$ is called orthogonal (w.r.t. a given integral I), if $I[fg] = 0$ whenever $\deg(g) < \deg(f)$.

The polynomials R_i that characterize a cubature formula of degree d are d-orthogonal.

In contrast with the one-dimensional case, in the n-dimensional case more than one orthogonal polynomial of a given degree d exists. Sequences of orthogonal polynomials can be constructed with $\dim \mathcal{V}_d^{n-1}$ linearly independent polynomials of degree d and many such sequences exist.

The trigonometric case

For the integral with region $C_n^\star$,

$$I[f] = \int_{[0,1)^n} f(\mathbf{x})\, \mathrm{d}\mathbf{x},$$

any trigonometric monomial is orthogonal to every trigonometric monomial of a lower degree. Hence, these are the obvious choice when $w(\mathbf{x}) \equiv 1$.

We have only found other weight functions in theoretical results where orthogonal polynomials are only used implicitly.

The algebraic case
It is a generalization of a result of Jackson (1936) that there exist $\dim \mathcal{P}_d^{n-1}$ unique orthogonal polynomials of degree d of the form

$$P^{\alpha_1,\alpha_2,\ldots,\alpha_n}(x_1,x_2,\ldots,x_n) = x_1^{\alpha_1}x_2^{\alpha_2}\ldots x_n^{\alpha_n} + Q \tag{6.3}$$

with $\sum_{i=1}^n \alpha_i = d$ and $Q \in \mathcal{P}_{d-1}^n$. The polynomials of the form (6.3) are called *basic orthogonal polynomials.*

For so-called *product regions*, that is, when the region of integration is a product of intervals and the weight function is a product of univariate functions, so that

$$I[f] = \int_{a_1}^{b_1} w_1(x_1)\ldots \int_{a_n}^{b_n} w_n(x_n) f(\mathbf{x})\,\mathrm{d}x_n \ldots \mathrm{d}x_1,$$

the basic invariant polynomials are products of monic univariate orthogonal polynomials. For example, in C_2, we have $P^{k,l}(x,y) = P_k(x)P_l(y)$, where $P_i(x)$ is the monic Legendre polynomial of degree i in x. The regions C_n and $E_n^{r^2}$ are product regions and their basic invariant polynomials are the product of monic Legendre and Hermite polynomials, respectively.

As the explicit expressions for the basic orthogonal polynomials for S_n and T_n are not well known, we list them here.

S_n: Let $\alpha \in \mathbb{N}^n$ and $\beta \le \alpha/2$ (that is, $0 \le \beta_i \le \alpha_i/2$ for $i = 1,\ldots,n$). Then,

$$P^\alpha(\mathbf{x}) = \sum_{\beta\le\alpha/2} (-1)^{|\beta|} \frac{\Gamma(|\alpha|-|\beta|+n/2)}{\Gamma(|\alpha|+n/2)2^{2|\beta|}} \left(\prod_{j=1}^n \frac{\alpha_j!}{(\alpha_j-2\beta_j)!\beta_j!}\right)\mathbf{x}^{\alpha-2\beta}.$$

See Appell and Kampé de Fériet (1926).

T_n: Let $\alpha \in \mathbb{N}^n$ and $\beta \le \alpha$ (that is, $0 \le \beta_i \le \alpha_i$ for $i = 1,\ldots,n$). Then,

$$P^\alpha(\mathbf{x}) = \sum_{\beta\le\alpha} (-1)^{|\alpha|+|\beta|} \frac{(|\alpha|+|\beta|+n-1)!}{(2|\alpha|+n-1)!} \left(\prod_{i=1}^n \binom{\alpha_i}{\beta_i}\frac{\alpha_i!}{\beta_i!}\right)\mathbf{x}^{\beta}.$$

See Appell and Kampé de Fériet (1926) for $n = 2$ and Grundmann and Möller (1978) for $n \in \mathbb{N}$.

E_n^r: An explicit expression for the basic invariant polynomials has not yet been shown.

The basic orthogonal polynomials reflect the symmetry of the integral. If the integral is centrally symmetric then the basic orthogonal polynomials of even (odd) degree consist of even (odd) degree monomials only. If the integral is fully symmetric then the basic orthogonal polynomial $P^\alpha(\mathbf{x})$ with all α_i even (odd) consists only of monomials with even (odd) powers of x_i, for all $i \in \{1, \ldots, n\}$. Furthermore, $P^{p(\alpha)}(\mathbf{x}) = P^\alpha(p(\mathbf{x}))$ where p performs a permutation on the components of its vector argument.

The structure of basic invariant polynomials motivates the following.

Definition 6.4 A set of polynomials S is called *fundamental of degree d* whenever $\dim \mathcal{V}_d^{n-1} (= \dim \mathcal{V}_d^n - \dim \mathcal{V}_{d-1}^n)$ linearly independent polynomials of the form $x_1^{\alpha_1} \ldots x_n^{\alpha_n} + Q_\alpha$, $Q_\alpha \in \mathcal{V}_{d-1}^n, |\alpha| = d$, belong to span S.

6.3. Polynomial ideals

The polynomials U_i and R_i are not uniquely determined. The direct sum of the vector spaces generated by these polynomials is

$$\operatorname{span}\{U_i\} \oplus \operatorname{span}\{R_i\} = \mathcal{V}_d^n.$$

$\operatorname{span}\{R_i\}$ is more than simply a vector space. Indeed, if one multiplies a polynomial that vanishes at all points of the cubature formula by an arbitrary polynomial, the product also vanishes at all points. And that brings us to ideals.

Definition 6.5 A *polynomial ideal* $\mathfrak{A}$ is a subset of the ring of polynomials in n variables $\mathcal{V}^n$ such that if $f_1, f_2 \in \mathfrak{A}$ and $g_1, g_2 \in \mathcal{V}^n$, then $f_1 g_1 + f_2 g_2 \in \mathfrak{A}$.

The genesis of ideal theory is described in Edwards (1980). In this section we describe the part of ideal theory needed in this paper.

Definition 6.6 If $\mathfrak{A}$ is a polynomial ideal, then the set of polynomials $\{f_1, \ldots, f_s\} \subset \mathfrak{A}$ form a basis for $\mathfrak{A}$ if each $f \in \mathfrak{A}$ can be written in the form

$$f = \sum_{j=1}^{s} g_j f_j \quad \text{where} \quad g_j \in \mathcal{V}^n.$$

The ideal generated by $\{f_1, \ldots, f_s\}$ is

$$(f_1, \ldots, f_s) := \left\{ f = \sum_{j=1}^{s} g_j f_j : g_j \in \mathcal{V}^n \right\}.$$

The polynomials R_i that characterize a cubature formula generate an ideal, denoted by $(R_1, \ldots, R_t)$.

Theorem 6.1 For any polynomial ideal there exists a finite basis.

Proof. See Hilbert (1890). □

There are several types of bases for ideals. For our purposes, H-bases and G-bases are important. H-bases are important as a theoretical tool. Their power will be shown by the short proof of Theorem 6.7. G-bases are important because algorithms exist to construct them and to derive properties of the ideal. It is thus very convenient for us that with some restrictions a G-basis is also an H-basis (Buchberger 1985, Möller and Mora 1986, Sturmfels 1996).

Definition 6.7 Let $\mathfrak{A}$ be a polynomial ideal. The set $\{f_1, \ldots, f_s\} \subset \mathfrak{A}$ is an H-basis for $\mathfrak{A}$ if for all $f \in \mathfrak{A}$ there exist polynomials $g_1, \ldots, g_s$ such that

$$f = \sum_{j=1}^{s} g_j f_j \quad \text{and} \quad \deg(g_j f_j) \leq \deg(f), \quad j = 1, \ldots, s.$$

Theorem 6.2 For any polynomial ideal an H-basis exists.

Proof. See Möller (1973). □

Other names for an H-basis are canonical basis and Macaulay basis.

Before defining G-bases, also called Gröbner-bases, we have to introduce some notation. Let the set of monomials $M = \{\mathbf{x}^\alpha : \alpha \in \mathbb{N}^n\}$ be ordered by $<$ such that, for any $f, f_1, f_2 \in M$, $1 \leq f$ and $f_1 \leq f_2$ imply $ff_1 \leq ff_2$. Let $f = \sum_{i=1}^{m} c_i f_i$ with $f_i \in M$ and $c_i \in \mathbb{R}_0$. Then the *headterm* of $f = \mathrm{Hterm}(f) := f_m$, and the *maximal part* of $f = M(f) := c_m f_m$. For $f, g \in \mathcal{P}^n \backslash \{0\}$ let

$$H(f,g) := \mathrm{lcm}\{\mathrm{Hterm}(f), \mathrm{Hterm}(g)\}. \tag{6.4}$$

Let $F \subset \mathcal{P}^n \backslash \{0\}$ be a finite set. We write $f \longrightarrow_F g$ if $f, g \in \mathcal{P}^n$ and there exist $h \in \mathcal{P}^n$, $f_i \in F$ such that $f = g + hf_i$, $\mathrm{Hterm}(g) < \mathrm{Hterm}(f)$ or $g = 0$. The map $\underset{F}{\longrightarrow}$ is called a reduction modulo F. By $\underset{F}{\longrightarrow}^{+}$ we denote the reflexive transitive closure of $\underset{F}{\longrightarrow}$.

Definition 6.8 A set $F := \{f_1, \ldots, f_l\}$ is a Gröbner basis (G-basis) for the ideal $\mathfrak{A}$ generated by F if

$$f \in \mathfrak{A} \text{ implies } f \underset{F}{\longrightarrow}^{+} 0.$$

Theorem 6.3 Let $F := \{f_1, \ldots, f_s\} \subset \mathcal{P}^n \backslash \{0\}$ and let $\mathfrak{A}$ be the ideal generated by F. Then the following conditions are equivalent.

- F is a Gröbner basis of $\mathfrak{A}$.
- For all (i,j) with $1 \leq i < j \leq s$,

$$SP(f_i, f_j) = \frac{H(f_i, f_j)}{M(f_i)} f_i - \frac{H(f_i, f_j)}{M(f_j)} f_j \underset{F}{\longrightarrow}^{+} 0.$$

Proof. See Möller and Mora (1986). □

Theorem 6.3 provides an algorithmic way to verify if a given set is a G-basis. Practical implementations incorporate several shortcuts. For example, according to Gebauer and Möller (1988), a pair (f_i, f_j) is superfluous if $H(f_l, f_j)$ divides properly $H(f_i, f_j)$ and $l < j$.

Theorem 6.4 If $<$ is compatible with the partial ordering by degrees, that is, $\deg(f) < \deg(g)$ implies $\text{Hterm}(f) < \text{Hterm}(g)$, then a G-basis with respect to $<$ is also an H-basis.

Proof. See Möller and Mora (1986). □

Definition 6.9. (Nullstellengebilde) The *zero set* of an ideal $\mathfrak{A}$ is

$$\mathsf{NG}(\mathfrak{A}) := \{\mathbf{y} \in \mathbb{C}^n : f(\mathbf{y}) = 0 \quad \text{for all } f \in \mathfrak{A}\}$$

If $\mathsf{NG}(\mathfrak{A})$ is a finite set of points, then the ideal is called *zero-dimensional*, and obviously any basis for $\mathfrak{A}$ consists of at least n polynomials.

An important function of a polynomial ideal is the Hilbert function (Hilbert 1890). It is useful to count the number of elements of $\mathsf{NG}(\mathfrak{A})$.

Definition 6.10 The *Hilbert function* $\mathcal{H}$ is defined as

$$\mathcal{H}(k; \mathfrak{A}) := \begin{cases} \dim \mathcal{P}_k^n - \dim(\mathfrak{A} \cap \mathcal{P}_k^n), & k \in \mathbb{N}, \\ 0, & -k \in \mathbb{N}_0. \end{cases}$$

Theorem 6.5 If $\mathcal{H}(k; \mathfrak{A}) = \mathcal{H}(K; \mathfrak{A})$ for all $k \geq K$ holds for a sufficiently large K, then the polynomials in $\mathfrak{A}$ have exactly $\mathcal{H}(K; \mathfrak{A})$ (complex) common zeros if these are counted with multiplicities.

Proof. See Gröbner (1949). □

Definition 6.11 An ideal $\mathfrak{A}$ is a real ideal if all polynomials vanishing at $\mathsf{NG}(\mathfrak{A}) \cap \mathbb{R}^n$ belong to $\mathfrak{A}$, that is,

$$f \in \mathfrak{A} \quad \text{if and only if} \quad f(\mathbf{y}) = 0, \qquad \text{for all } \mathbf{y} \in \mathsf{NG}(\mathfrak{A}) \cap \mathbb{R}^n.$$

Note that the theorems given in this subsection are proven only for algebraic polynomials in the literature. We do not see any problem in their application to ideals of invariant algebraic polynomials or trigonometric polynomials.

Within the ideal theoretical framework we can rephrase Theorem 3.1.

Theorem 6.6 Let I be an integral over an n-dimensional region. Let $\{\mathbf{y}^{(1)}, \ldots, \mathbf{y}^{(N)}\} \subset \mathbb{C}^n$ and $\mathfrak{A} := \{f \in \mathcal{V}^n : f(\mathbf{y}^{(i)}) = 0, i = 1, \ldots, N\}$. Then the following statements are equivalent.

- $f \in \mathfrak{A} \cap \mathcal{V}_d^n$ implies $I[f] = 0$.
- There exists a cubature formula Q (3.7) such that $I[f] = Q[f]$, for all $f \in \mathcal{V}_d^n$, with at most $\mathcal{H}(d; \mathfrak{A})$ (complex) weights different from zero.

Proof. This theorem is proven by Möller (1973) for the case $\mathcal{V}_d^n = P_d^n$. □

The role of H-bases is illustrated by the following theorem by Möller (1973).

Theorem 6.7 If $\{f_1, \ldots, f_s\}$ is an H-basis of a polynomial ideal $\mathfrak{A}$ and if the set of common zeros of $f_1, \ldots, f_s$ is finite and nonempty, then the following statements are equivalent.

- There is a cubature formula of degree d for the integral I which has as points the common zeros of $f_1, \ldots, f_s$. (These zeros may be multiple, leading to the use of function derivatives in the cubature formula.)
- f_i is d-orthogonal for $I, i = 1, 2, \ldots, s$.

Proof.

'$\Rightarrow$': Let $gf_i \in \mathcal{P}_d^n$. Then $I[gf_i] = \sum_{j=1}^{N} w_j g(\mathbf{y}^{(j)}) f_i(\mathbf{y}^{(j)}) = 0$, since $f_i \in \mathfrak{A}$.
'$\Leftarrow$': Let $f \in \mathfrak{A} \cup \mathcal{P}_d^n$. Then, with g_i as given in the definition of H-basis, $I[f] = I[\sum_{i=1}^{s} g_i f_i] = 0$.

□

Schmid managed to give a characterization of cubature formulae with real points and positive weights using real ideals.

Theorem 6.8 Let $\{R_1, \ldots, R_t\} \subset \mathcal{P}_{d+1}^n$ be a set of linearly independent d-orthogonal polynomials that is fundamental of degree $d+1$. Let $\mathfrak{A} := (R_1, \ldots, R_t)$ and $V := \text{span}\{R_1, \ldots, R_t\}$. Let $N + t = \dim \mathcal{P}_{d+1}^n$ and U an arbitrary but fixed vector space such that $\mathcal{P}_{d+1}^n = V \oplus U$. Then the following statements are equivalent.

- There exists an interpolatory cubature formula of degree d
$$Q[f] := \sum_{j=1}^{N} w_j f(\mathbf{y}^{(j)}), \quad \mathbf{y}^{(j)} \in \mathbb{R}^n, \quad w_j > 0$$
with $\{\mathbf{y}^{(1)}, \ldots, \mathbf{y}^{(N)}\} \subset \mathsf{NG}(\mathfrak{A})$.
- $\mathfrak{A}$ and U are characterized by:
  (i) $\mathfrak{A} \cap U = \{0\}$
  (ii) $I[f^2 - R^+] > 0$ for all $f \in U$, where $R^+ \in \mathfrak{A}$ is chosen such that $f^2 - R^+ \in \mathcal{P}_d^n$.
- $\mathfrak{A}$ is a real ideal and $|\mathsf{NG}(\mathfrak{A}) \cap \mathbb{R}^n| = N$. The points of the cubature formula are the elements of $\mathsf{NG}(\mathfrak{A}) \cap \mathbb{R}^n$.

Proof. See Schmid (1980*a*). □

This characterization was used to develop the T-method for constructing cubature formulae; see Section 9.2.

6.4. Tchakaloff's upper bound

To conclude this section we will prove an upper bound for the number of points in an interpolatory cubature formula using the concepts from ideal theory just introduced. We mentioned this result at the beginning of this section.

Corollary 6.1 If an interpolatory cubature formula of degree d for an integral over an n-dimensional region has N points, then $N \le \dim \mathcal{V}_d^n$.

Proof. Suppose a given cubature formula of degree d has M points. Let $\mathfrak{A}$ be the ideal of all polynomials that vanish at these points. According to Theorem 6.6, there exists a cubature formula with $N \le \mathcal{H}(d;\mathfrak{A}) \le M$ of these points, and from the definition of the Hilbert function, it follows immediately that $N \le \mathcal{H}(d;\mathfrak{A}) \le \dim \mathcal{V}_d^n$. A basis of any complement of $\mathfrak{A} \cap \mathcal{V}_d^n$ in $\mathcal{V}_d^n$ can be used to construct a set of linear equations to determine the weights. □

The above corollary is an elementary version of Tchakaloff's theorem.

Theorem 6.9 (Tchakaloff's theorem) Let I be an integral over an n-dimensional region Ω with a weight function that is nonnegative in Ω and for which the integrals of all monomials exist. Then a cubature formula of degree d with $N \le \dim \mathcal{V}_d^n$ points exists with all points inside Ω and all weights positive.

Proof. This theorem was proven by Tchakaloff (1957) for bounded regions and by Mysovskikh (1975) for unbounded regions for $\mathcal{V}_d^n = \mathcal{P}_d^n$. □

We will now prove, along the lines of Mysovskikh (1981), that this is the smallest general upper bound. We will construct an n-dimensional region for which a cubature formula of degree d with fewer points than $\dim \mathcal{V}_d^n$ does not exist.

Let $\mu := \dim \mathcal{V}_d^n$ and choose distinct points $\mathbf{a}^{(1)}, \ldots, \mathbf{a}^{(\mu)} \in \mathbb{R}^n$ that do not lie on a curve of order d. Let C_i be a cube with centre $\mathbf{a}^{(i)}$ and side ρ, such that the μ cubes do not intersect. We will now show that, for sufficiently small ρ, no cubature formula of degree d exists for $\Omega = C_1 \cup \ldots \cup C_\mu$ and $w(\mathbf{x}) \equiv 1$ with all points inside Ω and all weights positive.

Assume that such a cubature formula exists and let C_1 be the subregion containing no point of the cubature formula. Let $p(\mathbf{x}) \in \mathcal{V}_d^n$ satisfy

$$p(\mathbf{a}^{(1)}) = 1, \quad p(\mathbf{a}^{(i)}) = 0 \quad \text{for} \quad i = 2, \ldots, \mu,$$

and σ a number such that

$$0 < \sigma < \frac{1}{2\mu}. \tag{6.5}$$

Take ρ small enough such that $p(\mathbf{x}) \geq 1-\sigma$, for $\mathbf{x} \in C_1$, and $|p(\mathbf{x})| \leq \sigma$, for $\mathbf{x} \in \cup_{i=2}^{\mu} C_i$. Then

$$\begin{aligned}
\left|\int_\Omega p(\mathbf{x})\,\mathrm{d}\mathbf{x}\right| &= \left|\int_{C_1} p(\mathbf{x})\,\mathrm{d}\mathbf{x} + \sum_{i=2}^{\mu}\int_{C_i} p(\mathbf{x})\,\mathrm{d}\mathbf{x}\right| \\
&\geq \left|\int_{C_1} p(\mathbf{x})\,\mathrm{d}\mathbf{x}\right| - \sum_{i=2}^{\mu}\left|\int_{C_i} p(\mathbf{x})\,\mathrm{d}\mathbf{x}\right| \\
&\geq (1-\sigma)\rho^n - (\mu-1)\sigma\rho^n \\
&= \rho^n(1-\mu\sigma).
\end{aligned}$$

On the other hand,

$$\sum_{j=1}^{\mu} w_j p(\mathbf{y}^{(j)}) \leq \sigma \sum_{j=1}^{\mu} w_j = \sigma\mu\rho^n.$$

From the exactness of the cubature formula it follows that

$$(1-\mu\sigma)\rho^n \leq \sigma\mu\rho^n \quad \text{if and only if} \quad 1 \leq 2\mu\sigma,$$

which contradicts (6.5).

7. In search of minimal formulae

7.1. A general lower bound

We consider cubature formulae of the form

$$Q[f] = \sum_{j=1}^{N} w_j f(\mathbf{y}^{(j)}), \quad w_j \in \mathbb{R}, \tag{7.1}$$

for the approximation of the integral (3.1). In this section we identify the polynomials which are identical on the integration region Ω, and we restrict our attention to cubature formulae with all points inside Ω. This identification leaves the polynomial space unchanged if and only if Ω contains inner points.

Example 7.1 Consider the surface of the unit ball: $\Omega = \{\mathbf{x} : \sum_{i=1}^{n} x_i^2 = 1\}$. Then the polynomials $(\sum_{i=1}^{n} x_i^2)^p$, $p \in \mathbb{N}$, are all identified with the constant polynomial 1. So,

$$\dim \mathcal{P}_d^n|_\Omega = \binom{n+d}{n} - \binom{n+d-2}{n}.$$

Theorem 3.1 can be used to derive a very general lower bound. Good lower bounds are important because any method to construct cubature formulae (implicitly or explicitly) depends on a bound or estimate of the number of points. If a lower bound is known, then a method to construct cubature formulae attaining this bound is usually known.

Theorem 7.1 If the cubature formula (7.1) is exact for all polynomials of $\mathcal{V}_{2k}^n$, then the number of points $N \geq \dim \mathcal{V}_{k|\Omega}^n$.

Proof. Let $F = \mathcal{V}^n|_\Omega$, $F_1 = \mathcal{V}_k^n|_\Omega$ and

$$F_0 = \{f \in F_1 : f(\mathbf{y}^{(j)}) = 0, j = 1, \dots, N\}.$$

If $f \in F_0$, then $\deg(f) \leq k$ and $f(\mathbf{y}^{(j)}) = 0$, $j = 1, \dots, N$. Because f^2 is of degree at most $2k$, the cubature formula is exact and $I[f^2] = 0$. Hence, on Ω, $f \equiv 0$. So far, we have proved that

$$f \in F_0 \quad \text{implies} \quad f \equiv 0.$$

Let Q be a linear functional defined on $\mathcal{V}_k^n|_\Omega$. Then

$$f \in F_0 \Rightarrow f \equiv 0 \Rightarrow Q[f] = 0.$$

From Theorem 3.1 it follows that weights w_j can be found such that

$$Q[f] = \sum_{j=1}^{N} w_j f(\mathbf{y}^{(j)}), \quad \text{for all } f \in \mathcal{V}_k^n.$$

So, the vector space spanned by the functionals $L_j[f] = f(\mathbf{y}^{(j)})$ is equal to the space of all linear functionals defined on $\mathcal{V}_k^n|_\Omega$. Its dimension is also $\dim \mathcal{V}_k^n|_\Omega$. Hence $N \geq \dim \mathcal{V}_k^n|_\Omega$. □

For regions with interior points and algebraic degree, Theorem 7.1 is given by Radon (1948) for $n = 2$, and for general n by Stroud (1960). It should be noted that the well-known proof of the Radon–Stroud lower bound does not assume all points are inside the region. This restriction plays a role if one includes regions such as the surface of the n-ball, without interior points. For the surface of the n-ball, this result was given by Mysovskikh (1977). Table 1 lists all known formulae that attain the lower bound, for the regions we mentioned in Section 5.

For trigonometric degree, this theorem was probably first mentioned by Mysovskikh (1988). Table 2 lists all known formulae that attain the lower bound. Cools and Reztsov (1997) proved it for other spaces of trigonometric polynomials.

For regions with interior points and product algebraic degree, this theorem was presented by Gout and Guessab (1986). The bound is attained by Gauss product formulae. For other spaces of algebraic polynomials it was presented by Guessab (1986). The general formulation we gave is from Möller (1979) with a proof due to Mysovskikh (1981).

Because $\dim \overline{\mathcal{P}}_d^n = (\dim \overline{\mathcal{P}}_d^1)^n$ and $\dim \overline{\mathcal{T}}_d^n = (\dim \overline{\mathcal{T}}_d^1)^n$, this bound is attained for the overall degree case by the product rules based on minimal quadrature rules. Hence in the rest of this paper, not much attention is paid to this case. As Tables 1 and 2 illustrate, the ordinary degree case is totally

Table 1. *Minimal formulae of algebraic degree*

n	d	N	regions	references
n	2	$n+1$	C_n, S_n, T_n, U_n	see Stroud (1971), Mysovskikh (1981)
2	d	$d+1$	U_n	see Stroud (1971)
	$2k$	$\frac{k^2+3k+2}{2}$	$C_2^{0.5}$	Morrow and Patterson (1978)
	4	6	$C_2, S_2, T_2, E_2^{r^2}$	see Stroud (1971), Cools and Rabinowitz (1993)
	6	10	C_2	Schmid (1983)
			T_2	Rasputin (1983*a*)
			S_2	Rasputin (1986), Wissman and Becker (1986)
	8	15	C_2	Morrow and Patterson (1978)
			T_2	Cools and Haegemans (1987*c*)
3	4	10	C_3	Weiß (1991)

Table 2. *Minimal formulae of trigonometric degree for* C_n^*

n	d	N	references
n	2	$2n+1$	Noskov (1988*b*)
2	$2k$	$2k^2+2k+1$	Noskov (1988*b*)

different: odd degree formulae do not appear in these tables (except for U_2) and the known even degree formulae are rare.

The following theorem teaches us something about the weights. It generalizes a theorem from Mysovskikh (1981).

Theorem 7.2 If the cubature formula (7.1) is exact for all polynomials of degree $d > 0$ and has only real points and weights, then it has at least $\dim \mathcal{V}_k^n$ positive weights, $k = \lfloor \frac{d}{2} \rfloor$.

Proof. According to Theorem 7.1, $N \geq \dim \mathcal{V}_k^n = \kappa$. Because $d > 0$, the cubature formula is exact for $f \equiv 1$, that is, $\sum_{j=1}^{N} w_j = I[1] > 0$. Hence there must be positive weights. If $d = 1$, then $\kappa = 1$ and the theorem holds.

We now consider $d \geq 2$ and assume the theorem does not hold. Let the number of positive weights $\nu < \kappa$ and order the points of the cubature formula such that these positive weights correspond to $\mathbf{y}^{(1)}, \ldots, \mathbf{y}^{(\nu)}$. Then one can find a polynomial $p \in \mathcal{V}_k^n$ such that $p(\mathbf{y}^{(j)}) = 0$, $j = 1, \ldots, \nu$.

The cubature formula is exact for p^2, hence

$$I[p^2] = \sum_{j=\nu+1}^{N} w_j p^2(\mathbf{y}^{(j)}).$$

Because $I[p^2] > 0$, $p^2(\mathbf{y}^{(j)}) \geq 0$ and $w_j < 0$ we obtain a contradiction, hence our assumption was wrong. □

Corollary 7.1 If a cubature formula attains the lower bound of Theorem 7.1, then all its weights are positive.

Theorem 7.1 gives the same lower bound for cubature formulae of degree $2k$ and $2k+1$.

7.2. The characterization of minimal formulae and the reproducing kernel

For even degrees, I am unaware of any greater lower bound than that given in Theorem 7.1. The fact that not many formulae that attain this bound exist for the ordinary algebraic or trigonometric degree case has to do with the practical problems one encounters while attempting to construct these formulae. In this section, the reproducing kernel approach to construct cubature formulae is explained.

The concept of 'reproducing kernel' was first used for the construction of cubature formulae of algebraic degree by Mysovskikh (1968). For the trigonometric degree case it was first used by Mysovskikh (1990).

Choose the polynomials $\phi_1(\mathbf{x}), \phi_2(\mathbf{x}), \ldots \in \mathcal{V}^n$ such that $\phi_i(\mathbf{x})$ is orthogonal to $\phi_j(\mathbf{x})$, for all $j < i$, and $I[\phi_i \bar{\phi}_i] = 1$. This means that $\{\phi_i(\mathbf{x})\}_{i=1}^{\infty}$ is an orthonormal basis of $\mathcal{V}^n$. For a given $k \in \mathbb{N}$ we set $\kappa := \dim \mathcal{V}_k^n$ and

$$K(\mathbf{x}, \mathbf{y}) := \sum_{i=1}^{\kappa} \bar{\phi}_i(\mathbf{x}) \phi_i(\mathbf{y}).$$

$K(\mathbf{x}, \mathbf{y})$ is a reproducing kernel in the space $\mathcal{V}_k^n$: if $f \in \mathcal{V}_k^n$, then f coincides with its expansion in ϕ_i, so that for $a \in \mathbb{C}^n$ fixed,

$$f(\mathbf{a}) = I[f(\mathbf{x}) K(\mathbf{x}, \mathbf{a})] = \sum_{i=1}^{\kappa} I[f(\mathbf{x}) \bar{\phi}_i(\mathbf{x})] \phi_i(\mathbf{a}).$$

The reproducing kernel $K(\mathbf{x}, \mathbf{y})$ plays an important role in connection with Theorem 7.1, as the next theorem illustrates.

Theorem 7.3 A necessary and sufficient condition for the points $\mathbf{y}^{(j)}$, $j = 1, \ldots, N = \dim \mathcal{V}_k^n$, to be the points of a cubature formula that is exact for $\mathcal{V}_{2k}^n$ is that

$$K(\mathbf{y}^{(r)}, \mathbf{y}^{(s)}) = b_r \delta_{rs}, \tag{7.2}$$

with $b_r \neq 0$ and δ_{rs} the Kronecker symbol.

Proof. To prove the necessity, assume a cubature formula (7.1) exists that is exact on $\mathcal{V}_{2k}^n$. Hence, it is exact for

$$\phi_l(\mathbf{x})\phi_m(\mathbf{x}), \quad l,m = 1,\ldots,N,$$

and due to the orthonormality of the ϕ_i, we obtain

$$\sum_{j=1}^{N} w_j \bar{\phi}_l(\mathbf{y}^{(j)})\phi_m(\mathbf{y}^{(j)}) = \delta_{lm}. \tag{7.3}$$

Let $W := \mathrm{diag}(w_1,\ldots,w_N)$, E the unit matrix, and let A be

$$A = \begin{pmatrix} \phi_1(\mathbf{y}^{(1)}) & \ldots & \phi_1(\mathbf{y}^{(N)}) \\ \vdots & & \vdots \\ \phi_N(\mathbf{y}^{(1)}) & \ldots & \phi_N(\mathbf{y}^{(N)}) \end{pmatrix}.$$

Then (7.3) can be written as

$$AWA^\star = E,$$

where $A^\star$ denotes the Hermitian conjugate of A. A is non-singular, for otherwise there is an element of $\mathcal{V}_k^n$ that vanishes at all points of the cubature formula, which is impossible.

W is also non-singular, so we obtain

$$W = A^{-1}(A^\star)^{-1}$$

or

$$A^\star A = W^{-1}.$$

We deduce (7.2) with

$$b_r = 1/w_r = \sum_{i=1}^{N} |\phi_1(\mathbf{y}^{(r)})|^2 > 0.$$

(Remember Corollary 7.1!)

Sufficiency remains to be proven. Conditions (7.2) can be written as

$$A^\star A = B,$$

where $B := \mathrm{diag}(b_1,\ldots,b_N)$ is non-singular. This is equivalent to

$$AB^{-1}A^\star = E,$$

which in turn is equivalent to saying that the cubature formula with points $\mathbf{y}^{(j)}$, $j = 1,\ldots,N$, and weights $w_j = 1/b_j$ is exact for

$$\phi_l(\mathbf{x})\phi_m(\mathbf{x}), \quad l,m = 1,\ldots,N,$$

and thus for all elements of $\mathcal{V}_{2k}^n$. (The final step of this proof motivated the warning at the end of Section 3.) □

For the algebraic-degree case, the reproducing kernel approach has not been very successful in constructing minimal cubature formulae. It can, however, also be used to construct non-minimal formulae; see Möller (1973) and Mysovskikh (1980). Möller (1973) also gave a modified reproducing kernel method to construct centrally symmetric cubature formulae of odd algebraic degree. This modification is based on the same idea we use in the following section to derive a lower bound for such formulae. Cools and Sloan (1996) used a similar modified method to construct minimal shift symmetric cubature formulae of odd trigonometric degree. In this case an infinite number of minimal cubature formulae for each odd degree was obtained in the two-dimensional case.

The reproducing kernel approach has also led to interesting results on the weights of cubature formulae (Cools and Haegemans 1988*c*, Cools 1989, Beckers and Cools 1993). Such results have also led to the following theorem.

Theorem 7.4 A cubature formula of degree $d = 2k$ with $N = \dim \mathcal{P}_d^n$ points does not exist for U_n if $n > 2$ and $k > 2$.

Proof. See Taylor (1995). □

For other characterizations of cubature formulae attaining the bound of Theorem 7.1, see, for instance, Morrow and Patterson (1978), Schmid (1978), and Schmid (1995).

7.3. The general lower bound for some invariant formulae

Although Theorem 7.1 has already shown many of its faces in the literature, it has not yet unveiled all. We will now show what it can teach us about centrally symmetric cubature formulae. In combination with Theorem 5.1, Theorem 7.1 gives a lower bound for the number of G-orbits in a G-invariant cubature formula. This can be translated into a lower bound for the number of points by multiplying it with the highest possible cardinality of a G-orbit, but one expects this will not usually give strict bounds. There is, however, an interesting exception ...

Consider centrally symmetric cubature formulae of algebraic degree $2k+1$ with k even. According to Theorem 7.1, the number of orbits of this cubature formula, K, satisfies

$$K \geq \dim \mathcal{P}_k^n(G_{cs}) = \sum_{i=0}^{k/2} \binom{n-1+2i}{n-1}.$$

A G_{cs}-orbit has one or two points and there can be only one orbit with one point. Hence the above bound for K implies a bound for the number of points:

$$N \geq 2 \dim \mathcal{P}_k^n(G_{cs}) - 1. \tag{7.4}$$

For example, for $n = 2$, we obtain

$$N \geq 2\sum_{i=0}^{k/2}(2i+1) - 1 = \frac{k^2}{2} + 2k + 1 = \binom{k+2}{2} + \frac{k}{2}.$$

For C_2, for instance, this bound is now known to be sharp for degrees 1,5 and 9. Consider shift symmetric cubature formulae of trigonometric degree $2k+1$. A G_{ss}-orbit always has two points. Using the same arguments as in the previous paragraph, we obtain the lower bound

$$N \geq 2 \dim \mathcal{T}_k^n(G_{ss}).$$

for the number of points.

These results, which are derived under the restriction of central symmetry and shift symmetry, will appear again in Section 8.3.

The following questions have probably already occurred in the reader's mind while reading this section.

- Under what conditions is the lower bound of Theorem 7.1 sharp?
- What is the minimum number of points for a cubature formula for a given region?
- Is the symmetry of the region somehow reflected in the structure of minimal formulae?

These questions have kept researchers busy for approximately 50 years now, and are still only partially answered. We return to them in the next section.

8. In search of better bounds for odd degree formulae

8.1. The need for a better bound

In Section 7.1 we obtained a lower bound for the number of points N of a cubature formula that is exact on a vector space of functions $\mathcal{V}_d^n$. This bound, presented in Theorem 7.1, depends only on $\mathcal{V}_d^n$, restricted to Ω. In this section we will see that this bound is in general too low for odd degrees d. Higher lower bounds have to take into account more information on the region Ω and weight function $w(\mathbf{x})$.

Suppose we have a cubature formula of algebraic degree $d = 2k+1$ that attains the lower bound of Theorem 7.1, and let $\mathfrak{A}$ be the corresponding ideal. Then

$$\mathcal{H}(k;\mathfrak{A}) = \dim \mathcal{P}_k^n = N = \mathcal{H}(d;\mathfrak{A}).$$

Hence the ideal contains $\dim \mathcal{P}_{k+1}^n - \dim \mathcal{P}_k^n$ linearly independent polynomials of degree $k+1$. These polynomials must be d-orthogonal and thus, because of their degree, simply orthogonal. So we have in fact proved the following theorem.

Theorem 8.1 A necessary condition for the existence of a cubature formula of algebraic degree $2k+1$ with $N = \dim \mathcal{P}_k^n$ points is that the basic orthogonal polynomials of degree $k+1$ have N common zeros.

The condition of Theorem 8.1 does not hold for standard regions such as C_n, T_n, S_n, $E_n^{r^2}$ and E_n^r. Radon (1948) discovered that no cubature formula of degree 5 with 6 points exists for C_2, T_2, and S_2.

8.2. The quest for exceptional regions

Fritsch (1970) searched for an n-dimensional region for which a formula of degree 3 with $n+1$ points exists. He defined a region $S_n(d)$ as follows. Let S_n be the n-simplex with vertices $v_0, v_1, \ldots, v_n$ and centroid c. Let F_k be the face of S_n that does not contain the vertex v_k and let c_k be the centroid of F_k. Let $d > 0$ and define the points $u_k(d)$ by

$$u_k(d) = dc_k + (1-d)c \ , \ k = 0, 1, \ldots, n.$$

Let $S_{nk}(d)$ be the simplex with base F_k and vertex $u_k(d)$. Define

$$S_n(d) = \begin{cases} S_n \cup (\bigcup_{k=0}^n S_{nk}(d)) \ , & d \geq 1, \\ S_n - (\bigcup_{k=0}^n S_{nk}(d)) \ , & 0 < d < 1. \end{cases}$$

Fritsch constructed a cubature formula of degree 3 with $n+2$ points depending on n and d for the region $S_n(d)$. He proved that there exists a $d_n > 1$, a zero of a known polynomial, such that his formula has a zero weight, and thus uses only $n+1$ points. For two dimensions he found two such regions, as shown in Figure 2. He also proved that there exists one $d_n^\star$ for which a formula of the form he looked for does not exist. For two and three dimensions the region $S_n(d_n^\star)$ is centrally symmetric (that is, the region and weight function remain invariant after reflection through the centre) and we will see later that the minimal number of points in a formula of degree 3 for such a region requires $2n$ points.

Mysovskikh and Černicina (1971) constructed a region $\Omega = \Omega_1 \cup \Omega_2$ with

$$\begin{aligned} \Omega_1 &= \{(x,y) : -\tau \leq x \leq \tau \ , \ 0 \leq y \leq e^{-|x|}\}, \\ \Omega_2 &= \{(x,y) : -\sigma \leq x \leq \sigma \ , \ -\epsilon \leq y \leq 0\}, \end{aligned}$$

$$\tau = 3 \ , \ \epsilon \simeq 0.048 \ , \ \sigma \simeq 1.266,$$

for which there exists a cubature formula of degree 5 with 6 points.

Recently, Schmid and Xu (1994) found a two-dimensional region for which formulae with $\dim \mathcal{P}_k^2$ points exist for each degree $2k+1$.

Theorem 8.2 Let $W(u,v) := w(x)w(y)$ with $w(t) := (1-t)^\alpha(1+t)^\beta$ and let

$$\Omega := \{(u,v) : (x,y) \in [-1,1]^2 \ , \ x < y \ , \ u = x+y \ , \ v = xy\}.$$

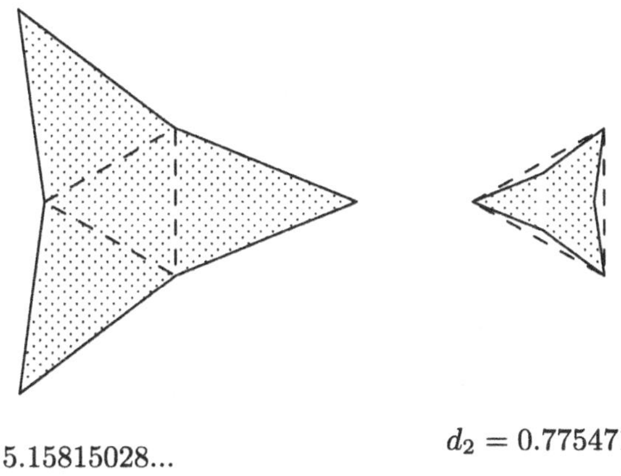

Fig. 2. Regions $S_2(d_2)$ for which a formula of degree 3 with 3 points exists

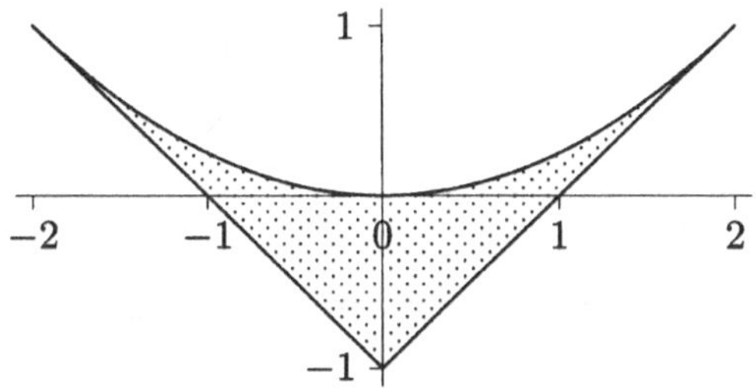

Fig. 3. Region for which a formula of degree $2k+1$ attains the bound of Theorem 7.1

Then there exists an infinite number of minimal cubature formulae of degree $2k$ and one (uniquely determined) minimal formula of degree $2k+1$ (both with $\dim \mathcal{P}_k^2$ points) for the following two classes of integrals,

$$\int_\Omega f(u,v)W(u,v)(u^2-4v)^\gamma \, du \, dv \quad \text{with} \quad \alpha, \beta > -1, \ \gamma = \pm\tfrac{1}{2}.$$

Proof. See Schmid and Xu (1994). □

Figure 3 displays Ω. Berens, Schmid and Xu (1995) obtained a similar result for arbitrary dimensions.

8.3. Improved bounds for centrally symmetric formulae

In Example 5.2 and at the end of Section 5 we encountered the pleasant effect of central symmetry on cubature formulae of algebraic degree. Mysovskikh (1966) showed that for centrally symmetric n-dimensional regions, the minimal number of points in a cubature formula of algebraic degree 3 is $2n$. The construction of such formulae is summarized by Stroud (1971). Möller (1973) generalized this improved lower bound for all odd degrees.

Theorem 8.3 Let R_{2k} denote the vector space of even polynomials of $\mathcal{P}^n_{2k+1}|_\Omega$ and R_{2k+1} denote the vector space of odd polynomials of $\mathcal{P}^n_{2k+1}|_\Omega$, $k \in \mathbb{N}_0$. If the algebraic degree of the cubature formula (7.1) for a centrally symmetric integral is $d = 2k+1$, then

$$\begin{aligned} N &\geq 2 \dim R_k - 1, && \text{if } k \text{ even and } 0 \text{ is a point,} \\ N &\geq 2 \dim R_k, && \text{otherwise.} \end{aligned}$$

A cubature formula that attains this bound is centrally symmetric and has all weights positive.

Proof. See Möller (1973) for the case where Ω has interior points and Möller (1979) for the general case. □

A similar result holds for cubature formulae of trigonometric degree.

Theorem 8.4 Let $R_k \subset \mathcal{T}^n_k$ denote the vector space of polynomials whose degree has the same parity as k. If the trigonometric degree of the cubature formula (7.1) for an integral over C^*_n is $d = 2k+1$, then

$$N \geq 2 \dim R_k.$$

Proof. See Mysovskikh (1988). □

A nice result about the weights was obtained using the reproducing kernel.

Theorem 8.5 A cubature formula that attains the bound of Theorem 8.4 has all weights equal to $1/N$.

Proof. See Beckers and Cools (1993). □

To illustrate my belief in the similarities between the algebraic degree case and the trigonometric degree case, as well as the similarities between central-symmetry and shift symmetry, I dare to pose the following conjecture.

Conjecture 8.1 Any cubature formula attaining the bound of Theorem 8.4 is shift symmetric.

How good are the lower bounds of Theorems 8.3 and 8.4? For two dimensions it is now known that these bounds are the best possible if further information on the integral is not available.

For the regions $C_2^{0.5}$ and $C_2^{-0.5}$, cubature formulae attaining the lower bound of Theorem 8.3 exist for arbitrary odd degree (Cools and Schmid 1989). In Table 3, we list the known minimal formulae for standard regions. For some regions, for instance S_2 and $E_2^{r^2}$, it has been proved that the bound of Theorem 8.3 cannot be attained for degrees $4k+1$, $k > 1$. For C_2, it is known that a cubature formula of degree 13 with 31 points cannot exist. For these regions at least one additional point is required (Verlinden and Cools 1992, Cools and Schmid 1993).

Table 3. *Minimal formulae of odd algebraic degree*

n	d	N	references C_n	S_n	$E_n^{r^2}$	E_n^r
2	3	4	[1]	[1]		
	5	7	[1]	[1]		
	7	12	[1]*	[1]	[1]	[1,4]
	9	17(18)	[3]	[2]	[4]	
	11	24	[5]			
3	3	6	[1]			
	5	13	[1]*	[1]	[1]	[1]

[1] = Stroud (1971), [2] = Piessens and Haegemans (1975), [3] = Möller (1976), [4] = Haegemans and Piessens (1977), [5] = Cools and Haegemans (1988*a*),
* = Many known formulae; see also Cools and Rabinowitz (1993).

Table 4. *Minimal formulae of odd trigonometric degree for* $C_n^\star$

n	d	N	references
n	1	2	Mysovskikh (1988)
	3	$4n$	Noskov (1988*a*)
2	d	$\frac{(d+1)^2}{2}$	Beckers and Cools (1993)
			Cools and Sloan (1996)
3	5	38	Frolov (1977)

For $C_2^\star$, cubature formulae attaining the lower bound of Theorem 8.4 exist for arbitrary odd degree (Cools and Sloan 1996). In Table 4, we list the known minimal formulae for $C_n^\star$.

8.4. An improved general bound for odd degrees

We will now present a lower bound especially derived for odd algebraic degrees, $d = 2k + 1$, without any assumptions on the symmetry of the region. Let

$$O_{k+1} := \{f \in \mathcal{P}_{k+1}^n : g \in \mathcal{P}_k^n \Rightarrow I[fg] = 0\},$$

Table 5. *Minimal formulae of odd algebraic degree for T_n*

n	d	N	references
2	3	4	Stroud (1971), Hillion (1977)
	5	7	Stroud (1971)
	7	12	Gatermann (1988), Becker (1987)
n	3	$n+2$	Stroud (1971)
4	3	6	Stroud (1971), Grundmann and Möller (1978), de Doncker (1979)

and define for arbitrary $l \in \{2, \ldots, n\}$

$$\begin{aligned}\gamma_l &:= \dim\left\{(f_1,\ldots,f_l) \in O_{k+1}^l \,\middle|\, \sum_{i=1}^{l} x_i f_i \in \mathcal{P}_{k+1}^n\right\}, \\ &- \dim\left\{(f_1,\ldots,f_l) \in O_{k+1}^l \,\middle|\, \sum_{i=1}^{l} x_i f_i \in O_{k+1}\right\}.\end{aligned}$$

Theorem 8.6 If a cubature formula has algebraic degree $2k+1$, then $N \geq \dim \mathcal{P}_k^n + \frac{\gamma_l}{l}$.

Proof. See Möller (1976) ($l = 2$, Ω with interior points), and Möller (1979). □

For two dimensions, the bounds of Theorems 8.3 and 8.6 coincide for centrally symmetric integrals:

$$N \geq \frac{(k+1)(k+2)}{2} + \left\lfloor \frac{k+1}{2} \right\rfloor. \tag{8.1}$$

For more than two dimensions, Theorem 8.3 gives a higher lower bound than Theorem 8.6 for centrally symmetric integrals. Theorem 8.6 was applied by Möller to the triangle T_2. He obtained (8.1) for $0 \leq k \leq 5$. Rasputin (1983*b*) generalized this to all k. Berens and Schmid (1992) proved that the same lower bound is obtained for some non-constant weight functions. In addition, Möller (1976) obtained the following results for T_n: for $k = 1$, $\gamma_2 = 2$ and for $k = 2$, $\gamma_2 = 2n - 2$. In Table 5 the known minimal formulae for this region are listed.

8.5. The quality of lower bounds

In this section we have presented the best known lower bounds for the number of points in cubature formulae of odd degree. We gave examples showing that these bounds can be attained for some regions. If one looks at Tables 3, 4 and 5, the results for standard regions look meagre: minimal formulae, that is, formulae attaining a known lower bound, are known only for low dimensions and low degrees. It is likely, but not certain, that these bounds are too low for standard regions and for higher degrees or dimensions. This uncertainty is one of the main problems in the construction of cubature formulae and the construction methods based on characterizing polynomials suffer from it, as we shall see in the following section.

9. Constructing cubature formulae using ideal theory

9.1. A bird's-eye view

Orthogonal polynomials were already being used by Appell (1890) and Radon (1948) to construct cubature formulae of algebraic degree for two-dimensional integrals. Radon tried without success to construct cubature formulae of degree 5 with 6 points. He constructed formulae of degree 5 with 7 points using the common zeros of three orthogonal polynomials. His work marked a starting point of a theory. During the 1960s, Stroud and Mysovskikh studied the relation between orthogonal polynomials and cubature formulae for n-dimensional integrals. In the mid-1970s, many new, mainly symmetric, cubature formulae were obtained using the common zeros of three orthogonal polynomials in two and three variables; see, for instance, Piessens and Haegemans (1975), Haegemans and Piessens (1976), Haegemans and Piessens (1977), Haegemans (1982). The theoretical results were put in the framework of ideal theory by Möller (1973). Methods to construct cubature formulae based on these and other theoretical achievements were derived by Morrow and Patterson (1978), Schmid (1980*b*) and Cools and Haegemans (1987*b*), amongst others.

We mentioned that one can also work with ideals of invariant polynomials. Gatermann (1992) combined ideal theory with the theory of linear representations of finite groups.

We will now present two successful methods to construct cubature formulae of algebraic degree. In order not to over-complicate everything, we restrict this to two dimensions.

9.2. The T-method

A starting point in Theorem 6.8 is that the ideal $\mathfrak{A}$ is fundamental of degree $d+1$. In general, $\mathfrak{A}$ will be fundamental of degree $l, l+1, \ldots$ where $\lfloor d/2 \rfloor + 1 \leq l \leq d+1$. Let m be such that $\mathfrak{A}$ is fundamental of degree m, but is not fundamental of degree $m-1$. One can try to determine a set of polynomials of degree m that form a basis of an ideal satisfying the conditions of Theorem 6.8. This idea was first suggested by Morrow and Patterson (1978) and Schmid (1978) for two-dimensional regions. It was further developed by Schmid (1980*a*); see also Schmid (1980*b*) and Schmid (1995).

Consider the case where the ideal $\mathfrak{A}$ associated with a cubature formula of degree $2k-1$ is fundamental of degree $k+1$. Let $R_0, \ldots, R_{k+1}$ be linearly independent polynomials of degree $k+1$ in two variables, x and y. These polynomials are orthogonal to all polynomials of degree $k-2$ if they vanish at the points of a cubature formula of degree $2k-1$. Thus the R_is can be written as

$$R_i = P^{k+1-i,i} + \sum_{j=0}^{k} \beta_{ij} P^{k-j,j} + \sum_{j=0}^{k-1} \gamma_{ij} P^{k-1-j,j}, \quad i = 0, \ldots, k+1,$$

where the $P^{a,b}$ are the basic orthogonal polynomials (6.3). The β_{ij} and γ_{ij} are parameters which have to be determined such that the R_is belong to an ideal $\mathfrak{A}$ satisfying the conditions of Theorem 6.8. When the integral is centrally symmetric, the basic orthogonal polynomials have a special form and the β_{ij} vanish.

The construction is based on the following observations.

- Let $Q_i := yR_i - xR_{i+1}$, $i = 0, \ldots, k$. Then Q_i is a polynomial of degree k and Q_i has to be orthogonal.
- The polynomials $xQ_i, yQ_i, i = 0, \ldots, k$, are of degree $k+1$ and they belong to $\mathfrak{A}$. Thus $xQ_i, yQ_i \in \operatorname{span}\{R_0, \ldots, R_{k+1}\}$.

Both conditions lead to necessary conditions: linear and quadratic equations in the γ_{ij}s. Starting from the explicit expressions for the basic orthogonal polynomials, a computer algebra system can be programmed to derive these equations. The linear equations can then be used to reduce the number of unknowns in the system of quadratic equations. In the resulting system the number of equations and unknowns is usually different. More recently, Schmid (1995) worked this out in detail using matrix equations.

The inequality in Theorem 6.8 translates into inequalities for the γ_{ij}. These inequalities together with the linear and quadratic equations give necessary and sufficient conditions for the γ_{ij}s so that all conditions of Theorem 6.8 are satisfied. Schmid (1983) used this method to construct cubature formulae of degree ≤ 9 for C_2^{α}. Cools and Schmid (1989) used it to construct formulae of arbitrary odd degree for $C_2^{-0.5}$ and $C_2^{0.5}$.

We will now prove, using G-bases, that the above method works. A similar proof for the n-dimensional case is given by Möller (1987).

Theorem 9.1 Let

$$\begin{aligned} R_i &= P^{k+1-i,i} + \textstyle\sum_{j=0}^{k-1} \gamma_{ij} P^{k-1-j,j}, \qquad & j &= 0, \ldots, k+1, \\ Q_i &= yR_i - xR_{i+1}, & i &= 0, \ldots, k. \end{aligned} \tag{9.1}$$

If the polynomials Q_i are $(2k-1)$-orthogonal and if all polynomials xQ_i, yQ_i are elements of $\operatorname{span}\{R_0, \ldots, R_{k+1}\}$, then $F := \{R_0, \ldots, R_{k+1}, Q_0, \ldots, Q_k\}$ is a G-basis.

Proof. We use the term ordering $1 < y < x < y^2 < xy < x^2 < \ldots$, apply Theorem 6.3 and distinguish three cases.

Case 1: (R_i, R_j), $i, j = 0, \ldots, k+1$.
$H(R_l, R_j) = x^{k+1-l}y^j$. This is a divisor of $H(R_i, R_j) = x^{k+1-i}y^j$ for $i < j$, if $i < l$. Hence the pair (R_i, R_j) is superfluous if there exists a l such that $i < l < j$. Therefore we only have to check pairs (R_i, R_{i+1}). But

$$\begin{aligned} SP(R_i, R_{i+1}) &= \frac{H(R_i, R_{i+1})}{M(R_i)} R_i - \frac{H(R_i, R_{i+1})}{M(R_{i+1})} R_{i+1} \\ &= \frac{x^{k+1-i}y^{i+1}}{x^{k+1-i}y^i} R_i - \frac{x^{k+1-i}y^{i+1}}{x^{k-i}y^{i+1}} R_{i+1} \\ &= yR_i - xR_{i+1} \\ &= Q_i, \end{aligned}$$

and thus $SP(R_i, R_{i+1}) \xrightarrow[F]{}^{+} 0$.

Case 2: (Q_i, Q_j), $i, j = 0, \ldots, k$.
If $\mathrm{Hterm}(Q_i) = \mathrm{Hterm}(Q_j)$ then $SP(Q_i, Q_j) \in \mathrm{span}\{Q_i\}$ and thus $SP(Q_i, Q_j) \xrightarrow[F]{}^{+} 0$.
If $\mathrm{Hterm}(Q_i) \neq \mathrm{Hterm}(Q_j)$ then there exist $u, v \in \{x, y\}$ for which $SP(Q_i, Q_j) = SP(uQ_i, vQ_j)$.
Because $xQ_i, yQ_i \in \mathrm{span}\{R_0, \ldots R_{k+1}\}$ this reduces to Case 1.

Case 3: (R_i, Q_j), $i = 0, \ldots, k+1$, $j = 0, \ldots, k$.
One can always find a $u \in \{x, y\}$ such that $SP(R_i, Q_j) = SP(R_i, uQ_j)$. Since x, Q_i, yQ_i are in $\mathrm{span}\{R_0, \ldots, R_{k+1}\}$, this reduces to Case 1.

□

Theorem 9.2 Let F be as defined in Theorem 9.1. If the common zeros of the polynomials in F are real and simple, then there exists a cubature formula of degree $2k-1$ with the elements of $\mathsf{NG}(F)$ as points. The number of points $N \leq \frac{k(k+3)}{2}$.

Proof. The ordering used in the proof of Theorem 9.1 is compatible with the partial ordering by degree. According to Theorem 6.4, F is thus an H-basis. Theorem 6.7 then guarantees the existence of the cubature formula.

An upper bound for the number of points in the cubature formula is given by the Hilbert function. Because F is fundamental of degree $k+1$,

$$\begin{aligned} \mathcal{H}(2k-1, F) &= \mathcal{H}(k, F) \\ &= \dim \mathcal{P}_k^2 - \dim(\mathcal{P}_k^2 \cap F) \\ &= \dim \mathcal{P}_{k-1}^2 + \operatorname{codim}(F \cap \mathcal{P}_k^2). \end{aligned}$$

There will be at least one polynomial Q_i, hence $\operatorname{codim}(F \cap \mathcal{P}_k^2) \le k$. Thus $N \le \frac{k(k+1)}{2} + k$. □

The upper bound of Theorem 9.2 is very weak. A tighter result is known.

Theorem 9.3 If the ideal of all polynomials that vanish at the N points of a cubature formula of degree $2k-1$ contains a fundamental set of degree $k+1$, then

$$\frac{k(k+1)}{2} + \left\lfloor \frac{k}{2} \right\rfloor \le N \le \frac{k(k+1)}{2} + \left\lfloor \frac{k}{2} \right\rfloor + 1.$$

Proof. See Cools (1989) or Schmid (1995). □

This clearly shows that the success of this method strongly depends on the quality of the lower bound (8.1) for the particular integral for which a cubature formula is wanted.

If the lower bound (8.1) underestimates the real minimal number of points by more than one, then the method is useless. At the moment it looks as if this is the case for $d \ge 15$ for the regions C_2, S_2 $E_2^{r^2}$ and E_2^r. The known exceptions are $C_2^{0.5}$ and $C_2^{-0.5}$.

9.3. The S-method

The S-method was suggested by Cools and Haegemans (1987*b*) in an attempt to find a method that is less dependent on the lower bound (8.1) than the T-method. If the T-method is used to construct symmetric cubature formulae for a two-dimensional symmetric integral, then $\gamma_{ij} = 0$ if $i+j$ is odd, in the polynomials R_i (9.1). The polynomials R_i can be divided into two sets: $A := \{R_i : i \text{ is even}\}$ and $B := \{R_i : i \text{ is odd}\}$. Instead of demanding that $(A \cup B) \subset \mathfrak{A}$, as in the T-method, we demand that $A \subset \mathfrak{A}$ or $B \subset \mathfrak{A}$. We assign $C := A$ and $q := 0$ if we want to investigate the case $A \subset \mathfrak{A}$. We assign $C := B$ and $q := 1$ if we want to investigate the case $B \subset \mathfrak{A}$. The S-method is based on the following observations.

- Let $S_i := y^2 R_i - x^2 R_{i+2}$, $i = q, q+2, \ldots, k-1$. Then S_i is a polynomial of degree $k+1$ and S_i must be orthogonal to all polynomials of degree $k-2$.
- Because S_i has degree $k+1$, $S_i \in \operatorname{span}(C)$.

Both conditions lead to necessary conditions for γ_{ij}: linear and quadratic equations in the γ_{ij}s. In Cools and Haegemans (1988*b*), necessary and sufficient conditions are given for this method, with proofs along the lines of the proof of Theorem 9.1.

The S-method has been used to construct cubature formulae of degree 13 with 36, 35 and 34 points for C_2, S_2, $E_2^{r^2}$, and of degree 17 with 57 points for C_2.

9.4. Evaluation

Orthogonal polynomials and ideal theory are powerful tools for theoretical investigations of cubature formulae. The most complex concepts of ideal theory have only been used to develop construction methods and to prove theorems about cubature formulae. The reader has probably noticed that we do not need the most sophisticated part of ideal theory to construct formulae: operations on vector spaces of polynomials suffice. This is one of the beautiful aspects of ideal theory. The construction methods described require the solution of systems of linear and quadratic equations. These systems are in general smaller than the systems that determine the formulae. One problem for these methods is that they stand or fall with the quality of the lower bounds given in Sections 7 and 8.

10. Constructing cubature formulae using invariant theory

10.1. A bird's-eye view

In this section we will describe how one tries to construct cubature formulae by solving the associated system of nonlinear equations (4.4). Sobolev's theorem plays a very important role: it is essential to limit the size of the nonlinear system by imposing structure on the cubature formulae. It suggests that we look for invariant cubature formulae, that is, solutions of the equations

$$Q[\phi_i] = I[\phi_i], \quad i = 1, \ldots, \dim \mathcal{P}_d^n(G), \tag{10.1}$$

where the ϕ_i form a basis for the space of G-invariant polynomials $\mathcal{P}_d^n(G)$.

The idea of demanding that a cubature formula has the same symmetries as the given integral is as old as the construction of cubature formulae itself. Indeed, when Maxwell (1877) constructed cubature formulae for the square and the cube, he considered only cubature formulae that are invariant with respect to the groups of symmetries of these regions, that is, G_{FS}.

There is no reason why a cubature formula should have the same structure as the integral. (What should a formula for a circle look like?) Cubature formulae that are invariant with respect to a subgroup of the symmetry group of the integral were already obtained by Radon (1948). His formula for C_2 is symmetric, that is, G_s-invariant, and his formula for S_2 has the origin and the vertices of a regular hexagon as points, that is, H_2^6-invariant.

Russian researchers, aware of Sobolev's result, applied the tools of invariant theory to construct cubature formulae invariant with respect to the symmetry groups of regular polytopes A_n, B_n and I_3 and the extension group $A_n^\star$. Notable results are those of Lebedev (1976) for U_n (see also Lebedev and Skorokhodov (1992) and Lebedev (1995)) and Konjaev (1977) for S_3, $E_2^{r^2}$ and E_2^r.

Table 6. *Different types of H_2^4-orbit*

type	generator	number of unknowns	number of points in an orbit	unknowns
0	(0,0)	1	1	weight
1	(a,0)	2	4	a, weight
2	(a,a)	2	4	a, weight
3	(a,b)	3	8	a, b, weight

Western researchers also considered subgroups without using the general theory. They realized that imposing too much structure on a formula prohibits attaining the minimal number of points. For instance, a fully symmetric formula for C_2 of degree 9 requires 20 points, a symmetric formula 18, but a rotational invariant (R_4) cubature formula requires 17, and this is minimal. Humans seem to have a preference for certain symmetries. Symmetry with respect to the axes (G_s) is studied regularly but symmetry with respect to the diagonals has been used only recently. The symmetry groups are nevertheless isomorphic. Rotational symmetries turned up unexpectedly in Möller (1976) and were later used to construct some other minimal formulae (Cools and Haegemans 1988*a*).

We will now present the consistency conditions approach to constructing fully symmetric cubature formulae. For simplicity, we again restrict ourselves to two dimensions.

10.2. Consistency conditions and fully symmetric regions

In this section, fully symmetric cubature formulae for two-dimensional integrals will be considered. The symmetry group is the dihedral group $H_2^4 = B_2 = G_{FS}$. In Example 5.1 it was shown that not all orbits have the same number of points. Each orbit in an invariant cubature formula introduces a number of unknowns in the nonlinear equations (5.2) and gives a number of points in the cubature formula (5.1). The role of the different types of orbit is described in Table 6.

Let K_i be the number of orbits of type i in an invariant cubature formula. One does not expect a solution of a system of nonlinear equations if there are more equations than unknowns. The previous sentence is the foundation upon which all work in this area is based. It sounds very reasonable but it also incorporates the weakness of this approach.

Rabinowitz and Richter (1969) introduced the notion of *consistency conditions*. A consistency condition is an inequality for the K_i that must be satisfied in order to obtain a system of nonlinear equations where the number of unknowns is greater than or equal to the number of equations in each subsystem. Cubature formulae that do not satisfy the consistency conditions are called 'fortuitous' and are thought to be rare.

We encountered basic invariant polynomials for H_2^4 in Example 5.3:

$$\sigma_2 := x^2 + y^2 \quad \text{and} \quad \sigma_4 := x^4 - 6x^2y^2 + y^4.$$

For this particular group it is more common to use $\phi_1 := x^2 + y^2$ and $\phi_2 := x^2y^2$ as basic invariant polynomials.

Demanding that the number of unknowns exceeds the number of equations gives a first consistency condition:

$$K_0 + 2K_1 + 2K_2 + 3K_3 \geq \dim \mathcal{P}_d^2(G_{FS}). \tag{10.2}$$

For $d = 2k+1$, $\dim \mathcal{P}_d^2(G_{FS}) = 1 + k + \lfloor \frac{k^2}{4} \rfloor$.

A cubature formula that is exact for ϕ_2 cannot use orbits of types 0 and 1 only, because such orbits have a zero contribution in a G_{FS}-invariant cubature formula. Thus, to integrate the polynomials

$$\phi_2(\phi_1^i \phi_2^j) \quad \text{for all } i, j : 0 \leq 2i + 4j \leq d - 4,$$

orbits of types 2 and 3 are needed. So we obtain the second consistency condition:

$$2K_2 + 3K_3 \geq \dim \mathcal{P}_{d-4}^2(G_{FS}). \tag{10.3}$$

A cubature formula that is exact for $(x-y)^2(x+y)^2 = \phi_1^2 - 4\phi_2$ cannot use orbits of types 0 and 2 only, for the same reasons as in the previous case. Analogously, the third consistency condition is obtained:

$$2K_1 + 3K_3 \geq \dim \mathcal{P}_{d-4}^2(G_{FS}). \tag{10.4}$$

A cubature formula that is exact for $x^2y^2(x-y)^2(x+y)^2 = \phi_1^2\phi_2 - 4\phi_2^2$ must use orbits of type 3 because all other orbits have a zero contribution. Thus, to integrate the polynomials

$$(\phi_1^2\phi_2 - 4\phi_2^2)(\phi_1^i \phi_2^j) \quad \text{for all } i, j : 0 \leq 2i + 4j \leq d - 8,$$

orbits of type 3 are needed. From this follows the fourth consistency condition:

$$3K_3 \geq \dim \mathcal{P}_{d-8}^2(G_{FS}). \tag{10.5}$$

The final consistency condition is that there can be only one orbit of type 0:

$$K_0 \leq 1. \tag{10.6}$$

The above consistency conditions were first derived by Mantel and Rabinowitz (1977).

If the structure of a cubature formula with $N = K_0 + 4K_1 + 4K_2 + 8K_3$ points satisfies the consistency conditions (10.2), (10.3), (10.4), (10.5) and

(10.6), the system of nonlinear equations (10.1), as well as each subsystem, has a number of unknowns that exceeds the number of equations and that looks promising to those interested in a solution of such a system. However, appearances can be deceptive.

The construction of a cubature formula with the lowest possible number of points requires two steps.

(1) Solve the integer programming problem:

$$\text{minimize } N(K_i : i = 0, 1, \ldots),$$

where the integers K_i satisfy the consistency conditions.

(2) Solve the system of polynomial equations (10.1). If no solution of the polynomial equations is found, then another (non-optimal) solution of the consistency conditions must be tried.

Example 10.1 For a fully symmetric formula of degree 7, the consistency conditions become

$$\begin{aligned} K_0 + 2K_1 + 2K_2 + 3K_3 &\geq 6, \\ 2K_2 + 3K_3 &\geq 2, \\ 2K_1 + 3K_3 &\geq 2, \\ 3K_3 &\geq 0, \\ K_0 &\leq 1. \end{aligned}$$

Optimal solutions are $[K_0, K_1, K_2, K_3] = [0, 1, 2, 0]$, and $[0, 2, 1, 0]$. (Optimal solutions are not necessarily unique!) This second structure corresponds to a cubature formula of the form

$$\begin{aligned} Q[f] = {} & w_1(f(x_1, 0) + f(-x_1, 0) + f(0, x_1) + f(0, -x_1)) \\ & + w_2(f(x_2, 0) + f(-x_2, 0) + f(0, x_2) + f(0, -x_2)) \\ & + w_3(f(x_3, x_3) + f(-x_3, x_3) + f(x_3, -x_3) + f(-x_3, -x_3)). \end{aligned}$$

The system of nonlinear equations (5.2) for this case is

$$\begin{cases} 4w_3\phi_2(x_3, x_3) & = 4w_3x_3^4 & = I[\phi_2], \\ 4w_3\phi_1(x_3, x_3)\phi_2(x_3, x_3) & = 8w_3x_3^6 & = I[\phi_1\phi_2] \end{cases}$$

$$\begin{cases} 4w_1 + 4w_2 & = I[0] - 4w_3, \\ 4w_1\phi_1(x_1, 0) + 4w_2\phi_1(x_2, 0) & = I[\phi_1] - 4w_3\phi_1(x_3, x_3), \\ 4w_1\phi_1^2(x_1, 0) + 4w_2\phi_1^2(x_2, 0) & = I[\phi_1^2] - 4w_3\phi_1^2(x_3, x_3), \\ 4w_1\phi_1^3(x_1, 0) + 4w_2\phi_1^3(x_2, 0) & = I[\phi_1^3] - 4w_3\phi_1^3(x_3, x_3). \end{cases}$$

From the first two equations one determines w_3 and x_3. Then w_1, x_1, w_2, and x_2 follow from the remaining four equations. Both systems have the familiar form of systems that determine a Gauss quadrature problem.

10.3. How to exploit symmetries

Invariant theory is very useful for constructing a system of nonlinear equations that determines a cubature formula with a particular structure. One advantage of imposing a structure is that the number of nonlinear equations is reduced. For instance, a cubature formula of degree 7 for a two-dimensional region is a solution of a system of $\dim \mathcal{P}_7^2 = 36$ equations. A fully symmetric formula of the same degree is determined by 6 equations.

A second advantage is that one can often find a basis for the invariant polynomials such that the equations are easy to solve. Typically, the system of nonlinear equations is split into several smaller subsystems which can be solved sequentially.

A third advantage is that, if the basis is chosen carefully, then each of these subsystems of nonlinear equations can be solved easily, because they have the same form as the systems that determine a quadrature formula.

The success of this approach depends on the selection of a proper basis for the invariant polynomials, and that is definitely more of an art than a science. This is clearly illustrated in Example 10.1. Other nice examples are given by Cools and Haegemans (1987*a*) and Beckers and Haegemans (1991).

10.4. Some critical notes

Consistency conditions can be derived for every structure and dimension. They can help to set up a system of nonlinear equations where in each subsystem the number of unknowns is larger than or equal to the number of equations. See, for example, Lyness and Jespersen (1975), Mantel and Rabinowitz (1977), Keast and Lyness (1979), Cools (1992), and Maeztu and Sainz de la Maza (1995).

In general, the system of nonlinear equations is still too large to be solved completely with currently available tools. One usually has to use an iterative zero finder and must provide very good starting values.

It must be emphasized that consistency conditions are neither sufficient nor necessary conditions. Even if a system of equations has more unknowns than equations, it might not have a real solution. Furthermore, fortuitous cubature formulae are known, for instance the minimal formulae for $C_2^{0.5}$ and $C_2^{-0.5}$.

The success of this approach depends on the quality of the lower bound for the number of points provided by the integer programming problem. For higher degrees and dimensions, many solutions of the consistency conditions exist for which no solutions of the nonlinear equations are known.

Most researchers have studied consistency conditions without worrying about the associated cubature formula. It is often easier to derive these conditions and, at the same time, obtain a system with a special structure that makes it easier to solve it, using the tools from invariant theory. See, for example, Beckers and Haegemans (1991).

Although the foundations of consistency conditions are built on quicksand, it must be said that most known cubature formulae of algebraic degree are obtained this way. In fact, for higher dimensions and higher degrees, only this approach has so far delivered cubature formulae.

11. A never-ending story

Let those patient readers who have borne with me thus far now join with me in looking back. We started from a solid, general theoretical foundation. Almost immediately we restricted our attention to the most common vector spaces, hence limiting consideration to cubature formulae of algebraic and trigonometric degree. We paid attention to lower bounds for the number of points and saw that they can easily be attained in the overall algebraic or trigonometric degree case. Following that, we searched for better bounds that, at least in the two-dimensional case, are attained for the trigonometric degree case. The rest of our time we spent on the most difficult and interesting algebraic degree case and ended with two approaches to constructing such formulae.

From the above, it is clear that solving systems of polynomial equations is very near to our heart. We therefore welcome the survey of Li (1997) in this volume.

Our list of references may seem long, yet it is incomplete. And there is much more to say: see also Engels (1980) and Davis and Rabinowitz (1984), and, if you can wait, Davis, Rabinowitz and Cools (199x). Let me whet your appetite.

A cubature formula is meant to be used to approximate integrals. Users want to have an indication of the accuracy of the approximation. A classical way to obtain an error estimate is to compare several approximations of different degrees of precision. Sequences of embedded cubature formulae help to reduce the burden. Indeed, these have already been investigated. As Cools (1992) incorporates a survey of some of the obtained results, I resist the temptation to elaborate on this subject.

Acknowledgements

The author gratefully acknowledges helpful comments by Professors Ann Haegemans, Dirk Laurie, Philip Rabinowitz and Hans Joachim Schmid on an earlier version of this paper. To the author alone belongs the responsibility for the selection of the material included.

REFERENCES

P. Appell (1890), 'Sur une classe de polynômes à deux variables et le calcul approché des intégrales double', *Ann. Fac. Sci. Univ. Toulouse* **4**, H1–H20.

P. Appell and J. Kampé de Fériet (1926), *Fonctions Hypergéométriques et Hypersphériques - Polynomes d'Hermite*, Gauthier-Villars, Paris.

T. Becker (1987), Konstruktion von interpolatorischen Kubaturformeln mit Anwendungen in der Finit-Element-Methode, PhD thesis, Technische Hochschule Darmstadt.

M. Beckers and R. Cools (1993), A relation between cubature formulae of trigonometric degree and lattice rules, in *Numerical Integration IV* (H. Brass and G. Hämmerlin, eds), Birkhäuser, Basel, pp. 13–24.

M. Beckers and A. Haegemans (1991), 'The construction of three-dimensional invariant cubature formulae', *J. Comput. Appl. Math.* **35**, 109–118.

H. Berens and H. J. Schmid (1992), On the number of nodes of odd degree cubature formulae for integrals with Jacobi weights on a simplex, in *Numerical Integration - Recent Developments, Software and Applications* (T. Espelid and A. Genz, eds), Vol. 357 of *NATO ASI Series C: Math. and Phys. Sciences*, Kluwer, Dordrecht, pp. 37–44.

H. Berens, H. J. Schmid and Y. Xu (1995), 'Multivariate Gaussian cubature formulae', *Arch. Math.* **64**, 26–32.

B. Buchberger (1985), Gröbner bases: An algorithmic method in polynomial ideal theory, in *Progress, Directions and Open Problems in Multidimensional Systems Theory* (N. Bose, ed.), Reidel, Dordrecht, pp. 184–232.

R. Cools (1989), The construction of cubature formulae using invariant theory and ideal theory, PhD thesis, Katholieke Universiteit Leuven.

R. Cools (1992), A survey of methods for constructing cubature formulae, in *Numerical Integration - Recent Developments, Software and Applications* (T. Espelid and A. Genz, eds), Vol. 357 of *NATO ASI Series C: Math. and Phys. Sciences*, Kluwer, Dordrecht, pp. 1–24.

R. Cools and A. Haegemans (1987*a*), 'Automatic computation of knots and weights of cubature formulae for circular symmetric planar regions', *J. Comput. Appl. Math.* **20**, 153–158.

R. Cools and A. Haegemans (1987*b*), 'Construction of fully symmetric cubature formulae of degree $4k-3$ for fully symmetric planar regions', *J. Comput. Appl. Math.* **17**, 173–180.

R. Cools and A. Haegemans (1987*c*), Construction of minimal cubature formulae for the square and the triangle using invariant theory, Report TW 96, Dept. of Computer Science, Katholieke Universiteit Leuven.

R. Cools and A. Haegemans (1988*a*), 'Another step forward in searching for cubature formulae with a minimal number of knots for the square', *Computing* **40**, 139–146.

R. Cools and A. Haegemans (1988*b*), Construction of symmetric cubature formulae with the number of knots (almost) equal to Möller's lower bound, in *Numerical Integration III* (H. Brass and G. Hämmerlin, eds), Birkhäuser, Basel, pp. 25–36.

R. Cools and A. Haegemans (1988*c*), 'Why do so many cubature formulae have so many positive weights?', *BIT* **28**, 792–802.

R. Cools and P. Rabinowitz (1993), 'Monomial cubature rules since 'Stroud': A compilation', *J. Comput. Appl. Math.* **48**, 309–326.

R. Cools and A. Reztsov (1997), 'Different quality indexes for lattice rules', *J. Complexity*. To appear.

R. Cools and H. J. Schmid (1989), 'Minimal cubature formulae of degree $2k-1$ for two classical functionals', *Computing* **43**, 141–157.

R. Cools and H. J. Schmid (1993), A new lower bound for the number of nodes in cubature formulae of degree $4n+1$ for some circularly symmetric integrals, in *Numerical Integration IV* (H. Brass and G. Hämmerlin, eds), Birkhäuser, Basel, pp. 57–66.

R. Cools and I. H. Sloan (1996), 'Minimal cubature formulae of trigonometric degree', *Math. Comp.* **65**, 1583–1600.

P. J. Davis (1967), 'A construction of nonnegative approximate quadratures', *Math. Comp.* **21**, 578–582.

P. J. Davis and P. Rabinowitz (1984), *Methods of Numerical Integration*, Academic, London.

P. J. Davis, P. Rabinowitz and R. Cools (199x), *Methods of Numerical Integration*. Work in progress.

E. de Doncker (1979), 'New Euler–Maclaurin expansions and their application to quadrature over the *s*-dimensional simplex', *Math. Comp.* **33**, 1003–1018.

M. G. Duffy (1982), 'Quadrature over a pyramid or cube of integrands with a singularity at a vertex', *SIAM J. Numer. Anal.* **19**, 1260–1262.

H. M. Edwards (1980), 'The genesis of ideal theory', *Arch. Hist. Exact Sci.* **23**, 321–378.

H. Engels (1980), *Numerical Quadrature and Cubature*, Academic, London.

C. S. Fisher (1967), 'The death of a mathematical theory: a study in the sociology of knowledge', *Arch. Hist. Exact Sci.* **3**, 136–159.

L. Flatto (1978), 'Invariants of finite reflection groups', *Enseign. Math.* **24**, 237–292.

F. N. Fritsch (1970), 'On the existence of regions with minimal third degree integration formulas', *Math. Comp.* **24**, 855–861.

K. K. Frolov (1977), 'On the connection between quadrature formulas and sublattices of the lattice of integral vectors', *Soviet Math. Dokl.* **18**, 37–41.

K. Gatermann (1988), 'The construction of symmetric cubature formulas for the square and the triangle', *Computing* **40**, 229–240.

K. Gatermann (1992), Linear representations of finite groups and the ideal theoretical construction of *G*-invariant cubature formulas, in *Numerical Integration – Recent Developments, Software and Applications* (T. Espelid and A. Genz, eds), Vol. 357 of *NATO ASI Series C: Math. and Phys. Sciences*, Kluwer, Dordrecht, pp. 25–35.

R. Gebauer and H. M. Möller (1988), 'On an installation of Buchberger's algorithm', *J. Symb. Computation* **6**, 275–286.

J. L. Gout and A. Guessab (1986), 'Sur les formules de quadrature numérique à nombre minimal de noeuds d'intégration', *Numer. Math.* **49**, 439–455.

W. Gröbner (1949), *Moderne Algebraische Geometrie*, Springer, Wien.

A. Grundmann and H. M. Möller (1978), 'Invariant integration formulas for the n-simplex by combinatorial methods', *SIAM J. Numer. Anal.* **15**, 282–290.

A. Guessab (1986), 'Cubature formulae which are exact on spaces P, intermediate between P_k and Q_k', *Numer. Math.* **49**, 561–576.

A. Haegemans (1982), Construction of known and new cubature formulas of degree five for three-dimensional symmetric regions, using orthogonal polynomials, in *Numerical Integration*, Birkhäuser, Basel, pp. 119–127.

A. Haegemans and R. Piessens (1976), 'Construction of cubature formulas of degree eleven for symmetric planar regions, using orthogonal polynomials', *Numer. Math.* **25**, 139–148.

A. Haegemans and R. Piessens (1977), 'Construction of cubature formulas of degree seven and nine symmetric planar regions, using orthogonal polynomials', *SIAM J. Numer. Anal.* **14**, 492–508.

D. Hilbert (1890), 'Über die Theorie der algebraischen Formen', *Math. Ann.* **36**, 473–534.

P. Hillion (1977), 'Numerical integration on a triangle', *Internat. J. Numer. Methods Engrg.* **11**, 797–815.

D. Jackson (1936), 'Formal properties of orthogonal polynomials in two variables', *Duke Math. J.* **2**, 423–434.

P. Keast and J. N. Lyness (1979), 'On the structure of fully symmetric multidimensional quadrature rules', *SIAM. J. Numer. Anal.* **16**, 11–29.

J. Kepler (1615), *Nova stereometria doliorum vinariorum, in primis Austriaci, figuræ omnium aptissimæ*, Authore Ioanne Kepplero, imp. Cæs. Matthiæ I. ejusq; fidd. Ordd. Austriæ supra Anasum Mathematico, Lincii, Anno MDCXV.

S. I. Konjaev (1977), 'Ninth-order quadrature formulas invariant with respect to the icosahedral group', *Soviet Math. Dokl.* **18**, 497–501.

N. M. Korobov (1959), 'On approximate calculation of multiple integrals', *Dokl. Akad. Nauk SSSR* **124**, 1207–1210. Russian.

V. I. Lebedev (1976), 'Quadrature on a sphere', *USSR Comput. Math. and Math. Phys.* **16**, 10–24.

V. I. Lebedev (1995), 'A quadrature formula for the sphere of 59th algebraic order of accuracy', *Dokl. Math.* **50**, 283–286.

V. I. Lebedev and A. L. Skorokhodov (1992), 'Quadrature formulas of orders 41, 47, and 53 for the sphere', *Dokl. Math.* **45**, 587–592.

T. Y. Li (1997), Numerical solution of multivariate polynomial systems by homotopy continuation methods, in *Acta Numerica*, Vol. 6, Cambridge University Press, pp. 399–436.

J. N. Lyness (1976), 'An error functional expansion for N-dimensional quadrature with an integrand function singular at a point', *Math. Comp.* **30**, 1–23.

J. N. Lyness (1992), On handling singularities in finite elements, in *Numerical Integration - Recent Developments, Software and Applications* (T. Espelid and A. Genz, eds), Vol. 357 of *NATO ASI Series C: Math. and Phys. Sciences*, Kluwer, Dordrecht, pp. 219–233.

J. N. Lyness and R. Cools (1994), A survey of numerical cubature over triangles, in *Mathematics of Computation 1943–1993: A Half-Century of Computational Mathematics* (W. Gautschi, ed.), Vol. 48 of *Proceedings of Symposia in Applied Mathematics*, AMS, Providence, RI, pp. 127–150.

J. N. Lyness and E. de Doncker (1993), 'Quadrature error expansions, II. The full corner singularity', *Numer. Math.* **64**, 355–370.

J. N. Lyness and E. de Doncker-Kapenga (1987), 'On quadrature error expansions, part I', *J. Comput. Appl. Math.* **17**, 131–149.

J. N. Lyness and D. Jespersen (1975), 'Moderate degree symmetric quadrature rules for the triangle', *J. Inst. Math. Appl.* **15**, 19–32.

J. N. Lyness and B. J. J. McHugh (1970), 'On the remainder term in the N-dimensional Euler–Maclaurin expansion', *Numer. Math.* **15**, 333–344.

J. N. Lyness and G. Monegato (1980), 'Quadrature error functional expansion for the simplex when the integrand has singularities at vertices', *Math. Comp.* **34**, 213–225.

J. N. Lyness and K. K. Puri (1973), 'The Euler–Maclaurin expansion for the simplex', *Math. Comp.* **27**, 273–293.

J. I. Maeztu and E. Sainz de la Maza (1995), 'Consistent structures of invariant quadrature rules for the n-simplex', *Math. Comp.* **64**, 1171–1192.

F. Mantel and P. Rabinowitz (1977), 'The application of integer programming to the computation of fully symmetric integration formulas in two and three dimensions', *SIAM J. Numer. Anal.* **14**, 391–425.

J. C. Maxwell (1877), 'On approximate multiple integration between limits of summation', *Proc. Cambridge Philos. Soc.* **3**, 39–47.

H. M. Möller (1973), Polynomideale und Kubaturformeln, PhD thesis, Universität Dortmund.

H. M. Möller (1976), 'Kubaturformeln mit minimaler Knotenzahl', *Numer. Math.* **25**, 185–200.

H. M. Möller (1979), Lower bounds for the number of nodes in cubature formulae, in *Numerische Integration*, Vol. 45 of *ISNM*, Birkhäuser, Basel, pp. 221–230.

H. M. Möller (1987), On the construction of cubature formulae with few nodes using Gröbner bases, in *Numerical Integration* (P. Keast and G. Fairweather, eds), Reidel, Dordrecht, pp. 177–192.

H. M. Möller and F. Mora (1986), 'New constructive methods in classical ideal theory', *J. Algebra* **100**, 138–178.

C. R. Morrow and T. N. L. Patterson (1978), 'Construction of algebraic cubature rules using polynomial ideal theory', *SIAM J. Numer. Anal.* **15**, 953–976.

I. P. Mysovskikh (1966), 'A proof of minimality of the number of nodes of a cubature formula for a hypersphere', *Zh. Vychisl. Mat. Mat. Fiz.* **6**, 621–630. Russian. Published as I. P. Mysovskih 1966.

I. P. Mysovskikh (1968), 'On the construction of cubature formulas with fewest nodes', *Soviet Math. Dokl.* **9**, 277–280.

I. P. Mysovskikh (1975), 'On Chakalov's theorem', *USSR Comput. Math. and Math. Phys.* **15**, 221–227.

I. P. Mysovskikh (1977), 'On the evaluation of integrals over the surface of a sphere', *Soviet Math. Dokl.* **18**, 925–929. Published as I. P. Mysovskih 1977.

I. P. Mysovskikh (1980), The approximation of multiple integrals by using interpolatory cubature formulae, in *Quantitative Approximation* (R. D. Vore and K. Scherer, eds), Academic, New York, pp. 217–243.

I. P. Mysovskikh (1981), *Interpolatory Cubature Formulas*, Izdat. 'Nauka', Moscow-Leningrad. Russian. See I. P. Mysovskikh 1992.

I. P. Mysovskikh (1988), 'Cubature formulas that are exact for trigonometric polynomials', *Metody Vychisl.* **15**, 7–19. Russian.

I. P. Mysovskikh (1990), On the construction of cubature formulas that are exact for trigonometric polynomials, in *Numerical Analysis and Mathematical Modelling* (A. Wakulicz, ed.), Vol. 24 of *Banach Center Publications*, PWN – Polish Scientific Publishers, Warsaw, pp. 29–38. Russian.

I. P. Mysovskikh (1992), Interpolatorische Kubaturformeln, Bericht Nr. 74, Institut für Geometrie und Praktische Mathematik der RWTH Aachen. Translated from the Russian by I. Dietrich and H. Engels. Published as J. P. Mysovskih 1992.

I. P. Mysovskikh and V. Ja Černicina (1971), 'The answer to a question of Radon', *Soviet Math. Dokl.* **12**, 852–854. Published as I. P. Mysovskih and V. Ja Černicina 1971.

H. Niederreiter (1992), *Random Number Generation and Quasi-Monte Carlo Methods*, Vol. 63 of *CBMS-NSF regional conference series in applied mathematics*, SIAM, Philadelphia.

M. V. Noskov (1988*a*), 'Cubature formulae for the approximate integration of functions of three variables', *USSR Comput. Math. and Math. Phys.* **28**, 200–202.

M. V. Noskov (1988*b*), 'Formulas for the approximate integration of periodic functions', *Metody Vychisl.* **15**, 19–22. Russian.

R. Piessens and A. Haegemans (1975), 'Cubature formulas of degree nine for symmetric planar regions', *Math. Comp.* **29**, 810–815.

P. Rabinowitz and N. Richter (1969), 'Perfectly symmetric two-dimensional integration formulas with minimal number of points', *Math. Comp.* **23**, 765–799.

J. Radon (1948), 'Zur mechanischen Kubatur', *Monatsh. Math.* **52**, 286–300.

G. G. Rasputin (1983*a*), 'On the construction of cubature formulas containing prespecified knots', *Metody Vychisl.* **13**, 122–128. Russian.

G. G. Rasputin (1983*b*), 'On the question of numerical characteristics for orthogonal polynomials of two variables', *Metody Vychisl.* **13**, 145–154. Russian.

G. G. Rasputin (1986), 'Construction of cubature formulas containing preassigned nodes', *Soviet Math. (Iz. VUZ)* **30**, 58–67.

L. F. Richardson (1927), 'The deferred approach to the limit', *Philos. Trans. Royal Soc. London* **226**, 261–299.

R. D. Richtmyer (1952), The evaluation of definite integrals, and a quasi-Monte-Carlo method based on the properties of algebraic numbers, Report LA-1342, Los Alamos Scientific Laboratory.

H. J. Schmid (1978), 'On cubature formulae with a minimal number of knots', *Numer. Math.* **31**, 281–297.

H. J. Schmid (1980*a*), 'Interpolatorische Kubaturformeln und reelle Ideale', *Math. Z.* **170**, 267–282.

H. J. Schmid (1980*b*), Interpolatory cubature formulae and real ideals, in *Quantitative Approximation* (R. D. Vore and K. Scherer, eds), Academic, New York, pp. 245–254.

H. J. Schmid (1983), *Interpolatorische Kubaturformeln*, Vol. CCXX of *Dissertationes Math.*, Polish Scientific Publishers, Warsaw.

H. J. Schmid (1995), 'Two-dimensional minimal cubature formulas and matrix equations', *SIAM J. Matrix Anal.* **16**(3), 898–921.

H. J. Schmid and Y. Xu (1994), 'On bivariate Gaussian cubature formulae', *Proc. Amer. Math. Soc.* **122**, 833–842.

I. H. Sloan and S. Joe (1994), *Lattice Methods for Multiple Integration*, Oxford University Press.

I. H. Sloan and P. J. Kachoyan (1987), 'Lattice mathods for multiple integration: theory, error analysis and examples', *SIAM J. Numer. Anal.* **24**, 116–128.

S. L. Sobolev (1962), 'The formulas of mechanical cubature on the surface of a sphere', *Sibirsk. Mat. Ž.* **3**, 769–796. Russian.

A. H. Stroud (1960), 'Quadrature methods for functions of more than one variable', *New York Acad. Sci.* **86**, 776–791.

A. H. Stroud (1971), *Approximate calculation of multiple integrals*, Prentice-Hall, Englewood Cliffs, NJ.

B. Sturmfels (1996), *Gröbner Bases and Convex Polytopes*, Vol. 8 of *University Lecture Series*, AMS, Providence, RI.

M. Taylor (1995), 'Cubature for the sphere and the discrete spherical harmonic transform', *SIAM J. Numer. Anal.* **32**(2), 667–670.

V. Tchakaloff (1957), 'Formules de cubatures mécaniques à coefficients non négatifs', *Bull. des Sciences Math., 2ᵉ série* **81**, 123–134.

P. Verlinden and R. Cools (1992), 'On cubature formulae of degree $4k+1$ attaining Möller's lower bound for integrals with circular symmetry', *Numer. Math.* **61**, 395–407.

P. Verlinden and A. Haegemans (1993), 'An error expansion for cubature with an integrand with homogeneous boundary singularities', *Numer. Math.* **65**, 383–406.

S. Weiß (1991), Über Kubaturformeln vom Grad $2k-2$, Master's thesis, Universität Erlangen.

J. W. Wissman and T. Becker (1986), 'Partially symmetric cubature formulas for even degrees of exactness', *SIAM J. Numer. Anal.* **23**, 676–685.

Acta Numerica (1997), *pp.* 55–228

Wavelet and multiscale methods for operator equations

Wolfgang Dahmen
Institut für Geometrie und Praktische Mathematik
RWTH Aachen
Templergraben 55, 52056 Aachen, Germany
E-mail: dahmen@igpm.rwth-aachen.de

CONTENTS

1. Introduction

More than anything else, the increase of computing power seems to stimulate the greed for tackling ever larger problems involving large-scale numerical simulation. As a consequence, the need for understanding something like the *intrinsic complexity* of a problem occupies a more and more pivotal position. Moreover, computability often only becomes feasible if an algorithm can be found that is *asymptotically optimal.* This means that storage and the number of floating point operations needed to resolve the problem with desired accuracy remain *proportional* to the problem size when the resolution of the discretization is refined. A significant reduction of complexity is indeed often possible, when the underlying problem admits a continuous model in terms of differential or integral equations. The physical phenomena behind such a model usually exhibit characteristic features over a wide range of scales. Accordingly, the most successful numerical schemes exploit in one way or another the interaction of different scales of discretization. A very prominent representative is the *multigrid* methodology; see, for instance, Hackbusch (1985) and Bramble (1993). In a way it has caused a breakthrough in numerical analysis since, in an important range of cases, it does indeed provide asymptotically optimal schemes. For closely related multilevel techniques and a unified treatment of several variants, such as *multiplicative* or *additive subspace correction* methods, see Bramble, Pasciak and Xu (1990), Oswald (1994), Xu (1992), and Yserentant (1993). Although there remain many unresolved problems, multigrid or multilevel schemes in the classical framework of finite difference and finite element discretizations exhibit by now a comparatively clear profile. They are particularly powerful for elliptic and parabolic problems.

1.1. Is there any vision?

Much more recently, the wavelet concept has (at least initially) raised high expectations. Traditional primary applications of wavelets have been signal analysis/processing, image processing/compression, *etc.* What are the reasons for the recent explosion of activities centred upon wavelets in connection with the numerical treatment of PDEs? Of course, anything that looks new inspires curiosity: there is certainly a bandwagon effect. Also, mathematical beauty plays a role. Perhaps it is just a fashionable new wave that will soon come to rest. In any case, comparisons of wavelet methods with conventional schemes should help in finding an answer. However, it is not that simple. First of all, the picture of wavelet concepts appears to be still quite fuzzy for several reasons. On one hand, at the present stage there simply do not yet exist complete software packages for complex real life problems, which would admit fair performance comparisons. On the other hand, the development of concepts and ideas is still far from steady state.

To find a reasonable path through the jungle, it is therefore worth spending some time on what could actually be expected.

First of all, is there any need to look at alternatives to multigrid? Of course, curiosity is a perfect reason. But looking again more closely at the multigrid methodology, its performance is best understood with respect to *uniform mesh refinements* and asymptotic optimality in the above sense refers to such settings. However, a fully refined mesh may not be necessary to resolve sufficiently the desired solution. To avoid this potential waste, *adaptive* techniques have to be and are indeed employed. There are many possibilities ranging from *a priori* local mesh refinements to fully self-adaptive schemes. In this regard, problems of a different type are encountered. On one hand, we need a thorough analysis to control the local refinement steps. The corresponding local information is usually *implicit*: it is derived by comparing different discretizations. At this point some heuristics usually enter. On the other hand, the mesh refinements cause *geometrical* problems that have nothing to do with the underlying problem. Overall, these matters appear to be somewhat better understood for the *additive* version (Oswald 1994) (which by the way is closer to the wavelet concept), while otherwise the multiplicative version is often more efficient. So there still appears to be a strong need for a better understanding of adaptivity in this context, both with regard to the underlying analytical concepts and to the corresponding data structures.

On the other hand, by their very nature, wavelet representations have a naturally built-in adaptivity through their ability to directly express and separate components living on different scales. This, combined with the fact that many operators and their inverses have (nearly) sparse representations in wavelet coordinates, may eventually lead to competitive or even superior schemes with regard to the following goal: keep the computational work proportional to the number of *significant* terms in the wavelet expansion of the searched object, which in some sense should reflect its intrinsic complexity; see, for example, Beylkin and Keiser (1997) and Dahlke, Dahmen, Hochmuth and Schneider (1997*b*).

The potential of this point of view will be one of the main themes of subsequent discussions. Wavelets are in some sense much more sophisticated tools than conventional discretizations. It will be seen that this also facilitates a refined analysis. One central objective of this paper is to highlight some of the underlying driving analytical mechanisms.

The price of a powerful tool is the effort required to construct and understand it. Its successful application hinges on the realization of a number of requirements. Some space has to be reserved for a clear identification of these requirements as well as for their realization. This is also particularly important for understanding the severe *obstructions* that keep us at present from readily materializing all the principally promising perspectives.

These obstructions are to a great extent related to constraints imposed by domain geometries. There may be a good chance to reduce many problems to a *periodized* one (by an additional separate treatment of boundary conditions). In the periodic case *ideal wavelets* are available. Nevertheless, there will still remain important problem classes for which this strategy does not work. Therefore I will deviate from the usual way of motivating and developing wavelet concepts by means of Fourier analysis. Instead, some effort will be spent on formulating a sufficiently general and flexible framework of multiresolution decompositions that can host a variety of specializations. Moreover, appropriate substitutes for the Fourier tools have to be developed.

Wavelets are traditionally associated with *orthonormal* bases. A closer look reveals that orthogonality is often convenient but *not essential.* In presenting the material I will deviate sometimes from the original sources by formulating things in the more flexible context of *biorthogonality.* While this often supports *locality* and helps to bring out what really *is* essential, it will also be seen sometimes to be simply better, and even offers interesting new combinations of different concepts.

Of course, the acceptance of new concepts increases with their practical success. Somehow the measure is set by the existing modern multigrid techniques. The competition between different methodologies can be very stimulating. It should not be the primary point of view though. I personally believe that the additional insight gained from different, yet related, concepts will be mutually beneficial. Perhaps at a later stage, a marriage of complementary components and an enriched supply of tools will lead to true improvements.

As mentioned before, the presentation of material will necessarily be very selective. The selection criteria will not include optimal performance in existing algorithms, but will instead attempt to bring out ideas and concepts that bear some potential for future developments or, on the other hand, explain inherent limitations. Last but not least, my ignorance is to blame. I apologize to all those whose contributions do not get a proper share.

I shall next briefly discuss some simple examples in connection with admittedly trivial problems. Their purpose is only to help in identifying a few characteristic features that will then serve as a guideline for subsequent developments.

1.2. The Haar basis

The scaled shifts

$$\phi_{j,k} = 2^{j/2}\phi\left(2^j \cdot -k\right), \quad k = 0, \dots 2^j - 1,$$

of the *box function*

$$\phi(x) = \begin{cases} 1, & 0 \le x \le 1, \\ 0, & \text{else}, \end{cases} \tag{1.1}$$

form an *orthonormal basis* of their linear span S_j relative to the standard inner product $\langle f, g\rangle = \langle f, g\rangle_{[0,1]} := \int_0^1 f(x)g(x)\,dx$. Since

$$\phi(x) = \phi(2x) + \phi(2x-1),$$

so that

$$\phi_{j,k} = \frac{1}{\sqrt{2}}(\phi_{j+1,2k} + \phi_{j+1,2k+1}), \tag{1.2}$$

the S_j are nested and the closure of their union relative to $\|\cdot\|_{L_2([0,1])} := \langle\cdot,\cdot\rangle^{1/2}$ is all of $L_2([0,1])$ (the space of square integrable functions on $[0,1]$). Thus, denoting by $P_j f$ the orthogonal projection of f onto S_j, one has the representation

$$f = P_0 f + \sum_{j=1}^{\infty}(P_j - P_{j-1})f.$$

The components $(P_j - P_{j-1})f$ represent the *'detail'* added to a given approximation when progressing to the next higher scale of discretization. In the present situation it can be conveniently encoded by the functions

$$\psi_{j,k} := \frac{1}{\sqrt{2}}(\phi_{j+1,2k} - \phi_{j+1,2k+1}), \quad k = 0,\ldots,2^j - 1, \tag{1.3}$$

where, as before, $\psi_{j,k} := 2^{j/2}\psi(2^j\cdot -k)$, and $\psi(x) := \phi(2x) - \phi(2x-1)$. In fact, one easily verifies that

$$\langle\phi(\cdot-k),\psi(\cdot-l)\rangle = 0, \quad \langle\psi(\cdot-k),\psi(\cdot-l)\rangle = \delta_{k,l}, \quad k,l = 0,\ldots,2^j-1, \tag{1.4}$$

so that

$$\langle\psi_{j,k},\psi_{n,l}\rangle = \delta_{j,n}\delta_{k,l}. \tag{1.5}$$

Thus

$$\Psi := \{\phi\} \cup \{\psi_{j,k} : k = 0,\ldots,2^j - 1,\ j = 0,1,2,\ldots\} \tag{1.6}$$

constitutes an *orthonormal basis* for $L_2([0,1])$ and every $f \in L_2([0,1])$ has a unique expansion

$$f = \sum_{\psi\in\Psi}\langle f,\psi\rangle\psi, \qquad \|f\|^2_{L_2([0,1])} = \sum_{\psi\in\Psi}|\langle f,\psi\rangle|^2. \tag{1.7}$$

The equivalence between *continuous* and *discrete* norms will frequently play a pivotal role in subsequent discussions.

As a first instance, relation (1.7) suggests the following simple strategy for approximating a given f by a piecewise constant with *possibly few* pieces. Suppose that all the wavelet coefficients $\langle f,\psi\rangle, \psi \in \Psi$, were known, and that the set $\Lambda \subset \Psi$, such that $\#\Lambda \le N$, contains the N *largest terms* $|\langle f,\psi\rangle|$, then the function $P_\Lambda f = \sum_{\psi\in\Lambda}\langle f,\psi\rangle\psi$ would, on account of (1.7), minimize the error among all piecewise constants on dyadic partitions with at most N pieces.

The selection of the N biggest terms is of course a nonlinear process. This aspect has been thoroughly discussed, for instance by DeVore and Lucier (1992) and DeVore, Jawerth and Popov (1992), and will be taken up in more detail again later in connection with adaptive methods. Here we add only a few comments, which are similar in spirit. Suppose that $g_J \in S_J$ is some approximation of g. It therefore has a representation

$$g_J = \sum_{k=0}^{2^J-1} c_k \phi_{J,k}$$

in terms of the *single scale* basis functions $\phi_{J,k}$ on the highest scale J. Note that although g_J may have a very simple structure, such as a constant throughout a large part or even all of $[0,1]$, all the 2^J coefficients could be *significant* in that they are needed to preserve accuracy. On the other hand, g_J has a wavelet or *multiscale* representation

$$g_J = \langle g, \phi_{0,0}\rangle \phi_{0,0} + \sum_{j=0}^{J-1} \sum_{k=0}^{2^j-1} d_{j,k}\, \psi_{j,k}.$$

If g_J were a constant, *all* the $d_{j,k}$ would vanish.

In general, one expects the $d_{j,k}$ to be very small where g_J does not vary much. In fact, if f were differentiable on the support of $\psi_{j,k}$, then, since

$$\langle \phi_{0,0}, \psi_{j,k}\rangle = \int_0^1 \psi_{j,k}(x)\, \mathrm{d}x = 0, \quad k = 0, \dots, 2^j - 1, \quad j \geq 0, \tag{1.8}$$

one obtains

$$\begin{aligned} |\langle f, \psi_{j,k}\rangle| &= \inf_{c\in\mathbb{R}} |\langle f - c, \psi_{j,k}\rangle| \leq \inf_{c\in\mathbb{R}} \|f - c\|_{L_2([2^{-j}k, 2^{-j}(k+1)])} \\ &\leq 2^{-j} \|f'\|_{L_2([2^{-j}k, 2^{-j}(k+1)])}. \end{aligned} \tag{1.9}$$

Thus, discarding wavelet coefficients that stay below a given threshold may compress the representation significantly, while the accuracy is, in view of (1.7), still controllable. The key is (1.8), which is often referred to as *moment conditions.* Obviously, the vanishing of moments of even higher polynomial order would increase the compression effect.

Of course, to exploit these facts practically requires switching back and forth between single- and multiscale representation. This issue will be addressed later in more generality.

1.3. The Hilbert transform

The compression of functions has a counterpart for operators. The fact that differential operators admit sparse representations is not surprising. Instead, consider the Hilbert transform

$$(\mathcal{H}f)(x) := \frac{1}{\pi} \mathrm{p.v.} \int_{\mathbb{R}} \frac{f(y)}{x - y}\, \mathrm{d}y \tag{1.10}$$

as an example of a typical singular integral operator. Here *p.v.* means 'principal value', that is, $\text{p.v.}\int_{\mathbb{R}} f(x)\,\mathrm{d}x = \lim_{\epsilon\to 0+} \int_{\mathbb{R}\setminus(-\epsilon,\epsilon)} f(x)\,\mathrm{d}x$. Its representation relative to the Haar basis requires the entries $H_{(j,k),(l,m)} := \langle \mathcal{H}\psi_{l,m}, \psi_{j,k}\rangle$. Suppose now that $2^{-l}(m+1) < 2^{-j}k$, and $l > j$, that is, the supports of $\psi_{l,m}$ and $\psi_{j,k}$ are disjoint. Then, by (1.8) and Taylor's expansion around $y = 2^{-l}m$, one obtains

$$\begin{aligned}
\pi\left|H_{(j,k),(l,m)}\right| &= \left|\int_{2^{-j}k}^{2^{-j}(k+1)} \left\{\int_{2^{-l}m}^{2^{-l}(m+1)} \left(\frac{1}{x-y} - \frac{1}{x-2^{-l}m}\right)\psi_{l,m}(y)\,\mathrm{d}y\right\}\psi_{j,k}(x)\,\mathrm{d}x\right| \\
&= \left|\int_{2^{-l}m}^{2^{-l}(m+1)} \left\{\int_{2^{-j}k}^{2^{-j}(k+1)} \frac{(y-2^{-l}m)}{(x-y_{l,m})^2}\psi_{j,k}(x)\,\mathrm{d}x\right\}\psi_{l,m}(y)\,\mathrm{d}y\right|,
\end{aligned}$$

for some $y_{l,m}$ in the support $[2^{-l}m, 2^{-l}(m+1)]$ of $\psi_{l,m}$. Repeating the same argument, one can subtract a constant in x which yields

$$\pi\left|H_{(j,k),(l,m)}\right| = \left|\int_{2^{-l}m}^{2^{-l}(m+1)} \left\{\int_{2^{-j}k}^{2^{-j}(k+1)} \left(\frac{(y-2^{-l}m)}{(x-y_{l,m})^2} - \frac{(y-2^{-l}m)}{(2^{-j}k-y_{l,m})^2}\right)\psi_{j,k}(x)\,\mathrm{d}x\right\}\psi_{l,m}(y)\,\mathrm{d}y\right|.$$

On account of Taylor's expansion around $x = 2^{-j}k$, the factor in front of $\psi_{j,k}(x)$ can be written as $-2(y-2^{-l}m)(x-2^{-j}k)/(x_{j,k}-y_{l,m})^3$, where $x_{j,k}$ is some point in the support $[2^{-j}k, 2^{-j}(k+1)]$ of the wavelet $\psi_{j,k}$. Noting that

$$\int_{\mathbb{R}} |\psi_{j,k}(x)|\,\mathrm{d}x \le 2^{-j/2},$$

a straightforward estimate provides

$$\pi|H_{(j,k),(l,m)}| \le 2^{-(n+j)\frac{3}{2}}|2^{-j}k - 2^{-l}m|^{-3} = \frac{2^{-|j-l|\frac{3}{2}}}{|k-2^{j-l}m|^3}. \tag{1.11}$$

Thus the entries $H_{(j,k),(m,l)}$ exhibit a decay with increasing distance of the supports of the wavelets as well as with increasing distance of scales. In essence this behaviour persists for a large class of integral operators and is the key to *sparsify* the discretization of such operators.

1.4. A two-point boundary value problem

Consider

$$-u'' = f \quad \text{on } [0,1], \quad u(0) = u(1) = 0, \tag{1.12}$$

as a simple model for an elliptic second-order boundary value problem. Although there are, of course, much simpler ways of solving (1.12), we start from the standard *weak formulation*

$$\langle u', v' \rangle = \langle f, v \rangle, \quad v \in H_0^1([0,1]). \tag{1.13}$$

Here $H_0^1([0,1])$ is the closure of all C^∞ functions with compact support in $(0,1)$ relative to the norm $\|f\|_{H^1([0,1])} = (\|f\|^2_{L_2([0,1])} + \|f'\|^2_{L_2([0,1])})^{1/2}$. To make our point, we use a standard Galerkin approach and solve (1.13) on finite-dimensional spaces $S_j \subset H_0^1([0,1])$. The simplest conforming choice of the trial spaces S_j are the spans of scaled *tent functions*

$$\phi_{j,k}(x) = 2^{j/2}\phi(2^j \cdot -k), \quad k = 0, \ldots, 2^j, \tag{1.14}$$

where

$$\phi(x) = \begin{cases} 1+x, & -1 \le x \le 0, \\ 1-x, & 0 \le x \le 1, \\ 0, & \text{otherwise.} \end{cases} \tag{1.15}$$

Choosing the $\phi_{j,k}$ as basis functions for S_J, the Galerkin conditions

$$\langle u_j, v \rangle = \langle f, v \rangle, \quad v \in S_j, \tag{1.16}$$

give rise to a linear system of equations

$$\mathbf{A}_J \mathbf{u} = \mathbf{f}, \tag{1.17}$$

where $\mathbf{A}_J$ is the stiffness matrix relative to the basis functions $\phi_{J,k}$ and $\mathbf{u}$, $\mathbf{f}$ are corresponding vectors with $f_k = \langle f, \phi_{J,k} \rangle$. Clearly $\mathbf{A}_J$ is tridiagonal. Hence (1.17) is very efficiently solvable. However, for higher-dimensional analogues the matrix would no longer have such a narrow bandwidth and one has to resort to *iterative methods* to preserve sparseness.

On the other hand, recalling the min-max characterization of the smallest and largest eigenvalue of a symmetric positive definite matrix, it is easy to see that the condition numbers of $\mathbf{A}_J$ grow like 2^{2J}, which renders classical iterative methods prohibitively inefficient.

To remedy this, one has to *precondition* the linear systems. One way is to exploit suitable multiscale decompositions of the trial spaces S_J. First note that, since

$$\phi(x) = \frac{1}{2}\phi(2x+1) + \phi(2x) + \frac{1}{2}\phi(2x-1), \tag{1.18}$$

that is,

$$\phi_{j,k} = \frac{1}{2\sqrt{2}}\phi_{j+1,2k-1} + \frac{1}{\sqrt{2}}\phi_{j+1,2k} + \frac{1}{2\sqrt{2}}\phi_{j+1,2k+1}, \tag{1.19}$$

the S_j are nested and, of course, their union is dense in $L_2([0,1])$.

In order to successively update solutions from coarser grids, we consider the following *hierarchical decomposition* of the trial spaces (Yserentant 1986). Instead of using orthogonal projections as in Section 1.2, we consider the *Lagrange projectors*

$$L_j f := \sum_{k=0}^{2^j} 2^{-j/2} f(2^{-j}k)\phi_{j,k}, \tag{1.20}$$

and note that the complements

$$W_j := (L_{j+1} - L_j)S_{j+1} \tag{1.21}$$

are simply spanned by the tent functions on new grid points on the next higher scale

$$\Psi_j := \{\psi_{j,k} := \phi_{j+1,2k+1} : k = 0, \ldots, 2^j - 1\}. \tag{1.22}$$

Note that neither the $\phi_{j,k}$ nor the $\psi_{j,k}$ are orthogonal but it is not hard to show that they satisfy the *stability condition*

$$c_1 \left(\sum_{k=0}^{2^j} |c_k|^2 \right)^{1/2} \leq \left\| \sum_{k=0}^{2^j} c_k \phi_{j,k} \right\|_{L_2([0,1])} \leq c_2 \left(\sum_{k=0}^{2^j} |c_k|^2 \right)^{1/2} \tag{1.23}$$

for some constants c_1, c_2 independent of the sequence $\{c_k\}_{k=0}^{2^j}$. Keeping this in mind, we now consider stiffness matrices relative to the *hierarchical bases* composed of the bases Ψ_j, and note that

$$\frac{d}{dx}\psi_{j,k}(x) = \frac{d}{dx}\phi_{j+1,2k+1}(x) = 2^{j+\frac{3}{2}}\psi^H_{j,k}(x), \tag{1.24}$$

where $\psi^H_{j,k}$ are the Haar wavelets from (1.3). Therefore one obtains from (1.5)

$$\left\langle \frac{d}{dx}\psi_{j,k}, \frac{d}{dx}\psi_{n,l} \right\rangle = 2^{j+n+3} \left\langle \psi^H_{j,k}, \psi^H_{n,l} \right\rangle = 2^{-2j+3}\delta_{j,n}\delta_{k,l}.$$

Hence $\mathbf{A}_J$ is, up to a 2×2 upper left block stemming from the coarse grid space S_0, a diagonal matrix, which is trivially *preconditioned* by symmetric *diagonal scalings*.

Now, one has to be somewhat careful when extrapolating from this observation. The fact that the hierarchical basis functions $\psi_{j,k}$ are actually orthogonal relative to the *energy inner product* is an artefact. In two dimensions this is no longer the case but it turns out that the hierarchical stiffness matrices can still be preconditioned by diagonal scaling to efficiently reduce the growth of the condition numbers to logarithmic behaviour. Moreover, it has suggested similar strategies involving other multiscale bases which do better. Corresponding preconditioning techniques are a central theme in subsequent discussions.

1.5. Some basic ideas

Next, I would like to blend these examples into a general picture to provide some orientation and unifying structure for the subsequent discussions of a diversity of ideas.

To this end, suppose that

$$\Psi = \{\psi_\lambda : \lambda \in \nabla\} \tag{1.25}$$

is a countable *basis* of some Hilbert space H. Thus every $v \in H$ has a unique convergent expansion in terms of elements of Ψ

$$v = \sum_{\lambda \in \nabla} d_\lambda \psi_\lambda. \tag{1.26}$$

The dependence of the coefficients $\{d_\lambda\}$ on v can be expressed via the *dual basis*. This is a collection of functionals

$$\tilde{\Psi} = \{\tilde{\psi}_\lambda : \lambda \in \nabla\},$$

such that

$$\langle \psi_\lambda, \tilde{\psi}_{\lambda'} \rangle = \delta_{\lambda,\lambda'}, \quad \lambda, \lambda' \in \nabla, \tag{1.27}$$

where $\langle \cdot, \cdot \rangle$ denotes the inner product on H. When H is infinite-dimensional, the notion of basis has to be further specified, but we will defer this issue for the moment. The collection of Haar functions $\psi_{j,k}^H$ forms such a basis for $H = L_2([0,1])$. In this case the indices $\lambda = (j,k)$ encode the information about scale and location. Of course, in the case of the Haar basis $\Psi = \Psi^H$, one has $\Psi = \tilde{\Psi}$. Equation (1.27) means that the coefficients d_λ in the expansion of v relative to Ψ are given by $d_\lambda = \langle v, \tilde{\psi}_\lambda \rangle$.

To simplify further exposition, I now introduce a compact notation for bases and their transforms that will be consistently used throughout the rest of the paper. Formally, let us view a given (countable) collection of functions Φ in H as a (column) *vector* (of possibly infinite length), so that an expansion with coefficients $c_\phi, \phi \in \Phi$ can be formally treated as an 'inner product'

$$\mathbf{c}^T \Phi := \sum_{\phi \in \Phi} c_\phi \phi.$$

The sum is always understood to converge in the norm of the underlying space, and the superscript T denotes 'transpose'. Likewise, for any $v \in H$, the quantities $\langle \Phi, v \rangle$ and $\langle v, \Phi \rangle$ mean the column-, respectively row-vector, of coefficients $\langle\phi, v\rangle$, $\langle v, \phi\rangle$, $\phi \in \Phi$. Thus (1.26) can be written for short as $\mathbf{d}^T\Psi$. Boldface lower case or capital letters will always denote sequences or matrices acting on sequences, respectively.

To push this a little further, for any two countable collections Φ, Ξ of functions, we consider the (possibly infinite) matrix

$$\langle\Phi, \Xi\rangle = (\langle\phi, \xi\rangle)_{\phi\in\Phi,\xi\in\Xi}.$$

Specifically, the above biorthogonality relations (1.27) then become

$$\langle\Psi, \tilde{\Psi}\rangle = \mathbf{I}, \tag{1.28}$$

where $\mathbf{I}$ denotes the identity matrix (whose dimension should be clear from the context).

The examples in Sections 1.3 and 1.4 can be viewed as special cases of the following situation. Suppose that H_1 and H_2 are *Hilbert* spaces such that either one of the continuous embeddings

$$H_1 \subseteq H \subseteq H_2 \quad \text{or} \quad H_2 \subseteq H \subseteq H_1$$

holds. In many cases of interest, H_2 is the *dual* of H_1, that is, the space of bounded linear functionals on H_1 relative to the duality pairing induced by the inner product $\langle\cdot,\cdot\rangle$ on the (intermediate) space H. Furthermore, suppose that $\mathcal{L}$ is a bounded linear bijection that maps H_1 onto H_2, that is,

$$\|\mathcal{L}v\|_{H_2} \sim \|v\|_{H_1}, \qquad v \in H_1, \tag{1.29}$$

where here and below $a \sim b$ means $a \lesssim b$ and $b \lesssim a$. The latter relations express that b can be bounded by some constant times a uniformly in any parameters on which a and b may depend. Hence the equation

$$\mathcal{L}u = f \tag{1.30}$$

has a unique solution $u \in H_1$ for every $f \in H_2$. In Section 1.3 we had $\mathcal{L} = H$, $H_1 = H_2 = L_2(\mathbb{R})$, while in Section 1.4 $\mathcal{L} = -\frac{d^2}{dx^2}$, $H_1 = H_0^1([0,1])$, $H_2 = H^{-1}([0,1])$, the dual of $H_0^1([0,1])$.

The basic idea is to transform the (continuous) equation (1.30) into an infinite *discrete* system of equations. This can be done with the aid of suitable *bases* for the spaces under consideration.

Given such bases, seeking the solution u of (1.29) is equivalent to finding the expansion sequence $\mathbf{d}$ of $u = \mathbf{d}^T\Psi$. Inserting this into (1.30) yields $(\mathcal{L}\Psi)^T\mathbf{d} = f$. Now suppose that $\Theta = \{\theta_\lambda : \lambda \in \nabla\}$ is *total* over H_2, that is, $\langle v, \Theta\rangle = 0$ implies $v = 0$ for $v \in H_2$. Then $(\mathcal{L}\Psi)^T\mathbf{d} = f$ becomes the (infinite) system

$$\langle\mathcal{L}\Psi, \Theta\rangle^{\mathrm{T}}\mathbf{d} = \langle f, \Theta\rangle^{\mathrm{T}}. \tag{1.31}$$

The objective now is to find collections Ψ and Θ for which the system (1.31) is efficiently solvable. This can be approached from several different angles.

(a) *Diagonalization*
The ideal case would be to know a complete system Ψ of *eigenfunctions* so that the choice $\Theta = \Psi$ would diagonalize (1.31). Of course, in practice this is usually not feasible. However, when Ψ and $\tilde{\Psi}$ are regular enough in the sense that the collections

$$\Theta := (\mathcal{L}^{-1})^*\tilde{\Psi} \subset H_1, \quad \tilde{\Theta} := \mathcal{L}\Psi \subset H_2, \tag{1.32}$$

then *biorthogonality* (1.28) implies

$$\langle \Theta, \tilde{\Theta} \rangle = \mathbf{I}, \tag{1.33}$$

that is, biorthogonality of the pair $\Theta, \tilde{\Theta}$. Here $\mathcal{L}^*$ denotes the *dual* or *adjoint* of $\mathcal{L}$ defined by $\langle \mathcal{L}u, v\rangle = \langle u, \mathcal{L}^*v\rangle$. In this case the solution $u = \mathbf{d}^T\Psi$ is given by

$$\mathbf{d} = \langle \Theta, f\rangle, \quad u = \langle f, \Theta\rangle\Psi. \tag{1.34}$$

When Ψ is a wavelet basis, it will be seen that under certain assumptions on $\mathcal{L}$ (defined on $\mathbb{R}^n$ or the torus), the elements of Θ share several properties with wavelets. The θ_λ are then called *vaguelettes.* Truncation of $\langle f, \Theta\rangle\Psi$ would readily yield an approximation to u. Note that this can be viewed as a *Petrov–Galerkin* scheme.

(b) *Preconditioning*
One expects that vaguelettes are numerically accessible only in special cases such as for constant coefficient differential elliptic operators on $\mathbb{R}^n$ or the torus. However, these cases may be in some sense *close* to more realistic cases, which opens possibilities for preconditioning.

Alternatively, one could relax the requirements on the bases Ψ and Θ. Again one could view the eigensystem as the ideal choice. A simple diagonal scaling would then transform $\langle \mathcal{L}\Psi, \Psi\rangle^T$ into $\mathbf{I}$. Thus one could ask for bases Ψ such that for a suitable *diagonal* matrix $\mathbf{D}$,

$$\mathbf{B} := \mathbf{D}\langle \mathcal{L}\Psi, \Psi\rangle^T\mathbf{D} \asymp \mathbf{I} \tag{1.35}$$

is spectrally equivalent to the identity, in the sense that $\mathbf{B}$ and its inverse $\mathbf{B}^{-1}$ are bounded in the Euclidean norm $\|\mathbf{d}\|^2_{\ell_2(\nabla)} := \mathbf{d}^*\mathbf{d}$, where $\mathbf{d}^* := \overline{\mathbf{d}}^T$ is the usual complex conjugate transpose.

Note that the principal sections of the infinite matrix $\mathbf{B}$ correspond to the *stiffness matrices* arising from a Galerkin scheme applied to (1.30) based on trial spaces spanned by subsets of Ψ. Relation (1.35) means that these linear systems are uniformly well conditioned. Such a Ψ would be in some sense sufficiently close to the eigensystem of $\mathcal{L}$. It will be seen that for a wide class of operators wavelet bases have that property. The precise choice of $\mathbf{D}$ depends on $\mathcal{L}$ or, more precisely, on the pair of spaces H_1, H_2 in (1.29). For instance, in Section 1.4 the diagonal entries of $\mathbf{D}$ would be 2^j for $\lambda := (j, k)$. In this context Sobolev spaces play a central role and the question of preconditioning will be seen to be intimately connected with the characterization of Sobolev spaces in terms of certain *discrete norms* induced by wavelet expansions.

(c) *Sparse representations*
The similarity between wavelet bases and eigensystems extends beyond the preconditioning effect. Indeed, for many operators the matrices $\mathbf{B}$ in (1.35), as well as their inverses, are *nearly sparse.* This means that replacing entries below a given threshold by zero yields a sparse matrix. When $\mathcal{L}$ is a differential operator and the wavelets have compact support this may not be too surprising (although the mixing of different levels creates, in general, a less sparse structure than shape functions with small support on the highest discretization level). However, it even remains true for certain integral operators as indicated by the estimate (1.11) for $\langle \mathcal{H}\Psi^H, \Psi^H \rangle$. Quantifying this sparsification will depend on $\mathcal{L}$ and on certain properties of the wavelet basis that will have to be clarified.

(d) *Significant coefficients and adaptivity*
Once you can track the wavelets in Ψ needed to represent the solution u of (1.30) accurately, one can, in principle, restrict the computations to the corresponding subspaces. Combining this with the sparse representation of operators is perhaps one of the most promising perspectives of wavelet concepts. A significant part of subsequent discussions will be initiated by this issue.

1.6. The structure of the paper

Here is a short overview of the material and the way it is organized. **Section 2** outlines the scope of problems to be treated and indicates corresponding basic obstructions to an efficient numerical solution. It is clear from the preceding discussion that, for each problem, the properties of underlying function spaces, in particular *Sobolev spaces*, have to be taken into account. A few preliminaries of this sort will therefore be collected first.

The objective of this paper is by no means the construction of wavelets. However, the performance of a wavelet scheme relies on very specific properties of the wavelet basis. I find it unsatisfactory to simply assume these properties without indicating to what extent and at what cost these properties may actually be realized. Therefore the construction of the tools also provides the necessary understanding for its limitations. Consequently some space has to be reserved for discussing properties of multiscale bases. Guided by the examples in Sections 1.2 and 1.4, **Section 3** begins by describing a general framework of multiresolution decompositions: this is to provide a uniform platform for all the subsequent specifications, in particular, those which involve more complex domain geometries. The simple but useful concepts of *stable completions* is emphasized as a construction device that can still be used under circumstances where, for instance, classical arguments based on Fourier techniques no longer apply.

Section 4 outlines some examples of multiscale decompositions and wavelet bases, which will later be referred to frequently. So-called *hierarchical bases* on bivariate triangulations (as a straightforward generalization of the construction in Section 1.4) will serve later as a bridge to developments in the finite element context. Wavelets defined on all of Euclidean space, their periodized versions, and wavelets on cubes are by far best understood. A few facts are recorded here which are important for further extensions needed later on.

Multiresolution originates from the classical setting concerning the full Euclidean space. The shift- and scale-invariance of its ingredients provide a comfortable basis for constructions and admit in combination with Fourier techniques best computational efficiency. While wavelets are usually associated with *orthogonal bases*, the concept of biorthogonal wavelets is emphasized, because it offers much more flexibility and localization (in physical space). I will try to indicate later that this actually pays dividends in several applications.

Much of the comfort of shift- and dilation-invariance can still be retained when dealing with wavelets on the *interval* (and hence on cubes). This still looks very restrictive, but it will turn out later to be an important ingredient for extending the application of wavelet schemes, for instance to closed surfaces or other manifolds. I have collected these construction issues in one section, so that those who are familiar with this material can easily skip over this section.

Section 5 addresses the heart of the matter. Once one is willing to dispense with orthogonality, one has to understand which type of decompositions are actually suitable. A classical theme in functional analysis is to characterize function spaces through isomorphisms into sequence spaces. The discussion in Section 1.5 has already stressed this point as a basic vehicle for developing *discretizations*. Orthonormal bases naturally induce such isomorphisms. When deviating from orthogonality, the leeway is easily seen to be set by the concept of *Riesz bases*, which in turn brings in the notion of *biorthogonal* bases. Whereas biorthogonality is necessary, it is not quite sufficient for establishing the desired norm equivalences. The objective of Section 5 is to bring out what is needed in addition. In order to be able to apply this to several cases, this is formulated for a general Hilbert space setting (see Section 5.2). It should be stressed that the additional stability criteria concern properties of the underlying multiresolution spaces *not* of the particular bases. Therefore things are kept in a basis-free form. Again, to guarantee flexible applicability, these criteria do *not* resort to Fourier techniques but are based on a pair of inequalities, describing *regularity* and *approximation* properties of the underlying multiresolution spaces (see Section 5.1). The most important application for the present purpose is the characterization of Sobolev spaces in Section 5.3 for all relevant versions of underlying domains, including manifolds such as closed surfaces. These facts will later play a crucial role in three different contexts, namely *preconditioning* (recall Section 1.5, (b)), thresholding strategies for *matrix compression* (see Section 1.5, (c)) and the analysis of *adaptive* schemes (see Section 1.5, (d)).

A first major application of the results in Section 5 is presented in **Section 6**. It is shown that (1.35), namely the transformation of a *continuous* problem into a *discrete* one, which is well-posed in the Euclidean metric, is realized for a wide class of elliptic *differential* and *integral* operators, described in Section 2.3. The entries of the diagonal matrix $\mathbf{D}$ depend on the *order* of the operator $\mathcal{L}$. Preconditioning is seen here to be an immediate consequence of the validity of norm equivalences for Sobolev spaces. It simply means that the shift in Sobolev scale caused by the operator $\mathcal{L}$ in (1.30) can be undone by a proper weighting of wavelet coefficients. Diagonal matrices act in some sense like differential or integral operators much like classical Bessel potentials.

To bring out the essential mechanism, this is formulated for a possibly abstract setting. One should look at the examples in Section 2.2 to see what it means in concrete cases. On the other hand, it is important to note that the full strength of wavelet bases is actually *not* always needed. When the order of the operator $\mathcal{L}$ is *positive*, the weaker concept of *frame* suffices. This establishes a strong link to recent, essentially independent, developments of *multilevel preconditioning techniques* in a finite element context. Both lines of development have largely ignored each other. Although the present discussion is primarily seen from the viewpoint of wavelet analysis, I will briefly discuss both schools and their interrelation.

While the concepts in Section 6 can also be realized in a finite element setting, **Section 7** confines the discussion to what will be called the *ideal setting*, meaning problems formulated on $\mathbb{R}^n$ or the torus. As detailed in Section 4.2, an extensive machinery of wavelet tools is available and much more refined properties can be exploited. Last but not least, through marriage with Fourier techniques such as FFT, this could be a tremendous support of computational efficiency. Some of the insight into local phenomena gained in this way can also be expected to help under more general circumstances. Of course, all these properties are preserved under periodization, so that wavelets still unfold their full potential for *periodic* problems.

One further reason for reserving some room for this admittedly restricted setting is to think of a *two-step approach*. Exploiting all the benefits of the ideal setting, one aims at developing highly efficient techniques that are to cope with the bulk of computation determined by, say, the spatial dimension of the domain. This should then justify efforts to treat geometric constraints entering through boundary conditions separately, hopefully at the expense of lower-order complexity.

The common ground for all the techniques mentioned in Section 7 is that the *inverse* of an elliptic operator is fairly well accessible in wavelet coordinates. This concerns *vaguelette* techniques, which in the spirit of Section 1.5 (a), aim at *diagonalizing* the operator $\mathcal{L}$ (see Section 7.2). Several issues such as the (adaptive) *evaluation* of vaguelette coefficients, *freezing coefficient* techniques for operators with variable coefficients and relaxed notions like *energy-pre-wavelets* are discussed.

The next step is to consider a class of univariate (periodic) nonlinear evolution model equations (Section 7.7). Several different approaches such as vaguelette schemes and *best bases* methods will be discussed. The so-called *pseudo-wavelet approach* aims at a systematic development of techniques for an adaptively controlled accurate application of evolution operators and nonlinear terms. An important vehicle in this context is the so-called *non-standard form* of operators. I will try to point out the difference between several approaches which are based on a number of very interesting and fairly unconventional concepts.

These evolution equations are to be viewed as simplified models of more complex systems like the *Stokes* and *Navier–Stokes* equations. In Section 7.11 some ways of dealing with corresponding additional difficulties are discussed. It seems that biorthogonal vaguelette versions combined with (biorthogonal) compactly supported divergence-free wavelet bases offer an interesting option, which has not been explored yet.

As mentioned before, a major motivation for the developments in Section 7 was to embed problems defined on more general domains into the ideal setting and then treat boundary conditions separately. **Section 8** is devoted to a brief discussion of several such embedding strategies. I will focus on three options. The first is to use extension techniques in conjunction with the multilevel Schwarz schemes described in Section 6.5. This is particularly tailored to variational formulations of problems involving self-adjoint operators. An alternative is to correct boundary values by solving a *boundary integral* equation. Finally, one can append boundary conditions with the aid of *Lagrange multipliers*.

Section 9 deals with pseudo-differential and integral operators. As an important case, this covers *boundary integral equations*. This type of problem is interesting for several reasons. First of all, it naturally came up in Section 8 in connection with partial differential equations. Second, it poses several challenges. On one hand, boundary integral formulations frequently offer physically more adequate formulations and reduce, in the case of exterior boundary value problems, the discretization of an unbounded domain to a discretization of a lower-dimensional compact domain. On the other hand, they have the serious drawback that (in some cases in addition to preconditioning issues) the resulting matrices are *dense*. However, as indicated by the example in Section 1.3, such operators have a *nearly* sparse representation relative to *appropriately* chosen wavelet bases. What exactly 'appropriate' means, and some ingredients of a rigorous analysis of corresponding *compression* techniques will be explained in this section. The issue here is twofold, namely reducing a matrix to a sparse matrix without losing *asymptotical* accuracy of the solution, and the efficient computation of the compressed matrices at costs that remain proportional to their size. Moreover, when the operator has *negative* order (see Section 2.2) preconditioning does require the full strength of wavelet decompositions. So, in principle, wavelets seem to be particularly promising for this type of problem. One expects that they offer a common platform for (i) efficiently applying operators that are otherwise dense, (ii) preconditioning the linear systems and (iii) facilitating adaptive strategies for further reducing complexity.

However, the embedding strategies from Section 8 do not apply to closed surfaces. So appropriate notions of wavelets on manifolds have to be developed. Discontinuous *multi-wavelets* have been employed so far. But according to the results in Section 6, they are not optimal for preconditioning operators of order -1. Therefore **Section 10** is devoted to the construction of wavelet bases on manifolds that have all the properties required by the analysis in Section 9. This rests on two pillars: the characterization of Sobolev spaces with respect to a *partition* of the manifold into parametric images of the unit cube (recall that the classical definition of Sobolev spaces on manifolds is based on open coverings), and certain biorthogonal wavelet bases on the unit cube that satisfy special boundary conditions. The construction of such bases, in turn, can be based on the ingredients presented in Section 4.4. This refers partly to work in progress. Some consequences with regard to domain decomposition are briefly indicated.

In **Section 11** we take up again the issue of adaptivity. The main objective is to outline a rigorous analysis for a possibly general setting that covers the previously discussed special cases. Some comments about relating this to adaptive strategies in the finite element context are included. In addition, this part should complement the intriguing algorithmic concepts discussed before. The section concludes with a brief discussion of the relation between the efficiency of adaptive approximation and *Besov regularity* of the solutions of elliptic equations.

Finally, in **Section 12** some further interesting directions of current and perhaps future research are indicated.

2. The scope of problems

The objective of this section is to put some meat on the skeleton of ideas in Section 1.5 by identifying first a list of concrete model problems satisfying (1.29) and (1.30). This requires some preparation.

2.1. Function spaces and other preliminaries

It is clear from the discussion in Section 1.5 that certain functional analytic concepts related to Sobolev spaces play an important role. This section contains corresponding relevant definitions, notation and conventions.

For any normed linear space S the norm is always denoted by $\|\cdot\|_S$. The *adjoint* or *dual* of an operator $\mathcal{L}$ is denoted by $\mathcal{L}^*$.

Important examples are L_p spaces. For $1 \le p \le \infty$ (with the usual sup-norm interpretation for $p = \infty$) and for any measure space $(\Omega, d\mu)$, the space $L_p(\Omega)$ consists of those measurable functions v such that

$$\|v\|_{L_p(\Omega)} := \left(\int_\Omega |v(x)|^p \,\mathrm{d}\mu(x) \right)^{1/p} < \infty.$$

For simplicity, we usually write $\mathrm{d}x$ instead of $\mathrm{d}\mu(x)$, since only the Lebesgue measure will matter. The case $p = 2$ is used most often. In this case $\|\cdot\|^2_{L_2(\Omega)} = \langle\cdot,\cdot\rangle_\Omega$, where

$$\langle u, v\rangle_\Omega := \int_\Omega u(x)\overline{v(x)}\,\mathrm{d}x$$

denotes the corresponding standard inner product. Here Ω may be $\mathbb{R}^n$ or a domain in $\mathbb{R}^n$ or, more generally, a manifold such as a closed surface. The latter interpretation is needed when dealing with boundary integral equations.

Partial derivatives are denoted by ∂, or ∂_x if it is stressed with respect to which variable it applies. Common multi-index notation is used, that is, $x^\alpha = x_1^{\alpha_1}\cdots x_n^{\alpha_n}$, $|\alpha| = |\alpha_1| + \cdots + |\alpha_n|$, for $\alpha \in \mathbb{N}_0$, $\mathbb{N}_0 := \{0, 1, 2, 3, \ldots\}$.

Suppose now that $\Omega \subset \mathbb{R}^n$ is a domain. We shall always assume that Ω is a bounded, open, and connected (at least) Lipschitz domain. This covers all cases of practical interest. If m is a positive integer, the *Sobolev* space $W^{m,p}((\Omega))$ consists of all functions $f \in L_p(\Omega)$, whose distributional derivatives $\partial^\nu f$, $|\nu| = m$, satisfy

$$|f|^p_{W^{m,p}(\Omega)} := \sum_{|\nu|=k} \|\partial^\nu f\|^p_{L_p(\Omega)} < \infty; \tag{2.1}$$

see, for example, Adams (1978). The pth root of (2.1) is the semi-norm for $W^{m,p}(\Omega)$, and adding to it $\|f\|_{L_p(\Omega)}$ gives the norm $\|f\|_{W^{m,p}(\Omega)}$ in $W^{m,p}(\Omega)$.

For the present purposes the most important case is again $p = 2$, which is denoted for short as $H^m(\Omega) := W^{m,2}(\Omega)$. Furthermore, Sobolev spaces with noninteger index $s \in \mathbb{R}$ are needed. There are several ways to define them. For $\Omega = \mathbb{R}^n$ one can use Fourier transforms

$$\hat{f}(y) := \int_{\mathbb{R}^n} f(x)e^{-ix\cdot y}\,dy,$$

and set

$$H^s(\mathbb{R}^n) = \{f \in L_2(\mathbb{R}^n) : \int_{\mathbb{R}^n} (1+|y|^2)^s |\hat{f}(y)|^2\,dy < \infty\},$$

where $|\cdot|$ is the Euclidean norm on $\mathbb{R}^n$. When $\Omega \neq \mathbb{R}^n$, the Lipschitz property implies that there exist extension operators E that are bounded in H^m for any $m \in \mathbb{N}$. For $s > 0$ one can define $\|f\|_{H^s(\Omega)} := \inf\{\|g\|_{H^s(\mathbb{R}^n)} : g\,|_\Omega = f\}$. Alternatively, $H^s(\Omega)$ can be defined by interpolation between $L_2(\Omega)$ and $H^m(\Omega)$, $m > s$; see Bergh and Löfström (1976), DeVore and Popov (1988*a*) and Triebel (1978). When $s < 0$ one can use duality. For any normed linear space V, the dual space, consisting of all bounded linear functionals on V, is denoted by V^*. It is a Banach space under the norm $\|w\|_{V^*} := \sup_{\|v\|_V=1} |w(v)|$. Specifically, when Ω is a closed manifold $(H^s(\Omega))^* = H^{-s}(\Omega)$.

We will briefly encounter *Besov* spaces $B^s_q(L_p(\Omega))$; see again Bergh and Löfström (1976), DeVore and Popov (1988*a*), DeVore and Sharpley (1993) and Triebel (1978). They arise by interpolation between $L_p(\Omega)$ and $W^{m,p}(\Omega)$. Recall that $H^s(\Omega) = B^s_2(L_2(\Omega))$.

As mentioned before, lower case boldface letters such as $\mathbf{c}, \mathbf{d}$ will always denote sequences over some (finite or infinite) index set Δ. As usual, for the same range of p as above, we set

$$\|\mathbf{c}\|_{\ell_p(\Delta)} := \left(\sum_{k\in\Delta} |c_k|^p\right)^{1/p}.$$

By convention, the elements of $\ell_p(\Delta)$ will always be viewed as *column* vectors, that is, $\mathbf{c}^T, \mathbf{c}^*$ are *rows*, the latter indicating complex conjugates when using the complex field. Analogously, for a matrix $\mathbf{M}$ the transpose is $\mathbf{M}^T$, while $\mathbf{M}^*$ denotes its complex conjugate transpose.

When there is no risk of confusion the reference to the domain or index set will sometimes be dropped, that is, we write $\langle\cdot,\cdot\rangle$, H^s, ℓ_2, *etc.*

2.2. A general class of elliptic problems

(a) *Scalar elliptic boundary value problems*
For $\Omega \subseteq \mathbb{R}^n$, an example of $\mathcal{L}$ in (1.30) is

$$\mathcal{L}u := i^{2m} \sum_{|\alpha|,|\beta|\le m} a_{\alpha,\beta}(x)\partial^\alpha\partial^\beta u = f \quad \text{on } \Omega, \quad \mathcal{B}u = 0 \quad \text{on } \partial\Omega, \tag{2.2}$$

where $\mathcal{B}$ is a suitable trace operator, and the polynomial

$$P(\xi) := \sum_{|\alpha|,|\beta|=m} a_{\alpha,\beta}(x)\xi^{\alpha+\beta}$$

satisfies

$$P(\xi) \ge \delta > 0, \quad \xi \in \mathbb{R}^n, \ x \in \Omega. \tag{2.3}$$

Depending on the regularity of the domain, (1.29) holds with $H_1 = H^s(\Omega)$, $H_2 = H^{s-2m}(\Omega)$ for a certain range of s. An important special case is

$$-\operatorname{div}(A(x)\nabla u) + a(x)u = f \text{ on } \Omega, \quad u = 0 \text{ on } \partial\Omega, \tag{2.4}$$

where $A(x)$ is uniformly positive definite and symmetric on Ω and for a vector field v the divergence operator is defined by $\operatorname{div} v := \sum_{j=1}^n \frac{\partial}{\partial x_j} v_j$. Clearly $A = I$, $a(x) = 0$, gives Poisson's equation with Dirichlet boundary conditions. Here $H_1 = H_0^1(\Omega)$ and $H_2 = H^{-1}(\Omega) = (H_0^1(\Omega))^*$. Likewise, one could take the Helmholtz equation $\mathcal{L} = -\Delta + aI$ for $a > 0$, or $\mathcal{L} = -\Delta + \beta\cdot\nabla$. Similarly with $\mathcal{L} = \Delta^2$, $H_1 = H_0^2(\Omega)$, $H_2 = H^{-2}(\Omega)$, fourth-order problems are covered as well.

The special case that $\mathcal{L}$ is positive definite and selfadjoint is of particular interest, that is,

$$a(u, v) = \langle \mathcal{L}u, v\rangle \tag{2.5}$$

is a *symmetric bilinear* form. *Ellipticity* here means that

$$a(\cdot,\cdot) \sim \|\cdot\|_{H_1}^2, \tag{2.6}$$

which implies (1.29). Clearly (2.4) falls into this category.

Such problems can be solved approximately with the aid of finite element-based Galerkin schemes. There are several different problems that arise. For $n \ge 2$ one obtains *large* linear systems, usually with *sparse* matrices which, for instance in the case (2.4), are symmetric positive definite. Thus a major challenge lies simply in the size of such problems. Since direct solvers based on matrix factorizations would cause a significant *fill-in* of nonzero entries in the factors, and therefore prohibitively limit storage and computing time, one has to resort to iterative solvers for large problem sizes. Unfortunately, the *condition numbers* of the system matrices grow with their size N like $N^{2m/n}$. It is therefore of vital importance to *precondition* these sytems.

In fact, an asymptotically optimal scheme would require *uniformly bounded* condition numbers.

When $\mathcal{L}$ is not selfadjoint, efficient schemes such as preconditioned conjugate gradient (PCG) iterations have to be replaced by more expensive ones, whose performance is no longer a simple function of the spectral condition numbers.

Finally, the coefficients in $A(x)$ or $a(x)$ may vary rapidly. On one hand, this may adversely affect the constants in (1.29). On the other hand, the resolution of such fluctuations may require too small mesh sizes, so that questions of *homogenization* arise. In the following we will primarily address the first two issues.

(b) *Saddle point equations*
An important example for a *system* of partial differential equations is the *Stokes* problem

$$\begin{aligned} -\Delta u + \nabla p &= f, \quad \text{on } \Omega, \quad u = 0 \text{ on } \partial\Omega, \\ \operatorname{div} u &= 0, \end{aligned} \tag{2.7}$$

as a simple model for viscous incompressible flow. The vector valued function u and the scalar field p represent velocity and pressure of the fluid, respectively. Obviously, one has to factor the constants from p, for instance by requiring $\int_\Omega p(x)\,\mathrm{d}x = 0$.

The *weak formulation* of (2.7) requires finding $(u, p) \in V \times M$, where

$$V := \left(H_0^1(\Omega)\right)^n, \quad M = L_{2,0}(\Omega) = \left\{f \in L_2(\Omega) : \int_\Omega f(x)\,\mathrm{d}x = 0\right\}, \tag{2.8}$$

such that

$$\begin{aligned} a(u, v) + b(v, p) &= \langle f, v\rangle_\Omega, \quad v \in V \\ b(u, \mu) &= 0, \quad \mu \in M, \end{aligned} \tag{2.9}$$

with

$$a(u, v) = \langle \nabla u, \nabla v\rangle_\Omega, \quad b(v, \mu) = \langle \operatorname{div} v, \mu\rangle_\Omega. \tag{2.10}$$

So-called *mixed formulations* of (2.4) for $a(x) = 0$ arise when introducing the *flux* $\sigma := -A\nabla u$ as a new variable, so that $-\operatorname{div}(A\nabla u) = f$ yields a coupled system of first-order equations

$$A\nabla u = -\sigma, \quad \operatorname{div} \sigma = f,$$

whose weak formulation is

$$\begin{aligned} a(\sigma, \tau) - b(\tau, u) &= 0, & v \in V &:= H(\operatorname{div}, \Omega), \\ -b(\sigma, v) &= -\langle f, v\rangle_\Omega, & v \in M &:= L_2(\Omega). \end{aligned} \tag{2.11}$$

Here $a(\cdot,\cdot) = \langle\cdot,\cdot\rangle_\Omega$, $b(\cdot,\cdot)$ is defined as before in the Stokes problem, and

$$H(\mathrm{div},\Omega) := \{\tau \in (L_2(\Omega))^n : \mathrm{div}\,\tau \in L_2(\Omega)\},$$

endowed with the graph norm $\|\tau\|_{H(\mathrm{div},\Omega)} = (\|\tau\|^2_{L_2(\Omega)} + \|\mathrm{div}\,\tau\|^2_{L_2(\Omega)})^{1/2}$.

Both cases (2.9) and (2.11) can be viewed as an operator equation of the form (1.30) with

$$\mathcal{L} = \begin{pmatrix} \mathcal{A} & \mathcal{B}^* \\ \mathcal{B} & 0 \end{pmatrix}, \tag{2.12}$$

and $\mathcal{A} : V \to V^*$, $\mathcal{B} : V \to M^*$ are defined by

$$\langle \mathcal{A}u, v\rangle_\Omega = a(u,v), \quad v \in V, \quad b(v,\mu) = \langle \mathcal{B}v, \mu\rangle_\Omega, \quad \mu \in M.$$

It is well known that in both cases $\mathcal{L}$ is an isomorphism from $H_1 := V \times M$ onto $H_2 := V^* \times M^*$, that is, (1.29) is valid (Braess 1997, Brezzi and Fortin 1991, Girault and Raviart 1986), which in this case means that

$$\inf_{\mu\in M} \sup_{v\in V} \frac{b(v,\mu)}{\|v\|_V \|\mu\|_M} \geq \beta > 0. \tag{2.13}$$

Note that in the case (2.11) the Galerkin approximation of $\mathcal{A}$ is a *mass matrix*. Introducing suitable weighted inner products on a high discretization level would precondition this part well, which is one possible strategy for dealing with fluctuating coefficients.

However, the numerical solution of (2.9) or (2.11) now poses additional difficulties. The operator $\mathcal{L}$ is no longer definite. Preconditioning therefore requires additional care. Furthermore, the discretizations of V and M must be compatible, that is, (2.13) has to hold *uniformly* in the family of trial spaces under consideration. Both issues, preconditioning as well as the construction of compatible trial spaces, will be discussed below.

(c) *Time-dependent problems*

Once elliptic problems of the above type can be handled, the next step is to consider problems of the form

$$\begin{aligned} \frac{\partial u}{\partial t} + \mathcal{L}u + \mathcal{G}(u) &= 0, \quad u(k) = u(k+l), \quad k,l \in \mathbb{Z}^n, \\ u(\cdot,0) &= u_0, \end{aligned} \tag{2.14}$$

where $\mathcal{L}$ is an elliptic operator of the form (2.4), and $\mathcal{G}$ is a possibly nonlinear function of u or a first-order derivative of u. Prominent examples are *reaction diffusion* equations

$$\frac{\partial u}{\partial t} = \nu \frac{\partial^2}{\partial x^2} u + u^p, \quad p > 1, \quad \nu > 0, \tag{2.15}$$

or the *viscous Burgers equation*

$$\frac{\partial u}{\partial t} + u\frac{\partial u}{\partial x} = \nu \frac{\partial^2}{\partial x^2} u, \tag{2.16}$$

which describes the formation of shocks. Several wavelet schemes for this type of equation will be discussed. Some will also apply to problems such as the *Korteweg–de Vries* equations

$$\frac{\partial u}{\partial t} + \alpha u\frac{\partial u}{\partial x} + \beta \frac{\partial^3}{\partial x^3} u = 0, \tag{2.17}$$

α, β constant, having special soliton solutions (Fornberg and Whitham 1978).

(d) *Boundary integral equations*

Many classical partial differential equations can be transformed into *boundary integral equations*. This includes the *Lamé–Navier equations* of linearized, three-dimensional elasticity, (Wendland 1987), the *oblique derivative problem* (Michlin 1965), arising in physical geodesy, the *exterior Stokes flow* (Ladyshenskaya 1969); see Schneider (1995) for a brief overview. Here it suffices to describe a simple example that exhibits the principal features of this class of problem. Consider the boundary value problem

$$\Delta U = 0, \text{ on } \Omega, \quad \partial_\nu U = f, \text{ on } \Gamma := \partial\Omega, \tag{2.18}$$

where Ω is a bounded domain in $\mathbb{R}^3$ and ∂_ν denotes the derivative in the direction of the outer normal to Γ. It is well known that this boundary value problem, which arises, for instance, in the computation of electrostatic fields, is equivalent to the following integral equation of second kind provided by the so-called *indirect method*

$$\mathcal{L}u = f, \tag{2.19}$$

where $\mathcal{L} = \frac{1}{2}\mathcal{I} + \mathcal{K}$ and

$$(\mathcal{K}u)(x) = \frac{1}{4\pi} \int_\Gamma \frac{\nu_y^{\mathrm{T}}(x-y)}{|x-y|^3} u(y)\,\mathrm{d}s_y. \tag{2.20}$$

Here ν_y denotes the exterior normal of Γ at y. $\mathcal{K}$ is called *double layer potential.* For smooth Γ the operator $\mathcal{K}$ is compact on $L_2(\Gamma)$ so that the principal symbol of $\mathcal{L}$ is $1/2$. Thus (1.29) holds with $H_1 = H_2 = L_2(\Gamma)$ and K is a *zero order* operator. Clearly, denoting by $G(x-y) := \frac{1}{4\pi|x-y|}$ the fundamental solution of (2.18), one has $\mathcal{K}u(x) = \int_\Gamma \partial_{\nu,y} G(x-y)u(y)\,\mathrm{d}S_y$ and the solution U of (2.18) can be obtained by evaluating $U(x) = \int_\Gamma G(x-y)u(y)\,\mathrm{d}s_y$, where u is the solution of (2.19).

This approach is particularly tempting when (2.19) is to be solved on the *exterior* $\mathbb{R}^3 \setminus \Omega$ of some bounded domain. In this case one has to append certain *radiation conditions* at infinity to determine the solution uniquely.

The so-called *direct method* arises in connection with *transmission problems* and is well suited to dealing with other boundary conditions. Problem (2.18) subject to Dirichlet conditions $U = f$ on Γ is known to be equivalent to

$$\mathcal{L}u = \mathcal{V}u = \left(\frac{1}{2}\mathcal{I} - \mathcal{K}\right) f, \tag{2.21}$$

where

$$(\mathcal{V}u)(x) = \int_{\Gamma} \frac{u(y)}{4\pi|x-y|} \, dy \tag{2.22}$$

is the *single layer* potential. In this case (1.29) can be shown to hold for $H_1 = H^{-1/2}(\Gamma)$ and $H_2 = H^{1/2}(\Gamma)$, and $\mathcal{L}$ has order *minus one.*

In both cases the unique solvability of (1.30) and (1.29) can be established along the following lines, which work for a much wider class of *pseudo-differential operators.* In fact, for smooth Γ these operators are classical pseudo-differential operators characterized by their *symbol*; see Hildebrandt and Wienholtz (1964), Kumano-go (1981). Equation (1.29) follows from the boundedness of $\mathcal{L}$, its injectivity on H_1, and *coercivity* of the principal part of its symbol.

The advantages of the approach are obvious. A 3D discretization of a possibly unbounded domain is reduced to a 2D discretization of a compact domain. One can also argue that in many cases the integral formulation is physically more adequate.

On the other hand, there are serious drawbacks. If the order of the operator $\mathcal{L}$ is different from zero, as in the case of the single layer potential operator, the need for preconditioning remains. In addition, conventional discretizations of the integral operators lead to dense matrices, which is perhaps the most severe obstruction to the use of these concepts for realistic problem sizes N. *Appropriate* wavelet bases will be seen to realize both desired effects (b), (c) in Section 1.5 for this class of problem.

2.3. A reference class of problems

The examples in Section 2.2 illustrate the variety of problems that will be discussed in this paper. To get some structure into the diversity of existing studies of various special cases, I stress the fact that certain results, mainly concerned with (b) in Section 1.5, actually hold in remarkable generality. Presenting them in this generality will help to bring out what really matters. In all the above examples the operator $\mathcal{L}$ satisfies (1.29) where H_1, H_2 are Sobolev spaces or products of such. In order to keep the discussion homogeneous, we will confine the formulation of a model class of problems to the scalar case. So we assume that there exist some positive constants $c_1, c_2 < \infty$ such that

$$c_1\|\mathcal{L}u\|_{H^{-t}} \leq \|u\|_{H^t} \leq c_2\|\mathcal{L}u\|_{H^{-t}}, \tag{2.23}$$

where H^s stands for a suitable (subspace of a) Sobolev space (for instance, determined by homogeneous boundary conditions) and H^{-s} for its dual space. The underlaying domain may be a bounded domain in $\mathbb{R}^n$, $\mathbb{R}^n$ itself or a more general manifold such as a closed surface according to the above examples. Thus the problem

$$\mathcal{L}u = f \tag{2.24}$$

has for every $f \in H^{-t}$ a unique solution.

The analysis that follows will also cover operators with global Schwartz kernel

$$\mathcal{L}u = \int_\Gamma K(\cdot, x)u(x)\,\mathrm{d}x,$$

as considered in Section 2.2. As in the above examples, K will always be assumed to be smooth off the diagonal $x = y$. Moreover, it is to satisfy the following asymptotic estimates, which obviously hold in the above cases as well,

$$\left|\partial_x^\alpha \partial_y^\beta K(x,y)\right| \;\lesssim\; \mathrm{dist}(x,y)^{-(n+2t+|\alpha|+|\beta|)}, \tag{2.25}$$

where $r = 2t$ is the *order* of the operator.

3. Multiscale decompositions of refinable spaces

In Section 1.5 the *transform point of view* has been stressed. As indicated there the corresponding numerical schemes can be viewed as *Galerkin* or, more generally, (generalized) *Petrov–Galerkin* schemes. The point is that these schemes are always seen in connection with a whole ascending sequence of trial spaces, often referred to as *multiresolution analysis.* This permits the interaction of different scales of discretizations. In *basis* or *transform* oriented methods this is effected with the aid of appropriate *multiscale bases* of hierarchical type. Following Carnicer, Dahmen and Peña (1996), Dahmen (1994), Dahmen (1996) and Dahmen (1995) a general framework of multiresolution and multiscale decompositions of trial spaces is described next in a form which will later host all the required specializations. The examples in Sections 1.2 and 1.4 can be used as a conceptual as well as a notational orientation.

3.1. Multiresolution

The concept of *multiresolution* analysis plays a central role in the context of classical wavelets on $\mathbb{R}^n$. The anticipated applications here require a suitable generalization. In the spirit of Section 1.5, let H be a Hilbert space with inner product $\langle\cdot,\cdot\rangle$ and associated norm $\|\cdot\| = \|\cdot\|_H = \langle\cdot,\cdot\rangle^{1/2}$. A multiresolution sequence $\mathcal{S} = \{S_j\}_{j\in\mathbb{N}_0}$ consists of nested closed subspaces $S_j \subset H$ whose union is dense in H

$$S_j \subset S_{j+1}, \quad \operatorname{clos}_H\left(\bigcup_{j\in\mathbb{N}_0} S_j\right) = H. \tag{3.1}$$

Define for any countable subset $\Phi \subset H$

$$S(\Phi) := \operatorname{clos}_H(\operatorname{span}\{\Phi\}),$$

the closure of the linear span of Φ. In all cases of practical interest the spaces S_j have the form

$$S_j := S(\Phi_j), \quad \Phi_j = \{\phi_{j,k} : k \in \Delta_j\} \tag{3.2}$$

for some (possibly infinite) index set Δ_j, where $\{\Phi_j\} = \{\Phi_j\}_{j\in\mathbb{N}_0}$ is *uniformly stable* in the sense that (see (1.23))

$$\|\mathbf{c}\|_{\ell_2(\Delta_j)} \sim \|\mathbf{c}^{\mathrm{T}}\Phi_j\|_H. \tag{3.3}$$

The Φ_j will sometimes be called *generator bases* or *single-scale bases.* The elements $\phi_{j,k}$ typically have good localization properties such as compact supports whose size depends on the scale j.

An arbitrary but fixed highest level of discretization will usually be denoted by J, and

$$N_J := \#\Delta_J$$

abbreviates the dimension of the corresponding space $S(\Phi_J)$.

Examples are $H = L_2([0,1])$ and $\phi_{j,k}$ the box or tent functions (see Sections 1.2 and 1.4) with $\Delta_j = \{0,\ldots,2^j-1\}$ or $\Delta_j = \{0,\ldots,2^j\}$, respectively.

Two-scale relations

Nestedness of the spaces $S(\Phi_j)$ combined with (3.3) means that every $\phi_{j,k} \in S(\Phi_j)$ possesses an expansion

$$\phi_{j,k} = \sum_{l\in\Delta_{j+1}} m^j_{l,k}\phi_{j+1,l}$$

with a *mask* or *filter sequence* $\mathbf{m}^j_k = \{m^j_{l,k}\}_{l\in\Delta_{j+1}} \in \ell_2(\Delta_{j+1})$; recall (1.2) and (1.19). In our compact notation this can be rewritten as

$$\Phi_j^{\mathrm{T}} = \Phi_{j+1}^{\mathrm{T}}\mathbf{M}_{j,0}, \tag{3.4}$$

where the *refinement* matrix $\mathbf{M}_{j,0}$ contains the $\mathbf{m}_k^j$ as columns.

I will make frequent use of this notation, for two reasons. First, it saves several layers of indices. Second, it clearly brings out the conceptual similarities shared by all the technically different subsequent specializations. On the other hand, a word of warning is also appropriate. The special features of the actual implementation remain somewhat obscure. For instance, it will by no means always be necessary to assemble the complete matrices $\mathbf{M}_{j,0}$. In most cases its application to a vector amounts to applying local filters. Keeping this in mind, I still grant priority to convenience.

To illustrate (3.4), recall from (1.2) that the refinement matrix for the box functions is the $2^{j+1} \times 2^j$ matrix

$$\mathbf{M}_{j,0} = \begin{pmatrix} \frac{1}{\sqrt{2}} & 0 & 0 & \dots & & & 0 \\ \frac{1}{\sqrt{2}} & 0 & 0 & & & & \\ 0 & \frac{1}{\sqrt{2}} & 0 & & & & \\ \vdots & \vdots & \vdots & & & & \vdots \\ & & & & 0 & \frac{1}{\sqrt{2}} & 0 \\ & & & & 0 & \frac{1}{\sqrt{2}} & 0 \\ & & & & 0 & 0 & \frac{1}{\sqrt{2}} \\ 0 & \dots & & & 0 & 0 & \frac{1}{\sqrt{2}} \end{pmatrix}, \tag{3.5}$$

whose dependence on j concerns only its size. Likewise, (1.19) gives the $(2^{j+1}-1) \times (2^j - 1)$ matrix

$$\mathbf{M}_{j,0} = \begin{pmatrix} \frac{1}{2\sqrt{2}} & 0 & 0 & \dots & \\ \frac{1}{\sqrt{2}} & 0 & 0 & \dots & \\ \frac{1}{2\sqrt{2}} & \frac{1}{2\sqrt{2}} & 0 & \dots & \\ 0 & \frac{1}{\sqrt{2}} & 0 & \dots & \\ 0 & \frac{1}{2\sqrt{2}} & \frac{1}{2\sqrt{2}} & 0 & \dots \\ 0 & 0 & \dots & & \\ \vdots & \vdots & \vdots & & \vdots \\ & & & & 0 \\ & & & \frac{1}{2\sqrt{2}} & \frac{1}{2\sqrt{2}} \\ 0 & \dots & & 0 & \frac{1}{\sqrt{2}} \\ 0 & \dots & & 0 & \frac{1}{2\sqrt{2}} \end{pmatrix}. \tag{3.6}$$

3.2. Stable completions

Since the union of $\mathcal{S}$ is dense in H, a basis for H can be assembled from functions which span complements between any two successive trial spaces. One may think of orthogonal complements as in Section 1.2 or of the hierarchical complements in Section 1.4 induced by Lagrange interpolation (1.21). Depending on the case at hand, different choices will be seen to be preferable. So at this point we follow Carnicer et al. (1996) and keep the specific choices open. Thus one looks for collections $\Psi_j = \{\psi_{j,k} : k \in \nabla_j\} \subset S(\Phi_{j+1})$, such that

$$S(\Phi_{j+1}) = S(\Phi_j) \oplus S(\Psi_j), \tag{3.7}$$

and $\{\Phi_j \cup \Psi_j\}$ is still uniformly stable in the sense of (3.3). Like refinability, such decompositions may be expressed equivalently in terms of *matrix relations* that will provide a convenient algebraic platform for a unified treatment of subsequent specializations. As above, (3.7) implies that there exists some matrix $\mathbf{M}_{j,1}$ such that

$$\Psi_j^{\mathrm{T}} = \Phi_{j+1}^{\mathrm{T}} \mathbf{M}_{j,1}. \tag{3.8}$$

It is easy to see that (3.7) is equivalent to the fact that the operator

$$\mathbf{M}_j := (\mathbf{M}_{j,0}, \mathbf{M}_{j,1}),$$

defined by $\mathbf{M}_j \binom{\mathbf{c}}{\mathbf{d}} := \mathbf{M}_{j,0}\mathbf{c} + \mathbf{M}_{j,1}\mathbf{d}$, for $\mathbf{c} \in \ell_2(\Delta_j)$, $\mathbf{d} \in \ell_2(\nabla_j)$, is *invertible* as a mapping from $\ell_2(\Delta_j) \times \ell_2(\nabla_j)$ onto $\ell_2(\Delta_{j+1})$. Moreover, $\{\Phi_j \cup \Psi_j\}$ is uniformly stable if and only if

$$\|\mathbf{M}_j\|, \|\mathbf{M}_j^{-1}\| = O(1), \quad j \in \mathbb{N}, \tag{3.9}$$

where $\|\cdot\|$ is the spectral norm (Carnicer et al. 1996).

It is convenient to block $\mathbf{M}_j^{-1}$ as

$$\mathbf{M}_j^{-1} =: \mathbf{G}_j = \begin{pmatrix} \mathbf{G}_{j,0} \\ \mathbf{G}_{j,1} \end{pmatrix}, \tag{3.10}$$

so that

$$\mathbf{I} = \mathbf{M}_j \mathbf{G}_j = \mathbf{M}_{j,0}\mathbf{G}_{j,0} + \mathbf{M}_{j,1}\mathbf{G}_{j,1} \tag{3.11}$$

and

$$\mathbf{G}_{j,e}\mathbf{M}_{j,e'} = \delta_{e,e'}\mathbf{I}, \quad e, e' \in \{0,1\}. \tag{3.12}$$

Of course, those who are familiar with wavelets recognize in (3.11) the classical filter relations. The matrix $\mathbf{M}_j$ describes a *change of bases* and hence the reverse change $\mathbf{G}_j$, that is, Φ_{j+1} can be expressed in terms of the coarse scale basis Φ_j and the complement basis Ψ_j. One readily concludes from (3.4), (3.8) and (3.11) the *reconstruction* relation

$$\Phi_{j+1}^{\mathrm{T}} = \Phi_j^{\mathrm{T}} \mathbf{G}_{j,0} + \Psi_j^{\mathrm{T}} \mathbf{G}_{j,1}. \tag{3.13}$$

In general it may be difficult to identify the inverse $\mathbf{G}_j$, or, better, to arrange $\mathbf{M}_{j,1}$ in such a way that also $\mathbf{G}_j$ has a nice structure such as sparseness. One rather expects that when $\mathbf{M}_j$ is sparse, $\mathbf{G}_j$ will be full. In some sense, the art of wavelet construction can be viewed as finding the exceptions.

It is again instructive to recall the examples in Section 1. The relation (1.3), defining the Haar wavelet, corresponds to the $2^{j+1} \times 2^j$ matrix

$$\mathbf{M}_{j,1} = \begin{pmatrix} \frac{1}{\sqrt{2}} & 0 & 0 & \dots & & & 0 \\ -\frac{1}{\sqrt{2}} & 0 & 0 & & & & \\ 0 & \frac{1}{\sqrt{2}} & 0 & & & & \\ \vdots & \vdots & \vdots & & & & \vdots \\ & & & & 0 & \frac{1}{\sqrt{2}} & 0 \\ & & & & 0 & -\frac{1}{\sqrt{2}} & 0 \\ & & & & 0 & 0 & \frac{1}{\sqrt{2}} \\ 0 & \dots & & & 0 & 0 & -\frac{1}{\sqrt{2}} \end{pmatrix}. \tag{3.14}$$

Since the Haar system is orthonormal, one simply has in this case

$$\mathbf{G}_j = \mathbf{M}_j^{\mathrm{T}}, \quad \|\mathbf{M}_j\| = \|\mathbf{M}_j^{-1}\| = 1. \tag{3.15}$$

Adding and subtracting (1.2) and (1.3), one could also deduce directly that

$$\phi_{j+1,2k} = \frac{1}{\sqrt{2}}(\phi_{j,k} + \psi^H_{j,k}), \quad \phi_{j+1,2k+1} = \frac{1}{\sqrt{2}}(\phi_{j,k} - \psi^H_{j,k}).$$

For the hierarchical basis from Section 1.4 one obtains the $(2^{j+1}-1) \times 2^j$ matrix

$$\mathbf{M}_{j,1} = \begin{pmatrix} 1 & 0 & 0 & & & & \\ 0 & 0 & 0 & & & & \\ 0 & 1 & 0 & & & & \\ \vdots & \vdots & \vdots & & & & \vdots \\ & & & & 0 & 1 & 0 \\ & & & & 0 & 0 & 0 \\ & & & & 0 & 0 & 1 \end{pmatrix}. \tag{3.16}$$

Moreover, since by (1.19) and (1.22),

$$\begin{aligned} \phi_{j+1,2k} &= \sqrt{2}\,\phi_{j,k} - \frac{1}{2}\left(\phi_{j+1,2k-1} + \phi_{j+1,2k+1}\right) \\ &= \sqrt{2}\,\phi_{j,k} - \frac{1}{2}\left(\psi_{j,k-1} + \psi_{j,k}\right), \quad k = 1, \dots, 2^j - 1, \\ \phi_{j+1,0} &= \sqrt{2}\,\phi_{j,0} - \frac{1}{2}\psi_{j,0}, \quad \phi_{j+1,2^{j+1}} = \sqrt{2}\,\phi_{j,2^j} - \frac{1}{2}\psi_{j,2^j-1}, \end{aligned}$$

while

$$\phi_{j+1,2k+1} = \psi_{j,k}, \quad k = 0, \ldots, 2^j - 1,$$

one readily identifies, in view of (3.13), the inverse $\mathbf{G}_j$ as

$$\mathbf{G}_{j,0} = \begin{pmatrix} 0 & \sqrt{2} & 0 & 0 & 0 & & \ldots & 0 & 0 \\ 0 & 0 & 0 & \sqrt{2} & 0 & & \vdots & 0 & 0 \\ \vdots & \vdots & & & & & \vdots & & \vdots \\ & & & & & \sqrt{2} & 0 & 0 & 0 \\ 0 & 0 & \ldots & & & 0 & 0 & \sqrt{2} & 0 \end{pmatrix}, \tag{3.17}$$

and

$$\mathbf{G}_{j,1} = \begin{pmatrix} 1 & -\frac{1}{2} & 0 & 0 & \ldots & & \\ 0 & -\frac{1}{2} & 1 & -\frac{1}{2} & & & \\ & & & & & & \\ & & & & & -\frac{1}{2} & 0 \\ \ldots & & & & & -\frac{1}{2} & 1 \end{pmatrix}. \tag{3.18}$$

Again one trivially has $\|\mathbf{M}_j\|, \|\mathbf{G}_j\| = O(1)$, $j \in \mathbb{N}_0$, so that the hierarchical complement bases are also uniformly stable in the above sense.

Remark 3.1 Evidently, the identification of a complement basis (3.7) is equivalent to *completing* a given refinement matrix $\mathbf{M}_{j,0}$ to an invertible mapping. Any $\mathbf{M}_{j,1}$ for which the completed matrix $\mathbf{M}_j$ satisfies (3.9) will be called *stable completion* of $\mathbf{M}_{j,0}$.

3.3. Multiscale bases

Repeating the decomposition (3.7), one can write each space $S(\Phi_J)$ as a sum of complement spaces

$$S(\Phi_J) = S(\Phi_0) \bigoplus_{j=0}^{J-1} S(\Psi_j).$$

Accordingly, $g_J \in S(\Phi_J)$ can be expanded in *single-scale* form with respect to Φ_J as

$$g_J = \Phi_J^T \mathbf{c}^J, \tag{3.19}$$

as well as in *multiscale form* as

$$g_J = \Phi_0^T \mathbf{c}^0 + \Psi_0^T \mathbf{d}^0 + \ldots + \Psi_{J-1}^T \mathbf{d}^{J-1}, \tag{3.20}$$

with respect to the multiscale basis

$$\Psi^J := \Phi_0 \bigcup_{j=0}^{J-1} \Psi_j. \tag{3.21}$$

Hence, by the denseness of $\mathcal{S}$ (3.1), the union

$$\Psi := \Phi_{j_0} \cup \bigcup_{j=j_0}^{\infty} \Psi_j =: \{\psi_\lambda : \lambda \in \nabla\} \tag{3.22}$$

is a candidate for a *basis* for the whole space H. Here j_0 is some fixed *coarsest level* (which, for simplicity, will usually be assumed to be $j_0 = 0$). We will always use the convention

$$\nabla := \Delta_+ \cup \nabla_-, \tag{3.23}$$

where

$$\Delta_+ := \Delta_{j_0}, \quad \psi_\lambda = \phi_{j_0,k}, \; \lambda := (j_0, k), \quad \nabla_- := \{(j,k) : k \in \nabla_j, j \in \mathbb{N}_0\}.$$

In principle, there is no need to consider only subsets Ψ^J of Ψ defined by *levelwise* truncation. Instead one can select *arbitrary* subsets $\Lambda \subset \nabla$ to form trial spaces

$$S_\Lambda := \mathcal{S}(\Psi_\Lambda), \quad \Psi_\Lambda := \{\psi_\lambda : \lambda \in \Lambda\},$$

to discretize (1.30), say. According to (d) in Section 1.5, the selection of Λ, depending on a particular problem at hand, is a very natural way of steering *adaptivity*. This is perhaps one of the most promising aspects of multiscale basis-oriented methods in comparison with conventional discretizations.

3.4. Multiscale transformations

On the other hand, working with arbitrary subsets $\Lambda \subset \nabla$ will be seen to cause practical problems that should not be underestimated. Adequate data structures have yet to be developed. Things are much simpler for the special case

$$\Lambda_J := \{\lambda \in \nabla : |\lambda| < J\}, \tag{3.24}$$

where

$$|\lambda| = \begin{cases} j & \text{if} \;\; \psi_\lambda \in \Psi_j, \\ j_0 - 1 & \text{if} \;\; \lambda \in \Delta_+, \end{cases}$$

which deserves some special attention.

To this end, both coefficient vectors $\mathbf{c}$ and $\mathbf{d}$ appearing in (3.19), (3.20), respectively, convey different information. While $\mathbf{c}^J$ in (3.19) indicates in many cases, for instance, the geometrical location of the graph of g_J, the $\mathbf{d}^j$ in (3.20) have the character of *differences*. While usually all the entries of $\mathbf{c}^J$ are needed to represent g_J accurately, many of the entries in $\mathbf{d}$ may be small, and replacing some of them by zero may still permit a sufficiently accurate approximation to g_J (recall (1.9) in Section 1.2). On the other hand, the pointwise evaluation of g_J is much simpler in the single-scale form (3.19). These questions will be encountered repeatedly in the course of subsequent developments.

To exploit the benefits of both representations, one needs a mechanism to convert one into the other. These transformations all have a common *pyramid structure*, which is explained next. Since by (3.4) and (3.8),

$$\Phi_j^{\mathrm{T}}\mathbf{c}^j + \Psi_j^{\mathrm{T}}\mathbf{d}^j = \Phi_{j+1}^{\mathrm{T}}\left(\mathbf{M}_{j,0}\mathbf{c}^j + \mathbf{M}_{j,1}\mathbf{d}^j\right),$$

the transformation

$$\mathbf{T}_J : \mathbf{d} \to \mathbf{c} \tag{3.25}$$

is schematically given by

$$\begin{array}{ccccccccc} & \mathbf{M}_{0,0} & & \mathbf{M}_{1,0} & & & \mathbf{M}_{J-1,0} & & \\ \mathbf{c}^0 & \to & \mathbf{c}^1 & \to & \mathbf{c}^2 & \to \cdots & \to & \mathbf{c}^J & \\ & \mathbf{M}_{0,1} & & \mathbf{M}_{1,1} & & & \mathbf{M}_{J-1,1} & & \\ & \nearrow & & \nearrow & & \nearrow \cdots & \nearrow & & \\ \mathbf{d}^0 & & \mathbf{d}^1 & & \mathbf{d}^2 & & \mathbf{d}^{J-1} & & . \end{array} \tag{3.26}$$

To express this in terms of matrix multiplications, define for $j < J$ the $\#\Phi_J \times \#\Phi_J$ matrix

$$\mathbf{T}_{J,j} := \begin{pmatrix} \mathbf{M}_j & \mathbf{0} \\ \mathbf{0} & \mathbf{I} \end{pmatrix},$$

where $\mathbf{I}$ is the identity block of size $\#\Phi_J - \#\Phi_{j+1}$. Then (3.26) becomes

$$\mathbf{T}_J = \mathbf{T}_{J,J-1} \cdots \mathbf{T}_{J,0}. \tag{3.27}$$

As for the inverse transformation, since, by (3.13),

$$\Phi_{j+1}^{\mathrm{T}}\mathbf{c}^{j+1} = \Phi_j^{\mathrm{T}}(\mathbf{G}_{j,0}\mathbf{c}^{j+1}) + \Psi_j^{\mathrm{T}}(\mathbf{G}_{j,1}\mathbf{c}^{j+1}) = \Phi_j^{\mathrm{T}}\mathbf{c}^j + \Psi_j^{\mathrm{T}}\mathbf{d}^j,$$

$\mathbf{T}_J^{-1}$ is realized by

$$\begin{array}{cccccccc} & \mathbf{G}_{J-1,0} & & \mathbf{G}_{J-2,0} & & & \mathbf{G}_{0,0} & \\ \mathbf{c}^J & \to & \mathbf{c}^{J-1} & \to & \mathbf{c}^{J-2} & \to \cdots & \to & \mathbf{c}^0 \\ & \mathbf{G}_{J-1,1} & & \mathbf{G}_{J-2,1} & & & \mathbf{G}_{0,1} & \\ & \searrow & & \searrow & & \searrow \cdots & \searrow & \\ & & \mathbf{d}^{J-1} & & \mathbf{d}^{J-2} & & & \mathbf{d}^0, \end{array} \tag{3.28}$$

which, of course, has a similar product structure as (3.27) involving the blocks $\mathbf{G}_j$.

Complexity of multiscale transformations

Let us comment first on the complexity of the transformations $\mathbf{T}_J$, $\mathbf{T}_J^{-1}$. In the above two examples (see (3.5), (3.6), (3.14), (3.16)) the matrices $\mathbf{M}_j$ and $\mathbf{G}_j$ have only finitely many nonzero entries in each column and row. Thus the operations that take $\mathbf{c}^j, \mathbf{d}^j$ into $\mathbf{c}^{j+1}$ as well as $\mathbf{c}^{j+1}$ into $\mathbf{c}^j, \mathbf{d}^j$ require the order of $\#\Delta_{j+1}$ operations uniformly in j. Since in both cases $\#\Delta_{j+1}/\#\Delta_j \sim \varrho > 1$ (here $\varrho = 2$), one concludes that the execution of $\mathbf{T}_J$ and $\mathbf{T}_J^{-1}$ requires the order of $\#\Delta_J = \dim S(\Phi_J)$ operations uniformly in $J \in \mathbb{N}$. Note that one need not assemble the global transformation $\mathbf{T}_J$ but rather apply local filters like (1.2) and (1.19) which correspond to the successive application of the factors $\mathbf{T}_{J,j}$. This pattern holds in much greater generality, as long as $\#\Delta_j/\#\Delta_{j-1} \geq \varrho > 1$, and the matrices $\mathbf{M}_j$, $\mathbf{G}_j$ stay *uniformly* sparse. By this we mean that the columns (rows) of $\mathbf{M}_j$, $(\mathbf{G}_j)$ contain only a uniformly bounded number of nonzero entries. Thus one may record the following for later use.

Remark 3.2 When all $\mathbf{M}_j$ are uniformly sparse and the cardinality of Φ_j grows geometrically, then the application of $\mathbf{T}_J$ requires $\mathcal{O}(\#\Delta_J)$ operations. Under the same assumptions on the $\mathbf{G}_j$ an analogous statement holds for the inverse transformation $\mathbf{T}_J^{-1}$.

Let us see next how the transformation $\mathbf{T}_J$ may enter a numerical scheme for the approximate solution of (1.30). Suppose one wants to employ a Galerkin scheme based on $S(\Phi_J)$, that is, one has to compute $u_J \in S(\Phi_J)$ satisfying

$$\langle \mathcal{L}u_J, v\rangle = \langle f, v\rangle, \quad v \in S(\Phi_J). \tag{3.29}$$

If u_J is to be represented in single-scale form $u_J = (\mathbf{c}^J)^T\Phi_J$, this amounts to solving the linear system

$$\langle \mathcal{L}\Phi_J, \Phi_J\rangle^T \mathbf{c}^J = \langle f, \Phi_J\rangle^T \tag{3.30}$$

for the unknown coefficient vector $\mathbf{c}^J$. As pointed out in Section 2.2, the matrix $\mathbf{A}_{\Phi_J} := \langle \mathcal{L}\Phi_J, \Phi_J\rangle^T$ may be sparse but increasingly ill-conditioned when J grows. In the special situation of Section 1.4 it has been observed that the stiffness matrix relative to the hierarchical basis has more favourable properties. One readily checks that, in general, the stiffness matrix $\mathbf{A}_{\Psi^J} := \langle \mathcal{L}\Psi^J, \Psi^J\rangle^T$ relative to the multiscale basis Ψ^J (see (3.21)) has the form

$$\mathbf{A}_{\Psi^J} = \mathbf{T}_J^T \mathbf{A}_{\Phi_J} \mathbf{T}_J. \tag{3.31}$$

Hence $\mathbf{A}_{\Psi^J}$ is a *principal section* of the (infinite) matrix

$$\mathbf{A}_\Psi := \langle \mathcal{L}\Psi, \Psi\rangle^T, \tag{3.32}$$

which is often called *standard representation* of $\mathcal{L}$.

Let us assume that $\mathcal{L}$ is a differential operator, so that, when Φ_J consists of compactly supported functions, $\mathbf{A}_{\Phi_J}$ is sparse and has only $\mathcal{O}(N_J)$ nonvanishing entries, where, as before, $N_J = \#\Delta_J$. Hence its accurate computation requires only the order of N_J operations and storage. Since the basis Ψ^J contains functions defined on coarse levels, basis functions from different scales will generally interact, so that $\mathbf{A}_{\Psi^J}$ will generally be much denser. However, in the context of iterative schemes, only the *application* of a matrix to a vector matters. By (3.31), the application of $\mathbf{A}_{\Psi^J}$ to a vector reduces to applying successively $\mathbf{T}_J, \mathbf{A}_{\Phi_j}$ and $\mathbf{T}_J^T$, each requiring, on account of Remark 3.2, the order of N_J operations.

In the above form these multiscale transformations are very efficient relative to the complexity of the *full* space $S(\Phi_J)$. At this point, though, it is not clear how to deal with spaces $S(\Psi_\Lambda)$ spanned by subsets of Ψ^J.

Stability and biorthogonality

There is obviously a continuum of possible complement bases Ψ_j that yield decompositions (3.7), and the question arises whether they are all equally suitable. The Haar basis corresponds to taking orthogonal complements relative to the ℓ_2-inner product, while the hierarchical basis spans orthogonal complements relative to the inner product $a(u,v) = \langle u', v'\rangle_{[0,1]}$ in $H_0^1([0,1])$. Thus orthogonal complements appear to be a canonical choice. However, they are frequently not easy to realize. For instance, any stable completion for (3.6), which induces orthogonal complements, is either dense or gives rise to dense inverses $\mathbf{G}_j$. Moreover, we will encounter situations where orthogonal complements are actually not the best choice.

At any rate, the qualification of the complement bases Ψ_j will be seen to depend crucially on the topological properties of their union Ψ. Aside from efficiency, a first reasonable constraint on the choice of the Ψ_j is the *stability* of the multiscale transformations; see, for example, Dahmen (1994, 1996).

Theorem 3.3 The transformations $\mathbf{T}_J$ are well conditioned in the sense that

$$\|\mathbf{T}_J\|, \|\mathbf{T}_J^{-1}\| = \mathcal{O}(1), \quad J \in \mathbb{N}, \tag{3.33}$$

if and only if the collection Ψ, defined by (3.22), is a *Riesz basis* of H. This means that every $f \in H$ has unique expansions

$$f = \sum_{\lambda\in\nabla} \langle f, \tilde{\psi}_\lambda\rangle \psi_\lambda = \sum_{\lambda\in\nabla} \langle f, \psi_\lambda\rangle \tilde{\psi}_\lambda, \tag{3.34}$$

where $\tilde{\Psi} \subset H$ is a *biorthogonal Riesz basis*, that is,

$$\langle \Psi, \tilde{\Psi}\rangle = \mathbf{I}, \tag{3.35}$$

such that

$$\|f\|_H \sim \|\langle f, \Psi\rangle\|_{\ell_2(\nabla)} \sim \|\langle f, \tilde{\Psi}\rangle\|_{\ell_2(\nabla)}. \tag{3.36}$$

Thus *biorthogonality* is as far as one can deviate from orthogonality. It will be seen that the framework of biorthogonal bases offers a much more flexible setting for constructing multiscale bases such that the matrices $\mathbf{M}_j$ as well as their inverses $\mathbf{G}_j$ are uniformly sparse and give rise to well-conditioned multiscale transformations. Moreover, several schemes that have originally been formulated for orthogonal wavelets (at the expense of infinite although decaying filters) can be adapted to the biorthogonal setting with better localization in physical space.

Remark 3.4 Biorthogonality came out as a necessary condition. In general, it is not quite a sufficient condition for the Riesz basis property (3.36). In fact, as observed by Meyer (1994), not every Schauder basis in a separable Hilbert space is a Riesz basis. Additional conditions ensuring (3.36) will be discussed later.

3.5. Stable completions continued

Constructing a stable completion in the sense of Section 3.2 does not yet guarantee that a collection Ψ of the form (3.22) is a Riesz basis in H. Since in general we cannot resort to Fourier techniques, other tools are needed. As we have seen in Sections 1.2 and 1.4, sometimes *certain* stable completions can be found that may not yet have the desired form. For instance, the hierarchical bases in Sections 1.4 and 4.1 are *not* Riesz bases. In such cases a simple device will help, that allows one to modify the complement bases (Carnicer et al. 1996). It will have several applications later. The first important observation is that, once *some* stable completion is known, *all* others can be *parametrized* as follows.

Proposition 3.5 Suppose that Φ_j are uniformly stable with refinement matrices $\mathbf{M}_{j,0}$ and let $\check{\mathbf{M}}_{j,1}$ be some (uniformly) stable completion of $\mathbf{M}_{j,0}$. Let $\check{\mathbf{G}}_j := \binom{\check{\mathbf{G}}_{j,0}}{\check{\mathbf{G}}_{j,1}}$ denote the inverse of $\check{\mathbf{M}}_j = (\mathbf{M}_{j,0}, \check{\mathbf{M}}_{j,1})$. Then $\mathbf{M}_{j,1}$ is also a stable completion of $\mathbf{M}_{j,0}$, if and only if there exist

$$\mathbf{L}_j : \ell_2(\nabla_j) \to \ell_2(\Delta_j), \quad \mathbf{K}_j : \ell_2(\nabla_j) \to \ell_2(\nabla_j)$$

such that $\mathbf{L}_j$, $\mathbf{K}_j$, $\mathbf{K}_j^{-1}$ are uniformly bounded as operators and

$$\mathbf{M}_{j,1} = \mathbf{M}_{j,0}\mathbf{L}_j + \check{\mathbf{M}}_{j,1}\mathbf{K}_j. \tag{3.37}$$

Moreover, the inverse $\mathbf{G}_j$ of $\mathbf{M}_j = (\mathbf{M}_{j,0}, \mathbf{M}_{j,1})$ is given by

$$\mathbf{G}_{j,0} = \check{\mathbf{G}}_{j,0} - \mathbf{L}_j\mathbf{K}_j^{-1}\check{\mathbf{G}}_{j,1}, \quad \mathbf{G}_{j,1} = \mathbf{K}_j^{-1}\check{\mathbf{G}}_{j,1}. \tag{3.38}$$

Thus, given $\check{\mathbf{M}}_{j,1}$, varying $\mathbf{L}_j$ and $\mathbf{K}_j$ produces a whole family of further stable completions and corresponding decompositions of the spaces $S(\Phi_j)$. The special case $\mathbf{K}_j = \mathbf{I}$ covers the *lifting scheme* proposed by

Sweldens (1996, 1997). In this case one has

$$\Psi_j^{\mathrm{T}} = \Phi_{j+1}^{\mathrm{T}}\mathbf{M}_{j,1} = \Phi_{j+1}^{\mathrm{T}}\mathbf{M}_{j,0}\mathbf{L}_j + \Phi_{j+1}^{\mathrm{T}}\check{\mathbf{M}}_{j,1} = \Phi_j^{\mathrm{T}}\mathbf{L} + \check{\Psi}_j,$$

that is, in terms of individual functions, one has

$$\psi_{j,k} = \sum_{l\in\Delta_j} (\mathbf{L}_j)_{l,k}\phi_{j,l} + \check{\psi}_{j,k}. \tag{3.39}$$

Thus the new wavelet $\psi_{j,k}$ is obtained from the initial wavelet $\check{\psi}_{j,k}$ by adding a linear combination of coarse scale generating functions.

Now the task remains to pick from the above family of stable completions a certain desired one. Specifically, we will have to identify stable completions associated with linear projectors of the form $\langle\cdot,\Xi_j\rangle\Phi_j$ where $\langle\Phi_j,\Xi_j\rangle = \mathbf{I}$.

In fact, Carnicer et al. (1996) have shown that

$$\mathbf{M}_{j,1} = \left(\mathbf{I} - \mathbf{M}_{j,0}\langle\Phi_{j+1},\Xi_j\rangle^{\mathrm{T}}\right)\check{\mathbf{M}}_{j,1} \tag{3.40}$$

are also stable completions with

$$\mathbf{G}_{j,0} = \check{\mathbf{G}}_{j,0} + \langle\Phi_{j+1},\Xi_j\rangle^{\mathrm{T}}\check{\mathbf{M}}_{j,1}\check{\mathbf{G}}_{j,1}, \quad \mathbf{G}_{j,1} = \check{\mathbf{G}}_{j,1}. \tag{3.41}$$

This obviously corresponds to the case $\mathbf{K}_j = \mathbf{I}$ and

$$\mathbf{L}_j = -\langle\Phi_{j+1},\Xi_j\rangle^{\mathrm{T}}\check{\mathbf{M}}_{j,1}. \tag{3.42}$$

To see the relevance of this latter observation in the present context, let for any $\Lambda \subset \nabla$

$$\Psi_\Lambda := \{\psi_\lambda : \lambda \in \Lambda\}. \tag{3.43}$$

If Ψ and $\tilde{\Psi}$ are biorthogonal collections (3.35), then

$$Q_\Lambda v := \langle v, \tilde{\Psi}_\Lambda\rangle\Psi_\Lambda, \quad Q_\Lambda^* v := \langle v, \Psi_\Lambda\rangle\tilde{\Psi}_\Lambda, \tag{3.44}$$

are *projectors* onto the spaces $S(\Psi_\Lambda)$, $S(\tilde{\Psi}_\Lambda)$, respectively, which are adjoints of each other. In particular, for $\Lambda = \Lambda_j$ we simply write $Q_j = Q_{\Lambda_j}$.

Remark 3.6 If Ψ and $\tilde{\Psi}$ are biorthogonal, then

$$Q_\Lambda Q_{\hat{\Lambda}} = Q_\Lambda \quad \text{when } \Lambda \subseteq \hat{\Lambda} \subset \nabla. \tag{3.45}$$

If in addition (3.36) holds, then the Q_Λ, Q_Λ^* are uniformly bounded in H, $\Lambda \subset \nabla$.

Suppose now that the desired biorthogonal multiscale bases are not yet known. Projectors can be also represented with respect to the basis Φ_j of $S(\Phi_j) = S(\Psi^j)$. So let

$$Q_j v = \langle v, \tilde{\Phi}_j\rangle\Phi_j, \tag{3.46}$$

where

$$\langle\Phi_j, \tilde{\Phi}_j\rangle = \mathbf{I}, \tag{3.47}$$

for some $\tilde{\Phi}_j \subset S(\tilde{\Psi}^j)$. We will see next what (3.45) means for the $\tilde{\Phi}_j$.

Remark 3.7 The Q_j defined by (3.46) satisfy (3.45), if and only if the collection $\tilde{\Phi}_j$ is refinable, that is, there exists a matrix $\tilde{\mathbf{M}}_{j,0}$ such that

$$\tilde{\Phi}_j^T = \tilde{\Phi}_{j+1}^T \tilde{\mathbf{M}}_{j,0}, \tag{3.48}$$

and

$$\tilde{\mathbf{M}}_{j,0}^* \mathbf{M}_{j,0} = \mathbf{I}. \tag{3.49}$$

The key to constructing biorthogonal wavelet bases is the following observation (Carnicer et al. 1996). The point is that if dual pairs of refinable generator bases $\Phi_j, \tilde{\Phi}_j$ satisfying (3.47) are given and *some initial* stable completion is known, then biorthogonal wavelets can easily be obtained as follows. One infers from (3.40) and (3.41) the following.

Proposition 3.8 Under the assumptions of Proposition 3.5,

$$\mathbf{M}_{j,1} = \left(\mathbf{I} - \mathbf{M}_{j,0}\tilde{\mathbf{M}}_{j,0}^*\right) \check{\mathbf{M}}_{j,1} \tag{3.50}$$

are also stable completions with

$$\mathbf{G}_{j,0} = \tilde{\mathbf{M}}_{j,0}^*, \quad \mathbf{G}_{j,1} = \check{\mathbf{G}}_{j,1}. \tag{3.51}$$

Moreover, $\tilde{\mathbf{M}}_{j,1} := \mathbf{G}_{j,1}^*$ is a stable completion of $\tilde{\mathbf{M}}_{j,0}$ and the collections Ψ, $\tilde{\Psi}$ obtained from

$$\Psi_j^T := \Phi_{j+1}^T \mathbf{M}_{j,1}, \quad \tilde{\Psi}_j^T := \tilde{\Phi}_{j+1}^T \tilde{\mathbf{M}}_{j,1}, \tag{3.52}$$

by (3.22), are biorthogonal.

Note that when $\check{\mathbf{M}}_j, \check{\mathbf{G}}_j$ and $\tilde{\mathbf{M}}_{j,0}$ are sparse, then the biorthogonal wavelets in Ψ and $\tilde{\Psi}$ have compact support.

4. Examples

The objective of this section is to identify several specializations of the setting described in Section 3, which will be needed later.

4.1. Hierarchical bases

The first example concerns the bivariate counterpart to the construction in Section 1.4. It has attracted considerable attention in connection with the *hierarchical bases preconditioner* (Yserentant 1986).

Suppose Ω is a bounded polygonal domain in $\mathbb{R}^2$ and $\mathcal{T}_0$ is some triangulation of Ω. This means the union of triangles in $\mathcal{T}_0$ agrees with $\overline{\Omega}$ and the intersection of any two different triangles $\tau, \tau' \in \mathcal{T}_0$ is either empty or a common vertex or a common edge. A sequence of triangulations $\mathcal{T}_j$ is then obtained by subdividing each $\tau \in \mathcal{T}_{j-1}$ into four congruent triangles. With each $\mathcal{T}_j$ we associate the space S_j of continuous piecewise linear functions on Ω. Thus, as in the univariate case (see Section 1.4), tent functions form a basis for S_j. In fact, denoting by $\phi_{j,k}$ the unique piecewise linear function which has the value 2^j at the vertex k of $\mathcal{T}_j$ while vanishing at all other vertices, one can show that the corresponding collections Φ_j are uniformly stable (3.3); see, for instance, Oswald (1990). It is clear that the union of the $S(\Phi_j)$ is dense in $H = L_2(\Omega)$.

The *hierarchical bases* are obtained by adding to Φ_j just those basis functions on the next level that correspond to the *new* vertices at the midpoints of the edges in $\mathcal{T}_j$. Thus, denoting by Δ_j the vertices in $\mathcal{T}_j$ and by ∇_j the midpoints of the edges in $\mathcal{T}_j$ or, equivalently, $\nabla_j = \Delta_{j+1} \setminus \Delta_j$, and calling $\mathcal{N}_{j+1,k}$ for $k \in \Delta_j$ the set of neighbouring vertices of k in Δ_{j+1}, one has

$$\phi_{j,k} = \sum_{m\in\{k\}\cup\mathcal{N}_{j+1,k}} 2^{-j-1}\phi_{j,k}(m)\phi_{j+1,m}, \quad k \in \Delta_j, \tag{4.1}$$

that is, the entries of $\mathbf{M}_{j,0}$ are given by

$$(\mathbf{M}_{j,0})_{m,k} = 2^{-j-1}\phi_{j,k}(m) = \begin{cases} \frac{1}{2}, & m = k, \\ \frac{1}{4}, & m \in \mathcal{N}_{j+1,k}, \\ 0, & \text{else.} \end{cases} \tag{4.2}$$

Since

$$\psi_{j,k} := \phi_{j+1,k}, \quad k \in \nabla_j \tag{4.3}$$

one has the completion

$$(\mathbf{M}_{j,1})_{m,k} = \delta_{m,k}, \quad m \in \Delta_{j+1}, \quad k \in \nabla_j. \tag{4.4}$$

On the other hand, since also for $m \in \Delta_j$ one has $\mathcal{N}_{j+1,m} \subseteq \nabla_j$, (4.1) and (4.3) imply

$$\phi_{j+1,m} = 2\phi_{j,m} - \sum_{k\in\mathcal{N}_{j+1,m}} \frac{1}{2}\psi_{j,k}, \tag{4.5}$$

so that in this case we infer from (3.13)

$$(\mathbf{G}_{j,0})_{k,m} = 2\delta_{k,m}, \quad m \in \Delta_{j+1}, \quad k \in \Delta_j, \tag{4.6}$$

and

$$(\mathbf{G}_{j,1})_{k,m} = \begin{cases} -\frac{1}{2}, & m \in \Delta_j, k \in \mathcal{N}_{j+1,m}, \\ \delta_{k,m}, & k,m \in \nabla_j, \\ 0, & \text{else.} \end{cases} \tag{4.7}$$

Since obviously $\|\mathbf{M}_j\|$, $\|\mathbf{G}_j\| = O(1)$, $j \in \mathbb{N}$, the $\mathbf{M}_{j,1}$, defined by (4.4), are indeed uniformly stable completions.

However, note that

$$S(\Psi_j) = (L_{j+1} - L_j)S(\Phi_{j+1}) \tag{4.8}$$

where the L_j are the *interpolation* projectors defined by

$$L_j f := \sum_{k\in\Delta_j} 2^{-j} f(k)\phi_{j,k}.$$

Hence the basis Ψ obtained in this way (see (3.22)) has no dual in $L_2(\Omega)$ and is therefore *not* a Riesz basis. However, it will serve as a convenient initial stable completion in the sense of Proposition 3.8. Some consequences of these facts will be discussed later in connection with preconditioning.

4.2. Wavelets on $\mathbb{R}^n$

The construction of wavelet bases is best understood for $H = L_2(\mathbb{R})$, where the notion of *multiresolution analysis* has originated from Mallat (1989), Meyer (1990) and Daubechies (1988).

Stationary multiresolution

Let us first consider the univariate case $n = 1$. Suppose that $\phi \in L_2(\mathbb{R})$ has *stable shifts*

$$\|\mathbf{c}\|_{\ell_2(\mathbb{Z})} \sim \left\| \sum_{k\in\mathbb{Z}} c_k \phi(\cdot - k) \right\|_{L_2(\mathbb{R})} \tag{4.9}$$

and is *refinable*, that is, there exists a *mask* $\mathbf{a} \in \ell_2(\mathbb{Z})$ such that

$$\phi(x) = \sum_{k\in\mathbb{Z}} a_k \phi(2x - k), \quad x \in \mathbb{R}, \text{ almost everywhere.} \tag{4.10}$$

Hence the collections

$$\Phi_j := \left\{ \phi_{j,k} := 2^{j/2}\phi(2^j \cdot -k) : k \in \mathbb{Z} \right\}$$

are *uniformly stable* (3.3) and satisfy (3.4) with $\mathbf{M}_{j,0} = \mathbf{M}_0 = (a_{l-2k})_{l,k\in\mathbb{Z}}$. Thus the refinement matrices are stationary, that is, they are independent of the scale j and the spatial location k. The examples from Sections 1.2 and 1.4 are obviously obtained by restricting collections Φ_j of this type to $[0, 1]$. The function ϕ is often called *scaling function* or *generator* of the multiresolution sequence $\mathcal{S} = \{S(\Phi_j)\}_{j\in\mathbb{Z}}$, which is known to be dense in $L_2(\mathbb{R})$; see, for example, de Boor, DeVore and Ron (1993) and Jia and Micchelli (1991).

Time-frequency analysis and Fourier techniques have been an indispensible source of construction tools. It is well known (de Boor et al. 1993,

Daubechies 1992, Mallat 1989, Jia and Micchelli 1991) that, in terms of the Fourier transform, stability (4.9) is equivalent to

$$\sum_{k\in\mathbb{Z}} |\hat{\phi}(y+2\pi k)|^2 \geq c > 0, \tag{4.11}$$

while the refinement relation (4.10) reads

$$\hat{\phi}(y) = 2^{-1}a(e^{-iy/2})\hat{\phi}(y/2). \tag{4.12}$$

The Laurent polynomial

$$a(z) = \sum_{k\in\mathbb{Z}} a_k z^k$$

is called the *symbol* of the mask **a**. Since under the present assumptions $\hat{\phi}$ is continuous, reiteration of (4.12) yields

$$\hat{\phi}(y) = \left\{\prod_{j=1}^{\infty}\left(2^{-1}a(e^{-i2^{-j}y})\right)\right\}\hat{\phi}(0), \tag{4.13}$$

where the product converges uniformly on compact sets so that we always have $\hat{\phi}(0) \neq 0$. Thus we may assume that ϕ is normalized to $\hat{\phi}(0) = 1$.

An important special case arises when the shifts $\phi(\cdot - k)$ are *orthonormal* so that (4.9) becomes an equality. An example is the scaling function (see Daubechies (1992, page 137))

$$\hat{\phi}(y) := \begin{cases} 1, & |y| \leq 2\pi/3, \\ \cos\left(\frac{\pi}{2}\nu(\frac{3}{2\pi}|y| - 1)\right), & 2\pi/3 \leq |y| \leq 4\pi/3, \\ 0, & \text{otherwise}, \end{cases} \tag{4.14}$$

where ν is a smooth function satisfying

$$\nu(x) = \begin{cases} 0, & x \leq 0, \\ 1, & x \geq 1. \end{cases}$$

Another interesting example is

$$\hat{\phi}(y) = 2 + (1 - e^{-iy})q(y), \tag{4.15}$$

where the trigonometric polynomial q is chosen, so that the shifts $\phi(\cdot - k), k \in \mathbb{Z}$ are orthonormal and

$$\frac{d^l}{dy^l}\hat{\phi}\,|_{y=0} = \delta_{0,l}, \quad l = 0, \ldots, d-1.$$

This latter condition means that

$$\int_{\mathbb{R}} x^l \phi(x)\,\mathrm{d}x = \delta_{0,l}, \quad l = 0, \ldots, d-1, \tag{4.16}$$

that is, the scaling function ϕ also has certain vanishing moments. Using essentially the same Taylor expansion argument as in Section 1.3, condition (4.16) implies that, for smooth f, one has $\langle f, \phi_{J,k}\rangle_{\mathbb{R}} \approx 2^{J/2} f(2^{-J}k)$. In fact, one can show that, for instance,

$$\left|\sum_{k\in\mathbb{Z}} f(2^{-J}k)\phi(2^J l - k) - f(2^{-J}l)\right| \lesssim 2^{-Jd}\|f\|_{W^{d,\infty}(\mathbb{R})}, \tag{4.17}$$

so that the expansion $\sum_{k\in\mathbb{Z}} f(2^{-J}k)\phi(2^J x - k)$ *almost interpolates* f. While ϕ from (4.14) has global support, the support width of ϕ from (4.15) is $3d-1$ (Daubechies 1992, page 258).

The now famous scaling functions ϕ with orthonormal shifts of smaller support (of width $2d-1$) have been constructed by Daubechies (1988, 1992).

When the shifts of ϕ are orthonormal, it can be verified that the shifts of

$$\psi(x) := \sum_{k\in\mathbb{Z}} (-1)^k a_{1-k}\phi(2x - k) \tag{4.18}$$

form an orthonormal basis of the orthogonal complement of $S(\Phi_0)$ in $S(\Phi_1)$, so that the corresponding $\psi_{j,k}$ constitute an orthonormal basis for $L_2(\mathbb{R})$. The function ϕ from (4.14) gives rise to the *Meyer wavelet* (Meyer 1990) which has extremely good localization in Fourier space but has rather slow decay in physical space. The wavelets for (4.15) are called *coiflets* and will be referred to later again.

In general one can say that $\hat{\phi}$ and $\hat{\psi}$ act like *low pass* and *band pass* filters. For an extensive discussion of this background see Daubechies (1992).

However, orthonormality will merely be viewed as a special case of the more flexible concept of biorthogonality that came up in Section 3.4; see Cohen, Daubechies and Feauveau (1992).

Dual pairs

The scaling functions $\phi, \tilde{\phi}$ are said to form a *dual pair* if

$$\langle \phi, \tilde{\phi}(\cdot - k)\rangle_{\mathbb{R}} := \int_{\mathbb{R}} \phi(x)\overline{\tilde{\phi}(x-k)}\,\mathrm{d}x = \delta_{0,k}, \quad k \in \mathbb{Z}. \tag{4.19}$$

We will sometimes refer to ϕ and $\tilde{\phi}$ as *primal* and *dual* generator, respectively. It is easy to see that compact support of ϕ and $\tilde{\phi}$ implies that the masks $\mathbf{a}$ and $\tilde{\mathbf{a}}$ have *finite support* and that (4.19) implies stability (4.9). Moreover, it is known that the functions

$$\psi(x) := \sum_{k\in\mathbb{Z}} (-1)^k \tilde{a}_{1-k}\phi(2x-k), \quad \tilde{\psi}(x) := \sum_{k\in\mathbb{Z}} (-1)^k a_{1-k}\tilde{\phi}(2x-k) \tag{4.20}$$

satisfy

$$\langle \phi, \tilde{\psi}(\cdot - k)\rangle_{\mathbb{R}} = \langle \tilde{\phi}, \psi(\cdot - k)\rangle_{\mathbb{R}} = 0, \quad \langle \psi, \tilde{\psi}(\cdot - k)\rangle_{\mathbb{R}} = \delta_{0,k}, \quad k \in \mathbb{Z}, \tag{4.21}$$

which obviously covers (4.18) as a special case. Observe next that straightforward computations confirm that the relations (4.19) and (4.21) are equivalent to

$$\begin{pmatrix} a(z) & a(-z) \\ b(z) & b(-z) \end{pmatrix} \begin{pmatrix} \overline{\tilde{a}(z)} & \overline{\tilde{b}(z)} \\ \overline{\tilde{a}(-z)} & \overline{\tilde{b}(-z)} \end{pmatrix} = \begin{pmatrix} 4 & 0 \\ 0 & 4 \end{pmatrix}. \tag{4.22}$$

This can be used for the construction of the dual generator $\tilde{\phi}$. Given the mask $\mathbf{a}$ one can determine $\tilde{\mathbf{a}}$ satisfying the first relation in (4.22) and then show that the product (4.13) with $\tilde{\mathbf{a}}$, instead of $\mathbf{a}$, is the Fourier transform of an L_2-function.

One easily deduces from (4.10) and (4.21) that for $\Psi_j := \{\psi_{j,k} : k \in \mathbb{Z}\}$, $\tilde{\Psi}_j := \{\tilde{\psi}_{j,k} : k \in \mathbb{Z}\}$ the collections

$$\Psi := \Phi_0 \bigcup_{j \geq 0} \Psi_j, \quad \tilde{\Psi} := \tilde{\Phi}_0 \bigcup_{j \geq 0} \tilde{\Psi}_j \tag{4.23}$$

are biorthogonal.

To relate this to the discussion in Section 3.2, note that with $b_k := (-1)^k \tilde{a}_{1-k}$, $\tilde{b}_k := (-1)^k a_{1-k}$ the bi-infinite matrix $\mathbf{M}_{j,1} = \mathbf{M}_1 := (b_{l-2k})_{l,k \in \mathbb{Z}}$ is a stable completion of $\mathbf{M}_0$ above and that in this case (see Proposition 3.8),

$$\mathbf{G}_0 = \tilde{\mathbf{M}}_0^* = \left(\bar{\tilde{a}}_{l-2k}\right)_{k,l \in \mathbb{Z}}, \quad \mathbf{G}_1 = \tilde{\mathbf{M}}_1^* = \left(\bar{\tilde{b}}_{l-2k}\right)_{k,l \in \mathbb{Z}}. \tag{4.24}$$

B-splines as primal generators give rise to an important class of dual pairs where both generators have compact support. Let $\lfloor x \rfloor$ ($\lceil x \rceil$) denote the largest (smallest) integer less (greater) than or equal to x, and define $N_d = \chi_{[0,1)} * \cdots * \chi_{[0,1)}$ as the d-fold convolution of the box function (1.1). Then, for

$$\phi = {}_d\phi := N_d\left(\cdot + \left\lfloor \tfrac{d}{2} \right\rfloor\right), \quad \hat{N}_d(y) = \left(\frac{1 - e^{-iy}}{iy}\right)^d, \tag{4.25}$$

(4.10) becomes

$${}_d\phi(x) = \sum_{k=-\lfloor \frac{d}{2} \rfloor}^{\lceil \frac{d}{2} \rceil} 2^{1-d} \binom{d}{k + \lceil \frac{d}{2} \rceil} {}_d\phi(2x - k). \tag{4.26}$$

Cohen et al. (1992) have shown that for every $d, \tilde{d} \in \mathbb{N}$, $\tilde{d} \geq d$, $d + \tilde{d}$ even, there exists a compactly supported scaling function ${}_{d,\tilde{d}}\tilde{\phi}$ such that $({}_d\phi, {}_{d,\tilde{d}}\tilde{\phi})$ form a dual pair. The role of the parameters $d, \tilde{d}$ will be pointed out below.

Polynomial exactness
It is remarkable that in the present *stationary setting* the refinement equation (4.12) has further important consequences. In fact, since, by (4.12),

$$\hat{\phi}(2\pi k 2^n) = \left\{ \prod_{j=1}^{n} \left(2^{-1} a(e^{-i2^{n-j}2\pi}) \right) \right\} \hat{\phi}(2\pi k),$$

letting n tend to infinity and applying the Riemann–Lebesgue lemma yields

$$\hat{\phi}(2\pi k) = 0, \quad k \in \mathbb{Z} \setminus \{0\}. \tag{4.27}$$

By the Poisson summation formula, this means that (Cavaretta, Dahmen and Micchelli 1991)

$$1 = \sum_{k \in \mathbb{Z}} \phi(x - k), \quad x \in \mathbb{R}. \tag{4.28}$$

Similarly, a somewhat refined argument shows that $\phi \in H^r(\mathbb{R})$ implies

$$\hat{\phi}^{(l)}(2\pi k) = 0, \quad k \in \mathbb{Z} \setminus \{0\}, \quad l = 0, \ldots, r, \tag{4.29}$$

(Cavaretta et al. 1991) so that Poisson's summation formula again implies that, for any polynomial p of degree at most r, there exists some polynomial q of lower degree such that

$$p(x) = \sum_{k \in \mathbb{Z}} p(k)\phi(x - k) + q(x). \tag{4.30}$$

In particular, when the scaling function ϕ also has vanishing moments (4.16), then the polynomial q can be shown to vanish. Combining this polynomial reproduction property with arguments from the proof of Proposition 5.1 below yields estimates of the form (4.17) above. The fact that shifts of ϕ represent polynomials of degree r exactly is reflected by the fact that the symbol $a(z)$ contains a power of $(1 + z)$, that is

$$a(z) = (1 + z)^{r+1} q(z) \tag{4.31}$$

where $q(1) = 2^{-r}$ (Daubechies 1992).

Returning to the above family $({}_d\phi, {}_{d,\tilde{d}}\tilde{\phi})$ of dual pairs, the parameters $d, \tilde{d}$ are exactly the respective orders of polynomial reproduction. Thus (4.19) yields

$$\begin{aligned} x^r &= \sum_{k \in \mathbb{Z}} \langle (\cdot)^r, {}_{d,\tilde{d}}\tilde{\phi}(\cdot - k) \rangle_{\mathbb{R}} \, {}_d\phi(x - k), \quad r = 0, \ldots, d - 1, \\ x^r &= \sum_{k \in \mathbb{Z}} \langle (\cdot)^r, {}_d\phi(\cdot - k) \rangle_{\mathbb{R}} \, {}_{d,\tilde{d}}\tilde{\phi}(x - k), \quad r = 0, \ldots, \tilde{d} - 1, \end{aligned} \tag{4.32}$$

which has two important consequences. On one hand, as indicated above, the order of polynomial reproduction governs the approximation power of the spaces $S(\Phi_j)$; see, for instance, Cavaretta et al. (1991). This will be established later in somewhat greater generality. Here we mention first the following important further implication.

Moment conditions
As an immediate consequence of (4.21) and (4.32), we state that

$$\int_{\mathbb{R}} x^r \psi(x)\,\mathrm{d}x = 0, \quad r = 0,\ldots,\tilde{d}-1, \qquad \int_{\mathbb{R}} x^r \tilde{\psi}(x)\,\mathrm{d}x = 0, \quad r = 0,\ldots,d-1, \tag{4.33}$$

when $\psi, \tilde{\psi}$ are the wavelets (4.20) relative to the dual pair $({}_d\phi, {}_{d,\tilde{d}}\tilde{\phi})$ from above. The wavelets $\psi, \tilde{\psi}$ are said to have *vanishing moments* of order $\tilde{d}, d$, respectively. Recall from Section 1.3 that the order of vanishing moments governs the compression capacity of a wavelet. The fact that in connection with biorthogonal wavelets the order of vanishing moments can be chosen *independently* of the order of exactness will play an important role later.

Integration by parts
There is an important trick for generating a dual pair from another one, essentially by *integrating up* and *differentiating down* (Dahmen, Kunoth and Urban 1996*c*, Lemarié-Rieusset 1992, Urban 1995*a*). To this end, suppose that $(\phi, \tilde{\phi})$ is a dual pair and $\phi \in H^{1+\varepsilon}(\mathbb{R})$. By the previous remarks, its symbol $a(z)$ is divisible by $(1+z)$. The new symbols

$$a^-(z) := \frac{2}{1+z}a(z), \quad \tilde{a}^+(z) := \frac{1+\bar{z}}{z}\tilde{a}(z) \tag{4.34}$$

obviously still satisfy the first relation in (4.22). Moreover, the refinement relations (4.11) relative to the masks $\mathbf{a}^-, \tilde{\mathbf{a}}^+$ can be shown still to possess solutions $\phi^-, \tilde{\phi}^+ \in L_2(\mathbb{R})$ with compact support, which are related by

$$\frac{d}{dx}\phi(x) = \phi^-(x) - \phi^-(x-1), \quad \frac{d}{dx}\tilde{\phi}^+(x) = \tilde{\phi}(x+1) - \tilde{\phi}(x). \tag{4.35}$$

Since one still has $a^-(z)\overline{\tilde{a}^+(z)} + a^-(-z)\overline{\tilde{a}^+(-z)} = 4$, $(\phi^-, \tilde{\phi}^+)$ is still a dual pair. Moreover, the corresponding wavelets $\psi^-, \tilde{\psi}^+$, defined by (4.20), are related to $\psi, \tilde{\psi}$ by

$$\frac{d}{dx}\psi(x) = 4\psi^-(x), \quad \frac{d}{dx}\tilde{\psi}^+(x) = -4\tilde{\psi}(x). \tag{4.36}$$

We will have several opportunities to make use of these facts later.

The multivariate case
The simplest way of generating orthogonal or biorthogonal wavelets on $\mathbb{R}^n$ is via tensor products. Given any dual pair $(\varphi, \tilde{\varphi})$ of univariate scaling

functions, the products

$$\phi(x) := \varphi(x_1)\cdots\varphi(x_n), \quad \tilde{\phi}(x) := \tilde{\varphi}(x_1)\cdots\tilde{\varphi}(x_n) \tag{4.37}$$

obviously form a dual pair in $L_2(\mathbb{R}^n)$.

The corresponding masks are obtained from the univariate ones in a straightforward fashion. One should note that for scalings by powers of two one now needs $2^n - 1$ different wavelets whose shifts span the complement spaces. Setting $E := \{0,1\}^n$, $E_* := E \setminus \{0\}$, it is convenient to index these *mother wavelets* as follows,

$$\psi_e(x) = \psi_{e_1}(x_1)\cdots\psi_{e_n}(x_n), \quad e \in E_*, \tag{4.38}$$

where we sometimes denote for convenience $\psi_0 := \varphi$. The $\tilde{\psi}_e$ are defined analogously. Thus, while associating the functions

$$\phi_{j,k} := 2^{nj/2}\phi(2^j \cdot -k), \quad k \in \mathbb{Z}^n, \tag{4.39}$$

with the index set or grid $\Delta_j := 2^{-j}\mathbb{Z}^n$, the wavelets $\psi_{e,j,k}$, $\tilde{\psi}_{e,j,k}$ correspond to $\nabla_{e,j} := 2^{-j}\left(\frac{e}{2} + \mathbb{Z}^n\right)$, so that $\Delta_{j+1} = \Delta_j \cup \left(\bigcup_{e\in E_*} \nabla_{e,j}\right)$.

Several alternatives have been studied. First, one might look for *genuinely* multivariate scaling functions and wavelets. The practical relevance in terms of small masks and locality seems to be confined to a few special cases; see, for instance, Cohen and Schlenker (1993). On the other hand, the tensor product structure offers numerous advantages with regard to computational efficiency, via reduction to univariate problems, and data structures, as long as the underlying grid structure is regular. However, to reduce the number of mother wavelets, one might employ scalings by suitable integer matrices M with all eigenvalues strictly greater than one. One then needs $|\det M| - 1$ mother wavelets (Gröchenich and Madych 1992, Cohen and Daubechies 1993, Dahlke, Dahmen and Latour 1995). Again, much less machinery is available in this case. Finally, instead of considering spaces generated by a single scaling function, one can use a fixed finite collection of generators. In summary, however, since none of these approaches overcomes the obstructions posed by more complex domain geometries, it is fair to say they do not offer any significant advantages for the problems considered here.

Computational issues

Obviously, the stationary setting offers a variety of computational advantages. One need not assemble any level dependent refinement or completion matrices. The multiscale transformations (3.26) and (3.28) reduce to local applications of finite filter masks which are fixed once and for all; see Barsch, Kunoth and Urban (1997) for a discussion of these issues. The main point of this section is to present some computational techniques for basic tasks like evaluating function values, derivatives and integrals of wavelets, which have no counterpart in conventional discretization settings.

Even though many scaling functions and hence corresponding wavelets possess no closed analytic representation, all essential information can be drawn from the masks. We will briefly exemplify this fact for the computation of integrals of products of scaling functions and wavelets or their derivatives. More details of the following facts can be found in Dahmen and Micchelli (1993) and Latto, Resnikoff and Tenenbaum (1992), and corresponding implementations are documented in Kunoth (1995).

Due to the two scale relations (4.20), integrals involving wavelets can be reduced to integrals involving only scaling functions. Thus Galerkin discretization of a partial differential equation requires evaluating terms like

$$\int_{\Omega} a(x)\partial^{\alpha}\phi_{j,k}(x)\partial^{\beta}\phi_{j,l}(x)\,\mathrm{d}x, \tag{4.40}$$

where a successive application of (4.10) has been used when wavelets on different levels j, j' are involved. Assuming for simplicity that Ω is a union of rectangular domains, the above integral can be written as

$$\sum_{m\in\mathbb{Z}^n}\int_{R^n} \chi_{j,m}(x)a(x)\partial^{\alpha}\phi_{j,k}(x)\partial^{\beta}\phi_{j,l}(x)\,\mathrm{d}x, \tag{4.41}$$

where $\chi = \chi_{\Box}$ is the characteristic function of the unit cube $\Box = [0,1]^n$. Due to the compact support of ϕ, the sum is actually finite and involves at most $|\operatorname{supp}\phi|$ terms.

Applying quadrature to quantities like (4.40) may not always be advisable, since although $a(x)$ may be very regular the accuracy of the quadrature is limited by the factors $\partial^{\alpha}\phi_{j,k}$, which may have very low regularity. Let us therefore point out how to evaluate (4.40) up to an accuracy that *only* depends on $a(x)$. To this end, let θ be any other scaling function such as a (tensor product) B-spline. Replacing $a(x)$ by some approximation $\sum_{l\in\mathbb{Z}^n} a_l\theta_{j,l}(x) =: a_j(x)$ which could, for instance, be obtained by interpolation, the compact support of θ again ensures that, when replacing $a(x)$ by $a_j(x)$ in (4.41), the sum over $l \in \mathbb{Z}^n$ is again finite, so that one ultimately has to compute after rescaling the quantities

$$\int_{\mathbb{R}} \chi_{\Box}(x)\theta(x-k^1)\partial^{\alpha}\phi(x-k^2)\partial^{\beta}\phi(x-k^3)\,\mathrm{d}x. \tag{4.42}$$

Similar expressions arise when discretizing nonlinear terms such as those appearing in Burgers equation.

Here a new idea enters. The point is now that, given *any* finite number of (possibly different) scaling functions ϕ_i, (with finitely supported masks),

$i = 0, \ldots, m$, with $\phi_i \in C^r(\mathbb{R}^n)$ say, then expressions of the form

$$I(k^1, \ldots, k^m, \mu^1, \ldots, \mu^m) := \int_{\mathbb{R}^n} \phi_0(x) \prod_{i=1}^{m} \partial^{\mu^i} \phi_i(x - k^i)\, \mathrm{d}x \tag{4.43}$$

can be computed *exactly* (up to round-off). Thus the accuracy of the quantities in (4.40) depends *only* on the approximability of the coefficient $a(x)$.

This is essentially a consequence of refinability and its close connection with *subdivision techniques*; see Cavaretta et al. (1991), Dahmen and Micchelli (1993) and Latto et al. (1992). The main ideas are now sketched. Suppose ϕ is a scaling function. Differentiating and evaluating (4.10) at (multi-)integers, yields

$$2^{-|\mu|} \partial^\mu \phi(k) = \sum_{l \in \mathbb{Z}^n} a_{2k-l} \partial^\mu \phi(l). \tag{4.44}$$

Clearly $(\partial^\mu \phi(k) : k \in \mathbb{Z}^n)$ is finitely supported. Thus (4.44) may be seen as an eigenvector relation, that is, the vector $\mathbf{V}^\mu = (\partial^\mu \phi(k) : k \in \operatorname{supp} \phi)$ is an eigenvector of a finite section of the transpose of the refinement matrix for the eigenvalue $2^{-|\mu|}$. When $n > 1$, that is, μ is a multi-integer, *every* $\mathbf{V}^\mu$ with $|\mu| = r$ is an eigenvector with eigenvalue 2^{-r}.

To exploit these relations for evaluating $\partial^\mu \phi(k)$, $k \in \mathbb{Z}^n$, one therefore has to find suitable additional conditions and show that they actually identify each $\mathbf{V}^\mu$ *uniquely*. Before we describe such conditions we point out that,

(i) once $\partial^\mu \phi|_{\mathbb{Z}^n}$ is known, successive use of (4.10) yields $\partial^\mu \phi|_{2^{-j}\mathbb{Z}^n}, j \in \mathbb{N}$
(ii) this can be used to determine the integrals (4.43).

To explain this latter fact, let us catenate $(k^1, \ldots, k^m)$, $(\mu^1, \ldots, \mu^m)$ to vectors k, μ in $\mathbb{Z}^s$, $\mathbb{Z}^s_+$, respectively, where $s = mn$. Note that

$$I(k, \mu) = (-1)^\mu \partial^\mu F(k), \tag{4.45}$$

where

$$F(y) := \int_{\mathbb{R}^n} \phi_0(x) \phi_1(x - y^1) \cdots \phi_m(x - y^m)\, \mathrm{d}x.$$

The point is that F is again a refinable function with mask coefficients

$$c_k = 2^{-n} \sum_{l \in \mathbb{Z}^n} a^0_l \prod_{i=1}^{m} a^i_{l-k^i}, \tag{4.46}$$

where $\mathbf{a}^i$ is the mask of ϕ_i.

Theorem 4.1 (Dahmen and Micchelli 1993) Suppose that all ϕ_i are stable in the sense of (4.9) and $\phi_i \in C^r(\mathbb{R}^n)$, $i = 1, \ldots, m$. Then for any $\mu \in \mathbb{Z}^{mn}_+$, $|\mu| \le r$, there exists a unique sequence $\mathbf{V}^\mu$ of finite support in

$\mathbb{Z}^{mn}$, satisfying

$$2^{-|\mu|}V_k^\mu = \sum_{l\in\mathbb{Z}^{mn}} c_{2k-l}V_l^\mu, \quad k \in \mathbb{Z}^{mn}, \tag{4.47}$$

and

$$\sum_{k\in\mathbb{Z}^{mn}} (-k)^\nu V_k^\mu = \mu!\,\delta_{\nu,\mu}, \quad |\nu| \le |\mu|, \quad \nu,\mu \in \mathbb{N}_0^{mn}, \tag{4.48}$$

where $\mathbf{c}$ is defined by (4.46). Moreover, one has

$$\mathbf{V}_k^\mu = \partial^\mu F(k) = (-1)^\mu I(k,\mu), \quad k \in \mathbb{Z}^{mn}. \tag{4.49}$$

The moment conditions (4.48) are implied by the polynomial reproduction (4.30), (4.32). The proof that these conditions determine the $\mathbf{V}^\mu$ uniquely, employs the concept of subdivision algorithms (Dahmen and Micchelli 1993, Cavaretta et al. 1991).

Other variants of similar nature can be found in Dahmen and Micchelli (1993) and Sweldens and Piessens (1994), among them recursions for evaluating moments like $\int_{\mathbb{R}^n} x^\beta \phi(x-\alpha)\,dx$.

Remark 4.2 The efficiency of this concept deteriorates when the factors in the integrals (4.40) or (4.41) involve functions on different scales, since this requires correspondingly many prior applications of refinement matrices. This problem does not arise when working with the so-called non-standard representation, which will be introduced later. Likewise, when $\mathcal{L}$ is a differential operator and ϕ has compact support, the above scheme can be used to compute the stiffness matrix $\mathbf{A}_{\Phi_J} := \langle \mathcal{L}\Phi_J, \Phi_J\rangle^{\mathrm{T}}$ accurately and efficiently. The multiscale transformation $\mathbf{T}_J$ (3.26) can then be employed to generate the stiffness matrix $\mathbf{A}_{\Psi^J}$ (3.32) at the expense of $\mathcal{O}(N_J)$ operations. Again this may not be the best strategy for dealing with matrices $\mathbf{A}_{\Psi_\Lambda}$ for arbitrary $\Lambda \subset \nabla$.

4.3. Periodization

The above setting is clearly not suitable yet for the treatment of operator equations which are usually defined on bounded domains.

A very special but nevertheless important framework is the *periodic* setting (Meyer 1990). It essentially retains all the structural and computational advantages of the stationary shift-invariant case considered above. There are at least two reasons for addressing this case with great care. First, many effects will be seen to be local in nature and hence also provide important insight for more general situations. Second, one might aim at a *two-stage* process, trying to carry out the bulk of computation via the full spatial dimension relative to a periodized problem, while treating domain-related effects like boundary conditions separately.

The simple trick is to replace the meaning of $g_{j,k} := 2^{nj/2} g(2^j \cdot - k)$, $k \in \mathbb{Z}^n$, for compactly supported or rapidly decaying $g \in L_2(\mathbb{R}^n)$ by its *periodized* counterpart

$$g_{j,k}(x) := 2^{nj/2} \sum_{l \in \mathbb{Z}^n} g\left(2^j(x+l) - k\right). \tag{4.50}$$

Given any dual pair $(\phi, \tilde{\phi})$ on $\mathbb{R}^n$, and setting $\Delta_j := \mathbb{Z}^n / 2^j \mathbb{Z}^n$, the corresponding sets

$$\Phi_j := \{\phi_{j,k} : k \in \Delta_j\}, \quad \Psi_{e,j} := \{\psi_{e,j,k} : k \in \Delta_j\}, \quad e \in E_*, \tag{4.51}$$

and likewise $\tilde{\Phi}_j, \tilde{\Psi}_{e,j}$, have finite cardinality 2^{nj} and consist of functions which are *one-periodic* in each variable. Note that this preserves orthogonality relations. One easily checks that (4.19) and (4.21) still imply that

$$\langle \phi_{j,k}, \tilde{\phi}_{j,l} \rangle_\square = \int_\square \phi_{j,k}(x) \overline{\tilde{\phi}_{j,l}(x)} \, \mathrm{d}x = \delta_{k,l}, \quad k, l \in \Delta_j, \tag{4.52}$$

and that the collections

$$\Psi := \Phi_0 \cup \bigcup_{j=0}^{\infty} \left(\bigcup_{e \in E_*} \Psi_{e,j} \right), \quad \tilde{\Psi} := \tilde{\Phi}_0 \cup \bigcup_{j=0}^{\infty} \left(\bigcup_{e \in E_*} \tilde{\Psi}_{e,j} \right), \tag{4.53}$$

are biorthogonal

$$\langle \Psi, \tilde{\Psi} \rangle_\square = \mathbf{I}. \tag{4.54}$$

Hence the $\mathcal{S} = \{S(\Phi_j)\}_{j \in \mathbb{N}_0}$, $\tilde{\mathcal{S}} = \{S(\tilde{\Phi}_j)\}_{j \in \mathbb{N}_0}$ form two biorthogonal multiresolution sequences fitting into the framework of Section 3 for $H = L_2(\mathbb{R}^n / \mathbb{Z}^n)$. One readily verifies that

$$\phi_{j,k}(x) = \sum_{l \in \Delta_{j+1}} \left(\sum_{m \in \mathbb{Z}^n} 2^{-n/2} a_{l-2k+2^{j+1}m} \right) \phi_{j+1,l}, \tag{4.55}$$

that is, the new masks are obtained by 2^{j+1}-periodization. Thus the refinement matrices $\mathbf{M}_{j,0}$ have *circulant* structure and analogously the completion $\mathbf{M}_{j,1}$ (as well as $\tilde{\mathbf{M}}_{j,0}$, $\tilde{\mathbf{M}}_{j,1}$).

Defining the discrete Fourier coefficients

$$\mathcal{F}_k(f) := \int_\square f(x) e^{-2\pi i x \cdot k} \, \mathrm{d}x, \quad k \in \mathbb{Z}^n,$$

it is clear that for the periodization $[g] := \sum_{m \in \mathbb{Z}^n} g(\cdot + m)$ one has

$$\mathcal{F}_k([g]) = \hat{g}(k), \quad k \in \mathbb{Z}^n.$$

Hence any results relative to $\mathbb{R}^n$ are readily related to corresponding results for $\mathbb{R}^n / \mathbb{Z}^n$; see also Fröhlich and Schneider (1995).

4.4. Wavelets on the interval

There is one further extension beyond $\mathbb{R}^n$ or $\mathbb{R}^n/\mathbb{Z}^n$ that is worth mentioning, namely wavelet-like bases on $[0,1]$. This still seems to be awfully restrictive. However, it will later be seen to be a key ingredient for the construction of wavelets on any domain that can be represented as a disjoint union of parametric images of cubes. This includes closed surfaces arising in connection with boundary integral equations (see Section 2.2).

Wavelets on the interval have been discussed in several papers; see, for instance Andersson, Hall, Jawerth and Peters (1994), Cohen, Daubechies and Vial (1993), Chui and Quak (1992) and Dahmen, Kunoth and Urban (1996*b*). The basic idea common to all these approaches is to construct multiresolution sequences $\mathcal{S}$ on $[0,1]$, which, up to local boundary effects, agree with the restriction of the stationary spaces defined on all of $\mathbb{R}$. Thus one retains possibly many translates $2^{j/2}\phi(2^j\cdot -k)$ whose support is strictly inside $(0,1)$. In addition, one takes fixed linear combinations of those translates interfering with the boundaries in such a way that the original order of polynomial exactness is preserved. The following discussion is based on Dahmen et al. (1996*b*), which differs somewhat from the other sources but seems to be tailored best to the needs of subsequent applications.

For any dual pair $(\phi,\tilde\phi)$ from the spline family (4.25), that is, $\phi = {}_d\phi$, $\tilde\phi = {}_{d,\tilde d}\tilde\phi$, $\tilde d \ge d$, $d+\tilde d$ even, define

$$\alpha^L_{j,m,r} := 2^{j/2}\langle (2^j\cdot)^r, 2^{j/2}\phi(2^j\cdot -m)\rangle_{\mathbb{R}} = \langle (\cdot)^r, \phi(\cdot - m)\rangle_{\mathbb{R}} =: \alpha_{m,r}, \quad (4.56)$$

and

$$\alpha^R_{j,m,r} := 2^{j/2}\langle (2^j(1-\cdot))^r, 2^{j/2}\phi(2^j\cdot -m)\rangle_{\mathbb{R}} = \langle (2^j-\cdot)^r, \phi(\cdot - m)\rangle_{\mathbb{R}}, \quad (4.57)$$

for $r=0,\ldots,\tilde d-1$. Likewise $\tilde\alpha^L_{j,m,r}$, $\tilde\alpha^R_{j,m,r}$, $r=0,\ldots,d-1$, are defined by replacing ϕ by $\tilde\phi$. It is known that the support of $\tilde\phi$ always contains $\operatorname{supp}\phi$. It turns out that things depend somewhat on the parity $l(d) := d \bmod 2$. So fix $\tilde l \in \mathbb{N}$, such that for $j \ge j_0$, $\operatorname{supp}\tilde\phi(2^j\cdot -m) \subset (0,1)$, for $l \le m \le 2^j - \tilde l - l(d)$. Define left ($L$) and right ($R$) boundary functions by

$$\begin{aligned}
\tilde\phi^L_{j,\tilde l-\tilde d+r} &:= \sum_{m=-\tilde l_2+1}^{\tilde l-1} \alpha_{m,r}2^{j/2}\tilde\phi(2^j\cdot -m)\Big|_{[0,1]}, \\
\tilde\phi^R_{j,\tilde l-2^j-l(d)+\tilde d-r} &:= \sum_{m=2^j-\tilde l-l(d)+1}^{2^j-\tilde l_1-1} \alpha^R_{j,m,r}2^{j/2}\tilde\phi(2^j\cdot -m)\Big|_{[0,1]}, \quad r=0,\ldots,\tilde d-1,
\end{aligned} \quad (4.58)$$

where $\operatorname{supp}\tilde\phi = [\tilde l_1, \tilde l_2]$. Since by (4.56), (4.57), the functions $\tilde\phi^L_{j,k}, \tilde\phi^R_{j,k}$ are simply truncations of the polynomial representations (4.32), it is easy to see that the collections of left and right *boundary functions*

$$\begin{aligned}\tilde{\Phi}_j^L &= \left\{\tilde{\phi}^L_{j,\tilde{l}-\tilde{d}+r} : r = 0,\dots,\tilde{d}-1\right\},\\ \tilde{\Phi}_j^R &= \left\{\tilde{\phi}^R_{j,2^j-\tilde{l}-l(d)+\tilde{d}-r} : r = 0,\dots,\tilde{d}-1\right\}\end{aligned} \tag{4.59}$$

together with the *interior* translates

$$\tilde{\Phi}_j^I := \left\{2^{j/2}\tilde{\phi}(2^j\cdot -m) : m = \tilde{l},\dots,2^j-\tilde{l}-l(d)\right\}$$

span all polynomials $\Pi_{\tilde{d}}$ of degree $\leq \tilde{d}-1$ on $[0,1]$, that is,

$$\Pi_{\tilde{d}} \subseteq S\left(\tilde{\Phi}_j^L \cup \tilde{\Phi}_j^I \cup \tilde{\Phi}_j^R\right). \tag{4.60}$$

Setting

$$l := \tilde{l} - (\tilde{d} - d),$$

the functions $\phi^L_{j,k}$, $\phi^R_{j,k}$ are defined in exactly the same way with all tildes removed, providing

$$\Pi_d \subseteq S(\Phi_j^L \cup \Phi_j^I \cup \Phi_j^R). \tag{4.61}$$

Also, by construction,

$$\#\left(\Phi_j^L \cup \Phi_j^I \cup \Phi_j^R\right) = \#\left(\tilde{\Phi}_j^L \cup \tilde{\Phi}_j^I \cup \tilde{\Phi}_j^R\right).$$

However, while the interior functions in Φ_j^I, $\tilde{\Phi}_j^I$ are still biorthogonal, the boundary modifications have certainly destroyed biorthogonality of the elements in Φ_j^X, $\tilde{\Phi}_j^X$, $X \in \{L,R\}$. Nevertheless, it can be shown that these collections can always be *biorthogonalized.* Moreover, this is a completely local process, which need be done only once. In this and in several other respects it is very fortunate that things have been set up to exploit symmetry as much as possible. In fact, using the fact that ϕ and $\tilde{\phi}$ are symmetric around $l(d)/2$, one can show that

$$\phi^R_{j,2^j-l-l(d)+d-r}(1-x) = \phi^L_{j,l-d+r}(x), \quad r = 0,\dots,d-1, \tag{4.62}$$

and likewise for $\tilde{\phi}^R_{j,k}$, $\tilde{\phi}^L_{j,k}$. Thus one ends up with pairs of collections

$$\Phi_j = \{\phi_{j,k} : k \in \Delta_j\}, \quad \tilde{\Phi}_j = \{\tilde{\phi}_{j,k} : k \in \Delta_j\},$$

where $\Delta_j := \{l-d,\dots,2^j-l-l(d)+d\}$, with the following properties (Dahmen et al. 1996*b*).

(i) The functions in $\Phi_j, \tilde{\Phi}_j$ have small support, that is,

$$\operatorname{diam}(\operatorname{supp}\phi_{j,k}), \quad \operatorname{diam}(\operatorname{supp}\tilde{\phi}_{j,k}) \sim 2^{-j}. \tag{4.63}$$

(ii) The $\Phi_j, \tilde{\Phi}_j$ are biorthogonal

$$\langle \Phi_j, \tilde{\Phi}_j^{\mathrm{T}}\rangle = \mathbf{I}.$$

(iii) The spaces $S(\Phi_j)$, $S(\tilde{\Phi}_j)$ are exact of order $d, \tilde{d}$, respectively, that is,

$$\Pi_d \subset S(\Phi_j), \quad \Pi_{\tilde{d}} \subset S(\tilde{\Phi}_j). \tag{4.64}$$

(iv) The spaces $S(\Phi_j)$, $S(\tilde{\Phi}_j)$ are nested. This can be verified by exploiting trivial scaling properties of polynomials and refinability of the interior translates.

It is worth commenting on the structure of the corresponding refinement matrices in

$$\Phi_j^{\mathrm{T}} = \Phi_{j+1}^{\mathrm{T}} \mathbf{M}_{j,0}, \quad \tilde{\Phi}_j^{\mathrm{T}} = \tilde{\Phi}_{j+1}^{\mathrm{T}} \tilde{\mathbf{M}}_{j,0}. \tag{4.65}$$

Each $\mathbf{M}_{j,0}$, $\tilde{\mathbf{M}}_{j,0}$ consists of a stationary interior block, whose size grows like 2^j, as well as an upper left and lower right block, which are completely independent of j and of fixed size. The interior blocks are just finite sections of the bi-infinite refinement matrices $(a_{k-2m})_{k,m\in\mathbb{Z}}$, $(\tilde{a}_{k-2m})_{k,m\in\mathbb{Z}}$. Moreover, symmetry surfaces again. Denoting for a given matrix $\mathbf{M}$ by $\mathbf{M}^{\updownarrow}$ the matrix obtained from $\mathbf{M}$ by reversing the order of rows and columns, one can show (see also (4.62)) that

$$\mathbf{M}_{j,0}^{\updownarrow} = \mathbf{M}_{j,0}, \quad \tilde{\mathbf{M}}_{j,0}^{\updownarrow} = \tilde{\mathbf{M}}_{j,0}. \tag{4.66}$$

The next step, namely to construct corresponding biorthogonal bases, is somewhat more involved. Using tools from spline theory, one can first construct suitable *initial* stable completions. Then Proposition 3.8 can be applied providing new (sparse) stable completions $\mathbf{M}_{j,1}$, $\tilde{\mathbf{M}}_{j,1}$ of the above refinement matrices, which have completely analogous structure and satisfy

$$\mathbf{M}_{j,0}\tilde{\mathbf{M}}_{j,0}^{\mathrm{T}} + \mathbf{M}_{j,1}\tilde{\mathbf{M}}_{j,1}^{\mathrm{T}} = \mathbf{I}, \quad \tilde{\mathbf{M}}_{j,e}^{\mathrm{T}}\mathbf{M}_{j,e'} = \delta_{e,e'}\mathbf{I}, \quad e, e' \in \{0,1\}. \tag{4.67}$$

Thus the wavelet bases

$$\Psi_j^{\mathrm{T}} = \Phi_{j+1}^{\mathrm{T}} \mathbf{M}_{j,1}, \quad \tilde{\Psi}_j^{\mathrm{T}} := \tilde{\Phi}_{j+1}^{\mathrm{T}} \tilde{\mathbf{M}}_{j,1} \tag{4.68}$$

satisfy $\langle \Psi_j, \tilde{\Psi}_{j'} \rangle_{[0,1]} = \delta_{j,j'}\mathbf{I}$ and hence

$$\langle \Psi, \tilde{\Psi} \rangle_{[0,1]} = \mathbf{I}, \quad \Psi := \Phi_{j_0} \cup \bigcup_{j\geq j_0} \Psi_j, \quad \tilde{\Psi} := \tilde{\Phi}_{j_0} \cup \bigcup_{j\geq j_0} \tilde{\Psi}_j, \tag{4.69}$$

are biorthogonal.

All filters have finite length so that the $\psi_{j,k}, \tilde{\psi}_{j,k}$ also satisfy (4.63). The filters are stationary in the above sense. Thus the multiscale transformations $\mathbf{T}_J$ (3.26) and $\mathbf{T}_J^{-1}$ (3.28) are still efficient and require the order of N_J operations. Finally, observe that the techniques described in Section 4.2 still apply, since all operations ultimately reduce to restrictions of $\phi(2^j \cdot -k)$ to $[0,1]$ which can be realized by choosing $\chi_{j,k}$ as an additional factor in (4.43).

5. Norm equivalences and function spaces

One of the most important properties of wavelets is that they can be used to characterize function spaces; see, for example, (DeVore et al. 1992, Meyer 1990). The Riesz basis property (3.36) which came up in connection with the stability of multiscale transformations (Theorem 3.3) is a special case in a whole scale of similar relations. This will be seen to play a vital role for preconditioning, matrix compression and adaptive techniques (recall (b), (c) and (d) in Section 3.1).

In the classical stationary shift invariant or periodic setting such results are established by making heavy use of Fourier techniques. They no longer apply in a straightforward manner for other domains such as the interval or more complex cases such as closed surfaces yet to come. Recall from Section 3.4 that the Riesz basis property as one instance of such norm equivalences naturally leads to the concept of biorthogonal bases. When these bases correspond to *orthogonal* complements between successive spaces $S(\Phi_j)$, $S(\Phi_{j+1})$, the Riesz basis property reduces to the Pythagorean theorem, once the complement bases Ψ_j are uniformly stable relative to each level. However, orthogonal decompositions are often difficult to realize, lead to dense matrices $\mathbf{G}_j$, and in some cases are not optimal for the application at hand. Thus understanding the general class of biorthogonal multiscale bases is vital. However, while being *necessary*, biorthogonality by itself is *not* quite *sufficient* to imply the Riesz basis property (Meyer 1994). The developments in this section are therefore guided by the following point:

- find criteria for the validity of the Riesz basis property and other norm equivalences for biorthogonal bases, which can still be employed in situations where Fourier techniques no longer work.

A key ingredient is a pair of direct and inverse estimates; these are also known to play an important role in convergence theory of multigrid algorithms.

5.1. Direct and inverse estimates

The type of estimate we are aiming at is rooted in approximation theory, concerning approximation and regularity properties of the trial spaces. To formulate versions suitable for the present purpose, suppose that $\Omega \subseteq \mathbb{R}^n$ is an open connected domain (the case $\Omega = \mathbb{R}^n$ included). If Ω has a boundary we assume that it has some minimal regularity such as the uniform cone condition; see, for instance, DeVore and Sharpley (1993) and Johnen and Scherer (1977). Thus there exists an extension operator $E : L_p(\Omega) \to L_p(\mathbb{R}^n)$ that is bounded in $W_p^m(\Omega)$ for any $m \in \mathbb{N}$. The estimates we require are

$$\inf_{v \in S(\Phi_j)} \|f - v\|_{L_p(\Omega)} \lesssim 2^{-dj} \|f\|_{W_p^d(\Omega)}, \quad f \in W_p^d(\Omega). \tag{5.1}$$

We will refer to such estimates as *direct* or *Jackson* estimates. By interpolation, one derives from (5.1) a scale of similar estimates with the right-hand side replaced by $2^{-sj}\|f\|_{B^s_q(L_p(\Omega))}$, $s < m$, $1 \le q \le \infty$, where $B^s_q(L_p(\Omega))$ are corresponding Besov spaces (see Section 2.1).

There is often a counterpart called *inverse* or *Bernstein* estimate

$$\|v\|_{B^s_q(L_p(\Omega))} \lesssim 2^{sj}\|v\|_{L_p(\Omega)}, \quad v \in S(\Phi_j). \tag{5.2}$$

We next give a simple criterion for verifying (5.1) that will apply in all cases of interest.

Proposition 5.1 Let $\Phi_j \subset L_p(\Omega)$ and $\Xi_j \subset L_{p'}(\Omega)$ with $\frac{1}{p} + \frac{1}{p'} = 1$ have the following properties:

(i) Ξ_j and Φ_j are biorthogonal,

$$\langle \Phi_j, \Xi_j \rangle_\Omega = \mathbf{I} \tag{5.3}$$

where $\langle \cdot, \cdot \rangle_\Omega$ denotes the dual pairing for $L_p(\Omega) \times L_{p'}(\Omega)$.

(ii) The elements of Φ_j and Ξ_j are uniformly bounded, that is,

$$\|\phi_{j,k}\|_{L_p}, \|\xi_{j,k}\|_{L_{p'}(\Omega)} = O(1), \quad j \in \mathbb{N}, \quad k \in \Delta_j. \tag{5.4}$$

(iii) The collections Φ_j, Ξ_j are locally finite, that is, there exists a constant $C < \infty$ such that

$$\#\{k' : \square_{j,k} \cap \square_{j,k'} \neq \emptyset\} \le C, \quad \operatorname{diam} \square_{j,k} \lesssim 2^{-j} \tag{5.5}$$

where $\square_{j,k}$ is the smallest cube containing $\operatorname{supp} \phi_{j,k}$ and $\operatorname{supp} \xi_{j,k}$.

(iv) The spaces $S(\Phi_j)$ contain all polynomials of order d (degree $\le d-1$) on Ω,

$$\Pi_d \subseteq S(\Phi_j). \tag{5.6}$$

Then one has

$$\|f - \langle f, \Xi_j \rangle_\Omega \Phi_j\|_{L_p(\Omega)} \lesssim 2^{-dj}\|f\|_{W^d_p(\Omega)}. \tag{5.7}$$

The type of argument needed here is essentially folklore. Since it plays a central role we sketch a proof. By (5.6), one has for any $P \in \Pi_d$

$$\|f - \langle f, \Xi_j \rangle_\Omega\|^p_{L_p(\square_{j,k})} \lesssim \|f-P\|^p_{L_p(\square_{j,k})} + \sum_{\square_{j,k'} \cap \square_{j,k} \neq \emptyset} |\langle f-P, \xi_{j,k'} \rangle_\Omega|^p \|\phi_{j,k'}\|^p_{L_p(\Omega)}. \tag{5.8}$$

On account of (5.5), the sum involves a uniformly bounded number of summands. Using (5.4) gives

$$|\langle f-P, \xi_{j,k'} \rangle_\Omega|^p \|\phi_{j,k'}\|^p_{L_p(\Omega)} \lesssim \|f-P\|^p_{L_p(\square_{j,k'})} \lesssim 2^{-dpj}\|f\|^p_{W^d_p(\square_{j,k'})}, \tag{5.9}$$

where a Bramble–Hilbert-type argument has been used in the last step. A little care has to be taken near the boundary. In order to employ the scaling argument needed for the Bramble–Hilbert argument, one can employ extension techniques; see Oswald (1997) for details. Bearing (5.5) in mind, and summing over $k \in \Delta_j$, yields (5.7). □

Estimates of the type (5.1) readily lead to estimates without *regularity* assumptions. Consider the *K-functional* (see, for instance, Bergh and Löfström (1976))

$$K_d(f,t) = K(f,t,L_p,W_p^d) := \inf_{g\in W_p^d(\Omega)} \left\{\|f-g\|_{L_p(\Omega)} + t^d\|g\|_{W_p^d(\Omega)}\right\}. \quad (5.10)$$

One immediately infers from (5.1) that

$$\inf_{v\in S_j} \|f-v\|_{L_p(\Omega)} \lesssim K_d(f,2^{-j}). \quad (5.11)$$

Remark 5.2 Under assumptions (5.4) and (5.5), the Φ_j are uniformly stable (relative to $\|\cdot\|_{L_p(\Omega)}$ and $\|\cdot\|_{\ell_p(\Delta_j)}$).

Remark 5.3 Obviously Proposition 5.1 applies to all the above examples of biorthogonal multiresolution sequences (with $\Xi_j = \tilde{\Phi}_j$). In fact, for wavelets on $\mathbb{R}^n$ recall (4.32), the multiresolution on $[0,1]$ was constructed in Section 4.4 so that (5.6) holds, while all other conditions are obviously satisfied. Thus we will assume from now on that the direct estimate (5.1) is valid for the order d of polynomial exactness.

Remark 5.4 Suppose $\phi \in L_2(\mathbb{R}^n)$ is a (compactly supported stable) scaling function. Let

$$\gamma := \sup\{s : \phi \in H^s(\mathbb{R}^n)\}.$$

Then

$$\|v\|_{H^s(R^n)} \lesssim 2^{sj}\|v\|_{L_2(\mathbb{R}^n)}, \quad v \in S(\Phi_j), \quad (5.12)$$

holds for any $s < \gamma$. It is also known that $\phi \in L_2(\mathbb{R}^n)$ implies $\phi \in H^s(\mathbb{R}^n)$ for some $s > 0$ (Villemoes 1993).

For a proof see, for instance, Dahmen (1995). One can show that when $\phi, \tilde{\phi}$ is a dual pair of compactly supported generators, then their integer shifts are *locally linearly independent*. Then $\|\cdot\|_{H^s(\square)}$ and $\|\cdot\|_{L_2(\square)}$ are equivalent norms on $S(\Phi_0)$ and the claim for integer s follows from summing the local norms and rescaling.

Remark 5.5 (Dahmen 1995, Dahmen et al. 1996*b*) Let $\Phi_j \subset L_2([0,1])$ denote the generator bases constructed in Section 4.4. Then

$$\|v\|_{H^s([0,1])} \lesssim 2^{sj}\|v\|_{L_2([0,1])}, \quad v \in S(\Phi_j), \quad s < \gamma. \quad (5.13)$$

Finally, the inverse inequalities can be expressed in terms of the K-functional as well. In fact, from (5.2), one can deduce that

$$K_d(v,t) \lesssim (\min\{1, t2^j\})^\gamma \|v\|_{L_p(\Omega)}, \quad v \in S(\Phi_j). \quad (5.14)$$

The form (5.10) and (5.14) of the direct and inverse estimates will guide the subsequent discussion.

5.2. The Riesz basis property

As pointed out in Section 3.4, the stability of multiscale transformations (3.26), (3.28) is equivalent to the Riesz basis property of the basis Ψ (and $\tilde{\Psi}$). It turns out that sufficient conditions which apply in our cases of interest can be formulated in a general Hilbert space setting. This will shed some light on the essential mechanisms. An important point is that, as one will see, once a biorthogonal pair $\Psi, \tilde{\Psi}$ is given, additional conditions implying the Riesz basis property *only* concern properties of the *spaces* spanned by subsets of Ψ and $\tilde{\Psi}$, *not* of the particular bases. These properties can be formulated in terms of the estimates (5.10), (5.14).

In order to stress this point, we will first reformulate the problem somewhat, which, by the way, corresponds also to the strategy of constructing Riesz bases employed in Section 4.4. First of all, it is usually not so difficult to assure stability of a complement basis $\Psi_j = \{\psi_{j,k} : k \in \nabla_j\}$ in the space Φ_{j+1}. We will therefore assume in the following that

$$\|\mathbf{d}^{\mathrm{T}}\Psi_j\|_H \sim \|\mathbf{d}\|_{\ell_2(\nabla_j)}. \tag{5.15}$$

Moreover, recall from (3.45) that biorthogonality is equivalent to

$$Q_j Q_l = Q_j \quad \text{for} \quad j \le l, \tag{5.16}$$

where the Q_j are the projectors $Q_j v = \langle v, \tilde{\Phi}_j \rangle \Phi_j = \langle v, \tilde{\Psi}^j \rangle \Psi^j$ of (3.44), (3.46), which, by Remark 3.6, have to be uniformly bounded when Ψ and $\tilde{\Psi}$ are Riesz bases. Then, by (5.15), the norm equivalence (3.36) can be equivalently expressed as

$$\|f\|_H \sim N_{\mathcal{Q}}(f) \sim N_{\mathcal{Q}^*}(f), \tag{5.17}$$

where, for $Q_{-1} := 0$,

$$N_{\mathcal{Q}}(f)^2 := \sum_{j=0}^{\infty} \|(Q_j - Q_{j-1})f\|_H^2. \tag{5.18}$$

The objective now is to establish the validity of (5.18) for a given sequence $\mathcal{Q}$ of projectors satisfying the necessary conditions of uniform boundedness and (5.16). It is important to note that in this form the result applies when the Q_j are given only in the form $Q_j v = \langle v, \tilde{\Phi}_j \rangle \Phi_j$, that is, without explicit knowledge of the *right* complement bases Ψ_j yet. Note also that the condition (5.16) implies that the ranges $\tilde{S}_j$ of the adjoints Q_j^* are also *nested.* Moreover, these spaces are also dense in H (Dahmen 1994, 1996). Let us denote the corresponding sequence by $\tilde{\mathcal{S}}$. The following result says that the Riesz basis property holds when, in addition to biorthogonality, the primal and dual multiresolution sequences $\mathcal{S}, \tilde{\mathcal{S}}$ both have some approximation and regularity properties expressed in terms of pairs of direct and inverse estimates (Dahmen 1996).

Theorem 5.6 Let $\mathcal{S}$ be an ascending dense sequence of closed subspaces of H and let $\mathcal{Q}$ be a sequence of uniformly H-bounded projectors with ranges $\mathcal{S}$ satisfying (5.16). Let $\tilde{\mathcal{S}}$ be the ranges of the adjoint sequence $\mathcal{Q}^*$. Suppose there exists a family of uniformly bounded subadditive functionals $\omega(\cdot,t) : H \to \mathbb{R}_+$, $t > 0$, such that $\lim_{t\to 0^+} \omega(f,t) = 0$ for each $f \in H$ and that the pair of estimates

$$\inf_{v\in V_j} \|f - v\|_H \lesssim \omega(f, 2^{-j}), \tag{5.19}$$

and

$$\omega(v_j, t) \lesssim (\min\{1, t2^j\})^\gamma \|v_j\|_H, \quad v_j \in V_j, \tag{5.20}$$

holds for $\mathcal{V} = \mathcal{S}$ and $\mathcal{V} = \tilde{\mathcal{S}}$ with some $\gamma, \tilde{\gamma} > 0$, respectively. Then

$$\|\cdot\|_H \sim N_{\mathcal{Q}}(\cdot) \sim N_{\mathcal{Q}^*}(\cdot). \tag{5.21}$$

Here is an immediate consequence of Theorem 5.6.

Remark 5.7 Note that the K-functional $K_d(\cdot,t)$ defined by (5.10) has, by (5.11), (5.14),(5.13), (5.12), all the properties of $\omega(\cdot,t)$ required above. Thus the biorthogonal bases constructed in Sections 4.2 and 4.4 are indeed Riesz bases.

A few comments on the proof of Theorem 5.6 are in order; see Dahmen (1996) for details. First one observes that (Cohen 1994)

$$N_{\mathcal{Q}}(\cdot) \lesssim \|\cdot\|_H \quad \text{if and only if} \quad \|\cdot\|_H \lesssim N_{\mathcal{Q}^*}(\cdot).$$

Thus it suffices to prove that

$$\|\cdot\|_H \lesssim N_{\mathcal{Q}}(\cdot) \quad \text{and} \quad \|\cdot\|_H \lesssim N_{\mathcal{Q}^*}(\cdot), \tag{5.22}$$

or the corresponding pair of opposite inequalities. To prove estimates of the form (5.22) one can employ a technique which is also familiar in the analysis of multilevel preconditioners.

Strengthened Cauchy inequalities

To this end, suppose there is a (dense) subspace $U \subset H$ with a (stronger) norm $\|\cdot\|_U$ such that, for some $\varepsilon > 0$,

$$\|f - Q_j f\|_{U^*} \lesssim 2^{-j\varepsilon}\|f\|_H, \quad \|f - Q_j f\|_H \lesssim 2^{-j\varepsilon}\|f\|_U, \tag{5.23}$$

and

$$\|v_j\|_U \lesssim 2^{j\varepsilon}\|v_j\|_H, \quad \|v_j\|_H \lesssim 2^{j\varepsilon}\|v_j\|_{U^*}, \quad v_j \in S_j. \tag{5.24}$$

Then one can estimate

$$\begin{aligned}&\langle (Q_j - Q_{j-1})f, (Q_i - Q_{i-1})f\rangle \\ &\leq \begin{cases} \|(Q_j - Q_{j-1})f\|_{U^*}\|(Q_i - Q_{i-1})f\|_U & \text{if } i \leq j, \\ \|(Q_j - Q_{j-1})f\|_U \; \|(Q_i - Q_{i-1})f\|_{U^*} & \text{if } i > j. \end{cases}\end{aligned}$$

Thus by (5.23), (5.24),

$$\begin{aligned}\|f\|_H^2 &= \sum_{i,j=0}^{\infty} \langle (Q_j - Q_{j-1})f, (Q_i - Q_{i-1})f\rangle \\ &\lesssim \sum_{i,j=0}^{\infty} 2^{-\varepsilon|i-j|}\|(Q_j - Q_{j-1})f\|_H\|(Q_i - Q_{i-1})f\|_H \lesssim N_{\mathcal{Q}}(f)^2.\end{aligned}$$

When $\mathcal{Q}$ is uniformly bounded on U^* the estimates (5.23) and (5.24) can be shown to hold by duality also for $\mathcal{Q}^*$. Thus the same argument also yields $\|f\|_H^2 \lesssim N_{\mathcal{Q}^*}(f)^2$.

A scale of interpolation spaces
So, it remains to find such a subspace U. Natural candidates are the spaces $\mathcal{A}^s_{\mathcal{Q}}$ which are defined for $s > 0$ as the collection of those $f \in H$ for which

$$\|f\|^s_{\mathcal{A}^s_{\mathcal{Q}}} = \sum_{j=0}^{\infty} 2^{2sj}\|(Q_j - Q_{j-1})f\|_H^2 < \infty.$$

They are dense reflexive subspaces of H, and, with a proper understanding of continuously extended projectors, one has a representation of their duals in terms of the dual projectors $\mathcal{Q}^*$ (Dahmen 1996)

$$(\mathcal{A}^s_{\mathcal{Q}})^* = \mathcal{A}^{-s}_{\mathcal{Q}^*}. \tag{5.25}$$

Moreover, these spaces are defined so that, again under assumption (5.16), a pair of direct and inverse inequalities hold, namely

$$\|f - Q_j f\|_H \lesssim 2^{-js}\|f\|_{\mathcal{A}^s_{\mathcal{Q}}}, \quad \|f - Q_j f\|_{(\mathcal{A}^s_{\mathcal{Q}^*})^*} \lesssim 2^{-js}\|f\|_H, \tag{5.26}$$

and

$$\|v_j\|_{\mathcal{A}^s_{\mathcal{Q}}} \lesssim 2^{js}\|v_j\|_H, \quad \|v_j\|_H \lesssim 2^{js}\|v_j\|_{(\mathcal{A}^s_{\mathcal{Q}^*})^*}. \tag{5.27}$$

Finally, by (5.16), $\mathcal{Q}$ is trivially uniformly bounded on $\mathcal{A}^s_{\mathcal{Q}}$ for all s, that is,

$$\|Q_j\|_{\mathcal{A}^s_{\mathcal{Q}}} = 1, \quad j \in \mathbb{N}_0, s \in \mathbb{R}. \tag{5.28}$$

Thus one *almost* has the pair of inequalities (5.23), (5.24) without any assumption on $\mathcal{Q}$ beyond (5.16) and uniform boundedness. What is missing is the relation between the spaces $(\mathcal{A}^s_{\mathcal{Q}^*})^*$ and $(\mathcal{A}^s_{\mathcal{Q}})^*$. If they were equivalent, (5.26) and (5.27), together with the strengthened Cauchy inequality argument, would confirm (5.22) and hence the claim of Theorem 5.6. This is where the direct and inverse inequalities (5.19), (5.20) come into play. In fact, with the aid of these inequalities one can prove that

$$\|\cdot\|_{\mathcal{A}^s_Q} \sim \|\cdot\|_{\mathcal{B}^s_\omega} \sim \|\cdot\|_{\mathcal{A}^s_{Q^*}} \quad \text{for } 0 < s < \min\{\gamma, \tilde{\gamma}\}, \tag{5.29}$$

where

$$\|f\|^s_{\mathcal{B}^s_\omega} := \|f\|^2_H + \sum_{j=0}^{\infty} 2^{2sj}\omega(f, 2^{-j})^2, \tag{5.30}$$

which closes the gap.

These results are closely related to interpolation theory. In fact, the $\mathcal{A}^s_Q$ are *interpolation spaces* obtained by the *real method*; see, for instance, Bergh and Löfström (1976), DeVore and Popov (1988*a*), DeVore and Sharpley (1993) and Peetre (1978). A detailed discussion of this point of view can be found in Dahmen (1995).

As mentioned before, the role of $\omega(\cdot, t)$ is typically played by a K-functional or a modulus of smoothness, which under our assumptions on the underlying domain are equivalent seminorms (Johnen and Scherer 1977). In that sense the spaces $\mathcal{B}^s_\omega$ can be viewed as generalized *Besov* spaces. Thus, in addition to the Riesz basis property, the above criteria *automatically* establish norm equivalences for a whole *scale* of spaces. The equivalence of the artificial spaces $\mathcal{A}^s_Q$ with the Besov-type spaces $\mathcal{B}^s_\omega$ in some range of s immediately yields norm equivalences for these (classical) function spaces, which will be addressed next.

5.3. Characterization of Sobolev spaces

When Ω is a domain in $\mathbb{R}^n$ as above and $\omega(\cdot, t)$ is an L_2 modulus of smoothness

$$\omega(f,t) = \omega_d(f,t)_{L_2(\Omega)} := \sup_{|h|\le t} \|\Delta^d_h f\|_{L_2(\Omega_{d,h})}$$

where $\Delta^d_h = \Delta_h \Delta^{d-1}_h$, $\Delta_h f = f(\cdot + h) - f(\cdot)$ and $\Omega_{d,h} = \{x : x + lh \in \Omega, l = 0, \dots, d\}$, or when $\omega(\cdot, t)$ is the K-functional from (5.10), the norm in (5.30) is equivalent to $\|\cdot\|_{H^s(\Omega)}$ for $0 < s < d$, and

$$H^s(\Omega) \sim B^s_2(L_2(\Omega)). \tag{5.31}$$

We will now apply the above results for $H = H^s(\Omega)$. For simplicity we focus on $H^0(\Omega) = L_2(\Omega)$. Moreover, let us denote for $s > 0$ by H^s some closed subspace of $H^s(\Omega)$ (or $H^s(\Omega)$ itself) which is, for instance, determined by some homogeneous boundary conditions. The key role is again played by a pair of direct and inverse inequalities

$$\inf_{v_j \in V_j} \|v - v_j\|_{L_2(\Omega)} \lesssim 2^{-sj}\|v\|_{H^s(\Omega)}, \quad v \in H^s, \ 0 \le s \le d_\mathcal{V}, \tag{5.32}$$

and

$$\|v_j\|_{H^s(\Omega)} \lesssim 2^{sj}\|v_j\|_{L_2(\Omega)}, \quad v_j \in V_j, \quad s < \gamma_{\mathcal{V}}. \tag{5.33}$$

Recall from Section 5.1 under which circumstances such inequalities hold. The direct inequality may be affected by homogeneous boundary conditions incorporated in $\mathcal{S}$. If this is done properly the argument stays essentially the same, since near the boundary not all polynomials are needed.

From (5.29), (5.28) and duality (5.25) one infers the following fact.

Theorem 5.8 Let $\mathcal{Q}$ be uniformly bounded with range $\mathcal{S}$ and suppose that (5.16) holds. Moreover, assume that $\mathcal{S}$ and the range $\tilde{\mathcal{S}}$ of $\mathcal{Q}^*$ satisfy (5.32) and (5.33) for some $d := d_{\mathcal{S}}, \tilde{d} := d_{\tilde{\mathcal{S}}}$, $0 < \gamma := \min\{\gamma_{\mathcal{S}}, d\}$ and $0 < \tilde{\gamma} := \min\{\gamma_{\tilde{\mathcal{S}}}, \tilde{d}\}$, respectively. Then

$$\left(\sum_{j=0}^{\infty} 2^{2sj}\|(Q_j - Q_{j-1})f\|^2_{L_2(\Omega)}\right)^{1/2} \sim \|f\|_{H^s(\Omega)}, \quad s \in (-\tilde{\gamma}, \gamma), \tag{5.34}$$

where it is to be understood that $H^s(\Omega) = (H^{-s}(\Omega))^*$ for $s < 0$. Moreover, $\mathcal{Q}$ is uniformly bounded in H^s for that range

$$\|Q_j v\|_{H^s(\Omega)} \lesssim \|v\|_{H^s(\Omega)}, \quad v \in H^s(\Omega). \tag{5.35}$$

When $H^s = H^s(\Omega)$ and both sequences $\mathcal{S}, \tilde{\mathcal{S}}$ have a high order of exactness $d, \tilde{d}$, respectively, the above range may have a significant part for $s < 0$. There is, however, always some $\tilde{\gamma} > 0$ reaching into the negative range. It could be small if H^s is a true subspace of $H^s(\Omega)$ and the corresponding boundary conditions are incorporated in $\mathcal{S}$. What matters, though, is that, by (5.25), (5.31) and the above result applied to $s \geq 0$, one still has

$$\left(\sum_{j=0}^{\infty} 2^{-2sj}\|(Q_j^* - Q_{j-1}^*)f\|^2_{L_2(\Omega)}\right)^{1/2} \sim \|f\|_{H^{-s}}, \quad s \in [0, \gamma). \tag{5.36}$$

It is convenient to express these relations in terms of the operators

$$\Sigma_s f := \sum_{j=0}^{\infty} 2^{js}(Q_j - Q_{j-1})f, \tag{5.37}$$

which act as a *shift* in the Sobolev scale

$$\|\Sigma_s f\|_{H^t(\Omega)} \sim \|f\|_{H^{t+s}(\Omega)}, \quad t + s \in (-\tilde{\gamma}, \gamma), \tag{5.38}$$

just like classical Bessel potential operators in harmonic analysis. Due to (5.16), one has

$$\Sigma_s^{-1} = \Sigma_{-s}, \quad \Sigma_s^* = \sum_{j=0}^{\infty} 2^{js}\left(Q_j^* - Q_{j-1}^*\right). \tag{5.39}$$

It is important to note that one-sided estimates of type (5.34) hold for a wider range of s. In fact, the uniform boundedness of the Q_j ensures that

$$\|Q_j f - f\|_{L_2(\Omega)} \lesssim \inf_{v \in S(\Phi_j)} \|f - v\|_{L_2(\Omega)}.$$

Thus by Proposition 5.1 and (5.11) one obtains

$$\|(Q_j - Q_{j-1})f\|_{L_2(\Omega)} \lesssim K\left(f, 2^{-j+1}, L_2, H^d\right).$$

Since

$$\left(\|f\|^2_{L_2(\Omega)} + \sum_{J=0}^{\infty} 2^{2sj} K\left(f, 2^{-j+1}, L_2, H^d\right)^2\right)^{1/2}$$

is known to be a norm for the Besov space $B^s_2(L_2(\Omega)) = H^s(\Omega)$, one obtains, for instance,

$$\sum_{j=0}^{\infty} 2^{2js} \|(Q_j - Q_{j-1})f\|^2_{L_2(\Omega)} \lesssim \|f\|^2_{H^s(\Omega)}, \quad -\tilde{\gamma} < s < d. \tag{5.40}$$

If corresponding wavelet bases $\Psi, \tilde{\Psi}$ are known, Σ_s can be written as

$$\Sigma_s f = \sum_{\lambda \in \nabla} 2^{s|\lambda|} \langle f, \tilde{\psi}_\lambda \rangle_\Omega \psi_\lambda, \tag{5.41}$$

and (5.34) becomes

$$\|f\|_{H^s} \sim \|\mathbf{D}^s \langle f, \tilde{\Psi} \rangle^{\mathrm{T}}_\Omega\|_{\ell_2(\nabla)}, \quad s \in (-\tilde{\gamma}, \gamma), \tag{5.42}$$

where $\mathbf{D}^s$ denotes the diagonal matrix

$$(\mathbf{D}^s)_{\lambda,\lambda'} = 2^{s|\lambda|} \delta_{\lambda,\lambda'}. \tag{5.43}$$

Remark 5.9 In view of Proposition 5.1, Remark 5.4, Remark 5.5, Theorem 5.8 implies that the wavelet bases constructed in (4.23) for $L_2(\mathbb{R})$, in (4.53) for the periodic case and (4.69) for the interval $[0, 1]$ all satisfy (5.42) for $s \in (-\tilde{\gamma}, \gamma)$.

Frames

It is important to note that norm equivalences of the type (5.34) for $s > 0$ do *not* require knowlege of concrete bases for decompositions $(Q_j - Q_{j-1})S_j$. Instead one can prove that (Dahmen 1995, Oswald 1994, Oswald 1992, Oswald 1990)

$$\|f\|^2_{H^s(\Omega)} \sim \inf \left\{ \sum_{j=0}^{\infty} 2^{2sj} \|f_j\|^2_{L_2(\Omega)} : f = \sum_{j=0}^{\infty} f_j \right\}. \tag{5.44}$$

In terms of interpolation theory, norms of this type correspond to the J-method (Bergh and Löfström 1976, Peetre 1978). Such norm equivalences will play a crucial role in preconditioning.

These results have further natural extensions to other (reflexive) Banach spaces like L_p-spaces ($1 < p < \infty$). Interpolation between $L_p(\Omega)$ and $W_p^m(\Omega)$, say, leads to Besov spaces $B_q^s(L_p(\Omega))$ endowed with the norms

$$\|f\|_{B_q^s(L_p(\Omega))}^q = \|f\|_{L_p(\Omega)}^q + \sum_{j=0}^{\infty} 2^{qsj} K\left(f, 2^{-j}, L_p, W_p^d\right)^q \tag{5.45}$$

for $d \geq s > 0$. Assuming that Ψ, $\tilde{\Psi}$ are biorthogonal wavelet bases (in $L_2(\Omega)$) one still obtains norm equivalences of the form

$$\|f\|_{B_q^s(L_p(\Omega))} \sim \left(\|\langle f, \tilde{\Phi}_0\rangle_\Omega\|_{\ell_p(\Delta_{j_0})}^q + \sum_{j=j_0}^{\infty} 2^{jq(s+\frac{n}{2}-\frac{n}{p})} \|\langle f, \tilde{\Psi}_j\rangle_\Omega\|_{\ell_p(\nabla_j)}^q \right)^{1/q} \tag{5.46}$$

which, of course, reduce to (5.42) for $s > 0$, $p = q = 2$. These norm equivalences play an important role in *nonlinear approximation* (DeVore and Lucier 1992, DeVore et al. 1992). This, in turn, will be of interest in connection with adaptive schemes (see Section 11).

6. Preconditioning

This section is only concerned with preconditioning systems arising from discretizations of operator equations, which in a loose sense may be termed *elliptic*. In particular, all the examples in Section 2 are covered (see also (b) in Section 1.5). I would like to stress the following points.

- Once the norm equivalences discussed in Section 5.3 are available, the principal argument is rather simple and applies to a relatively wide range of cases, represented by the reference problem in Section 2.3. To bring out the basic mechanism, I will address it first in this generality, which will cover various special cases treated in the literature.
- The strongest interrelation between rather independent developments in the area of wavelets on one hand and finite element discretizations on the other hand occurs in connection with preconditioning. Since these developments usually ignore each other, I will comment on both. In view of the existing excellent treatments of *multilevel subspace correction methods* seen through the finite element eye, the main focus here will be on the wavelet or basis oriented point of view.
- In the present generality the results are purely asymptotical. The actual performance of corresponding schemes depends very much on the concrete case at hand. In general, it is hard to say which concept is best able do cope with near degeneracies or strong isotropies.

6.1. Discretization and projection methods

Consider problem (2.24) for the spaces H^t, H^{-t}, L_2 as described in Section 2.3. Throughout the following we will assume that $\Psi = \{\psi_\lambda : \lambda \in \nabla\}$, $\tilde{\Psi} = \{\tilde{\psi}_\lambda : \lambda \in \nabla\}$, with $\nabla = \Delta_+ \cup \nabla_-$, are biorthogonal wavelet bases in L_2 such that the norm equivalences

$$\|f\|_{H^s} \sim \|\mathbf{D}^s\langle f, \tilde{\Psi}\rangle^{\mathrm{T}}\|_{\ell_2(\nabla)}, \quad s \in (-\tilde{\gamma}, \gamma), \tag{6.1}$$

hold.

The numerical schemes to be used for the solution of (2.24) may be viewed as *generalized Petrov–Galerkin schemes.* To describe this, we adhere to notation (3.43) and suppose that Θ_Λ is a collection of functionals that is defined and total over $S(\mathcal{L}\Psi_\Lambda)$. To solve (2.24), the objective is to determine $u_\Lambda \in S(\Psi_\Lambda)$ such that

$$\langle \mathcal{L}u_\Lambda, \Theta_\Lambda\rangle = \langle f, \Theta_\Lambda\rangle. \tag{6.2}$$

Of course, $\Theta = \Psi$ gives rise to a classical Galerkin scheme, while collocation is obtained when Θ involves Dirac functionals. In the latter case the right-hand side has to be taken from a sufficiently smooth space. This is appropriate when $\mathcal{L}$ is also known to be boundedly invertible as an operator from H^s into H^{s-2t} for some larger $s \in \mathbb{R}$. To explain what is meant by *stability* of the scheme, it is convenient to reinterpret (6.2) as a *projection method.* Suppose that $\tilde{\Theta}$ is a sufficiently regular dual set for Θ and let $P_\Lambda := \langle \cdot, \Theta_\Lambda\rangle\tilde{\Theta}_\Lambda$ be an associated projector. Then (6.2) is equivalent to

$$P_\Lambda \mathcal{L} Q_\Lambda u = P_\Lambda f. \tag{6.3}$$

The scheme (6.2) is said to be $(s, 2t)$-*stable* if for $\#\Lambda$ large

$$\|P_\Lambda \mathcal{L} v\|_{H^{s-2t}} \sim \|v\|_{H^s}, \quad v \in S(\Psi_\Lambda), \tag{6.4}$$

that is, the finite-dimensional operators $\mathcal{L}_\Lambda := P_\Lambda \mathcal{L} Q_\Lambda$ are uniformly bounded invertible mappings from $H^s \cap S(\Psi_\Lambda)$ onto $H^{s-2t} \cap S(\tilde{\Theta}_\Lambda)$. In terms of linear systems, substituting $u_\Lambda = \mathbf{d}^T \Psi_\Lambda$ into (6.2) yields the linear system

$$\mathbf{d}^T\langle \mathcal{L}\Psi_\Lambda, \Theta_\Lambda\rangle = \langle f, \Theta_\Lambda\rangle. \tag{6.5}$$

In particular, for the Galerkin case, (6.3) becomes

$$Q_\Lambda^* \mathcal{L} Q_\Lambda u_\Lambda = Q_\Lambda^* f. \tag{6.6}$$

The most important case for the subsequent discussion is $(t, 2t)$-stability, in brief *stability*, which then means

$$\|Q_\Lambda^* \mathcal{L} v\|_{H^{-t}} \sim \|v\|_{H^t}, \quad u \in S_\Lambda. \tag{6.7}$$

In general not much is known about stability for the above general class of Petrov–Galerkin schemes. For (nonconstant coefficient) pseudo-differential operators on the torus, stability conditions are established in Dahmen,

Prößdorf and Schneider (1994*c*); see also Dahmen, Kleemann, Prößdorf and Schneider (1996*a*) for an application to collocation.

When $\mathcal{L}$ is a pseudo-differential operator, its injectivity, boundedness and coercivity of the principal part of its symbol also imply stability (6.7) of the Galerkin scheme (Dahmen et al. 1994*c*, Dahmen, Prößdorf and Schneider 1994*b*, Hildebrandt and Wienholtz 1964). Of course, when $\mathcal{L}$ is *selfadjoint* in the sense that

$$a(u,v) := \langle \mathcal{L}u, v\rangle \tag{6.8}$$

is a *symmetric bilinear form*, ellipticity (2.23) means that

$$\|\cdot\|^2 := a(\cdot,\cdot)^2 \sim \|\cdot\|_{H^t}, \tag{6.9}$$

and the Galerkin scheme is trivially stable.

We think of the trial spaces having large dimension so that direct solvers based on factorization techniques are prohibitively expensive in storage and computing time. On the other hand, in the symmetric case (6.9), for instance, the speed of convergence of *iterative methods* is known to be governed by the condition numbers

$$\kappa_2(\mathcal{L}_\Lambda) := \lambda_{\max}(\mathcal{L}_\Lambda)/\lambda_{\min}(\mathcal{L}_\Lambda), \tag{6.10}$$

where

$$\lambda_{\max}(\mathcal{L}_\Lambda) := \sup_{v\in S(\Psi_\Lambda)} \frac{\langle \mathcal{L}v, v\rangle}{\langle v, v\rangle}, \quad \lambda_{\min}(\mathcal{L}_\Lambda) := \inf_{v\in S(\Psi_\Lambda)} \frac{\langle \mathcal{L}v, v\rangle}{\langle v, v\rangle}. \tag{6.11}$$

Note that when $t \neq 0$, the condition numbers grow with increasing $\#\Lambda$. In fact, on account of the norm equivalence (6.1) and (6.9), one obtains

$$\lambda_{\min}(\mathcal{L}_\Lambda) \leq \langle \mathcal{L}\psi_\lambda, \psi_\lambda\rangle / \|\psi_\lambda\|_{L_2} \sim 2^{2t|\lambda|},$$

while

$$\lambda_{\max}(\mathcal{L}_\Lambda) \geq \langle \mathcal{L}\psi_\lambda, \psi_\lambda\rangle / \|\psi_\lambda\|_{L_2} \sim 2^{2t|\lambda|}.$$

Thus choosing $|\lambda|$ as the lowest or highest level in Λ, depending on the sign of t, it is clear that

$$\kappa_2(\mathcal{L}_\Lambda) \gtrsim 2^{2|t||\Lambda|}, \tag{6.12}$$

where $|\Lambda| := \max\{|\lambda| - |\lambda'| : \lambda, \lambda' \in \Lambda\}$.

Thus, in such cases the objective is to find a symmetric positive definite operator $\mathcal{C}_\Lambda$ such that $\kappa_2(\mathcal{C}_\Lambda\mathcal{L}_\Lambda)$ remains possibly uniformly bounded, so that schemes like

$$u_\Lambda^{l+1} := u_\Lambda^l + \mathcal{C}_\Lambda(\mathcal{L}_\Lambda u_\Lambda^l - f),$$

or, better, correspondingly preconditioned conjugate gradient iterations, would converge rapidly.

6.2. An application of norm equivalences

With the results of Section 5.3 at hand, the task of preconditioning has become relatively easy. Since, under the assumption (2.23), $\mathcal{L}$ acts as a shift in the Sobolev scale, it is reasonable to exploit the fact that Σ_s from (5.37) does that too. Hence Σ_s should have the capability of undoing the effect of $\mathcal{L}$. By the previous remarks, the stiffness matrix

$$\mathbf{A}_\Lambda := \langle \mathcal{L}\Psi_\Lambda, \Psi_\Lambda \rangle^{\mathrm{T}} \tag{6.13}$$

relative to the wavelet basis Ψ_Λ is ill conditioned for $t \neq 0$ and large $\#\Lambda$. However, a diagonal symmetric scaling suffices to remedy this. This observation has been made on various different levels of generality in several papers; see, for instance Beylkin (1993), Dahmen and Kunoth (1992), Dahmen et al. (1996*c*), Dahmen et al. (1994*b*), Jaffard (1992) and Oswald (1992).

Theorem 6.1 (Dahmen et al. 1994*b*) Suppose that the Galerkin scheme (6.6) is stable (6.7) and that the parameters $\gamma, \tilde{\gamma}$ in (6.1) satisfy

$$|t| < \gamma, \tilde{\gamma}. \tag{6.14}$$

Let $\mathbf{D}_\Lambda^s$ be the diagonal matrix defined by (5.43). Then the matrices

$$\mathbf{B}_\Lambda := \mathbf{D}_\Lambda^{-t}\mathbf{A}_\Lambda\mathbf{D}_\Lambda^{-t} \tag{6.15}$$

have uniformly bounded spectral condition numbers

$$\|\mathbf{B}_\Lambda\|\,\|\mathbf{B}_\Lambda^{-1}\| = O(1), \quad \Lambda \subset \nabla. \tag{6.16}$$

Proof. Consider any $v \in S_\Lambda$ and set $w := \Sigma_t v$ (see (5.37)). Thus, by (6.14) and (5.38), one obtains

$$\|w\|_{L_2} = \|\Sigma_t v\|_{L_2} \sim \|v\|_{H^t} \sim \|Q_\Lambda^* \mathcal{L} Q_\Lambda v\|_{H^{-t}},$$

where we have used the stability (6.7) in the last step. Employing the norm equivalence (5.38), now relative to the dual basis, and bearing (5.39) in mind, yields

$$\|w\|_{L_2} \sim \|\Sigma_{-t}^* Q_\Lambda^* \mathcal{L} Q_\Lambda \Sigma_{-t} w\|_{L_2}.$$

This means that the operators

$$\mathcal{L}_{t,\Lambda} := \Sigma_{-t}^* Q_\Lambda^* \mathcal{L} Q_\Lambda \Sigma_{-t} : S_\Lambda \to \tilde{S}_\Lambda$$

are uniformly boundedly invertible, that is,

$$\|\mathcal{L}_{t,\Lambda}\|\,\|\mathcal{L}_{t,\Lambda}^{-1}\| = O(1), \quad \#\Lambda \to \infty. \tag{6.17}$$

It is now a matter of straightforward calculation to verify that the matrix representation of $\mathcal{L}_{t,\Lambda}$ relative to Ψ_Λ is

$$\langle \mathcal{L}_{t,\Lambda}\Psi_\Lambda, \Psi_\Lambda \rangle^{\mathrm{T}} = \mathbf{D}_\Lambda^{-t}\mathbf{A}_\Lambda\mathbf{D}_\Lambda^{-t}, \tag{6.18}$$

which proves the claim. □

Letting $\#\Lambda$ tend to infinity, the original equation $\mathcal{L}u = f$ can be viewed as an infinite discrete system (recall (1.35) in Section 1.5)

$$\mathbf{D}^{-t}\mathbf{A}\mathbf{D}^{-t}\mathbf{d} = \mathbf{D}^{-t}\mathbf{f} \tag{6.19}$$

where $\mathbf{D}^{-t}$ and $\mathbf{A}$ are the infinite counterparts of $\mathbf{D}_\Lambda^{-t}$, $\mathbf{A}_\Lambda$, respectively, and $\mathbf{f} := \langle f, \Psi\rangle^{\mathrm{T}}$ is the coefficient sequence of f expanded relative to the *dual* basis $\tilde{\Psi}$. The sequence $\mathbf{d}$ then consists of the wavelet coefficients (relative to Ψ) of the solution

$$u = \mathbf{d}^{\mathrm{T}}\Psi$$

of (2.24). The infinite matrix $\mathbf{B} := \mathbf{D}^{-t}\mathbf{A}\mathbf{D}^{-t}$ is, on account of Theorem 6.1, a boundedly invertible mapping from $\ell_2(\nabla)$ onto $\ell_2(\nabla)$.

It is remarkable that similar techniques also lead to preconditioners for *collocation* matrices (Schneider 1995). In brief, recall that (6.3) defines a collocation scheme, when the P_Λ in (6.3) are *interpolation* projectors. Let us consider full sets Λ_J, defined in (3.24), which means $S(\Phi_J) = S_{\Lambda_J}$, and assume that for a suitable mesh of points $\{x_{J,k}\}_{k\in\Delta_J}$, the corresponding projectors have the form

$$P_{\Lambda_J} f = L_J f = \sum_{k\in\Delta_J} 2^{-nJ/2} f(x_{J,k})\tilde{\theta}_{J,k} =: \langle f, \delta_J\rangle\tilde{\theta}_J,$$

that is, $\delta_{J,k}\theta_{J,m} = \theta_{J,m}(x_{J,k}) = 2^{Jd/2}\delta_{k,m}$, $k, m \in \Delta_J$. For instance, $\tilde{\theta}$ could be a spline function interpolating the Kronecker sequence. Moreover, assume that

$$(L_{j+1} - L_j)\, f = \langle f, \vartheta_j\rangle\gamma_j, \tag{6.20}$$

where

$$\vartheta_j^{\mathrm{T}} = \delta_{j+1}^{\mathrm{T}}\mathbf{M}_{j,1}^{\delta}, \quad \gamma_j = \tilde{\theta}_{j+1}^{\mathrm{T}}\tilde{\mathbf{M}}_{j,1}^{\delta}, \tag{6.21}$$

are corresponding stable completions.

Theorem 6.2 (Schneider 1995) Suppose that the collocation method (6.3) relative to $P_{\Lambda_J} = L_J$ is $(s, 2t)$-stable in the sense of (6.4), and assume that

$$\frac{n}{2} < s - 2t, \quad \frac{n}{2} < \gamma_{\tilde{\theta}}, \quad \tilde{\gamma} > 0, \quad s < \gamma. \tag{6.22}$$

Then the matrices

$$\mathbf{D}_J^{s-2t}\left\langle \mathcal{L}\Psi^J, \vartheta^J\right\rangle^{\mathrm{T}}\mathbf{D}_J^{-s} \tag{6.23}$$

have uniformly bounded spectral condition numbers.

Details of the proof can be found in Schneider (1995). It uses continuity of $\mathcal{L}$ as a mapping on Sobolev spaces, the stability (6.4) and the fact that

$$\|f\|_{H^\tau}^2 \sim \sum_{j=0}^{\infty} 2^{2\tau j} \|(L_{j+1} - L_j)f\|_{L_2}^2$$

for $n/2 < \tau < \gamma_{\tilde{\theta}}$. As for this norm equivalence, note that the L_j do satisfy (5.16). But they are only bounded in higher Sobolev spaces, which causes conditions (6.22). Keeping this in mind, the above equivalence can be deduced from the general results in Sections 5.2 and 5.3. For details see Dahmen (1996).

The above simple argument is designed to show the qualitative role of norm equivalences in connection with preconditioning. In practice, the constants involved will matter. However, in principle, it should be noted that, in Theorem 6.1, neither

- selfadjointness of $\mathcal{L}$, nor
- positive order $2t > 0$

is required for the validity of (6.16).

For unsymmetric problems, (6.16) alone is not sufficient to imply the efficiency of corresponding variants of the preconditioned conjugate gradient method, such as GMRES. But behind the validity of (1.29) or (2.23) there is usually a symmetric principal part of the operator $\mathcal{L}$, in which case GMRES will perform well, provided that the condition numbers stay small. Alternatively, if the constant in (6.16) stays moderate, one can square the preconditioned system and the conjugate gradient scheme works well. We can summarize this under the following purely asymptotic result.

Remark 6.3 Suppose that every matrix vector multiplication with $\mathbf{A}_\Lambda$ can be carried out in $O(\#\Lambda)$ operations uniformly in $\Lambda \subset \nabla$, and assume that the (exact) Galerkin solution u_Λ of (6.6) in S_Λ satisfies

$$\|u - u_\Lambda\|_{H^t} = \varepsilon_\Lambda.$$

Then an approximate solution $\hat{u}_\Lambda$ of (6.6) satisfying $\|u - \hat{u}_\Lambda\|_{H^t} = O(\varepsilon_\Lambda)$ uniformly in Λ can be computed at the expense of $O(\#\Lambda)$ operations.

The argument is based on standard *nested iteration*. Solving first on a small $\Lambda_0 \subset \nabla$, then doubling $\#\Lambda_0$ to Λ_1, say, and noting that $\varepsilon_{\Lambda_0}/\varepsilon_{\Lambda_1} \leq C$, only $O(\Lambda_1)$ iterations on the preconditioned system are needed to reduce the error from ε_{Λ_0} to ε_{Λ_1}, when using $\hat{u}_{\Lambda_0}$ as a starting solution. Repeating this argument confirms the assertion.

Next, let us address some algorithmic issues. When $\mathcal{L}$ is a differential operator, the stiffness matrices

$$\mathbf{A}_{\Phi_J} = \langle \mathcal{L}\Phi_J, \Phi_J \rangle^{\mathrm{T}}$$

relative to the fine scale (nodal) bases Φ_J are sparse under the usual assumption (5.5). Due to the larger supports of wavelets from coarser scales the corresponding stiffness matrices $\mathbf{A}_{\Lambda_J}$ relative to the wavelet bases of $S(\Phi_J)$ are less sparse. Thus, assembling the wavelet stiffness matrix exactly would increase computational and storage complexity. However, when working with the fully refined sequence of spaces $S(\Phi_j)$, this can be remedied as follows. All that is needed in an iterative scheme is the *application* of the preconditioned matrix. Since by (3.31),

$$\mathbf{A}_{\Lambda_J} = \mathbf{T}_J^{\mathrm{T}} \mathbf{A}_{\Phi_J} \mathbf{T}_J, \tag{6.24}$$

where $\mathbf{T}_J$ is the multiscale transformation from (3.25), (3.26), the application of the preconditioned matrix $\mathbf{B}_J := \mathbf{D}_{\Lambda_J}^{-t} \mathbf{A}_{\Lambda_J} \mathbf{D}_{\Lambda_J}^{-t}$ to a vector $\mathbf{v}$ can be carried out as follows.

ALGORITHM 1 (CB: CHANGE OF BASES)

(1) Compute $\mathbf{w} = \mathbf{T}_J \mathbf{D}_{\Lambda_J}^{-t} \mathbf{v}$. Due to the pyramid structure of $\mathbf{T}_J$ (3.26) and the geometrical increase of $\#\Phi_j$, this requires $O(\#\Phi_J)$ operations, where the constant depends on the length of the masks in $\mathbf{M}_j$.
(2) Compute $\mathbf{z} := \mathbf{A}_{\Phi_J} \mathbf{w}$, which, due to the sparseness of $\mathbf{A}_{\Phi_J}$ is again a $O(\#\Phi_J)$ process.
(3) Compute $\mathbf{D}_{\Lambda_J}^{-t} \mathbf{T}_J^{\mathrm{T}} \mathbf{z}$, which corresponds to the first step.

Remark 6.4 When $\mathcal{L}$ is a differential operator, the application of the preconditioned matrix $\mathbf{D}_{\Lambda_J}^{-t} \mathbf{A}_{\Lambda_J} \mathbf{D}_{\Lambda_J}^{-t}$ relative to the full spaces $S(\Phi_J)$ to a vector requires the amount of $O(\#\Phi_J) = \mathcal{O}(N_J)$ operations and storage.

Remark 6.5 In the periodic case, or when working on the interval, $\mathbf{A}_{\Phi_J}$ can be computed very efficiently (even for variable coefficients) by the methods described in Section 4.2.

Remark 6.6 It is also important to note that the above preconditioner only requires knowlege of the transformation $\mathbf{T}_J$ in (3.26) *not* of the inverse $\mathbf{T}_J^{-1}$ (see Section 3.4). Recall that $\mathbf{T}_J$ involves the refinement matrices for Φ_j and the stable completions $\mathbf{M}_{j,1}$, $j < J$, that is, the masks of the wavelets (see (3.27)). Hence this method can still be used in the present context with the same efficiency when only the matrices $\mathbf{M}_j$ are sparse while the inverses $\mathbf{G}_j$ are fully populated. This is the case for many *pre-wavelets*, that is, for stable complement bases Ψ_j, which span the *orthogonal* complement of $S(\Phi_j)$ in $S(\Phi_{j+1})$.

So far this strategy refers to *fully refined* spaces $S(\Phi_J)$. Things change when the trial spaces are to be adapted *during* the solution process. This means that one actually wants to compute a solution from spaces S_Λ where Λ is a much smaller *lacunary* subset of Λ_J, $J = \max\{|\lambda| + 1 : \lambda \in \Lambda\}$. To take full advantage of the corresponding principal reduction of complexity, any steps requiring the computational complexity of the full space $S(\Phi_J)$ should be avoided. This suggests building up the matrix $\mathbf{A}_\Lambda$ directly but only relative to the elements in Λ. How to do this efficiently depends very much on the particular operator $\mathcal{L}$. We will comment on this issue later in more detail for operators satisfying (2.25). In this case one can exploit certain decay properties of the entries of $\mathbf{A}_\Lambda$ to compute $\mathbf{A}_\Lambda$ approximately to any desired accuracy.

6.3. Hierarchical bases preconditioner

The change of bases preconditioner has already been employed in connection with hierarchical bases (Yserentant 1986). The corresponding setting of piecewise linear bivariate finite element and associated hierarchical bases was described in Section 4.1. Due to the simple and very sparse structure of the matrices $\mathbf{M}_j$ (see (4.2) and (4.4)), Algorithm 1 above is very efficient. However, the hierarchical bases are invariant under the application of $L_j - L_{j-1}$ as in (6.20), where the L_j are Lagrange interpolation operators relative to the triangulation $\mathcal{T}_j$. Hence they are *not* bounded in $L_2(\Omega)$ and the collection $\Phi_0 \cup \bigcup_{j=0}^{\infty} \Psi_j$ is *not* a Riesz basis for $L_2(\Omega)$. Moreover, $\|\cdot\|_{H^1}$ is *not* equivalent to the discrete norm $(L_{-1} := 0)$ $\left(\sum_{j=0}^{\infty} 2^{2j} \|(L_{j+1} - L_j)f\|_{L_2}^2\right)^{1/2}$ for $n \geq 2$. Hence the *hierarchical* basis preconditioner, based on Algorithm 1, is *not* asymptotically optimal. For $n = 2$, the condition numbers grow like the square of the number of levels, while for $n = 3$ they already exhibit an exponential growth. Nevertheless, its extreme simplicity accounts for its attractiveness for $n = 2$. Ways of *stabilizing* it, for instance with the aid of the techniques in Section 3.5, will be presented later.

6.4. BPX scheme

Although, as it stands, the simple hierarchical complement bases do not provide an asymptotically optimal scheme with regard to preconditioning, it turns out that the full power of wavelet decompositions is needed only for operators of *non-positive* order. Throughout this section we will assume that $\mathcal{L}$ is selfadjoint positive definite (6.8), (6.9) and that the order $2t$ of $\mathcal{L}$ is *positive.* In this case one gets away with much less. So, suppose the bases Ψ_J are stable in the sense of (3.3) and give rise to a hierarchy of nested spaces $S_J = S(\Phi_J) \subset H^t$ as before. The following discussion reflects an approach to multilevel preconditioners developed in the context of finite element discretizations (Bramble et al. 1990, Oswald 1992, Yserentant 1990, Xu 1992, Zhang 1992). The objective is to find a positive definite selfadjoint operator C_J on S_J such that

$$\langle C_J^{-1} v, v\rangle \sim \langle \mathcal{L}_J v, v\rangle = a(v,v), \quad v \in S_J, \tag{6.25}$$

which means that $\mathcal{C}_J$ and $\mathcal{L}_J$ are *spectrally equivalent.* In fact, the uniformity of (6.25) in J implies, in view of the min-max characterization of eigenvalues, that

$$\frac{\lambda_{\max}\left(\mathcal{C}_J^{1/2}\mathcal{L}_J\mathcal{C}_J^{1/2}\right)}{\lambda_{\min}\left(\mathcal{C}_J^{1/2}\mathcal{L}_J\mathcal{C}_J^{1/2}\right)} \sim 1. \tag{6.26}$$

To describe a candidate for $\mathcal{C}_J$, let P_J denote the orthogonal projector onto S_J. Clearly the P_J satisfy (5.16) and $P_J^* = P_J$. Thus Theorem 5.8 applies and (5.38) means that

$$\hat{\mathcal{C}}_J^{-1} = \sum_{j=0}^{J} 2^{2tj}(P_j - P_{j-1}), \quad P_{-1} := 0,$$

satisfies

$$\langle \hat{\mathcal{C}}_J^{-1} v, v\rangle = \langle \hat{\mathcal{C}}_J^{-1/2} v, \hat{\mathcal{C}}_J^{-1/2} v\rangle = \|\Sigma_t v\|_{L_2} \sim \|v\|_{H^t}. \tag{6.27}$$

Hence, by ellipticity,

$$\langle \hat{\mathcal{C}}_J^{-1} v, v\rangle \sim a(v,v), \quad v \in S_J, \tag{6.28}$$

so that $\hat{\mathcal{C}}_J P_J \mathcal{L} P_J$ have uniformly bounded condition numbers. This corresponds to the situation assumed in Theorem 6.1, since the evaluation of $\hat{\mathcal{C}}_J$ seems to require knowledge of explicit bases for the orthogonal complements. However, since, clearly, by (5.39),

$$\hat{\mathcal{C}}_J = \sum_{j=0}^{J} 2^{-2tj}(P_j - P_{j-1}),$$

and $t > 0$, $\hat{\mathcal{C}}_J$ is easily seen to be spectrally equivalent to $\overline{\mathcal{C}}_J := \sum_{j=0}^{J} 2^{-2tj} P_j$, which, by the uniform stability of the Φ_j, is spectrally equivalent to

$$\mathcal{C}_J v := \sum_{j=0}^{J} 2^{-2tj} \sum_{k\in\Delta_j} \langle v, \phi_{j,k}\rangle \phi_{j,k}. \tag{6.29}$$

Combining the spectral equivalence of $\mathcal{C}_J$ and $\hat{\mathcal{C}}_J$ with (6.27) and (6.29) yields

$$\langle \mathcal{C}_J P_J \mathcal{L} P_J v, v\rangle \sim \langle v, v\rangle. \tag{6.30}$$

Hence (Dahmen and Kunoth 1992, Oswald 1992, Zhang 1992),

$$\kappa_2\left(\mathcal{C}_J P_J \mathcal{L} P_J\right) = O(1), \quad j \in \mathbb{N}. \tag{6.31}$$

Note that application of $\mathcal{C}_J$ does *not* require explicit knowledge of any complement basis. It also requires only the order of $\#\Phi_J$ operations. For more details about the actual implementation, the reader is referred to Bramble et al. (1990) and Xu (1992).

Exact decompositions in terms of complement bases have been replaced by *redundant* spanning sets, which consist here of properly weighted nodal basis functions on *each* level. In brief, the collections $\{2^{-2jt}\phi_{j,k} : k \in \Delta_j, j = 0, \dots, J\}$ form *frames* for $H^t(\Omega)$. Here $\{g_j\}$ is called a frame for H, if

$$\|v\|_H^2 \sim \sum_j |\langle v, g_j \rangle_H|^2, \quad v \in H. \tag{6.32}$$

It is perhaps worth stressing the relation to the wavelet transforms.

Remark 6.7 The corresponding wavelet preconditioner looks like

$$C_J^w v = \sum_{j=-1}^{J-1} 2^{-2jt} \sum_{k \in \nabla_j} \langle v, \psi_{j,k} \rangle \psi_{j,k}.$$

Since the $\psi_{j,k}$ are linear combinations of the $\phi_{j+1,m}$, its evaluation always seems to be *more* expensive than that of (6.29). The cost of each iteration increases with the lengths of the masks of the wavelets (Griebel and Oswald 1995*b*, Oswald 1994).

Adaptive grids

Theorem 6.1 has been formulated for *arbitrary* subsets $\Lambda \subset \nabla$. Thus adaptivity can be based, in principle, on adapting the choice of Λ to the problem at hand. It will be explained later how to arrange that. Roughly speaking, the behaviour of the wavelet coefficients themselves is an indication for the selection of relevant indices. The point is that this kind of adaptation essentially requires managing *index sets*.

So far, in a finite element context, the above discussion of the BPX scheme refers to spaces generated by *uniform* refinements. Adaptivity usually requires *mesh refinement* strategies based on monitoring the current solution through additional local comparisons. It is interesting to see how preconditioning is affected when working with adaptively refined meshes. Employing *hanging* or *slave* nodes, that is, adding locally further nodal basis functions, corresponds, roughly speaking, to considering submatrices of those stemming from uniform refinements. Since the convex hull of the spectrum of the latter matrices contains the spectrum of the submatrices, the BPX scheme is trivially adapted to nonuniform refinements and the condition numbers remain bounded.

Slave nodes require a little care retrieving stable bases for the resulting finite element spaces. If one wants to avoid slave nodes, the nonuniform refinements have to be *closed* by introducing suitable transition elements; see, for instance, Bank, Sherman and Weiser (1983). In this case, the submatrix argument does not work in a strict sense. Nevertheless, one can prove that for such adaptive refinements resulting in highly nonuniform meshes, the BPX scheme still produces uniformly bounded condition numbers. This has been shown first in Dahmen and Kunoth (1992), where further details can be found; see also Bornemann and Yserentant (1993).

An analogous result holds for fourth-order problems. As a model case, one could consider $\mathcal{L} = \Delta^2$ with homogeneous Dirichlet boundary conditions. A convenient conforming finite element discretization can be based on certain piecewise cubic macro patches generated by suitable subdivisions of (non rectangular) quadrilaterals. These are obtained by connecting the intersection of diagonals with the midpoint of the edges of the quadrilateral. The nodal basis functions are fundamental interpolators relative to point values and gradients at the corners of the quadrilaterals and normal derivatives at the midpoints of edges. The resulting spaces are nested and the underlying mesh refinements stay regular, in the sense that smallest angles are bounded away from zero. See Dahmen, Oswald and Shi (1993*a*) for more details. Adaptive refinements analogous to the piecewise linear case are discussed by Kunoth (1994), where the corresponding result about uniformly bounded condition numbers is also established. One should note that the classical cubic Clough–Tocher macro element is not suited for refinements. Since the quintic C^1-Argyris element requires higher smoothness at the vertices, its refinement leads to nonnested trial spaces.

The hierarchical basis and BPX preconditioner are special instances of the following more general class of schemes that have a long tradition in the finite element context.

6.5. Multilevel Schwarz schemes

We will briefly indicate how the above material ties into the more general setting of *Schwarz schemes* and *stable splittings*, which is also a convenient framework for incorporating domain decomposition and multigrid techniques. For a more extensive treatment of these issues, as well as further details concerning the following discussion, we refer, for example, to Griebel and Oswald (1995*a*), Oswald (1994), Xu (1992) and Yserentant (1993). As above, $\mathcal{L}$ will be selfadjoint positive definite on some separable Hilbert space $H = H_1$, that is, $a(u,v) := \langle \mathcal{L}u, v\rangle$ is a symmetric bilinear form and we assume that (6.9) holds with H^t replaced by H. We wish to find $u \in H$ such that

$$a(u,v) = f(v), \quad v \in H, \tag{6.33}$$

where f is a linear functional on H. In fact, at this point one can think of H being some Sobolev space H^t as above but also of the finite dimensional trial space $S(\Phi_J)$ of highest resolution. Let $\{V_j\}$ be an at most countable collection of closed nested subspaces of H such that every $v \in H$ has at least one expansion

$$v = \sum_j v_j,$$

which converges in H, in brief $H = \sum_j V_j$.

The basic idea is to solve for each V_j the problem restricted to V_j and then add these solutions up. This corresponds to (block) Gauss–Seidel or Jacobi relaxation. The solution of the subproblems will be based on auxiliary inner products $b_j(\cdot,\cdot)$ on V_j which approximate $a(\cdot,\cdot)$. Following Oswald (1994), we write $\{H; a\}$, $\{V_j; b_j\}$ to express that each V_j, H, are Hilbertian relative to the scalar products b_j, a. The *subspace splitting*

$$\{H; a\} = \sum_j \{V_j; b_j\}$$

is called *stable* if

$$\|v\|^2_{\{b_j\}} := \inf\left\{\sum_j b_j(v_j, v_j) : v_j \in V_j, v = \sum_j v_j\right\} \sim a(v, v). \tag{6.34}$$

Taking $V_j = S(\Phi_j)$, $b_j(v,v) := 2^{2sj}\langle v, v\rangle$, $H = H^s(\Omega)$, we see that the norm equivalence (5.44) is a special case of (6.34). Alternatively, setting $V_{-1} := S(\Phi_0)$, $V_j = S(\Psi_j)$, $j \geq 0$, $v_j = (Q_j - Q_{j-1})v$, $b_j(v,v)$ as before, it is clear that $\|v\|_{\{b_j\}} \lesssim \|v\|_{\mathcal{A}^s_Q}$. Now consider the following Riesz operators $\mathcal{T}_j$, g_j which interrelate the scalar products.

$$\begin{aligned} \mathcal{T}_j : H \to V_j : \quad & b_j(\mathcal{T}_j v, v_j) = a(v, v_j), \\ g_j \in V_j : \quad & b_j(g_j, v_j) = f(v_j), \quad v_j \in V_j. \end{aligned} \tag{6.35}$$

Defining $B_j : H \to V_j$ by $b(v, v_j) = \langle B_j v, v_j\rangle, v_j \in V_j$, and recalling that $a(u,v) = \langle \mathcal{L}u, v\rangle$, we infer that $\mathcal{T}_j = B_j^{-1}\mathcal{L}$, so that each application of $\mathcal{T}_j$ corresponds to the solution of a restricted (typically small) problem. A central observation in this context is the following result; see Oswald (1994) for the background and further references.

Theorem 6.8 Let

$$\mathcal{T} := \sum_j \mathcal{T}_j, \quad g = \sum_j g_j. \tag{6.36}$$

Then equation (6.33) is equivalent to the operator equation

$$\mathcal{T}u = g, \tag{6.37}$$

which is called the *additive* Schwarz formulation of (6.33). Moreover, $\mathcal{T}$ is selfadjoint positive definite and when H is finite-dimensional its smallest and largest eigenvalue are $\lambda_{\min}(\mathcal{T})$, $\lambda_{\max}(\mathcal{T})$, where

$$\lambda_{\max}(\mathcal{T}) = \sup_{v\in H} \frac{a(v,v)}{\|v\|^2_{\{b_j\}}}, \quad \lambda_{\min}(\mathcal{T}) = \inf_{v\in H} \frac{a(v,v)}{\|v\|^2_{\{b_j\}}},$$

respectively.

Since

$$\kappa_2(\mathcal{T}) := \frac{\lambda_{\max}(\mathcal{T})}{\lambda_{\min}(\mathcal{T})}, \tag{6.38}$$

the condition of (6.37) is bounded by the ratio of the constants in the upper and lower bound in (6.34). A corresponding asymptotic statement requires the ratio of the upper and lower bound to remain uniformly bounded when the dimension of H increases. The quantitative relevance of such statements again depends, of course, on the problem at hand.

Theorem 6.8 can be deduced from the following result, which is interesting in its own right and has several further applications mentioned below.

Theorem 6.9 (Nepomnyaschikh 1990) Let $H, \tilde{H}$ be two Hilbert spaces with scalar products $\langle \cdot, \cdot \rangle_H$, $\langle \cdot, \cdot \rangle_{\tilde{H}}$, respectively, and with bilinear forms $a, \tilde{a}$ induced by symmetric positive definite operators $\mathcal{L} : H \to H$, $\tilde{\mathcal{L}} : \tilde{H} \to \tilde{H}$. Suppose that there exists a surjective bounded linear operator $\mathcal{R} : \tilde{H} \to H$ such that

$$a(v, v) \sim \inf_{\tilde{v} \in \tilde{H}, v = \mathcal{R}\tilde{v}} \tilde{a}(\tilde{v}, \tilde{v}) := \|v\|, \quad v \in H.$$

Then $\mathcal{P} := \mathcal{R}\tilde{\mathcal{L}}^{-1}\mathcal{R}^*\mathcal{L} : H \to H$ is symmetric positive definite with

$$\lambda_{\max}(\mathcal{P}) = \sup_{v \in H} \frac{a(v, v)}{\|v\|^2}, \quad \lambda_{\min}(\mathcal{P}) = \inf_{v \in H} \frac{a(v, v)}{\|v\|^2}.$$

Of course, Theorem 6.8 is obtained by taking $\tilde{H} := \{\tilde{v} = \{v_i\} : v_i \in V_i, \sum_i b_i(v_i, v_i) < \infty\}$; see, for example, Oswald (1994).

An interesting case is, for example, $H = S(\Phi_J)$ for some (large J) where the splitting consists of *one-dimensional* subspaces $V_{j,k} := S(\{\phi_{j,k}\})$. Thus the corresponding Riesz operator $\mathcal{T}_{j,k}$ has the form $\mathcal{T}_{j,k}v = c_{j,k}\phi_{j,k}$ where

$$c_{j,k} = a(v, \phi_{j,k}) / b_{j,k}(\phi_{j,k}, \phi_{j,k}). \tag{6.39}$$

Thus

$$\mathcal{T}v = \sum_{j=0}^{J} \sum_{k \in \Delta_j} \frac{a(v, \phi_{j,k})}{b_{j,k}(\phi_{j,k}, \phi_{j,k})} \phi_{j,k}. \tag{6.40}$$

Hence (6.34) means that $\{\phi_{j,k}/b_{j,k}(\phi_{j,k}, \phi_{j,k})\}$ forms a frame for $\{H_j a\}$.

The BPX scheme and the hierarchical basis preconditioner are examples of this type. To see this, let $f(v) = \langle f, v \rangle$, where $\langle \cdot, \cdot \rangle$ is induced by the L_2 inner product. Let $B_{j,k}$ be defined by $b_{j,k}(u_{j,k}, v_{j,k}) = \langle B_{j,k}u_{j,k}, v_{j,k} \rangle$, so that $\mathcal{T}_{j,k} = B_{j,k}^{-1}P_{j,k}\mathcal{L}$, $P_{j,k}$ the orthogonal projection onto $V_{j,k}$. Then one can write

$$\mathcal{T} = \left(\sum_{j=0}^{J} \sum_{k \in \Delta_j} B_{j,k}^{-1} P_{j,k} \right) \mathcal{L} =: \mathcal{C}^{-1}\mathcal{L}.$$

Thus, choosing $b_{j,k}(\cdot,\cdot) := 2^{2jt}\langle\cdot,\cdot\rangle$ and $V_{j,k} = S(\{\phi_{j,k}\}), k \in \Delta_j$, $V_{j,k} = S(\{\phi_{j,k}\}), k \in \Delta_j \setminus \Delta_{j-1}$, yields the BPX and hierarchical basis preconditioner, respectively (see (6.29)).

There is obviously great flexibility in choosing the subspaces V_j. In general, realization of the operators $\mathcal{T}_j$ requires solving linear systems of the size determined by V_j. In the context of parallel computing one might accept spaces V_j of growing dimension. For instance, the V_j could be chosen as subspaces of H associated with a decomposition of the underlying domain. A survey of applications of this type can be found in Chan and Mathew (1994).

The proofs establishing the critical equivalence (6.34) are, of course, related to the concepts discussed in Section 5.3. The norm $|\!|\!|\cdot|\!|\!|_{\{b_j\}}$ can in certain cases be evaluated exactly, which greatly simplifies the analysis; compare Griebel and Oswald (1995*a*), and Nießen (1995). The bounds $|\!|\!|\cdot|\!|\!|^2_{\{b_j\}} \lesssim a(\cdot,\cdot)$ typically involve approximation theory tools. The converse estimates are often reduced to the validity of a *strengthened Cauchy–Schwarz inequality* (recall Section 5.2), which in this context has the form

$$a(v_j, v_k) \lesssim \gamma_{j,k} b_j(v_j, v_j) b_k(v_k, v_k), \quad v_j \in V_j, v_k \in V_k, \tag{6.41}$$

where $(\gamma_{j,k})$ should be bounded in ℓ_2. For a detailed discussion of these issues, compare Griebel and Oswald (1995*a*), Yserentant (1993).

Again following Oswald (1997), we mention an interesting extension of the splitting concept which aims at relaxing the assumption of nestedness $V_j \subset V_{j+1}$, as well as of conformity $V_j \subset H$. This, again, is an application of Theorem 6.9. It requires introducing mappings $R_j : V_j \to H$ such that $R := \sum_j R_j : \prod_j V_j \to H$ is *onto* and

$$|\!|\!|v|\!|\!|_{\{b_j;R_j\}} := \inf\Big\{\sum_j b_j(v_j, v_j) : v_j \in V_j, v = \sum_j R_j v_j\Big\} \sim a(v,v), \quad v \in H. \tag{6.42}$$

In this case, $\mathcal{T}$ in (6.37) has to be replaced by $\mathcal{T}' := \sum_j (R_j B_j^{-1} R_j^*)\mathcal{L}$, where $B_j : V_j \to V_j$ is now defined by $b(u_j, v_j) = \langle B_j u_j, v_j\rangle_{V_j}$, $v_j \in V_j$. Equivalently, one can write $\mathcal{T}' = \sum_j R_j \mathcal{T}_j'$, where $\mathcal{T}_j' : H \to V_j$ is given by $b(\mathcal{T}_j' v, v_j) = a(v, R_j v_j)$, $v_j \in V_j$. See Oswald (1994), Griebel and Oswald (1995*a*).

We now indicate a typical iteration based on (6.35) and (6.36). The *additive version* **A** creates a sequence of approximations $\{u^l\}$ given by

$$u^{l+1} = u^l + \omega \sum_{j=0}^{J} (g_j - \mathcal{T}_j u^l). \tag{6.43}$$

Here ω plays the same role of a relaxation parameter as in the Jacobi or Richardson iteration. Recall that each iteration requires the solution of variational problems on the spaces V_j. Perhaps the structure of the iteration (6.43) becomes more transparent on recalling that $\mathcal{T}_j = B_j^{-1}\mathcal{L}$, so that $g_j - \mathcal{T}_j u^l = B_j^{-1} r^l$, where $r^l = f - \mathcal{L}u^l$ is the residual from the last step. Likewise, in the more general version (6.42), $g_j - \mathcal{T}_j u^l$ has to be replaced by $R_j B_j^{-1} R_j^* r^l$.

The *multiplicative* version **M** reads

$$\begin{aligned} v^0 &:= u^l, \\ v^{j+1} &= v^j + \omega(g_{J-j} - \mathcal{T}_{J-j} v^j), \quad j = 0, \ldots, J, \\ u^{l+1} &= v^{J+1}, \end{aligned} \tag{6.44}$$

generalizing SOR. The corresponding iteration operators are

$$M_{\mathbf{A}} = I - \omega\mathcal{T}, \quad M_{\mathbf{M}} = (I - \omega\mathcal{T}_0)(I - \omega\mathcal{T}_1)\cdots(I - \omega\mathcal{T}_J).$$

The convergence theory is given by Bramble (1993), Griebel and Oswald (1995*a*), Xu (1992) and Yserentant (1993). For the interpretation of these schemes in the multigrid context, see Bramble (1993) and Griebel (1994). Here we quote the following result from Griebel and Oswald (1995*a*).

Theorem 6.10 Assume that H is finite-dimensional and the algorithms **M** and **A** are given by (6.43) and (6.44).

(i) **A** converges for $0 < \omega < 2/\lambda_{\max}(\mathcal{T})$. The optimal rate is achieved for $\omega^* = 2/(\lambda_{\max}(\mathcal{T}) + \lambda_{\min}(\mathcal{T}))$ and equals

$$\rho_{\mathbf{A}} = \min_{0<\omega<2/\lambda_{\max}} \|M_{\mathbf{A}}\|_a = 1 - \frac{2}{1 + \kappa_2(\mathcal{T})}.$$

(ii) Suppose (6.41) holds with $\gamma_{jj} = 1$. Then **M** converges for $0 < w < 2$. The optimal rate is bounded by

$$\rho_M \leq 1 - \frac{\lambda_{\min}(\mathcal{T})}{2\lambda_{\max}(\mathcal{T}) + 1}.$$

For various modifications see Oswald (1997) and the literature cited there.

6.6. Finite element-based wavelets

The previous discussion shows that preconditioning matrices stemming from Galerkin discretizations of elliptic operators of positive order does *not* require explicit knowledge of wavelet bases. Nevertheless, a number of recent studies have addressed the construction and application of wavelets in a finite element context, to obtain wavelet-based stable splittings for Schwarz schemes. Let us briefly postpone giving reasons why the additional effort might still pay in this context, and first outline some ingredients of the various approaches.

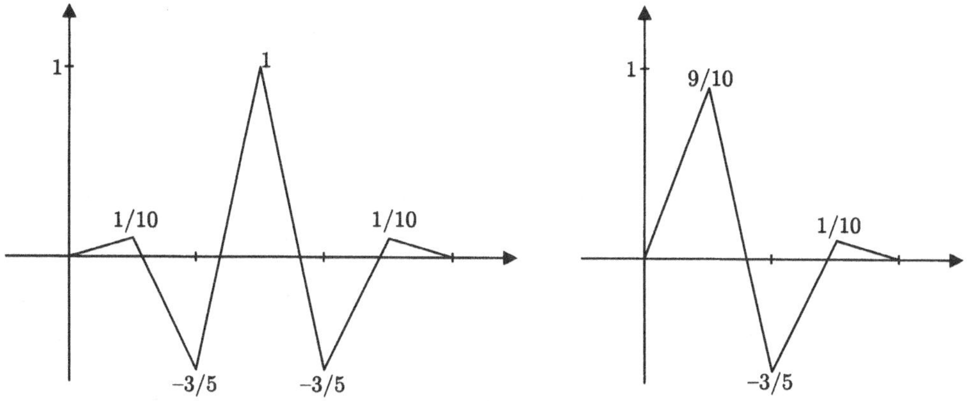

Fig. 1. Pre-wavelets for $H_0^1([0,1])$

Pre-wavelets
Recall from (6.24) that preconditioning based on a change of basis does *not* require the transform $\mathbf{T}_J^{-1}$. Thus it suffices to make sure that the scheme (3.26) is efficient, which means that the masks of the wavelets have possibly small support. Most of the presently known FE-based wavelets still refer to an underlying uniform grid structure for multilinear finite elements on regular lattices $h\mathbb{Z}^d$ (type-1 mesh) or to regular triangulations of the plane which are generated from the standard uniform rectangular mesh by inserting in each square element the southwest–northeast diagonal (type-2 mesh).

When dealing with meshes of type 1 restricted to the unit square $\square = [0,1]^2$, say, one can employ tensor products of biorthogonal wavelets on $[0,1]$ discussed in Section 4.4. For $k \in \mathbb{Z}$, $e \in \{0,1\}^2$, $j \geq j_0$, they have the form

$$\psi_{e,j,k}(x) = \psi_{e_1,j,k_1}(x_1)\psi_{e_2,j,k_2}(x_2)$$

where $\psi_{0,j,k} = \phi_{j,k}$, $\psi_{1,j,k} = \psi_{j,k}$ are the corresponding univariate generator and wavelet functions. When ϕ is the standard piecewise linear tent function (1.15) and the dual bases are exact of order 2 as well, the mask coefficients can be found in Dahmen et al. (1996*b*), and Dahmen and Schneider (1997*a*).

In most cases, however, so-called piecewise linear pre-wavelets have been used; see, for example, Griebel and Oswald (1995*b*). Interior and boundary wavelets are shown below in Figure 1.

Here, pre-wavelet means that these wavelets form uniformly L_2-stable bases for *orthogonal* complements between two successive trial spaces. Hence they also form a Riesz basis for $L_2([0,1])$. In this case the masks in the inverse transformations are not local but, as mentioned before, this is harmless here. Obviously the masks for the tensor product wavelets have 15 or 25 nonzero coefficients.

Piecewise linear pre-wavelets for meshes of type 2 have been constructed by Kotyczka and Oswald (1996). Those of smallest possible support have 13 nonzero coefficients. The construction principle is to make an ansatz of a linear combination of tent functions on the fine scale so that, for possibly few nonzero coefficients, orthogonality to the tent functions on the coarser scale holds. Usually, the difficult part is to verify that three such linear combinations form a stable basis on each given level. Also, the adaptation to the boundary is in this case more difficult than in the tensor product case.

The resulting pre-wavelets still have relatively large support. Therefore several alternatives have been proposed resulting in complement spaces that are no longer orthogonal but are spanned by functions of smaller support, while still exhibiting better stability properties across levels than the hierarchical bases.

For instance, the discretization of the double layer potential equation on a polyhedron in Dahmen, Kleemann, Prößdorf and Schneider (1994*a*) involves piecewise linear wavelet type functions of the form shown in Figure 2.

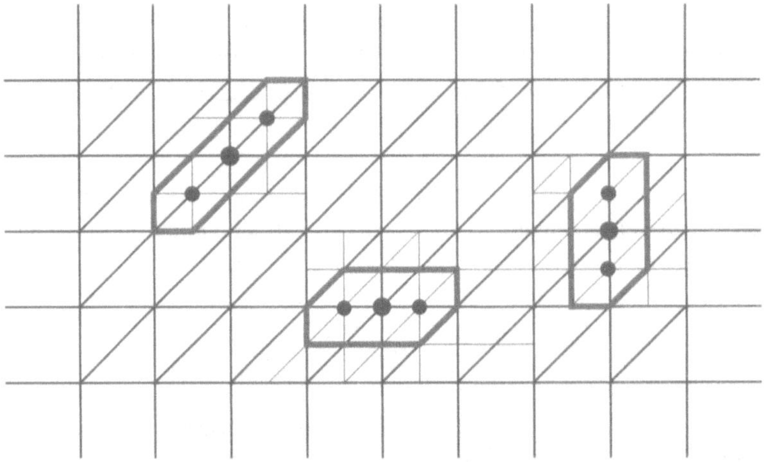

Fig. 2. Short support wavelets

The stencils in this case are

$$\begin{pmatrix} 0 & 0 & -1/2 \\ 0 & 1 & 0 \\ -1/2 & 0 & 0 \end{pmatrix}, \begin{pmatrix} 0 & 0 & 0 \\ -1/2 & 1 & -1/2 \\ 0 & 0 & 0 \end{pmatrix}, \begin{pmatrix} 0 & -1/2 & 0 \\ 0 & 1 & 0 \\ 0 & -1/2 & 0 \end{pmatrix}.$$

Here the central coefficient refers to a point in the coarse mesh, while all neighbours refer to points in the next finer mesh. Its univariate counterpart for ϕ defined by (1.15) is

$$\psi(x) = -\frac{1}{2}\phi(2x+1) + \phi(2x) - \frac{1}{2}\phi(2x-1), \tag{6.45}$$

used by Dahmen et al. (1996*a*) to discretize the Helmholtz equation on a closed curve. The fact that these functions actually span Riesz bases in L_2 was shown by Stevenson (1995*b*). More precisely, as pointed out by Lorentz and Oswald (1996, 1997), the functions give rise to Sobolev norm equivalences (6.1) for $n \leq 3$ in the range $s \in (-0.992036, 3/2)$.

Motivated by earlier work by Hackbusch (1989) about *frequency filtering*, an interesting systematic approach to constructing L_2-stable finite element wavelet bases was proposed by Stevenson (1995*b*, 1996, 1995*a*). Again let S_j denote the space of piecewise linear finite elements on meshes of type 2 with scale $2^{-j} = h$. The central idea is to employ level *dependent discrete* scalar products which on those spaces are uniformly equivalent to the standard L_2-inner product. For instance, writing

$$\int_\Omega f(x)g(x)\,\mathrm{d}x = \langle f,g\rangle_\Omega = \sum_{\tau\in\mathcal{T}_j} \langle f,g\rangle_\tau, \tag{6.46}$$

where $\mathcal{T}_j$ is the triangulation of level j of Ω, the terms $\langle f,g\rangle_\tau$ are replaced by a *quadrature rule*. The rationale is that orthogonality with respect to discrete inner products is often easier to realize and corresponding masks are shorter, which gives rise to functions with smaller support. Thus, when $\tau = [x^1, x^2, x^3] \in \mathcal{T}_j$ has vertices x^1, x^2, x^3 one can set

$$\langle f,g\rangle_{\tau,j} = \frac{1}{|\tau|}\sum_{i=1}^{3} f(x^i)g(x^i). \tag{6.47}$$

Now, given the usual tent functions $\phi_{j,k}$ from (4.1) as generators for S_j, one then seeks for a biorthogonal collection $\Xi_j \subset S_{j+1}$ of linear combinations on the next higher level, that is,

$$\langle \Phi_j, \Xi_j\rangle = \mathbf{I}, \tag{6.48}$$

where the $\zeta_{j,k} \in \Xi_j$ have possibly small support. These auxiliary collections Ξ_j are then used to construct complement functions in S_{j+1} which are orthogonal to S_j relative to the level dependent inner product. As mentioned before, one exploits the fact that orthogonality with respect to the discrete inner products is much easier to realize than for the standard inner product. Details and concrete examples can be found in Stevenson (1995*b*, 1996, 1995*a*). In light of Section 5.2, the discrete inner products have been used here to construct a Riesz basis in L_2 without identifying the dual basis relative to the standard inner product. Compared with orthogonal splittings, one takes advantage of significantly smaller filters.

Stabilization of hierarchical bases

A further alternative has been proposed by Carnicer et al. (1996) for refinements $\mathcal{T}_j$ of *arbitrary* triangulations described in Section 4.1. Again denoting by Φ_j the L_2-normalized tent functions relative to $\mathcal{T}_j$, a biorthogonal collection Ξ_j is constructed which consists of (discontinuous) piecewise linear functions. To describe this briefly, let Δ_j again denote the set of vertices of triangulation $\mathcal{T}_j$ and let $\tau = [k, m, p]$ be a triangle in $\mathcal{T}_j$. Then there exist unique affine functions $\sigma^{\tau}_{j,q}$ such that $\int_{\tau} \phi_{j,q}(x)\sigma^{\tau}_{j,q'}(x)\,\mathrm{d}x = \delta_{q,q'}$, $q, q' \in \{k, m, p\}$. Set

$$\zeta_{j,k}(x) := \begin{cases} \frac{1}{n_k}\sigma^{\tau}_{j,k}(x), & x \in \tau, \tau \subset \operatorname{supp}\phi_{j,k}, \\ 0, & x \notin \operatorname{supp}\phi_{j,k}, \end{cases} \tag{6.49}$$

where n_k is the number of triangles having k as a vertex. Thus (6.48) $\langle \Phi_j, \Xi_j \rangle = \mathbf{I}$ again holds, and the question arises of how to identify a stable basis Ψ_j for the complement space

$$W_j := \{\langle g, \Xi_{j+1}\rangle_\Omega \Phi_{j+1} - \langle g, \Xi_j\rangle_\Omega \Phi_j : g \in S_{j+1}\}, \tag{6.50}$$

induced by the projectors $\langle \cdot, \Xi_j\rangle_\Omega \Phi_j$. As in the previously discussed case, this can be done by exploiting the fact that some *initial* complement space is available, namely the one spanned by the hierarchical basis described in Section 4.1. To distinguish it, it will be denoted here as

$$\check{\Psi}_j := \{\phi_{j+1,k} : k \in \Delta_{j+1} \setminus \Delta_j\}. \tag{6.51}$$

At this point the techniques described in Section 3.5 come into play. In particular, (3.40) applies. In fact, $\mathbf{M}_{j,0}$, $\check{\mathbf{M}}_{j,1}$, $\check{\mathbf{G}}_{j,0}$, $\check{\mathbf{G}}_{j,1}$ are given by (4.2), (4.4), (4.6), (4.7) respectively. Thus, with Ξ_j defined by (6.49), the new stable completions $\mathbf{M}_{j,1}$ defined by (3.40) are readily computable. In view of the form (3.39) of the new basis functions, this process may be viewed as a *coarse grid stabilization.* The construction is *not* restricted to regular triangulations. In the case $n = 2$ and the special case of regular triangulations of type 2 (see above), the stabilized complement basis functions have the form

$$\psi_{j,k} = \phi_{j+1,k} - \sum_{l=1}^{4} a_l \phi_{j,k(l)}, \quad k \in \Delta_{j+1} \setminus \Delta_j, \tag{6.52}$$

where k are the midpoints of the edges in the triangulation $\mathcal{T}_j$, and the $k(l)$ denote the vertices of the parallelepiped having the edge associated with k as a diagonal. The construction obtained above through (6.49) is a special case of a whole family of stabilizations (6.52) of the form (Lorentz and Oswald 1996, Lorentz and Oswald 1997)

$$a_1 = a_2 = a, \quad a_3 = a_4 = \frac{1}{8} - a, \quad a \in \mathbb{R}, \tag{6.53}$$

namely for $a = 1/6$. The corresponding basis for the uniform setting is shown by Lorentz and Oswald (1996, 1997) to satisfy (6.1) for $s \in (-0.35768, 3/2)$. The choice $a = 3/16$ from Lorentz and Oswald (1996) gives a somewhat larger range $s \in (-0.440765, 3/2)$, which is maximal in this class.

There is a closely related but slightly different approach to such coarse grid stabilizations, proposed by Vassilevski and Wang (1997*a*). Recall that the multiscale transformations $\check{\mathbf{T}}_j$ associated with the hierarchical bases $\check{\Psi}_j$ from (6.51) are extremely efficient. Since it only involves nodal basis functions, not with respect to the full bases on each level but with respect to the complements only, it is even more efficient than BPX. The objective is to stabilize the hierarchical basis while retaining as much of its efficiency as possible. When switching to another stable completion of the form (3.40), the efficiency of $\check{\mathbf{T}}_j$ can still be exploited. In fact, the multiscale transformation $\mathbf{T}_j$ can be performed in two stages. First perform a step of $\check{\mathbf{T}}_j$ and then correct it, on account of (3.39) or (6.52), by terms involving only the coarse scale generator basis functions. Relevant algorithmic details are given in Sweldens (1996, 1997). The idea is to construct complement functions, that are *close* to functions which span the *orthogonal* complement between two successive trial spaces. I would like to deviate from the original approach and phrase this here in terms of the stable completions described in Section 3.5. Again, straightforward computations show that, given $\check{\mathbf{M}}_{j,1}$ as above,

$$\mathbf{M}_{j,1} = \left(\mathbf{I} - \left(\mathbf{M}_{j,0}\langle\Phi_j, \Phi_j\rangle^{-1}\langle\Phi_j, \Phi_{j+1}\rangle\right)\right)\check{\mathbf{M}}_{j,1} \tag{6.54}$$

gives rise to a basis $\Psi_j^{\mathrm{T}} = \Phi_{j+1}^{\mathrm{T}}\mathbf{M}_{j,1}$ spanning the orthogonal complement of $S(\Phi_j)$ in $S(\Phi_{j+1})$. In other words,

$$\mathbf{L}_j = -\langle\Phi_j, \Phi_j\rangle^{-1}\langle\Phi_j, \Phi_{j+1}\rangle\check{\mathbf{M}}_{j,1}, \quad \mathbf{K}_j = \mathbf{I}, \tag{6.55}$$

(see (3.37)) yield a suitable new stable completion. Note that, by (3.4), $\langle\Phi_j, \Phi_{j+1}\rangle = \mathbf{M}_{j,0}\langle\Phi_{j+1}, \Phi_{j+1}\rangle$. Of course, the matrix $\langle\Phi_j, \Phi_j\rangle^{-1}$ is dense and so is $\mathbf{M}_{j,1}$. However, to compute $\mathbf{M}_{j,1}\mathbf{d}$ for any coefficient vector $\mathbf{d}$ it suffices to compute $\check{\mathbf{M}}_{j,1}\mathbf{d} =: \hat{\mathbf{d}}$ and $\langle\Phi_j, \Phi_{j+1}\rangle\hat{\mathbf{d}} =: \mathbf{b}$. Next, instead of computing $\langle\Phi_j, \Phi_j\rangle^{-1}\mathbf{b}$ exactly, one performs only a few relaxation sweeps for the linear system

$$\langle\Phi_j, \Phi_j\rangle\mathbf{y} = \mathbf{b},$$

followed by $\hat{\mathbf{d}} - \mathbf{M}_{j,0}\mathbf{y}$. Note that $\langle\Phi_j, \Phi_j\rangle$ is positive definite and uniformly well conditioned, since the Φ_j are uniformly stable. Further details are found in Vassilevski and Wang (1997*a*), Vassilevski and Wang (1997*b*). Special cases again lead to (6.52) with coefficients a_l as in (6.53) with $a = 5/48$ and $s \in (0.248994, 3/2)$ (Lorentz and Oswald 1996).

Efficiency and robustness
Which of these options should be preferred? There is, of course, no uniform answer. The decision would depend on the precise problem, on the mesh and on many other side constraints. As soon as one fixes a particular model problem, some aspects become prominent. Nevertheless, there have been some recent comparisons that provide interesting information. These comparisons address two important issues, namely *efficiency* and *robustness.*

Firstly, efficiency comparisons are discussed by Ko, Kurdila and Oswald (1997) for the model problem

$$-\Delta u + qu = f \text{ on } \Omega, \quad u = 0 \text{ on } \partial\Omega, \tag{6.56}$$

where $\Omega \subset \mathbb{R}^2$ is a simple domain such as a rectangle, so that wavelet-based preconditioners can compete without struggling too much with technicalities.

Poisson-like problems
The tests in Ko et al. (1997), Lorentz and Oswald (1996) and Lorentz and Oswald (1997) indicate that, for the Poisson problem $q = 0$ in (6.56), the BPX scheme is superior to the wavelet-based methods, both with regard to the number of iterations needed to ensure a desired accuracy and to the cost of each iteration. As for the cost, this is obvious (recall Remark 6.7). Several types of wavelets, such as Daubechies wavelets, the above finite element-based wavelets, and so-called multi-wavelets, were included in the comparisons. For this kind of problem, the scheme based on Daubechies wavelets appears to be the weakest, since the cost per iteration is higher due to larger masks, while a careful study of corresponding condition numbers(Lorentz and Oswald 1997) shows that the frame bounds for the H_0^1-frames behind the BPX scheme are tighter than those of all wavelet bases. However, the finite element-based wavelets with small support come quite close. The condition numbers produced by the BPX- and by the coarse grid corrections (6.52) with $a = 1/6$ and $a = 3/16$ are reported to stay below 11 (Lorentz and Oswald 1997). The fact that the wavelets also form a Riesz basis in L_2 is not crucial in this case.

Helmholtz problems
The situation changes when $q > 0$ in (6.56) is increased. Now the additional zero-order term starts to affect stability. The efficiency of the BPX scheme in its original form starts to deteriorate. However, a suitable (inexpensive) modification, namely including a properly weighted zero-order term in the auxiliary form $b_{j,k}(\cdot,\cdot)$ has been observed to stabilize it (Oswald 1994). The condition numbers for the wavelet schemes are now smaller and the finite element-based wavelets with small support do quite well (Stevenson 1996, 1995*a*. These schemes are in that sense more *robust* for the class of prob-

lems (6.56). The reason is that, in contrast to the H_0^1-frames behind the BPX scheme, the wavelets form a Riesz basis for a larger range of Sobolev spaces including L_2, so that the zero-order term $q\,u$ is handled better. In particular, when q gets very large, the condition numbers for orthonormal wavelets tend to one, simply because the operator approaches (a multiple of) the identity. Eventually this starts to offset the higher cost per iteration. This is of particular importance for implicit discretizations of parabolic problems

$$\frac{\partial}{\partial t}u = \Delta u, \tag{6.57}$$

where for each time step an elliptic problem $\mathcal{L} = I - \Delta t\,\Delta$ has to be solved. Here wavelet preconditioners work well for a wide range of time steps without additional tuning.

The same robustness issue is also treated by Stevenson (1995*b*), using the wavelet (6.45) derived from the frequency decomposition approach mentioned in Section 6.6. It is shown to be superior to the BPX scheme for this type of problem with regard to efficiency and robustness.

Of course, when the solution is very smooth, the higher cost of a higher-order wavelet scheme per iteration may well be offset by the better approximation. Also, the effect of adaptive refinements has not been taken into account in the above comparisons.

Anisotropies

A similar observation can be made for problems of the type

$$-\varepsilon\frac{\partial^2 u}{\partial {x_1}^2} - \frac{\partial^2 u}{\partial {x_2}^2} + qu = f \text{ on } \Omega, \quad u = 0 \text{ on } \partial\Omega, \tag{6.58}$$

where Ω is again a rectangle for simplicity and ε is small. Such anisotropies aligned with coordinated lines arise, for instance, when employing boundary-adapted grids with high-aspect ratios in flow computations. Griebel and Oswald (1995*b*) compare multilevel Schwarz preconditioners based on tensor product pre-wavelets (see Figure 1) with nodal basis oriented splittings for problems of the type (6.58). Again, the latter method is typically twice as efficient as the wavelet scheme when using proper tuning, while the wavelet scheme is clearly more robust relative to varying ε and q in (6.58). Moreover, in the 3D case it still works in combination with *sparse grid* techniques.

The same issue is treated by Stevenson (1996) (for $q = 0$) with the aid of the modified frequency decomposition multilevel schemes discussed in Section 6.6. Moreover, triangular-based wavelets constructed via discrete inner products are applied by Stevenson (1995*a*) to several types of second-order elliptic boundary value problems with leading term div (A grad). In particular, the case where A is a piecewise constant diagonal matrix with large jumps is considered. The main result is to show that the proposed wavelet-based multilevel scheme is robust for the class of problem where A exhibits possible anisotropies along the three directions of a type-2 grid, when this mesh is used for the discretization.

More detailed information about comparisons can be found in Ko et al. (1997), Lorentz and Oswald (1996) and Lorentz and Oswald (1997). In summary, it seems that the robustness issue is in favour of wavelet-based discretizations. One should not forget, though, that the comparisons did not include multiplicative multigrid schemes, which are usually more efficient than additive counterparts such as the BPX method. Of course, the story changes again when nonuniform grids and complicated domain geometries are considered. Classical wavelets no longer apply directly (see below for a two-stage approach), while the above coarse grid corrected wavelets are still defined.

Finally, a more interesting question occurs when adaptivity is employed, for instance, for a domain with reentrant corners. To my knowledge, direct comparisons have not yet been made. It will be seen later that wavelets seem to have a great potential in this regard.

7. The ideal setting

7.1. Preliminary remarks

Preconditioning is only *one* aspect of wavelet schemes. At least for positive order operators it does not require the full sophistication of wavelets, since simpler suitable frames are seen to work as well, often even better, and for more flexible meshes. In this sense, preconditioning puts only weak demands on the wavelet as a discretization tool. To exploit the full potential of wavelets one is led to ask for more. Two possible directions are, firstly, to consider operators such as integral operators whose conventional discretization gives rise to dense matrices, or, secondly, when dealing with differential operators, to try to *diagonalize* $\mathcal{L}$ in the sense of Section 1.5 (a). We will first address the latter issue. Of course, one cannot expect such an objective to be feasible under the most general circumstances. The basic rationale is to develop a two-stage process. First design highly efficient schemes for an *ideal setting* (ideal with regard to highest efficiency and availability of the tools) and then try to reduce realistic problems to the ideal case at an additional expense which, however, should be of lower order. As indicated in Sections 4.2 and 4.3, wavelets unfold their full potential when working on the whole Euclidean space $\mathbb{R}^n$ or on the n-torus $\mathbb{R}^n/\mathbb{Z}^n$. I will refer to this as the *ideal setting.*

A great deal of effort has been spent on studying elliptic operators $\mathcal{L}$ of the form (2.4), where $a > 0$ and $A(x)$ is a symmetric matrix satisfying

$$\sup_{x\in\mathbb{R}^n} \|A(x)\| < \infty, \quad \xi^{\mathrm{T}} A(x)\xi \geq \delta|\xi|^2, \quad x,\xi \in \mathbb{R}^n, \tag{7.1}$$

for some $\delta > 0$, and whose coefficients satisfy certain (weak) regularity properties. The subsequent discussion mainly reflects related work by Angeletti, Mazet and Tchamitchian (1997), Liandrat and Tchamitchian (1997) and Tchamitchian (1996). Typical questions to be studied are

(i) the continuity of $\mathcal{L}^{-1}$ in Sobolev scales $W^{s,p}$ depending on the coefficients of $A(x)$
(ii) the boundedness of associated Galerkin projections in this scale
(iii) efficient numerical procedures for the approximate solution of

$$\mathcal{L}u = f \quad \text{on } \mathbb{R}^n \text{ or } \mathbb{R}^n/\mathbb{Z}^n \tag{7.2}$$

with the aid of wavelet bases.

These questions are clearly closely interrelated. Our main concern here will be (iii), while recent studies of (i) and (ii) are given by Angeletti et al. (1997).

Recall from Section 4.2 the format of the biorthogonal wavelet bases $\Psi = \{\psi_\lambda : \lambda \in \nabla\}$ and likewise $\tilde{\Psi}$ to be used when $\Omega = \mathbb{R}^n$. By (4.38), one has to deal with $2^n - 1$ mother wavelets ψ_e, $e \in E_* := \{0,1\}^n \setminus \{0\}$. In this case it is convenient to take $j_0 = -\infty$ as the coarsest level so that $\nabla = \nabla_-$. When working on the torus it will always be tacitly assumed that $j_0 \geq 0$ is fixed. In either case, the indices $\lambda \in \nabla$ have the geometric interpretation $\lambda = 2^{-j}(k + \frac{e}{2})$, $k \in \mathbb{Z}^n$, $j \in \mathbb{Z}, j \geq j_0$, $e \in E_*$. We continue denoting by $d, \tilde{d}$ the order of polynomial exactness of the spaces $S(\Phi_j)$, $S(\tilde{\Phi}_j)$, so that the direct inequalities (5.1) hold with $d, \tilde{d}$, respectively.

The principal goal is to *diagonalize* the operator $\mathcal{L}$ in (7.2) as indicated in Section 1.5 (a). The relevant theoretical background is the theory of Calderón–Zygmund operators. To support the understanding of the subsequent developments I include some brief comments in this regard, mainly following Tchamitchian (1996).

7.2. Vaguelettes and Calderón–Zygmund operators

The subsequent discussion first follows the original development that has been tailored to orthonormal wavelets and therefore admits wavelets with global support as long as there is enough decay. Orthonormality is not crucial, though, and analogous statements can be made for the biorthogonal case as well. So assume that the ψ_e decay rapidly along with their derivatives up to some order $r_0 \geq 3$. For $|\alpha| = r_0$, $\partial^\alpha \psi_e$ is supposed to be defined almost everywhere while $\psi_e \in C^{r-1}(\mathbb{R}^n)$, $e \in \{0,1\}^n$.

In addition to the wavelet basis Ψ we consider another family $\Theta = \{\theta_\lambda : \lambda \in \nabla\}$ which is related to Ψ by

$$\mathcal{T}\Psi = \Theta, \tag{7.3}$$

for some linear operator $\mathcal{T}$. Ultimately, we will be interested in $\mathcal{T} = (\mathcal{L}^{-1})^*$. Unfortunately, the θ_λ will *not* arise from finitely many mother functions by means of dilation and translation. However, in cases of interest to us, the θ_λ still share the following properties with the ψ_λ, $\lambda \in \nabla$: there exist positive numbers C, q, r and a non-negative integer $\tilde{d} \in [0, q)$, such that for $\lambda \in \nabla_-$,

$$|\theta_\lambda(x)| \le C 2^{jn/2} \left(1 + 2^j |x - \lambda|\right)^{-n-q}, \quad x \in \mathbb{R}^n. \tag{7.4}$$

Furthermore, $\theta_\lambda \in C^{\lfloor r \rfloor}(\mathbb{R}^n)$, ($\lfloor a \rfloor$ being the largest integer less than or equal to a), and for all $\alpha \in \mathbb{Z}_+^n$, $|\alpha| \le \lfloor r \rfloor$ one has

$$|\partial^\alpha \theta_\lambda(x)| \le C 2^{j(\frac{n}{2} + |\alpha|)} \left(1 + 2^j |x - \lambda|\right)^{-n-q}, \quad x \in \mathbb{R}^n. \tag{7.5}$$

Moreover, for $|\alpha| = \lfloor r \rfloor$ the $\partial^\alpha \theta_\lambda$ are Hölder continuous, that is,

$$|\partial^\alpha \theta_\lambda(x + h) - \partial^\alpha \theta_\lambda(x)| \le C 2^{j(\frac{n}{2} + r)} |h|^{r - \lfloor r \rfloor} \left(1 + 2^j |x - \lambda|\right)^{-n-q}, \quad x \in \mathbb{R}^n, \tag{7.6}$$

and for every polynomial P of order at most $\tilde{d}$ one has

$$\int_{\mathbb{R}^n} P(x) \theta_\lambda(x) \, dx = 0, \quad \lambda \in \nabla_-. \tag{7.7}$$

The set Θ is called a family of *vaguelettes* with index $(\tilde{d}, q, r)$. Note that the kernel of the operator $\mathcal{T}$ defined by (7.3) has the form

$$K(x; y) = \sum_{\lambda \in \nabla} \theta_\lambda(x) \overline{\psi_\lambda(y)}. \tag{7.8}$$

One can then show (see, for instance, Tchamitchian (1996)) that the estimates (7.4), combined with corresponding standard estimates for the wavelets, imply that there exist constants $C, \delta > 0$, such that

$$|K(x, y)| \le \frac{C}{|x - y|}, \quad x \ne y, \tag{7.9}$$

and

$$|K(x, y) - K(x', y)| + |K(y, x) - K(y, x')| \le C \frac{|x - x'|^\delta}{|x - y|^{1+\delta}}, \tag{7.10}$$

when $|x - x'| \le |x - y|/2$. An operator $\mathcal{T}$, such that for any two test functions f, g with disjoint compact supports

$$\langle \mathcal{T} f, g \rangle = \iint K(x, y) f(y) \overline{g(x)} \, dy \, dx,$$

is called a *Calderón–Zygmund operator* (CZO) if $\mathcal{T}$ is continuous on L_2 and K satisfies the so-called *standard estimates* (7.9), (7.10). The above notions are now interrelated by the following result, whose proof can be found in Meyer (1990).

Theorem 7.1 Suppose Θ is a family of vaguelettes with index $(\tilde{d}, q, r)$. Then $\mathcal{T}$ defined by (7.3) extends to a CZO. It is continuous on $W^{s,p}(\mathbb{R}^n)$ for $1 < p < \infty$ and $|s| < \inf(r_0, r, \tilde{d} + 1, q)$. The corresponding operator norms depend on the parameters and on the constant C in (7.4)–(7.7).

Let us denote by $\mathcal{LMA}$ the collection of operators $\mathcal{T}$ such that $\Theta = \mathcal{T}\Psi$ is a family of vaguelettes.

Theorem 7.2

(i) $\mathcal{LMA}$ is independent of the choice of Ψ (provided that $r_0 > 0$).
(ii) $\mathcal{LMA}$ is an algebra which is stable under taking adjoints.
(iii) $\mathcal{LMA}$ is exactly the set of CZOs $\mathcal{T}$ such that $\mathcal{T}(1) = \mathcal{T}^*(1) = 0$.
(iv) $\mathcal{LMA}$ is not stable under taking inverses.

Assertions (i) to (iii) are due to Lemarié (1984); (iii) is related to the T1-Theorem by G. David and J. L. Journé, which characterizes the continuity of an operator $\mathcal{T}$ satisfying the estimates (7.9), (7.10). This result will be mentioned later again. As for (iv), we refer the reader to Tchamitchian (1996), and its implications will become clearer later on. Since a CZO takes $L_\infty(\mathbb{R}^n)$ into the space of functions of bounded mean oscillation (BMO), $\mathcal{T}(1)$ and $\mathcal{T}^*(1)$ are indeed defined. Recall that a locally integrable function f belongs to BMO if and only if, for any cube C, there exists a constant a such that

$$\frac{1}{|C|}\int_C |f(x) - a|\,\mathrm{d}x \lesssim 1,$$

where $|C|$ denotes the volume of C.

The following characterization of $\mathcal{LMA}$ in terms of matrices will be important in the present context. Let $\mathcal{M}$ denote the set of matrices $\mathbf{A}$ such that for some $\gamma > 0$ one has

$$|\mathbf{A}_{\lambda,\lambda'}| \lesssim \frac{2^{-||\lambda|-|\lambda'||(\gamma+\frac{n}{2})}}{\left(1 + 2^{\min\{|\lambda|,|\lambda'|\}}|\lambda - \lambda'|\right)^{1+\gamma}}. \tag{7.11}$$

Thus the entries of $\mathbf{A}$ decay with increasing difference in scale and spatial location of the indices λ, λ'. Recall that this estimate is of the type (1.11). In fact, essentially the same argument can be employed to show that, when $\int \theta_\lambda = 0, \quad \lambda \in \nabla$, then $\langle \Theta, \Theta \rangle \in \mathcal{M}$. Moreover, the following result can be found in Angeletti et al. (1997).

Theorem 7.3 $\mathcal{T}$ belongs to $\mathcal{LMA}$ if and only if the matrix $\langle \Psi, \mathcal{T}\Psi \rangle$ belongs to $\mathcal{M}$.

Next, we will describe an approach, initiated by Tchamitchian (1987), which is based on the above concepts and aims at avoiding the solution of linear systems essentially by *diagonalizing* the operator.

7.3. Constant coefficient operators

The basic strategy is to proceed again in two steps: first treat carefully the case of *constant coefficient* operators and then use a *freezing coefficient* technique. We will specify the operator $\mathcal{L}$ from (2.2) first to the special case where $a_{\alpha\beta}(x) = a_{\alpha\beta}$ are constant so that

$$\mathcal{L}u = - \sum_{|\alpha|,|\beta|=1} a_{\alpha\beta}\partial^\alpha\partial^\beta + a. \tag{7.12}$$

The operator $\mathcal{L}$ will always be assumed to be elliptic, which here means that the principal part $\sigma_0(y)$ of its *symbol* σ is strictly positive on $\mathbb{R}^n$, that is,

$$\sigma_0(y) := \sum_{|\alpha|,|\beta|=1} a_{\alpha\beta}y^{\alpha+\beta} \geq \delta > 0, \quad y \in \mathbb{R}^n, \quad \sigma(y) = \sigma_0(y) + a. \tag{7.13}$$

We follow Liandrat and Tchamitchian (1997) and try to solve $\mathcal{L}u = f$ conceptually by applying the inverse $\mathcal{L}^{-1}$ to the right-hand side f. Although at first glance this may contradict basic principles in numerical analysis, it does have tempting aspects, as shown next.

Suppose that $\Psi, \tilde{\Psi}$ are biorthogonal Riesz bases in $L_2(\mathbb{R}^n)$. Then the solution u of (7.2) has the form $u = \mathbf{d}^T\Psi$ with unknown coefficient sequence $\mathbf{d}^T = \langle u, \tilde{\Psi}\rangle$. Inserting $u = \mathcal{L}^{-1}f$, one obtains

$$\mathbf{d}^T = \langle u, \tilde{\Psi}\rangle = \langle \mathcal{L}^{-1}f, \tilde{\Psi}\rangle = \langle f, (\mathcal{L}^{-1})^*\tilde{\Psi}\rangle = \langle f, \Theta\rangle, \tag{7.14}$$

that is, the roles of $\mathcal{T}$ and Ψ in (7.3) are played here by $(\mathcal{L}^{-1})^*$ and $\tilde{\Psi}$, respectively. Thus the solution u of $\mathcal{L}u = f$ is formally given as

$$u = \langle f, \Theta\rangle\Psi. \tag{7.15}$$

Proposition 7.4 (Angeletti et al. 1997) The collection Θ, defined by $\Theta = (\mathcal{L}^{-1})^*\tilde{\Psi}$, is a family of vaguelettes.The constant C in (7.4)–(7.7) depends on the $a_{\alpha\beta}$ and $\Psi, \tilde{\Psi}$ but *not* on a.

Thus the image of $\tilde{\Psi}$ under $(\mathcal{L}^{-1})^*$ still has nice localization properties reflected by estimates (7.4)–(7.7). This suggests the following approach (Liandrat and Tchamitchian 1997).

A projection scheme

A natural idea is to compute an approximate solution of (7.2) by truncating $\langle f, \Theta\rangle$. Fixing *any* finite $\Lambda \subset \nabla$, this corresponds to projecting u into the finite-dimensional space $S_\Lambda = S(\Psi_\Lambda)$ $\Psi_\Lambda = \{\psi_\lambda : \lambda \in \Lambda\}$, (see (3.43)), that is,

$$u_\Lambda = \langle u, \tilde{\Psi}_\Lambda\rangle\Psi_\Lambda = \langle f, \Theta_\Lambda\rangle\Psi_\Lambda. \tag{7.16}$$

Note that this is a Petrov–Galerkin approximation (6.2).

Convergence

Under assumptions (7.13), $\mathcal{L}$ is also a boundedly invertible mapping from $H^{2+s}(\mathbb{R}^n)$ to $H^s(\mathbb{R}^n)$, $s \in \mathbb{R}$. Thus, noting that, by (7.15), $u_\Lambda = Q_\Lambda u = Q_\Lambda \mathcal{L}^{-1} f$, one has $u_\Lambda - u = (Q_\Lambda - I)\mathcal{L}^{-1} f$. Employing our direct estimates (5.1) (for $p = 2$), we obtain

$$\|u_\Lambda - u\|_{L_2} = \left\|(Q_\Lambda - I)\mathcal{L}^{-1} f\right\|_{L_2} \lesssim 2^{-dj(\Lambda)} \|\mathcal{L}^{-1} f\|_{H^d}, \tag{7.17}$$

where $j(\Lambda) = \max\{j : |\lambda| < j \Rightarrow \lambda \in \Lambda\}$. Continuity of $\mathcal{L}^{-1}$ in the Sobolev scale gives

$$\|u_\Lambda - u\|_{L_2} \lesssim 2^{-dj(\Lambda)} \|f\|_{H^{d-2}}. \tag{7.18}$$

Analogous estimates for the spaces $W^{s,p}$ are obtained in exactly the same way as long as one has continuity of $\mathcal{L}^{-1}$ (see question (ii) in the Section 7.1). Moreover, standard interpolation arguments yield

$$\|u_\Lambda - u\|_{L_2} \lesssim 2^{-sj(\Lambda)} \|f\|_{H^{s-2}}, \quad d \geq s \geq 2. \tag{7.19}$$

Estimates of this type are very crude. They guarantee convergence as long as the spaces S_Λ include sufficiently many low frequencies, that is, $j(\Lambda)$ grows with $\#\Lambda$. The interesting part, of course, concerns the adaptation of Λ to the problem at hand, which may result in a selection of highly nonuniformly shaped subsets $\Lambda \subset \nabla$. In view of (7.16), this is closely related to the next point.

7.4. Evaluation of $\langle f, \Theta_\Lambda \rangle$

By (7.15), the success of the approach hinges on identifying and computing the *significant* coefficients of $\langle f, \Theta \rangle$, represented here by the finite array $\langle f, \Theta_\Lambda \rangle$. The important point is that, by Proposition 7.4, $\mathcal{L}^{-*}$ is a CZO; here and elsewhere, we shall write $\mathcal{L}^{-*}$ instead of $(\mathcal{L}^*)^{-1}$. Hence, according to (7.11), $\mathcal{L}^{-*}$ has a quasi sparse matrix representation.

Noting that $(\mathcal{L}u)\hat{}(y) = \sigma(y)\hat{u}(y)$, where σ is by (7.13) strictly positive on $\mathbb{R}^n$, the definition (7.14) of Θ means that

$$\langle f, \theta_\lambda \rangle = \frac{1}{(2\pi)^n} \langle \hat{f}, \hat{\theta}_\lambda \rangle = \frac{1}{(2\pi)^n} \langle \hat{f}, \bar{\sigma}^{-1} \hat{\tilde{\psi}}_\lambda \rangle.$$

Since the wavelets are well localized in Fourier space, one could employ quadrature to compute $\langle \hat{f}, \bar{\sigma}^{-1} \hat{\tilde{\psi}}_\lambda \rangle$ up to any desired precision.

Projections into S and convolutions

We now describe an alternative approach proposed by Liandrat and Tchamitchian (1997). Due to the vaguelette estimates, θ_λ belongs up to a desired tolerance to $S(\tilde{\Phi}_{|\lambda|+p})$ for some $p \in \mathbb{N}$, which depends on $\mathcal{L}$, Ψ and $\tilde{\Psi}$ but *not* on $|\lambda|$. This suggests projecting f into the space $S(\Phi_{|\lambda|+p})$. As before, we will always denote by Q_j or Q_Λ the canonical projectors onto the spaces $S(\Phi_j)$, $S(\Psi_\Lambda)$, respectively. This suggests replacing $\langle f, \theta_\lambda\rangle$ by $\langle Q_{|\lambda|+p} f, \theta_\lambda\rangle$. In fact, since

$$|\langle f, \theta_\lambda\rangle - \langle Q_j f, \theta_\lambda\rangle| \le \|f\|_{L_2} \|(I - Q_j^*)\theta_\lambda\|_{L_2},$$

the precision depends on the approximation properties of the spaces $S(\tilde{\Phi}_j)$ and on the regularity of the θ_λ and hence of the ψ_λ. Now suppose $|\lambda| = j$ and write

$$Q_{j+p} f = Q_{j+1} f + \sum_{l=1}^{p-1} (Q_{j+l+1} - Q_{j+l}) f = \mathbf{c}_{j+1}^{\mathrm{T}} \Phi_{j+1} + \sum_{l=1}^{p-1} \mathbf{d}_{j+l}^{\mathrm{T}} \Psi_{j+l},$$

where $\mathbf{c}_{j+1}^T = \langle f, \tilde{\Phi}_{j+1}\rangle$, $\mathbf{d}_{j+l}^T = \langle f, \tilde{\Psi}_{j+l}\rangle$, so that

$$\langle Q_{j+p} f, \theta_\lambda\rangle = \mathbf{c}_{j+1}^{\mathrm{T}} \langle \Phi_{j+1}, \theta_\lambda\rangle + \sum_{l=1}^{p-1} \mathbf{d}_{j+l}^{\mathrm{T}} \langle \Psi_{j+l}, \theta_\lambda\rangle. \tag{7.20}$$

This amounts to discrete convolution of the wavelet coefficients of f with filters that depend only on Ψ and $\mathcal{L}$. To compute these filters once and for all, one can again resort to Fourier transforms.

Moreover, if the right-hand side f is smooth except at isolated points, then only a small number of the coefficients $d_\lambda = \langle f, \tilde{\psi}_\lambda\rangle$ exceed a given threshold in magnitude. The sequences $\langle \Psi_{j+l}, \theta_\lambda\rangle$, $l = 1, \ldots, p-1$, describe how the wavelet coefficients d_λ, $|\lambda| = j + l$, are *smeared* by the application of $\mathcal{L}^{-*}$. Thus (7.20) is to be applied to the *compressed* arrays of wavelet coefficients, which result from thresholding.

Let us first add a few comments on the structure of the sequences $\langle \Psi_{j+l}, \theta_\lambda\rangle$ which can be viewed as one column of the matrix $\langle \Psi_{j+l}, \Theta_j\rangle$ where $\Theta_j := \{\theta_\lambda : |\lambda| = j\}$. Recalling the two-scale relations (3.4) and (3.8), $\Psi_j^{\mathrm{T}} = \Phi_{j+1}^{\mathrm{T}} \mathbf{M}_{j,1}$, (which is here stationary in j but would sizewise depend on j in the periodic case), one obtains

$$\begin{aligned} \langle \Psi_{j+l}, \Theta_j\rangle &= \langle \Psi_{j+l}, \mathcal{L}^{-*}\tilde{\Psi}_j\rangle = \mathbf{M}_{j+l,1}^{\mathrm{T}} \langle \Phi_{j+l+1}, \mathcal{L}^{-*}\tilde{\Phi}_{j+1}\rangle \tilde{\mathbf{M}}_{j,1} \\ &= \mathbf{M}_{j+l,1}^{\mathrm{T}} \langle \Phi_{j+l+1}, \mathcal{L}^{-*}\tilde{\Phi}_{j+1+l}\rangle \tilde{\mathbf{M}}_{j+l,0} \cdots \tilde{\mathbf{M}}_{j+1,0} \tilde{\mathbf{M}}_{j,1}. \end{aligned} \tag{7.21}$$

Thus, once the arrays $\mathbf{F}_q := \langle \Phi_q, \mathcal{L}^{-*}\tilde{\Phi}_q\rangle$ are known, the matrices $\langle \Psi_{j+l}, \Theta_j\rangle$ are obtained with the aid of pyramid-type schemes like (3.26). Moreover, a typical entry of $\mathbf{F}_q$ has the form

$$\begin{aligned} \langle \phi_{q,k}, \mathcal{L}^{-*}\tilde{\phi}_{q,l}\rangle &= (2\pi)^{-n} \langle \hat{\phi}_{q,k}, \bar{\sigma}^{-1} \hat{\tilde{\phi}}_{q,l}\rangle \\ &= (2\pi)^{-n} \int_{\mathbb{R}^n} \hat{\phi}(y) \overline{\hat{\tilde{\phi}}(y) \sigma(2^q y)^{-1}} e^{-iy\cdot(k-l)}\, \mathrm{d}y. \end{aligned} \tag{7.22}$$

As for the cost of these operations, let us consider the periodic case (see Section 4.3), where J denotes the highest discretization level and p as above is fixed. Suppose that the arrays $\mathbf{F}_q$ have been (approximately) computed once and for all in an *initialization step*. Note that the fast decay of $\sigma(2^q y)^{-1}$ for large q implies that $\mathbf{F}_q$ can be ever better approximated by a bounded matrix with small bandwidth. Hence the computation of all vaguelette coefficients on level r boils down to

$$\langle Q_{r+p}f, \Theta_r\rangle = \mathbf{c}_{r+1}^{\mathrm{T}}\mathbf{F}_{r+1}\tilde{\mathbf{M}}_{r,1} + \sum_{l=1}^{p-1} \mathbf{d}_{r+l}^{\mathrm{T}}\mathbf{M}_{r+l,1}^{\mathrm{T}}\mathbf{F}_{r+l+1}\tilde{\mathbf{M}}_{r+l,0}\cdots\tilde{\mathbf{M}}_{r+1,0}\tilde{\mathbf{M}}_{r,1}. \tag{7.23}$$

Thus, when the wavelets have compact support and each $\mathbf{F}_{r+l+1}$ is replaced by a sparse matrix, this requires the order of $2^{(r+p)n}$ (n being the spatial dimension) operations, where the constant depends on $\Psi, \tilde{\Psi}$ and the accuracy of $\mathbf{F}_{r+l+1}$. Consequently, the computation of u_Λ is of the order $2^{(p+|\Lambda|)n} \sim N_\Lambda$, $N_\Lambda := \dim S(\Phi_{|\Lambda|})$, where $|\Lambda| := \max\{|\lambda| + 1 : \lambda \in \Lambda\}$, with a constant depending on p. Note that when rapidly decaying wavelets with global support (but very good localization in Fourier space – see Section 4.2) are used, the matrices $\tilde{\mathbf{M}}_{j,0}$, $\tilde{\mathbf{M}}_{j,1}$ are no longer sparse but, in the periodic case, are circulants, so that FFT can be employed to limit the order of operations to $N_\Lambda \log N_\Lambda$.

On the other hand, the above work estimate has been very crude. In fact, when the right-hand side f is smooth except at isolated points, only very few of its wavelet coefficients d_λ are expected to exceed a given tolerance $\varepsilon > 0$. Due to the localization and cancellation properties of the vaguelettes θ_λ, the coefficients $\langle f, \theta_\lambda\rangle$ are expected to exhibit similar behaviour. In fact, since $\mathcal{L}^{-*}$ is a CZO the decay of the entries in $\langle \Psi_{j+l}, \Theta_j\rangle$ is governed by estimates of the form (7.11), and the *spread* of the wavelet coefficients of f due to $\mathcal{L}^{-*}$ can be seen from (7.20). This suggests computing $\langle f, \theta_\lambda\rangle$ or, better, $\langle Q_{|\lambda|+p}f, \theta_\lambda\rangle$ only for those λ in a certain neighbourhood of the significant coefficients of f. The number of these coefficients may, of course, be much smaller than $\dim S(\Phi_{|\Lambda|})$. A more formal treatment of this issue in Liandrat and Tchamitchian (1997) is based on the notion of (ε, s)-*adapted spaces*.

Nevertheless, it does not appear to be completely obvious how to carry out all computations without requiring the full complexity of the highest discretization level at some point. In fact, while thresholding the arrays $\mathbf{d}_{r+l}$ on the right-hand side of (7.23) facilitates the successive multiplication with possibly very short vectors, the first summand involving $\mathbf{c}_{r+1}$ does not seem to be compressible in this form.

Remark 7.5 Another point concerns the various tolerances in the above procedure. Uniformity of the work estimates in $\#\Lambda$ are ultimately of limited value when the involved tolerances and thresholds are kept *fixed*. In fact, increasing $\#\Lambda$ should produce better overall accuracy. Correspondingly tighter thresholds, in turn, are expected to require a larger p, and hence a higher computational cost that may no longer stay proportional to $\#\Lambda$. Questions of this form will be encountered in similar contexts several times.

A hybrid scheme
There exist several variants of the above scheme (see, for instance, Ponenti (1994)) among which I would like to mention the *hybrid* scheme proposed by Fröhlich and Schneider (1995), which differs from the above procedure in an essential way. The main point in Fröhlich and Schneider (1995) is to economize the evaluation of the vaguelette coefficients $\langle f, \theta_\lambda\rangle$ by incorporating *interpolation techniques.* Again it is designed for the periodic case. For simplicity, we consider bases on $\mathbb{R}$ and refer the reader to Section 4.3 for standard periodization (see also Fröhlich and Schneider (1995)) and tensor product versions for the bivariate situation.

Let $\mathcal{L} = \sum_{m=0}^{s} a_m \left(\frac{d}{dx}\right)^m$ be an elliptic operator; that is, as before, its symbol $\sigma(y) := \sum_{m=0}^{s} a_m(iy)^m$ is strictly positive on $\mathbb{R}$. By construction, the family $\tilde{\Theta} := \mathcal{L}\Psi$ is biorthogonal to $\Theta = (\mathcal{L}^{-1})^*\tilde{\Psi}$. Hence one has

$$f = \langle f, \Theta\rangle\tilde{\Theta}. \tag{7.24}$$

Thus, instead of approximating f first by projecting into the spaces $S(\Phi_j)$, as in the previous approach, one could try to expand f approximately with respect to $\tilde{\Theta}$. Thus, consider the spaces

$$S_{\mathcal{L},J} := S(\{\tilde{\theta}_\lambda \in \tilde{\Theta} : |\lambda| < J\}) = S(\mathcal{L}\Phi_J). \tag{7.25}$$

The idea is now to employ *Lagrange interpolation* to efficiently obtain an approximation to f in $S_{\mathcal{L},J}$, say. Therefore one is interested in finding the *fundamental Lagrange functions*

$$L_j(x) = \sum_{k\in\mathbb{Z}} g_k\left(\mathcal{L}\phi_{j,k}\right)(x) = \mathbf{g}^{\mathrm{T}}\mathcal{L}\Phi_j(x), \tag{7.26}$$

such that

$$L_j(2^{-j}k) = \delta_{0,k}, \quad k \in \mathbb{Z}. \tag{7.27}$$

This is equivalent to saying that

$$1 = \sum_{k\in\mathbb{Z}} \langle L_j(2^{-j}\cdot), \delta(\cdot - k)\rangle e^{iky} = 2^j \sum_{k\in\mathbb{Z}} \hat{L}_j(2^j(y + 2\pi k)).$$

Standard arguments (Dahmen et al. 1994*c*, Fröhlich and Schneider 1995) yield

$$\hat{L}_j(y) = \frac{\sigma(y)\hat{\phi}(2^{-j}y)}{\sum_{k\in\mathbb{Z}} \sigma(2^j y + 2^j 2\pi k)\hat{\phi}(y + 2\pi k)}, \tag{7.28}$$

which, of course, requires the sum in the denominator to be nonzero.

Recall that, by (7.14), the solution u of $\mathcal{L}u = f$ is given by

$$u = \langle f, \mathcal{L}^{-*}\tilde{\Phi}_0\rangle\Phi_0 + \langle f, \Theta\rangle\Psi. \tag{7.29}$$

In order to project the right-hand side f first into $S_{\mathcal{L},J}$, one can use the samples of f at $2^{-J}k$. In view of (7.27),

$$f_J(x) := \sum_{k\in\mathbb{Z}} f(2^{-J}k)L_J(x - 2^{-J}k) \tag{7.30}$$

interpolates f in $S_{\mathcal{L},J}$. Thus, to obtain an approximation to the coefficients in $\langle f, \Theta^J\rangle$, say, one has to rewrite f_J in (7.30), in view of (7.24), in terms of $\tilde{\Theta}^J = \mathcal{L}\Psi^J$. Since this is the central point, we describe this change of bases in a little more detail. Let $\mathbf{L}_j = (L_j(\cdot - 2^{-j}k) : k \in \mathbb{Z})$ and define

$$\begin{aligned} \mathbf{D}_j &:= \langle \mathbf{L}_j, \Theta_{j-1}\rangle, \quad j = 1,\ldots,J, \quad \mathbf{D}_{-1} := \langle \mathbf{L}_0, (\mathcal{L}^{-1})^*\tilde{\Phi}_0\rangle, \\ \mathbf{f}_j &:= (f_j(2^{-j}k) : k \in \mathbb{Z}). \end{aligned} \tag{7.31}$$

Since $f_J \in S_{\mathcal{L},J}$, it can be expanded as

$$f_J = \mathbf{f}_J^{\mathrm{T}}\mathbf{L}_J = \langle \mathbf{f}_J^{\mathrm{T}}\,\mathbf{L}_J, \mathcal{L}^{-*}\tilde{\Phi}_J\rangle\mathcal{L}\Phi_J = \mathbf{f}_J^{\mathrm{T}}\,\langle \mathbf{L}_J, \mathcal{L}^{-*}\tilde{\Phi}_J\rangle\mathcal{L}\Phi_J.$$

Now, combining (3.11) and (3.13) with Proposition 3.8, one obtains $\tilde{\Phi}_J^T = \tilde{\Phi}_{J-1}^T\mathbf{M}_{J-1,0}^* + \tilde{\Psi}_{J-1}^T\mathbf{M}_{J-1,1}^*$. Substituting this into the above relation and using (3.12) yields

$$f_J = \mathbf{f}_J^{\mathrm{T}}\,\langle \mathbf{L}_J, \mathcal{L}^{-*}\Phi_{J-1}\rangle\mathcal{L}\tilde{\Phi}_{J-1} + \mathbf{f}^{\mathrm{T}}\,\langle \mathbf{L}_J, \Theta_{J-1}\rangle\tilde{\Theta}_{J-1}, \tag{7.32}$$

where

$$f_{J-1} := \mathbf{f}_J^{\mathrm{T}}\langle \mathbf{L}_J, \mathcal{L}^{-*}\tilde{\Phi}_{J-1}\rangle\mathcal{L}\Phi_{J-1} \in S_{\mathcal{L},J-1}. \tag{7.33}$$

Thus we have determined the vaguelette coefficients $\langle f, \Theta_{J-1}\rangle \approx \langle f_J, \Theta_{J-1}\rangle$ of f relative to $\tilde{\Theta}$ as

$$\mathbf{d}_{J-1} = \mathbf{f}_J^{\mathrm{T}}\,\mathbf{D}_J, \tag{7.34}$$

where $\mathbf{D}_J$ is given by (7.31). To continue this process, one only has to determine the samples of f_{J-1} defined in (7.33) on the coarse grid, that is,

$$f_{J-1}(2^{-J+1}k) = f_J(2^{-J}2k) - \mathbf{d}_{J-1}^{\mathrm{T}}\tilde{\Theta}_{J-1}(2^{-J}k), \tag{7.35}$$

once one has computed $\tilde{\Theta}_{J-1} = \mathcal{L}\Psi_{J-1}$. Instead of performing (7.35) exactly, one can discard entries in $\mathbf{d}_{J-1}$ that stay below a certain threshold, to generate step-by-step compressed vectors $\mathbf{d}_j$, $j < J$, such that

$$u_J := \mathbf{c}_0^{\mathrm{T}}\Phi_0 + \sum_{0\le j<J} \mathbf{d}_j^{\mathrm{T}}\Psi_j$$

approximates u in (7.29). The following algorithm, from Fröhlich and Schneider (1995), does exactly that.

ALGORITHM 2 (FS)
Initialization: Compute the filters $\mathbf{D}_j$ and $\mathcal{L}\phi_{0,0}, \mathcal{L}\psi_{j,0}$, $j = 0, \ldots, J-1$.

(1) Set $j = J$ and determine $f_J(2^{-J}k) = f(2^{-J}k)$, $k \in \mathbb{Z}$.
(2) For $j = J-1, J-2, \ldots, 0$,

$$\begin{aligned} \mathbf{d}_j &= \mathbf{f}_{j+1}^{\mathrm{T}} \mathbf{D}_{j+1}, \\ \mathbf{f}_j &= (f_j(2^{-j}k) : k \in \mathbb{Z}), \quad f_j(2^{-j}k) = f_{j+1}(2^{-j}k) - \mathbf{d}_j^{\mathrm{T}} \tilde{\Theta}_j(2^{-j-1}k). \end{aligned}$$

(3) Compute $\mathbf{c}_0^{\mathrm{T}}$ with the aid of the filter in $\langle \mathbf{L}_0, \mathcal{L}^{-*}\tilde{\Phi}_0 \rangle$.

A few comments on this scheme are in order.

- When the change of bases is done exactly and no thresholding is applied, the above scheme is a collocation scheme.

- Instead of starting with a set Φ_j of orthonormal scaling functions as in Fröhlich and Schneider (1995), we have kept the flexibility of using biorthogonal pairs $\Phi_j, \tilde{\Phi}_j$. In Fröhlich and Schneider (1995), orthogonality was paid for by infinite masks which require additional truncation. Starting with biorthogonal spline wavelets (4.25), the collection $\tilde{\Theta}$ still consists of compactly supported functions. Likewise the representation of u_J involves only the compactly supported functions in Ψ and Φ_0. This might favour embedding techniques for more general domains. It is clear that the L_j have typically global support but decay exponentially. Here the actual computation requires a truncation. Of course, the $\mathbf{D}_j$ are obtained by computing only *one* mask, which involves truncation of the vaguelettes too. The matrix formulation for the periodic case is identical once $\mathbb{Z}$ is replaced by $\mathbb{Z}/2\mathbb{Z}$.

- In Fröhlich and Schneider (1995), it is assumed that a reduced set $\Lambda \subset \{\lambda : |\lambda| < J\}$ is given from the start. The above algorithm is formulated there in a way that takes advantage of this data reduction. This requires *a priori* knowledge about the solution u. Such information is often available when dealing with *time-dependent problems* and an initial guess of Λ can be obtained from the approximation on the previous time level. In this case the samples of f are not required on the full grid of level J. This can be incorporated above as well by requiring samples only at places determined by significant vaguelette coefficients.

The scheme is applied by Fröhlich and Schneider (1996) to Helmholz-type problems as well as to nonlinear parabolic PDEs and to the computation of flame fronts. The experiments indicate dramatic savings if the computation can be fully confined to the significant wavelet contributions.

7.5. Freezing coefficients

The numerical feasibility of the above vaguelette schemes hinges in an essential way on the constant coefficient model problem. Let us now sketch some ideas from Lazaar, Liandrat and Tchamitchian (1994) about how to extend these techniques to the case of non-constant coefficients. Roughly speaking, an exact solver on a coarse scale is employed in conjunction with a *freezing coefficient/vaguelette* scheme on higher scales. For simplicity, we consider the univariate case $n = 1$ only, that is, $\mathcal{L} = \mathcal{I} - \frac{\partial}{\partial x}\left(\nu(x)\frac{\partial}{\partial x}\right)$ where $\nu(x) \geq \nu > 0$ is Lipschitz continuous. Here we have $\mathcal{L} = \mathcal{L}^*$.

The objective is to evaluate a projection of the inverse $\mathcal{L}^{-1}$. Consider the Galerkin projection of the low-frequency part

$$\mathcal{A}_q := Q_q(Q_q^*\mathcal{L}Q_q)^{-1}Q_q^* \tag{7.36}$$

of the inverse, where $Q_q f = \langle f, \tilde{\Phi}_q\rangle\Phi_q$. Due to the variable coefficient $\nu(x)$, the evaluation of $\mathcal{L}^{-1}\tilde{\psi}_\lambda$ is not feasible where, as before, $\lambda = 2^{-j}\left(k + \frac{1}{2}\right)$. Instead one defines functions θ_λ by

$$-\nu(\lambda)\frac{d^2}{dx^2}\theta_\lambda = \tilde{\psi}_\lambda, \quad \lambda \in \nabla, \tag{7.37}$$

which, according to the preceding discussion, are vaguelettes (see Tchamitchian (1997)), so that the operator $\mathcal{P}_q$ defined by

$$\mathcal{P}_q(\tilde{\psi}_\lambda) = \begin{cases} \theta_\lambda, & |\lambda| \geq q, \\ 0, & |\lambda| < q, \end{cases} \tag{7.38}$$

is a bounded mapping from $L_2(\mathbb{R})$ to $H^2(\mathbb{R})$. Here, $\mathcal{P}_q$ is often termed a *parametrix* of $\mathcal{L}$, that is, the *exact* inverse of an approximation to $\mathcal{L}$ at high frequencies. In fact, by definition, one has, for $|\lambda| \geq q$,

$$\begin{aligned} \mathcal{L}\mathcal{P}_q(\tilde{\psi}_\lambda) &= \Theta_\lambda - \frac{d}{dx}(\nu(\cdot) - \nu(\lambda))\frac{d}{dx}\Theta_\lambda - \nu(\lambda)\frac{d^2}{dx^2}\Theta_\lambda \\ &=: \mathcal{R}_q(\tilde{\psi}_\lambda) + \tilde{\psi}_\lambda, \end{aligned}$$

while $\mathcal{L}\mathcal{P}_q(\tilde{\psi}_\lambda) = 0$ for $|\lambda| < q$. Hence one obtains

$$\mathcal{L}\mathcal{P}_q = (\mathcal{I} - Q_q) + \mathcal{R}_q, \tag{7.39}$$

and one can show that (Lazaar et al. 1994)

$$\|\mathcal{R}_q g\|_{L_2} \lesssim 2^{-q}\|g\|_{L_2}. \tag{7.40}$$

Now $\mathcal{A}_q + \mathcal{P}_g$ is expected to approximate $\mathcal{L}^{-1}$ well. In fact, a von Neumann series argument yields the following theorem.

Theorem 7.6 (Lazaar et al. 1994) Let $\mathcal{U}_q := \mathcal{I} - \mathcal{L}(\mathcal{A}_q + \mathcal{P}_q)$. Then $\|\mathcal{U}_q^2\|_{L_2} \lesssim 2^{-q}$ so that, for q sufficiently large (depending on $\Psi, \tilde{\Psi}$ and ν),

$$\mathcal{L}^{-1} = (\mathcal{A}_q + \mathcal{P}_q) \sum_{l=0}^{\infty} \mathcal{U}^l. \tag{7.41}$$

As for the numerical realization, choose some $J > q$. The idea is to replace the role of L_2 in the above scheme by $S_J = S(\Phi_J)$, that is, let $\mathcal{L}_J := Q_J^* \mathcal{L} Q_J$, and denote by $\mathcal{A}_{J,q}$ the Galerkin projection obtained by replacing $\mathcal{L}$ in (7.36) by $\mathcal{L}_J$.

The next step is to approximate $\mathcal{L}_J^{-1}$ in the neighbourhood of each wavelet on high scales by $\mathcal{P}_{J,q}$ defined by

$$\mathcal{P}_{J,q} g := \sum_{q \leq |\lambda| < J} \langle g, \psi_\lambda \rangle \tau_\lambda,$$

where the τ_λ are here defined by

$$-\nu(\lambda) Q_J^* \partial_x^2 Q_J \tau_\lambda = \tilde{\psi}_\lambda, \quad |\lambda| \geq q,$$

As above, $\mathcal{A}_{J,q} + \mathcal{P}_{J,q}$ approximates $\mathcal{L}_J^{-1}$ on S_J. One can now formulate an analogue to Theorem 7.6, setting $\mathcal{U}_{J,q} = \mathcal{I} - \mathcal{L}_p(\mathcal{A}_{J,q} + \mathcal{P}_{J,q})$, so that

$$\mathcal{L}_J^{-1} = (\mathcal{A}_{J,q} + \mathcal{P}_{J,q}) \sum_{l \geq 0} \mathcal{U}_{J,q}^l.$$

Thus the solution of $\mathcal{L}u = f$ in S_J is given by

$$u_J = \sum_{k \geq 1} f_k, \tag{7.42}$$

where

$$f_k := (\mathcal{I} - (\mathcal{A}_{J,q} + \mathcal{P}_{J,q})\mathcal{L}) f_{k-1}, \quad k \geq 1,$$

and $f_1 := (\mathcal{A}_{J,q} + \mathcal{P}_{J,q}) f$. An approximate solution in S_J is obtained by truncating the series (7.42).

Note that, in view of (5.16), $\mathcal{A}_{J,q}$ is defined by

$$Q_q^*(Q_J^* \mathcal{L} Q_J) Q_q \mathcal{A}_{J,q} = Q_q^* \mathcal{L} Q_q \mathcal{A}_{J,q} = \mathcal{I},$$

so that the application of $\mathcal{A}_{J,q}$ requires solving the (small) linear system in $S_q \subset S_J$. However, the discretization of $\mathcal{L}$ that involves inner products with non-constant coefficients has to have the accuracy of the highest discretization level J in order not to spoil the overall accuracy. The application of $\mathcal{P}_{J,q}$ again requires a sufficiently accurate evaluation of the vaguelettes τ_λ. Here the remarks of the preceding discussion apply. Again, in the periodic setting wavelets with global support are usually admitted at the expense of an additional log term introduced by FFT. Employing compactly supported wavelets and approximating the vaguelettes as indicated before, one may still hope to keep the computational work proportional to $\dim S_J$. Of course, the practical realization involves several approximation and truncation steps depending on the choice of J, q and $\Psi, \tilde{\Psi}$, and these have to be carefully balanced. One has to keep in mind that the general philosophy is to spend quite some effort on initialization and precomputation in order to reduce the solution to rapid evaluation schemes. It should be interesting to compare the scheme with conventional preconditioning schemes. Numerical tests for a periodic model problem show rapid convergence of the scheme (Lazaar et al. 1994). Details of the numerical schemes, their analysis and numerical experiments are presented in Lazaar (1995).

7.6. Energy pre-wavelets

One drawback of the above vaguelette schemes is that even when biorthogonal pairs $\Psi, \tilde{\Psi}$ of compactly supported wavelets are used, the collections Θ generally involve globally supported functions. This can be remedied in certain cases at the expense of exact diagonalization. In fact, consider again a constant coefficient elliptic operator $\mathcal{L} = \sum_{|\alpha|,|\beta|\leq s} a_{\alpha\beta}\partial^\alpha\partial^\beta$ with strictly positive symbol σ. Suppose that $\phi \in L_2(\mathbb{R}^n)$, $n \leq 3$, is a stable generator (see (4.9)), which is smooth enough to satisfy

$$\sum_{k\in\mathbb{Z}^n} \left|\phi\left(\frac{y+4\pi k}{2}\right)\right|^2 \sigma(y+4\pi k) \sim 1, \tag{7.43}$$

and $\int_{[0,1]^n} \left(\sum_{k\in\mathbb{Z}^n} |\phi(x-k)|\right)^{-2} dx < \infty$. Moreover, assume that ϕ is skew-symmetric about some point $\alpha \in \mathbb{R}^n$, that is, $\phi(\alpha + x) = \overline{\phi(\alpha - x)}$, $x \in \mathbb{R}^n$. It was shown by Dahlke and Weinreich (1994) (see also Dahlke (1996) and Dahlke and Weinreich (1993)) that there exist $\psi_e \in S(\Phi_1)$, $e \in E_* = \{0,1\}^n \setminus \{0\}$, such that

$$S(\Phi_1) = S(\Phi_0) \oplus S\Big(\{\psi_e(\cdot - k) : k \in \mathbb{Z}^n, e \in E_*\}\Big),$$

and

$$\langle \mathcal{L}\Phi_0, \Psi_e \rangle = 0, \quad e \in E_*. \tag{7.44}$$

Thus the ψ_e generate complement spaces that are *orthogonal* relative to the energy inner product $a(u,v) = \langle \mathcal{L}u, v\rangle$ (when $\mathcal{L}$ is symmetric, recall Section 1.4). One should note that this also covers *genuinely* multivariate generators ϕ not obtained by tensor products of univariate ones. The restriction to spatial dimensions $n \leq 3$ arises from the fact that in these cases the masks for the wavelets can be retrieved from the mask of the generators in an explicit way, which plays a central role in the construction.

The above result (7.44) concerns the decomposition for *one* level. Due to the appearance of the symbol σ, the adapted wavelets are (as in the vaguelette case) *scale-dependent*. To obtain a complete wavelet basis, one has to demand that (7.43) holds for all symbols $\sigma_j := \sigma(2^j\cdot)$. Let $\{\psi_{j,e}\}_{e\in E_*}$ be the wavelet family constructed above relative to σ_j. Then $\{\phi(\cdot - k) : k \in \mathbb{Z}^n\} \cup \{\psi_{j,e}(2^j \cdot -k) : k \in \mathbb{Z}^n,\ e \in E_*,\ j = 0, 1, 2, \ldots\}$ forms a wavelet basis satisfying

$$\langle \mathcal{L}\psi_{j,e}(\cdot - k), \psi_{j',e'}(\cdot - k')\rangle = 0, \quad e, e' \in E_*, \quad k, k' \in \mathbb{Z}^n, \quad j \neq j', \quad (7.45)$$

(Dahlke and Weinreich 1994, Dahlke 1996).

Returning to the periodic case, the stiffness matrices relative to this basis is therefore *block-diagonal.* In particular, the case $\mathcal{L} = -\Delta + a$, $a > 0$ is covered. In this case the wavelets can be chosen to have *compact support.* Therefore the diagonal blocks are sparse. Properly scaled, each block is *well conditioned.* Thus such matrices are easily inverted, which suggests using them for preconditioning purposes, when $\mathcal{L}$ has a more complicated form.

7.7. *Evolution equations*

The next step is to consider problems of the form (2.14) described in Section 2.2. A common approach to such problems is to fix a time discretization that is *implicit* in the leading second-order term $\mathcal{L}u$ and *explicit* in $\mathcal{G}(u)$. The simplest example is the Euler scheme

$$\frac{u^{(l+1)} - u^{(l)}}{\Delta t} + \mathcal{L}u^{(l+1)} + \mathcal{G}(u^{(l)}) = 0, \quad (7.46)$$

where the upper index l denotes the time level, that is, $u^{(l)}$ is an approximation to $u(\cdot, t_l)$, $t_l = t_{l-1} + \Delta t$. Thus for each time step one has to solve an elliptic problem

$$(I + \Delta t\mathcal{L})u^{(l+1)} = u^{(l)} - \Delta t\mathcal{G}(u^{(l)}), \quad (7.47)$$

of the form discussed in previous sections. Of course, any elliptic solver such as the robust FE-based wavelet preconditioners discussed in Section 6 can be used for that purpose too. Once space discretizations $\mathcal{L}_j$, $\mathcal{G}_j$ for $\mathcal{L}$ and $\mathcal{G}$ relative to the spaces $S(\Phi_j) =: S_j$, say, have been chosen, one has to solve *linear* problems

$$(I + \Delta t\mathcal{L}_j)u_j^{(l+1)} = u_j^{(l)} - \Delta t\mathcal{G}_j(u_j^{(l)}).$$

In particular, when $\mathcal{G} \equiv 0$ one formally obtains

$$u_j^{(l+1)} = (I + \Delta t\mathcal{L}_j)^{-l}u_j^{(0)}. \quad (7.48)$$

Note that the projection scheme from the previous section, with respect to orthonormal wavelet bases, gives rise to a conceptually somewhat different scheme of the form

$$u_j^{(l+1)} = (Q_j(I + \Delta t\mathcal{L})^{-1})^l u_j^{(0)}, \tag{7.49}$$

where one can use the fact that

$$(Q_j(I + \Delta t\mathcal{L})^{-1})^l = (Q_j(I + \Delta t\mathcal{L})^{-1}Q_j)^{l-1}(Q_j(I + \Delta t\mathcal{L})^{-1}). \tag{7.50}$$

Liandrat and Tchamitchian (1997) have pointed out that there exists *no* discretization $\mathcal{L}_j$ of $\mathcal{L}$, independent of Δt, such that $(I + \Delta t\mathcal{L}_j)^{-1} = Q_j(I + \Delta t\mathcal{L})^{-1}Q_j$.

An algorithm based on (7.49) is proposed and analysed in Liandrat and Tchamitchian (1997). This *time-dependent vaguelette scheme* looks schematically as follows.

ALGORITHM 3 (TIME-DEPENDENT V-SCHEME)

(1) Choice of Ψ, $\tilde{\Psi}$, Δt.
(2) Initialization:

- When globally supported wavelets are used, fix truncated versions of the filter matrices $\mathbf{M}_{j,0}$, $\mathbf{M}_{j,1}$ from (3.4), (3.8).
- Approximate the filters

$$\mathbf{F}_q := \langle \Phi_q, (I + \Delta t\mathcal{L})^{-1}\tilde{\Phi}_q \rangle.$$

(3) Compute the scalar products

$$\langle u_j^{(l)} - \Delta t\mathcal{G}(u_j^{(l)}), \theta_\lambda \rangle$$

according to (7.23) (or the hybrid evaluation scheme in Section 7.4).

(4) The representation of $u_j^{(l)}$ in terms of Φ_j is obtained with the aid of (3.26).

To ensure that this scheme is competitive with finite element schemes, an efficient evaluation of the nonlinear terms $\mathcal{G}(u_j^{(l)})$ is needed. This is a nontrivial task when working in the wavelet representation. Some proposals on how to deal with this task can be found in Liandrat and Tchamitchian (1997). We will address this issue again later.

Wavelet representation of evolution operators

We will now briefly describe an alternative approach pursued by a number of researchers; see, for example, Beylkin and Keiser (1997), Dorobantu (1995), Enquist, Osher and Zhong (1994) and Perrier (1996). The order of time and space discretization is now reversed. The basic ideas will be explained again for the model case of univariate evolution equations on $[0, 1]$ with periodic boundary conditions of the form

$$\frac{\partial u}{\partial t} = \mathcal{L}u + \mathcal{G}(u), \quad u(\cdot, t_0) = u_0, \quad u(x, t) = u(x + 1, t), \tag{7.51}$$

where, as before, $\mathcal{G}$ is a possibly nonlinear operator and $\mathcal{L}$ is a constant coefficient second-order operator $\mathcal{L}u = \nu\frac{\partial^2}{\partial x^2}$, $\nu > 0$. Again the examples (2.15) and (2.16) are covered.

A key role is played by the classical *semi-group* approach. In fact, by *Duhamel's principle* the solution $u(x,t)$ to (7.51) is given by

$$u(x,t) = e^{(t-t_0)\mathcal{L}}u_0(x) + \int_{t_0}^{t} e^{(t-\tau)\mathcal{L}}\mathcal{G}(u(x,\tau))\,\mathrm{d}\tau. \tag{7.52}$$

In particular, for the heat equation where $\mathcal{G} = 0$ and $\nu = 1$, the solution operator $e^{t\mathcal{L}}$ has the form

$$(e^{t\mathcal{L}}v)(x) = \frac{1}{\sqrt{2\pi t}}\int_{\mathbb{R}} e^{\frac{(x-y)^2}{4t}}v(y)\,\mathrm{d}y. \tag{7.53}$$

The reason why the use of this representation has been mainly confined to theoretical purposes is that conventional discretizations of the involved operators are *not sparse.* The main thrust of the above mentioned papers is that this is different when employing wavelet-based discretizations (recall Section 1.5 (c)).

An example of this type is the following proposal from Enquist et al. (1994) concerning *long time solutions.* It begins with a conventional discretization by the method of lines

$$\frac{d}{dt}\mathbf{U} = \mathcal{L}_j\mathbf{U} + \mathbf{F} \tag{7.54}$$

of $\frac{\partial u}{\partial t} = \mathcal{L}u + f$, where $\mathbf{U} = (U_k(t) \approx u(k2^{-j},t))_{k=0}^{2^j-1}$, $\mathbf{F} = (F_k(t))_{k=0}^{2^j-1}$, $F_k(t) = f(k2^{-j},t)$ and

$$(\mathcal{L}_j\mathbf{U})_k = \nu 2^{2j}(U_{k-1}(t) - 2U_k(t) + U_{k+1}(t))$$

is the classical second divided difference operator. Now Duhamel's principle applied to (7.54) yields

$$\mathbf{U}(t+\Delta t) = e^{\Delta t\mathcal{L}_j}\mathbf{U}(t) + \int_{t}^{t+\Delta t} e^{(t+\Delta t-s)\mathcal{L}_j}\mathbf{F}(s)\,\mathrm{d}s. \tag{7.55}$$

Conventional numerical schemes are now obtained by expanding the evolution operator $e^{\Delta t\mathcal{L}_j}$. For instance, Taylor expansion and truncation yields explicit schemes with the usual stability constraints on the time step Δt relative to spatial mesh size $\Delta x = 2^{-j}$ in the present case. Any such approximation $\mathcal{E}$ to $e^{\Delta t\mathcal{L}_j}$ provides

$$\mathbf{U}(n\Delta t) \approx \mathbf{U}^n = \mathcal{E}^n\mathbf{U}^0 + \sum_{i=1}^{n}\mathcal{E}^{n-i}\mathbf{F} \tag{7.56}$$

as a discrete counterpart of (7.55). The simplest examples are $\mathcal{E} = (I + \Delta t \mathcal{L}_j)$ or $\mathcal{E} = (I - \Delta t \mathcal{L}_j)^{-1}$ for the *explicit* and *implicit Euler* schemes, respectively. Alternatively, $\mathcal{E} = (I - \frac{\Delta t}{2}\mathcal{L}_j)^{-1}(I + \frac{\Delta t}{2}\mathcal{L}_j)$ corresponds to the *Crank–Nicholson scheme.*

Now, suppose one is interested in long time solutions of the heat equation. This requires high powers of $\mathcal{E}$. In particular, the powers $\mathcal{E}^{2^l}$ can be obtained by repeated squaring. Setting $\mathcal{S}_m := \mathcal{E}^{2^m}$, $\mathcal{C}_m := \sum_{i=0}^{2^m-1} \mathcal{E}^i \mathbf{F}$, and noting that

$$\begin{aligned}\sum_{i=0}^{2^m-1} \mathcal{E}^i &= I + \mathcal{E} + \mathcal{E}^2(I+\mathcal{E}) + \mathcal{E}^4(I + \mathcal{E} + \mathcal{E}^2 + \mathcal{E}^3) + \cdots \\ &\quad + \mathcal{E}^{2^{m-1}}(I + \mathcal{E} + \cdots + \mathcal{E}^{2^{m-1}-1}), \end{aligned} \tag{7.57}$$

the following algorithm approximates the solution at time $t = 2^m \Delta t$ after m steps.

ALGORITHM 4
Set $\mathcal{S}_0 := \mathcal{E}$, $\mathcal{C}_0 = \mathbf{F}$.

(1) For $i = 1, 2, \ldots, m$:

$$\begin{aligned}\mathcal{S}_i &= \mathcal{S}_{i-1}^2, \\ \mathcal{C}_i &= (I + \mathcal{S}_{i-1})\mathcal{C}_{i-1}.\end{aligned}$$

(2) Then $\mathbf{U}^{(2^m)} = \mathcal{S}_m \mathbf{U}^{(0)} + \mathcal{C}_m$ is the approximate solution of (7.54) at time $2^m \Delta t$.

The conceptual advantage is that time is rapidly advanced by a few applications of powers of $\mathcal{E}$. However, in this form Algorithm 4 cannot be applied in practice since the corresponding matrices fill up after a few squarings, so that each step becomes too costly. The basic idea of Enquist et al. (1994) is to transform Algorithm 4 in such a way that the $\mathcal{S}_i$ become sparse (within some tolerance). One exploits the fact that wavelet representations of CZO (and their powers) are nearly sparse (recall Section 7.2). Similar ideas are used by Perrier (1996). Consider again the 1-periodic case and a corresponding dual pair of periodized generator bases Φ_j, $\tilde{\Phi}_j$. Let $N = 2^j$. Each $\mathbf{c} \in \mathbb{R}^N$ can then be indentified with $\mathbf{c}^{\mathrm{T}}\Phi_j \in S(\Phi_j)$ and the transform $\mathbf{T}_j^{-1}$ defined by (3.28) transforms $\mathbf{c}$ into the corresponding wavelet coefficient vector $\mathbf{d} = \mathbf{T}_j^{-1}\mathbf{c}$. If Φ_j consists of pairwise orthogonal functions, for instance periodized Daubechies scaling functions, so that $\Phi_j = \tilde{\Phi}_j$, the transformation $\mathbf{T}_j$ is orthogonal and $\mathbf{T}_j^{-1} = \mathbf{T}_j^{\mathrm{T}}$. Hence the application of $\mathcal{S}_i^2$ in wavelet representation becomes

$$\mathbf{T}_j^{\mathrm{T}}\mathcal{S}_i^2\mathbf{c} = \mathbf{T}_j^{\mathrm{T}}\mathcal{S}_i^2\mathbf{T}_j\mathbf{d} = (\mathbf{T}_j^{\mathrm{T}}\mathcal{S}_i\mathbf{T}_j)^2\mathbf{d}.$$

Replacing $\mathcal{S}_0$ in Algorithm 4 by $\mathbf{T}_j^{\mathrm{T}}\mathcal{E}\mathbf{T}_j$ produces an equivalent scheme.

The gain lies in the fact that the iterates $\mathcal{S}_i$ now become sparse (*cf.* (7.11), (6.24)). To increase the efficiency of the above scheme, one can introduce the operation $\mathrm{trunc}(\mathcal{S}_i, \varepsilon)$, which sets all entries to zero whose absolute value stays below the threshold $\varepsilon > 0$. This leads to the following.

ALGORITHM 5
Set $\mathcal{S}_o = \mathrm{trunc}(\mathbf{T}_j^{\mathrm{T}} \mathcal{E} \mathbf{T}_j, \varepsilon)$ and $\mathcal{C}_0 = \mathbf{T}_j^{\mathrm{T}} \mathbf{F}$.

(1) For $i = 1, 2, \ldots, m$:

$$\begin{aligned} \mathcal{S}_i &= \mathrm{trunc}(\mathcal{S}_{i-1}^2, \varepsilon), \\ \mathcal{C}_i &= (I + \mathcal{S}_{i-1})\mathcal{C}_{i-1}, \end{aligned}$$

$$\mathbf{U}^{(2^m)} = \mathbf{T}_j(\mathcal{S}_m \mathbf{T}_j^{\mathrm{T}} \mathbf{U}^{(0)} + \mathcal{C}_m).$$

Of course, the threshold ε has to be chosen appropriately. Also, modifications and additional assumptions are necessary when f depends on t explicitly. The error analysis carried out by Dorobantu (1995) indicates that $\varepsilon \leq \Delta t$ is a reasonable choice. Although an explicit scheme constrains Δt relative to $\Delta x = 2^{-j}$, the experiments in Dorobantu (1995) suggest that a simple explicit Euler scheme $\mathcal{E} = (I + \Delta t \mathcal{L}_j)$ is in this context superior to an implicit scheme, although the solution of corresponding systems benefits from the preconditioning effects of the wavelet bases. One should also note that the choice of the wavelet basis is not necessarily related to the space discretization, which above was just finite differences. On one hand, this increases flexibility and reminds us of algebraic multigrid. On the other hand, the scope for rigorous analysis of the scheme certainly decreases.

The non-standard form
The key idea of the above scheme is that, as soon as sparse representations of evolution operators are available, discretizations of the integral representation (7.52) reduce to matrix-vector multiplications with (nearly) sparse matrices. Therefore, the efficiency of this operation is crucial (just as in the context of iterative solvers). So far, we have primarily exploited the (near) sparseness of the matrices $\langle \mathcal{T}\Psi, \Psi\rangle^{\mathrm{T}}$, which are often referred to as the standard form of the operator $\mathcal{T}$. In particular, in the context of periodic problems the following alternate representation has been propagated by several researches. It is called *non-standard (NS) form*; see, for example, Beylkin, Coifman and Rokhlin (1991), Beylkin and Keiser (1997) and Dorobantu (1995). While $\langle \mathcal{T}\Psi, \Psi\rangle^{\mathrm{T}}$ arises from the formal expansion (see (5.37))

$$\mathcal{T} = \Sigma_0^* \mathcal{T} \Sigma_0 = \sum_{j,l=0}^{\infty} (Q_j^* - Q_{j-1}^*)\mathcal{T}(Q_l - Q_{l-1}) \tag{7.58}$$

setting $Q_{-1} = 0$, the alternative telescoping expansion gives

$$\mathcal{T} = \sum_{j=0}^{\infty} (Q_{j+1}^* \mathcal{T} Q_{j+1} - Q_j^* \mathcal{T} Q_j) + Q_0^* \mathcal{T} Q_0,$$

where as before $Q_j f = \langle f, \tilde{\Phi}_j \rangle \Phi_j = \langle f, \tilde{\Phi}_0 \rangle \Phi_0 + \sum_{l=0}^{j-1} \langle f, \tilde{\Psi}_l \rangle \Psi_l$. One readily checks that

$$\mathcal{T} = \sum_{j=0}^{\infty} \left\{ (Q_{j+1}^* - Q_j^*) \mathcal{T} (Q_{j+1} - Q_j) + Q_j^* \mathcal{T} (Q_{j+1} - Q_j) + (Q_{j+1}^* - Q_j^*) \mathcal{T} Q_j \right\} + Q_0^* \mathcal{T} Q_0. \tag{7.59}$$

Of course, one can start the expansion at any other fixed coarsest level j_0 instead of $j_0 = 0$. Another way of looking at the NS form is to expand the kernel of $\mathcal{T}$ relative to the dilates and translates of the bivariate wavelets

$$\psi(x)\psi(y), \quad \phi(x)\psi(y), \quad \psi(x)\phi(y).$$

Since

$$\begin{aligned}
(Q_{j+1}^* - Q_j^*) \mathcal{T} (Q_{j+1} - Q_j) v &= \langle v, \tilde{\Psi}_j \rangle \langle \mathcal{T} \Psi_j, \Psi_j \rangle \tilde{\Psi}_j, \\
(Q_{j+1}^* - Q_j^*) \mathcal{T} Q_j v &= \langle v, \tilde{\Phi}_j \rangle \langle \mathcal{T} \Phi_j, \Psi_j \rangle \tilde{\Psi}_j, \\
Q_j^* \mathcal{T} (Q_{j+1} - Q_j) v &= \langle v, \tilde{\Psi}_j \rangle \langle \mathcal{T} \Psi_j, \Phi_j \rangle \tilde{\Phi}_j,
\end{aligned}$$

the matrix representations of the block operators are

$$\begin{aligned}
\mathbf{A}_j &;= \langle \mathcal{T} \Psi_j, \Psi_j \rangle = \mathbf{M}_{j,1}^{\mathrm{T}} \langle \mathcal{T} \Phi_{j+1}, \Phi_{j+1} \rangle \mathbf{M}_{j,1}, \\
\mathbf{B}_j &:= \langle \mathcal{T} \Phi_j, \Psi_j \rangle = \mathbf{M}_{j,0}^{\mathrm{T}} \langle \mathcal{T} \Phi_{j+1}, \Phi_{j+1} \rangle \mathbf{M}_{j,1}, \\
\mathbf{C}_j &:= \langle \mathcal{T} \Psi_j, \Phi_j \rangle = \mathbf{M}_{j,1}^{\mathrm{T}} \langle \mathcal{T} \Phi_{j+1}, \Phi_{j+1} \rangle \mathbf{M}_{j,0},
\end{aligned} \tag{7.60}$$

and $\mathbf{H}_0 := \langle \mathcal{T} \Phi_0, \Phi_0 \rangle$ for the coarse level contribution. Thus these blocks involve the three types of scalar products

$$\alpha_{k,l}^j = \langle \mathcal{T} \psi_{j,k}, \psi_{j,l} \rangle, \quad \beta_{k,l}^j = \langle \mathcal{T} \phi_{j,k}, \psi_{j,l} \rangle, \quad \gamma_{k,l}^j = \langle \mathcal{T} \psi_{j,k}, \phi_{j,l} \rangle, \tag{7.61}$$

which, in contrast to the standard form, involve *only* functions on the *same* level j in each block.

As a consequence, several practical advantages can be attributed to the NS form. In contrast to the standard form, the NS form maintains the convolution structure of an operator. Thus FFT can be used to enhance further the efficiency of matrix vector multiplication in NS form. Moreover, since the scalar products only involve functions on the same level, the methods described in Section 4.2 can be used to calculate them efficiently. Only finitely different coefficients are needed to represent a constant coefficient

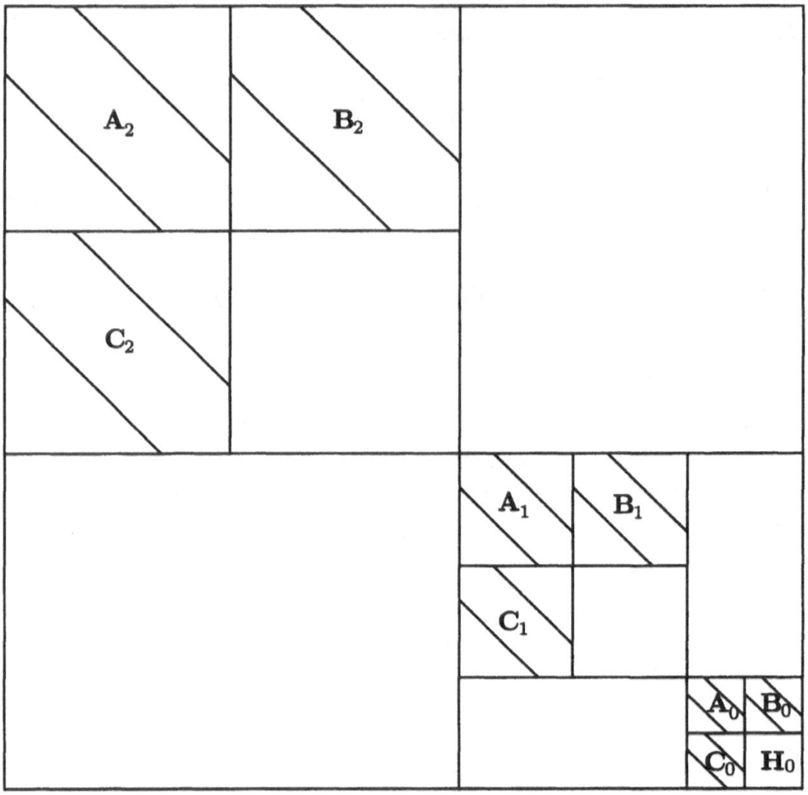

Fig. 3. Schematic view of the NS form

differential operator in NS form. For an extensive discussion of the representation of operators in wavelet bases, the reader is referred to Beylkin (1992). However, one has to stress that one consequence of uncoupling levels in the NS form is that the vectors it applies to are *not* representations of the original vector with respect to *any* basis. Instead they could be viewed as coefficient vectors relative to a redundant spanning set. Accordingly the size of the NS form is up to almost twice the size of the corresponding standard form; see, for example, Beylkin (1992) and Beylkin and Keiser (1997).

More precisely, the action of the truncated operator $\mathcal{T}_J := Q_J^* \mathcal{T} Q_J$ can be described as

$$\begin{aligned} \mathcal{T}_J v &= \mathcal{T}_0 v + \sum_{j=0}^{J-1} \left\{ \langle v, \tilde{\Psi}_j \rangle \mathbf{A}_j \tilde{\Psi}_j + \langle v, \tilde{\Phi}_j \rangle \mathbf{B}_j \tilde{\Psi}_j + \langle v, \tilde{\Psi}_j \rangle \mathbf{C}_j \tilde{\Phi}_j \right\} \\ &=: \sum_{j=0}^{J-1} \{ \tilde{\mathbf{d}}_j^{\mathrm{T}} \tilde{\Psi}_j + \tilde{\mathbf{c}}_j^{\mathrm{T}} \tilde{\Phi}_j \}, \end{aligned} \qquad (7.62)$$

where

$$\begin{aligned}
\tilde{\mathbf{d}}_j^{\mathrm{T}} &:= \langle v, \tilde{\Psi}_j\rangle \mathbf{A}_j + \langle v, \tilde{\Phi}_j\rangle \mathbf{B}_j, \quad j = 0, \dots, J-1, \\
\tilde{\mathbf{c}}_j^{\mathrm{T}} &:= \langle v, \tilde{\Psi}_j\rangle \mathbf{C}_j, \quad j = 1, \dots, J-1, \\
\tilde{\mathbf{c}}_0^{\mathrm{T}} &:= \langle v, \tilde{\Phi}_0\rangle \langle \mathcal{T}\Phi_0, \Phi_0\rangle + \langle v, \tilde{\Psi}_0\rangle \mathbf{B}_0.
\end{aligned} \tag{7.63}$$

Thus the application of $\mathcal{T}_J$ in the NS form requires, in addition to the wavelet coefficients $\langle v, \tilde{\Psi}_j\rangle$ of v in $S(\Psi_j)$, the scaling function coefficients $\langle v, \tilde{\Phi}_j\rangle$ on level j. Hence, when $v \in S(\Phi_J)$ is given in single-scale representation $v = \langle v, \tilde{\Phi}_J\rangle \Phi_J$, the array $\tilde{\mathbf{d}}_J^{\mathrm{NS}} := (\tilde{\mathbf{c}}_0, \tilde{\mathbf{d}}_0, \dots, \tilde{\mathbf{c}}_{J-1}, \tilde{\mathbf{d}}_{J-1})$ representing the application of $\mathcal{T}_j$ to v is obtained, according to (7.63), by applying the blocks $\mathbf{A}_j, \mathbf{B}_j, \mathbf{C}_j$ to the result of the pyramid scheme (3.28) for the corresponding level.

Conversely, to transform the output $\tilde{\mathbf{d}}_J^{\mathrm{NS}}$ back into a coefficient vector relative to a basis of $S(\tilde{\Phi}_J)$, one can proceed as follows. Since by (3.4), (3.8),

$$\tilde{\mathbf{c}}_j^{\mathrm{T}} \tilde{\Phi}_j = \tilde{\mathbf{c}}_j^{\mathrm{T}} \tilde{\mathbf{M}}_{j,0}^{\mathrm{T}} \tilde{\Phi}_{j+1}, \quad \tilde{\mathbf{d}}_j^{\mathrm{T}} \tilde{\Psi}_j = \tilde{\mathbf{d}}_j^{\mathrm{T}} \tilde{\mathbf{M}}_{j,1}^{\mathrm{T}} \tilde{\Phi}_{j+1},$$

the pyramid scheme

$$\begin{array}{ccccccccccc}
\tilde{\mathbf{c}}_0 & \longrightarrow & \hat{\mathbf{c}}_1 & \longrightarrow & \hat{\mathbf{c}}_2 & \longrightarrow & \cdots & \longrightarrow & \hat{\mathbf{c}}_{J-1} & \longrightarrow & \hat{\mathbf{c}}_J \\
\tilde{\mathbf{d}}_0, \tilde{\mathbf{c}}_1 & \nearrow & \tilde{\mathbf{d}}_1, \tilde{\mathbf{c}}_2 & \nearrow & \tilde{\mathbf{d}}_2, \tilde{\mathbf{c}}_3 & \nearrow & \cdots & \nearrow & \tilde{\mathbf{d}}_{J-1}, \tilde{\mathbf{c}}_J & \nearrow &
\end{array}, \tag{7.64}$$

where

$$\hat{\mathbf{c}}_0 := \tilde{\mathbf{c}}_0, \quad \hat{\mathbf{c}}_j := \tilde{\mathbf{M}}_{j-1,0} \hat{\mathbf{c}}_{j-1} + \tilde{\mathbf{M}}_{j-1,1} \tilde{\mathbf{d}}_{j-1} + \tilde{\mathbf{c}}_j, \quad j = 1, \dots, J,$$

similarly to (3.26) produces, in view of (7.62), the single-scale representation $\mathcal{T}_J v = \hat{\mathbf{c}}_J^{\mathrm{T}} \tilde{\Phi}_J$. Likewise, in view of

$$\tilde{\mathbf{c}}_j^{\mathrm{T}} \tilde{\Phi}_j = \tilde{\mathbf{c}}_j^{\mathrm{T}} (\overline{\mathbf{M}}_{j-1,0} \tilde{\Phi}_{j-1} + \overline{\mathbf{M}}_{j-1,1} \tilde{\Psi}_{j-1}),$$

(recall (3.13))

$$\begin{array}{cccccccccc}
\tilde{\mathbf{c}}_{J-1}, \tilde{\mathbf{d}}_{J-2} & \longrightarrow & \hat{\mathbf{c}}_{J-2}, \tilde{\mathbf{d}}_{J-3} & \longrightarrow & \hat{\mathbf{c}}_{J-3}, \tilde{\mathbf{d}}_{J-4} & \longrightarrow & \cdots & \longrightarrow & \hat{\mathbf{c}}_0 \\
\tilde{\mathbf{d}}_{J-1} & \searrow & \hat{\mathbf{d}}_{J-2} & \searrow & \hat{\mathbf{d}}_{J-3} & \searrow & & \searrow & \hat{\mathbf{d}}_0
\end{array}, \tag{7.65}$$

where

$$\hat{\mathbf{c}}_j := \tilde{\mathbf{c}}_j + \mathbf{M}_{j,0}^* \tilde{\mathbf{c}}_{j+1}, \quad \hat{\mathbf{d}}_j := \tilde{\mathbf{d}}_j + \mathbf{M}_{j,1}^* \tilde{\mathbf{c}}_{j+1}, \quad j = J-2, \dots, 0,$$

generates the wavelet representation

$$\mathcal{T}_J v = \hat{\mathbf{c}}_0^{\mathrm{T}} \tilde{\Phi}_0 + \hat{\mathbf{d}}_0^{\mathrm{T}} \tilde{\Psi}_0 + \dots + \hat{\mathbf{d}}_{J-2}^{\mathrm{T}} \tilde{\Psi}_{J-2} + \tilde{\mathbf{d}}_{J-1}^{\mathrm{T}} \tilde{\Psi}_{J-1}. \tag{7.66}$$

Computation of the blocks $\mathbf{A}_j, \mathbf{B}_j, \mathbf{C}_j$

When $\mathcal{T}$ is a convolution operator one only has to determine the filter coefficients $\alpha_l^j, \beta_l^j, \gamma_l^j$. Moreover, in view of (7.60), it suffices to determine the coefficients of $\langle \mathcal{T}\Phi_j, \Phi_j\rangle$ and then apply portions of the pyramid schemes (3.26). When $\mathcal{T}$ is a homogeneous operator of order p such as $\left(\frac{d}{dx}\right)^p$ this is not even necessary, and one puts

$$\langle \mathcal{T}\Phi_j, \Phi_j\rangle = 2^{-(J-j)p}\langle \mathcal{T}\Phi_J, \Phi_J\rangle. \tag{7.67}$$

Specifically, when $\mathcal{T} = \left(\frac{d}{dx}\right)^p$, the coefficients in $\langle \mathcal{T}\Phi_J, \Phi_J\rangle$ form finite difference approximations of $\mathcal{T}$ in $S(\tilde{\Phi}_J)$ of order $d + \tilde{d} - 1$, when $d, \tilde{d}$ are the respective orders of exactness of the $\Phi_j, \tilde{\Phi}_j$ and $\tilde{d} \geq d$. In fact, let $\square_{j,k} := 2^{-j}(k + [0,1])$ and $v \in C^\infty(\mathbb{R})$. Hence there exists a polynomial P of degree $d-1$ and a smooth remainder R so that

$$v|_{\square_{j,k}} = (P + 2^{-jd}R)|_{\square_{j,k}}.$$

Thus

$$\begin{aligned} Q_j^*\mathcal{T}Q_jv - \mathcal{T}v &= Q_j^*\mathcal{T}P + 2^{-jd}Q_j^*\mathcal{T}R - \mathcal{T}P - 2^{-jd}\mathcal{T}R \\ &= (Q_j^* - I)\mathcal{T}P + 2^{-jd}(Q_j^* - I)\mathcal{T}R. \end{aligned}$$

Since for $\mathcal{T} = \left(\frac{d}{dx}\right)^p$, $\mathcal{T}P$ is a polynomial of degree at most $d-1-p < \tilde{d}$, the first summand on the right-hand side vanishes. Moreover, locally $(Q_j^*-I)\mathcal{T}R$ behaves for smooth R in a neighbourhood of $\square_{j,k}$ like $2^{-\tilde{d}j}$ in the L_∞-norm, say (see Proposition 5.1).

Moment conditions of the **B**-*blocks*

By (7.62), the blocks $\mathbf{A}_j$ and $\mathbf{C}_j$ are multiplied by vectors that contain wavelet coefficients. Since possibly only a few of these coefficients exceed a given threshold, one expects that these multiplications can be carried out efficiently within a desired accuracy. Instead, the vectors multiplying the blocks $\mathbf{B}_j$ consist of scaling function coefficients representing averages. Therefore these arrays are generally dense. However, it is important to note that the matrices $\mathbf{B}_j = \langle \mathcal{T}\Phi_j, \Psi_j\rangle$ have *vanishing moments* when

$$\mathcal{T} = \mathcal{H} \quad \text{or } \mathcal{T} = f\left(\frac{\partial}{\partial x}\right), \tag{7.68}$$

where $\mathcal{H}$ is the Hilbert transform (see Section 1.3) and f is analytic. More precisely, for any $\mathbf{p} := (P(l))_{l\in\mathbb{Z}}$, P a polynomial of degree $\leq d-1$, one has

$$\mathbf{p}^{\mathrm{T}}\mathbf{B}_j = 0 \tag{7.69}$$

for any $\mathcal{T}$ from (7.68); see Beylkin and Keiser (1997). In fact, by (7.60), one has

$$\mathbf{p}^{\mathrm{T}}\mathbf{B}_j = \langle \mathcal{T}\mathbf{p}^{\mathrm{T}}\Phi_j, \Psi_j\rangle. \tag{7.70}$$

By (4.30), $\mathbf{p}^{\mathrm{T}}\Phi_j$ is a polynomial of degree $d-1$. Expanding $f(\frac{d}{dx})$ in powers of $\frac{d}{dx}$, it is clear that $f(\frac{d}{dx})\mathbf{p}^{\mathrm{T}}\Phi_j$ is a polynomial of degree $\leq d-1$. Since, by assumption $\tilde{d} \geq d$, (7.69) follows in this case from (4.33). When $\mathcal{T} = \mathcal{H}$ an argument similar to the one used in Section 1.3 also confirms (7.69); see Beylkin and Keiser (1997).

Some theoretical remarks

Due to the appearance of at least one wavelet in the scalar products (7.61) there is still a compression effect. In fact, if the kernel K of $\mathcal{T}$ satisfies

$$\left|\frac{\partial^r}{\partial x^r}K(x,y)\right| + \left|\frac{\partial^r}{\partial y^r}K(x,y)\right| \lesssim |x-y|^{-(r+1)} \tag{7.71}$$

for $x \neq y \bmod 1$, one can show that (Tchamitchian 1996)

$$\left|\alpha^j_{k,l}\right| + \left|\beta^j_{k,l}\right| + \left|\gamma^j_{k,l}\right| \lesssim |k-l|^{r+1}, \tag{7.72}$$

provided that the corresponding functions with indices k and l have disjoint supports. For the remaining cases the additional assumption

$$\left|\alpha^j_{k,l}\right| + \left|\beta^j_{k,l}\right| + \left|\gamma^j_{k,l}\right| \lesssim 1, \quad k,l,j \in \mathbb{Z}, \tag{7.73}$$

is needed, which is called the *weak boundedness property.* This condition is weaker than L_2-boundedness of the operator. It plays an important role in the following celebrated theorem due to G. David and J. L. Journé; see, for instance, Tchamitchian (1996).

Theorem 7.7 Suppose that the kernel of $\mathcal{T}$ satisfies (7.9), (7.10). Then $\mathcal{T}$ is continuous on L_2 if and only if it has the weak boundedness property (7.73), $\mathcal{T}(1) \in \mathrm{BMO}$ and $\mathcal{T}^*(1) \in \mathrm{BMO}$.

7.8. A pseudo-wavelet approach

The previous section contains major ingredients of an approach to solving periodic nonlinear equations of the form (7.51) proposed by Beylkin and Keiser (1997). There it is termed the *pseudo-wavelet approach.* It is a systematic attempt to compute an approximate solution to (7.51) at the expense of a number of arithmetic operations proportional to the number of wavelet coefficients required for representing the approximate solution to the desired accuracy. The central idea is to employ appropriate discretizations of (7.52) which ultimately reduce to the *adaptive* application of certain operators in the NS form to corresponding coefficient vectors.

One basic tool is a class of time discretization schemes presented by Beylkin, Keiser and Vozovoi (1996). For instance, in the case of Burgers' equation (2.16) the term

$$I(t,t_0) := \int_{t_0}^{t} e^{(t-\tau)\mathcal{L}} u(\cdot,\tau)\frac{\partial}{\partial x}u(\cdot,\tau)\,d\tau \tag{7.74}$$

is approximated by

$$\begin{aligned} &I(t+\Delta t, t) &(7.75)\\ &= \frac{1}{2}\mathcal{O}_{\mathcal{L},1}\left(u(\cdot,t_0)\frac{\partial}{\partial x}u(\cdot,t+\Delta t) + u(\cdot,t+\Delta t)\frac{\partial}{\partial x}u(\cdot,t)\right) + \mathcal{O}((\Delta t)^2), \end{aligned}$$

where

$$\mathcal{O}_{\mathcal{L},m} := \left(e^{m\Delta t\mathcal{L}} - \mathcal{I}\right)\mathcal{L}^{-1}. \tag{7.76}$$

So the idea is to discretize $\mathcal{G}(u(\cdot,\tau))$ in the time variable τ so that the *exact* application of $\int_t^{t+\Delta t} e^{(t+\Delta t-\tau)\mathcal{L}}\,\mathrm{d}\tau$ reduces to the application of $\mathcal{O}_{\mathcal{L},1}$. This is essentially different from the procedure in Section 7.7 where the space discretization was fixed before. For the derivation of higher-order schemes and a corresponding stability analysis see Beylkin et al. (1996). Here it is important that the operators $\mathcal{L}^{-1}$ or $e^{m\Delta t\mathcal{L}}$ can be evaluated *exactly* within any chosen accuracy. Again in the case of Burgers' equation one has to evaluate

$$u(x,t+\Delta t) = e^{\Delta t\mathcal{L}}u(x,t) - I(t+\Delta t,t), \tag{7.77}$$

where $\mathcal{L} = \nu\left(\frac{\partial}{\partial x}\right)^2$ and $\mathcal{O}_{\mathcal{L},1}$ is given by (7.76).

In order to apply the operator functions $e^{\Delta t\mathcal{L}}$ and $\mathcal{O}_{\mathcal{L},1}$ efficiently one is interested in computing their NS form. Therefore it is important to determine the NS form of $f\left(\frac{\partial}{\partial x}\right)$ when f is analytic. Beylkin and Keiser (1997) propose two approaches, namely to compute

$$Q_J^* f\left(\frac{\partial}{\partial x}\right)Q_J, \tag{7.78}$$

or

$$f\left(Q_J^*\frac{\partial}{\partial x}Q_J\right). \tag{7.79}$$

To compute the NS form of $f\left(\frac{\partial}{\partial x}\right)$ via (7.79) one can diagonalize $\frac{\partial}{\partial x}$ with the aid of the discrete Fourier transform and apply the spectral theorem (Beylkin and Keiser 1997).

Using (7.78), according to the discussion in the previous section (see (7.60)), one can first determine the arrays $\mathbf{c}^j := \langle f\left(\frac{\partial}{\partial x}\right)\Phi_j, \Phi_j\rangle$, consisting of the coefficients

$$c^j_{k,k'} = 2^j \int_{\mathbb{R}} \phi(2^j x - k) f\left(\frac{\partial}{\partial x}\right)\phi(2^j x - k')\,\mathrm{d}x = \mathbf{C}^j_{k-k'}.$$

Using Fourier transforms, one can show that (Beylkin and Keiser 1997)

$$c_l^j = \int_0^{2\pi} g_j(\xi) e^{i\xi l}\,\mathrm{d}\xi, \tag{7.80}$$

where

$$g_j(\xi) = \sum_{k\in\mathbb{Z}} f(-i2^j(\xi + 2\pi k))|\hat{\phi}(\xi + 2\pi k)|^2,$$

and exploit the fact that $|\hat{\phi}(\xi)|^2$ acts like a *cut-off function*. Thus $g_j(\xi)$ can be approximated arbitrarily well by a finite sum $\tilde{g}_j(\xi)$ which can be used to discretize (7.80). Recall the similar reasoning in the vaguelette approach.

Adaptive application of operators in NS form
According to (7.63), the application of an operator in NS form requires evaluating

$$\tilde{\mathbf{d}}^j = \langle v, \tilde{\Psi}_j\rangle \mathbf{A}_j + \langle v, \tilde{\Phi}_j\rangle \mathbf{B}_j, \quad \tilde{\mathbf{c}}_j = \langle v, \tilde{\Psi}_j\rangle \mathbf{C}_j.$$

To accomplish the goal of realizing an overall solution complexity, which is proportional to the number of significant wavelet coefficients of the solution relative to a given accuracy, each calculation of $\mathcal{T}_J v$ has to be realized within this order of complexity. A heuristic reasoning towards this goal can be summarized as follows. The solution to the differential equations under consideration are typically smooth except at isolated locations where singularities such as shocks can build up. Consequently, many wavelet coefficients of the solution can be expected to stay below a given threshold. Hence the arrays $\langle v, \tilde{\Psi}_j\rangle$ are typically *short*. However, the arrays $\langle v, \tilde{\Phi}_j\rangle$ consist of averages and may be dense in spite of the smoothness of v. At this point the vanishing moment property of the **B**-blocks established in the previous section is crucial. Exploiting this property, Beylkin and Keiser (1997) argue that, when a smooth vector is applied to $\mathbf{B}_j$, the result will be sparse. In fact, Beylkin and Keiser (1997) indicate how to use the wavelet coefficients of v to replace the dense array $\langle v, \tilde{\Phi}_j\rangle$ by a sparse vector $\mathbf{s}^j$ so as to realize an efficient application of the $\mathbf{B}_j$ block within a desired tolerance of accuracy. For a more detailed discussion of the components of such a scheme we refer to Beylkin and Keiser (1997) and the literature cited there.

In addition, some interesting numerical experiments are discussed by Beylkin and Keiser (1997). First a classical Crank–Nicholson scheme for the heat equation $\frac{\partial u}{\partial t} = \nu \frac{\partial^2}{\partial x^2} u$ is compared to the wavelet-based scheme, which consists in this case of a repeated application of the NS form of $e^{\Delta t \mathcal{L}}$ via

$$u(\cdot, t_{j+1}) = e^{\Delta t\mathcal{L}} u(\cdot, t_j), \quad u(\cdot, t_0) = u_0.$$

This is an explicit procedure, yet unconditionally stable, once the evaluation of $e^{\Delta t \mathcal{L}}$ is *accurate enough* to cope with the higher-order oscillations introduced by the nonlinear terms. In particular, the advantage of *higher-order schemes* is stressed. In fact, due to the higher number of vanishing moments, they result in better sparseness of the NS form of the operators. Subsequently, the scheme is tested on several versions of Burgers' equation and its generalizations. These experiments apparently confirm that the number of operations needed to update the solution in each time step remains proportional to the number of significant wavelet coefficients.

7.9. Wavelet packets and best bases

There is yet another technique for discretizing evolution equations of the form (2.14), which has been proposed by Joly, Maday and Perrier (1997), for instance. It aims at realizing best possible *compression* of the approximate solution by employing the concept of *wavelet packets* and *best bases* developed by Coifman, Meyer, Quake and Wickerhauser (1993) and Coifman, Meyer and Wickerhauser (1992). This technique is also used by Farge, Goirand, Meyer, Pascal and Wickerhauser (1992). Therefore we will briefly indicate some of the ideas in Joly et al. (1997), where further details and relevant references can be found.

To describe the concept of wavelet packets, we confine the discussion to scaling functions $\phi \in L_2(\mathbb{R})$, whose translates $\phi(\cdot - k)$, $k \in \mathbb{Z}$, are *orthonormal*, that is, $\phi = \tilde{\phi}$. Let us denote by $\mathbf{a}, \mathbf{b}$ the masks of ϕ and the wavelet ψ, that is, $b_k = (-1)^k a_{1-k}$, $k \in \mathbb{Z}$ (see (4.20)). One can use these masks to recursively generate further basis functions, defined with $\psi_0 = \phi$, $\psi_1 = \psi$, for $n \geq 1$ by

$$\psi_{2n}(x) = \sum_{k\in\mathbb{Z}} a_k \psi_n(2x-k), \quad \psi_{2n+1}(x) = \sum_{k\in\mathbb{Z}} b_k \psi_n(2x-k). \tag{7.81}$$

One can then show that

$$S(\Phi_j) = S\Big(\{\psi_n(\cdot - k) : 0 \leq n < 2^j, k \in \mathbb{Z}\}\Big),$$

so that a variety of orthonormal bases are available. Let $\mathcal{E}_J$, respectively $\mathcal{E}$, denote any subset of $\mathbb{N} \times (\mathbb{Z} \cap (-\infty, J])$, respectively $\mathbb{N} \times \mathbb{Z}$. With each $(n,j) \in \mathbb{N} \times \mathbb{Z}$ associate the interval $I_{n,j} := [2^j n, 2^j(n+1))$. Let

$$\psi_{n,j,k}(x) := 2^{j/2}\psi_n(2^j x - k).$$

Theorem 7.8 (Coifman et al. 1992) Any collection $\{\psi_{n,j,k} : (n,j,k) \in \mathcal{E}_J \times \mathbb{Z}$ resp. $\mathcal{E} \times \mathbb{Z}\}$ is an orthonormal basis of $S(\Phi_J)$, respectively $L_2(\mathbb{R})$, if and only if

(i) $\displaystyle\bigcup_{(n,j)\in\mathcal{E}_J} I_{n,j} = [0, 2^J)$, resp. $\overline{\displaystyle\bigcup_{(n,j)\in\mathcal{E}_J} I_{n,j}} = \mathbb{R}^+$

(ii) for all $(n, j), (n', j') \in \mathcal{E}_J$, resp. $\mathcal{E}$, one has $I_{n,j} \cap I_{n',j'} = \emptyset$ if $(n, j) \neq (n', j')$.

It is again convenient to abbreviate $\lambda = (n, j, k) \in \mathcal{E} \times \mathbb{Z}$. For a given problem, the point is now to select that basis which is *best* in a certain sense, rather in the spirit of signal analysis. The decision is based on an appropriate notion of *entropy*. For a given $v \in L_2(\mathbb{R})$, let

$$v = \sum_{\lambda \in \mathcal{E} \times \mathbb{Z}} c_\lambda \psi_\lambda.$$

For $\mathcal{D} \subset \mathcal{E}$ and any $\varepsilon > 0$, the quantity

$$H^{\varepsilon,\mathcal{D}}(v) := \#\{c_\lambda : |c_\lambda| \geq \varepsilon, \lambda \in \mathcal{D} \times \mathbb{Z}\} \tag{7.82}$$

is called the *cardinal entropy* of v. Note the difference from the following more familiar measure for the content of information,

$$H^{\mathcal{D}}(v) := - \sum_{\lambda \in \mathcal{D} \times \mathbb{Z}} |c_\lambda|^2 \ln |c_\lambda|, \tag{7.83}$$

which is called *Shannon entropy*. Minimizing the cardinal entropy over $\mathcal{D}$ corresponds to selecting a basis with respect to which the representation of v has possibly few coefficients above the threshold ε. The adaptation of these notions to the periodic case is again standard. Corresponding decomposition and reconstruction algorithms as well as the recursive determination of best bases are described in Joly et al. (1997). A comparison between the different notions of entropy favours the cardinal entropy for present purposes. Let us denote by $B_\varepsilon(v) \subset \mathcal{E}_J$ the index set for a best basis for v relative to (7.82) and set $\Lambda_\varepsilon(v) := \{\lambda = (n, j, k) : (n, j) \in B_\varepsilon(v), k \in \mathbb{Z}/2^j\mathbb{Z}\}$.

To make use of these concepts here, one associates with each λ the *centre* x_λ of the basis function ψ_λ. Moreover, one assigns to $\lambda = (n, j, k)$ an *influence rectangle* centred at $(x_\lambda, n2^j)$ in the position–frequency diagram, which symbolizes the time–frequency support of ψ_λ. Once the position of ψ_λ is determined, one can define the *neighbours* of each ψ_λ; see Joly et al. (1997) for the precise definition. Given the best basis, the reduced representation of v is given by

$$Q_\varepsilon v := \sum_{\lambda \in \tilde{\Lambda}_\varepsilon(v)} \langle v, \psi_\lambda \rangle \psi_\lambda, \tag{7.84}$$

where

$$\tilde{\Lambda}_\varepsilon(v) = \{\lambda \in \Lambda_\varepsilon(v) : |\langle v, \psi_\lambda \rangle| \geq \varepsilon\}. \tag{7.85}$$

The central step of the adaptive procedure proposed by Joly et al. (1997) is to add to $\tilde{\Lambda}_\varepsilon(v)$ the neighbours of its indices, which typically results in a set that is not much larger than $\tilde{\Lambda}_\varepsilon(v)$. In fact, for most elements in $\tilde{\Lambda}_\varepsilon(v)$, one expects that its neighbours already belong to $\tilde{\Lambda}_\varepsilon(v)$.

Of course, for non-stationary problems the solution will change in time. Small variations of a function may actually cause significant changes of the best basis. However, Joly et al. (1997) observe that the entropy does not change much when dealing with evolution equations like Burgers' equation. Thus the best basis does not have to be updated after each time step.

We now outline the algorithm from Joly et al. (1997) for Burgers' equation (2.16). We wish to approximate the solution $u(m\Delta t, \cdot)$ at time $m\Delta t$ by $u^m = u_J^m \in S(\Phi_J)$ as follows.

ALGORITHM 6 (JMP)
Let u_J^0 be the approximation of the initial value u^0 and let $B_\varepsilon(u_J^0)$ denote its best basis. Let

$$\Lambda^0 := \tilde{\Lambda}_\varepsilon(u_J^0)$$

be the reduced index set after thresholding.

$(m+1)$st step: Given

$$u^m = \sum_{\lambda \in \Lambda^m} c_\lambda(m)\psi_\lambda,$$

where $\Lambda^m \subseteq \Lambda(u_J^0)$.

(i) Form $\tilde{\Lambda}^m$ as in (7.85) and $\tilde{u}^m$ as in (7.84) (relative to $\tilde{\Lambda}^m$).
(ii) Form Λ^{m+1} by adding to $\tilde{\Lambda}^m$ those elements in $\Lambda_\varepsilon(u_J^0)$ that are neighbours of elements from $\tilde{\Lambda}^m$.
(iii) Determine

$$u^{m+1} = \sum_{\lambda \in \Lambda^{m+1}} c_\lambda(m+1)\psi_\lambda$$

by requiring that the following Galerkin conditions hold:

$$\left\langle \frac{1}{\Delta t}(u^{m+1} - \tilde{u}^m) + \frac{1}{2}\frac{\partial}{\partial x}(\tilde{u}_*^m)^2, \psi_\lambda \right\rangle = -\nu \left\langle \frac{\partial}{\partial x}\tilde{u}_*^m, \frac{\partial}{\partial x}\psi_\lambda \right\rangle, \tag{7.86}$$

for $\lambda \in \Lambda^{m+1}$, where $\tilde{u}_*^m := \frac{3}{2}\tilde{u}^m - \frac{1}{2}\tilde{u}^{m-1}$.

We conclude with a brief discussion of the implementation.

Since the sets Λ^m change, the stiffness matrices needed in (7.86) change as well. Therefore Joly et al. (1997) propose to precompute the *whole* stiffness matrix relative to the entire (periodized) basis Φ_J. To generate the possible wavelet packets, one then has to use the corresponding multiscale transformations, providing a matrix of size $2^J \times J2^J$.

Using orthonormal spline wavelets, the masks are no longer short, so that one has to resort to FFT in the multiscale transformations. This introduces additional log factors in the operation count. The decomposition of the columns in the extended stiffness matrix according to the chosen best basis requires the order of $J2^{2J}$ operations. It seems that this strategy at some stage requires computational work which is quadratic in the dimension of the uncompressed problem size corresponding to $S(\Phi_J)$. This may be a serious drawback when dealing with several spatial variables.

The evaluation of nonlinear terms requires special care, which will be discussed in the following section.

7.10. Evaluation of nonlinear terms

A critical role in all the above developments is played by the evaluation of nonlinear terms. It is perhaps worthwhile to comment briefly on the principal problems arising in this context. As before, let $\Psi_\Lambda = \{\psi_\lambda : \lambda \in \Lambda\}$. So far the discussion has stressed important advantages of multiscale representations of the form $u_\Lambda = \mathbf{d}_\Lambda^T \Psi_\Lambda$, where Λ selects only those wavelets that are needed to represent a function u to some given tolerance. However, note that at any point in the domain, wavelets from all levels appearing in Λ may contribute. Thus the cost of evaluating a function in multiscale representation at a single point could be proportional to the highest level J appearing in Λ. When frequent evaluations are necessary, this could of course significantly diminish efficiency. In contrast, evaluating a function in single-scale representation requires only a finite number of operations *independent* of the level J. On the other hand, $\#\Lambda$ could be very small compared to $\dim S(\Phi_J)$. Thus the transformation of u_Λ into single-scale representation in $S(\Phi_J)$ would produce a much larger array of coefficients which, due to their nature of representing averages, may all be significant. This would waste the significant reduction of complexity gained by the sparse representation of u_Λ in wavelet coordinates.

This problem is exacerbated when, instead of point evaluations, one has to compute *nonlinear* functionals of a function u_Λ given in multiscale representation. A typical example arises in connection with the elliptic problem (7.47). Suppose that the approximate solution $u^{(l)}$ from the previous time step l is given as $u^{(l)} = \mathbf{d}_\Lambda^T \Psi_\Lambda$, where Λ is a possibly small lacunary subset of ∇. If one uses a collocation scheme for solving (7.47), one has to evaluate the nonlinear term $\mathcal{G}(u^{(l)})$ on some grid. This requires the evaluation of $u^{(l)}$ on that grid, which is the task discussed above, followed by the application of $\mathcal{G}$. If the application of $\mathcal{G}$ is expensive, an alternative is to *approximate* $\mathcal{G}(u^{(l)})$ first and then evaluate this approximation. When J is the highest scale in Λ, depending on the nature of $\mathcal{G}$, one expects that $\mathcal{G}(\mathbf{d}_\Lambda^T \Psi_\Lambda)$ can be accurately resolved on a level $\hat{J} > J$. But, again, if the approximation were given in a single-scale form, its evaluation would be inexpensive, but the representation itself would possibly involve far more coefficients than those in the array $\mathbf{d}_\Lambda$. This suggests also seeking some lacunary multiscale representation $\hat{\mathbf{d}}_{\hat{\Lambda}}^T \Xi_{\hat{\Lambda}} \approx \mathcal{G}(\mathbf{d}_\Lambda^T \Psi_\Lambda)$ with respect to a suitable basis Ξ (not necessarily equal to Ψ). In fact, when (7.47) is to be solved by a Galerkin scheme, one would have to approximate the quantities $\langle \mathcal{G}(\mathbf{d}_\Lambda^T \Psi_\Lambda), \psi_\lambda \rangle$. Thus, for $\Xi = \tilde{\Psi}$, the array $\hat{\mathbf{d}}_{\hat{\Lambda}}^T$ would readily provide these quantities.

Overall, since in many cases one expects that the somewhat higher cost of evaluating a function in multiscale representation is by far offset by the sparseness of the representation, the central objective can be summarized as follows. If, for $\epsilon > 0$, the set $\Lambda \subset \nabla$ is needed to approximate the solution by $u_\Lambda = \mathbf{d}_\Lambda^T \Psi_\Lambda$ within a tolerance ϵ, find a possibly small set $\hat{\Lambda}$ and an approximation $\hat{\mathbf{d}}_{\hat{\Lambda}}^T \Xi_{\hat{\Lambda}}$ to $\mathcal{G}(\mathbf{d}_\Lambda^T \Psi_\Lambda)$ that is sufficiently accurate to preserve the overall precision of the solution scheme. Moreover, when ϵ tends to zero, so that the cardinality of $\Lambda = \Lambda_\epsilon$ increases, the *ideal* situation would be that the corresponding size of $\hat{\Lambda} = \hat{\Lambda}_\epsilon$ stays *proportional* to $\#\Lambda_\epsilon$ *uniformly* in ϵ. Likewise, the computational work needed to determine the approximation $\hat{\mathbf{d}}_{\hat{\Lambda}_\epsilon}^T \Xi_{\hat{\Lambda}_\epsilon}$ should also be of the order of $\#\Lambda_\epsilon$ (perhaps times a logarithmic term).

It seems that we are at present far from this goal, at least in the above strict asymptotic sense. Since this is currently a subject of intense research, the state of the art will probably change quickly in the near future. Giving a detailed account of the various existing approaches would certainly go beyond the scope of this paper. Nevertheless, sketching some ideas, at least, should be worthwhile.

A typical nonlinear term arising, for instance, in (2.16) and (2.17) is $u\frac{\partial}{\partial x}u$. Since products of functions can be obtained as differences of squares, it suffices to consider $f(u) = u^2$. The approach pursued by Beylkin and Keiser (1997) starts with the expansion

$$
\begin{aligned}
(Q_J v)^2 - (Q_0 v)^2 &= \sum_{j=0}^{J-1} \left((Q_{j+1} v)^2 - (Q_j v)^2 \right) \\
&= \sum_{j=0}^{J-1} \left(2(Q_j v)(R_j v) + (R_j v)^2 \right), \qquad (7.87)
\end{aligned}
$$

where we abbreviate $R_j := Q_{j+1} - Q_j$.

In fact, within a given tolerance, one has $v \approx Q_J v$ for J sufficiently large. This gives

$$
v^2 \approx (Q_0 v)^2 + \sum_{j=0}^{J-1} \left(2(Q_j v)(R_j v) + (R_j v)^2 \right). \qquad (7.88)
$$

The evaluation of $(Q_0 v)^2$ is inexpensive. The problem is that products in the summation will generally not belong to the same space as the factors. Since products correspond to convolutions in the Fourier domain, one can estimate the extent to which higher oscillations are introduced. To resolve them accurately enough one needs a higher level of resolution. However, estimating the spread in the Fourier domain, one can argue that, again within some tolerance, $(Q_j v)(R_j v)$ and $(R_j v)^2$ belong to $S(\Phi_{j+j_0})$ for some positive j_0. For any given tolerance a positive j_0 does indeed exist independent of j (Beylkin and Keiser 1997). Thus, by repeated application of the refinement matrices, one can determine the representation of $Q_j v$ and $R_j v$ in $S(\Phi_{j+j_0})$, for instance,

$$R_j v = (\mathbf{d}^{j+j_0})^{\mathrm{T}} \Phi_{j+j_0}.$$

If, in addition, the functions in Φ_j were *interpolatory*, that is, $\phi(\cdot - k) = \delta_{0,k}$, the assumption $v, f(v) \in S(\Phi_j)$ would give

$$f(v) = \sum_k f(c_k)\phi(\cdot - k), \quad v = \sum_k c_k \phi(\cdot - k). \tag{7.89}$$

In this case one would have $(R_j v) = ((\mathbf{d}^{j+j_0})^2)^{\mathrm{T}} \Phi_{j+j_0}$, where the square is to be understood componentwise. Thus the coefficients $\mathbf{c}^{j+j_0}(v^2)$ in $S(\Phi_{j+j_0})$ are computed (approximately) as

$$\mathbf{c}^{j+j_0}(v^2) = 2(\mathbf{c}^{j+j_0}(Q_j v))(\mathbf{d}^{j+j_0}(R_j v)) + (\mathbf{d}^{j+j_0}(R_j v))^2. \tag{7.90}$$

The justification for taking componentwise products of the coefficient sequences assumes the use of scaling functions whose shifts are orthogonal and which are *almost interpolatory* in the sense of (4.17). Recall that this is the case when the scaling function has sufficiently many vanishing moments (4.16); see Beylkin et al. (1991).

Note that at least one factor in each product on the right-hand side of (7.90) involves wavelet coefficients. These arrays are usually sparse, so that only significant products need be calculated. Accordingly, one should only compute those scaling function coefficients in $\mathbf{c}^{j+j_0}(Q_j v)$ affected by large wavelet coefficients. This requires suitably localized multiscale transformations. Depending on the context, the resulting (local) single-scale arrays $\mathbf{c}^{j+j_0}(v^2)$ can be used for point evaluations, or have to be transformed into wavelet representations. Therefore, the development of appropriate data structures is certainly an important issue.

A promising alternative is offered by an adaptation of Algorithm 2 from Section 7.4, by which an interpolating approximation is transformed into a wavelet representation. A different strategy is pursued in Joly et al. (1997); see also related work in Danchin (1997), Maday, Perrier and Ravel (1991). Joly et al. (1997) propose interpolating the reduced approximation $\tilde{u}^m$ at *all* the points x_λ corresponding to the entire basis. The values at these points are computed through a fast evaluation scheme (setting those coefficients to zero whose indices do not belong to Λ^m: see the algorithm in the previous section). The values of $(\tilde{u}^m)^2$ are then computed at each point and $(\tilde{u}^m)^2$ is interpolated with respect to the best basis. The overall cost is $\mathcal{O}(J2^J)$, which unfortunately exceeds the number of significant coefficients in Λ^m. If the best bases need not be changed, the collocation can be based on the centres x_λ, $\lambda \in \Lambda^{m+1}$ (Danchin 1997).

7.11. Stokes and Navier–Stokes equations

Of course, the above evolution equations can be viewed as simplified test cases for the next higher mathematical model, namely the Navier–Stokes equations for *incompressible* fluids, which, properly normalized, read

$$\begin{aligned} \frac{\partial u}{\partial t} - \nu \Delta u + (u \cdot \nabla) u + \nabla p &= f, \\ \nabla \cdot u &= 0; \end{aligned} \tag{7.91}$$

see, for example, Girault and Raviart (1986). Here $u : \mathbb{R}^n \times \mathbb{R}^+ \to \mathbb{R}^n$ represents the velocity of the fluid, and $p : \mathbb{R}^n \times \mathbb{R}^+ \to \mathbb{R}$ the pressure, and ν is a positive number called the *kinematic viscosity*. So when ν gets small the formally parabolic first system becomes hyperbolic. One usually looks for (u, p) satisfying (7.91) in some domain Ω subject to initial and boundary conditions

$$u(\cdot, 0) = u_0 \quad \text{in } \Omega, \quad u = 0 \text{ on } \partial\Omega \times [0, T], \tag{7.92}$$

and

$$\int_\Omega p(x,t)\,\mathrm{d}x = 0, \quad t \in (0,T), \tag{7.93}$$

since p is only determined up to a constant. In this section we outline some recent attempts that tackle the numerical solution of this kind of problem by means of wavelet discretizations.

Amongst the difficulties in treating (7.91) numerically is the constraint $\nabla \cdot u = 0$. One way to avoid this is to write (7.91) in the so-called *vorticity stream function formulation*

$$\begin{aligned} \frac{\partial u}{\partial t} + u \cdot \nabla \omega - \nu \Delta \omega &= 0, \\ \nabla \times u &= \omega, \end{aligned} \tag{7.94}$$

which is valid in this form for $n = 2$; see, for example, Quartapelle (1993).

For $\Omega = \mathbb{R}^2/\mathbb{Z}^2$ and periodic boundary conditions, Algorithm 2 (see Section 7.4) is applied to (7.94) by Fröhlich and Schneider (1996). Wavelet schemes based on the methodologies described in Section 7.7 are applied to (7.94), among other model problems, in Charton and Perrier (1996), accompanied by a complexity analysis which indicates that the complexity of the scheme is proportional to the dimension of the highest resolution. Although working with possibly highly lacunary sets Λ of wavelet indices, one has to employ at some stage the transformation between single and multiscale representation. So part of the principal efficiency is diminished again by this typical bottleneck. On the other hand, if this is the only place where the complexity of the full spaces enters, the constants appearing in the complexity estimates can be expected to be moderate.

To become competitive with the existing numerical methods, it is conceptually important to overcome the restriction to the vorticity stream function formulation, which is convenient only in the two-dimensional case. When working with the primitive variables u, p in (7.91), it is known that the constraint $\nabla \cdot u = 0$ imposes certain compatibility conditions on the trial spaces for velocity and pressure that are necessary for an asymptotically stable solution procedure Suitable families of such finite element spaces are known (Girault and Raviart 1986). However, for $n = 3$, they become quite involved when trying to raise the order of exactness. We will therefore briefly discuss what potential contributions of wavelet concepts in this regard can be expected.

Saddle point formulation
Suppose we fix trial spaces $V_h \subset (H_0^1(\Omega))^n$ and

$$M_h \subset L_{2,0}(\Omega) = \{g \in L_2(\Omega) : \int_\Omega g(x)\, dx = 0\}.$$

A semi-implicit discretization of (7.91) in conjunction with a weak formulation of the corresponding linear problem yields

$$\begin{aligned} \langle u_h^{m+1}, v_h\rangle + \Delta t\nu\langle \nabla u_h^{m+1}, \nabla v_h\rangle + \langle v_h, \nabla p\rangle &= \langle f - u_h^m \cdot \nabla u_h^m, v_h\rangle, \quad v_h \in V_h, \\ \langle \operatorname{div} u_h^{m+1}, \mu_h\rangle &= 0, \quad \mu_h \in M_h. \end{aligned} \tag{7.95}$$

One may question for which time steps and under which circumstances it is reasonable to use an explicit discretization of the transport term $u_h^m \cdot \nabla u_h^m$ and put it on the right-hand side. But for the time being we ignore this point and remark that (7.95) corresponds to the linear system of equations

$$\begin{pmatrix} \mathbf{A}_{h,\alpha} & \mathbf{B}_h^{\mathrm{T}} \\ \mathbf{B}_h & \mathbf{0} \end{pmatrix} \begin{pmatrix} \mathbf{u}_h \\ \mathbf{p}_h \end{pmatrix} = \begin{pmatrix} \mathbf{F}_h \\ \mathbf{0} \end{pmatrix}. \tag{7.96}$$

Here $\mathbf{A}_{h,\alpha}$ is the stiffness matrix of the operator $\mathcal{L} = I - \alpha\Delta$, $\alpha = \Delta t\nu$, and $\mathbf{B}_h^{\mathrm{T}}$ is the discretization of the gradient ∇ relative to the chosen bases in V_h, M_h. Recall from Section 2.2 (b) that the stationary Stokes problem (2.7) leads to an analogous system where $\mathbf{A}_{h,\alpha}$ is replaced by the discretization of $\mathcal{L} = \nu\Delta$. Both share the same operator $\mathbf{B}_h$, however.

To ensure that the discretizations are stable, that is, that the inverses of the discretized operators $\mathcal{L}_h$ are uniformly bounded, the pairs (V_h, M_h) have to satisfy the *Ladyšenskaya–Babuška–Brezzi condition* (LBB), which means that the *inf–sup condition* (2.13) holds uniformly in h

$$\inf_{\mu_h \in M_h} \sup_{v_h \in V_h} \frac{b(v_h, \mu_h)}{\|v_h\|_V \|\mu_h\|_M} \geq \beta > 0, \tag{7.97}$$

where as before $b(v, \mu) = \langle \text{div } v, \mu \rangle$. The following well-known fact characterizes the validity of (7.97); see Fortin (1977).

Proposition 7.9 Suppose (2.13) holds for the pairs (V, M) and some $\tilde{\beta} > 0$. Then the subspaces V_h, M_h satisfy (7.97) uniformly in h, if and only if there exist linear operators $Q_h : V \to V_h$ satisfying

$$\|Q_h v\|_V \lesssim \|v\|_V, \quad v \in V, \tag{7.98}$$

and

$$b(v - Q_h v, \mu_h) = 0, \quad v \in V, \quad \mu_h \in M_h. \tag{7.99}$$

While in the finite element context this observation is primarily of theoretical use, it does offer a constructive angle in the wavelet setting. We briefly sketch the approach of Dahmen et al. (1996*c*). In fact, (7.99) may be viewed as a biorthogonality condition with respect to $b(\cdot, \cdot)$. Again this is most conveniently explained first for the case $\Omega = \mathbb{R}^n$.

To describe this consider any dual pair of biorthogonal compactly supported generators $\phi, \tilde{\phi} \in L_2(\mathbb{R}^n)$ (see (4.19)). Assuming that $\phi \in H^{1+\varepsilon}(\mathbb{R})$, the procedure mentioned in Section 4.2 (see (4.34)) yields another dual pair $(\phi^-, \tilde{\phi}^+)$ of biorthogonal compactly supported generators. More generally, let us set

$$\psi_0^- := \phi^-, \quad \psi_1^- := \psi^-, \quad \tilde{\psi}_0^+ := \tilde{\phi}^+, \quad \tilde{\psi}_1^+ := \tilde{\psi}^+,$$

and likewise $\psi_i, \tilde{\psi}_i$, $i = 0, 1$. Here $\psi^-, \tilde{\psi}^+$ are the corresponding compactly supported new biorthogonal mother wavelets, which, by (4.36), arise from ψ and $\tilde{\psi}$ essentially by *differentiation* and *integration*, respectively. The trial spaces on $\mathbb{R}^n$ are again obtained by taking tensor products. In particular, the multivariate scaling functions and mother wavelets $\psi_e^{-,i}(x), \tilde{\psi}_e^{+,i}(x)$ are obtained for $e \in E = \{0, 1\}^n$, $i = 1, \ldots, n$, by replacing in (4.38) the ith factor by $\psi_{e_i}^-(x_i), \tilde{\psi}_{e_i}^+(x_i)$, respectively. Specifically, we set $\phi^{-,i} =: \psi_0^{-,i}$, $\tilde{\phi}^{+,i} =: \tilde{\psi}_0^{+,i}$.

Now let

$$V_j := \{v \in V : v_i \in S(\tilde{\Phi}_j^{+,i}), i = 1, \ldots, n\} \tag{7.100}$$

and

$$M_j := S(\Phi_j). \tag{7.101}$$

Thus we can also write

$$V_j = S(\tilde{\Phi}_{j,\otimes}^+), \quad \tilde{\Phi}_{j,\otimes}^+ := \tilde{\Phi}_j^{+,1} \times \cdots \times \tilde{\Phi}_j^{+,n}, \tag{7.102}$$

and the pairs $\tilde{\Phi}^{+}_{j,\otimes}, \Phi^{-}_{j,\otimes}$ are biorthogonal, where $\Phi^{-}_{j,\otimes}$ is defined analogously. Hence

$$Q_j v := \langle v, \Phi^{-}_{j,\otimes} \rangle \tilde{\Phi}^{+}_{j,\otimes} \tag{7.103}$$

are projectors from V onto V_j. Since

$$v - Q_j v = \sum_{l=j+1}^{\infty} (Q_{l+1} - Q_l) v \tag{7.104}$$

and

$$((Q_{l+1} - Q_l)v)_i = \langle v_i, \Psi_l^{-,i} \rangle \tilde{\Psi}_l^{+,i},$$

we conclude from (4.36) that

$$\frac{\partial}{\partial x_i} ((Q_{l+1} - Q_l)v)_i = -4 \langle v_i, \Psi_l^{-,i} \rangle \tilde{\Psi}_l. \tag{7.105}$$

But since biorthogonality of Ψ and $\tilde{\Psi}$ ensures that $\langle \Phi_j, \tilde{\Psi}_l \rangle = 0$, $l \geq j$, one immediately infers from (7.101), (7.104) and (7.105) that

$$\langle \text{div } (v - Q_j v), \mu_j \rangle = 0, \quad \text{for all } \mu_j \in M_j = S(\Phi_j). \tag{7.106}$$

Moreover, for $\tilde{\phi} \in H^{1+\varepsilon}$, Theorem 5.8 implies that the Q_j are uniformly bounded on V. Thus Proposition 7.9 applies and confirms that the spaces V_j, M_j defined by (7.100) and (7.101) do satisfy the LBB condition (7.97).

There is no difficulty in adapting this construction to the periodic case. It is perhaps more interesting to note that it can also be extended to $\Omega = \square = (0,1)^n$ and homogeneous boundary conditions (7.92). This is done in Dahmen et al. (1996*c*) by starting with a dual pair $\phi, \tilde{\phi}$ as above and constructing biorthogonal refinable bases $\Phi_j, \tilde{\Phi}_j$ adapted to $[0,1]$ as indicated in Section 4.4. The key is that the modified dual pair $\phi^{-}, \tilde{\phi}^{+}$ again gives rise to pairs of refinable biorthogonal bases $\Phi_j^{-}, \tilde{\Phi}_j^{+}$ where, however, we now have the inclusion $S(\tilde{\Phi}_j^{+}) \subset H_0^1(\square)$. Defining the collections $\Psi_j^{-,i}, \tilde{\Psi}_j^{+,i}$ in analogy to the previous construction, one can prove that one still has

$$S\left(\frac{\partial}{\partial x_i} \tilde{\Psi}_j^{+,i}\right) = S(\tilde{\Psi}_j). \tag{7.107}$$

Thus the same reasoning as before shows that for analogously defined projectors Q_j (7.106) is still valid. The validity of the LBB condition (7.97) also follows in this case from Proposition 7.9.

Solving the linear systems

Since the matrix in (7.96) is indefinite, the solution of (7.96) by iterative methods requires a bit more care; see Bramble and Pasciak (1988). The upshot of all the options is that, whenever a good preconditioner for the (positive definite) matrix $\mathbf{A}_{j,\alpha}$, as well as for the *Schur complement*

$\mathbf{K}_{j,\alpha} := \mathbf{B}_j \mathbf{A}_{j,\alpha}^{-1} \mathbf{B}_j^{\mathrm{T}}$, is available, one can combine both so as to obtain a correspondingly efficient iterative scheme for the treatment of (7.96). An example is mentioned below in Section 8.3 in a different context. In the case of the stationary Stokes problem, the block $\mathbf{A}_{j,\alpha}$ corresponds to a stiffness matrix for the Laplace operator Δ. Asymptotically optimal preconditioners for this component were discussed in Section 6. In this case the Schur complement is an operator of order zero and hence does not require any further preconditioning. In the time-dependent case, the roles are reversed: $\mathbf{A}_{j,\alpha}$ is a discretization of the Helmholtz operator $I - \alpha\Delta$, which, for small α, resembles the identity. Thus $\mathbf{A}_{j,\alpha}$ is already moderately well conditioned. Nevertheless, the robust wavelet-based preconditioners discussed in Section 6.6 would here cover the full range of possible values of α. Now, for small α the Schur complement tends more and more to a second-order operator, so that preconditioning becomes necessary; see Bramble and Pasciak (1994). Again, an asymptotically optimal preconditioner-based on the above wavelet bases, namely Algorithm 1 in Section 6.2, is proposed by Dahmen et al. (1996*c*). The concrete examples considered there are based on dual pairs $\phi, \tilde{\phi}$ where $\tilde{\phi}$ is chosen as a B-spline (see Section 4.4). All basis functions and wavelets for V_j and M_j have compact support. The construction allows one to realize any desired order of exactness for any spatial dimension. The numerical experiments in Dahmen et al. (1996*c*) for the linearized problem cover two- and three-dimensional examples and confirm the predicted asymptotic optimality, that is, iteration numbers are independent of the size of the problem.

Divergence-free wavelets

Instead of seeking pairs of trial spaces V_j, M_j satisfying the LBB conditions, one could try to find trial spaces V_j which satisfy the constraint div $v = 0$, $v \in V_j$ *weakly*, that is, $b(v, \mu) = 0$ for all $\mu \in M_j$, $v \in V_j$. This has been realized in the finite element context but corresponding constructions are rather involved, in particular for the 3D case. One could even go one step further and try to construct spaces $V_j^0 \subset V^0 := \{v \in (H_0^1(\Omega))^n : \text{div } v = 0\}$. Orthogonal *divergence-free* wavelets have been constructed by Battle and Federbusch. These wavelets have necessarily global support (Lemarié-Rieusset 1994) although they decay exponentially. A somewhat different line based on Section 4.2 has been pursued by Jouini (1992) and Lemarié-Rieusset (1992). Dispensing with orthogonality, one can construct *divergence-free biorthogonal wavelets* with *compact* support. This point was taken up by Urban (1995*a*, 1995*b*), where, in addition to tensor products, *genuinely multivariate* divergence-free wavelets are constructed.

Using such trial spaces, the weak formulation (7.95) reduces to

$$\langle u_j^{m+1}, v_j \rangle + \Delta t \nu \langle \nabla u_j^{m+1}, \nabla v_j \rangle = \langle F, v_j \rangle, \quad v_j \in V_j^0, \tag{7.108}$$

that is, to a Helmholtz problem on V^0. Here F collects the terms on the right-hand side of (7.95).

First, numerical experiences for the 3D case are reported by Urban (1995*c*, 1996). These experiments concern classical Galerkin schemes. On the other hand, the above reduction to the Helmholtz problem suggests the following interesting alternative. Because of the constraint $\operatorname{div} u = 0$, the fast vaguelette evaluation schemes based on orthonormal wavelets have been confined to the vorticity stream function formulation and thus to the bivariate case. Recalling that the vaguelette approach can be extended to biorthogonal wavelet bases, one can combine it with the above divergence-free wavelets, which also work in the three-dimensional case.

We conclude this section with some brief remarks on the construction. The key observation (Lemarié-Rieusset 1992, Urban 1995*a*, Urban 1995*b*) is the commutation property

$$\frac{\partial}{\partial x_i}\langle v, \tilde{\Phi}_j\rangle \Phi_j = \left\langle \frac{\partial}{\partial x_i} v, \tilde{\Phi}_j^{+,i}\right\rangle \Phi_j^{-,i}. \tag{7.109}$$

To make use of this fact for the construction of divergence-free wavelets, one has to iterate the modifications from Section 4.2 by setting

$$\phi^{-,(i,l)} = (\phi^{-,i})^{-,l},$$

and analogously for $\tilde{\phi}, \psi$ and $\tilde{\psi}$. For $i \in \{1,\ldots,n\}$ let $\mathcal{N}_i = \{1,\ldots,n\} \setminus \{i\}$ and define functions $\psi^{\nabla}_{e,\nu}$, $\nu \in \{1,\ldots,n\}$, by

$$\left(\psi^{\nabla}_{e,\nu}\right)_i = \begin{cases} 0, & i \notin \{\nu, i_e\}, \\ \psi_e^{-,\mathcal{N}_\nu}, & i = \nu, \\ -\frac{1}{4}\frac{\partial}{\partial x_\nu}\psi_e^{-,\mathcal{N}_\nu \setminus \{i_e\}}, & i = i_e, \end{cases}$$

where $i_e \in \{1,\ldots,n\}$ is any index such that $e_i = 1$. The collections

$$\Psi^{\nabla} := \{\psi^{\nabla}_{e,\nu,j,k} : e \in E_*, \nu \neq i_e, j \in \mathbb{Z}, k \in \mathbb{Z}^n\}$$

can then be shown to be a divergence-free wavelet basis (Urban 1995*a*, 1995*b*). Again, further analysis, implementations and numerical experiments can be found in Urban (1995*b*), (1995*c*) and (1996).

8. Extension to more general domains

Except for the extensions to wavelets on cubes (see Section 4.4 and the comments in the previous section) all approaches described so far rely in an essential way on the underlying stationary shift-invariant structure of the discretization. It has long been known in numerical analysis that, beyond mere asymptotic estimates, regular discretizations often support efficiency in many ways, reflected by superconvergence effects, for instance. Therefore it may pay in the end in many situations to exploit such advantages for the bulk of computation and treat boundary effects separately. The matrix capacitance method is an example of such a strategy which has been extensively studied in connection with finite difference schemes. Moreover, when dealing with problems where the geometry changes in time, such a concept may even be a necessity rather than an option. One possibility that may come to mind first is to enforce essential boundary conditions by means of penalty terms; see, for example, Glowinski, Pan, Wells and Zhou (1996) and Glowinski, Rieder, Wells and Zhou (1993). Aside from accuracy issues, a conceptual difficulty with this approach seems to be that it stiffens the problem significantly and thereby wastes previously gained advantages on the preconditioning side. A relatively simple alternative is to refine the spaces near the boundary. Incorporating additional basis functions on higher discretization levels whose supports are still inside the domain can compensate the loss of accuracy encountered otherwise. From a complexity point of view this works in the bivariate case but no longer for domains in $\mathbb{R}^3$ (Jaffard 1992, Oswald 1997). Therefore we will now concentrate on three alternative possibilities.

8.1. An extension technique

Throughout this subsection assume that $\mathcal{L}$ is a selfadjoint elliptic operator so that, for $a(\cdot,\cdot)$ defined by (2.5), with respect to natural boundary conditions, the problem in variational form is to find u in $H = H^t(\Omega)$ such that

$$a(u,v) = \langle f, v\rangle, \quad v \in H \tag{8.1}$$

for some $f \in H^*$ (*cf.* (6.9)). We briefly sketch some ideas from Oswald (1997) that fit into the multilevel Schwarz concepts described in Section 6.5.

The starting point is a nested sequence of finite element or spline spaces S_j, $j \in \mathbb{N}_0$, defined on regular meshes of types 1 or 2 (see Section 6.6). Now $\Omega \subset \mathbb{R}^n$ is supposed to be an arbitrary bounded domain with sufficiently regular boundary to admit the existence of extension operators E, which are bounded in an appropriate Sobolev scale. For instance, the validity of a *uniform cone condition* or Lipschitz boundaries would do (Johnen and Scherer 1977).

The first step is to construct collections $\Phi_{j,\Omega}$ consisting mainly of functions $\phi(2^j\cdot -k)$, $k \in \mathbb{Z}^n$, whose support does not intersect $\partial\Omega$, where ϕ is in this case a tensor product B-spline, say. In addition, one needs functions that are adapted to the boundary. Their restriction to Ω is supported in a margin of width $\sim 2^{-j}$ along the boundary. They consist of fixed linear combinations of $\phi(2^j\cdot -k)$ designed in such a way that the span of the entire collection Φ_j contains all polynomials up to some degree $d-1$ on Ω. This is similar to the ideas presented in Section 4.4 and to the recent developments in Cohen, Dahmen and DeVore (1995). However, in order to keep these boundary-near cluster functions as simple as possible, they are, in contrast to Cohen et al. (1995), *not* required to be refinable. Thus one generally has

$$S(\Phi_{j,\Omega}) \not\subset S(\Phi_{j+1,\Omega}). \tag{8.2}$$

However, along with $\Phi_{j,\Omega}$, a biorthogonal collection $\Xi_{j,\Omega}$ is constructed in such a way that the projectors

$$Q_j v := \langle v, \Xi_{j,\Omega} \rangle \Phi_{j,\Omega} \tag{8.3}$$

satisfy the direct estimates (5.7). Essential hypotheses are that the $\phi(\cdot - k)$ are *locally* linearly independent (that is, the vanishing of a linear combination on any neighbourhood implies that the coefficients of the overlapping translates $\phi(\cdot - k)$ are zero), as well as the availability of extension operators.

Due to the lack of nestedness (8.2), the techniques from Section 6.5 cannot yet be applied directly. To remedy this, Oswald (1997) shows how to construct another sequence of nested spaces $\hat{S}_j \subset H^t(\mathbb{R})$ spanned by suitably chosen B-splines (on all levels $l \le j$) which overlap Ω. In addition, appropriate restriction and extension operators

$$R_j : \hat{S}_j \to S(\Phi_{j,\Omega}), \quad E_j : S(\Phi_{j,\Omega}) \to \hat{S}_j,$$

are identified. In fact, $R_j v = \langle v, \Xi_{j,\Omega} \rangle \Phi_{j,\Omega}$. The E_j have the form

$$E_J v_J := \sum_{j=0}^{J} (P_j - P_{j-1}) v_J \in \hat{S}_J,$$

where the P_j are similar quasi-interpolant type operators as the Q_j in (8.3) above. In fact, $P_j v = \langle v_1 \Xi_j \rangle \Phi_j$, where the elements $\xi_{j,k} \in \Xi_j$ are either supported in Ω when $\operatorname{supp} \phi(2^j \cdot - k) \cap \Omega \ne \emptyset$, or zero otherwise. To establish suitable norm estimates for these operators, one needs certain additional requirements on the domain which, for instance, ensure that the margin of boundary affected basis functions has width $\sim 2^{-j}$ on level j. One can then prove that (Oswald 1997)

$$R_j E_j v_j = v_j, \quad v_j \in S(\Phi_{j,\Omega}),$$

and

$$\begin{aligned} a(R_j \hat{v}_j, R_j \hat{v}_j) &\lesssim \|\hat{v}_j\|^2_{H^t(\mathbb{R}^n)}, & \hat{v}_j \in \hat{S}_j, \\ \|E_j v_j\|^2_{H^t(\mathbb{R}^n)} &\lesssim a(v_j, v_j), & v_j \in S(\Phi_{j,\Omega}). \end{aligned} \tag{8.4}$$

One can then proceed as follows. Fix any $H^t(\mathbb{R}^n)$-elliptic form $\hat{a}(\cdot,\cdot)$ and determine a preconditioner $\hat{\mathcal{C}}_j$ on $\hat{S}_j$ for the operator $\hat{\mathcal{L}}_j$ defined by $\hat{a}(\hat{u}_j, \hat{v}_j) = \langle \hat{\mathcal{L}}_j \hat{u}_j, \hat{v}_j \rangle$, $\hat{u}_j, \hat{v}_j \in \hat{S}_j$, by the methods from Section 6.5. Then, defining

$$\mathcal{C}_j := R_j \hat{\mathcal{C}}_j R_j^* \tag{8.5}$$

and $\mathcal{L}_j$ by $\langle \mathcal{L}u_j, v_j\rangle_\Omega = a(u_j, v_j)$, $u_j, v_j \in S(\Phi_{j,\Omega})$, the operators $C_j\mathcal{L}_j$ satisfy

$$\kappa_2(C_j\mathcal{L}_j) \sim 1 \quad \text{if} \quad \kappa_2(\hat{C}_j\hat{\mathcal{L}}_j) \sim 1. \tag{8.6}$$

The proof relies on the fictitious space Theorem 6.9 (Nepomnyaschikh 1990); see Oswald (1997) for details. The treatment of essential boundary conditions and further extensions are also discussed in Oswald (1997).

In this form the scheme does *not* make explicit use of any wavelet basis or a corresponding exact representation of complement components. Hence it is tailored to the selfadjoint case but otherwise very flexible in connection with many standard discretizations.

8.2. Boundary value correction

Consider operators of the form (2.4), that is, $\mathcal{L} = -\text{div}(A(x)\nabla) + a(x)\mathcal{I}$. Suppose $\Omega \subset \mathbb{R}^n$ is a bounded domain. The following strategy for solving

$$\mathcal{L}u = f \quad \text{on } \Omega, \quad \mathcal{B}u|_{\partial\Omega} = g, \tag{8.7}$$

where $\mathcal{B}$ is some boundary value operator, has been proposed in Averbuch, Beylkin, Coifman and Israeli (1995). Without loss of generality one may assume that $\Omega \subset \square := (0,1)^n$.

(1) Determine a smooth extension f_{ext} and an operator $\mathcal{L}_{\text{ext}}$ of f and $\mathcal{L}$ respectively, from Ω to $\square$.

(2) Solve the problem

$$\mathcal{L}_{\text{ext}}u = f_{\text{ext}} \quad \text{on} \quad \square \tag{8.8}$$

with periodic boundary conditions.

(3) Given the solution u_{ext} of (8.8), solve the *homogeneous* problem

$$\mathcal{L}u = 0 \quad \text{on} \quad \Omega \tag{8.9}$$

subject to the boundary conditions

$$\mathcal{B}u|_{\partial\Omega} = g - \mathcal{B}u_{\text{ext}}|_{\partial\Omega}, \tag{8.10}$$

with the aid of a boundary integral method (see Section 2.2 (d)).

Averbuch et al. (1995) only address (8.8), arguing that efficient methods for (8.9), (8.10) are available. The rationale is that fast wavelet methods such as those described in Section 7 do a particularly efficient job on the bulk of the problem. In fact, in the periodic setting, the significant wavelet coefficients are indeed determined by the significant wavelet coefficients of the right-hand side in the following sense. Suppose that $\Lambda_{f,\varepsilon}$ is the subset of wavelet coefficients needed to represent f_{ext} on $\square$ with accuracy ε. Then the set $\Lambda_{u,\varepsilon}$ of coefficients needed to represent the solution u_{ext} with accuracy ε is contained in a certain *'neighbourhood'* of $\Lambda_{f,\varepsilon}$, that is, a somewhat larger set containing $\Lambda_{f,\varepsilon}$ where $\#\Lambda_{u,\varepsilon}$ is claimed to be proportional to $\#\Lambda_{f,\varepsilon}$. However, it is not apparent how this proportionality depends on ε and on the norm with respect to which accuracy is measured. Nevertheless, according to Theorem 6.2, a diagonally preconditioned conjugate gradient scheme constrained to the space $S(\Psi_{\Lambda_{u,\varepsilon}})$ would produce (perhaps combined with nested iteration) an approximate solution of accuracy ε at the expense of $\mathcal{O}(\#\Lambda_{u,\varepsilon})$ operations. For higher dimensions, in particular, this seems very tempting, since a high degree of adaptivity can be obtained without worrying about the substantial complications caused by mesh refinement strategies in conventional finite-difference or finite-element schemes.

On the other hand, there are still many points that need to be carefully addressed.

(1) If the boundary $\partial\Omega$ is fairly regular, a standard multilevel finite element scheme, at least in the 2D case, combined with the existing adaptive refinement schemes (see, for instance, Bornemann, Erdmann and Kornhuber (1996), Bank and Weiser (1985)) applied directly to the problem on Ω would realize at least the same favourable complexity.

(2) If the boundary has very little regularity, it is not clear how to properly balance the regularity of the extension to avoid introducing artificial singularities, and how to realize the extension numerically.

(3) For problems in $\mathbb{R}^n$ with $n > 2$, and nonconstant diffusion matrix $A(x)$ in (2.4), the treatment of the boundary integral equation arising from (8.9) and (8.10) may no longer be so trivial, let alone the extension problem.

Nevertheless, this approach offers a methodology for separating the bulk of computation in the highest spatial dimension from the boundary treatment.

8.3. Lagrange multipliers

The following alternative is in principle by no means new, but has been to some extent revived by the development of wavelet schemes; see, for example, Babuška (1973) and Brezzi and Fortin (1991). The idea of appending essential boundary conditions by means of *Lagrange multipliers* has been taken up again and analysed from the point of view of multilevel schemes, in Kunoth (1994) and Kunoth (1995). Suppose that $\hat{\Omega}$ is a cube containing Ω, let $a(u,v) = \langle \mathcal{L}_{\text{ext}} u, v\rangle_{\hat{\Omega}}$ and $M = \left(H^{s-\beta}(\partial\Omega)\right)^*$, when $\mathcal{B}$ maps $H^s(\Omega)$ onto $H^{s-\beta}(\partial\Omega)$. Choosing $H = H^s(\hat{\Omega})$ or $H = H_0^s(\hat{\Omega})$ or the subspace $H_p^s(\hat{\Omega})$ consisting of periodic functions in $H^s(\hat{\Omega})$ and defining

$$b(v,\mu) := \langle v, \mu\rangle_{\partial\Omega}, \tag{8.11}$$

the corresponding weak formulation of (8.7) requires finding $(u,p) \in H \times M$ such that

$$\begin{aligned} a(u,v) + b(v,p) &= \langle f, v\rangle_{\hat\Omega}, \quad v \in H, \\ b(u,\mu) &= g, \qquad\quad \mu \in M. \end{aligned} \tag{8.12}$$

The solution (u,p) of (8.12) solves the *saddle-point problem*

$$\inf_{v\in H}\sup_{\mu\in M}\left\{\frac{1}{2}a(v,v) + b(v,\mu) - \langle f, v\rangle_{\hat\Omega} - b(g,\mu)\right\} \tag{8.13}$$

(recall (7.91) and (7.95)). For general conditions under which (8.12) and (8.13) are equivalent see Brezzi and Fortin (1991). It is also known that, for instance for $\mathcal{L} = -\Delta + a$ and $\mathcal{B} = \mathcal{I}$, the Lagrange multiplier p in the solution of (8.12) agrees with $\frac{\partial u}{\partial \nu}$ on $\partial\Omega$ where $\partial\nu$ denotes the derivative in the direction of the outward normal of $\partial\Omega$.

Let us again denote by

$$\begin{pmatrix} \mathcal{A}_h & \mathcal{B}_h^* \\ \mathcal{B}_h & 0 \end{pmatrix}\begin{pmatrix} u \\ p \end{pmatrix} = \begin{pmatrix} f \\ g \end{pmatrix} \tag{8.14}$$

the operator equation corresponding to (8.12) projected on $S_h \times M_h \subseteq H \times M$. Recall that solving (8.14) (and hence solving (8.7) approximately for the above choice of $a(\cdot,\cdot)$ and $b(\cdot,\cdot)$), requires addressing the following two issues.

- Ensure that (S_h, M_h) satisfies the corresponding LBB condition.
- Find an efficient iteration scheme coping with the fact that the matrices in (8.14) are indefinite.

The first issue depends on the particular situation at hand; see Bramble (1981) and Glowinski et al. (1996). Following Bramble and Pasciak (1988), Kunoth (1994) and Kunoth (1995), the second task can be tackled for instance as follows.

Suppose that the selfadjoint positive definite operator $\mathcal{C}_h$ is a preconditioner for $\mathcal{A}_h$ satisfying

$$\langle \mathcal{C}_h^{-1}v, v\rangle \sim \langle \mathcal{A}_h v, v\rangle, \quad \langle(\mathcal{A}_h - \mathcal{C}_h)v, v\rangle \le \eta\langle \mathcal{A}_h v, v\rangle, \quad v \in S_h. \tag{8.15}$$

Moreover, assume that $\mathcal{K}_h$ is a preconditioner for the Schur complement, that is,

$$\langle \mathcal{K}_h^{-1}\mu, \mu\rangle_{\partial\Omega} \sim \langle \mathcal{B}_h\mathcal{A}_h^{-1}\mathcal{B}_h^*\mu, \mu\rangle_{\partial\Omega}, \quad \mu \in M_h. \tag{8.16}$$

According to Bramble and Pasciak (1988) (see also Kunoth (1995)), one can use the fact that (8.14) is equivalent to

$$\mathcal{M}_h\begin{pmatrix} u \\ p \end{pmatrix} := \begin{pmatrix} \mathcal{C}_h\mathcal{A}_h & \mathcal{C}_h\mathcal{B}_h^* \\ \mathcal{B}_h(\mathcal{C}_h\mathcal{A}_h - \mathcal{I}) & \mathcal{B}_h\mathcal{C}_h\mathcal{B}_h^* \end{pmatrix}\begin{pmatrix} u \\ p \end{pmatrix} = \begin{pmatrix} \mathcal{C}_h f \\ \mathcal{B}_h\mathcal{C}_h f - g \end{pmatrix}, \tag{8.17}$$

and then show that, under the assumption (8.15), $\mathcal{M}_h$ is positive definite relative to the inner product

$$\left[\begin{pmatrix} v \\ \mu \end{pmatrix}, \begin{pmatrix} w \\ \nu \end{pmatrix}\right] := \left\langle (\mathcal{A}_h - \mathcal{C}_h^{-1})v, w \right\rangle_{\hat{\Omega}} + \langle \mu, \nu \rangle_{\partial\Omega}.$$

Moreover, when in addition (8.16) holds, one can verify that

$$\kappa_2\left(\mathcal{G}_h^{1/2}\mathcal{M}_h\mathcal{G}_h^{1/2}\right) \sim 1, \quad h \to 0, \tag{8.18}$$

where

$$\mathcal{G}_h := \begin{pmatrix} \mathcal{I} & 0 \\ 0 & \mathcal{K}_h \end{pmatrix}.$$

Kunoth (1994, 1995) has shown how to construct preconditioners $\mathcal{C}_h$ and $\mathcal{K}_h$ based on multilevel decompositions of appropriate trial spaces $S_j = S_{h_j}$, $M_j = M_{h_j}$, $j \in \mathbb{N}_0$. For $\mathcal{C}_h$ one could use a multilevel Schwarz scheme or a wavelet-based preconditioner, as detailed in Sections 6.2 (Algorithm 1), 6.4 and 6.5. The operator $\mathcal{B}_h\mathcal{A}_h^{-1}\mathcal{B}_h^*$ takes $\left(H^{s-\beta}(\partial\Omega)\right)^* = H^{\beta-s}(\partial\Omega)$ into $H^{s-\beta}(\partial\Omega)$ and thus has typically *negative* order and BPX or Schwarz schemes do not apply directly. The following strategy is suggested by the results in Kunoth (1995). Let $\Gamma = \partial\Omega$ and suppose that Ψ^Γ, $\tilde{\Psi}^\Gamma$ are biorthogonal Riesz bases for $L_2(\Gamma)$ with corresponding single-scale bases Φ_j^Γ, $\tilde{\Phi}_j^\Gamma$. Let

$$Q_j^\Gamma \mu = \langle \mu, \tilde{\Phi}_j^\Gamma \rangle_\Gamma \Phi_j^\Gamma,$$

and define $\mathcal{B}_j$ by

$$\langle \mathcal{B}_j v, \mu \rangle_\Gamma = \langle Q_j^\Gamma \mathcal{B} v, \mu \rangle_\Gamma, \quad v \in S_{j+j_0}, \quad \mu \in S(\tilde{\Phi}_j^\Gamma),$$

that is, $M_j = S(\tilde{\Phi}_j^\Gamma)$. Here the choice of $j_0 \in \mathbb{Z}$ leaves some flexibility for satisfying the LBB condition. One can then realize an asymptotically optimal preconditioner $\mathcal{K}_j = \mathcal{K}_{h_j}$ for the Schur complement $\mathcal{B}_j\mathcal{A}_j^{-1}\mathcal{B}_j$ with the aid of the change of bases scheme (Algorithm 1) from Section 6.2. For further details see Kunoth (1995).

The tempting aspect of this strategy is that it has the potential to be extended to a wider class of problems. For instance, using divergence-free wavelets for discretizing the Stokes problem on a cube or torus and appending boundary conditions by Lagrange multipliers leads to the type of saddle-point problem considered above with $\mathcal{L} = -\Delta$. Moreover, in view of (8.17), one need not deal with the exact Schur complement but retain sparse representations of the zero-order operator on Γ.

On the other hand, one needs suitable multiscale bases Ψ^Γ, $\tilde{\Psi}^\Gamma$ on Γ. When $n = 2$, Γ is a curve and one can readily resort, at least for sufficiently smooth curves, to periodic univariate wavelets, or to composite wavelet bases of the type considered in Section 4.4. When $n > 2$, things are more complicated. However, due to the typically low order of the Schur complement, the bases $\Psi^\Gamma, \tilde{\Psi}^\Gamma$ generally need not be very regular. A more detailed discussion of constructing wavelet bases on manifolds such as closed surfaces, particularly in the context of boundary integral equations, is given in Sections 9 and 10.

9. Pseudo-differential and boundary integral equations

So far the discussion has been essentially confined to differential operators. Of course, the appearance of integral operators is also implicit in the vaguelette concept. Moreover, they occur explicitly in Section 8.2 as a solution component for treating partial differential equations (see also Section 2.2 (d)). In addition to the issue of preconditioning, the numerical treatment of integral operators or, more generally, of operators with global Schwartz kernel faces a further serious obstruction: conventional discretizations lead to *dense* matrices so that both assembling these matrices and solving the linear systems quickly become prohibitively expensive for realistic problems. In fact, direct solvers require the order of N^3 operations when N denotes the problem size, and each matrix vector multiplication in an iterative method is of the order N^2. A conceptual remedy is to perform the matrix vector multiplications only *approximately* within some tolerance. In many cases this indeed allows one to reduce the computational complexity to *almost linear* growth, if the analytical background of the problem is properly exploited. Examples of this type are *panel clustering* (Hackbusch and Nowak 1984, Hackbusch and Sauter 1993, Sauter 1992) or the closely related multipole expansions (Carrier, Greengard and Rokhlin 1988, Greengard and Rokhlin 1987, Rokhlin 1985). A similar finite difference-based approach is presented in Brandt and Venner (preprint) and Brandt and Lubrecht (1990).

Yet another direction has been initiated by the startling paper by Beylkin et al. (1991). As announced in Section 1.5 (c), the representation of certain integral operators in wavelet coordinates is nearly sparse (see Section 1.3). Roughly speaking, the idea is to replace the exact stiffness matrix $\mathbf{A}_J := \langle \mathcal{L}\Psi_J, \Psi_J \rangle^{\mathrm{T}}$ by a *compressed* matrix $\mathbf{A}_J^c$ arising from $\mathbf{A}_J$ by setting all entries below a given threshold to zero. Beylkin et al. (1991) have shown that the product $\mathbf{A}_J^c\mathbf{d}$, $\mathbf{d} \in \mathbb{R}^{N_J}$, $N_J = \dim S(\Phi_J)$, is still within accuracy ε from $\mathbf{A}_J\mathbf{d}$ if only the order of $N_J \log N_J$ entries in $\mathbf{A}_J^c$ are different from zero. This result has since started a number of investigations centred upon the following questions.

(1) How to deal with operators of *nonzero order*?
(2) What can be said about other schemes such as *collocation*?
(3) How to deal with *nonperiodic problems*, specifically with boundary integral equations on closed surfaces?
(4) What can be said about asymptotics, that is, how sparse can $\mathbf{A}_J^c$

be made while still guaranteeing that the solution exhibits the same asymptotic accuracy as the solution of the uncompressed system?

(i) Beylkin et al. (1991) consider a zero-order operator, so that no preconditioning is necessary. Important applications (see Section 2.2 (d)) involve operators of order different from zero, such as the single-layer potential operator. Aside from a possible need of regularization in such cases, preconditioning again becomes necessary. For operators of order minus one, a preconditioner-based on multigrid techniques was developed by Bramble, Leyk and Pasciak (1994), by introducing a suitable discrete norm for H^{-1}. One then obtains a fast method by combining this concept with any of the above mentioned fast matrix-vector multiplication schemes. It seems that, in the context of wavelet-based schemes, the preconditioning of operators of *any* order, as explained in Section 6.2, and its effect on matrix compression were first solved by Dahmen, Prößdorf and Schneider (1993*b*) and Dahmen et al. (1994*b*).

(ii) While Beylkin et al. (1991) only consider a 'classical Galerkin scheme', in practice collocation is often preferred to Galerkin schemes as a discretization tool for integral operators, because it reduces the dimension of numerical integration. Comparatively little is known about stability criteria for collocation schemes in that context. The class of classical periodic pseudo-differential operators

$$(\mathcal{L}u)(x) = \sum_{k\in\mathbb{Z}^n} \sigma(x,k)\hat{u}(k)e^{2\pi i k\cdot x}, \tag{9.1}$$

where $\hat{u}(k)$ are the Fourier coefficients of u, was chosen by Dahmen et al. (1994*c*) as a model setting for studying the following issues: stability criteria for various types of elliptic pseudo-differential operators and various types of generalized Petrov–Galerkin discretization in a multiresolution context, as well as an asymptotic analysis of fast solution by compression techniques. The schemes considered there are of the type (6.3) with projectors of the form

$$P_j v = \sum_{k\in\mathbb{Z}^n/2^j\mathbb{Z}^n} \eta_{j,k}(v)\phi_{j,k},$$

where Φ_j are periodized refinable single-scale bases and

$$\eta_{j,k}(v) := 2^{-nj/2}\eta\Big(v(2^{-j}(\cdot + k))\Big)$$

arise from some fixed functional η defined on $S(\mathcal{L}\Phi_j)$. Thus $\eta = \delta(\cdot - \alpha)$ corresponds to collocation while $\eta = \phi$ covers the Galerkin scheme. The operators $\mathcal{L}$ under consideration are assumed to be *elliptic* in the sense that the principal part $\sigma_0(x,\xi)$ of their symbol $\sigma(x,\xi)$ is *coercive*, that is, for some $\delta > 0$ one has

$$\operatorname{Re}\sigma_0(x,\xi) \gtrsim |\xi|^{2|t|}, \text{ for } |\xi| > \delta, \quad x \in \mathbb{R}^n/\mathbb{Z}^n. \tag{9.2}$$

The main result of Dahmen et al. (1994*c*) can be stated as follows. For any fixed $y \in \mathbb{R}^n/\mathbb{Z}^n$, let $\sigma_y(k) := \sigma_0(y,k)$ induce the constant coefficient operator $\mathcal{L}_y$. Let

$$\alpha_\eta(w,y) := \sum_{k\in\mathbb{Z}^n} \sigma_y(w+2\pi k)\hat{\phi}(w+2\pi k)\overline{\hat{\eta}(w+2\pi k)}$$

denote the *numerical symbol* relative to $\mathcal{L}_y$, which under suitable assumptions on ϕ is well defined. Here $\hat{\eta}$ is the Fourier transform of η in the distributional sense. The numerical symbol is called *elliptic* (see Wendland (1987)) if

$$|\alpha_\eta(w,y)| \gtrsim |w|^r, \quad \text{for } w \in \left[-\tfrac{1}{2},\tfrac{1}{2}\right]^n, \quad \text{and } y \in \mathbb{R}^n/\mathbb{Z}^n. \tag{9.3}$$

A freezing coefficient technique based on superconvergence results in connection with the so-called *discrete commutator property* is used by Dahmen et al. (1994*c*) to show that the generalized Petrov–Galerkin scheme (6.3) is (s,r)-*stable*, in the sense of (6.4), if and only if the scheme is (s,r)-stable for $\mathcal{L}_y$ for all $y \in \mathbb{R}^n/\mathbb{Z}^n$. This in turn finally yields that, under the above assumptions on $\mathcal{L}$, the scheme (6.3) is (s,r)-stable if and only if the numerical symbol α_η is elliptic in the sense of (9.3). Condition (9.3) naturally extends the stability condition (4.11), which refers to $\mathcal{L} = \mathcal{I}$.

This criterion is useful for verifying stability of collocation schemes where $\hat{\eta} = 1$ (compare with (7.28)); see, for instance, Dahmen et al. (1996*a*).

(iii) The above-mentioned results on periodic pseudo-differential equations immediately apply to boundary integral equations for two-dimensional domains with smooth boundary, which, via a smooth reparametrization, can be identified with the circle. Univariate periodic wavelets provide all the necessary tools for this case. Important contributions for Galerkin schemes are given by von Petersdorff and Schwab (1997*b*).

However, when the boundary integral equation lives on a surface of higher dimension, being able to treat periodic problems is ultimately *not* sufficient any longer. This puts conceptually new demands on the tools, that is, on the construction of appropriate wavelets. This issue will be addressed later in more detail.

To see how well the analysis of the periodic case predicts the right behaviour in more realistic situations, a multiscale collocation method for the double-layer potential equation on two-dimensional polyhedral surfaces in $\mathbb{R}^3$ was developed and tested by Dahmen et al. (1994*a*). The multiresolution spaces consist of continuous piecewise linear finite elements relative to uniform triangulations of the (triangular) faces of the polyhedron. The functions indicated in Figure 2 were used as wavelets. Since in this construction the order of vanishing moments decreases near face edges, this approach was still provisional. Although, in contrast to the torus, the surfaces considered by Dahmen et al. (1994*a*) were no longer smooth, by and large the same compression and convergence behaviour could be observed as predicted by the analysis of the idealized situation. However, full practical use of these findings requires computing the compressed matrices $\mathbf{A}_J^c$ at costs which are essentially of the order of nonvanishing entries. Meanwhile substantial progress has been made in this regard, which we sketch later; see Dahmen and Schneider (1997*b*) and von Petersdorff and Schwab (1997*a*).

(iv) While in Beylkin et al. (1991) the compression rate referred to a *fixed* accuracy ε, an asymptotic analysis was carried out in Dahmen et al. (1993*b*), Dahmen et al. (1994*b*), von Petersdorff and Schwab (1997*b*) and von Petersdorff, Schneider and Schwab (1997). In particular, Schneider (1995) has shown that, under certain assumptions on the domain and on the wavelet bases, $\mathbf{A}_J^c$ can be compressed to $\mathcal{O}(N_J)$ nonvanishing entries while still realizing the asymptotic accuracy of the unperturbed scheme. Recently, significant progress on a practicable realization in terms of a nearby asymptotically optimal fully discrete scheme for zero-order operators was accomplished by von Petersdorff and Schwab (1997*a*).

In summary, the practical success of such concepts requires handling the following central tasks.

(a) Construct appropriate wavelet bases $\Psi, \tilde{\Psi}$ defined on a manifold Γ such that the underlying operator can be preconditioned well and efficiently compressed.
(b) Develop a scheme for computing the compressed operator at an expense that stays proportional to the number of nonvanishing entries.
(c) Combine these techniques with adaptive space refinement strategies, that is, with identifying sets $\Lambda \subset \nabla$ adapted to the problem at hand. By Remark 6.3, these together would provide an asymptotically optimal scheme.

In principle, all three goals are in sight. We will first sketch some basic ingredients of several contributions to (a) and (b).

9.1. Geometry considerations

The numerical treatment of realistic boundary integral equations obviously requires more than periodized wavelets. A natural starting point is the representation of the boundary manifold $\Gamma = \partial\Omega$. In the context of boundary integral equations, one is primarily interested in spatial dimensions $n = 1, 2$ of Γ. However, the same ideas also apply in principle to other manifolds, such as bounded domains in $\mathbb{R}^3$, so it is worth keeping n arbitrary at this point.

Concrete *free-form surface* representations are generated by CAD packages. There, a surface Γ is usually *parametrically* defined, that is, Γ is a disjoint union of (open) *patches* Γ_i,

$$\Gamma = \overline{\bigcup_{i=1}^{M} \Gamma_i}, \quad \Gamma_i \cap \Gamma_l = \emptyset, \quad i \neq l. \tag{9.4}$$

The global regularity of Γ is usually described with the aid of an *atlas* $\{(\hat{\Gamma}_i, \kappa_i)\}_{i=1}^{M}$. This consists of a covering $\Gamma = \bigcup_{i=1}^{M} \hat{\Gamma}_i$ and associated *regular* mappings

$$\kappa_i : \widehat{\square}_i \to \hat{\Gamma}_i, \quad \widehat{\square}_i \subset \mathbb{R}^n, \quad i = 1, \ldots, M,$$

that is, κ_i and κ_i^{-1} are smooth mappings so that, in particular, the corresponding functional determinant $|\partial\kappa_i(x)|$ does not vanish on $\hat{\Gamma}_i$. Moreover, for $\square \subset \bigcap_{i=1}^{M} \widehat{\square}_i$, one has

$$\kappa_i|_{\square} = \Gamma_i, \quad i = 1, \ldots, M. \tag{9.5}$$

The set Γ is called a C^m manifold, respectively Lipschitz manifold, if the mappings are C^m, respectively Lipschitz. In practice, one does not work with coverings. Instead the global smoothness requirements are then translated into relations between the control parameters in the mappings κ_i corresponding to adjacent patches. Again, these considerations also apply to domain decompositions of domains in $\mathbb{R}^3$.

9.2. Function spaces on Γ

The discussion in Section 6 has made it very clear that the qualification of a wavelet basis Ψ for a given problem is closely related to the relevant function spaces. Thus one has to understand such function spaces defined on Γ. Denoting by $\mathrm{d}s$ the surface measure on Γ, the space $L_2(\Gamma)$ of square integrable functions on Γ is a Hilbert space with respect to the inner product

$$\langle u, v\rangle_\Gamma = \int_\Gamma u(x)\overline{v(x)}\,\mathrm{d}s_x. \tag{9.6}$$

With the aid of the above atlas, one can also define Sobolev spaces $H^s(\Gamma)$ on Γ. On the other hand, it would be extremely useful to relate the function space structure back to the parameter domain $\square$. Locally this is possible. Since the κ_i are smooth, it is easy to see that for $s \geq 0$

$$H^s(\Gamma_i) = \{v \in L_2(\Gamma_i) : v \circ \kappa_i \in H^s(\square)\}. \tag{9.7}$$

Moreover,

$$(u, v) := \sum_{i=1}^{M} (u, v)_i, \tag{9.8}$$

where

$$(u, v)_i = \int_{\square} (u \circ \kappa_i)(x)\overline{(v \circ \kappa_i)(x)}\, \mathrm{d}x, \tag{9.9}$$

defines an inner product on Γ and

$$(v, v) \sim \langle v, v \rangle_\Gamma, \quad v \in L_2(\Gamma). \tag{9.10}$$

Equations (9.7) and (9.8) suggest the norms

$$\|v\|_s^2 := \sum_{i=1}^{M} \|v\|_{H^s(\Gamma_i)}^2, \quad 0 \le s, \tag{9.11}$$

for the space $\prod_{i=1}^M H^s(\Gamma_i)$.

Since the properties of the operator equation are usually specified in terms of a global topology on Γ, such as the one induced by spaces $H^s(\Gamma)$, say, it is important to know how these spaces relate to each other. While $H^s(\Gamma)$ is generally a closed *subspace* of $\prod_{i=1}^M H^s(\Gamma_i)$ with respect to the norm (9.11), one even has

$$H^s(\Gamma) \cong \prod_{i=1}^{M} H^s(\Gamma_i), \quad -\frac{1}{2} < s < \frac{1}{2}, \tag{9.12}$$

that is, both spaces agree as sets and the norms are equivalent. However, there is, of course, the restriction $s < 1/2$, which will be seen later to be an unfortunate obstruction.

9.3. Multi-wavelets

The above geometric setting suggests the following natural concept; see Alpert (1993), Alpert, Beylkin, Coifman and Rokhlin (1993), von Petersdorff and Schwab (1997*a*) and von Petersdorff et al. (1997). Let Π_d be the set of polynomials of total degree less than d on $\mathbb{R}^n$ and let $P := \{P_\nu : |\nu| = \nu_1 + \ldots + \nu_n < d\}$ be an orthonormal basis of Π_d on $\square$, which can be generated by the Gram–Schmidt process from the monomial basis. For simplicity, let us now write $\square = (0,1)^n$. A similar variant of what follows can be developed for the standard simplex (and even more generally for *invariant sets* (Micchelli and Xu 1994)) as well. Let $\square$ be divided into 2^{jn} congruent cubes

$$\square_{j,\eta} := 2^{-j}(\eta + \square), \quad \eta \in \{0, \ldots, 2^j - 1\}^n =: E_j,$$

and let $\tau_{j,\eta}(x) := 2^j x - \eta$ denote the affine transformation that takes $\square_{j,\eta}$ onto $\square$. Now one easily generates spaces of (discontinuous) piecewise polynomials of degree $< d$ relative to the partition of $\square$ into $\square_{j,\eta}$, $\eta \in E_j$. Transporting these spaces to the patches Γ_i then creates 'piecewise polynomials' defined on Γ. Formally, this can be described as follows. Let

$$\Delta_j := \{k = (i, \eta, \nu) : \eta \in E_j, \nu \in \mathbb{Z}^n_+, |\nu| < d, i = 1, \ldots, M\},$$

and set for $k = (i, \eta, \nu)$

$$\begin{aligned} \phi_{j,k}(x) &:= \begin{cases} 2^{jn/2}(\varphi_{j,\eta,\nu} \circ \kappa_i^{-1})(x), & x \in \bar{\Gamma}_i, \\ 0, & x \notin \Gamma_i, \end{cases} \\ \varphi_{j,\eta,\nu}(y) &= \begin{cases} (p_\nu \circ \tau_{j,\eta})(y), & y \in \square_{j,\eta}, \\ 0, & \text{else}. \end{cases} \end{aligned} \tag{9.13}$$

Obviously the spaces $S(\Phi_j)$ are nested and their union is dense in $L_2(\Gamma)$.

The construction of orthogonal complements between adjacent trial spaces works as follows. Again using Gram–Schmidt, one can construct an orthogonal basis $\{r_l : l = 1, \ldots, (2^n - 1)\binom{n+d-1}{d-1}\}$ of the local space $S(P)$ in $S(\{\varphi_{1,\eta,\nu} : \eta \in E_1, |\nu| < d\})$. The complement basis Ψ_j in $S(\Phi_{j+1})$ is then obtained by (9.13) with p_ν replaced by r_l. The collection

$$\begin{aligned} \Psi &= \Phi_0 \cup \bigcup_{j=0}^{\infty} \Psi_j \\ &= \Phi_0 \cup \left\{\psi_\lambda : \lambda = (i, j, l), 1 \le i \le M, j \ge 0, 1 \le l \le (2^n - 1)\binom{n+d-1}{d-1}\right\}, \end{aligned}$$

is by construction orthonormal with respect to the inner product $(\cdot,\cdot)$ defined by (9.8), (9.9). Thus every $v \in L_2(\Gamma)$ has a unique expansion

$$v = (v, \Psi)\Psi, \quad \|v\|_{L_2(\Gamma)} \sim \|(v, \Psi)\|_{\ell_2}. \tag{9.14}$$

Moreover, for every $i \in \{1, \ldots, M\}$ and any polynomial $p \in \Pi_d$ the *generalized moment conditions* hold

$$(p \circ \kappa_i^{-1}, \psi) = 0, \quad \psi \in \Psi \setminus \Phi_0. \tag{9.15}$$

This relation implies that for any smooth function f on Γ one has for $\lambda = (i, j, l)$, $|\lambda| = j$,

$$|\langle f, \psi_\lambda \rangle_\Gamma| \lesssim 2^{-|\lambda|(d+\frac{n}{2})} \|f\|_{W^{\infty,d}(\operatorname{supp}\psi_\lambda)}. \tag{9.16}$$

In fact, setting $w_i(y) = |\partial\kappa_i(y)|$, $g(y) := w_i(y)(f \circ \kappa_i)(y)$, yields

$$\langle f, \psi_\lambda \rangle_\Gamma = \int_\square g(y)(\psi_\lambda \circ \kappa_i)(y)\, dy.$$

Since, by construction, $\int_\square p(x)(\psi_\lambda \circ \kappa_i)(x)\, dx = 0$, $p \in \Pi_d$, (9.16) follows from Taylor expansion of g around any point in $\operatorname{supp}\psi_\lambda \circ \kappa_i$, and the fact that $w_i(y)$ and κ_i are smooth.

Using Proposition 5.1, it is also standard to confirm the direct estimates (Dahmen and Schneider 1996)

$$\|v - (v, \Phi_J)\Phi_J\|_{L_2(\Gamma)} \lesssim 2^{-Jd}\|v\|_d, \tag{9.17}$$

provided that Γ is smooth enough to admit the definition of $H^d(\Gamma)$ in the above sense.

This approach can be extended to other parameter domains exhibiting a certain selfsimilarity; see Micchelli and Xu (1994). The following points should be kept in mind, however.

(i) Note that the order of moment conditions in (9.15) equals the order of accuracy in (9.17). It will be seen later that asymptotic optimality sometimes demands that the order of moment conditions is *higher* than the order of exactness.

(ii) The multi-wavelet basis is very flexible and relatively easy to implement. On the other hand, due to the discontinuous character of the trial functions, $\dim S(\Phi_j) = N2^{nj}\binom{n+d-1}{d-1}$. This effect could be damped by forming *composite wavelet* bases according to the following recipe (Dahmen and Schneider 1996).

- Construct biorthogonal wavelet bases $\Psi^\square, \tilde{\Psi}^\square$ on the parameter domain $\square$ by taking tensor products of the bases discussed in Section 4.4.
- Lift these bases with the aid of the parametric mappings κ_i as above to composite biorthogonal bases $\Psi, \tilde{\Psi}$ with respect to the inner product $(\cdot, \cdot)$ (9.8).

This alternative has the following attractive features.

- Although the same order d of exactness (9.17) is retained, one has $\dim S(\Phi_j) \leq N2^{nj}$, which is the fraction $\binom{n+d-1}{d-1}^{-1}$ of the dimension of the corresponding discontinuous space. Moreover, on each patch the trial functions are still $d-2$ times differentiable.
- The order $\tilde{d}$ of vanishing moments can be chosen as $\tilde{d} \geq d$ independently of the order d of accuracy, which will be seen to support compression.
- The Riesz basis property of the biorthogonal bases is not quite as straightforward as in the orthonormal case. However, it is still straightforward to verify the validity of direct and inverse estimates as in Section 5.1 (see Proposition 5.1), so that Theorem 5.8 applies and confirms, among other things, (9.14) in this case.
- Recall from (6.14) and Theorem 6.1 that optimal preconditioning depends on the validity of norm equivalences (6.1) in a range $(-\tilde{\gamma}, \gamma)$ containing t, where $2t = r$ is the order of $\mathcal{L}$. Thus, by (9.12), bases

of the above type are not optimal for the single-layer potential operator, which requires $-\frac{1}{2} \in (-\tilde{\gamma}, \gamma)$. Since discontinuities are confined to the patch boundaries, this adverse effect is expected to be milder than for a basis with increasingly dense discontinuities. The validity of norm equivalences of the form (6.1) for function spaces on manifolds will be seen later to be closely related to suitable characterizations of the function spaces with respect to *partitions*, not coverings, of the manifold (see Section 10.1 below).

Nevertheless, for operators of order zero, multi-wavelets are admissible. This program has been carried through by von Petersdorff and Schwab (1997*a*), arriving ultimately at a fully discrete scheme which solves the discretized boundary integral equation with matrices of size N_J at a cost of $\mathcal{O}(N_J(\log N_J)^4)$ operations and storage up to nearly asymptotically optimal accuracy. Some ingredients of the schemes in Dahmen et al. (1994*b*), Dahmen and Schneider (1997*b*) and Schneider (1995) will now be sketched, primarily from the point of view taken in Dahmen et al. (1994*b*), namely to identify the precise requirements on a pair of biorthogonal bases $\Psi^\Gamma, \tilde{\Psi}^\Gamma$ for $L_2(\Gamma)$ that gives rise to an symptotically optimal scheme. These findings, in turn, will then guide the construction of suitable bases for the general case. One can then also get rid of logarithmic factors in the work estimates.

9.4. A basic estimate

In the following we will assume that the operator

$$\mathcal{L}v = \int_\Gamma K(\cdot, x)v(x)\,\mathrm{d}s_x \tag{9.18}$$

satisfies the estimate (2.23) and that its Schwartz kernel K is smooth except on the diagonal, such that (2.25) holds (see Section 2.3). Γ is an (at least Lipschitz) manifold of dimension n. To solve the equation

$$\mathcal{L}u = f, \tag{9.19}$$

we wish to employ a pair of biorthogonal wavelet bases $\Psi = \{\psi_\lambda : \lambda \in \nabla\}$, $\tilde{\Psi} = \{\tilde{\psi}_\lambda : \lambda \in \nabla\}$ with $\nabla = \Delta_+ \cup \nabla_-$ as before. We will assume for the moment that this pair is *ideal* in the following sense.

Assumptions. For any order of accuracy d we have $\tilde{d}$th order of vanishing moments

$$(p \circ \kappa^{-1}, \psi_\lambda) = 0, \quad \lambda \in \nabla_-, \quad p \in \Pi_{\tilde{d}}, \tag{9.20}$$

where κ is a regular parametrization as above. Moreover, the pair of biorthogonal bases $\Psi, \tilde{\Psi}$ satisfy the norm equivalence (6.1) (or (5.42)) for the range

$s \in (-\tilde{\gamma}, \gamma)$. The regularity bounds $\gamma, \tilde{\gamma}$ are related to $\mathcal{L}$ by

$$|t| < \gamma, \tilde{\gamma}, \tag{9.21}$$

where again $r = 2t$ is the order of $\mathcal{L}$. Of course, we will also assume that the Galerkin method is stable, that is, (6.7) holds.

The first important step is to verify an estimate of the type (7.11) (see also (1.11)). Denote again by $|\lambda|$ the scale associated with ψ_λ and by Ω_λ the support of ψ_λ. In view of the moment condition (9.20), the argument leading to (9.16) can be applied to each variable consecutively (recall Section 1.3), which provides

$$|\langle \mathcal{L}\psi_{\lambda'}, \psi_\lambda \rangle| \lesssim \frac{2^{-(|\lambda|+|\lambda'|)(n/2+\tilde{d})}}{(\operatorname{dist}(\Omega_\lambda, \Omega_{\lambda'}))^{n+2\tilde{d}+2t}}, \tag{9.22}$$

whenever $\operatorname{dist}(\Omega_\lambda, \Omega_{\lambda'}) \gtrsim 2^{-\min(|\lambda|,|\lambda'|)}$ (Dahmen et al. 1994*b*, von Petersdorff and Schwab 1997*b*, von Petersdorff and Schwab 1997*a*). When the supports of ψ_λ and $\psi_{\lambda'}$ overlap, or more generally, when $\operatorname{dist}(\Omega_\lambda, \Omega_{\lambda'}) \lesssim 2^{-\min(|\lambda|,|\lambda'|)}$, one can use the *norm equivalence* (6.1) as follows (Dahlke et al. 1997*b*). To this end, suppose that $\mathcal{L}$ has the following additional continuity properties. There exists some $\tau > 0$ such that

$$\|Av\|_{H^{-t+s}} \lesssim \|v\|_{H^{t+s}}, \quad v \in H^{t+s}, 0 \le |s| \le \tau. \tag{9.23}$$

Without loss of generality one can assume that $|\lambda| > |\lambda'|$, that is, $\lambda \in \nabla_-, \lambda' \in \nabla$. Using Schwarz's inequality and the continuity of $\mathcal{L}$ (9.23) gives

$$|\langle \mathcal{L}\psi_{\lambda'}, \psi_\lambda \rangle| \le \|\mathcal{L}\psi_{\lambda'}\|_{H^{-t+\sigma}} \|\psi_\lambda\|_{H^{t-\sigma}} \lesssim \|\psi_{\lambda'}\|_{H^{t+\sigma}} \|\psi_\lambda\|_{H^{t-\sigma}}. \tag{9.24}$$

Thus, when

$$\sigma \le \tau, \quad t + \sigma < \gamma, \quad t - \sigma > -\tilde{\gamma},$$

one can apply now the norm equivalence (5.42) to each factor on the right-hand side of (9.24) which, upon using biorthogonality, yields

$$|\langle \mathcal{L}\psi_{\lambda'}, \psi_\lambda \rangle| \le 2^{t(|\lambda|+|\lambda'|)} 2^{\sigma(|\lambda'|-|\lambda|)}. \tag{9.25}$$

Combining (9.22) and (9.25) and assuming that

$$n/2 + \tilde{d} + t \ge \sigma, \tag{9.26}$$

one arrives at the following central estimate

$$2^{-(|\lambda'|+|\lambda|)t} |\langle \mathcal{L}\psi_{\lambda'}, \psi_\lambda \rangle| \lesssim \frac{2^{-||\lambda|-|\lambda'||\sigma}}{(1 + 2^{\min(|\lambda|,|\lambda'|)} \operatorname{dist}(\Omega_\lambda, \Omega_{\lambda'}))^{n+2\tilde{d}+2t}}. \tag{9.27}$$

This is exactly of the form (7.11). Note that the *preconditioning* has already been incorporated so that, in agreement with (7.11), the quantities on the left-hand side now represent a zero-order operator. Note also that the number of vanishing moments $\tilde{d}$ determines the decay on fixed levels. It was important above in (9.26) to be able to choose $\tilde{d}$ large enough.

As earlier, let $\Psi^J := \Phi_0 \cup_{j=0}^{J} \Psi_j$. The idea is to replace by zero those entries in the stiffness matrices

$$\mathbf{A}_{\Psi^J} := \langle \mathcal{L}\Psi^J, \Psi^J \rangle_\Gamma^T$$

which, according to the *a priori* estimates (9.27), are guaranteed to stay below a given threshold. However, that would leave the order of $J2^J$ entries for which no decay is predicted by (9.27) (Dahmen et al. 1994*b*, Dahmen and Schneider 1997*b*, von Petersdorff and Schwab 1997*b*, von Petersdorff and Schwab 1997*a*). A further reduction requires more subtle estimates developed by Schneider (1995). To this end, we will assume that the wavelets are, up to parametric transformation, piecewise polynomials, and we will denote the *singular support* of $\psi_{\lambda'}$ (Dahmen, Kunoth and Schneider 1997, Schneider 1995) by

$$\Omega_{\lambda'}^S := \text{sing supp}\, \psi_{\lambda'},$$

which consists of the boundaries of the subdomains in $\Omega_{\lambda'}$ whose parametric preimages in $\square$ are maximal regions where $\psi_{\lambda'} \circ \kappa_i$ is a polynomial (in this case of order d). If $|\lambda'| < |\lambda|$ and $\text{dist}\,(\Omega_\lambda, \Omega_{\lambda'}) \lesssim 2^{-|\lambda'|}$, then it is shown in Schneider (1995) that the estimate

$$|\langle \mathcal{L}\psi_{\lambda'}, \psi_\lambda \rangle_\Gamma| \lesssim \frac{2^{-|\lambda|(1+\tilde{d})} 2^{|\lambda'|}}{\left(\text{dist}\,(\Omega_\lambda, \Omega_{\lambda'}^S)\right)^{2t+\tilde{d}}} \tag{9.28}$$

holds.

9.5. *Matrix compression*

With the above estimates at hand, a *level dependent a priori truncation rule* can be designed in such a way that, on zeroing all entries which stay below the corresponding threshold, the resulting compressed matrix $\mathbf{A}_{\Psi^J}^c$ is sparse and contains only $\mathcal{O}(N_J)$ nonvanishing entries. As earlier, $N_J := \dim S(\Phi_J)$ is the dimension of the trial space of highest resolution. In addition to the above constraint (9.26) on $\tilde{d}$ it is important here to have

$$d < \tilde{d} + 2t. \tag{9.29}$$

Thus for operators of nonpositive order the order of vanishing moments should *exceed* the order of accuracy of the underlying scheme.

The compression proceeds in two steps. Fixing some $a > 0$ and $d' \in (d, \tilde{d} + 2t)$, let for $j = |\lambda|, j' = |\lambda'|$

$$b_{j,j'} \sim \max\left\{a\, 2^{-j}, a\, 2^{-j'}, a 2^{(J(2d'-2t) - j'(\tilde{d}+d') - j(\tilde{d}+d'))/(2\tilde{d}+2t)}\right\}, \tag{9.30}$$

and set

$$a^1_{\lambda,\lambda'} := \begin{cases} (\mathbf{A}_{\Psi^J})_{\lambda,\lambda'}, & \text{if } \operatorname{dist}(\Omega_\lambda, \Omega_{\lambda'}) \le b_{j,j'}, \\ 0, & \text{otherwise.} \end{cases} \tag{9.31}$$

Hence the bands get narrower when progressing to higher scales. In a second step, one sets

$$(\mathbf{A}^c_{\Psi^J})_{\lambda,\lambda'} := \begin{cases} a^1_{\lambda,\lambda'}, & j' \le j \text{ and } \operatorname{dist}(\Omega_\lambda, \Omega^S_{\lambda'}) \le b^S_{j,j'}, \\ & j \le j' \text{ and } \operatorname{dist}(\Omega^S_\lambda, \Omega_{\lambda'}) \le b^S_{j,j'}, \\ 0, & \text{otherwise.} \end{cases} \tag{9.32}$$

Here the truncation parameters $b^S_{j,j'}$ controlling the distance from the singular support are given by

$$b^S_{j,j'} \sim \max\left\{a'\,2^{-j}, a'\,2^{-j'}, a'2^{(J(2d'-2t)-\max\{j,j'\}\tilde{d}-(j+j')d')/(\tilde{d}+2t)}\right\}, \tag{9.33}$$

and the parameters a, a' are fixed constants independent of J. For instance, a determines the bandwidth in the block matrices $\mathbf{A}^c_{J,J} = (\mathbf{A}_{\psi^J})_{J,J} = ((\langle \mathcal{L}\psi_{\lambda'}, \psi_\lambda\rangle)_{|\lambda'|,|\lambda|=J}$. The choice of a, a' will be further specified later (Dahmen et al. 1994*b*, Schneider 1995).

Theorem 9.1 If the moment conditions (9.20) hold for $\tilde{d}$ satisfying (9.29), then under the above assumptions on $\mathcal{L}$ and $\Psi, \tilde{\Psi}$ the compression strategy (9.31), (9.32) generates matrices $\mathbf{A}^c_{\Psi^J}$ containing only $\mathcal{O}(N_J)$ non-vanishing entries.

9.6. Asymptotic estimates

The basic tool for estimating the effect of the above compression is a suitable version of a *weighted Schur lemma*. Recall that if for some matrix $\mathbf{A} = (a_{i,j})_{i,j\in I}$ there exists a positive constant c and a sequence $\mathbf{b}$ with $b_i > 0$, such that

$$\sum_{i\in I} |a_{i,j}| b_i \le c\, b_j \quad \text{for all } \ j \in I,$$

and

$$\sum_{j\in I} |a_{i,j}| b_j \le c\, b_i \quad \text{for all } \ j \in I,$$

then $\|\mathbf{A}\| \le c$, where $\|\cdot\|$ denotes the spectral norm. In the present context the b_j are chosen as 2^{-sj} for suitable choices of $s \ge 0$. Again denoting by $\mathbf{D}^s$ the diagonal matrix with entries $(\mathbf{D}^s)_{\lambda,\lambda'} = 2^{s|\lambda|}\delta_{\lambda,\lambda'}$, the Schur lemma can be used to show that

$$\|\mathbf{D}_J^{-s}(\mathbf{A}_{\Psi^J} - \mathbf{A}^c_{\Psi^J})\mathbf{D}_J^{-\tilde{s}}\| \ \lesssim \ J^{-1}a^{-2t-2\tilde{d}}2^{-J(s+\tilde{s}-2t)}.$$

At this point the *norm equivalences* enter again. In fact, one infers from the above estimate combined with (6.1) *consistency estimates* of the form

$$\|(\mathcal{L}_J - \mathcal{L}_J^c)\, u\|_{H^{s-2t}} \lesssim a^{-2t-2\tilde{d}} 2^{J(s-\tau)} \|u\|_{H^\tau}, \tag{9.34}$$

where $a > 1$ is fixed, $\mathcal{L}_J$, $\mathcal{L}_J^c$ are the finite-dimensional operators corresponding to $\mathbf{A}_{\Psi^J}$ and $\mathbf{A}_{\Psi^J}^c$, respectively, and the range of the parameters s and τ is $-d + 2t \le s < \gamma$, $-\gamma < \tau \le d$. As before, γ and d reflect the regularity and the order of accuracy of the trial functions. In particular, for any $\epsilon > 0$ one can choose $a > 1$ such that

$$\|(\mathcal{L}_J - \mathcal{L}_J^c)\, u\|_{H^{-t}} \lesssim \epsilon \|u\|_{H^t}. \tag{9.35}$$

A perturbation argument combined with these estimates ensures stability of the compressed operator in the energy norm and even for lower norms, we have

$$\|\mathcal{L}_J^c v_J\|_{H^{s-2t}} \gtrsim \|v_J\|_{H^s}, \quad v_J \in S(\Phi_J), \tag{9.36}$$

for $2t - d \le s \le t$; see, for example, Dahmen et al. (1994*b*).

These facts can then be combined to prove the following result (Dahmen et al. 1994*b*, Dahmen et al. 1993*b*, Dahmen and Schneider 1997*b*, Schneider 1995).

Theorem 9.2 Under the above circumstances the compressed system

$$\mathbf{A}_{\Psi^J}^c \mathbf{d}_J = \langle f, \Psi^J \rangle_\Gamma^{\mathrm{T}}$$

possesses a unique solution and $u_J^c := \mathbf{d}_J^T \Psi^J$ has asymptotically optimal accuracy

$$\|u - u_J^c\|_{H^\tau} \lesssim 2^{J(\tau - s)} \|u\|_{H^s}, \tag{9.37}$$

where $-d + 2t \le \tau < \gamma$, $\tau \le s$, $t \le s \le d$ and u is the exact solution of $\mathcal{L}u = f$. Moreover, the matrices $\mathbf{B}_J^c = \mathbf{D}^{-t}\mathbf{A}_{\Psi^J}^c \mathbf{D}^{-t}$ have the order of N_J nonvanishing entries and uniformly bounded condition numbers.

By Remark 6.3, one obtains a scheme that solves (9.19) with asymptotically optimal accuracy in linear time.

We summarize the required conditions on the wavelet basis. To realize an asymptotically optimal balance between accuracy and efficiency, the regularity γ of Ψ, the regularity $\tilde{\gamma}$ of the dual basis $\tilde{\Psi}$, the order of vanishing moments $\tilde{d}$ and the order of exactness d of the trial spaces $S(\Phi_J)$ should be related in the following way.

Regularity	$\gamma > t$ conformity	$\tilde{\gamma} > -t$ preconditioning
Order	d	convergence rate $2^{-J(2d+2-2t)}$
Vanishing moments	$\tilde{d} > d - 2t$	

9.7. Adaptive quadrature

In the above analysis it has been assumed that the matrix entries $\langle \mathcal{L}\psi_\lambda, \psi_{\lambda'}\rangle_\Gamma$ are given exactly. Of course, in general they have no closed analytical representation.

In principle, one can first compute the stiffness matrix $\langle \mathcal{L}\Phi_J, \Phi_J\rangle_\Gamma^{\mathrm{T}}$ relative to the single-scale basis Φ_J (for instance with the aid of the techniques described in Section 4.2) with sufficient accuracy to preserve the overall precision of the above scheme. In fact, the *multiscale transformation* $\mathbf{T}_J$ from (3.25) yields

$$\mathbf{A}_{\Psi^J} = \mathbf{T}_J^T \langle \mathcal{L}\Phi_J, \Phi_J\rangle_\Gamma^{\mathrm{T}} \mathbf{T}_J.$$

However, since $\langle \mathcal{L}\Phi_J, \Phi_J\rangle_\Gamma$ is a dense matrix, this process requires at least the order of N_J^2 operations and storage which would completely destroy the efficiency of the fully discrete scheme.

To find a more economic strategy, one has to bear the following points in mind.

- There is an *a priori* criterion to decide whether a matrix coefficient must be computed or can be neglected.
- Note that dist $(\Omega_\lambda, \Omega_{\lambda'}) > b_{|\lambda|,|\lambda'|}$ implies that dist $(\Omega_\nu, \Omega_{\nu'}) > b_{|\nu|,|\nu'|}$ holds for $\Omega_\nu \subset \Omega_\lambda$ and $\Omega_{\nu'} \subset \Omega_{\lambda'}$, $|\nu| \geq |\lambda|, |\nu'| \geq |\lambda'|$. Thus, one does not have to check condition (9.31) or (9.32) for all pairs λ, λ'. Exploiting the hierarchical structure of multiscale bases, one needs at most $\mathcal{O}(2^{nJ}) = \mathcal{O}(N_J)$ checks to decide whether or not an entry has to be computed.

An accurate computation of the remaining nonzero coefficients by numerical quadrature is a difficult task. Significant coefficients involving low-level wavelets have to be computed with accuracy determined by the discretization error of the scheme. We will see later that, based on the construction outlined in Section 4.4, wavelets with the above ideal properties can be constructed whose pullbacks to the parameter domain are piecewise polynomials. Hence the approximation of $\langle \mathcal{L}\psi_\lambda, \psi_{\lambda'}\rangle_\Gamma$ can be reduced to the evaluation of integrals of the form

$$\int_{\kappa_i(\Box_\lambda)} \int_{\kappa_l(\Box_{\lambda'})} K(\hat{x}, \hat{y}) \psi_\lambda(\hat{x}) \psi_{\lambda'}(\hat{y}) \, \mathrm{d}s_{\hat{x}} \, \mathrm{d}s_{\hat{y}}, \tag{9.38}$$

where $\Box_\lambda \subset \Box$ denotes a cube such that $\psi_\lambda \circ \kappa_i \,|_{\Box_\lambda}$ is a polynomial of degree $d-1$. Thus one ultimately has to compute expressions of the type

$$\int_{\Box_\lambda} \int_{\Box_{\lambda'}} H(x, y) p_\lambda(x) p_{\lambda'}(y) \, \mathrm{d}x \, \mathrm{d}y, \tag{9.39}$$

where $p_\lambda, p_{\lambda'}$ are polynomials of degree $d-1$ satisfying

$$\|p_\lambda\|_{W^{s,\infty}(\square_\lambda)} \lesssim 2^{(s+n/2)j}. \tag{9.40}$$

When $\kappa_i(\square_\lambda) \cap \kappa_l(\square_{\lambda'}) = \emptyset$, $H(x,y) = K(\kappa_i(x), \kappa_l(y))|\partial\kappa_i(x)||\partial\kappa_l(y)|$ is arbitrarily smooth. Thus high-order quadrature can be used to compute entries not discarded by the decay estimates. When integrating over pairs of domains that share an edge, a vertex or are identical, then in general the integral is singular. In this case some sort of *regularization* should be applied to reduce the integral to a weakly singular integral (Nedelec 1982, von Petersdorff and Schwab 1997*a*). Then one can use transformation techniques like Duffy's trick proposed by Sauter (1992) to end up with analytical integrals (von Petersdorff and Schwab 1997*a*, Schwab 1994).

The central objective is now to balance the error caused by quadrature with the desired overall accuracy of the scheme, while preserving efficiency. Employing adaptive quadrature in connection with a multi-wavelet discretization for zero-order operators, a fully discrete scheme has recently been developed in von Petersdorff and Schwab (1997*a*), where essential use is also made of the analyticity of the kernel K in a neighbourhood of the two-dimensional surface Γ in $\mathbb{R}^3$. The resulting fully discrete scheme requires $\mathcal{O}(N_J(\log N_J)^4)$ operations. A somewhat different approach is given by Dahmen and Schneider (1997*b*), ending up with a slightly more favourable complexity analysis.

The balancing of errors is guided by the following considerations. The problem of quadrature has to be seen in close connection with compression and the special features of multiscale bases. Basis functions from coarser scales introduce large domains of integration while requiring high accuracy. In particular, on the coarsest scale $\lambda, \lambda' \in \Delta_+$ the full accuracy $2^{-J(2d'-2t)}$ depending on J is required, while on the highest scale $|\lambda|, |\lambda'| = J$ the computation of each entry requires only a fixed number of quadrature points independent of J. In fact, diam supp $\psi_\lambda \sim 2^{-J}$ and $|\langle \mathcal{L}\psi_\lambda, \psi_{\lambda'}\rangle_\Gamma| \lesssim 2^{j2t}$ for $|\lambda| = |\lambda'| = j$. Thus, many entries only have to be computed with low accuracy, while high accuracy is merely required for a small portion of the matrix. Using the analysis of matrix compression as a guideline, a careful balancing of the various effects shows that most matrix entries $\langle \mathcal{L}\psi_\lambda, \psi_{\lambda'}\rangle_\Gamma$ must be computed with a precision

$$e_{\lambda,\lambda'} \lesssim 2^{-J(2d'-2t)} 2^{\max\{|\lambda|,\lambda'|\}(d'+1)} 2^{\min\{|\lambda|,|\lambda'|\}(d'+1)} 2^{-2\max\{|\lambda|,|\lambda'|\}}$$

for some $d' > d$ (Dahmen et al. 1997, Dahmen and Schneider 1997*b*).

The fully discretized Galerkin method in Dahmen and Schneider (1997*b*) is based on product-type Gaussian formulae of order D for approximating inner and outer integrals

$$\int_\tau \int_{\tau'} p(x)p'(y)\,\mathrm{d}x\,\mathrm{d}y = Q_x^D \otimes Q_y^D\,(p \cdot p'), \quad \text{for all} \quad p, p' \in \Pi_D, \tag{9.41}$$

where the domains τ and τ' are congruent to $\square$. According to the previous remarks, the error estimate for the quadrature method has much in common with estimating matrix coefficients relative to wavelet bases. The relevant estimates are summarized as follows.

Lemma 9.3 Let $Q_\tau^D \otimes Q_{\tau'}^D$ be a product-type Gaussian quadrature method of order D and $\tau \subset \square_\lambda$, $\tau' \subset \square_{\lambda'}$. Furthermore, suppose that $\mathcal{L}$ is a boundary integral operator with the above properties and Γ is a piecewise analytic boundary surface. In local parametrization let the kernel be denoted by $H(x,y)$ as above and set $G(x,y) := H(x,y)p_\lambda(x)p_{\lambda'}(y)$. If $\tau \cap \tau' = \emptyset$, then there exists a constant c such that the estimate

$$\begin{aligned} &|\int_\tau \int_{\tau'} G(x,y)\,\mathrm{d}x\,\mathrm{d}y - Q_\tau^D \otimes Q_{\tau'}^D(G)| \\ \leq\; & c\frac{2^{(|\lambda|+|\lambda'|)}(\max\{\text{diam}\,\tau, \text{diam}\,\tau'\})^{D-d}(\text{diam}\,\tau)^2(\text{diam}\,\tau')^2}{\text{dist}\,(\kappa_i(\tau'), \kappa_l(\tau))^{2+2t+D-d}} \end{aligned}$$

holds, provided that $2 + 2t + D - d > 0$.

The principal strategy is to choose the diameter of the subdomains proportional to the distance from the singularity while the degree D has to be adapted to maintain the desired accuracy taking the decay of the entries into account. Details can be found in Dahmen et al. (1997) and Dahmen and Schneider (1997*b*):

In summary the following result can be proved (Dahmen et al. 1997, Dahmen and Schneider 1997*b*).

Theorem 9.4 Under the above assumptions the fully discretized compressed system $\mathbf{A}_{\Psi^J}^{cq}\mathbf{d}_J^{cq} = \langle f, \Psi^J\rangle^{\mathrm{T}}$ possesses a unique solution and $u_J^{cq} := (\mathbf{d}_J^{cq})^T\Psi^J$ realizes asymptotically optimal accuracy

$$\|u - u_J^{cq}\|_{H^\tau} \;\lesssim\; 2^{j(\tau-s)}\,\|u\|_{H^s} \tag{9.42}$$

where $-d + 2t \leq \tau < \gamma$, $\tau \leq s$, $t \leq s \leq d$ and u is the exact solution of $\mathcal{L}u = f$. Moreover, the nonzero coefficients of the matrix $\mathbf{A}_{\Psi^J}^{cq}$ can be computed at the expense of $\mathcal{O}(N_J)$ floating point operations and storage.

10. Wavelets on manifolds and domain decomposition

The periodic case is certainly the most convenient setting for constructing wavelets and exploiting their full computational efficiency. On the other hand, the application of embedding techniques as described in Section 8 is certainly limited. For instance, problems defined on closed surfaces, as discussed in Section 9, cannot be treated in this way. Also, the resolution of boundary layer effects may cause difficulties.

This has motivated various attempts to extend wavelet-like tools to more general settings. This is, for instance, reflected by the general framework in Section 3, and concepts like stable completions; see Section 3.2 and Carnicer et al. (1996). The so-called *lifting scheme* (Sweldens 1996, Sweldens 1997) is very similar in spirit. Its applications, for instance in computer graphics, also demonstrate its versatility and efficiency in connection with unstructured grids (Schröder and Sweldens 1995). Unfortunately, the understanding of analytical properties like stability and norm equivalences in a more general setting still appears to be in its infancy. Attempts to develop stability criteria that work in sufficiently flexible settings have been sketched in Section 5; see Dahmen (1994) and (1996). Some recent consequences of these developments will be indicated next.

Many problem formulations suggest in a natural way a decomposition of the underlying domain into subdomains, which in turn are often representable as parametric images of *cubes*. As was indicated in Section 4.4, wavelet bases on cubes are well understood and much of the efficiency of wavelet bases in the ideal setting can be retained. This can readily be combined with the idea described in Section 9.2 to obtain wavelet bases with essentially the same nice properties on any domain Ω, as long as $\Omega = \kappa(\square)$ is a smooth regular parametric image of the unit cube, that is,

$$|\partial\kappa(x)| \neq 0, \quad x \in \square. \tag{10.1}$$

In fact, the canonical inner product $\langle\cdot,\cdot\rangle_\Omega$ can be replaced by the inner product (see (9.9))

$$(u,v) := \int_\square (u\circ\kappa)(x)\overline{(v\circ\kappa)(x)}\,\mathrm{d}x, \tag{10.2}$$

which induces an equivalent norm for $L_2(\Omega)$, say. Moreover, when $F \subseteq L_2(\Omega)$ denotes a Besov or Sobolev space on Ω, it can be pulled back to a corresponding space on $\square$ by

$$F(\Omega) = \{g\circ\kappa^{-1} : g \in F(\square)\}, \tag{10.3}$$

with

$$\|v\|_{F(\Omega)} \sim \|v\circ\kappa\|_{F(\square)}. \tag{10.4}$$

Any biorthogonal wavelet bases $\Psi, \tilde{\Psi}$ on $\square$ then induce collections $\Psi^\Omega := \Psi\circ\kappa^{-1}$, $\tilde{\Psi}^\Omega := \tilde{\Psi}\circ\kappa^{-1}$ which are biorthogonal Riesz bases on Ω relative to the inner product (10.2). On account of (10.4), they inherit all the norm equivalences satisfied by $\Psi, \tilde{\Psi}$. In this way, all computations are ultimately carried out on the standard domain $\square$.

Of course, the qualitative properties of the bases $\Psi^{\Omega}, \tilde{\Psi}^{\Omega}$ depend on the mapping κ, which in practice confines this approach to rather simple domains Ω. However, the next step, which was made, to some extent, in Section 9, is to consider domains that are *disjoint unions* of such simple domains. Modelling closed surfaces, as considered in Section 9.1, falls exactly into this category. Although the following facts are by no means restricted to this case, we will adopt the same notation and assumptions made in Section 9.1 but keep in mind that Γ may as well denote some bounded domain in Euclidean space.

The whole preceding development shows that the power of wavelet discretizations hinges on its relation to certain function spaces, in particular, on corresponding norm equivalences. However, this is exactly the point where one easily gets stuck. In fact, recall from (9.12) that managing norm equivalences on the individual spaces $F(\Gamma_i)$ with the aid of the transported bases $\Psi^{\Gamma_i}, \tilde{\Psi}^{\Gamma_i}$ does not generally imply corresponding relations with respect to the global space $F(\Gamma)$. The problem is that the norms $\|\cdot\|_{F(\Gamma)}$ and $(\sum_{i=1}^{M}\|\cdot\|^2_{F(\Gamma_i)})^{1/2}$ do *not* generally determine the same space. Below we indicate several attempts, mostly referring to work in progress, to overcome this difficulty.

10.1. Composite wavelet bases

The following comments are based on Dahmen and Schneider (1996), and related special cases considered in Jouini and Lemarié-Rieusset (1993). The basic idea is to glue the bases defined on each patch together so that the resulting global bases are at least continuous on all of Γ. One way to achieve this is to carefully inspect the construction of biorthogonal spline wavelets on $[0,1]$, described in Section 4.4. One can show that the biorthogonal generator bases Φ_j and $\tilde{\Phi}_j$ on $[0,1]$ can be arranged to have the following property. All but one basis function at each end of the interval vanish at 0 and 1. This fact can then be exploited to construct pairs of refinable biorthogonal generator bases $\Phi_j^{\Gamma}, \tilde{\Phi}_j^{\Gamma}$, which belong to $C(\Gamma)$. Unfortunately, the wavelets corresponding to these global generator bases *cannot* be easily obtained by stitching local wavelet bases together. The reason is that not all the wavelets for the local bases can be arranged to vanish at the patch boundaries. Nevertheless, one can employ the concept of stable completions from Section 3.2 to construct compactly supported biorthogonal wavelet bases $\Psi^{\Gamma}, \tilde{\Psi}^{\Gamma}$ on Γ, which also belong to $C(\Gamma)$ (Dahmen and Schneider 1996). The disadvantage of this construction is that, since some wavelets have support in more than one patch Γ_i, moment conditions of the form (9.20) no longer hold in full strength near the patch boundaries.

Nevertheless, since all basis functions are local and since the trial spaces $S(\Phi_j^{\Gamma})$ retain the same approximation properties as the local spaces transported from $\square$, these spaces can be used for conforming Galerkin discretizations for second-order problems, even in connection with *nonoverlapping domain decomposition strategies*, for instance. On the other hand, it is clear that such an approach is limited for principal reasons. For example, it does not provide the *ideal* bases in the sense of Section 9.6 for operators of order minus one. More generally, this approach is not suited for handling duality.

An alternative approach, which is interesting from several points of view, will be outlined next.

10.2. Characterization of function spaces via partitions of domains

In the following we will denote by $F(\Gamma')$ spaces of the form $H^s(\Gamma')$ or Besov spaces, where $\Gamma' \subseteq \Gamma$. The problem with (9.12) is that the spaces $F(\Gamma)$ are usually defined through *overlapping coverings* of Γ, not through a *partition* of Γ. Therefore a fundamental step towards overcoming limitations of the type (9.12) is first to derive a characterization of function spaces on Γ in terms of *partitions*. Such characterizations were developed by Ciesielski and Figiel (1983), in terms of mappings

$$T : F(\Gamma) \to \prod_{i=1}^{M} \chi_{\Gamma_i}(P_i(F(\Gamma))), \quad V : F(\Gamma) \to \prod_{i=1}^{M} \chi_{\Gamma_i}(P_i^*(F(\Gamma))), \tag{10.5}$$

defined by

$$Tv = (\chi_{\Gamma_i} P_i v)_{i=1}^M, \quad Vv = (\chi_{\Gamma_i} P_i^* v)_{i=1}^M. \tag{10.6}$$

Here χ_{Γ_i} denotes the characteristic function of Γ_i and the P_i are certain projectors on $L_2(\Gamma)$, constructed in such way that T and V are actually *topological isomorphisms* with respect to F, and the factors $\chi_{\Gamma_i}(P_i(F(\Gamma)))$ are closed subspaces of $F(\Gamma_i)$ determined by certain homogeneous *trace conditions*.

The main focus of Ciesielski and Figiel (1983) was the existence of unconditional bases of Sobolev and Besov spaces on compact C^∞-manifolds. The objective of Dahmen and Schneider (1997*a*) is to employ such concepts for the development of practicable schemes. This requires us to identify practically realizable projections P_i needed in (10.5) and to combine this with the recently developed technology of biorthogonal wavelet bases on $\square$. This provides practically feasible wavelet bases for the component spaces $\chi_{\Gamma_i}(P_i(F(\Gamma)))$, and hence through (10.5) also for Γ. The resulting bases can be shown to exhibit all the desired properties listed in Section 9.6. The main ingredients of this program can be outlined as follows.

Ordering of patches

First one *orders* the patches Γ_i in a certain fashion. If $\overline{\Gamma}_i \cap \overline{\Gamma}_l := \epsilon_{i,l}$ is a common face and $i < l$, then $\epsilon_{i,l}$ is called an *outflow* (*inflow*) face for Γ_i (Γ_l).

$\partial\Gamma_i^{\uparrow}$, $\partial\Gamma_i^{\downarrow}$ are called the *outflow* and *inflow* boundary of the patch Γ_i. Let $\Gamma_i^{\uparrow}$ denote an extension of Γ_i in Γ which contains the outflow boundary $\partial\Gamma_i^{\uparrow}$ in its relative interior and whose boundary contains the inflow boundary $\partial\Gamma_i^{\downarrow}$ of Γ_i. Thus $\Gamma_i^{\uparrow}$ could be taken as the union of Γ_i and those patches whose closure intersects the relative interior of the outflow boundary $\partial\Gamma_i^{\uparrow}$. Analogously one defines $\Gamma_i^{\downarrow}$ with respect to the reverse flow.

Extensions

Now suppose that E_i is an extension operator from the domain Γ_i to $\Gamma_i^{\uparrow}$. It turns out that the topological properties of the projectors P_i to be constructed for (10.5) hinge upon the following continuity properties of the extensions E_i. To describe this, the following notation is convenient. Let

$$f^{\uparrow}(x) := \begin{cases} f(x), & x \in \Gamma_i, \\ 0, & x \in \Gamma_i^{\uparrow} \setminus \Gamma_i \end{cases}$$

denote the trivial extension of $f \in F(\Gamma_i)$ to $\Gamma_i^{\uparrow}$ and define

$$F(\Gamma_i)^{\uparrow} := \{f \in F(\Gamma_i) : f^{\uparrow} \in F(\Gamma_i^{\uparrow})\}, \quad \|f\|_{F(\Gamma_i)^{\uparrow}} := \|f^{\uparrow}\|_{F(\Gamma_i^{\uparrow})}.$$

Thus $F(\Gamma_i)^{\uparrow}$ consists of those elements in the local space $F(\Gamma_i)$ whose trace vanishes on the outflow boundary $\partial\Gamma_i^{\uparrow}$. Again the spaces $F(\Gamma_i)^{\downarrow}$ are defined analogously.

Now suppose that the extensions E_i satisfy

$$\|E_i f\|_{F(\Gamma_i^{\uparrow})} \lesssim \|f\|_{F(\Gamma_i)}, \quad \|(E_i^* f)^{\uparrow}\|_{F(\Gamma_i)^{\uparrow}} \lesssim \|f\|_{F(\Gamma_i^{\uparrow})}. \tag{10.7}$$

Due to the simple form of the parameter domain $\square$, such extensions can be constructed explicitly as tensor products of *Hestenes-type extensions* (Ciesielski and Figiel 1983, Dahmen and Schneider 1997*a*). However, some deviations from the construction in Ciesielski and Figiel (1983), which are essential from a practical point of view, will be mentioned later.

Topological isomorphisms

Given E_i as above, one now defines

$$P_1 f := E_1(\chi_{\Gamma_1} f), \quad P_i f := E_i(\chi_{\Gamma_i}(f - \sum_{l<i} P_l f)), \quad i = 2, \ldots, M. \tag{10.8}$$

One can prove the following facts (Dahmen and Schneider 1997*a*).

Theorem 10.1 One has

$$\chi_{\Gamma_i}(P_i(F(\Gamma))) = F(\Gamma_i)^{\downarrow}, \quad \chi_{\Gamma_i}(P_i^*(F(\Gamma))) = F(\Gamma_i)^{\uparrow}. \tag{10.9}$$

The mappings

$$T : f \mapsto \{\chi_{\Gamma_i} P_i f\}_{i=1}^{M}, \quad V : f \mapsto \{\chi_{\Gamma_i} P_i^* f\}_{i=1}^{M} \tag{10.10}$$

define topological isomorphisms acting from $F(\Gamma)$ onto the product spaces $\Pi_{i=1}^{M} F(\Gamma_i)^{\downarrow}$, $\Pi_{i=1}^{M} F(\Gamma_i)^{\uparrow}$, respectively, whose inverses are given for $\mathbf{v} = (v_i)_{i=1}^{M} \in \prod_{i=1}^{M} L_2(\Gamma_i)$ by

$$S\mathbf{v} = \sum_{i=1}^{M} P_i \chi_{\Gamma_i} v_i, \quad U\mathbf{v} = \sum_{i=1}^{M} P_i^* \chi_{\Gamma_i} v_i, \tag{10.11}$$

respectively. Specifically, one has

$$F(\Gamma) \cong \prod_{i=1}^{M} F(\Gamma_i)^{\downarrow} \cong \prod_{i=1}^{M} F(\Gamma_i)^{\uparrow},$$

and

$$\|v\|_{F(\Gamma)} \sim \left(\sum_{i=1}^{M} \|P_i v\|^2_{F(\Gamma_i)^{\downarrow}}\right)^{\frac{1}{2}} \sim \left(\sum_{i=1}^{M} \|P_i^* v\|^2_{F(\Gamma_i)^{\uparrow}}\right)^{\frac{1}{2}}, \quad v \in F(\Gamma). \tag{10.12}$$

Moreover, the maps T, V extend to isomorphisms from $F^*(\Gamma)$ onto the spaces $\prod_{i=1}^{M} F^*(\Gamma_i)^{\downarrow}$ and $\prod_{i=1}^{M} F^*(\Gamma_i)^{\uparrow}$, respectively, and

$$\|v\|_{F^*(\Gamma)} \sim \left(\sum_{j=1}^{M} \|P_j v\|^2_{F^*(\Gamma_j)^{\downarrow}}\right)^{\frac{1}{2}}, \quad v \in F^*(\Gamma). \tag{10.13}$$

Note that duality is incorporated in a natural way.

10.3. Biorthogonal wavelets on Γ

With Theorem 10.1 at hand, one can now construct wavelet bases on Γ that give rise to the desired norm equivalences. The basic steps can be roughly sketched as follows.

First, for each i let $\hat{\kappa}_i$ be an extension of κ_i (with as much smoothness as permitted by the regularity of $\Gamma_i^{\downarrow}$) and $\square_i^{\downarrow}$ a hyperrectangle such that $\hat{\kappa}_i(\square_i^{\downarrow}) = \Gamma_i^{\downarrow}$ and $\hat{\kappa}_i \,|_{\square} = \kappa_i$. As above, the spaces $F(\square)^{\downarrow,i}$ then consist of those elements in $F(\square)$ whose trivial extension to $\square_i^{\downarrow}$ by zero belongs to $F(\square_i^{\downarrow})$.

- For each pair of *complementary* homogeneous boundary conditions in $F([0,1])$, construct biorthogonal wavelet bases on $[0,1]$ based on the schemes described in Section 4.4. By this we mean, for instance, that when the wavelets and generators on the primal side are to vanish up to some order at zero, there are no boundary constraints at zero for the functions in the dual system (and analogously for all possible combinations).

- Using tensor products, this leads to biorthogonal wavelet bases

$$\Psi^{\square,i} \subset F(\square)^{\downarrow,i}, \quad \tilde{\Psi}^{\square,i} \subset F(\square)^{\uparrow,i}.$$

- The bases

$$\Psi^{\Gamma_i} := \Psi^{\square,i} \circ \kappa_i^{-1} \subset F(\Gamma_i)^{\downarrow}, \quad \tilde{\Psi}^{\Gamma_i} := \tilde{\Psi}^{\square,i} \circ \kappa_i^{-1} \subset F(\Gamma_i)^{\uparrow},$$

are biorthogonal with respect to the inner product $(\cdot,\cdot)_i$ defined by (9.9).

- The collections

$$\Psi^{\Gamma} := S(\{\Psi^{\Gamma_i}\}_{i=1}^M), \quad \tilde{\Psi}^{\Gamma} := U(\{\tilde{\Psi}^{\Gamma_i}\}_{i=1}^M), \tag{10.14}$$

where S, U are defined in (10.11), are biorthogonal wavelet bases on Γ relative to the inner product (9.8). Moreover, from Theorem 10.1 and (10.3), (10.4) one infers that for $F = H^s$

$$\|\mathbf{d}^T\Psi^{\Gamma}\|_{H^s(\Gamma)} \sim \|\mathbf{D}^s\mathbf{d}\|_{\ell_2}. \tag{10.15}$$

The range of $s \in \mathbb{R}$ is constrained here by the regularity bounds $\gamma, \tilde{\gamma}$ of the bases $\Psi^{\square,i}$, $\tilde{\Psi}^{\square,i}$, respectively, and by the regularity of Γ, which restricts the range of Sobolev indices. As before, $\mathbf{D}^s$ denotes the diagonal matrix with entries $(\mathbf{D}^s)_{\lambda,\lambda'} = 2^{s|\lambda|}\delta_{\lambda,\lambda'}$.

10.4. Computational aspects

In practice one would *not* compute Ψ^{Γ} explicitly. To discuss this issue, consider the inner product

$$\langle \mathbf{v}, \mathbf{u}\rangle_{\Pi} := \sum_{i=1}^{M} \langle v_i, u_i\rangle_{\Gamma_i}$$

on $\Pi_{i=1}^M L_2(\Gamma_i)$, which is of course also equivalent to $(\cdot,\cdot)$ defined by (9.8). Formally the stiffness matrix relative to Ψ^{Γ} constructed above is given by

$$\langle \mathcal{L}\Psi^{\Gamma}, \Psi^{\Gamma}\rangle_{\Gamma} = \langle (S^*\mathcal{L}S)\{\Psi^{\Gamma_l}\}_l, \{\Psi^{\Gamma_i}\}_i\rangle_{\Pi},$$

where S is defined in (10.11). When $\mathcal{L}$ is an isomorphism from $F(\Gamma)$ into $F^*(\Gamma)$, Theorem 10.1 assures that $\mathcal{L}_{\Pi} := S^*\mathcal{L}S$ is an isomorphism from $\Pi_{\downarrow} := \prod_{i=1}^M F(\Gamma_i)^{\downarrow}$ into $\Pi_{\uparrow}^* := \prod_{i=1}^M F^*(\Gamma_i)^{\uparrow}$, that is

$$\|\mathcal{L}_{\Pi}\mathbf{v}\|_{\Pi_{\uparrow}^*} \sim \|\mathbf{v}\|_{\Pi_{\downarrow}}, \quad \mathbf{v} \in \Pi_{\downarrow}. \tag{10.16}$$

Thus the problem $\mathcal{L}u = f$ is equivalent to

$$\mathcal{L}_{\Pi}\mathbf{u} = \mathbf{f}, \tag{10.17}$$

where $\mathbf{f} = S^*f = Vf$. Of course, when $\mathbf{u}$ solves (10.17), then $u = S\mathbf{u}$ is the solution to the original problem. Straightforward calculation shows that (10.17), in turn, can be stated as

$$\sum_{l=1}^{M} \mathcal{L}_{i,l} u_l = f_i, \quad i = 1, \dots, M,$$

where

$$\mathcal{L}_{i,l} = \chi_{\Gamma_i} P_i^* \mathcal{L} P_l \chi_{\Gamma_l}, \quad f_i = \chi_{\Gamma_i} P_i^* f, \quad i, l = 1, \dots M. \tag{10.18}$$

If, in addition, $\mathcal{L}$ is selfadjoint, one infers from (10.16) that

$$\|\mathbf{v}\|_{\Pi_\downarrow}^2 \sim \langle \mathcal{L}_\Pi \mathbf{v}, \mathbf{v} \rangle_\Pi.$$

Thus, choosing $\mathbf{v} := \{v\delta_{i,l}\}_{l=1}^M$, this yields

$$\|\mathcal{L}_{i,i} v\|_{F^*(\Gamma_i)^\uparrow} \sim \|v\|_{F(\Gamma_i)^\downarrow}, \quad i = 1, \dots, M, \tag{10.19}$$

which suggests solving (10.17) by an iteration of the form

$$u_i^{j+1} = u_i^j + \omega \mathcal{L}_{i,i}^{-1} (f_i - \sum_{l=1}^{M} \mathcal{L}_{i,l} u_l^j), \quad i = 1, \dots, M. \tag{10.20}$$

In fact this fits into the framework of Schwarz-type iterations described in Section 6.5. Specifically, on account of Theorem 10.1, one can apply Theorem 6.9, where S, defined by (10.11), plays the role of the mapping $\mathcal{R}$ in Theorem 6.9, so that convergence of the iteration follows from Theorem 6.10.

Hence the solution of (10.17) has been reduced to the *parallel* solution of *local* problems of the form

$$\mathcal{L}_{i,i} u_i = g_i, \quad i = 1, \dots, M, \tag{10.21}$$

which may be viewed as a *domain decomposition* method. On account of the relation $F(\Gamma_i)^\downarrow = \{g \circ \kappa_i^{-1} : g \in F(\square)^{\downarrow,i}\}$ (*cf.* (9.7)) and the definition of the bases Ψ^{Γ_i}, each equation in (10.21) is in effect an elliptic problem defined on the *unit cube*. On the unit cube $\square$, wavelet bases with all the desired properties are available. In addition, full advantage can be taken of highly efficient tensor product grid structures. As will be shown in Section 11, the adaptive potential of wavelet bases for elliptic problems can be fully exploited to facilitate an economic solution of each equation (10.21).

Note that, in principle, the approach works for differential as well as integral operators $\mathcal{L}$. The practical realization of the pullback of $\mathcal{L}_{i,i}$ to $\square$ depends, of course, on the type of $\mathcal{L}$. Let us therefore briefly comment on the practical aspects. First observe that, on Γ_i,

$$\psi_\nu^{\Gamma_i} = \chi_{\Gamma_i} P_i \psi_\nu^{\Gamma_i} = \chi_{\Gamma_i} E_i \psi_\nu^{\Gamma_i}. \tag{10.22}$$

Now, if a wavelet $\psi_\nu^{\Gamma_i}$ is supported inside Γ_i, then its trivial extension $(\psi_\nu^{\Gamma_i})^\uparrow$ to $\Gamma_i^\uparrow$ (by zero) already belongs to $F(\Gamma_i^\uparrow)$. However, the extensions constructed by Ciesielski and Figiel (1983) may still give rise to a nontrivial extension $P_i\psi_\nu^{\Gamma_i} = E_i\psi_\nu^{\Gamma_i}$, which on $\Gamma_i^\uparrow \setminus \Gamma_i$ differs from zero, and hence from $(\psi_\nu^{\Gamma_i})^\uparrow$, even though the wavelet $\psi_\nu^{\Gamma_i}$ is not close to the outflow boundary. To suppress this strong coupling between adjacent patches, Dahmen and Schneider (1997*b*) have shown how to construct extensions with the required continuity properties for which all wavelets in Ψ^{Γ_i} that already belong to $F(\Gamma_i)^\uparrow$ are extended by zero. This is again done by exploiting properties of suitable local multiscale bases on $\square$. The nontrivial extension of the remaining (boundary-near) wavelets represent the (scale-dependent) *coupling conditions* for the domain decomposition method. Thus Lagrange multipliers are *not* necessary for coupling the subproblems so that indefinite sytems are avoided. Note also that the discretizations, particularly their respective order of exactness, can be chosen independently on each patch Γ_i.

Since domain decomposition is comparatively less developed for integral operators, we take a closer look at the case where $\mathcal{L}$ has a global kernel K. One can show (Dahmen and Schneider 1997*b*) that the entries of the stiffness matrices then take the form

$$\left\langle \mathcal{L}_{i,l}\psi_\nu^{\Gamma_l}, \psi_\lambda^{\Gamma_i} \right\rangle_{\Gamma_i} = \int_\square \int_\square K_{i,l}(x,y)\psi_\nu^{\square,l}(y)\psi_\lambda^{\square,i}(x)\,\mathrm{d}y\,\mathrm{d}x, \tag{10.23}$$

where the kernel $K_{i,l}$ depends on the indices ν, λ of the wavelets in the following way. When both wavelets are supported in the interior of the cube, one has $K_{i,l}(x,y) = |\partial\kappa_i(x)||\partial\kappa_l(y)|K(\kappa_i(x), \kappa_i(y))$, where $|\partial\kappa_i|$ denotes the functional determinant of the mapping κ_i. However, when both wavelets have nontrivial extensions, for instance, one has to set

$$K_{i,l}(x,y) = |\partial\kappa_i(x)||\partial\kappa_l(y)|((E_i^* \otimes E_l^*)K)(\kappa_i(x), \kappa_i(y)).$$

The remaining mixed cases are analogous. Hence, in this case the coupling conditions simply boil down to modifications of the kernel. Note that $(E_i^* \otimes E_l^*)$ are *restriction operators*. In particular, this enforces the appropriate boundary conditions. In fact, one (locally) has

$$K_{i,l}(\cdot, y) \in F(\square)^{\uparrow,i}, \quad H_{i,l}(x,\cdot) \in F(\square)^{\uparrow,l}, \tag{10.24}$$

as long as the parameters y, x stay away from the respective outflow boundaries. This has the following important consequences (Dahmen and Schneider 1997*b*).

Moment conditions

Due to the complementary boundary conditions satisfied by the pairs of bases $\Psi^{\square,i}$, $\tilde{\Psi}^{\square,i}$ on $\square$, the spaces $S(\tilde{\Phi}_j^{\square,i})$ generally do *not* contain all polynomials of order d on $\square$. Hence the wavelets $\psi^{\square,i}$ near the outflow boundary do not have vanishing moments of corresponding low orders, but annihilate only those polynomials locally contained in $F(\Box)^{\uparrow,i}$. Therefore the wavelets still satisfy the estimate (9.16) for any function $f \in F(\Box)^{\uparrow,i}$. In view of (10.24), the kernel $K_{i,l}$ satisfies these boundary conditions. Hence the same argument that led to (9.27) still applies to $K_{i,l}(x,y)$. Therefore the wavelets still give rise to estimates like (9.22), (9.27) and hence to optimal compression determined by the order $\tilde{d}$ of the dual multiresolution. In particular, the kernels $K_{i,l}$ become more and more negligible when Γ_i and Γ_l are far apart.

Norm equivalences

Since the wavelets on $\Box$ give rise to norm equivalences of the form (6.1), the individual equations (10.21) are easily preconditioned. Moreover, the analysis of corresponding adaptive schemes described in the next section applies to the situation at hand.

11. Analysis of adaptive schemes for elliptic problems

11.1. Some preliminary remarks

The motivation for the following discussion is twofold. On one hand, the inherent potential of wavelet discretizations for adaptivity has been stated often above. However, as natural as it appears, a closer look reveals that on a rigorous and on a conceptual level a number of questions remain open. The discretizations typically involve several types of truncation that often remain unspecified. It is not always clear how corresponding errors propagate in the global scheme and how the tolerances have to be chosen to guarantee a specified overall accuracy. Moreover, thresholding arguments are often not clearly related to the *norm*, that is to measure global accuracy.

In many studies, some *a priori* assumptions are made about the type of singularity, for instance, in terms of the distribution of significant wavelet coefficients. For periodic problems the singularities of the solution are determined by the right-hand side alone (when the coefficients are smooth). This is no longer the case when essential boundary conditions for more complex geometries are imposed. Finally, what is the preferred strategy? In the spirit of image compression, a *fine-to-coarse* approach would aim at discarding insignificant wavelet coefficients, starting from a discretization for a fixed highest level of resolution. The obvious disadvantage is that such an approach accepts the complexity of a fully refined discretization at some stage. Alternatively, in a coarse-to-fine approach, one would try to track the significant wavelets needed to realize the desired accuracy, starting from a coarse discretization. The risk of missing important information along the way is perhaps even higher in this approach. However, the analysis outlined below indicates ways of dealing with this problem. So the subsequent discussion can be viewed as an attempt to address such questions on a rigorous level, and thereby complement the intriguing adaptive algorithmic developments discussed before.

On the other hand, adaptive techniques have been extensively studied in the context of finite element discretizations of (primarily) elliptic differential equations; see, for instance, Babuška and Miller (1987), Babuška and Rheinboldt (1978), Bank and Weiser (1985), Bornemann et al. (1996), Eriksson, Estep, Hansbo and Johnson (1995) and Verfürth (1994). These methods are based on *a posteriori error indicators* or *estimators*. In practice they have been proven to be quite successful. However, the analysis and the schemes are rather dependent on the particular problem at hand and on the particular type of finite element discretization. The geometrical problems caused by suitable mesh refinements become nontrivial for 3D problems. From a principal point of view, it is furthermore unsatisfactory that the proof of the overall convergence of such schemes usually requires making an *a priori assumption* on the unknown solution, as explained below in more detail.

The adaptive treatment of integral equations in the context of classical finite element discretizations is comparatively less developed. The global nature of the operator makes a local analysis harder. Typical *a posteriori* strategies therefore constrain the structure of admissible meshes (Carstensen 1996), which certainly interferes with the essence of adaptive methods.

These considerations have motivated recent investigations by Dahlke et al. (1997*b*), which substantiate that the main potential of wavelet discretizations lies in adaptivity. Some of the ingredients of the analysis will be outlined next. As in the context of preconditioning, a wide range of problems, including differential as well as integral operators, can be treated in a unified way. A convergence proof is only based on assumptions on the (accessible) data rather than on the (unknown) solution. Furthermore, there is no restriction on the emerging index sets.

Again, a key role is played by the validity of *norm equivalences* of the form (6.1) in combination with *compression* arguments based on the estimates (9.27) or (11.2) below.

To bring out the essential mechanisms, we will refer to the general problem in Section 2.3. Thus we will assume throughout the rest of this section that $\mathcal{L}$ satisfies (2.23) and (2.25). We consider stationary elliptic problems because they also arise in timestepping schemes. In fact, time-dependent problems are in some sense even easier, because information from the preceding time step can be used. Likewise the present formulation can be viewed as an ingredient of an iteration in nonlinear problems.

Moreover, in view of the developments in preceding sections, it is justified to assume that Ψ and $\tilde{\Psi}$ are biorthogonal wavelet bases satisfying the norm equivalences (6.1). Their range of validity is to satisfy (6.14). Specifically, there then exist finite positive constants c_3, c_4 such that

$$c_3\|\mathbf{D}^{-t}\mathbf{d}\|_{\ell_2} \le \|\mathbf{d}^T\tilde{\Psi}\|_{H^{-t}} \le c_4\|\mathbf{D}^{-t}\mathbf{d}\|_{\ell_2}. \tag{11.1}$$

Moreover, the corresponding spaces $S(\Phi_j)$ are assumed to be exact of order d and the wavelets $\psi_\lambda, \lambda \in \nabla_-$, satisfy a suitable version of moment conditions of order $\tilde{d}$ (see, for instance, (9.15)) when $\mathcal{L}$ is an integral operator, or are regular enough when $\mathcal{L}$ is a differential operator, so that in either case the estimate

$$2^{-(|\lambda'|+|\lambda|)t}|\langle \mathcal{L}\psi_{\lambda'}, \psi_\lambda\rangle| \lesssim \frac{2^{-||\lambda|-|\lambda'||\sigma}}{(1+2^{\min(|\lambda|,|\lambda'|)}\operatorname{dist}(\Omega_\lambda,\Omega_{\lambda'}))^{n+2\tilde{d}+2t}} \tag{11.2}$$

holds (see (9.27)). Finally, we will assume that the Galerkin scheme is stable (6.7) (recall the comments in Section 9).

11.2. The saturation property

Suppose for a moment that $\mathcal{L}$ is selfadjoint, in which case (2.23) means that the bilinear form

$$a(u,v) := \langle \mathcal{L}u, v\rangle \tag{11.3}$$

induces a norm which is equivalent to $\|\cdot\|_{H^t}$

$$\|\cdot\|^2 := a(\cdot,\cdot) \sim \|\cdot\|^2_{H^t}. \tag{11.4}$$

In this case a well-known starting point for finite element-based adaptive schemes is the following observation concerning the equivalence between the validity of two-sided error estimates and the so-called *saturation* property (Bornemann et al. 1996). The basic reasoning can be sketched as follows. Suppose that $S \subset V \subset H^t$ are two trial spaces with respective Galerkin solutions u_S, u_V. By orthogonality one has

$$\|u_V - u_S\| \le \|u - u_S\|.$$

Moreover, one easily checks that

$$\|u - u_V\| \le \beta\|u - u_S\| \tag{11.5}$$

holds for some $\beta < 1$, if and only if

$$(1-\beta^2)^{1/2}\|u - u_S\| \le \|u_V - u_S\|. \tag{11.6}$$

Here and elsewhere u denotes the exact solution to $\mathcal{L}u = f$. Thus, if the refined solution u_V captures a sufficiently large portion of the remainder (11.6) the global energy error is guaranteed to decrease by a factor β when passing to the refined solution u_V. Moreover, one has the bounds

$$\|u_V - u_S\| \le \|u - u_S\| \le (1-\beta^2)^{-1/2}\|u_V - u_S\|, \tag{11.7}$$

which are computable. In practice one controls the local behaviour of $u_V - u_S$ and refines the mesh at places where (an estimate for) this difference is largest. This results in balancing the error bounds. Although this has been observed to work well in many cases, the principal problem remains that, to prove convergence of the overall adaptive algorithm, something like (11.6) has to be *assumed* about the unknown solution.

Dahlke et al. (1997*b*) pursue a similar updating strategy. Let some current solution space S_Λ and a Galerkin solution u_Λ be given. The objective is to find for a fixed *decay rate* $\beta < 1$, a possibly small $\tilde{\Lambda} \subset \nabla = \Delta_+ \cup \nabla_-$, $\Lambda \subset \tilde{\Lambda}$ such that

$$\|u - u_{\tilde{\Lambda}}\|_{H^t} \leq \beta \|u - u_\Lambda\|_{H^t},$$

which implies convergence.

11.3. A posteriori *error estimates*

It is well known that for elliptic problems the error in energy norm can be estimated by the residual in a dual norm which, at least in principle, can be evaluated. In fact, since

$$r_\Lambda := \mathcal{L}u_\Lambda - f = \mathcal{L}(u_\Lambda - u),$$

the bounded invertibility of $\mathcal{L}$ (2.23) yields

$$c_1 \|r_\Lambda\|_{H^{-t}} \leq \|u - u_\Lambda\|_{H^t} \leq c_2 \|r_\Lambda\|_{H^{-t}}. \tag{11.8}$$

Expanding the residual r_Λ relative to the dual basis $\tilde{\Psi}$ and taking the Galerkin conditions into account, the norm equivalence (11.1) and (11.8) provide

$$c_1 c_3 \left(\sum_{\lambda \in \nabla \setminus \Lambda} \delta_\lambda(\Lambda)^2 \right)^{1/2} \leq \|u - u_\Lambda\|_{H^t} \leq c_2 c_4 \left(\sum_{\lambda \in \nabla \setminus \Lambda} \delta_\lambda(\Lambda)^2 \right)^{1/2}, \tag{11.9}$$

where the quantities

$$\delta_\lambda = \delta_\lambda(\Lambda) := 2^{-t|\lambda|} \left|\langle r_\Lambda, \psi_\lambda \rangle\right|, \quad \lambda \in \nabla \setminus \Lambda,$$

are, in principle, local quantities bounding the error $\|u - u_\Lambda\|_{H^t}$ from below and above. They indicate which wavelets are significant in the representation of u. However, since these quantities involve infinitely many (unknown) terms, (11.9) is in its present form of no practical use.

The objective of the following considerations is to replace the quantities $\delta_\lambda(\Lambda)$ in (11.9) by finitely many computable ones which, up to a given tolerance depending only on the data, still provide lower and upper bounds.

Denoting by $u_{\lambda'} = \langle u_\Lambda, \tilde{\psi}_{\lambda'} \rangle$, $f_\lambda := \langle f, \psi_\lambda \rangle$ the wavelet coefficients of the current approximation u_Λ and the right-hand side f with respect to Ψ and

$\tilde{\Psi}$, respectively, it is helpful to rewrite

$$\delta_\lambda(\Lambda) = 2^{-t|\lambda|} \left| f_\lambda - \sum_{\lambda' \in \Lambda} \langle \mathcal{L}\psi_{\lambda'}, \psi_\lambda \rangle u_{\lambda'} \right| . \tag{11.10}$$

This shows that the size of $\delta_\lambda(\Lambda)$ is influenced by two quantities. First, if the right-hand side f itself has singularities, this will result in large wavelet coefficients f_λ. Second, the sum $\sum_{\lambda' \in \Lambda} \langle \mathcal{L}\psi_{\lambda'}, \psi_\lambda \rangle u_{\lambda'}$ gives the contribution of the current solution which, for instance, could reflect the influence of the boundary. Thus, to estimate the $\delta_\lambda(\Lambda)$ one needs

(a) estimates on the smearing effect of $\mathcal{L}$
(b) some *a priori* knowledge about f.

So far we have only used the ellipticity (2.23) of $\mathcal{L}$ and the norm equivalence (11.1). To deal with (a) one has to make essential use of the decay estimates (11.2). We now describe their use. Let $\delta < \sigma - n/2$, where $\sigma > n/2$ is the constant in (11.2). Choose for any $\epsilon > 0$, positive numbers ϵ_1, ϵ_2 such that

$$\epsilon_1^{2\tilde{d}+2t} + 2^{-\frac{\delta}{\epsilon_2}} \leq \epsilon.$$

For each $\lambda \in \nabla$, define the *influence sets*

$$\nabla_{\lambda,\epsilon} := \{\lambda' \in \nabla : \left| |\lambda| - |\lambda'| \right| \leq \epsilon_2^{-1} \text{ and } 2^{\min\{|\lambda|,|\lambda'|\}} \operatorname{dist}(\Omega_\lambda, \Omega_{\lambda'}) \leq \epsilon_1^{-1}\},$$

where Ω_λ again denotes the support of ψ_λ. The sets $\nabla_{\lambda,\epsilon}$ describe the significant portion of $\langle \mathcal{L}u_\Lambda, \psi_\lambda \rangle$ appearing in the residual weights $\delta_\lambda(\Lambda)$ (11.10). In fact, using the estimate of (9.35), one can show the existence of a constant c_5 independent of f and Λ, such that the remainder

$$e_\lambda := \sum_{\lambda' \in \Lambda \setminus \nabla_{\lambda,\epsilon}} \langle \mathcal{L}\psi_{\lambda'}, \psi_\lambda \rangle u_{\lambda'}$$

can be estimated by

$$\Big(\sum_{\lambda \in \nabla \setminus \Lambda} 2^{-|\lambda| 2t} \left| e_\lambda \right|^2 \Big)^{\frac{1}{2}} \leq c_5 \epsilon \left\| u_\Lambda \right\| ; \tag{11.11}$$

see also Dahlke et al. (1997*b*), Dahmen et al. (1993*b*) and Dahmen et al. (1994*b*). Note that, again by (6.1),

$$\|u_\Lambda\| \sim \|u_\Lambda\|_{H^t} \sim \|\mathbf{D}^t \langle u_\Lambda, \tilde{\Psi}_\Lambda \rangle\|_{\ell_2},$$

so that the right-hand side in (11.11) can be evaluated by means of the wavelet coefficients of the current solution u_Λ. Moreover, one can even give an *a priori* bound. In fact, the stability of the Galerkin scheme (6.7) states, on account of the uniform boundedness of the Q_Λ^* in H^{-t} (see Theorem 5.8), that

$$\|u_\Lambda\| \lesssim \|Q^*_\Lambda f\|_{H^{-t}} \leq c'_5 \|f\|_{H^{-t}}. \tag{11.12}$$

As for (b) above, by construction, the *significant neighbourhood* of Λ in $\nabla \setminus \Lambda$

$$N_{\Lambda,\epsilon} := \{\lambda \in \nabla \setminus \Lambda : \Lambda \cap \nabla_{\lambda,\epsilon} \neq \emptyset\} \tag{11.13}$$

is finite

$$\# N_{\Lambda,\epsilon} < \infty.$$

Outside $N_{\Lambda,\epsilon}$, the quantities $\delta_\lambda(\Lambda)$ in (11.10) are essentially influenced by wavelet coefficients of f. But this portion is a remainder of f. In fact, by (6.1),

$$\begin{aligned}\Big(\sum_{\lambda\in\nabla\setminus(\Lambda\cup N_{\Lambda,\epsilon})} 2^{-2t|\lambda|}\,|f_\lambda|^2\Big)^{\frac12} &\leq c_2\|f - Q^*_{\Lambda\cup N_{\Lambda,\epsilon}}\|_{H^{-t}}\\ &\leq c_6 \inf_{v\in\tilde S_{\Lambda\cup N_{\Lambda,\epsilon}}} \|f-v\|_{H^{-t}} \leq c_6 \inf_{v\in\tilde S_\Lambda} \|f-v\|_{H^{-t}},\end{aligned}$$

for some $c_6 < \infty$. This suggests defining

$$d_\lambda(\Lambda,\epsilon) := 2^{-t|\lambda|}\Big|\sum_{\lambda'\in\Lambda\cap\nabla_{\lambda,\epsilon}} \langle \mathcal{L}\psi_{\lambda'}, \psi_\lambda\rangle u_{\lambda'}\Big|, \quad \lambda\in\nabla\setminus\Lambda.$$

Note that, in view of (11.13),

$$d_\lambda(\Lambda,\epsilon) = 0, \quad \lambda\in\nabla\setminus\Lambda, \quad \lambda\notin N_{\lambda,\epsilon}. \tag{11.14}$$

The main result can now be formulated as follows (Dahlke et al. 1997*b*).

Theorem 11.1 Under the above assumptions, one has

$$\|u-u_\Lambda\|_{H^t} \leq c_2c_4\Big(\Big(\sum_{\lambda\in N_{\Lambda,\epsilon}} d_\lambda(\Lambda,\epsilon)^2\Big)^{\frac12} + c'_5\epsilon\,\|f\|_{H^{-t}} + c_6 \inf_{v\in\tilde S_\Lambda}\|f-v\|_{H^{-t}}\Big)$$

as well as

$$\Big(\sum_{\lambda\in N_{\Lambda,\epsilon}} d_\lambda(\Lambda,\epsilon)^2\Big)^{\frac12} \leq \frac{1}{c_1c_3}\|u-u_\Lambda\|_{H^t} + c'_5\epsilon\,\|f\|_{H^{-t}} + c_6\inf_{v\in\tilde S_\Lambda}\|f-v\|_{H^{-t}}.$$

Moreover, for any $\tilde\Lambda\subset\nabla$, $\Lambda\subset\tilde\Lambda$, one has

$$\Big(\sum_{\lambda\in\tilde\Lambda\cap N_{\Lambda,\epsilon}} d_\lambda(\Lambda,\epsilon)^2\Big)^{\frac12} \leq \frac{1}{c_1c_3}\|u_{\tilde\Lambda}-u_\Lambda\|_{H^t} + c'_5\epsilon\,\|f\|_{H^{-t}} + c_6\inf_{v\in\tilde S_\Lambda}\|f-v\|_{H^{-t}}.$$

This result provides, up to the controllable tolerance

$$\tau(\Lambda,\epsilon) := c'_5\mathrm{eps}\,\|f\|_{H^{-t}} + c_6\inf_{v\in\tilde S_\Lambda}\|f-v\|_{H^{-t}},$$

computable lower and upper bounds for the error $\|u - u_\Lambda\|_{H^t}$. For second-order two-point boundary value problems, estimates of the above type were first obtained by Bertoluzza (1994). Under much more specialized assumptions, results of similar nature have also been established in the finite element context; see, for example, Dörfler (1996).

11.4. Convergence of an adaptive refinement scheme

In the present setting, it can be shown with the aid of Theorem 11.1 that, under mild assumptions on the right-hand side f, a suitable adaptive choice of $\tilde{\Lambda}$ *enforces* the validity of the saturation property (11.6). We continue with the notation of Section 11.3. However, for simplicity we confine the discussion to the selfadjoint case (11.3), (11.4), that is, the norm $\|\cdot\|_{H^t}$ is replaced by the *energy norm* $\|\cdot\|$. The constants c_i have to be properly adjusted. The following theorem was proved by Dahlke et al. (1997*b*).

Theorem 11.2 Let tol > 0 be a given tolerance and fix $\theta \in (0,1)$. Define

$$C^* := \left(\frac{1}{c_1 c_3} + \frac{1-\theta}{2c_2c_4}\right), \tag{11.15}$$

choose $\mu > 0$ such that

$$\mu C^* \leq \frac{1-\theta}{2(2-\theta)c_2c_4}, \tag{11.16}$$

and set

$$\epsilon := \frac{\mu \text{ tol}}{2c_5' \, \|f\|_{H^{-t}}}. \tag{11.17}$$

Suppose that for $\Lambda \subset \nabla$, one has

$$c_6 \inf_{v \in \tilde{S}_\Lambda} \|f - v\|_{H^{-t}} < \frac{1}{2}\mu \text{ tol}.$$

Then, whenever $\tilde{\Lambda} \subset \nabla$, $\Lambda \subset \tilde{\Lambda}$ is chosen so that

$$\Big(\sum_{\lambda \in \tilde{\Lambda} \cap N_{\Lambda,\epsilon}} d_\lambda(\Lambda,\epsilon)^2\Big)^{\frac{1}{2}} \geq (1-\theta)\Big(\sum_{\lambda \in N_{\Lambda,\epsilon}} d_\lambda(\Lambda,\epsilon)^2\Big)^{\frac{1}{2}},$$

there exists a constant $\beta \in (0,1)$ depending only on the constants μ, θ, c_i, $i = 1, \ldots, 6$, such that either

$$\|u - u_{\tilde{\Lambda}}\| \leq \beta \, \|u - u_\Lambda\|$$

or

$$\Big(\sum_{\lambda \in N_{\lambda,\epsilon}} d_\lambda(\Lambda,\epsilon)^2\Big)^{\frac{1}{2}} = \Big(\sum_{\lambda \in \nabla \setminus \Lambda} d_\lambda(\Lambda,\epsilon)^2\Big)^{\frac{1}{2}} < \text{tol}.$$

For the discussion of unsymmetric problems see Dahlke et al. (1997*b*) and Hochmuth (1996).

Of course, the idea is to choose $\tilde{\Lambda} \supset \Lambda$ as small as possible, that is, in any case $\tilde{\Lambda} \setminus \Lambda \subset N_{\Lambda,\epsilon}$. This leads to the following.

ALGORITHM 7

Choose $\Lambda_0 = \emptyset$, eps > 0, tol $>$ eps, $\theta \in (0,1)$.

(1) Compute C^*, μ according to (11.15), (11.16).
(2) Compute $\epsilon = \epsilon(\mu, \mathrm{tol})$ by (11.17).
(3) Determine $\Lambda \subset \nabla$, $\Lambda_0 \subset \Lambda$ such that

$$c_6 \inf_{v \in \tilde{S}_\Lambda} \|f - v\|_{H^{-t}} < \frac{1}{2}\mu \ \mathrm{tol}.$$

(4) Solve

$$\langle \mathcal{L} u_\Lambda, v\rangle = \langle f, v\rangle, \quad \text{for all } v \in S_\Lambda.$$

(5) Compute

$$\eta_{\Lambda,\epsilon} := \Big(\sum_{\lambda \in N_{\Lambda,\epsilon}} d_\lambda(\Lambda,\epsilon)^2\Big)^{\frac{1}{2}}.$$

If $\eta_{\Lambda,\epsilon} <$ tol:

- If tol $\le$ eps, accept u_Λ as solution and stop.
- Otherwise set $\Lambda \to \Lambda_0$, $\frac{\mathrm{tol}}{2} \to$ tol, and go to (2).

Otherwise, go to (6).

(6) Determine $\tilde{\Lambda}$ with $\Lambda \subset \tilde{\Lambda} \subset \Lambda \cup N_{\Lambda,\epsilon}$ such that

$$\Big(\sum_{\lambda \in \tilde{\Lambda}} d_\lambda(\Lambda,\epsilon)^2\Big)^{\frac{1}{2}} \ge (1-\theta)\eta_{\Lambda,\epsilon}.$$

Set $\tilde{\Lambda} \to \Lambda$ and go to (4).

Although quite different with regard to its technical ingredients, the above algorithm is very similar in spirit to the adaptive scheme proposed by Dörfler (1996) for bivariate piecewise linear finite element discretizations of Poisson's equation. As above, Dörfler (1996) chooses the coarsest grid in such a way that all errors stemming from data are kept below any desired tolerance.

A brief comment on step (4) in Algorithm 7 is in order. By Theorem 6.1, the principal sections of the matrix $\mathbf{B}_\Lambda := \mathbf{D}^{-t}\langle \mathcal{L}\Psi_\Lambda, \Psi_\Lambda\rangle^{\mathrm{T}}\mathbf{D}^{-t}$ are well conditioned. This can be exploited to update a current Galerkin approximation u_Λ, as follows. Let $\mathbf{u}_\Lambda := \langle u_\Lambda, \tilde{\Psi}_\Lambda\rangle^T$ be the vector of wavelet coefficients of u_Λ. To compute the coefficient vector $\mathbf{u}_{\tilde{\Lambda}}$ of $u_{\tilde{\Lambda}}$ we choose an initial approximation $\mathbf{v}$ according to

$$v_\lambda = \begin{cases} u_\lambda, & \lambda \in \Lambda\,, \\ w_\lambda, & \lambda \in \tilde{\Lambda} \setminus \Lambda, \end{cases} \tag{11.18}$$

where $\mathbf{w}_{\tilde{\Lambda}\setminus\Lambda}$ are the coefficients of the Galerkin solution $w_{\tilde{\Lambda}\setminus\Lambda}$ of the complement system

$$\langle \mathcal{L}w_{\tilde{\Lambda}\setminus\Lambda}, v\rangle = \langle f, v\rangle, \quad v \in S_{\tilde{\Lambda}\setminus\Lambda} := \operatorname{span}\{\psi_\lambda : \lambda \in \tilde{\Lambda} \setminus \Lambda\}.$$

The corresponding matrix entries have to be determined anyway for the adaptive refinement. Since, by (6.16), the corresponding section $\mathbf{B}_{\tilde{\Lambda}\setminus\Lambda}$ of $\mathbf{B}_{\tilde{\Lambda}}$ is well conditioned, only a few conjugate gradient iterations are expected to be necessary to approximate $\mathbf{w}_{\tilde{\Lambda}\setminus\Lambda}$ well enough to provide a good starting approximation of the form (11.18). This will then have to be improved by (a few) further iterations on the system matrix $\mathbf{B}_{\tilde{\Lambda}}$.

11.5. Besov regularity

The results of the previous section imply convergence of the adaptic scheme but do not provide any concrete information about the efficiency, for instance by relating the final accuracy to the number $\#\Lambda$ needed to realize it by the scheme. The ideal case would be that the scheme picks at each stage the *minimal* number of additional indices needed to reduce the current error by a fixed fraction. This cannot be concluded, since the scheme selects the indices with respect to bounds, not with respect to the true error. Nevertheless, since these bounds are lower and upper ones, one expects that the selected index sets are close to minimal ones. Given this assumption, the question of for which circumstances the above adaptive scheme is significantly more efficient than working simply with uniform refinements is closely related to characterizing the efficiency of so-called *best N-term approximation*, or *nonlinear approximation*. A beautiful theory for these issues has been developed in a number of papers; see, for instance, DeVore and Popov (1988*a*), DeVore et al. (1992) and DeVore and Lucier (1992). Here we indicate very briefly some typical facts suited to the present context. To this end, consider

$$\sigma_{N,t}(g) := \inf\left\{ \|g - \sum_{\lambda\in\Lambda} d_\lambda \psi_\lambda\|_{H^t} : d_\lambda \in \mathbb{R},\ \lambda \in \Lambda \subset \nabla,\ \#\Lambda = N \right\}.$$

Employing the norm equivalence (5.38) yields

$$\sigma_{N,t}(v) \sim \sigma_{N,0}(\Sigma_t v) := \sigma_N(\Sigma_t v), \tag{11.19}$$

which in turn leads to the following (Dahlke, Dahmen and DeVore 1997*a*).

Remark 11.3 Let $v \in H^t$. We take Λ_N to be a set of N indices λ for which $2^{t|\lambda|}|\langle v, \tilde{\psi}_\lambda\rangle|$ is largest. Then one has

$$\sigma_{N,t}(v) \sim \|v - Q_{\Lambda_N} v\|_{H^t}, \quad N \in \mathbb{N}. \tag{11.20}$$

Thus, picking the N first largest *weighted* coefficients realizes asymptotically the best N-term approximation relative to the norm $\|\cdot\|_{H^t}$ and hence, in case (11.4), also relative to the energy norm $\|\cdot\|$.

Combining (11.20) with analogous results about $\sigma_{N,t}$ for $t = 0$, the best N-term approximation of a function v relative to $\|\cdot\|_{H^t}$ can be characterized in terms of its *Besov regularity* (Dahlke et al. 1997*a*).

Proposition 11.4 Assume that $\alpha - t < \gamma$ and let for $t \leq \alpha$

$$\frac{1}{\tau^*} := \frac{\alpha - t}{n} + \frac{1}{2}. \tag{11.21}$$

Then one has

$$\sum_{N=1}^{\infty} \left(N^{(\alpha-t)/n}\sigma_{N,t}(v)\right)^{\tau^*} < \infty \tag{11.22}$$

(where n is again the spatial dimension of the underlying domain Ω), if and only if $v \in B^{\alpha}_{\tau^*}(L_{\tau^*}(\Omega))$. Recall the characterization of Besov norms (5.46).

Proposition 11.4 has an interesting application to the Poisson equation

$$-\Delta u = f \quad \text{in} \quad \Omega, \quad u = 0 \quad \text{on} \quad \partial\Omega, \tag{11.23}$$

when Ω is a bounded *Lipschitz domain* in $\mathbb{R}^n$. The efficiency of the best N-term approximation when applied to the solution of Laplace's equation has been studied by Dahlke and DeVore (1995). However, these results were formulated with respect to approximation in $L_2(\Omega)$. For elliptic equations, the energy norm is more natural. A combination of Proposition 11.4 and the results of Dahlke and DeVore (1995) provides the following result concerning approximation relative to $\|\cdot\|_{H^1}$ (Dahlke et al. 1997*a*).

Proposition 11.5 Let Ω be a bounded Lipschitz domain in $\mathbb{R}^n$, and let u denote the solution of (11.23) with $f \in B_2^{\alpha-1}(L_2(\Omega))$, $\alpha \geq 1$. Then

$$\sum_{N=1}^{\infty} \left(N^{s/n}\sigma_{N,1}(u)\right)^{\tau} < \infty \quad \text{for all} \quad 0 < s < s^*/3, \tag{11.24}$$

where $s^* = \min\{\frac{3n}{2(n-1)}, \alpha + 1\}$ and $\tau = (s-1)/n + 1/2$.

To illustrate this result, consider the example where $n = 2$. If $\alpha \geq 2$, then $s^* = 3$. Hence, in this case, the nonlinear method gives an H^1–approximation to u of order up to $N^{-1/2}$, whereas a linear method, that is, uniform refinements, using N terms could only give $N^{-1/4}$ in the worst case.

These facts indicate that adaptive refinements will generally perform significantly better. Establishing a closer connection between the adaptive scheme discussed in the preceding section and N-term approximation is an interesting question under current investigation.

12. What else?

Evidently a lot more than could be included in a survey. Therefore I would like to add only a few brief comments on further interesting directions.

Collocation
It has been pointed out that collocation plays an important role in connection with the fast evaluation of nonlinear terms. Bertoluzza (1997) has discussed some promising features of collocation in connection with highly accurate discretizations. There, interpolatory scaling functions are employed and corresponding analogues to hierarchical bases are established. Again, interpolatory representations are helpful with regard to evaluating nonlinear terms.

Transport problems
The above concepts are more or less tailored to *elliptic* problems. It is less clear how to treat transport terms. Typical model problems are *convection–diffusion* problems of the form

$$-\Delta u + \beta(x) \cdot \nabla u = f \quad \text{in} \quad \Omega, \quad u = 0 \quad \text{on} \quad \partial\Omega, \tag{12.1}$$

where the convection term is strongly dominant. Canuto and Cravero (1996) have proposed discretizing (12.1) with a conventional finite element method and use wavelet expansions of the current solution to determine successive mesh refinements at locations where wavelet coefficients are large. First results by Dahmen, Müller and Schlinkmann (199x) indicate that the concept of stable completions can be successfully employed to design *level-dependent* Petrov–Galerkin discretizations, which in a multigrid context recover the usual multigrid efficiency for elliptic problems even in the case of strong convection terms.

Discrete multiresolution concepts have been developed by Harten (1995), with special emphasis on the treatment of *hyperbolic conservation* laws. It is well known that such systems can be viewed as evolution equations for *cell averages.* This fact serves as the basis for *finite volume discretizations.* However, when advancing in time, these schemes require at some stage the computation of fluxes across cell boundaries, which in turn need pointwise values of the conservative variables. To realize high accuracy, one therefore has to design highly accurate reconstruction schemes to recover the pointwise values from the cell averages, which is actually the only place where a discretization error is introduced. Unfortunately, in realistic problems the evaluation of corresponding numerical fluxes is very costly. The main thrust of Harten's concept therefore aims at reducing the cost of numerical flux computations according to the following idea. Fluxes are initially computed only on a very coarse grid (using, however, data corresponding to the highest level of resolution). The flux values on successively finer grids are then determined either by cheap interpolation schemes from those on coarser levels, whenever they are smooth, or otherwise by expensive accurate schemes. This decision is based on a suitable multiscale representation of the data. This is highly reminiscent of data compression techniques. The underlying multiscale decomposition concept proposed by Harten is very flexible and has to be made concrete in each application. The multiscale transformations have the format (3.26) and (3.28), although, in principle, no explicit knowledge of underlying bases Φ_j, Ψ_j is required. Nevertheless, many technical as well as conceptual problems arise when applying this methodology to concrete problems, in particular, when dealing with several space variables. Some recent contributions can be found in Gottschlich-Müller and Müller (1996), Sjögreen (1995) and Sonar (1995), for instance.

Software

To unfold the full efficiency of most of the concepts discussed so far, rather new data structures are needed. It does not seem to be possible to simply hook wavelet components to existing software for conventional discretization schemes. Existing codes still seem to be confined to model problems. The beginnings of a systematic software development for wavelet schemes in a PDE context are discussed by Barsch et al. (1997), for example.

Wavelets as analysis tools

The primary objective of the developments detailed in this paper is the understanding and design of highly efficient solvers for large-scale problems. I find the variety of contributions very promising and interesting. However, because of the state of the software development, and for conceptual reasons mainly in connection with geometry constraints, it is fair to say that wavelet schemes have not yet become quite competitive with well tuned multigrid codes for realistic problems. On the other hand, the discussion also indicates that the potential of wavelets has not yet been exhausted, and that the results that have been achieved so far provide a highly stimulating source of ideas and further progress. In fact, the above comments on the convection–diffusion problem suggest that true benefit for future generations of multiscale techniques may result from a marriage of different methodologies. I would be very pleased if the present paper could be of some help in this regard.

On the other hand, it has already been indicated that, aside from algorithmic developments, wavelets offer powerful analysis tools. An example, namely investigating the *boundedness of Galerkin projectors in L_p-Sobolev spaces* has already been mentioned by Angeletti et al. (1997). The determination of *Besov regularity* of solutions to elliptic boundary value problems (Dahlke and DeVore 1995, Dahlke 1996) is another intriguing instance which is important for the understanding of adaptivity. The multiresolution approach to *homogenization* (Brewster and Beylkin 1995, Dorobantu 1995) opens further startling perspectives. Wavelets have recently been employed in the study of turbulence and multiscale interaction of flow phenomena (Elezgaray, Berkooz, Dankowicz, Holmes and Myers 1997, Wickerhauser, Farge and Goirand 1997, Farge et al. 1992).

In summary, it seems that wavelets have become indispensible as a conceptual source for understanding multiscale phenomena and corresponding solution schemes.

Acknowledgements

I would like to thank A. Kunoth and S. Müller for their assistance and valuable suggestions during the preparation of the manuscript. I am also indebted to T. Bronger for turning my terrible handwriting into the above result.

REFERENCES

R. A. Adams (1978), *Sobolev Spaces*, Academic Press.

A. Alpert, G. Beylkin, R. Coifman and V. Rokhlin (1993), 'Wavelet-like bases for the fast solution of second-kind integral equations', *SIAM J. Sci. Statist. Comput.* **14**, 159–184.

B. Alpert (1993), 'A class of bases in L_2 for sparse representation of integral operators', *SIAM J. Math. Anal.* **24**, 246–262.

L. Andersson, N. Hall, B. Jawerth and G. Peters (1994), Wavelets on closed subsets of the real line, in *Topics in the Theory and Applications of Wavelets* (L. L. Schumaker and G. Webb, eds), Academic Press, Boston, pp. 1–61.

J. M. Angeletti, S. Mazet and P. Tchamitchian (1997), Analysis of second order elliptic operators without boundary conditions and with VMO or Hölderian coefficients, in *Multiscale Wavelet Methods for PDEs* (W. Dahmen, A. J. Kurdila and P. Oswald, eds), Academic Press. To appear.

A. Averbuch, G. Beylkin, R. Coifman and M. Israeli (1995), Multiscale inversion of elliptic operators, in *Signal and Image Representation in Combined Spaces* (J. Zeevi and R. Coifman, eds), Academic Press, pp. 1–16.

I. Babuška (1973), 'The finite element method with Lagrange multipliers', *Numer. Math.* **20**, 179–192.

I. Babuška and A. Miller (1987), 'A feedback finite element method with *a posteriori* error estimation: Part I. The finite element method and some basic properties of the *a posteriori* error estimator', *Comput. Meth. Appl. Mech. Eng.* **61**, 1–40.

I. Babuška and W. C. Rheinboldt (1978), 'Error estimates for adaptive finite element computations', *SIAM J. Numer. Anal.* **15**, 736–754.

R. E. Bank and A. Weiser (1985), 'Some *a posteriori* error estimates for elliptic partial differential equations', *Math. Comput.* **44**, 283–301.

R. E. Bank, A. H. Sherman and A. Weiser (1983), Refinement algorithms and data structures for regular local mesh refinement, in *Scientific Computing* (R. Stepleman and et al., eds), IMACS, North-Holland, Amsterdam, pp. 3–17.

T. Barsch, A. Kunoth and K. Urban (1997), Towards object oriented software tools for numerical multiscale methods for PDEs using wavelets, in *Multiscale Wavelet Methods for PDEs* (W. Dahmen, A. J. Kurdila and P. Oswald, eds), Academic Press. To appear.

J. Bergh and J. Löfström (1976), *Interpolation Spaces: An Introduction*, Springer.

G. Berkooz, J. Elezgaray and P. Holmes (1993), Wavelet analysis of the motion of coherent structures, in *Progress in Wavelet Analysis and Applications* (Y. Meyer and S. Roques, eds), Editions Frontières, pp. 471–476.

S. Bertoluzza (1994), '*A posteriori* error estimates for wavelet Galerkin methods', Istituto di Analisi Numerica, Pavia. Preprint Nr. 935.

S. Bertoluzza (1997), An adaptive collocation method based on interpolating wavelets, in *Multiscale Wavelet Methods for PDEs* (W. Dahmen, A. J. Kurdila and P. Oswald, eds), Academic Press. To appear.

G. Beylkin (1992), 'On the representation of operators in bases of compactly supported wavelets', *SIAM J. Numer. Anal.* **29**, 1716–1740.

G. Beylkin (1993), Wavelets and fast numerical algorithms, in *Different Perspectives on Wavelets* (I. Daubechies, ed.), Vol. 47 of *Proc. Symp. Appl. Math.*, pp. 89–117.

G. Beylkin and J. M. Keiser (1997), An adaptive pseudo-wavelet approach for solving nonlinear partial differential equations, in *Multiscale Wavelet Methods for PDEs* (W. Dahmen, A. J. Kurdila and P. Oswald, eds), Academic Press. To appear.

G. Beylkin, R. R. Coifman and V. Rokhlin (1991), 'Fast wavelet transforms and numerical algorithms I', *Comm. Pure Appl. Math.* **44**, 141–183.

G. Beylkin, J. M. Keiser and L. Vozovoi (1996), 'A new class of stable time discretization schemes for the solution of nonlinear PDE's'. Preprint.

F. Bornemann and H. Yserentant (1993), 'A basic norm equivalence for the theory of multilevel methods', *Numer. Math.* **64**, 455–476.

F. Bornemann, B. Erdmann and R. Kornhuber (1996), '*A posteriori* error estimates for elliptic problems in two and three space dimensions', *SIAM J. Numer. Anal.* **33**, 1188–1204.

D. Braess (1997), *Finite Elements: Theory, Fast Solvers and Applications in Solid Mechanics*, Cambridge University Press.

J. H. Bramble (1981), 'The Lagrange multiplier method for Dirichlet's problem', *Math. Comput.* **37**, 1–11.

J. H. Bramble (1993), *Multigrid Methods*, Vol. 294 of *Pitman Research Notes in Mathematics*, Longman, London. Co-published in the USA with Wiley, New York.

J. H. Bramble and J. Pasciak (1988), 'A preconditioning technique for indefinite systems resulting from mixed approximations for elliptic problems', *Math. Comput.* **50**, 1–17.

J. H. Bramble and J. Pasciak (1994), Iterative techniques for time dependent Stokes problems. Preprint.

J. H. Bramble, Z. Leyk and J. E. Pasciak (1994), 'The analysis of multigrid algorithms for pseudo-differential operators of order minus one', *Math. Comput.* **63**, 461–478.

J. H. Bramble, J. E. Pasciak and J. Xu (1990), 'Parallel multilevel preconditioners', *Math. Comput.* **55**, 1–22.

A. Brandt and A. A. Lubrecht (1990), 'Multilevel matrix multiplication and the fast solution of integral equations', *J. Comput. Phys.* **90**, 348–370.

A. Brandt and K. Venner (preprint), Multilevel evaluation of integral transforms on adaptive grids.

M. E. Brewster and G. Beylkin (1995), 'A multiresolution strategy for numerical homogenization', *Appl. Comput. Harm. Anal.* **2**, 327–349.

F. Brezzi and M. Fortin (1991), *Mixed and Hybrid Finite Element Methods*, Springer, New York.

C. Canuto and I. Cravero (1996), Wavelet-based adaptive methods for advection–diffusion problems. Preprint, University of Torino.

J. M. Carnicer, W. Dahmen and J. M. Peña (1996), 'Local decomposition of refinable spaces', *Appl. Comput. Harm. Anal.* **3**, 127–153.

J. Carrier, L. Greengard and V. Rokhlin (1988), 'A fast adaptive multipole algorithm for particle simulations', *SIAM J. Sci. Statist. Comput.* **9**, 669–686.

C. Carstensen (1996), 'Efficiency of *a posteriori* BEM-error estimates for first-kind integral equations on quasi-uniform meshes', *Math. Comput.* **65**, 69–84.

A. S. Cavaretta, W. Dahmen and C. A. Micchelli (1991), *Stationary Subdivision.* Mem. Amer. Math. Soc., **93**, #453.

T. Chan and T. Mathew (1994), Domain decomposition algorithms, in *Acta Numerica*, Vol. 3, Cambridge University Press, pp. 61–143.

P. Charton and V. Perrier (1995), Towards a wavelet-based numerical scheme for the two-dimensional Navier–Stokes equations, in *Proceedings of the ICIAM*, Hamburg.

P. Charton and V. Perrier (1996), A pseudo-wavelet scheme for the two-dimensional Navier–Stokes equations. Preprint.

C. K. Chui and E. Quak (1992), Wavelets on a bounded interval, in *Numerical Methods of Approximation Theory* (D. Braess and L. Schumaker, eds), Birkhäuser, Basel, pp. 1–24.

Z. Ciesielski and T. Figiel (1983), 'Spline bases in classical function spaces on compact C^∞ manifolds, parts I & II', *Studia Math.* **76**, 1–58, 95–136.

A. Cohen (1994). Private communication.

A. Cohen and I. Daubechies (1993), 'Non-separable bidimensional wavelet bases', *Revista Mat. Iberoamericana* **9**, 51–137.

A. Cohen and J.-M. Schlenker (1993), 'Compactly supported bi-dimensional wavelet bases with hexagonal symmetry', *Constr. Appr.* **9**, 209–236.

A. Cohen, W. Dahmen and R. DeVore (1995), 'Multiscale decompositions on bounded domains'. IGPM-Report 113, RWTH Aachen. To appear in *Trans. Amer. Math. Soc.*

A. Cohen, I. Daubechies and J.-C. Feauveau (1992), 'Biorthogonal bases of compactly supported wavelets', *Comm. Pure Appl. Math.* **45**, 485–560.

A. Cohen, I. Daubechies and P. Vial (1993), 'Wavelets on the interval and fast wavelet transforms', *Appl. Comput. Harm. Anal.* **1**, 54–81.

R. R. Coifman, Y. Meyer and M. V. Wickerhauser (1992), Size properties of the wavelet packets, in *Wavelets and their Applications* (G. Beylkin, R. R. Coifman, I. Daubechies, S. Mallat, Y. Meyer, L. A. Raphael and M. B. Ruskai, eds), Jones and Bartlett, Cambridge, MA, pp. 453–470.

R. R. Coifman, Y. Meyer, S. R. Quake and M. V. Wickerhauser (1993), Signal processing and compression with wavelet packets, in *Progress in Wavelet Analysis and Applications* (Y. Meyer and S. Roques, eds), Editions Frontières, Paris, pp. 77–93.

S. Dahlke (1996), 'Wavelets: Construction principles and applications to the numerical treatment of operator equations'. Habilitationsschrift, Aachen.

S. Dahlke and R. DeVore (1995), 'Besov regularity for elliptic boundary value problems'. IGPM-Report 116, RWTH Aachen.

S. Dahlke and I. Weinreich (1993), 'Wavelet-Galerkin methods: An adapted biorthogonal wavelet basis', *Constr. Approx.* **9**, 237–262.

S. Dahlke and I. Weinreich (1994), 'Wavelet bases adapted to pseudo-differential operators', *Appl. Comput. Harm. Anal.* **1**, 267–283.

S. Dahlke, W. Dahmen and R. DeVore (1997*a*), Nonlinear approximation and adaptive techniques for solving elliptic operator equations, in *Multiscale Wavelet Methods for PDEs* (W. Dahmen, A. Kurdila and P. Oswald, eds), Academic Press. To appear.

S. Dahlke, W. Dahmen and V. Latour (1995), 'Smooth refinable functions and wavelets obtained by convolution products', *Appl. Comput. Harm. Anal.* **2**, 68–84.

S. Dahlke, W. Dahmen, R. Hochmuth and R. Schneider (1997*b*), 'Stable multiscale bases and local error estimation for elliptic problems', *Appl. Numer. Math.* **23**, 21–48.

W. Dahmen (1994), Some remarks on multiscale transformations, stability and biorthogonality, in *Wavelets, Images and Surface Fitting* (P. J. Laurent, A. le Méhauté and L. L. Schumaker, eds), A. K. Peters, Wellesley, MA, pp. 157–188.

W. Dahmen (1995), Multiscale analysis, approximation, and interpolation spaces, in *Approximation Theory VIII, Wavelets and Multilevel Approximation* (C. K. Chui and L. L. Schumaker, eds), World Scientific, pp. 47–88.

W. Dahmen (1996), 'Stability of multiscale transformations', *Journal of Fourier Analysis and Applications* **2**, 341–361.

W. Dahmen and A. Kunoth (1992), 'Multilevel preconditioning', *Numer. Math.* **63**, 315–344.

W. Dahmen and C. A. Micchelli (1993), 'Using the refinement equation for evaluating integrals of wavelets', *SIAM J. Numer. Anal.* **30**, 507–537.

W. Dahmen and R. Schneider (1996), 'Composite wavelet bases'. IGPM-Report 133, RWTH Aachen.

W. Dahmen and R. Schneider (1997*a*), Wavelets on manifolds I. Construction and domain decomposition. In preparation.

W. Dahmen and R. Schneider (1997*b*), Wavelets on manifolds II. Applications to boundary integral equations. In preparation.

W. Dahmen, B. Kleemann, S. Prößdorf and R. Schneider (1994*a*), A multiscale method for the double layer potential equation on a polyhedron, in *Advances in Computational Mathematics* (H. P. Dikshit and C. A. Micchelli, eds), World Scientific, pp. 15–57.

W. Dahmen, B. Kleemann, S. Prößdorf and R. Schneider (1996*a*), Multiscale methods for the solution of the Helmholtz and Laplace equation, in *Boundary*

Element Methods: Report from the Final Conference of the Priority Research Programme 1989–1995 of the German Research Foundation, Oct. 2–4, 1995 in Stuttgart (W. Wendland, ed.), Springer, pp. 180–211.

W. Dahmen, A. Kunoth and R. Schneider (1997), Operator equations, multiscale concepts and complexity, in *Mathematics of Numerical Analysis: Real Number Algorithms* (J. Renegar, M. Shub and S. Smale, eds), Vol. 32 of *Lectures in Applied Mathematics*, AMS, Providence, RI, pp. 225–261.

W. Dahmen, A. Kunoth and K. Urban (1996*b*), 'Biorthogonal spline-wavelets on the interval: Stability and moment conditions'. IGPM-Report 129, RWTH Aachen.

W. Dahmen, A. Kunoth and K. Urban (1996*c*), 'A wavelet Galerkin method for the Stokes problem', *Computing* **56**, 259–302.

W. Dahmen, S. Müller and T. Schlinkmann (199x), Multiscale techniques for convection-dominated problems. In preparation.

W. Dahmen, P. Oswald and X.-Q. Shi (1993*a*), 'C^1-hierarchical bases', *J. Comput. Appl. Math.* **9**, 263–281.

W. Dahmen, S. Prößdorf and R. Schneider (1993*b*), 'Wavelet approximation methods for pseudodifferential equations II: Matrix compression and fast solution', *Advances in Computational Mathematics* **1**, 259–335.

W. Dahmen, S. Prößdorf and R. Schneider (1994*b*), Multiscale methods for pseudo-differential equations on smooth manifolds, in *Proceedings of the International Conference on Wavelets: Theory, Algorithms, and Applications* (C. K. Chui, L. Montefusco and L. Puccio, eds), Academic Press, pp. 385–424.

W. Dahmen, S. Prößdorf and R. Schneider (1994*c*), 'Wavelet approximation methods for pseudodifferential equations I: Stability and convergence', *Math. Z.* **215**, 583–620.

R. Danchin (1997), PhD thesis, Université Paris VI. In preparation.

I. Daubechies (1988), 'Orthonormal bases of compactly supported wavelets', *Comm. Pure Appl. Math.* **41**, 909–996.

I. Daubechies (1992), *Ten Lectures on Wavelets*, Vol. 61 of *CBMS-NSF Regional Conference Series in Applied Math.*, SIAM, Philadelphia.

C. de Boor, R. DeVore and A. Ron (1993), 'On the construction of multivariate (pre-) wavelets', *Constr. Approx.* **9**(2), 123–166.

R. DeVore and B. Lucier (1992), Wavelets, in *Acta Numerica*, Vol. 1, Cambridge University Press, pp. 1–56.

R. DeVore and V. Popov (1988*a*), 'Interpolation of Besov spaces', *Trans. Amer. Math. Soc.* **305**, 397–414.

R. DeVore and V. Popov (1988*b*), Interpolation spaces and nonlinear approximation, in *Function Spaces and Approximation* (M. Cwikel, J. Peetre, Y. Sagher and H. Wallin, eds), Lecture Notes in Math., Springer, pp. 191–205.

R. DeVore and B. Sharpley (1993), 'Besov spaces on domains in $\mathbb{R}^d$', *Trans. Amer. Math. Soc.* **335**, 843–864.

R. DeVore, B. Jawerth and V. Popov (1992), 'Compression of wavelet decompositions', *Amer. J. Math.* **114**, 737–785.

W. Dörfler (1996), 'A convergent adaptive algorithm for Poisson's equation', *SIAM J. Numer. Anal.* **33**, 1106–1124.

M. Dorobantu (1995), Wavelet-Based Algorithms for Fast PDE Solvers, PhD thesis, Royal Institute of Technology, Stockholm University.

J. Elezgaray, G. Berkooz, H. Dankowicz, P. Holmes and M. Myers (1997), Local models and large scale statistics of the Kuramoto–Sivansky equation, in *Multiscale Wavelet Methods for PDEs* (W. Dahmen, A. J. Kurdila and P. Oswald, eds), Academic Press. To appear.

B. Enquist, S. Osher and S. Zhong (1994), 'Fast wavelet-based algorithms for linear evolution operators', *SIAM J. Sci. Comput.* **15**, 755–775.

K. Eriksson, D. Estep, P. Hansbo and C. Johnson (1995), Introduction to adaptive methods for differential equations, in *Acta Numerica*, Vol. 4, Cambridge University Press, pp. 105–158.

M. Farge, E. Goirand, Y. Meyer, F. Pascal and M. V. Wickerhauser (1992), 'Improved predictability of two-dimensional turbulent flows using wavelet packet compression', *Fluid Dynam. Res.* **10**, 229–250.

B. Fornberg and G. B. Whitham (1978), 'A numerical and theoretical study of certain nonlinear wave phenomena', *Philos. Trans. R. Soc. London, Ser. A* **289**, 373–404.

M. Fortin (1977), 'An analysis of convergence of mixed finite element methods', *R.A.I.R.O. Anal. Numer.* **11 R3**, 341–354.

J. Fröhlich and K. Schneider (1995), 'An adaptive wavelet-vaguelette algorithm for the solution of nonlinear PDEs'. Preprint SC 95-28, ZIB.

J. Fröhlich and K. Schneider (1996), 'Numerical simulation of decaying turbulence in an adaptive wavelet basis'. Preprint, Universität Kaiserslautern, Fachbereich Chemie.

V. Girault and P.-A. Raviart (1986), *Finite Element Methods for Navier–Stokes Equations: Theory and Algorithms*, Series in Computational Mathematics, Springer.

R. Glowinski, T. W. Pan, R. O. Wells and X. Zhou (1996), 'Wavelet and finite element solutions for the Neumann problem using fictitious domains', *J. Comput. Phys.* **126**, 40–51.

R. Glowinski, A. Rieder, R. O. Wells and X. Zhou (1993), A wavelet multigrid preconditioner for Dirichlet boundary value problems in general domains, Technical Report 93-06, Rice University, Houston.

B. Gottschlich-Müller and S. Müller (1996), 'Multiscale concept for conservation laws'. IGPM-Report 128, RWTH Aachen.

L. Greengard and V. Rokhlin (1987), 'A fast algorithm for particle simulations', *J. Comput. Phys.* **73**, 325–348.

M. Griebel (1994), *Multilevelmethoden als Iterationsverfahren über Erzeugendensystemen*, Teubner Skripten zur Numerik, Teubner, Stuttgart.

M. Griebel and P. Oswald (1995*a*), 'Remarks on the abstract theory of additive and multiplicative Schwarz algorithms', *Numer. Math.* **70**, 163–180.

M. Griebel and P. Oswald (1995*b*), 'Tensor product type subspace splittings and multilevel iterative methods for anisotropic problems', *Advances in Computational Mathematics* **4**, 171–206.

K. H. Gröchenich and W. R. Madych (1992), 'Haar bases and self-affine tilings of $\mathbb{R}^n$', *IEEE Trans. Inform. Theory* **38**, 556–568.

W. Hackbusch (1985), *Multigrid Methods and Applications*, Springer, New York.

W. Hackbusch (1989), 'The frequency-decomposition multigrid method I, Applications to anisotropic equations', *Numer. Math.* **56**, 229–245.

W. Hackbusch (1992), 'The frequency-decomposition multigrid method II, Convergence analysis based on the additive Schwarz method', *Numer. Math.* **63**, 433–453.

W. Hackbusch and Z. P. Nowak (1984), 'On the fast matrix multiplication in the boundary element method by panel clustering', *Numer. Math.* **54**, 463–491.

W. Hackbusch and S. Sauter (1993), 'On the efficient use of the Galerkin method to solve Fredholm integral equations', *Appl. Math.* **38**, 301–322.

A. Harten (1995), 'Multiresolution algorithms for the numerical solution of hyperbolic conservation laws', *Comm. Pure Appl. Math.* **48**, 1305–1342.

S. Hildebrandt and E. Wienholtz (1964), 'Constructive proofs of representation theorems in separable Hilbert spaces', *Comm. Pure Appl. Math.* **17**, 369–373.

R. Hochmuth (1996), '*A posteriori* estimates and adaptive schemes for transmission problems'. IGPM Report 131, RWTH Aachen.

S. Jaffard (1992), 'Wavelet methods for fast resolution of elliptic equations', *SIAM J. Numer. Anal.* **29**, 965–986.

R. Q. Jia and C. A. Micchelli (1991), Using the refinement equation for the construction of pre-wavelets II: Powers of two, in *Curves and Surfaces* (P. J. Laurent, A. le Méhauté and L. L. Schumaker, eds), Academic Press, pp. 209–246.

H. Johnen and K. Scherer (1977), On the equivalence of the K-functional and moduli of continuity and some applications, in *Constructive Theory of Functions of Several Variables*, Vol. 571 of *Lecture Notes in Math.*, Springer, pp. 119–140.

P. Joly, Y. Maday and V. Perrier (1994), 'Towards a method for solving partial differential equations by using wavelet packet bases', *Comput. Meth. Appl. Mech. Engrg.* **116**, 301–307.

P. Joly, Y. Maday and V. Perrier (1997), A dynamical adaptive concept based on wavelet packet best bases: Application to convection diffusion partial differential equations, in *Multiscale Wavelet Methods for PDEs* (W. Dahmen, A. Kurdila and P. Oswald, eds), Academic Press. To appear.

A. Jouini (1992), 'Constructions de bases d'ondelettes sur les variétés'. Dissertation, Université Paris Sud–Centre d'Orsay.

A. Jouini and P. G. Lemarié-Rieusset (1992), Ondelettes sur un ouvert borné du plan. Preprint.

A. Jouini and P. G. Lemarié-Rieusset (1993), 'Analyses multirésolutions biorthogonales et applications', *Ann. Inst. Henri Poincaré, Anal. Non Lineaire* **10**, 453–476.

J. Ko, A. J. Kurdila and P. Oswald (1997), Scaling function and wavelet preconditioners for second order elliptic problems, in *Multiscale Wavelet Methods for PDEs* (W. Dahmen, A. Kurdila and P. Oswald, eds), Academic Press. To appear.

U. Kotyczka and P. Oswald (1996), Piecewise linear pre-wavelets of small support, in *Approximation Theory VIII, vol. 2* (C. K. Chui and L. L. Schumaker, eds), World Scientific, Singapore, pp. 235–242.

H. Kumano-go (1981), *Pseudo-Differential Operators*, MIT Press, Boston.

A. Kunoth (1994), Multilevel Preconditioning, PhD thesis, FU Berlin. Shaker, Aachen.

A. Kunoth (1995), 'Multilevel preconditioning: Appending boundary conditions by Lagrange multipliers', *Advances in Computational Mathematics* **4**, 145–170.

O. A. Ladyshenskaya (1969), *The Mathematical Theory of Viscous Incompressible Flow*, 2nd edn, Gordon and Breach, New York.

A. Latto, H. L. Resnikoff and E. Tenenbaum (1992), The evaluation of connection coefficients of compactly supported wavelets, in *Proceedings of French–USA Workshop on Wavelets and Turbulence* (Y. Maday, ed.), Springer.

S. Lazaar (1995), Algorithmes à base d'ondelettes et résolution numérique de problèmes elliptiques à coefficients variables, PhD thesis, Université d'Aix-Marseille I.

S. Lazaar, J. Liandrat and P. Tchamitchian (1994), 'Algorithme à base d'ondelettes pour la résolution numérique d'équations aux dérivées partielles à coefficients variables', *C.R. Acad. Sci., Série I* **319**, 1101–1107.

P. G. Lemarié (1984), *Algèbre d'opérateurs et semi-groupes de Poisson sur un espace de nature homogène*, Publ. Math. d'Orsay.

P. G. Lemarié-Rieusset (1992), 'Analyses, multi-résolutions nonorthogonales, commutation entre projecteurs et derivation et ondelettes vecteurs à divergence nulle', *Revista Mat. Iberoamericana* **8**, 221–236.

P. G. Lemarié-Rieusset (1994), 'Un théorème d'inexistence pour des ondelettes vecteurs à divergence nulle', *C. R. Acad. Sci. Paris I* **319**, 811–813.

J. Liandrat and P. Tchamitchian (1997), 'Elliptic operators, adaptivity and wavelets', *SIAM J. Numer. Anal.* To appear.

R. Lorentz and P. Oswald (1996), 'Constructing 'economic' Riesz bases for Sobolev spaces', GMD-Birlinghoven. Preprint.

R. Lorentz and P. Oswald (1997), Multilevel finite element Riesz bases in Sobolev spaces, in *DD9 Proceedings* (P. Bjorstad, M. Espedal and D. Keyes, eds), Wiley. To appear.

Y. Maday, V. Perrier and J. C. Ravel (1991), 'Adaptivité dynamique sur base d'ondelettes pour l'approximation d'équations aux dérivées partielles', *C. R. Acad. Sci. Paris Sér. I Math.* **312**, 405–410.

S. Mallat (1989), 'Multiresolution approximations and wavelet orthonormal bases of $L_2(\mathbb{R})$', *Trans. Amer. Math. Soc.* **315**, 69–87.

Y. Meyer (1990), *Ondelettes et opérateurs 1–3: Ondelettes*, Hermann, Paris.

Y. Meyer (1994). Private communication.

C. A. Micchelli and Y. Xu (1994), 'Using the matrix refinement equation for the construction of wavelets on invariant set', *Appl. Comput. Harm. Anal.* **1**, 391–401.

S. G. Michlin (1965), *Multidimensional Singular Integral Equations*, Pergamon Press, Oxford.

J. C. Nedelec (1982), 'Integral equations with non-integrable kernels', *Integral Equations Operator Theory* **5**, 562–572.

S. V. Nepomnyaschikh (1990), 'Fictitious components and subdomain alternating methods', *Sov. J. Numer. Anal. Math. Modelling* **5**, 53–68.

G. Nießen (1995), 'An explicit norm representation for the analysis of multilevel methods'. IGPM-Preprint 115, RWTH Aachen.

P. Oswald (1990), 'On function spaces related to finite element approximation theory', *Z. Anal. Anwendungen* **9**, 43–64.

P. Oswald (1992), On discrete norm estimates related to multilevel preconditioners in the finite element method, in *Constructive Theory of Functions (Proc. Int. Conf. Varna, 1991)* (K. G. Ivanov, P. Petrushev and B. Sendov, eds), Bulg. Acad. Sci., Sofia, pp. 203–214.

P. Oswald (1994), *Multilevel Finite Element Approximations*, Teubner Skripten zur Numerik, Teubner, Stuttgart.

P. Oswald (1997), Multilevel solvers for elliptic problems on domains, in *Multiscale Wavelet Methods for PDEs* (W. Dahmen, A. Kurdila and P. Oswald, eds), Academic Press. To appear.

J. Peetre (1978), *New Thoughts on Besov Spaces*, Duke University Press, Durham, NC.

V. Perrier (1996), Numerical schemes for 2D-Navier–Stokes equations using wavelet bases. Preprint.

P. J. Ponenti (1994), Algorithmes en ondelettes pour la résolution d'équations aux dérivées partielles, PhD thesis, Université Aix-Marseille I.

L. Quartapelle (1993), *Numerical Solution of the Incompressible Navier–Stokes Equations*, Vol. 113 of *International Series of Numerical Mathematics*, Birkhäuser.

V. Rokhlin (1985), 'Rapid solution of integral equations of classical potential theory', *J. Comput. Phys.* **60**, 187–207.

S. Sauter (1992), Über die effiziente Verwendung des Galerkinverfahrens zur Lösung Fredholmscher Integralgleichungen, PhD thesis, Universität Kiel.

R. Schneider (1995), 'Multiskalen- und Wavelet-Matrixkompression: Analysisbasierte Methoden zur effizienten Lösung großer vollbesetzter Gleichungssysteme'. Habilitationsschrift, Technische Hochschule Darmstadt.

P. Schröder and W. Sweldens (1995), Spherical wavelets: Efficiently representing functions on the sphere, in *Computer Graphics Proceedings (SIGGRAPH 95)*, ACM SIGGRAPH, pp. 161–172.

C. Schwab (1994), 'Variable order composite quadrature of singular and nearly singular integrals', *Computing* **53**, 173–194.

B. Sjögreen (1995), 'Numerical experiments with the multiresolution scheme for the compressible Euler equations', *J. Comput. Phys.* **117**, 251–261.

T. Sonar (1995), 'Multivariate Rekonstruktionsverfahren zur numerischen Berechnung hyperbolischer Erhaltungsgleichungen'. Habilitationsschrift, Technische Hochschule Darmstadt.

R. P. Stevenson (1995*a*), A robust hierarchical basis preconditioner on general meshes. Preprint, University of Nijmegen.

R. P. Stevenson (1995*b*), 'Robustness of the additive and multiplicative frequency decomposition multilevel method', *Computing* **54**, 331–346.

R. P. Stevenson (1996), 'The frequency decomposition multilevel method: A robust additive hierarchical basis preconditioner', *Math. Comput.* **65**, 983–997.

W. Sweldens (1996), 'The lifting scheme: A custom-design construction of biorthogonal wavelets', *Appl. Comput. Harm. Anal.* **3**, 186–200.

W. Sweldens (1997), 'The lifting scheme: A construction of second generation wavelets', *SIAM J. Math. Anal.* To appear.

W. Sweldens and R. Piessens (1994), 'Quadrature formulae and asymptotic error expansions for wavelet approximations of smooth functions', *SIAM J. Num. Anal.* **31**, 2140–2164.

P. Tchamitchian (1987), 'Biorthogonalité et théorie des opérateurs', *Revista Mat. Iberoamericana* **3**, 163–189.

P. Tchamitchian (1996), Wavelets, functions, and operators, in *Wavelets: Theory and Applications* (G. Erlebacher, M. Y. Hussaini and L. Jameson, eds), ICASE/LaRC Series in Computational Science and Engineering, Oxford University Press, pp. 83–181.

P. Tchamitchian (1997), 'Inversion de certains opérateurs elliptiques à coefficients variables', *SIAM J. Math. Anal.* To appear.

H. Triebel (1978), *Interpolation Theory, Function Spaces, and Differential Operators*, North-Holland, Amsterdam.

K. Urban (1995*a*), Multiskalenverfahren für das Stokes-Problem und angepaßte Wavelet-Basen, PhD thesis, RWTH Aachen. Aachener Beiträge zur Mathematik.

K. Urban (1995*b*), 'On divergence free wavelets', *Advances in Computational Mathematics* **4**, 51–82.

K. Urban (1995*c*), A wavelet-Galerkin algorithm for the driven cavity Stokes problem in two space dimension, in *Numerical Modelling in Continuum Mechanics* (M. Feistauer, R. Rannacher and K. Korzel, eds), Charles University, Prague, pp. 278–289.

K. Urban (1996), Using divergence free wavelets for the numerical solution of the Stokes problem, in *AMLI '96: Proceedings of the Conference on Algebraic Multilevel Iteration Methods with Applications* (O. Axelsson and B. Polman, eds), University of Nijmegen, pp. 261–278.

P. S. Vassilevski and J. Wang (1997*a*), 'Stabilizing the hierarchical basis by approximate wavelets, I: Theory', *Numer. Lin. Alg. Appl.* To appear.

P. S. Vassilevski and J. Wang (1997*b*), 'Stabilizing the hierarchical basis by approximate wavelets, II: Implementation', *SIAM J. Sci. Comput.* To appear.

R. Verfürth (1994), '*A posteriori* error estimation and adaptive mesh refinement techniques', *J. Comput. Appl. Math.* **50**, 67–83.

L. F. Villemoes (1993), Sobolev regularity of wavelets and stability of iterated filter banks, in *Progress in Wavelet Analysis and Applications* (Y. Meyer and S. Roques, eds), Editions Frontières, Paris, pp. 243–251.

T. von Petersdorff and C. Schwab (1997*a*), Fully discrete multiscale Galerkin BEM, in *Multiscale Wavelet Methods for PDEs* (W. Dahmen, A. Kurdila and P. Oswald, eds), Academic Press. To appear.

T. von Petersdorff and C. Schwab (1997*b*), 'Wavelet approximation for first kind integral equations on polygons', *Numer. Math.* To appear.

T. von Petersdorff, R. Schneider and C. Schwab (1997), 'Multiwavelets for second kind integral equations', *SIAM J. Numer. Anal.* To appear.

W. L. Wendland (1987), Strongly elliptic boundary integral equations, in *The State of the Art in Numerical Analysis* (A. Iserles and M. J. D. Powell, eds), Clarendon Press, Oxford, pp. 511–561.

M. V. Wickerhauser, M. Farge and E. Goirand (1997), Theoretical dimension and the complexity of simulated turbulence, in *Multiscale Wavelet Methods for*

PDEs (W. Dahmen, A. J. Kurdila and P. Oswald, eds), Academic Press. To appear.

J. Xu (1992), 'Iterative methods by space decomposition and subspace correction', *SIAM Review* **34**, 581–613.

H. Yserentant (1986), 'On the multilevel splitting of finite element spaces', *Numer. Math.* **49**, 379–412.

H. Yserentant (1990), 'Two preconditioners based on the multi-level splitting of finite element spaces', *Numer. Math.* **58**, 163–184.

H. Yserentant (1993), Old and new proofs for multigrid algorithms, in *Acta Numerica*, Vol. 2, Cambridge University Press, pp. 285–326.

X. Zhang (1992), 'Multilevel Schwarz methods', *Numer. Math.* **63**, 521–539.

Acta Numerica (1997), *pp.* 229–269

A new version of the Fast Multipole Method for the Laplace equation in three dimensions

Leslie Greengard*
Courant Institute of Mathematical Sciences,
New York University,
New York, NY 10012, USA

Vladimir Rokhlin†
Departments of Mathematics and Computer Science,
Yale University,
New Haven, CT 06520, USA

We introduce a new version of the Fast Multipole Method for the evaluation of potential fields in three dimensions. It is based on a new diagonal form for translation operators and yields high accuracy at a reasonable cost.

CONTENTS

* The work of this author was supported by DARPA/AFOSR under Contract F49620-95-C-0075 and by the US Department of Energy under Contract DEFGO288ER25053.

† The work of this author was supported by DARPA/AFOSR under Contract F49620-95-C-0075 and by ONR under Grant N00014-96-1-0188.

1. Introduction

In this paper, we introduce a new version of the Fast Multipole Method (FMM) for the evaluation of potential fields in three dimensions. The scheme evaluates all pairwise interactions in large ensembles of particles, *i.e.* expressions of the form

$$\Phi(x_j) = \sum_{\substack{i=1 \\ i \neq j}}^{n} \frac{q_i}{\|x_j - x_i\|} \tag{1.1}$$

for the gravitational or electrostatic potential, and

$$E(x_j) = \sum_{\substack{i=1 \\ i \neq j}}^{n} q_i \frac{x_j - x_i}{\|x_j - x_i\|^3} \tag{1.2}$$

for the field, where $x_1, x_2, \ldots, x_n$ are points in $\mathbb{R}^3$, and $q_1, q_2, \ldots, q_n$ are a set of (real) coefficients. Here $\| \cdot \|$ denotes the Euclidean norm.

The evaluation of expressions of the form (1.1) is closely related to a number of important problems in applied mathematics, physics, chemistry and biology. Molecular dynamics and Hartree–Fock calculations in chemistry, the evolution of large-scale gravitational systems in astrophysics, capacitance extraction in electrical engineering, and vortex methods in fluid dynamics are all examples of areas where simulations require rapid and accurate evaluation of sums of the form (1.1) and (1.2). When certain closely related interactions are considered as well, involving expressions of the form

$$\Phi(x_j) = \sum_{\substack{i=1 \\ i \neq j}}^{n} q_i \frac{e^{ik\|x_j - x_i\|}}{\|x_j - x_i\|}, \tag{1.3}$$

the list of applications becomes even more extensive.

This paper is a continuation (after an interval of several years) of a sequence of joint papers by the authors, starting with Greengard and Rokhlin (1987) and Carrier, Greengard and Rokhlin (1988) which introduced the Fast Multipole Method in two dimensions. Subsequent work extended the method to three dimensions (Greengard 1988, Greengard and Rokhlin 1988*a*, 1988*b*), and there followed a number of versions of the scheme, both by the present authors and by other researchers; see, for example, Ånderson (1992), Nabors, Korsmeyer, Leighton and White (1994), Berman (1995), Epton and Dembart (1995), Elliott and Board (1996). After about ten years of research, however, a somewhat unsatisfactory picture has emerged. In short, there now exist extremely efficient algorithms for the evaluation of the two-dimensional analogues of (1.1), (1.2) with (practically) arbitrarily high precision, as well as very efficient and accurate algorithms for a host of related problems (Rokhlin 1988, Alpert and Rokhlin 1991, Beylkin, Coifman and Rokhlin 1991, Coifman and Meyer 1991, Greengard and Strain 1991,

Strain 1991, Alpert, Beylkin, Coifman and Rokhlin 1993). However, for the sums (1.1) and (1.2), there are few practical schemes, and these provide only limited accuracy. Since most real-world problems are three-dimensional, it can be said that analysis-based 'fast' methods are a promising group of techniques, but that they have not yet lived up to all their expectations.

In the present paper, we try to remedy this situation. We describe a version of the Fast Multipole Method in three dimensions that produces high accuracy at an acceptable computational cost. As will be seen from the numerical examples in Section 9, the new scheme has a break-even point of $n \sim 2000$ when compared with direct calculation in single precision; with 10-digit accuracy, the break-even point is $n \sim 5000$; with 3-digit accuracy, it is $n \sim 500$. The approach uses a considerably more involved mathematical (and numerical) apparatus than is customary in the design of fast multipole-type algorithms. This apparatus is based on a new diagonal form for translation operators acting on harmonic functions, extending the two-dimensional version introduced by Hrycak and Rokhlin (1995). The overall approach bears some resemblance to that used in Fast Multipole Methods for high-frequency scattering problems, which are based on diagonal forms of translation operators for the Helmholtz equation (Rokhlin 1990*b*, 1995, Epton and Dembart 1995).

2. Philosophical preliminaries

We begin with an overview of analysis-based 'fast' numerical algorithms, concentrating on the evaluation of expressions of the form (1.1). Where possible, we summarize the current 'state of the art' in the field.

If we define the $n \times n$ matrix A by the formula

$$A_{ij} = \frac{1}{\|x_j - x_i\|}, \tag{2.1}$$

we can rewrite (1.1) in the form

$$\Phi = Aq, \tag{2.2}$$

with $\Phi, q \in \mathbb{R}^n$ (the expression (1.2) can be rewritten in a similar fashion). Obviously, straightforward evaluation of either of the expressions (1.1), (1.2) requires $O(n^2)$ operations (evaluating n potentials at n points), and for large-scale problems this estimate is prohibitively large. On the other hand, the evaluation of expressions of the forms (1.1), (1.2) is an integral part of the numerical solution of many important problems in applied mathematics, and during the last decade, several 'fast' schemes have been proposed for this purpose, that is, schemes whose computational cost is less than $O(n^2)$. Typically, such methods require $O(n)$ or $O(n \log n)$ operations (Rokhlin 1985, Anderson 1986, Greengard and Rokhlin 1987, Carrier et al. 1988, Rokhlin 1988, 1990*b*, Brandt and Lubrecht 1990, Brandt 1991, Alpert and Rokhlin

1991, Beylkin et al. 1991, Coifman and Meyer 1991, Greengard and Strain 1991, Strain 1991, 1992, Epton and Dembart 1995). All of them are based on the straightforward observation that the potentials are smooth functions in $\mathbb{R}^3$, except when x_i is near x_j, and as a result, large submatrices of A are well approximated by low-rank matrices. Clearly, applying a matrix of dimension $n \times n$ but rank J to an arbitrary vector requires only nJ operations (as opposed to n^2); this simple observation leads directly to a variety of asymptotically 'fast' schemes for the evaluation of (1.1); below, we illustrate the construction of such schemes with a simple example.

Suppose that, in the expression (1.1), the points $x_1, x_2, \ldots, x_n$ are equispaced and lie on the interval $[-1, 1]$, so that

$$x_1 = -1, \quad x_2 = -1 + h, \quad \ldots, \quad x_{n-1} = 1 - h, \quad x_n = 1, \tag{2.3}$$

where $h = 2/(n-1)$. Given three integers l, m, k such that

$$\begin{aligned} 1 &\leq l \leq n, \\ 1 &\leq m \leq n, \\ 1 &\leq k \leq n - l, \\ 1 &\leq k \leq n - m, \end{aligned} \tag{2.4}$$

we will denote by $A_{l,m,k}$ the submatrix of A consisting of such elements A_{ij} that

$$\begin{aligned} l &\leq i \leq l + k - 1, \\ m &\leq j \leq m + k - 1, \end{aligned} \tag{2.5}$$

and say that $A_{l,m,k}$ is separated from the diagonal if

$$| \, l - (m + k - 1) \, | > k, \tag{2.6}$$

and

$$| \, m - (l + k - 1) \, | > k. \tag{2.7}$$

In other words, we will say that the submatrix $A_{l,m,k}$ of the matrix A is separated from the diagonal if its distance from the diagonal of A is greater than or equal to its own size (Figure 1). We will construct a rudimentary 'fast' algorithm for the application of the matrix A to an arbitrary vector by means of the following lemma; its proof is based on several well-known facts, all of which can be found in Dahlquist and Bjork (1974).

Lemma 2.1 For any integer $p \leq 1$, and any l, m, k satisfying the conditions (2.4), there exists a matrix $B_{l,m,k}$ of dimension $k \times k$ and rank J, such that

$$\| A_{l,m,k} - B_{l,m,k} \| \leq \frac{1}{4^J}. \tag{2.8}$$

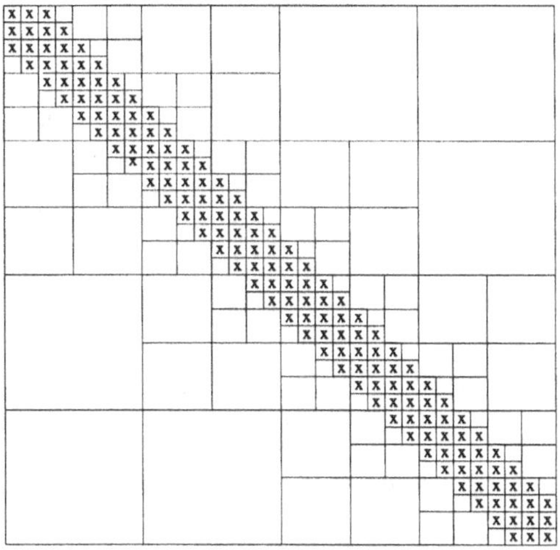

Fig. 1. Subdivision of matrix into well separated blocks. The submatrices marked by an X are not well separated from the diagonal

In other words, any submatrix of A separated from the diagonal is of rank J, to the precision $1/4^J$.

Outline of proof. We start by defining the function $f : \mathbb{R}^2 \to \mathbb{R}^1$ by the formula

$$f(x,y) = \frac{1}{\|x-y\|}, \tag{2.9}$$

and observing that f is smooth everywhere in $\mathbb{R}^2$, except when $x = y$. We will say that the square $[a, a+c] \times [b, b+c] \subset \mathbb{R}^2$ is separated from the diagonal if

$$\mid a+c-b \mid > c, \tag{2.10}$$

and

$$\mid b+c-a \mid > c, \tag{2.11}$$

and observe that on any such square, the function f can be expanded in a two-dimensional Chebychev series, that is, represented in the form

$$f(x,y) = \sum_{p,q=0}^{\infty} \alpha_{pq} T_p \left(\frac{2x}{c} - \frac{c+2a}{c} \right) T_q \left(\frac{2y}{c} - \frac{c+2b}{c} \right), \tag{2.12}$$

with T_j denoting the jth Chebychev polynomial. Finally, we observe that for any a, b, c satisfying the conditions (2.10), (2.11), the convergence of the expansion (2.12) is given by the formula

$$\left| f(x,y) - \sum_{p,q=0}^{J} \alpha_{pq} T_p\left(\frac{2x}{c} - \frac{c+2a}{c}\right) T_q\left(\frac{2y}{c} - \frac{c+2b}{c}\right) \right| < \frac{1}{4^J}. \tag{2.13}$$

In other words, for any square separated from the diagonal, the expansion (2.12) converges to accuracy ε after no more than $\log_4(\varepsilon)$ terms. Combining (2.6), (2.7) and (1.1) with (2.12) and (2.13), we observe that, for any i, j satisfying the inequalities (2.5),

$$\left| A_{ij} - \sum_{p,q=0}^{J} \alpha_{pq} T_p\left(\frac{2x_i}{c} - \frac{c+2a}{c}\right) T_q\left(\frac{2y_j}{c} - \frac{c+2b}{c}\right) \right| < \frac{1}{4^J}, \tag{2.14}$$

with $a = (2l)/n - 1$, $b = (2m)/n - 1$, $c = (2k)/n$. The matrix $B_{l,m,k}$ defined by

$$(B_{l,m,k})_{ij} = \sum_{p,q=0}^{J} \alpha_{pq} T_p\left(\frac{2x_i}{c} - \frac{c+2a}{c}\right) T_q\left(\frac{2y_j}{c} - \frac{c+2b}{c}\right) \tag{2.15}$$

clearly satisfies the desired condition (2.8). □

In order to develop a fast algorithm, we first subdivide the matrix A into a collection of submatrices, as depicted in Figure 1. Each of the submatrices in this structure is separated from the diagonal, except the submatrices near the diagonal whose ranks are small simply because their dimensionality is small. By virtue of Lemma 2.1, each of the separated submatrices is of rank J, to the accuracy 4^{-J}. In order to apply A to an arbitrary vector with fixed but finite accuracy (which is always the case in numerical computations), we can apply each of the submatrices to the appropriate part of the vector for a cost proportional to kJ, where k is the size of the submatrix. Adding up the costs for all such submatrices, we obtain the operation count of

$$Jn \log n \sim \log\left(\frac{1}{\varepsilon}\right) n \log n, \tag{2.16}$$

instead of n^2.

The scheme outlined above is extremely simple, but representative of the current approach to the design of 'fast' summation algorithms. Several comments are in order.

1. It is easy to see that the matrix A defined in (2.1) with the spacing defined by (2.3) is in fact a Toeplitz matrix that can be applied to an arbitrary vector for a cost proportional to $n \log n$ via the Fast Fourier Transform. This situation occurs sometimes, both in one and higher dimensions. However, the Toeplitz nature of the matrix A is lost when the points are not distributed on a uniform grid, and direct application

of the FFT becomes impossible. For 'somewhat uniformly' distributed points x_i, various types of local corrections have been successfully utilized. When the points are not distributed uniformly (for example, on a curve or surface), FFT-based methods become ineffective.

2. As described, the scheme is only applicable to one-dimensional problems, and under very limited conditions. In most situations, the subdivision of the matrix has to be modified, taking into account the geometric distribution of points in order to locate submatrices whose 'numerical rank' is low. Examples of such subdivisions can be found in Carrier et al. (1988), Van Dommelen and Rundensteiner (1989), Beylkin et al. (1991) and Nabors et al. (1994).

3. The scheme is extremely simple and general. It is entirely unrelated to the detailed nature of the matrix A, needing only some inequality like (2.13). In other words, so long as the entries of the matrix A are smooth functions of their indices away from the diagonal, a scheme of the type outlined above will work. In fact, even that is not necessary; the elements of the matrix have only to be *sufficiently smooth functions of their indices on a sufficiently large part of the matrix.*

4. The scheme admits a large number of modifications; the most obvious ones replace the Chebychev expansion in (2.12) with other approximations; one should be careful in doing so, since under many conditions the Chebychev approximation is optimal (among polynomial approximations), or nearly so. Some of the special-purpose approximation schemes that have been used successfully employ wavelets and related bases (Beylkin et al. 1991, Alpert et al. 1993).

Another obvious modification is a change in the choice of submatrices of low rank; the use of rectangular submatrices (as opposed to the square ones in Figure 1) permits coarser subdivisions and tends to result in more efficient algorithms.

5. Algorithms of the type described above usually do not work for problems where the matrix A is a discretization of an integral operator with an oscillatory kernel, since such discretizations (normally) have a more or less constant number of nodes per wavelength of the dominant oscillation. As a result, the rank of each submatrix is proportional to its size, and the resulting algorithms have CPU time estimates of the order $O(n^2)$. Sometimes, the calculation can be accelerated by reducing the size of the constant (Wagner and Chew 1994), but the asymptotic complexity in such cases is the same as for the direct approach. For certain classes of oscillatory problems (such as Helmholtz and Schrödinger equations at high frequency), there exist asymptotically 'fast' schemes based on a different (and considerably more involved) analytical apparatus; see, for example, Rokhlin (1988, 1990*b*, 1993), Canning (1989, 1992, 1993), Coifman and Meyer (1991), Bradie, Coifman and Grossmann (1993), Coifman, Rokhlin and Wandzura (1993, 1994), Wagner and Chew (1994), Epton and Dembart (1995). As noted in the introduction, these schemes are related to the scheme we will present below. They are, however, outside the scope of this paper.

3. Mathematical preliminaries I

In this section, we briefly derive the multipole expansion of a charge distribution and refer the reader to Kellogg (1953), Jackson (1975), Wallace (1984), and Greengard (1988) for more detailed discussions.

If a point charge of strength q is located at $P_0 = (x_0, y_0, z_0)$, then the potential and electrostatic field due to this charge at a distinct point $P = (x, y, z)$ are given by

$$\Phi = \frac{1}{R} \tag{3.1}$$

and

$$\mathbf{E} = -\nabla\Phi = \left(\frac{x - x_0}{R^3}, \frac{y - y_0}{R^3}, \frac{z - z_0}{R^3}\right), \tag{3.2}$$

respectively, where R denotes the distance between points P_0 and P.

We would like to derive a series expansion for the potential at P in terms of its distance from the origin r. For this, let the spherical coordinates of P be (r, θ, ϕ) and of P_0 be (ρ, α, β). Letting γ be the angle between the vectors P and P_0, we have from the cosine rule

$$R^2 = r^2 + \rho^2 - 2r\rho\cos\gamma, \tag{3.3}$$

with

$$\cos\gamma = \cos\theta\cos\alpha + \sin\theta\sin\alpha\cos(\phi - \beta). \tag{3.4}$$

Thus,

$$\frac{1}{R} = \frac{1}{r\sqrt{1 - 2\frac{\rho}{r}\cos\gamma + \frac{\rho^2}{r^2}}} = \frac{1}{r\sqrt{1 - 2u\mu + \mu^2}}, \tag{3.5}$$

having set

$$\mu = \frac{\rho}{r} \quad \text{and} \quad u = \cos\gamma. \tag{3.6}$$

For $\mu < 1$, we may expand the inverse square root in powers of μ, resulting in a series of the form

$$\frac{1}{\sqrt{1 - 2u\mu + \mu^2}} = \sum_{n=0}^{\infty} P_n(u)\mu^n \tag{3.7}$$

where

$$P_0(u) = 1, \quad P_1(u) = u, \quad P_2(u) = \frac{3}{2}\left(u^2 - \frac{1}{3}\right), \quad \ldots \tag{3.8}$$

and, in general, $P_n(u)$ is the Legendre polynomial of degree n. Our expression for the field now takes the form

$$\frac{1}{R} = \sum_{n=0}^{\infty} \frac{\rho^n}{r^{n+1}} P_n(u). \tag{3.9}$$

The angular parameter u, however, depends on both the source and the target locations. A more general representation will require the introduction of spherical harmonics, which are solutions of the Laplace equation obtained by separation of variables in spherical coordinates. Any harmonic function Φ can be expanded in the form

$$\Phi = \sum_{n=0}^{\infty} \sum_{m=-n}^{n} \left(L_n^m r^n + \frac{M_n^m}{r^{n+1}} \right) Y_n^m(\theta, \phi). \tag{3.10}$$

The terms $Y_n^m(\theta, \phi) r^n$ are referred to as spherical harmonics of degree n or *solid harmonics*, the terms $Y_n^m(\theta, \phi)/r^{n+1}$ are called spherical harmonics of degree $-n-1$ or *multipoles*, and the coefficients L_n^m and M_n^m are known as the moments of the expansion.

The spherical harmonics can be expressed in terms of partial derivatives of $1/r$ (Wallace 1984) as

$$\frac{Y_n^0(\theta, \phi)}{r^{n+1}} = A_n^0 \frac{\partial^n}{\partial z^n} \left(\frac{1}{r}\right). \tag{3.11}$$

For $m > 0$, we have

$$\frac{Y_n^m(\theta, \phi)}{r^{n+1}} = A_n^m \left(\frac{\partial}{\partial x} + i\frac{\partial}{\partial y}\right)^m \left(\frac{\partial}{\partial z}\right)^{n-m} \left(\frac{1}{r}\right), \tag{3.12}$$

and

$$\frac{Y_n^{-m}(\theta, \phi)}{r^{n+1}} = A_n^m \left(\frac{\partial}{\partial x} - i\frac{\partial}{\partial y}\right)^m \left(\frac{\partial}{\partial z}\right)^{n-m} \left(\frac{1}{r}\right), \tag{3.13}$$

where

$$A_n^m = \frac{(-1)^n}{\sqrt{(n-m)!(n+m)!}}. \tag{3.14}$$

They also satisfy the relation

$$Y_n^m(\theta, \phi) \equiv \sqrt{\frac{(n-|m|)!}{(n+|m|)!}} P_n^{|m|}(\cos\theta) e^{im\phi}, \tag{3.15}$$

where we have omitted the normalization factor of $\sqrt{(2n+1)/4\pi}$, to match the definitions (3.11)–(3.13) given above. The special functions P_n^m are called associated Legendre functions and can be defined by Rodrigues' formula

$$P_n^m(x) = (-1)^m \left(1 - x^2\right)^{m/2} \frac{d^m}{dx^m} P_n(x).$$

Theorem 3.1 (Addition theorem for Legendre polynomials) Let P and Q be points with spherical coordinates (r, θ, ϕ) and (ρ, α, β), respectively, and let γ be the angle subtended between them. Then

$$P_n(\cos\gamma) = \sum_{m=-n}^{n} Y_n^{-m}(\alpha, \beta) Y_n^m(\theta, \phi). \tag{3.16}$$

Combining Theorem 3.1 and equation (3.9), we have

$$\frac{1}{R} = \sum_{n=0}^{\infty} \sum_{m=-n}^{n} \rho^n Y_n^{-m}(\alpha, \beta) \frac{Y_n^m(\theta, \phi)}{r^{n+1}}. \tag{3.17}$$

It is now straightforward to expand the field due to a collection of sources in terms of multipoles.

Theorem 3.2 (Multipole expansion) Suppose that k charges of strengths $\{q_i,\ i = 1, \ldots, k\}$ are located at the points $\{Q_i = (\rho_i, \alpha_i, \beta_i),\ i = 1, \ldots, k\}$, with $|\rho_i| < a$. Then for any $P = (r, \theta, \phi) \in \mathbb{R}^3$ with $r > a$, the potential $\Phi(P)$ is given by

$$\Phi(P) = \sum_{n=0}^{\infty} \sum_{m=-n}^{n} \frac{M_n^m}{r^{n+1}} Y_n^m(\theta, \phi), \tag{3.18}$$

where

$$M_n^m = \sum_{i=1}^{k} q_i \rho_i^n Y_n^{-m}(\alpha_i, \beta_i). \tag{3.19}$$

Furthermore, for any $p \geq 1$,

$$\left| \Phi(P) - \sum_{n=0}^{p} \sum_{m=-n}^{n} \frac{M_n^m}{r^{n+1}} Y_n^m(\theta, \phi) \right| \leq \frac{A}{r-a} \left(\frac{a}{r}\right)^{p+1}, \tag{3.20}$$

where

$$A = \sum_{i=1}^{k} |q_i|. \tag{3.21}$$

Proof. The formula (3.19) follows from equation (3.17) and superposition. The error bound is obtained from the triangle inequality and the fact that the ratios ρ_i/r are bounded from above by a/r. □

Suppose now that $r = 2a$ in the context of the preceding theorem. Then the error bound (3.20) becomes

$$\left|\Phi(P) - \sum_{n=0}^{p}\sum_{m=-n}^{n} \frac{M_n^m}{r^{n+1}} Y_n^m(\theta,\phi)\right| \le \frac{A}{a}\left(\frac{1}{2}\right)^{p+1}, \tag{3.22}$$

and setting $p = \log_2(1/\varepsilon)$ yields a precision ε relative to the ratio A/a.

4. An $N \log N$ algorithm

Theorem 3.2 is all that is required to construct a simple fast algorithm of arbitrary precision. To reduce the number of issues addressed, we assume that the particles are fairly homogeneously distributed in a square so that adaptive refinement is not required.

In order to make systematic use of multipole expansions, we introduce a hierarchy of boxes which refine the computational domain into smaller and smaller regions. At refinement level 0, we have the entire computational domain. Refinement level $l+1$ is obtained recursively from level l by subdivision of each box into eight equal parts. This yields a natural tree structure, where the eight boxes at level $l+1$ obtained by subdivision of a box at level l are considered its children.

Definition 4.1 Two boxes are said to be *near neighbours* if they are at the same refinement level and share a boundary point (a box is a near neighbour of itself).

Definition 4.2 Two boxes are said to be *well separated* if they are at the same refinement level and are not near neighbours.

Definition 4.3 With each box i we associate an *interaction list*, consisting of the children of the near neighbours of i's parent which are well separated from box i (Figure 4).

Definition 4.4 With each box i at level l we associate a multipole expansion $\Phi_{l,i}$ about the box centre, which describes the far field induced by the particles contained inside the box.

The basic idea is to consider clusters of particles at successive levels of spatial refinement, and to compute interactions between distant clusters by means of multipole expansions when possible. It is clear that at levels 0 and 1, there are no pairs of boxes that are well separated. At level 2, on the other hand, sixty-four boxes have been created and there is a number of well separated pairs. Multipole expansions can then be used to compute

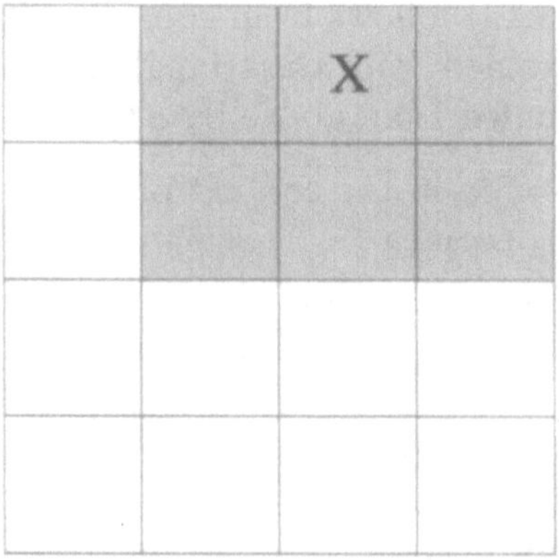

Fig. 2. The first step of the algorithm, depicted in two space dimensions for clarity. Interactions between particles in box X and its near neighbours (grey) are not computed. Interactions between well separated boxes are computed via multipole expansions

interactions between these well separated pairs (Figure 2) with rigorous bounds on the error. In fact, it is easy to see that the bound (3.20) applies with the ratio $a/r < 1/\sqrt{3}$. Thus, to achieve a given precision ε, we need to use $p = \log_{\sqrt{3}}(1/\varepsilon)$ terms.

It remains to compute the interactions between particles contained in each box with those contained in the box's near neighbours, and this is done recursively. We first refine each level 2 box to create level 3. For a given level 3 box, we then seek to determine which other level 3 boxes can be interacted with by means of multipole expansions. Since those boxes outside the region of the *parent*'s nearest neighbours are already accounted for (at level 2), they can be ignored. Since interactions with near neighbours cannot be accounted for accurately by means of an expansion, they can also be ignored for the moment. The remaining boxes correspond exactly to the interaction list defined above (Figure 3).

The nature of the recursion is now clear. At every level, the multipole expansion is formed for each box due to the particles it contains. The resulting expansion is then evaluated for each particle in the region covered by its interaction list (Figure 4).

We halt the recursive process after roughly $\log_8 N$ levels of refinement. The amount of work done at each level is of the order $O(N)$. To see this, note first that approximately $N p^2$ operations are needed to create all expansions, since each particle contributes to p^2 expansion coefficients. Secondly, from the point of view of a single particle, there are at most 189 boxes (the maximum size of the interaction list) whose expansions are computed, so that $189 N p^2$ operations are needed for all evaluations.

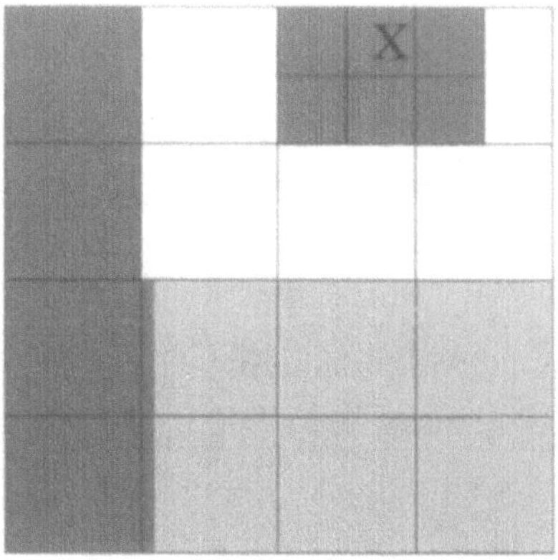

Fig. 3. The second step of the algorithm, depicted in two space dimensions. After refinement, note that the particles in the box marked X have already interacted with the most distant particles (light grey). They are now well separated from the particles in the white boxes, so that these interactions can be computed via multipole expansions. The near neighbour interactions (dark grey) are not computed

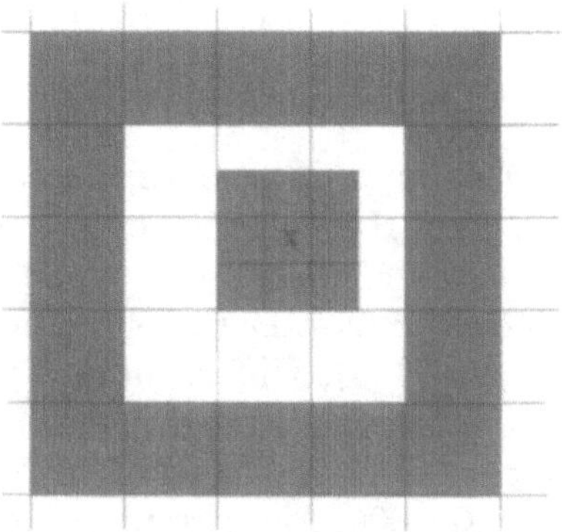

Fig. 4. Subsequent steps of the algorithm. The interaction list for box X is indicated in white. In three dimensions, it contains up to 189 boxes

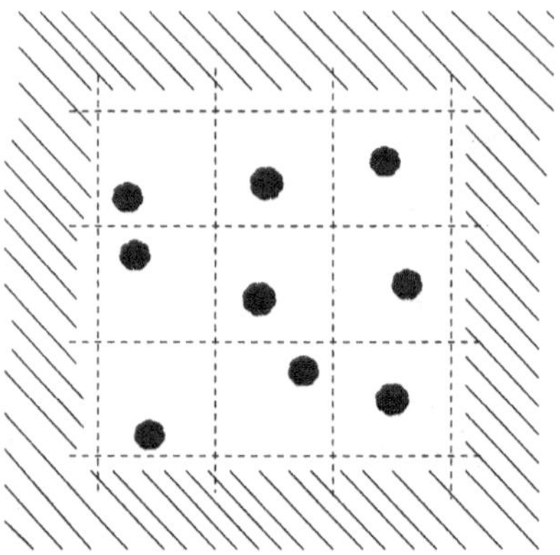

Fig. 5. At the finest level, interactions with near neighbours are computed directly. In three dimensions, there are up to 27 near neighbours

At the finest level, we have created roughly $8^{\log_8 N} = N$ boxes and it remains only to compute interactions between nearest neighbours. By the assumption of homogeneity, there are $O(1)$ particles per box, so that this last step requires about $27\,N$ operations (Figure 5). The total cost is approximately

$$189Np^2 \log_8 N + 27N. \tag{4.1}$$

The algorithm just described is, in essence, a nonadaptive version of the one proposed by Barnes and Hut (1986), except that it achieves arbitrary precision through the use of high order expansions. Two-dimensional schemes of this type are due to Van Dommelen and Rundensteiner (1989) and Odlyzko and Schönhage (1988). Unfortunately, while such schemes have good asymptotic work estimates, the three-dimensional versions provide only modest speedups at high precision for the values of N encountered in present day applications. At $N = 100,000$, for example, seven digits of accuracy require $p \approx 20$, and the $N \log N$ scheme is only two to three times faster than the direct $O(N^2)$ method. In order to accelerate the calculation significantly, we need some further analytic machinery.

5. Mathematical preliminaries II

The FMM relies on three translation operators, acting on either multipole (far field) or solid harmonic (local) expansions. They are described in the next three theorems (Greengard and Rokhlin 1988*a*, Greengard 1988).

Theorem 5.1 (Translation of a multipole expansion) Suppose that l charges of strengths $q_1, q_2, \ldots, q_l$ are located inside the sphere D of radius a with centre at $Q = (\rho, \alpha, \beta)$, and that for points $P = (r, \theta, \phi)$ outside D, the potential due to these charges is given by the multipole expansion

$$\Phi(P) = \sum_{n=0}^{\infty} \sum_{m=-n}^{n} \frac{O_n^m}{r'^{\,n+1}} Y_n^m(\theta', \phi'), \tag{5.1}$$

where $P - Q = (r', \theta', \phi')$. Then for any point $P = (r, \theta, \phi)$ outside the sphere D_1 of radius $(a + \rho)$,

$$\Phi(P) = \sum_{j=0}^{\infty} \sum_{k=-j}^{j} \frac{M_j^k}{r^{j+1}} Y_j^k(\theta, \phi), \tag{5.2}$$

where

$$M_j^k = \sum_{n=0}^{j} \sum_{m=-n}^{n} \frac{O_{j-n}^{k-m} i^{|k|-|m|-|k-m|} A_n^m A_{j-n}^{k-m} \rho^n Y_n^{-m}(\alpha, \beta)}{A_j^k}, \tag{5.3}$$

with A_n^m defined by equation (3.14). Furthermore, for any $p \geq 1$,

$$\left| \Phi(P) - \sum_{j=0}^{p} \sum_{k=-j}^{j} \frac{M_j^k}{r^{j+1}} Y_j^k(\theta, \phi) \right| \leq \left(\frac{\sum_{i=1}^{l} |q_i|}{r - (a + \rho)} \right) \left(\frac{a + \rho}{r} \right)^{p+1}. \tag{5.4}$$

Definition 5.1 The linear operator mapping old multipole coefficients $\{O_j^k : 0 \leq n \leq p, \ -n \leq m \leq n\}$, to new multipole coefficients $\{M_j^k : 0 \leq n \leq p, \ -n \leq m \leq n\}$ according to equation (5.3) will be denoted by $\mathcal{T}_{MM}$.

Theorem 5.2 (Conversion of a multipole expansion into a local expansion) Suppose that l charges of strengths $q_1, q_2, \ldots, q_l$ are located inside the sphere D_Q of radius a with centre at $Q = (\rho, \alpha, \beta)$, and that $\rho > (c + 1)a$ with $c > 1$. Then the corresponding multipole expansion (5.1) converges inside the sphere D_0 of radius a centred at the origin. Inside D_0, the potential due to the charges $q_1, q_2, \ldots, q_l$ is described by a local expansion:

$$\Phi(P) = \sum_{j=0}^{\infty} \sum_{k=-j}^{j} L_j^k Y_j^k(\theta, \phi) r^j, \tag{5.5}$$

where

$$L_j^k = \sum_{n=0}^{\infty} \sum_{m=-n}^{n} \frac{O_n^m i^{|k-m|-|k|-|m|} A_n^m A_j^k Y_{j+n}^{m-k}(\alpha, \beta)}{(-1)^n A_{j+n}^{m-k} \rho^{j+n+1}}, \tag{5.6}$$

with A_r^s defined by equation (3.14). Furthermore, for any $p \geq 1$,

$$\left| \Phi(P) - \sum_{j=0}^{p} \sum_{k=-j}^{j} L_j^k Y_j^k(\theta, \phi) r^{j+1} \right| \leq \left(\frac{\sum_{i=1}^{l} |q_i|}{ca - a} \right) \left(\frac{1}{c} \right)^{p+1}. \tag{5.7}$$

Definition 5.2 The linear operator mapping truncated multipole expansion coefficients $\{O_j^k : 0 \le j \le p,\ -j \le k \le j\}$ to local coefficients $\{L_j^k : 0 \le j \le p,\ -j \le k \le j\}$ according to equation (5.6) will be denoted by $\mathcal{T}_{ML}$.

Theorem 5.3 (Translation of a local expansion) Let $Q = (\rho, \alpha, \beta)$ be the origin of a local expansion

$$\Phi(P) = \sum_{n=0}^{p} \sum_{m=-n}^{n} O_n^m Y_n^m(\theta', \phi') r'^n, \tag{5.8}$$

where $P = (r, \theta, \phi)$ and $P - Q = (r', \theta', \phi')$. Then

$$\Phi(P) = \sum_{j=0}^{p} \sum_{k=-j}^{j} L_j^k Y_j^k(\theta, \phi) r^j, \tag{5.9}$$

where

$$L_j^k = \sum_{n=j}^{p} \sum_{m=-n}^{n} \frac{O_n^m i^{|m|-|m-k|-|k|} A_{n-j}^{m-k} A_j^k Y_{n-j}^{m-k}(\alpha, \beta) \rho^{n-j}}{(-1)^{n+j} A_n^m}, \tag{5.10}$$

with A_r^s defined by equation (3.14).

Definition 5.3 The linear operator mapping old local expansion coefficients $\{O_n^m : 0 \le n \le p,\ -n \le m \le n\}$ to new local expansion coefficients $\{L_n^m : 0 \le n \le p,\ -n \le m \le n\}$ according to equation (5.10) will be denoted by $\mathcal{T}_{LL}$.

6. The original FMM

We can now construct a scheme with cost proportional to N, by using Theorem 5.2 to convert the far field expansion of a source box into a local expansion inside a target box, rather than by direct evaluation of the far field expansion at individual target positions.

Definition 6.1 With each box i at level l we associate a local expansion $\Psi_{l,i}$ about the box centre, which describes the potential field induced by all particles outside box i's near neighbours.

Definition 6.2 With each box i at level l we associate a local expansion $\tilde{\Psi}_{l,i}$ about the box centre, which describes the potential field induced by all particles outside the near neighbours of i's *parent*.

ALGORITHM 1

The parent of a box j will be denoted by $p(j)$. The list of children of a box j will be denoted by $c(j)$. The interaction list of a box j will be denoted by *ilist*(j).

Upward pass

Initialization

Choose the number of refinement levels $n \approx \log_8 N$, and the order of the multipole expansion desired p. The number of boxes at the finest level is then 8^n, and the average number of particles per box is $s = N/(8^n)$.

Step 1

Form multipole expansions $\Phi_{n,i}$ of potential field due to particles in each box about the box centre at the finest mesh level, via Theorem 3.2.

Step 2

For levels $l = n-1, \ldots, 2$,

Form multipole expansion $\Phi_{l,j}$ about the centre of each box at level l by merging expansions from its eight children via Theorem 5.1.

$$\Phi_{l,j} = \textstyle\sum_{k \in c(j)} \mathcal{T}_{MM} \Phi_{l+1,k}.$$

Downward pass

Initialization

Set $\Psi_{1,1} = \Psi_{1,2} = \cdots = \Psi_{1,8} = (0, 0, \ldots, 0)$.

Step 3

For levels $l = 2, \ldots, n$,

Form the expansion $\tilde{\Psi}_{l,j}$ for each box j at level l, by using Theorem 5.3 to shift the local Ψ expansion of j's parent to j itself.

$$\tilde{\Psi}_{l,j} = \mathcal{T}_{LL} \Psi_{l-1,p(j)}.$$

Form $\Psi_{l,j}$ by using Theorem 5.2 to convert the multipole expansion $\Phi_{l,k}$ of each box k in the *interaction list* of box j to a local expansion about the centre of box j, adding these local expansions together, and adding the result to $\tilde{\Psi}_{l,j}$.

$$\Psi_{l,j} = \tilde{\Psi}_{l,j} + \textstyle\sum_{k \in \text{ilist}(j)} \mathcal{T}_{ML} \Phi_{l,k}.$$

Step 4

For each particle in each box j at the finest level n,

evaluate $\Psi_{n,j}$ at the particle position.

Step 5

For each particle in each box j at the finest level n,

compute interactions with particles in near neighbour boxes directly.

Since s is the average number of particles per box at the finest level, there are approximately N/s boxes in the tree hierarchy. Therefore, Step 1 requires approximately Np^2 work, Step 2 requires $(N/s)\,p^4$ work, Step 3 requires $189(N/s)\,p^4$ work, Step 4 requires $N\,p^2$ work, and Step 5 requires $27N\,s$ work. Thus, a reasonable estimate for the total operation count is

$$191\left(\frac{N}{s}\right)p^4 + 2Np^2 + 27Ns. \tag{6.1}$$

With $s = 2p^2$, the operation count becomes approximately

$$150N\,p^2. \tag{6.2}$$

This would appear to beat the estimate (4.1) for any N, but there is a subtle catch. The number of terms p needed for a fixed precision in the $N \log N$ scheme is smaller than the number of terms needed in the FMM described above. To see why, consider two interacting cubes A and B of unit volume, with sources in A and targets in B. The worst-case multipole error decays like $(\sqrt{3}/3)^p$, since $\sqrt{3}/2$ is the radius of the smallest sphere enclosing cube A and $3/2$ is the shortest distance to a target in B. The conversion of a multipole expansion in A to a local expansion in B, however, satisfies an error bound which depends on the smallest sphere enclosing B as well as the smallest sphere enclosing A. From equation (5.7), the worst case error is less than $(0.76)^p$, although with more detailed analysis, one can show that the error is bounded by $(0.75)^p$ (Petersen, Smith and Soelvason 1995).

In the original FMM (Greengard and Rokhlin 1988*a*, Greengard 1988), it was suggested that one redefine the nearest neighbour list to include 'second nearest neighbours,' so that boxes which interact via multipole expansions are separated by at least two intervening boxes of the same size. The error can then be shown to decay approximately like $(0.4)^p$. However, the number of near neighbours increases from 27 to 125 and the size of the interaction list increases from 189 to 875.

It is clear that the major obstacle to achieving reasonable efficiency at high precision is the cost of the multipole to local translations ($189p^4$ operations per box). There are several schemes that have been suggested for reducing the cost of applying translation operators. The simplest is based on rotating the coordinate system so that the vector connecting the source box B and the target box C lies along the z-axis, shifting the expansion along the z-axis, and then rotating back to the original coordinate system.

6.1. The FMM using rotation matrices

We begin with the following obvious result.

Lemma 6.1 Consider a harmonic function given by

$$\Phi(P) = \sum_{n=0}^{\infty} \sum_{m=-n}^{n} \left(L_n^m r^n + \frac{M_n^m}{r^{n+1}} \right) Y_n^m(\theta, \phi),$$

where (r, θ, ϕ) are the spherical coordinates of the point P. If we rotate the coordinate system through an angle β in the positive sense about the z-axis, then

$$\Phi(P) = \sum_{n=0}^{\infty} \sum_{m=-n}^{n} \left(\tilde{L}_n^m r^n + \frac{\tilde{M}_n^m}{r^{n+1}} \right) Y_n^m(\theta, \phi'),$$

where (r, θ, ϕ') are the new coordinates of P,

$$\tilde{L}_n^m = L_n^m \, e^{im\beta}, \quad \text{and} \quad \tilde{M}_n^m = M_n^m \, e^{im\beta}.$$

Definition 6.3 Given a rotation angle β, the diagonal operator mapping old multipole coefficients to rotated multipole coefficients ($O_n^m \to O_n^m \, e^{im\beta}$) will be denoted by $\mathcal{R}_z(\beta)$.

We also need to be able to rotate the coordinate system about the y-axis.

Lemma 6.2 Consider a harmonic function given by

$$\Phi(P) = \sum_{n=0}^{\infty} \sum_{m=-n}^{n} \left(L_n^m r^n + \frac{M_n^m}{r^{n+1}} \right) Y_n^m(\theta, \phi),$$

where (r, θ, ϕ) are the spherical coordinates of the point P. If we rotate the coordinate system through an angle α in the positive sense about the y-axis, then there exist coefficients $R(n, m, m', \alpha)$ such that

$$\Phi(P) = \sum_{n=0}^{\infty} \sum_{m'=-n}^{n} \left(\tilde{L}_n^{m'} r^n + \frac{\tilde{M}_n^{m'}}{r^{n+1}} \right) Y_n^{m'}(\theta', \phi'),$$

where (r, θ, ϕ') are the new coordinates of P,

$$\tilde{L}_n^{m'} = \sum_{m=-n}^{n} R(n, m, m', \alpha) L_n^m \tag{6.3}$$

and

$$\tilde{M}_n^{m'} = \sum_{m=-n}^{n} R(n, m, m', \alpha) M_n^m. \tag{6.4}$$

Proof. See Biedenharn and Louck (1981) for a complete discussion and for a variety of methods that can be used to compute the coefficients $R(n, m, m', \alpha)$. □

Lemma 6.3 In order to shift a multipole expansion a distance ρ along the z-axis, one can replace equation (5.3) with the simpler formula

$$M_j^k = \sum_{n=0}^{j} \frac{O_{j-n}^k A_n^0 A_{j-n}^k \rho^n Y_n^0(1,0)}{A_j^k}. \tag{6.5}$$

In order to convert a multipole expansion centred at the origin into a local expansion centred at $(0,0,\rho)$, one can replace equation (5.6) with the simpler formula

$$L_j^k = \sum_{n=0}^{\infty} \frac{O_n^m A_n^k A_j^k Y_{j+n}^0(1,0)}{(-1)^n A_{j+n}^0 \rho^{j+n+1}}, \tag{6.6}$$

In order to translate the centre of a local expansion from the origin to the point $(0,0,\rho)$, one can replace equation (5.10) with the simpler formula

$$L_j^k = \sum_{n=j}^{p} \frac{O_n^m A_{n-j}^0 A_j^k Y_{n-j}^0(1,0)\rho^{n-j}}{(-1)^{n+j} A_n^k}, \tag{6.7}$$

Definition 6.4 Given a rotation angle α, the diagonal operator mapping old multipole coefficients to rotated multipole coefficients according to formula (6.3) or (6.4) will be denoted by $\mathcal{R}_y(\alpha)$. The special cases of the linear operators $\mathcal{T}_{MM}$, $\mathcal{T}_{ML}$, and $\mathcal{T}_{LL}$ which shift a distance ρ in the z-direction according to the formulae (6.5), (6.6), and (6.7) will be denoted by $\mathcal{T}_{MM}^z(\rho)$, $\mathcal{T}_{ML}^z(\rho)$, and $\mathcal{T}_{LL}^z(\rho)$.

We can now combine Lemmas 6.1, 6.2 and 6.3 to obtain the desired factorizations of $\mathcal{T}_{MM}$, $\mathcal{T}_{ML}$, $\mathcal{T}_{LL}$.

Lemma 6.4

$$\begin{aligned}
\mathcal{T}_{MM} &= \mathcal{R}_z(-\beta)\mathcal{R}_y(-\alpha)\mathcal{T}_{MM}^z(\rho)\mathcal{R}_y(\alpha)\mathcal{R}_z(\beta), \\
\mathcal{T}_{ML} &= \mathcal{R}_z(-\beta)\mathcal{R}_y(-\alpha)\mathcal{T}_{ML}^z(\rho)\mathcal{R}_y(\alpha)\mathcal{R}_z(\beta), \\
\mathcal{T}_{LL} &= \mathcal{R}_z(-\beta)\mathcal{R}_y(-\alpha)\mathcal{T}_{LL}^z(\rho)\mathcal{R}_y(\alpha)\mathcal{R}_z(\beta),
\end{aligned}$$

where (ρ, α, β) is the desired shifting vector.

Clearly, the cost of applying $\mathcal{T}_{MM}$, $\mathcal{T}_{ML}$, or $\mathcal{T}_{LL}$ by means of the preceding factorization is

$$O(p^2) + O(p^3) + O(p^3) + O(p^3) + O(p^2).$$

Thus, the total computational cost of the FMM can be reduced to approximately

$$191\left(\frac{N}{s}\right)3p^3 + 2Np^2 + 27Ns.$$

With $s = 3p^{3/2}$, the operation count becomes

$$270N\,p^{3/2} + 2N\,p^2. \tag{6.8}$$

7. Mathematical preliminaries III

Over the last few years, a number of 'fast' or diagonal translation schemes have been developed that require $O(p^2)$ work (Greengard and Rokhlin 1988*b*, Berman 1995, Elliott and Board 1996). Unfortunately, they are all subject to certain numerical instabilities. The instabilities can be overcome, but at additional cost, the details of which we leave to the cited papers.

The latest generation of fast algorithms is based on combining multipole expansions with exponential or 'plane wave' expansions. The reason for using exponentials is that translation corresponds to multiplication and, like the earlier fast schemes, requires only $O(p^2)$ work. Unlike in the earlier diagonal schemes, however, no numerical instabilities are encountered. The two-dimensional theory is described in Hrycak and Rokhlin (1995), and we present the three-dimensional theory here.

Remark 7.1 A complicating feature of the new approach is that *six* plane wave expansions will be associated with each box, one emanating from each face of the cube. To fix notation, we will refer to the $+z$ direction as *up*, to the $-z$ direction as *down*; to the $+y$ direction as *north*, to the $-y$ direction as *south*; to the $+x$ direction as *east*, and to the $-x$ direction as *west*. The interaction list for each box will be subdivided into six lists, one associated with each direction.

Definition 7.1 The *Uplist* for a box B consists of elements of the interaction list that lie *above* B and are separated by at least one box in the $+z$ direction (Figure 6). The *Downlist* for a box B consists of elements of the interaction list that lie *below* B and are separated by at least one box in the $-z$ direction. The *Northlist* for a box B consists of elements of the interaction list that lie *north* of B, are separated by at least one box in the $+y$ direction, and are not contained in the Up- or Downlists. The *Southlist* for a box B consists of elements of the interaction list that lie *south* of B, are separated by at least one box in the $-y$ direction, and are not contained in the Up- or Downlists. The *Eastlist* for a box B consists of elements of the interaction list that lie *east* of B, are separated by at least one box in the $+x$ direction, and are not contained in the Up-, Down-, North-, or Southlists. The *Westlist* for a box B consists of elements of the interaction list that lie *west* of B, are separated by at least one box in the $-x$ direction, and are not contained in the Up-, Down-, North-, or Southlists.

It is easy to verify that the original interaction list is the union of the Up-, Down-, North-, South-, East- and Westlists. It is also easy to verify that

$$\begin{aligned} C \in \text{Uplist}(B) &\Leftrightarrow B \in \text{Downlist}(C) \\ C \in \text{Northlist}(B) &\Leftrightarrow B \in \text{Southlist}(C) \\ C \in \text{Eastlist}(B) &\Leftrightarrow B \in \text{Westlist}(C). \end{aligned} \tag{7.1}$$

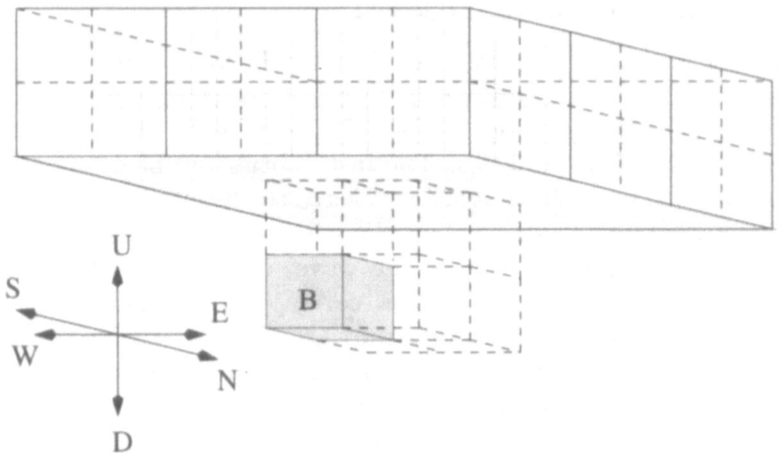

Fig. 6. The Uplist for the box B (see Definition 7.1)

Given a source location $P = (x_0, y_0, z_0)$ and a target location $Q = (x, y, z)$, our starting point is the well-known integral representation (Morse and Feshbach 1953, p. 1256)

$$\begin{aligned} &\frac{1}{\sqrt{(x-x_0)^2+(y-y_0)^2+(z-z_0)^2}} \\ &\quad = \frac{1}{2\pi}\int_0^\infty e^{-\lambda(z-z_0)}\int_0^{2\pi} e^{i\lambda((x-x_0)\cos\alpha+(y-y_0)\sin\alpha)}\,\mathrm{d}\alpha\,\mathrm{d}\lambda \\ &\quad = \int_0^\infty e^{-\lambda(z-z_0)} J_0(\lambda\sqrt{(x-x_0)^2+(y-y_0)^2})\,\mathrm{d}\lambda, \qquad (7.2) \end{aligned}$$

valid for $z > z_0$.

To get a discrete representation, we must use an appropriate quadrature formula. The inner integral, with respect to α, is easily handled by the trapezoidal rule (which achieves spectral accuracy for periodic functions), but the outer integral requires more care. Laguerre quadrature is an appropriate choice here, but even better performance can be obtained using generalized Gaussian quadrature rules (Yarvin and Rokhlin 1996). These have been designed with the geometry of the interaction list in mind.

Because of the restriction that $z > z_0$, we will assume, for the moment, that the source P is contained in a box B and that the target Q lies in a box $C \in \text{Uplist}(B)$. The following lemma describes several discrete approximations of the double integral in (7.2) as double sums.

Lemma 7.1 Let $P \in B$ and $Q \in C \in \text{Uplist}(B)$, where B is a box of unit volume. Then

$$\left| \frac{1}{r_{PQ}} - \sum_{k=1}^{9} \frac{w_k}{M(k)} \sum_{j=1}^{M(k)} e^{-\lambda_k[(z-z_0)-i(x-x_0)\cos\alpha_j-(y-y_0)\sin\alpha_j]} \right| < 10^{-3}, \quad (7.3)$$

where $\alpha_j = 2\pi j/M(k)$, and the weights $w_1, \ldots, w_9$, nodes $\lambda_1, \ldots, \lambda_9$, and values $M(1), \ldots, M(9)$ are given in Section 12, Table 5. (The total number of exponentials required is 109.)

$$\left| \frac{1}{r_{PQ}} - \sum_{k=1}^{18} \frac{w_k}{M(k)} \sum_{j=1}^{M(k)} e^{-\lambda_k[(z-z_0)-i(x-x_0)\cos\alpha_j-(y-y_0)\sin\alpha_j]} \right| < 10^{-6}, \quad (7.4)$$

where $\alpha_j = 2\pi j/M(k)$, and the weights $w_1, \ldots, w_{18}$, nodes $\lambda_1, \ldots, \lambda_{18}$, and values $M(1), \ldots, M(18)$ are given in Section 12, Table 6. (The total number of exponentials required is 558.)

$$\left| \frac{1}{r_{PQ}} - \sum_{k=1}^{30} \frac{w_k}{M(k)} \sum_{j=1}^{M(k)} e^{-\lambda_k[(z-z_0)-i(x-x_0)\cos\alpha_j-(y-y_0)\sin\alpha_j]} \right| < 5 \times 10^{-11}, \quad (7.5)$$

where $\alpha_j = 2\pi j/M(k)$, and the weights $w_1, \ldots, w_{30}$, nodes $\lambda_1, \ldots, \lambda_{30}$, and values $M(1), \ldots, M(30)$ are given in Section 12, Table 7. (The total number of exponentials required is 1751.)

Remark 7.2 The formulae (7.3)–(7.5) are somewhat complex, but have a simple interpretation. The outer sums use the generalized Gaussian weights and nodes $\{w_k, \lambda_k\}$ obtained in Yarvin and Rokhlin (1996) to approximate the outer integral (with respect to λ), while the inner sums use the trapezoidal rule to approximate the inner integral (with respect to α). The number of nodes in each inner integral depends on the value λ_k for which the integration is being performed, and is denoted by $M(k)$. These are derived from standard estimates concerning Bessel functions (Watson 1944, pp. 227, 255; Rokhlin 1995).

Remark 7.3 In the remainder of this paper, we will assume that the desired precision ε is clear from the context, and will write

$$\left| \frac{1}{r_{PQ}} - \sum_{k=1}^{s(\varepsilon)} \sum_{j=1}^{M(k)} \frac{w_k}{M(k)} e^{-\lambda_k(z-z_0)} e^{i\lambda_k((x-x_0)\cos\alpha_j+(y-y_0)\sin\alpha_j)} \right| < \varepsilon, \quad (7.6)$$

where $\alpha_j = 2\pi j/M(k)$. This is a mild abuse of notation, since the weights, nodes and values $M(k)$ depend on ε as well. The total number of exponential basis functions used will be denoted by $S_{\exp}$, so that

$$S_{\exp} = \sum_{k=1}^{s(\varepsilon)} M(k).$$

Corollary 7.1 Let B be a box of unit volume centred at the origin containing N charges of strengths $\{q_l,\ l = 1, \ldots, N\}$, located at the points $\{Q_l = (x_l, y_l, z_l),\ l = 1, \ldots, N\}$. Then, for any P contained in Uplist(B), the potential $\Phi(P)$ satisfies

$$\left| \Phi(P) - \sum_{k=1}^{s(\varepsilon)} \sum_{j=1}^{M(k)} W(k,j) e^{-\lambda_k z} e^{i\lambda_k (x \cos\alpha_j + y \sin\alpha_j)} \right| < A\varepsilon, \tag{7.7}$$

where $A = \sum_{l=1}^{N} |q_l|$ and

$$W(k,j) = \sum_{l=1}^{N} q_l e^{\lambda_k z_l} e^{-i\lambda_k (x_l \cos\alpha_j + y_l \sin\alpha_j)}. \tag{7.8}$$

Corollary 7.2 (Diagonal translation) Let B be a box of unit volume centred at the origin containing N charges of strengths $\{q_l : l = 1, \ldots, N\}$, located at the points $\{Q_l = (x_l, y_l, z_l) : l = 1, \ldots, N\}$ and let C be a box in Uplist(B) centred at (x_1, y_1, z_1). For $P \in C$, let the potential $\Phi(P)$ be approximated by the exponential expansion centred at the origin

$$\Phi(P) = \sum_{k=1}^{s(\varepsilon)} \sum_{j=1}^{M(k)} W(k,j) e^{-\lambda_k z} e^{i\lambda_k (x \cos\alpha_j + y \sin\alpha_j)} + O(\varepsilon). \tag{7.9}$$

Then

$$\Phi(P) = \sum_{k=1}^{s(\varepsilon)} \sum_{j=1}^{M(k)} V(k,j) e^{-\lambda_k (z - z_1)} e^{i\lambda_k ((x - x_1)\cos\alpha_j + (y - y_1)\sin\alpha_j)} + O(\varepsilon), \tag{7.10}$$

where

$$V(k,j) = W(k,j)\, e^{-\lambda_k z_1} e^{i\lambda_k (x_1 \cos\alpha_j + y_1 \sin\alpha_j)}. \tag{7.11}$$

Definition 7.2 The diagonal operator mapping the original set of exponential expansion coefficients $\{W(k,j)\}$ to the shifted exponential expansion coefficients $\{V(k,j)\}$ according to (7.11) will be denoted by $\mathcal{D}_{\vec{BC}}$, where $\vec{BC} = (x_1, y_1, z_1)$ is the vector from the centre of B to the centre of C.

In the FMM, we will be given the multipole expansion of a charge distribution for a box B rather than the charge distribution itself, and will need to convert it to an exponential expansion. This is accomplished by the following theorem.

Theorem 7.1 Let B be a box of unit volume centred at the origin containing N charges of strengths $\{q_l,\ l = 1,\ldots,N\}$, located at the points $\{Q_l = (x_l, y_l, z_l),\ l = 1,\ldots,N\}$. Let $P \in C \in \text{Uplist}(B)$ and suppose that the potential $\Phi(P)$ is given as the multipole expansion

$$\Phi(P) = \sum_{n=0}^{\infty} \sum_{m=-n}^{n} \frac{M_n^m}{r^{n+1}} Y_n^m(\theta, \phi). \tag{7.12}$$

Then

$$\left| \Phi(P) - \sum_{k=1}^{s(\varepsilon)} \sum_{j=1}^{M(k)} W(k,j) e^{-\lambda_k z} e^{i\lambda_k (x \cos\alpha_j + y \sin\alpha_j)} \right| < A\,\varepsilon, \tag{7.13}$$

where $A = \sum_{l=1}^{N} |q_l|$ and

$$W(k,j) = \frac{w_k}{M(k)} \sum_{m=-\infty}^{\infty} (-i)^{|m|} e^{im\alpha_j} \sum_{n=|m|}^{\infty} \frac{M_n^m}{\sqrt{(n-m)!(n+m)!}} \lambda_k^n. \tag{7.14}$$

Proof. The formula (7.14) follows from the definitions (3.11) (3.12) and (3.13). The estimate (7.13) follows from Corollary 7.1. □

Definition 7.3 The linear operator mapping a finite multipole expansion $\{M_n^m : 0 \le n \le p,\ -n \le m \le n\}$, to the corresponding set of coefficients in an exponential expansion $\{W(k,j)\}$ according to equation (7.14) will be denoted by $\mathcal{C}_{MX}$.

Once the multipole expansion for a source box has been converted into an exponential expansion (via Theorem 7.1) and translated to a target box centre (via Corollary 7.2), we will need to convert the exponential expansion back into a solid harmonic series. The following theorem provides the necessary machinery.

Theorem 7.2 Let B be a box of unit volume containing N charges of strengths $\{q_l,\ l = 1,\ldots,N\}$, located at the points $\{Q_l = (x_l, y_l, z_l),\ l = 1,\ldots,N\}$. Let P be contained in a box $C \in \text{Uplist}(B)$, centred at the origin, and suppose that the potential $\Phi(P)$ is given as the exponential expansion

$$\Phi(P) - \sum_{k=1}^{s(\varepsilon)} \sum_{j=1}^{M(k)} W(k,j) e^{-\lambda_k z} e^{i\lambda_k (x \cos\alpha_j + y \sin\alpha_j)}| < A\,\varepsilon, \tag{7.15}$$

where $A = \sum_{l=1}^{N} |q_l|$. Then

$$\left| \Phi(P) - \sum_{n=0}^{\infty} \sum_{m=-n}^{n} L_n^m Y_n^m(\theta, \phi) r^n \right| < A\,\varepsilon, \tag{7.16}$$

where

$$L_n^m = \frac{(-i)^{|m|}}{\sqrt{(n-m)!(n+m)!}} \sum_{k=1}^{s(\varepsilon)} (-\lambda_k)^n \sum_{j=1}^{M(k)} W(k,j) e^{im\alpha_j}. \tag{7.17}$$

Proof. Equation (7.17) follows easily from the formula in Hobson (1955, p. 123),

$$(z + ix\cos\alpha + iy\sin\alpha)^n =$$

$$r^n \left\{ P_n(\cos\theta) + 2\sum_{m=1}^{n} (i)^{-m} \frac{n!}{(n+m)!} (-1)^m P_n^m(\cos\theta)\cos m(\phi-\alpha) \right\},$$

where (r, θ, ϕ) are the spherical coordinates of the point with Cartesian coordinates (x, y, z). □

Definition 7.4 The linear operator mapping the set of coefficients in an exponential expansion $\{W(k,j)\}$ to the coefficients in the corresponding truncated solid harmonic expansion $\{L_n^m : 0 \le n \le p, \ -n \le m \le n\}$, according to equation (7.17) will be denoted by $\mathcal{C}_{XL}$.

Remark 7.4 Theorems 7.1 and 7.2, like Theorem 5.2, are not quite the right tools needed to obtain rigorous error estimates for the FMM. In both cases, we have ignored the fact that the multipole and local expansions are truncated. It is straightforward but tedious to derive precise estimates, and we ignore this issue in the present paper. We should note that the nature of such estimates depends on how the multipole-to-exponential, multipole-to-solid harmonic or exponential-to-solid harmonic conversion is carried out. Formulae (7.14), (7.17) and (5.6) are the easiest to derive, being the Taylor expansions of the potential Φ. However, each of these conversions is simply a linear mapping from one set of basis functions to another. The formulae (7.17), (7.14), and (5.6) can be shown to correspond to minimizing the L_2 error on the surface of a *sphere* enclosing the given source or target box. One could choose a variety of other possible projections, such as minimizing the L_2 or L_∞ error on the surface of the corresponding box itself.

Remark 7.5 By inspection of formula (7.14), it is clear that the cost of applying the operator $\mathcal{T}_{MX}$ is $p^2 s(\varepsilon) + p S_{\text{exp}}$. The same is true for the operator $\mathcal{T}_{XL}$. It is also worth noting that Fast Fourier Transforms can be used to reduce the cost of the outer sum in the truncated version of formula (7.14) and the inner sum in the truncated version of formula (7.17).

Corollary 7.3 (Multipole to local factorization) Let B be a box of unit volume and C a box in Uplist(B). If $\mathcal{T}_{ML}$ is the translation operator converting the multipole expansion centred in B to the local expansion centred in C, then

$$\mathcal{T}_{ML} = \mathcal{C}_{XL} \mathcal{D}_{\vec{BC}} \mathcal{C}_{MX}. \tag{7.18}$$

Remark 7.6 It is important to note that Lemma 7.1 provides a carefully designed quadrature formula which assumes that the source box B has unit volume and that the target is in B's Uplist. In order to use these quadrature weights and nodes, we need to rescale the multipole and local expansions so that the box dimension always has unit volume. To accomplish this, if

$$\Phi(P) = \sum_{n=0}^{\infty} \sum_{m=-n}^{n} \frac{M_n^m}{r^{n+1}} Y_n^m(\theta, \phi) \tag{7.19}$$

is the multipole expansion for a box B of volume d^3, we simply write

$$\Phi(P) = \sum_{n=0}^{\infty} \sum_{m=-n}^{n} \frac{(M_n^m / d^{n+1})}{(r/d)^{n+1}} Y_n^m(\theta, \phi). \tag{7.20}$$

The local expansion for a target box in B's interaction list is accumulated as

$$\Phi(P) = \sum_{j=0}^{\infty} \sum_{k=-j}^{j} L_j^k \, d^j Y_j^k(\theta, \phi) \left(\frac{r}{d}\right)^j . \tag{7.21}$$

Corollary 7.4 (Scaled multipole to local factorization) Let B be a box of volume d^3 and C a box in Uplist(B), with the vector from the centre of B to the centre of C given by (x_1, y_1, z_1). If $\mathcal{T}_{ML}$ is the translation operator converting the multipole expansion centred in B to the local expansion centred in C, then

$$\mathcal{T}_{ML} = \mathcal{D}_{d,L} \, \mathcal{C}_{XL} \mathcal{D}_{\overline{BC}} \mathcal{C}_{MX} \, \mathcal{D}_{d,M}, \tag{7.22}$$

where

$$\mathcal{D}_{d,M} M_n^m = M_n^m / d^{n+1}, \quad \mathcal{D}_{d,L} L_n^m = L_n^m / d^n,$$

and $\overline{BC} = \vec{BC}/d$.

The cost of a single multipole-to-local translation using the factorization of Corollary 7.4 is

$$2p^2 + 2p^2 s(\varepsilon) + 2pS_{\text{exp}} \approx 2p^3,$$

since $s \approx p$ and $S_{\text{exp}} \approx p^2$. If each translation were carried out in this manner, we would not improve on the rotation-based scheme discussed in Section 6.1. However, once the multipole expansion for a box B has been converted to an exponential expansion (via the application of $\mathcal{D}_{d,M}$ and $\mathcal{C}_{MX}$), it can be translated to each box in its Uplist at a cost of $S_{\text{exp}} \approx p^2$ operations. Conversely, once a box B has *accumulated* all the exponential expansions transmitted from its Downlist (see equation (7.1)), a single application of the operators $\mathcal{C}_{XL}$ and $\mathcal{D}_{d,L}$ yields the local harmonic expansion describing the field due to the sources in the Downlist of box B (Figure 7).

Up to this point, we have considered only the exponential representation

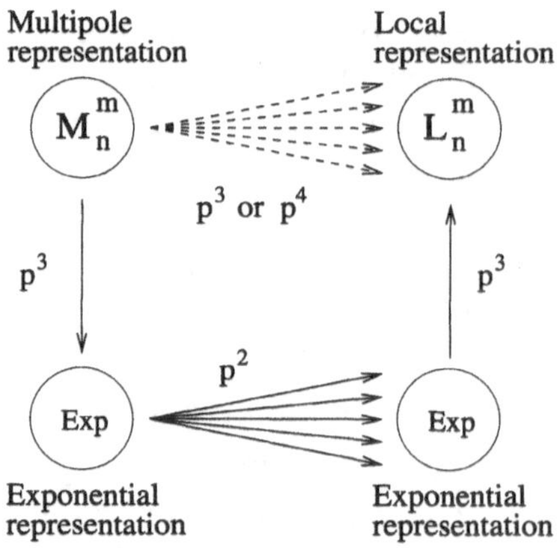

Fig. 7. In the new FMM, a large number of multipole-to-local translations, costing $O(p^3)$ or $O(p^4)$ work, can be replaced by a large number of exponential translations, costing $O(p^2)$ work

needed to shift information in the upward ($+z$) direction. As noted in the beginning of this section, however, there are six outgoing directions that need to be accounted for. The most straightforward way of generating the appropriate expansions is to rotate the coordinate system so that the z-axis points in the desired direction. The following lemma provides the necessary formulae.

Lemma 7.2 *Let B be a box of volume d^3 and C a 'target' box. Let $\mathcal{T}_{ML}$ be the translation operator converting the multipole expansion centred in B to the local expansion centred in C.*
If $C \in$ Downlist(B), then

$$\mathcal{T}_{ML}^{\text{Down}} = \mathcal{D}_{d,L}\,\mathcal{R}_y(-\pi)\,\mathcal{C}_{XL}\mathcal{D}_{\overline{BC}}\mathcal{C}_{MX}\,\mathcal{R}_y(\pi)\,\mathcal{D}_{d,M}.$$

If $C \in$ Eastlist(B), then

$$\mathcal{T}_{ML}^{\text{East}} = \mathcal{D}_{d,L}\,\mathcal{R}_y(-\pi/2)\,\mathcal{C}_{XL}\mathcal{D}_{\overline{BC}}\mathcal{C}_{MX}\,\mathcal{R}_y(\pi/2)\,\mathcal{D}_{d,M}.$$

If $C \in$ Westlist(B), then

$$\mathcal{T}_{ML}^{\text{West}} = \mathcal{D}_{d,L}\,\mathcal{R}_y(\pi/2)\,\mathcal{C}_{XL}\mathcal{D}_{\overline{BC}}\mathcal{C}_{MX}\,\mathcal{R}_y(-\pi/2)\,\mathcal{D}_{d,M}.$$

If $C \in$ Northlist(B), then

$$\mathcal{T}_{ML}^{\text{North}} = \mathcal{D}_{d,L}\,\mathcal{R}_y(-\pi/2)\,\mathcal{R}_z(-\pi/2)\,\mathcal{C}_{XL}\,\mathcal{D}_{\overline{BC}}\mathcal{C}_{MX}\,\mathcal{R}_y(\pi/2)\,\mathcal{R}_z(\pi/2)\,\mathcal{D}_{d,M}.$$

If $C \in$ Southlist(B), then

$$\mathcal{T}_{ML}^{\text{South}} = \mathcal{D}_{d,L}\,\mathcal{R}_y(\pi/2)\,\mathcal{R}_z(-\pi/2)\,\mathcal{C}_{XL}\,\mathcal{D}_{\overline{BC}}\mathcal{C}_{MX}\,\mathcal{R}_y(-\pi/2)\,\mathcal{R}_z(\pi/2)\,\mathcal{D}_{d,M},$$

where $\overline{BC}$ is the appropriately scaled vector from the centre of B to the centre of C in the rotated coordinate system. The operators $\mathcal{R}_z$ and $\mathcal{R}_y$ are defined in Section 6.1.

Definition 7.5 Let $\mathcal{T}_{ML}^{\mathrm{Up}}$ be given by the operator $\mathcal{T}_{ML}$ defined in equation (7.22). Then, for Dir $\in$ {Up, Down, East, West, North, South}, we will write

$$\mathcal{T}_{ML}^{\mathrm{Dir}} = \mathcal{Q}^{\mathrm{Dir}}\,\mathcal{D}_{\overline{BC}}\,\mathcal{P}^{\mathrm{Dir}},$$

so that

$$\begin{aligned}
\mathcal{Q}^{\mathrm{Up}} &= \mathcal{D}_{d,L}\,\mathcal{C}_{XL},\\
\mathcal{P}^{\mathrm{Up}} &= \mathcal{C}_{MX}\,\mathcal{D}_{d,M},\\
\mathcal{Q}^{\mathrm{Down}} &= \mathcal{D}_{d,L}\,\mathcal{R}_y(-\pi)\,\mathcal{C}_{XL},\\
\mathcal{P}^{\mathrm{Down}} &= \mathcal{C}_{MX}\,\mathcal{R}_y(\pi)\,\mathcal{D}_{d,M},
\end{aligned}$$

etc.

We are now in a position to describe the new FMM in detail.

8. The new FMM

ALGORITHM 2

The parent of a box j will be denoted by $p(j)$. The list of children of a box j will be denoted by $c(j)$. For each box j, the 'outgoing' exponential expansion with coefficients $\{W(n,m) : 1 \le n \le s(\varepsilon),\ 1 \le m \le M(n)\}$, will be denoted by W_j. We will also associate an 'incoming' exponential expansion with each box, denoted by V_j.

Upward pass

Initialization

Choose the number of refinement levels $n \approx \log_8 N$, and the order of the multipole expansion desired p. The number of boxes at the finest level is then 8^n, and the average number of particles per box is $s = N/(8^n)$.

Step 1

Form multipole expansions $\Phi_{n,i}$ of potential field due to particles in each box about the box centre at the finest mesh level, via Theorem 3.2.

Step 2

Do for levels $l = n-1, \ldots, 2$,

Form multipole expansion $\Phi_{l,j}$ about the centre of each box at level l by merging expansions from its eight children via Theorem 5.1.

$$\Phi_{l,j} = \textstyle\sum_{k \in c(j)} \mathcal{T}_{MM}\Phi_{l+1,k}.$$

(In applying $\mathcal{T}_{MM}$, use the factorization of Lemma 6.4.)

End do

Downward pass
Initialization
Set $\Psi_{1,1} = \Psi_{1,2} = \cdots = \Psi_{1,8} = (0, 0, \ldots, 0)$.

Step 3A
Do for levels $l = 2, \ldots, n$,
Form the expansion $\tilde{\Psi}_{l,j}$ for each box j at level l by using Theorem 5.3 to shift the local Ψ expansion of j's parent to j itself.

$$\tilde{\Psi}_{l,j} = \mathcal{T}_{LL} \Psi_{l-1,p(j)}.$$

(In applying $\mathcal{T}_{LL}$, use the factorization of Lemma 6.4.) Set $\Psi_{l.j} = \tilde{\Psi}_{l.j}$.

Step 3B
For each direction Dir = Up, Down, North, South, East, West, the opposite direction will be denoted by −Dir, so that −Up = Down, −Down = Up, *etc.* Thus, if a box B sends an outgoing expansion in direction Dir to Box C on its Dirlist, then C can be viewed as receiving the expansion from B which is an element of its −Dirlist (see equation (7.1)).

Do for Dir = Up, Down, North, South, East, West,
For each box j at level l, convert the multipole expansion $\Phi_{l,j}$ into the 'outgoing' exponential expansion for direction Dir.

$$W_j = \mathcal{P}^{\mathrm{Dir}} \Phi_{l,j}.$$

For each box j at level l, collect the 'outgoing' exponential expansions from the −Dirlist of box j as an 'incoming' exponential expansion

$$V_j = \textstyle\sum_{k \in -\mathrm{Dirlist}} \mathcal{D}_{\tilde{kj}} W_k,$$

where $\tilde{kj}$ is the appropriately scaled vector from the centre of box k to the centre of box j in the rotated coordinate system.

For each box j at level l, convert the accumulated 'incoming' exponential expansion V_j into a local harmonic expansion and add result to $\Psi_{l,j}$.

$$\Psi_{l,j} = \Psi_{l,j} + \mathcal{Q}^{\mathrm{Dir}} V_j.$$

End do
End do

Step 4
For each particle in each box j at the finest level n,
evaluate $\Psi_{n,j}$ at the particle position.

Step 5
For each particle in each box j at the finest level n,
compute interactions with particles in near neighbour boxes directly.

Since we are using the rotation scheme for applying $\mathcal{T}_{MM}$ and $\mathcal{T}_{LL}$ in Steps 2 and 3A, these now require a total of $3p^3\,(N/s)$ work, where s is the number of particles per box on the finest level. In Step 3B, the applications of the multipole to exponential operators $\mathcal{P}^{\rm Dir}$ and the exponential-to-local-operators $\mathcal{Q}^{\rm Dir}$ require a total of approximately $6\,p^3(N/s)$ work, while the exponential translations require approximately $189\,p^2\,(N/s)$ work. The total operation count is therefore of the order

$$189\,\frac{N}{s}\,p^2 + 2\,N\,p^2 + 27N\,s + 6\frac{N}{s}\,p^3.$$

With $s = 2p$, the total operation count is about

$$150\,N\,p + 5N\,p^2.$$

8.1. *Current improvements*

There are several ways in which the algorithm described above has been accelerated. Symmetry considerations, for example, allow the pairs of operators $\{\mathcal{P}^{\rm Up}, \mathcal{P}^{\rm Down}\}$, $\{\mathcal{P}^{\rm North}, \mathcal{P}^{\rm South}\}$, and $\{\mathcal{P}^{\rm East}, \mathcal{P}^{\rm West}\}$ to be applied simultaneously. The same is true for the adjoint pairs $\{\mathcal{Q}^{\rm Up}, \mathcal{Q}^{\rm Down}\}$, *etc.* Thus, the $6\,p^3(N/s)$ work needed in Step 3B can be replaced by $3\,p^3(N/s)$ work.

Even more significant is the fact that the number of translations per box can be reduced from 189 to less than 40. To see why, suppose that a box B at level l has eight children, denoted $B_1, \ldots, B_8$, and that boxes $C_1, \ldots, C_J$ lie in the Uplist of each child. In the new FMM described above, we accumulated an 'incoming' exponential expansion in each box C_j as

$$V_j = \sum_{k=1}^{8} \mathcal{D}_{\overline{B_k C_j}} W_k,$$

where W_k is the 'outgoing' exponential expansion for B_k. Repeating this for $j = 1, \ldots, J$ requires a total of $8\,J$ translations. Since all translations are diagonal, however, it is easy to verify that

$$\begin{aligned} V_j &= \sum_{k=1}^{8} \mathcal{D}_{\overline{BC_j}} \mathcal{D}_{\overline{B_k B}} W_k \\ &= \mathcal{D}_{\overline{BC_j}} \sum_{k=1}^{8} \mathcal{D}_{\overline{B_k B}} W_k. \end{aligned}$$

Thus, by first merging the 'outgoing' expansions, and then translating their sum to each target box C_j, only $8 + J$ translations are needed. It should be emphasized that this improvement relies on the diagonal form of the operators. One could try to merge expansions in this manner in the context of the original FMM, but the local expansion coefficients computed with and without merging would not be the same. There would be a significant loss of precision, consistent with the error bound (5.7).

8.2. Further improvements

There are several ways in which the scheme can be accelerated that have not been incorporated into the existing code. The most significant of these is probably a change in the choice of the translation operators $\mathcal{T}_{MM}$ and $\mathcal{T}_{LL}$, as well as the multipole-to-exponential and exponential-to-local conversion operators $\mathcal{C}_{MX}$ and $\mathcal{C}_{XL}$. As mentioned previously, the obvious formulae (5.3), (5.10), (7.14), and (7.17) are obtained via Taylor expansion and are clearly not optimal. Preliminary numerical experiments indicate that replacing them with more carefully optimized tools will reduce the cost of these calculations within the FMM by a factor of three. Furthermore, the improvement described in Remark 7.5 has not yet been implemented; we are using the explicit matrix form of the discrete Fourier transform in applying $\mathcal{C}_{MX}$ and $\mathcal{C}_{XL}$, rather than the FFT.

The incorporation of all these modifications is likely to reduce the overall cost by a factor of two.

9. Numerical results

The new FMM has been implemented in Fortran 77 and tested on uniform random distributions. The results of our experiments are summarized in Tables 1–4, with all times calculated in seconds using a Sun Ultra-1/140 workstation. In each table, the first column lists the number of particles, the second column lists the number of levels used in the multipole hierarchy, the third column lists the order of the multipole expansion used, and the fourth column lists the corresponding number of exponential basis functions. Columns five and six indicate the times required by the FMM and the direct calculation, respectively, and column seven lists the l^2 norm of the error in the FMM approximation

$$E = \left(\frac{\sum_{i=1}^{N} |\Phi(x_i) - \tilde{\Phi}(x_i)|^2}{\sum_{i=1}^{N} |\Phi(x_i)|^2} \right)^{1/2}. \tag{9.1}$$

For the largest simulations, with $N > 10000$, we have carried out the direct calculation on a subset of only 100 particles. The stated times, indicated in parentheses, are then computed by extrapolation and the errors are obtained by restricting the formula (9.1) to this subset.

Table 1. *Timing results for the FMM using fifth-order expansions and twenty-eight exponential basis functions*

N	Levels	p	S_{exp}	T_{FMM}	T_{dir}	Error
500	3	5	28	0.18	0.20	4.5×10^{-3}
5000	4	5	28	1.9	20.1	7.6×10^{-3}
40000	5	5	28	20	(1461)	7.0×10^{-3}
300000	6	5	28	175	(82475)	1.3×10^{-2}

Table 2. *Timing results for the FMM using ninth-order expansions and* 109 *exponential basis functions*

N	Levels	p	S_{exp}	T_{FMM}	T_{dir}	Error
2000	3	9	109	1.4	3.37	1.4×10^{-4}
10000	4	9	109	7.9	83	3.6×10^{-4}
80000	5	9	109	111	(5838)	4.1×10^{-4}

Table 3. *Timing results for the FMM using eighteenth-order expansions and* 558 *exponential basis functions*

N	Levels	p	S_{exp}	T_{FMM}	T_{dir}	Error
4000	3	18	558	8.3	13.4	1.1×10^{-7}
25000	4	18	558	68	(567)	1.5×10^{-7}
150000	5	18	558	495	(20100)	1.9×10^{-7}

Table 4. *Timing results for the FMM using thirtieth-order expansions and* 1751 *exponential basis functions*

N	Levels	p	S_{exp}	T_{FMM}	T_{dir}	Error
5000	3	30	1751	22	20.8	6.2×10^{-12}
50000	4	30	1751	316	(2280)	6.2×10^{-12}

10. Extensions and generalizations

The scheme presented in this paper is not adaptive and assumes that the distribution of points is reasonably uniform in space. In order to handle more general distributions, one needs to allow some regions to be subdivided into finer refinement levels than others. Adaptive structures of this type have been designed by several groups (Carrier et al. 1988, Van Dommelen and Rundensteiner 1989, Nabors et al. 1994) and we are in the process of incorporating these structures into the new FMM.

While a number of techniques now exist for high-frequency scattering problems (Rokhlin 1988, 1990, 1993, Canning 1989, 1992, 1993, Coifman and Meyer 1991, Bradie et al. 1993, Coifman et al. 1993, 1994, Wagner and Chew 1994, Epton and Dembart 1995), an important generalization of the algorithm of this paper is to the calculation of potentials governed by the Helmholtz equation at low frequency. By this we mean an environment in which the region of interest is no more than a few wavelengths in size, but contains a large number of discretization points (for example, due to the complexity of some structure being modelled). Algorithms for such problems are currently being designed.

11. Conclusions

A new version of the FMM has been developed. It is based on a new diagonal form for translation operators, and is significantly faster than previous implementations at any desired level of precision. Of particular interest is the fact that high precision calculations have been brought within practical reach.

12. Tables: quadrature weights and nodes

Table 5. *Columns 1 and 2 contain the nine weights and nodes needed for discretization of the outer integral in (7.2) at three-digit accuracy. Column 3 contains the number of discretization points needed in the inner integral, which we denote by* $M(k)$

Node	Weight	$M(k)$
0.09927399673971	0.24776441819008	4
0.47725674637049	0.49188566500464	7
1.05533661382183	0.65378749137677	11
1.76759343354008	0.76433038408784	15
2.57342629351471	0.84376180565628	20
3.44824339201583	0.90445883985098	20
4.37680983554726	0.95378613136833	24
5.34895757205460	0.99670261613218	7
6.35765785313375	1.10429422730252	1

Table 6. *Columns 1 and 2 contain the eighteen weights and nodes for discretization of the outer integral in (7.2) at six-digit accuracy. Column 3 contains the number of discretization points needed in the inner integral, which we denote by* $M(k)$

Node	Weight	$M(k)$
0.05278852766117	0.13438265914335	5
0.26949859838931	0.29457752727395	8
0.63220353174689	0.42607819361148	12
1.11307564277608	0.53189220776549	16
1.68939496140213	0.61787306245538	20
2.34376200469530	0.68863156078905	25
3.06269982907806	0.74749099381426	29
3.83562941265296	0.79699192718599	34
4.65424734321562	0.83917454386997	38
5.51209386593581	0.87570092283745	43
6.40421268377278	0.90792943590067	47
7.32688001906175	0.93698393742461	51
8.27740099258238	0.96382546688788	56
9.25397180602489	0.98932985769673	59
10.25560272374640	1.01438284597917	59
11.28208829787774	1.04003654374165	51
12.33406790967692	1.06815489269567	4
13.41492024017240	1.10907580975537	1

Table 7. *Columns 1 and 2 contain the thirty weights and nodes for discretization of the outer integral in (7.2) at ten-digit accuracy. Column 3 contains the number of discretization points needed in the inner integral, which we denote by $M(k)$*

Node	Weight	$M(k)$
0.03239542384523	0.08289159611006	7
0.16861844033714	0.18838810673274	10
0.40611377169029	0.28485143005306	14
0.73466473057596	0.37041553715895	18
1.14340561998398	0.44539043894975	22
1.62232408412252	0.51100452150290	26
2.16276138867422	0.56865283856139	30
2.75739199003682	0.61958013174010	35
3.40002470112078	0.66481004321965	39
4.08539104793552	0.70517204769960	43
4.80897515497095	0.74134967169016	48
5.56688915983444	0.77392103530415	53
6.35578243654166	0.80338600122756	57
7.17277232990713	0.83018277269650	62
8.01538803542112	0.85469824839953	66
8.88152313049502	0.87727539085565	71
9.76939480982937	0.89821948245755	76
10.67750922034750	0.91780416582368	80
11.60463289992789	0.93627766216629	85
12.54977061299652	0.95386940504388	89
13.51215012257297	0.97079739700556	94
14.49121482655196	0.98727684670885	97
15.48662587630224	1.00353112433459	103
16.49827659770404	1.01980697905712	107
17.52632405530625	1.03639774457222	110
18.57124579700721	1.05368191266322	112
19.63393428118300	1.07219343903929	108
20.71585163675095	1.09278318162014	84
21.81939113866225	1.11737373706779	4
22.95080495008893	1.15786184931141	1

REFERENCES

B. K. Alpert and V. Rokhlin (1991), 'A fast algorithm for the evaluation of Legendre expansions', *SIAM J. Sci. Statist. Comput.* **12**, 158–179.

B. K. Alpert, G. Beylkin, R. Coifman and V. Rokhlin (1993), 'Wavelet-like bases for the fast solution of second-kind integral equations', *SIAM J. Sci. Statist. Comput.* **14**, 159–184.

C. R. Anderson (1986), 'A method of local corrections for computing the velocity field due to a distribution of vortex blobs', *J. Comput. Phys.* **62**, 111–123.

C. R. Anderson (1992), 'An implementation of the fast multipole method without multipoles', *SIAM J. Sci. Statist. Comput.* **13**, 923–947.

A. W. Appel (1985), 'An efficient program for many-body simulation', *SIAM J. Sci. Statist. Comput.* **6**, 85–103.

J. Barnes and P. Hut (1986), 'A hierarchical $O(N \log N)$ force-calculation algorithm', *Nature* **324**, 446–449.

L. Berman (1995), 'Grid-multipole calculations', *SIAM J. Sci. Comput.* **16**, 1082–1091.

G. Beylkin, R. Coifman and V. Rokhlin (1991), 'Fast wavelet transforms and numerical algorithms I', *Comm. Pure Appl. Math.* **44**, 141–183.

L. C. Biedenharn and J. D. Louck (1981), *Angular Momentum in Quantum Physics: Theory and Application*, Addison Wesley, London.

J.A. Board, J. W. Causey, J. F. Leathrum, A. Windemuth and K. Schulten (1992), 'Accelerated molecular dynamics simulation with the parallel fast multipole method', *Chem. Phys. Let.* **198**, 89–94.

B. Bradie, R. Coifman and A. Grossmann (1993), 'Fast numerical computations of oscillatory integrals related to acoustic scattering, I', *Appl. Comput. Harm. Anal.* **1**, 94–99.

A. Brandt (1991), 'Multilevel computations of integral transforms and particle interactions with oscillatory kernels', *Comp. Phys. Comm.* **65**, 24–38.

A. Brandt and A. A. Lubrecht (1990), 'Multilevel matrix multiplication and fast solution of integral equations', *J. Comput. Phys.* **90**, 348–370.

F. X. Canning (1989), 'Reducing moment method storage from order N^2 to order N', *Electron. Let.* **25**, 1274–1275.

F. X. Canning (1992), 'Sparse approximation for solving integral equations with oscillatory kernels', *SIAM J. Sci. Statist. Comput.* **13**, 71–87.

F. X. Canning (1993), 'Improved impedance matrix localization method', *IEEE Trans. Antennas and Propagation* **41**, 658–667.

J. Carrier, L. Greengard and V. Rokhlin (1988), 'A fast adaptive multipole algorithm for particle simulations', *SIAM J. Sci. Statist. Comput.* **9**, 669–686.

R. Coifman and Y. Meyer (1991), 'Remarques sur l'analyse de Fourier à fenêtre', *C. R. Acad. Sci. Paris* **312**, Serie 1, 259–261.

R. Coifman, V. Rokhlin and S. Wandzura (1993), 'The fast multipole method for the wave equation: a pedestrian prescription', *IEEE Antennas and Propagation Mag.* **35**, 7–12.

R. Coifman, V. Rokhlin and S. Wandzura (1994), 'Faster single-stage Multipole Method for the wave equation', 10th Annual Review of Progress in Applied Computational Electromagnetics, Vol. 1, pp. 19-24, Monterey, CA, Applied Computational Electromagnetics Society.

G. Dahlquist and A. Bjork (1974), *Numerical Methods* Prentice-Hall, Englewood Cliffs, NJ.

H.-Q. Ding, N. Karasawa and W. A. Goddard, III (1992), 'Atomic level simulations on a million particles: The Cell Multipole Method for Coulomb and London nonbond interactions, *J. Chem. Phys.* **97**, 4309–4315.

W. D. Elliott and J.A. Board (1996), 'Fast Fourier Transform accelerated fast multipole algorithm', *SIAM J. Sci. Comput.* **17**, 398–415.

M. A. Epton and B. Dembart (1995), 'Multipole translation theory for three-dimensional Laplace and Helmholtz equations', *SIAM J. Sci. Comput.* **16**, 865–897.

A. Greenbaum, L. Greengard and G. B. McFadden (1993), 'Laplace's equation and the Dirichlet–Neumann map in multiply connected domains', *J. Comput. Phys.* **105**, 267–278.

L. Greengard (1988), *The Rapid Evaluation of Potential Fields in Particle Systems*, MIT Press, Cambridge, MA.

L. Greengard (1990), 'The numerical solution of the N-body problem', *Computers in Physics* **4**, 142–152.

L. Greengard (1994), 'Fast algorithms for classical physics', *Science* **265**, 909–914.

L. Greengard and J.-Y. Lee (1996), 'A direct adaptive Poisson solver of arbitrary order accuracy', *J. Comput. Phys.* **125**, 415–424.

L. Greengard and M. Moura (1994), 'On the numerical evaluation of electrostatic fields in composite materials', in *Acta Numerica*, Vol. 3, Cambridge University Press, pp. 379–410.

L. Greengard and V. Rokhlin (1987), 'A fast algorithm for particle simulations', *J. Comput. Phys.* **73**, 325–348.

L. Greengard and V. Rokhlin (1988*a*), 'Rapid evaluation of potential fields in three dimensions', in *Vortex Methods*, C. Anderson and C. Greengard (eds.), Lecture Notes in Mathematics, vol. 1360, Springer, 121–141.

L. Greengard and V. Rokhlin (1988*b*), 'On the efficient implementation of the fast multipole algorithm', *Department of Computer Science Research Report 602, Yale University.*

L. Greengard and V. Rokhlin (1989), 'On the evaluation of electrostatic interactions in molecular modeling', *Chemica Scripta* **29A**, 139–144.

L. Greengard and J. Strain (1991), 'The fast Gauss transform', *SIAM J. Sci. Statist. Comput.* **12**, 79–94.

L. Greengard, M. C. Kropinski and A. Mayo (1996), 'Integral equation methods for Stokes flow and isotropic elasticity', *J. Comput. Phys.* **125**, 403–414.

M. Gu and S. C. Eisenstat (1992), 'A divide-and-conquer algorithm for the symmetric tridiagonal eigenproblem', *Department of Computer Science Research Report 932, Yale University.*

W. Hackbusch and Z. P. Nowak (1989), 'On the fast matrix multiplication in the boundary element method by panel clustering', *Numer. Math.* **54**, 463–491.

E. W. Hobson (1955), *Spherical and Ellipsoidal Harmonics*, Dover, New York.

R. W. Hockney and J. W. Eastwood (1981), *Computer Simulation Using Particles*, McGraw-Hill, New York.

T. Hrycak and V. Rokhlin (1995), 'An improved fast multipole algorithm for potential fields', *Department of Computer Science Research Report 1089, Yale University.*

J. D. Jackson (1975), *Classical Electrodynamics*, Wiley, New York.

O. D. Kellogg (1953), *Foundations of Potential Theory*, Dover, New York.

P. M. Morse and H. Feshbach (1953), *Methods of Theoretical Physics*, McGraw-Hill, New York.

K. Nabors and J. White (1991), 'FastCap: a multipole accelerated 3-D capacitance extraction program', *IEEE Trans. Computer-Aided Design* **10**, 1447–1459.

K. Nabors and J. White (1992), 'Multipole-accelerated capacitance extraction algorithms for 3-D structures with multiple dielectrics', *IEEE Trans. Circuits and Systems* **39**, 946–954.

K. Nabors, F. T. Korsmeyer, F. T. Leighton and J. White (1994), 'Preconditioned, adaptive, multipole-accelerated iterative methods for three-dimensional first-kind integral equations of potential theory', *SIAM J. Sci. Statist. Comput.* **15**, 714–735.

A. M. Odlyzko and A. Schönhage (1988), 'Fast algorithms for multiple evaluations of the Riemann zeta function', *Trans. Amer. Math. Soc.* **309**, 797–809.

H. G. Petersen, E. R. Smith and D. Soelvason (1995), 'Error estimates for the fast multipole method. II. The three-dimensional case', *Proc. R. Soc. London, Series A* **448**, 401–418.

V. Rokhlin (1985), 'Rapid solution of integral equations of classical potential theory', *J. Comput. Phys.* **60**, 187–207.

V. Rokhlin (1988), 'A fast algorithm for the discrete Laplace transformation', *J. Complexity* **4**, 12–32.

V. Rokhlin (1990*a*), 'End-point corrected trapezoidal quadrature rules for singular functions', *Computers Math. Applic.* **20**, 51–62.

V. Rokhlin (1990*b*), 'Rapid solution of integral equations of scattering theory in two dimensions', *J. Comput. Phys.* **86**, 414–439.

V. Rokhlin (1993), 'Diagonal forms of translation operators for the Helmholtz equation in three dimensions', *Appl. Comput. Harm. Anal.* **1**, 82–93.

V. Rokhlin (1995), 'Sparse diagonal forms of translation operators for the Helmholtz equation in two dimensions, *Department of Computer Science Research Report 1095, Yale University.*

J. M. Song and W. C. Chew (1995), 'Multilevel fast multipole algorithm for solving combined field integral equations of electromagnetic scattering', *Microwave and Opt. Technol. Letters* **10** 14–19.

J. Strain (1991), 'The Fast Gauss Transform with variable scales', *SIAM J. Sci. Statist. Comput.* **12**, 1131–1139.

J. Strain (1992), 'The Fast Laplace Transform based on Laguerre functions', *Math. Comp.* **58**, 275–283.

L. Van Dommelen and E. A. Rundensteiner (1989), 'Fast, adaptive summation of point forces in the two-dimensional Poisson equation', *J. Comput. Phys.* **83**, 126–147.

P. R. Wallace (1984), *Mathematical Analysis of Physical Problems*, Dover, New York.

R. L. Wagner and W. C. Chew (1994), 'A ray-propagation fast multipole algorithm', *Microwave and Opt. Technol. Letters* **7** 348–351.

R. L. Wagner and W. C. Chew (1995), 'A study of wavelets for the solution of electromagnetic integral equations', *IEEE Antennas Propag.* **43** 802–810.

H. Y. Wang and R. LeSar (1995), 'An efficient fast-multipole algorithm based on an expansion in the solid harmonics', *J. Chem. Phys.* **104**, 4173–4179.

G. N. Watson (1944), *A Treatise on the Theory of Bessel Functions*, Cambridge University Press.

N. Yarvin and V. Rokhlin (1996), 'Generalized Gaussian quadratures and singular value decompositions of integral operators', *Department of Computer Science Research Report 1109, Yale University.*

Acta Numerica (1997), *pp.* 271–397

Lanczos-type solvers for nonsymmetric linear systems of equations

Martin H. Gutknecht
Swiss Center for Scientific Computing
ETH-Zentrum, CH-8092 Zürich, Switzerland
E-mail: mhg@scsc.ethz.ch

Among the iterative methods for solving large linear systems with a sparse (or, possibly, structured) nonsymmetric matrix, those that are based on the Lanczos process feature short recurrences for the generation of the Krylov space. This means low cost and low memory requirement. This review article introduces the reader not only to the basic forms of the Lanczos process and some of the related theory, but also describes in detail a number of solvers that are based on it, including those that are considered to be the most efficient ones. Possible breakdowns of the algorithms and ways to cure them by look-ahead are also discussed.

CONTENTS

1. Introduction

The task of solving huge sparse systems of linear equations comes up in many if not most large-scale problems of scientific computing. In fact, if tasks were judged according to hours spent on them on high-performance computers, the one of solving linear systems might be by far the most important one. There are two types of approach: direct methods, which are basically ingenious variations of Gaussian elimination, and iterative ones, which come in many flavours. The latter are the clear winners when we have to solve equations arising from the discretization of three-dimensional partial differential equations, while for two-dimensional problems none of the two approaches can claim to be superior in general.

Among the many existing iterative methods, those based on the Lanczos process – and we consider the conjugate gradient (CG) method to be included in this class – are definitely among the most effective ones. For symmetric positive definite systems, CG is normally the best choice, and arguments among users are restricted to which preconditioning technique to use and whether it is worthwhile, or even necessary, to combine the method with other techniques such as domain decomposition or multigrid.

For nonsymmetric (or, more correctly, non-Hermitian) systems it would be hard to make a generally accepted recommendation. There are dozens of algorithms that are generalizations of CG or are at least related to it. They fall basically into two classes: (i) methods based on orthogonalization, many of which feature a minimality property of the residuals with respect to some norm, but have to make use of long recurrences involving all previously found iterates and residuals (or direction vectors), unless truncated or restarted, in which case the optimality is lost; and (ii) methods based on biorthogonalization (or, duality) that feature short recurrences and a competitive speed of convergence. It is the latter class that is the topic of this article. The gain in memory requirement and computational effort that comes from short recurrences is often crucial for making a problem solvable. While computers get faster and memory cheaper, users turn to bigger problems, so that efficient methods become more rather than less important.

The application of recursive biorthogonalization to the numerical solution of eigenvalue problems and linear systems goes back to Lanczos (1950, 1952) and is therefore referred to as the Lanczos process. In its basic form, the process generates a pair of biorthogonal (or, dual) bases for a pair of Krylov spaces, one generated by the coefficient matrix $\mathbf{A}$ and the other by its Hermitian transpose or adjoint $\mathbf{A}^\star$. This process features a three-term recurrence and is here called Lanczos biorthogonalization or BIO algorithm (see Section 2). A variation of it, described in the second Lanczos paper, applies instead a pair of coupled two-term recurrences and is here referred to as BIOC algorithm, because it produces additionally a second pair of biconjugate bases (see Section 7). Both these algorithms can be applied for solving a linear system $\mathbf{Ax} = \mathbf{b}$ or for finding a part of the spectrum of $\mathbf{A}$. For the eigenvalue problem it has so far been standard to use the BIO algorithm, but there are indications that this may change in the future (see the comments in Section 7).

The emphasis here is neither on eigenvalue problems nor on symmetric linear systems, but on solving nonsymmetric systems. Although the determination of eigenvalues is based on the same process, its application to this problem has a different flavour and is well known for additional numerical difficulties. Moreover, for the eigenvalue problem, the spectrum of a tridiagonal matrix has to be determined in a postprocessing step. For generality, our formulations include complex systems, although the methods are mainly applied to real data.

For symmetric positive definite systems, the Lanczos process is equivalent to the conjugate gradient (CG) method of Hestenes and Stiefel (1952), which has been well understood for a long time and is widely treated in the literature. Related algorithms for indefinite symmetric systems are also well known. Therefore, we can concentrate on the nonsymmetric case.

For an annotated bibliography of the early work on the CG and the Lanczos methods we refer to Golub and O'Leary (1989). The two Lanczos papers are briefly reviewed in Stewart (1994).

There are several ways of solving linear systems iteratively with the Lanczos process. Like any other Krylov space method, the Lanczos process generates a nested sequence of Krylov subspaces[1] $\mathcal{K}_n$: at each step, the so far created basis $\{\mathbf{y}_0, \dots, \mathbf{y}_{n-1}\}$ is augmented by a new (right) Lanczos vector $\mathbf{y}_n$ that is a linear combination of $\mathbf{A}\mathbf{y}_{n-1}$ and the old basis. The starting vector $\mathbf{y}_0$ is the residual of some initial approximation $\mathbf{x}_0$, that is, $\mathbf{y}_0 := \mathbf{b} - \mathbf{A}\mathbf{x}_0$. The nth approximation $\mathbf{x}_n$ (the nth 'iterate') is then chosen to have a representation

$$\mathbf{x}_n = \mathbf{x}_0 + \sum_{j=0}^{n} \mathbf{y}_j \kappa_j \quad \text{so that } \mathbf{x}_n - \mathbf{x}_0 \in \mathcal{K}_n. \tag{1.1}$$

This is what characterizes a Krylov space solver. Note, however, that the algorithms to be discussed will not make use of this representation since it would require us to store the whole Krylov space basis. It is a feature of all competitive Lanczos-type solvers that the iterates can be obtained with short recurrences.

The Lanczos process is special in that it generates two nested sequences of Krylov spaces, one, $\{\mathcal{K}_n\}$, from $\mathbf{A}$ and some $\mathbf{y}_0$, the other, $\{\widetilde{\mathcal{K}}_n\}$ from $\mathbf{A}^\star$ and some $\widetilde{\mathbf{y}}_0$. The iterates of the classical Lanczos-type solver, the biconjugate gradient (BiCG) method, are then characterized by $\widetilde{\mathcal{K}}_n \perp \mathbf{r}_n$, where $\mathbf{r}_n := \mathbf{b} - \mathbf{A}\mathbf{x}_n$ is the nth residual.

In the 'symmetric case', when $\mathbf{A}$ is Hermitian and $\widetilde{\mathbf{y}}_0 = \mathbf{y}_0$, so that $\widetilde{\mathcal{K}}_n = \mathcal{K}_n$ (for all n), this orthogonality condition is a Galerkin condition, and the method reduces to the conjugate gradient (CG) method, which is known to minimize the error in the $\mathbf{A}$-norm when the matrix is positive definite. Of course, the minimization is subject to the condition (1.1). By replacing in the symmetric case the orthogonality by $\mathbf{A}$-orthogonality, that is, imposing $\widetilde{\mathcal{K}}_n \perp \mathbf{A}\mathbf{r}_n$, we obtain the conjugate residual (CR) or minimum residual method – a particular algorithm due to Paige and Saunders (1975) is called MINRES, see Section 5 – with the property that the residual is minimized. As a consequence, the norm of the residuals decreases monotonically.

For non-Hermitian matrices $\mathbf{A}$ the two Krylov spaces are different, and $\widetilde{\mathcal{K}}_n \perp \mathbf{r}_n$ becomes a Petrov–Galerkin condition. In the BiCG method the short recurrences of CG and CR survive, but, unfortunately, the error and residual minimization properties are lost. In fact, in practice the norm of

[1] Throughout the paper we refer for simplicity mostly to Krylov spaces, not subspaces, as do many other authors.

the residuals sometimes increases suddenly by orders of magnitude, but also reduces soon after to the previous level. This is referred to as a *peak* in the residual norm plot, and when this happens several times in an example, BiCG is said to have an erratic convergence behaviour.

There are two good ways to achieve a smoother residual norm plot of BiCG. First, one can pipe the iterates $\mathbf{x}_n$ and the residuals $\mathbf{r}_n$ of BiCG through a simple smoothing process that determines smoothed iterates $\widetilde{\mathbf{x}}_n$ and $\widetilde{\mathbf{r}}_n$ according to

$$\widetilde{\mathbf{x}}_n := \widetilde{\mathbf{x}}_{n-1}(1-\theta_n) + \mathbf{x}_n\theta_n, \quad \widetilde{\mathbf{r}}_n := \widetilde{\mathbf{r}}_{n-1}(1-\theta_n) + \mathbf{r}_n\theta_n,$$

where θ_n is chosen such that the 2-norm of $\widetilde{\mathbf{r}}_n$ is as small as possible. This simple recursive weighting process is very effective. It was proposed by Schönauer (1987) and further investigated by Weiss (1990). Now it is referred to as minimal residual (MR) smoothing; we will discuss it in Section 17.

An alternative is the quasi-minimal residual (QMR) method of Freund and Nachtigal (1991), which does not use the BiCG iterates directly, but only the basis $\{\mathbf{y}_j\}$ of $\mathcal{K}_n$ that is generated by the Lanczos process. The basic idea is to determine iterates $\mathbf{x}_n$ so that the coordinate vector of their residual with respect to the Lanczos basis has minimum length. This turns out to be a least squares problem with an $(n+1)\times n$ tridiagonal matrix, the same problem as in MinRes; see Section 5.

QMR can also be understood as a harmonic mean smoothing process for the residual norm plot, and therefore, theoretically, neither QMR nor MR smoothing can really speed up the convergence considerably. In practice, however, it often happens that straightforward implementations of the BiCG method converge more slowly than a carefully implemented QMR algorithm (or not at all).

At this point we should add that there are various ways to realize the BiCG method (see Sections 4, 8 9), and that they all allow us to combine it with a smoothing process. In theory, the various algorithms are mathematically nearly equivalent, but with respect to round-off they differ. And round-off is indeed a serious problem for all Lanczos process based algorithms with short recurrences: when building up the dual bases we only enforce the orthogonality to the two previous vectors; the orthogonality to the earlier vectors is inherited and, with time, is more and more lost due to round-off. We will discuss this and other issues of finite precision arithmetic briefly in Section 18.

In contrast to smoothing, an idea due to Sonneveld (1989) really increases the speed of convergence. At once, it eliminates the following two disadvantages of the nonsymmetric Lanczos process: first, in addition to the subroutine for the product $\mathbf{Ax}$ that is required by all Krylov space solvers, BiCG also needs one for $\mathbf{A}^\star\mathbf{x}$; second, each step of BiCG increases the dimension of both $\mathcal{K}_n$ and $\widetilde{\mathcal{K}}_n$ by one and, naturally, needs two matrix-vector multiplications to do so, but only one of them helps to reduce the approximation error by increasing the dimension of the approximation space. To explain Sonneveld's idea we note that every vector in the Krylov space $\mathcal{K}_{n+1}$ has a representation of the form $p_n(\mathbf{A})\mathbf{y}_0$ with a polynomial of degree at most n. In the standard version of the BiCG method, the basis vectors generated are linked by this polynomial:

$$\mathbf{y}_n = p_n(\mathbf{A})\mathbf{y}_0, \quad \widetilde{\mathbf{y}}_n = \overline{p_n}(\mathbf{A}^\star)\widetilde{\mathbf{y}}_0.$$

Here, the bar denotes complex conjugation of the coefficients. Since $\mathbf{y}_n$ happens to be the residual $\mathbf{r}_n$ of $\mathbf{x}_n$, p_n is actually the residual polynomial. Sonneveld found with his conjugate gradient squared (CGS) algorithm a method where the nth residual is instead given by $\mathbf{r}_n = p_n^2(\mathbf{A})\mathbf{y}_0 \in \mathcal{K}_{2n}$. Per step it increases the Krylov space by two dimensions. Moreover, the transpose or adjoint matrix is no longer needed. In practice, the convergence is indeed typically nearly twice as fast as for BiCG, and thus also in terms of matrix-vector multiplications the method is nearly twice as effective. However, it turns out that the convergence is even more erratic. We will refer to this method more appropriately as biconjugate gradient squared (BiCGS) method and treat various forms of it in Section 14.

To get smoother convergence, van der Vorst (1992) modified the approach in his BiCGStab algorithm by choosing residuals of the form $\mathbf{r}_n = p_n(\mathbf{A})t_n(\mathbf{A})\mathbf{y}_0 \in \mathcal{K}_{2n}$, where the polynomials t_n are built up in factored form, with a new zero being added in each step in such a way that the residual undergoes a one-dimensional minimization process. This method was soon enhanced further and became the germ of a group of methods that one might call the BiCGStab family. It includes BiCGStab2 with two-dimensional minimization every other step, and the more general BiCGStab(ℓ) with ℓ-dimensional minimization after a compound step that costs 2ℓ matrix-vector products and increases the dimension of the Krylov space by 2ℓ. The BiCGStab family is a subset of an even larger class, the Lanczos-type product methods (LTPMs), which are characterized by residual polynomials that are products of a Lanczos polynomial p_n and another polynomial t_n of the same degree. All LTPMs are transpose-free and gain one dimension of the Krylov space per matrix-vector multiplication. Basically an infinite number of such methods exist, but it is not so easy to find one that can outperform, say, BiCGStab2. One that seems to be slightly better is based on the same idea as BiCGStab2 and, in fact, requires only the modification of a single condition in the code. We call it BiCG$\times$MR2. It does a two-dimensional minimization in each step. LTPMs in general and the examples mentioned here are discussed in Section 16.

Another good idea is to apply the QMR approach suitably to BiCGS. The resulting transpose-free QMR (TFQMR) algorithm of Freund (1993) can compete in efficiency with the best LTPMs; we treat it in Section 15. It is also possible to apply the QMR approach to LTPMs; see the introduction of Section 16 for references.

A disadvantage of Krylov space methods based on the Lanczos process is that they can break down for several reasons (even in exact arithmetic). Particularly troublesome are the 'serious' Lanczos breakdowns that occur when

$$\langle \widetilde{\mathbf{y}}_n, \mathbf{y}_n \rangle = 0, \quad \text{but} \quad \widetilde{\mathbf{y}}_n \neq \mathbf{o}, \;\; \mathbf{y}_n \neq \mathbf{o}.$$

Though true ('exact') breakdowns are extremely unlikely (except in contrived or specially structured examples), near-breakdowns can be (but need not be) the cause for an interruption or a slow-down of the convergence. These breakdowns were surely among the reasons why the BiCG method was rarely used over decades. However, as problem sizes grew, it became more and more important to have a method with short recurrences. Fortunately, it turned out that there is a mathematically correct way to circumvent both exact and near-breakdowns. The first to come up with such a procedure for the BiO algorithm for eigenvalue computations were Parlett, Taylor and Liu (1985), who also coined the term 'look-ahead'. In their view, this was a generalization of the two-sided Gram–Schmidt algorithm. In 1988 look-ahead was rediscovered by Gutknecht from a completely different perspective: for him, look-ahead was a translation of general recurrences in the Padé table, and a realization of what Gragg (1974) had indicated in a few lines long before. Although classical Padé and continued fraction theory put no emphasis on singular cases, the recurrences for what corresponds to an exact breakdown had been known for a long time. Moreover, it was possible to generalize them in order to treat near-breakdowns. This theory and the application to several versions of the BiCG method are compiled in Gutknecht (1992, 1994*a*). The careful implementation and the numerical tests of Freund, Gutknecht and Nachtigal (1993) ultimately proved, that all this can be turned into robust and efficient algorithms. Many other authors have also contributed to look-ahead Lanczos algorithms; see our references in Section 19. Moreover, the basic idea can be adapted to many other related recurrences in numerical analysis.

Here, we describe in Section 3 in detail the various ways in which the BiO algorithm can break down or terminate. Throughout the manuscript we then point out under which conditions the other algorithms break down. Although breakdowns are rare, knowing where and why they occur is important for learning how to avoid them and for gaining a full understanding of Lanczos-type algorithms. Quality software should be able to cope with breakdowns, or at least indicate to the user when a critical situation occurs. Seemingly, we only indicate exact breakdowns, but, of course, to include near-breakdowns the conditions '$= \mathbf{o}$' and '$= 0$' just have to be replaced by other appropriate conditions. In practice, however, the question of what 'appropriate' means is not always so easy to answer. It is briefly touched on in Section 19.

The derivations of the look-ahead BIO and BIOC algorithms we present are based on the interpretation as two-sided Gram–Schmidt process. The algorithms themselves are improved versions with reduced memory requirement. The look-ahead BIO algorithm makes use of a trick hidden in Freund and Zha (1993) and pointed out in Hochbruck (1996). We establish this simplification in a few lines, making use of a result from Gutknecht (1992). For the look-ahead BIOC algorithm the simplification is due to Hochbruck (1996), but also proved here in a different way.

We give many pointers to the vast literature on Krylov space solvers based on the Lanczos process, but treating all the algorithms proposed is far beyond our scope. In fact, this literature has grown so much in the last few years that we have even had to limit the number of references. Moreover, there must exist papers we are unaware of. Our intention is mainly to give an easy introduction to the nonsymmetric Lanczos process and some of the linear system solvers based on it. We have tried in particular to explain the underlying ideas and to make clear the limitations and difficulties of this family of methods. We have chosen to treat those algorithms that we think are important for understanding, as well as those we think are among the most effective.

There are many aspects that are not covered or only briefly referred to. First of all, the important question of convergence and its deterioration due to numerical effects is touched on only superficially. Also, we do not give any numerical results, since giving a representative set would have required considerable time and space. In fact, while straightforward implementation of the algorithms requires little work, we believe that the production of quality software taking into account some of the possible enhancements we mention in Section 18 – not to speak of the look-ahead option that should be included – requires considerable effort. Testing and evaluating such software is not easy either, as simple examples do not show the effects of the enhancements, while complicated ones make it difficult to link causes and effects. To our knowledge, the best available software is Freund and Nachtigal's QMRPACK, which is freely available from NETLIB, but is restricted to various forms of the QMR and TFQMR methods; see Freund and Nachtigal (1996).

While we are aware that in practice preconditioning can improve the convergence of an iterative method dramatically and can reduce the overall costs, we describe only briefly in Section 11 how preconditioners can be integrated into the algorithms, but not how they are found. For a survey of preconditioning techniques we refer, for example, to Saad (1996).

Also among the things we skip are the adaptations of the Lanczos method to systems with multiple right-hand sides. Recent work includes Aliaga,

Hernandez and Boley (1994) and Freund and Malhotra (1997), which is an extension of Ruhe's band Lanczos algorithm (Ruhe 1979) blended with QMR.

The Lanczos process is closely linked to other topics with the same mathematical background, in particular, formal orthogonal polynomials, Padé approximation, continued fractions, the recursive solution of linear systems with Hankel matrix, and the partial realization problem of system theory. We discuss formal orthogonal polynomials (FOPs) briefly in Section 12, but the other topics are not touched on. For the Padé connection, which is very helpful for understanding breakdowns and the look-ahead approach to cure them, we refer to Gutknecht (1994*b*) for a simple treatment. The relationship is described in much more detail and capitalized upon in Gutknecht (1992, 1994*a*), where Lanczos look-ahead for both the BIO and the BIOC process was introduced based on this connection. The relations between various look-ahead recurrences in the Padé table and variations of look-ahead Lanczos algorithms are also a topic of Hochbruck (1996). A fast Hankel solver based on look-ahead Lanczos was worked out in detail in Freund and Zha (1993). For the connection to the partial realization problem; see Boley (1994), Boley and Golub (1991), Golub, Kågström and Van Dooren (1992), and Parlett (1992).

In the wider neighbourhood of the Lanczos process we also find the Euclidean algorithm, Gauss quadrature, the matrix moment problem and its modified form, certain extrapolation methods, and the interpretation of conjugate direction methods as a special form of Gauss elimination. But these subjects are not treated here.

A preliminary version of a few sections of this overview was presented as an invited talk at the Copper Mountain Conference on Iterative Methods 1990 and was printed in the *Preliminary Proceedings* distributed at the conference. The present Section 14 (plus some additional material) was made available on separate sheets at that conference. We refer here to these two sources as Gutknecht (1990).

Notation

Matrices and vectors are denoted by upper and lower case boldface letters. In particular, $\mathbf{O}$ and $\mathbf{o}$ are the zero matrix and vector, respectively. The transpose of $\mathbf{A}$ is $\mathbf{A}^\top$; its conjugate transpose is $\mathbf{A}^\star$. Blocks of block vectors and matrices are sometimes printed in roman instead of boldface. For coefficients and other scalars we normally choose lower case Greek letters. However, for polynomials (and some function values) we use lower case roman letters. Sets are denoted by calligraphic letters; for instance, $\mathcal{P}_n$ is the set of polynomials of degree at most n.

We write scalars on the right-hand side of vectors, for instance $\mathbf{x}\alpha$, so that this product becomes a special case of matrix multiplication[2]. The inner product in $\mathbb{C}^N$ is assumed to be defined by $\langle \mathbf{x}, \mathbf{y} \rangle := \mathbf{x}^\star \mathbf{y}$, so that $\langle \mathbf{x}\alpha, \mathbf{y}\beta \rangle = \overline{\alpha}\beta \langle \mathbf{x}, \mathbf{y} \rangle$. The symbol $:=$ is used both for definitions and for algorithmic assignments.

To achieve a uniform nomenclature for the methods and algorithms we consider, we have modified some of the denominations used by other authors (but we also refer to the original name if the identity is not apparent). We hope that this will help readers to find their way in the maze of Lanczos-type solvers.

2. The unsymmetric Lanczos or BIO algorithm

In this section we describe the basic nonsymmetric Lanczos process that generates a pair of biorthogonal vector sequences. These define a pair of nested sequences of Krylov spaces, one generated by the given matrix or operator $\mathbf{A}$, the other by its Hermitian transpose $\mathbf{A}^\star$. The Lanczos algorithm is based on a successive extension of these two spaces, coupled with a two-sided Gram–Schmidt process for the construction of dual bases for them. The recurrence coefficients are the elements of a tridiagonal matrix that is, in theory, similar to $\mathbf{A}$ if the algorithm does not stop early. In the next section we will discuss the various ways in which the process can terminate or break down.

2.1. Derivation of the BIO *algorithm*

Let $\mathbf{A} \in \mathbb{C}^{N \times N}$ be any real or complex $N \times N$ matrix, and let $\mathbf{B}$ be a nonsingular matrix that commutes with $\mathbf{A}$ and is used to define the formal (*i.e.*, not necessarily symmetric definite) inner product $\langle \widetilde{\mathbf{y}}, \mathbf{y} \rangle_\mathbf{B} := \widetilde{\mathbf{y}}^\star \mathbf{B} \mathbf{y}$ on $\mathbb{C}^N \times \mathbb{C}^N$. Orthogonality will usually be referred to with respect to this formal inner product. For simplicity, we do *not* call it $\mathbf{B}$-orthogonality, except when this inaccuracy becomes misleading. However, we are mostly interested in the case where $\mathbf{B}$ is the identity $\mathbf{I}$ and thus $\langle \widetilde{\mathbf{y}}, \mathbf{y} \rangle_\mathbf{B}$ is the ordinary Euclidean inner product of $\mathbb{C}^N$. The case $\mathbf{B} = \mathbf{A}$ will also play a role. Due to $\mathbf{AB} = \mathbf{BA}$ we will have $\langle \widetilde{\mathbf{y}}, \mathbf{A}\mathbf{y} \rangle_\mathbf{B} = \langle \mathbf{A}^\star \widetilde{\mathbf{y}}, \mathbf{y} \rangle_\mathbf{B}$. Finally, we let $\widetilde{\mathbf{y}}_0, \mathbf{y}_0 \in \mathbb{C}^N$ be two non-orthogonal initial vectors: $\langle \widetilde{\mathbf{y}}_0, \mathbf{y}_0 \rangle_\mathbf{B} \neq 0$.

The *Lanczos biorthogonalization* (BIO) *algorithm*, called *method of minimized iterations* by Lanczos (1950, 1952), but often referred to as the *unsymmetric* or *two-sided Lanczos algorithm*, is a process for generating two finite sequences $\{\mathbf{y}_n\}_{n=0}^{\nu-1}$ and $\{\widetilde{\mathbf{y}}_n\}_{n=0}^{\nu-1}$, whose length ν depends on $\mathbf{A}$, $\mathbf{B}$,

[2] To see the rationale, think of the case where α turns out to be an inner product or where we generalize to block algorithms and α becomes a block of a vector.

$\mathbf{y}_0$, and $\widetilde{\mathbf{y}}_0$, such that, for $m, n = 0, 1, \ldots, \nu - 1$,

$$\begin{aligned} \mathbf{y}_n &\in \mathcal{K}_{n+1} := \operatorname{span}(\mathbf{y}_0, \mathbf{A}\mathbf{y}_0, \ldots, \mathbf{A}^n \mathbf{y}_0), \\ \widetilde{\mathbf{y}}_m &\in \widetilde{\mathcal{K}}_{m+1} := \operatorname{span}(\widetilde{\mathbf{y}}_0, \mathbf{A}^\star \widetilde{\mathbf{y}}_0, \ldots, (\mathbf{A}^\star)^m \widetilde{\mathbf{y}}_0) \end{aligned} \tag{2.1}$$

and

$$\langle \widetilde{\mathbf{y}}_m, \mathbf{y}_n \rangle_{\mathbf{B}} = \begin{cases} 0 & \text{if } m \neq n, \\ \delta_n \neq 0 & \text{if } m = n. \end{cases} \tag{2.2}$$

$\mathcal{K}_n$ and $\widetilde{\mathcal{K}}_m$ are *Krylov (sub)spaces* of $\mathbf{A}$ and $\mathbf{A}^\star$, respectively. The condition (2.2) means that the sequences are *biorthogonal.* Their elements $\mathbf{y}_n$ and $\widetilde{\mathbf{y}}_m$ are called *right* and *left Lanczos vectors*, respectively. In view of (2.1) $\mathbf{y}_n$ and $\widetilde{\mathbf{y}}_n$ can be expressed as

$$\mathbf{y}_n = p_n(\mathbf{A})\mathbf{y}_0, \quad \widetilde{\mathbf{y}}_n = \widetilde{p}_n(\mathbf{A}^\star)\widetilde{\mathbf{y}}_0, \tag{2.3}$$

where p_n and $\widetilde{p}_n$ are polynomials of degree at most n into which $\mathbf{A}$ and $\mathbf{A}^\star$ are substituted for the variable. We call p_n the nth *Lanczos polynomial*[3]. We will see in a moment that p_n has exact degree n and that the sequence $\{\widetilde{\mathbf{y}}_m\}$ can be chosen such that $\widetilde{p}_n = \overline{p_n}$, where $\overline{p_n}$ is the polynomial with the complex conjugate coefficients. In the general case, $\widetilde{p}_n$ will be seen to be a scalar multiple of $\overline{p_n}$.

The biorthogonal sequences of Lanczos vectors are constructed by a two-sided version of the well-known Gram–Schmidt process, but the latter is not applied to the bases used in the definition (2.1), since these are normally very ill-conditioned, that is, close to linearly dependent. Instead, the vectors that are orthogonalized are of the form $\mathbf{A}\mathbf{y}_n$ and $\mathbf{A}^\star\widetilde{\mathbf{y}}_n$; that is, they are created in each step from the most recently constructed pair by multiplication with $\mathbf{A}$ and $\mathbf{A}^\star$, respectively.

The length ν of the sequences is determined by the impossibility of extending them such that the conditions (2.1) and (2.2) still hold with $\delta_n \neq 0$. We will discuss the various reasons for a termination or a breakdown of the process below.

Clearly $\mathcal{K}_{n+1} \supseteq \mathcal{K}_n$, $\widetilde{\mathcal{K}}_{n+1} \supseteq \widetilde{\mathcal{K}}_n$, and from (2.2) it follows that equality cannot hold when $n < \nu$, since $\langle \widetilde{\mathbf{y}}, \mathbf{y}_n \rangle_{\mathbf{B}} = 0$ for all $\widetilde{\mathbf{y}} \in \widetilde{\mathcal{K}}_n$ and $\langle \widetilde{\mathbf{y}}_n, \mathbf{y} \rangle_{\mathbf{B}} = 0$ for all $\mathbf{y} \in \mathcal{K}_n$, or, briefly,

$$\widetilde{\mathcal{K}}_n \perp_{\mathbf{B}} \mathbf{y}_n, \quad \widetilde{\mathbf{y}}_n \perp_{\mathbf{B}} \mathcal{K}_n, \tag{2.4}$$

but $\widetilde{\mathcal{K}}_{n+1} \not\perp_{\mathbf{B}} \mathbf{y}_n$, $\widetilde{\mathbf{y}}_n \not\perp_{\mathbf{B}} \mathcal{K}_{n+1}$. Consequently, for $n = 1, \ldots, \nu - 1$,

$$\mathbf{y}_n \in \mathcal{K}_{n+1} \backslash \mathcal{K}_n, \quad \widetilde{\mathbf{y}}_n \in \widetilde{\mathcal{K}}_{n+1} \backslash \widetilde{\mathcal{K}}_n \tag{2.5}$$

[3] In classical analysis, depending on the normalization, the Lanczos polynomials are called *Hankel polynomials* or *Hadamard polynomials*, the latter being monic (Henrici 1974).

(where the backslash denotes the set-theoretic difference), and thus

$$\mathcal{K}_{n+1} = \text{span}\,(\mathbf{y}_0, \mathbf{y}_1, \ldots, \mathbf{y}_n), \quad \widetilde{\mathcal{K}}_{n+1} = \text{span}\,(\widetilde{\mathbf{y}}_0, \widetilde{\mathbf{y}}_1, \ldots, \widetilde{\mathbf{y}}_n). \tag{2.6}$$

Note that the relation on the left of (2.4) is equivalent to $\mathbf{y}_n \perp_{\mathbf{B}^\star} \widetilde{\mathcal{K}}_n$, but not to $\mathbf{y}_n \perp_{\mathbf{B}} \widetilde{\mathcal{K}}_n$, in general.

From (2.5) and (2.6) we conclude that for some complex constants $\tau_{k,n-1}$, $\widetilde{\tau}_{k,n-1}$ ($k = 0, \ldots, n$; $n = 1, \ldots, \nu - 1$)

$$\begin{aligned} \mathbf{y}_n \tau_{n,n-1} &= \mathbf{A}\mathbf{y}_{n-1} - \mathbf{y}_{n-1}\tau_{n-1,n-1} - \cdots - \mathbf{y}_0\tau_{0,n-1}, \\ \widetilde{\mathbf{y}}_n \widetilde{\tau}_{n,n-1} &= \mathbf{A}^\star\widetilde{\mathbf{y}}_{n-1} - \widetilde{\mathbf{y}}_{n-1}\widetilde{\tau}_{n-1,n-1} - \cdots - \widetilde{\mathbf{y}}_0\widetilde{\tau}_{0,n-1}, \end{aligned} \tag{2.7}$$

where $\tau_{n,n-1} \neq 0$ and $\widetilde{\tau}_{n,n-1} \neq 0$ can be chosen arbitrarily, for instance, for normalizing $\mathbf{y}_n$ and $\widetilde{\mathbf{y}}_n$. The choice will of course affect the constants δ_m and the coefficients $\tau_{n,m}$ and $\widetilde{\tau}_{n,m}$ for $m \geq n$, but only in a transparent way.

For $n = \nu$, (2.7) can still be satisfied for any nonzero values of $\tau_{\nu,\nu-1}$ and $\widetilde{\tau}_{\nu,\nu-1}$ by some $\mathbf{y}_\nu \perp_{\mathbf{B}^\star} \widetilde{\mathcal{K}}_\nu$ and some $\widetilde{\mathbf{y}}_\nu \perp_{\mathbf{B}} \mathcal{K}_\nu$, but these two vectors may be orthogonal to each other, so that (2.2) does not hold. In particular, one or even both may be the zero vector $\mathbf{o}$. Nevertheless, (2.4) and (2.6)–(2.7) also hold for $n = \nu$.

For $n \leq \nu$, let us introduce the $N \times n$ matrices

$$\mathbf{Y}_n := [\, \mathbf{y}_0 \;\; \mathbf{y}_1 \;\; \cdots \;\; \mathbf{y}_{n-1} \,], \quad \widetilde{\mathbf{Y}}_n := [\, \widetilde{\mathbf{y}}_0 \;\; \widetilde{\mathbf{y}}_1 \;\; \cdots \;\; \widetilde{\mathbf{y}}_{n-1} \,], \tag{2.8}$$

the diagonal matrices

$$\mathbf{D}_{\delta;n} := \text{diag}(\delta_0, \delta_1, \ldots, \delta_{n-1}),$$

and the $n \times n$ Hessenberg matrices $\mathbf{T}_n := [\tau_{k,j}]_{k,j=0}^{n-1}$ and $\widetilde{\mathbf{T}}_n := [\widetilde{\tau}_{k,j}]_{k,j=0}^{n-1}$. Then (2.2) and (2.7) become

$$\widetilde{\mathbf{Y}}_\nu^\star \mathbf{B} \mathbf{Y}_\nu = \mathbf{D}_{\delta;\nu} \tag{2.9}$$

and

$$\mathbf{A}\mathbf{Y}_\nu = \mathbf{Y}_\nu \mathbf{T}_\nu + \mathbf{y}_\nu \gamma_{\nu-1} \mathbf{l}_\nu^\top, \quad \mathbf{A}^\star \widetilde{\mathbf{Y}}_\nu = \widetilde{\mathbf{Y}}_\nu \widetilde{\mathbf{T}}_\nu + \widetilde{\mathbf{y}}_\nu \widetilde{\gamma}_{\nu-1} \mathbf{l}_\nu^\top, \tag{2.10}$$

where $\gamma_{n-1} := \tau_{n,n-1}$ and $\widetilde{\gamma}_{n-1} := \widetilde{\tau}_{n,n-1}$, and where $\mathbf{l}_n^\top$ is the last row of the $n \times n$ identity matrix. $n \leq \nu$. Using (2.9) and $\widetilde{\mathcal{K}}_\nu \perp_{\mathbf{B}} \mathbf{y}_\nu$, $\widetilde{\mathbf{y}}_\nu \perp_{\mathbf{B}} \mathcal{K}_\nu$ we conclude further that

$$\widetilde{\mathbf{Y}}_\nu^\star \mathbf{B} \mathbf{A} \mathbf{Y}_\nu = \mathbf{D}_{\delta;\nu} \mathbf{T}_\nu, \quad \mathbf{Y}_\nu^\star \mathbf{B}^\star \mathbf{A}^\star \widetilde{\mathbf{Y}}_\nu = \overline{\mathbf{D}_{\delta;\nu}} \widetilde{\mathbf{T}}_\nu, \tag{2.11}$$

and hence

$$\mathbf{D}_{\delta;\nu} \mathbf{T}_\nu = \widetilde{\mathbf{T}}_\nu^\star \mathbf{D}_{\delta;\nu}. \tag{2.12}$$

Here, on the left-hand side we have a lower Hessenberg matrix, while on the right-hand side there is an upper Hessenberg matrix. Consequently, both

$\mathbf{T}_\nu$ and $\widetilde{\mathbf{T}}_\nu$ must be tridiagonal. We simplify the notation by letting

$$\mathbf{T}_n =: \begin{bmatrix} \alpha_0 & \beta_0 & & & \\ \gamma_0 & \alpha_1 & \beta_1 & & \\ & \gamma_1 & \alpha_2 & \ddots & \\ & & \ddots & \ddots & \beta_{n-2} \\ & & & \gamma_{n-2} & \alpha_{n-1} \end{bmatrix},$$

and naming the elements of $\widetilde{\mathbf{T}}_n$ analogously with tildes.

By comparing diagonal and off-diagonal elements on both sides of (2.12) we obtain

$$\alpha_n = \overline{\widetilde{\alpha}_n} \tag{2.13}$$

and

$$\delta_n \beta_n = \delta_{n+1} \overline{\widetilde{\gamma}_n}, \quad \delta_{n+1} \gamma_n = \delta_n \overline{\widetilde{\beta}_n}. \tag{2.14}$$

Hence,

$$\widetilde{\beta}_n \widetilde{\gamma}_n = \overline{\beta_n \gamma_n}, \tag{2.15}$$

which could be satisfied by setting[4]

$$\widetilde{\beta}_n := \overline{\beta_n}, \quad \widetilde{\gamma}_n := \overline{\gamma_n}, \quad i.e., \quad \widetilde{\mathbf{T}}_\nu := \overline{\mathbf{T}_\nu}. \tag{2.16}$$

This would further allow us to set $\gamma_n := 1$ (for all n) Another frequent choice is

$$\widetilde{\beta}_n := \overline{\gamma_n}, \quad \widetilde{\gamma}_n := \overline{\beta_n}, \quad i.e., \quad \widetilde{\mathbf{T}}_\nu := \mathbf{T}_\nu^\star, \tag{2.17}$$

which according to (2.14) yields $\delta_n = \delta_0$ for all n.

However, we want to keep the freedom to scale $\mathbf{y}_n$ and $\widetilde{\mathbf{y}}_n$, since most algorithms that apply the Lanczos process for solving linear systems make use of it. (As, for instance, the standard BiCG algorithm discussed in Section 9 and the QMR approach of Section 5). Moreover, $\gamma_n = 1$ (for all n) can cause overflow or underflow. In view of (2.14), we have in general

$$\beta_n = \overline{\widetilde{\gamma}_n} \delta_{n+1} / \delta_n, \quad \widetilde{\beta}_n = \overline{\gamma_n \delta_{n+1} / \delta_n} = \overline{\beta_n \gamma_n} / \widetilde{\gamma}_n, \tag{2.18}$$

in accordance with (2.15). The choice between (2.16) and (2.17), and more generally, any choice that can be made to satisfy (2.14)–(2.15) only affects the scaling (including the sign, or, in the complex case, the argument of all components) of $\mathbf{y}_n$ and $\widetilde{\mathbf{y}}_n$. As we will see, it just leads to diagonal similarity transformations of the matrices $\mathbf{T}_\nu$ and $\widetilde{\mathbf{T}}_\nu$.

[4] By replacing the Euclidean inner product $\langle \widetilde{\mathbf{y}}, \mathbf{y} \rangle = \widetilde{\mathbf{y}}^\star \mathbf{y}$ by the bilinear form $\widetilde{\mathbf{y}}^\top \mathbf{y}$, and replacing $\mathbf{A}^\star$ by $\mathbf{A}^\top$, we could avoid the complex conjugation of the coefficients of $\mathbf{A}$ and of the recurrence coefficients; see, for instance, Freund et al. (1993). We prefer here to use the standard inner product.

After shifting the index n by 1, (2.7) can now be written as

$$\begin{aligned} \mathbf{y}_{n+1}\gamma_n &= \mathbf{A}\mathbf{y}_n - \mathbf{y}_n\alpha_n - \mathbf{y}_{n-1}\beta_{n-1}, \\ \widetilde{\mathbf{y}}_{n+1}\widetilde{\gamma}_n &= \mathbf{A}^\star\widetilde{\mathbf{y}}_n - \widetilde{\mathbf{y}}_n\widetilde{\alpha}_n - \widetilde{\mathbf{y}}_{n-1}\widetilde{\beta}_{n-1}, \end{aligned} \tag{2.19}$$

$n = 0, \ldots, \nu - 1$, with $\mathbf{y}_{-1} := \widetilde{\mathbf{y}}_{-1} := \mathbf{o}$, $\beta_{-1} := \widetilde{\beta}_{-1} := 0$. Taking inner products of the first formula with $\widetilde{\mathbf{y}}_{n-1}$, $\widetilde{\mathbf{y}}_n$, and $\widetilde{\mathbf{y}}_{n+1}$ we get three relations involving δ_{n+1}, δ_n, δ_{n-1}, and the recurrence coefficients α_n, β_{n-1}, γ_n. In particular, $\alpha_n = \langle \widetilde{\mathbf{y}}_n, \mathbf{A}\mathbf{y}_n \rangle_\mathbf{B} / \delta_n$ and, as is seen by inserting $\mathbf{A}^\star\widetilde{\mathbf{y}}_{n-1}$ according to (2.19),

$$\beta_{n-1} = \langle \widetilde{\mathbf{y}}_{n-1}, \mathbf{A}\mathbf{y}_n \rangle_\mathbf{B} / \delta_{n-1} = \langle \mathbf{A}^\star\widetilde{\mathbf{y}}_{n-1}, \mathbf{y}_n \rangle_\mathbf{B} / \delta_{n-1} = \overline{\widetilde{\gamma}_{n-1}}\delta_n/\delta_{n-1} \tag{2.20}$$

as in (2.18). Since (2.13) and the second relation in (2.18) determine $\widetilde{\alpha}_n$ and $\widetilde{\beta}_{n-1}$ we are left with two degrees of freedom that can be used to fix two of the three coefficients γ_n, $\widetilde{\gamma}_n$, and δ_{n+1}. We just exploit here the fact that the relations (2.9)–(2.11) can be used to determine $\mathbf{Y}_\nu$, $\widetilde{\mathbf{Y}}_\nu$, $\mathbf{T}_\nu$, $\widetilde{\mathbf{T}}_\nu$, and $\mathbf{D}_{\delta;\nu}$ column by column. Altogether we get the following general version of the unsymmetric Lanczos or biorthogonalization (BIO) algorithm.

ALGORITHM 1. (BIO ALGORITHM)
Choose $\mathbf{y}_0$, $\widetilde{\mathbf{y}}_0 \in \mathbb{C}^N$ such that $\delta_0 := \langle \widetilde{\mathbf{y}}_0, \mathbf{y}_0 \rangle_\mathbf{B} \neq 0$, and set $\beta_{-1} := 0$. For $n = 0, 1, \ldots$ compute

$$\alpha_n := \langle \widetilde{\mathbf{y}}_n, \mathbf{A}\mathbf{y}_n \rangle_\mathbf{B} / \delta_n, \tag{2.21a}$$

$$\widetilde{\alpha}_n := \overline{\alpha_n}, \tag{2.21b}$$

$$\beta_{n-1} := \overline{\widetilde{\gamma}_{n-1}}\delta_n/\delta_{n-1} \quad (\text{if } n > 0), \tag{2.21c}$$

$$\widetilde{\beta}_{n-1} := \overline{\gamma_{n-1}\delta_n/\delta_{n-1}} = \overline{\beta_{n-1}\gamma_{n-1}/\widetilde{\gamma}_{n-1}} \quad (\text{if } n > 0), \tag{2.21d}$$

$$\mathbf{y}_{\text{temp}} := \mathbf{A}\mathbf{y}_n - \mathbf{y}_n\alpha_n - \mathbf{y}_{n-1}\beta_{n-1}, \tag{2.21e}$$

$$\widetilde{\mathbf{y}}_{\text{temp}} := \mathbf{A}^\star\widetilde{\mathbf{y}}_n - \widetilde{\mathbf{y}}_n\widetilde{\alpha}_n - \widetilde{\mathbf{y}}_{n-1}\widetilde{\beta}_{n-1}, \tag{2.21f}$$

$$\delta_{\text{temp}} := \langle \widetilde{\mathbf{y}}_{\text{temp}}, \mathbf{y}_{\text{temp}} \rangle_\mathbf{B}\,; \tag{2.21g}$$

if $\delta_{\text{temp}} = 0$, choose $\gamma_n \neq 0$ and $\widetilde{\gamma}_n \neq 0$, set

$$\nu := n + 1, \quad \mathbf{y}_\nu := \mathbf{y}_{\text{temp}}/\gamma_n, \quad \widetilde{\mathbf{y}}_\nu := \widetilde{\mathbf{y}}_{\text{temp}}/\widetilde{\gamma}_n, \quad \delta_{n+1} := 0, \tag{2.21h}$$

and stop; otherwise, choose $\gamma_n \neq 0$, $\widetilde{\gamma}_n \neq 0$, and δ_{n+1} such that

$$\gamma_n\overline{\widetilde{\gamma}_n}\delta_{n+1} = \delta_{\text{temp}}, \tag{2.21i}$$

set

$$\mathbf{y}_{n+1} := \mathbf{y}_{\text{temp}}/\gamma_n, \quad \widetilde{\mathbf{y}}_{n+1} := \widetilde{\mathbf{y}}_{\text{temp}}/\widetilde{\gamma}_n, \tag{2.21j}$$

and proceed with the next step.

The definitions in (2.21h) will guarantee that formulae we derive below remain valid for $n = \nu$.

In exact arithmetic, which we assume throughout most of the text, the following basic result holds.

Theorem 2.1 The two sequences $\{\mathbf{y}_n\}_{n=0}^{\nu-1}$ and $\{\widetilde{\mathbf{y}}_n\}_{n=0}^{\nu-1}$ generated by the BIO algorithm satisfy (2.1) and (2.2), and hence (2.4)–(2.5). Moreover, (2.1) and (2.2) also hold when $m = \nu$ or $n = \nu$, except that $\delta_\nu = 0$.

Conversely, if (2.1) and (2.2) hold for $m, n = 0, \ldots, \nu$, except that $\delta_\nu = 0$, and if the choice (2.16) for satisfying (2.15) is made, then the recurrence formulae of the BIO algorithm are valid for $n = 0, \ldots, \nu - 1$, but the algorithm will stop at $n = \nu - 1$ due to $\delta_{\text{temp}} = 0$. The conditions (2.1) and (2.2) determine $\{\mathbf{y}_n\}_{n=0}^{\nu}$ and $\{\widetilde{\mathbf{y}}_n\}_{n=0}^{\nu}$ uniquely up to scaling.

Proof. The second part is covered by our derivation. It remains to prove the first part by induction. Assume that the sequences $\{\mathbf{y}_n\}_{n=0}^{\nu-1}$ and $\{\widetilde{\mathbf{y}}_m\}_{m=0}^{\nu-1}$ have been generated by the BIO algorithm and that (2.1) and (2.2) hold for $m, n = 0, \ldots, k$ $(< \nu)$. (For $k = 0$ this is clearly the case.) Then the validity of (2.1) for $m = k + 1$ or $n = k + 1$ is obvious from (2.21e) and (2.21f). Moreover, by (2.21j), (2.21e), and (2.21f),

$$\begin{aligned}\langle \widetilde{\mathbf{y}}_m, \mathbf{y}_{k+1} \rangle_{\mathbf{B}} &= \langle \widetilde{\mathbf{y}}_m, \mathbf{A}\mathbf{y}_k - \mathbf{y}_k\alpha_k - \mathbf{y}_{k-1}\beta_{k-1} \rangle_{\mathbf{B}} / \gamma_k \qquad (2.22\text{a}) \\ &= \Big(\langle \widetilde{\gamma}_m \widetilde{\mathbf{y}}_{m+1} + \widetilde{\alpha}_m \widetilde{\mathbf{y}}_m + \widetilde{\beta}_{m-1} \widetilde{\mathbf{y}}_{m-1}, \mathbf{y}_k \rangle_{\mathbf{B}} \\ &\quad - \langle \widetilde{\mathbf{y}}_m, \mathbf{y}_k \alpha_k \rangle_{\mathbf{B}} - \langle \widetilde{\mathbf{y}}_m, \mathbf{y}_{k-1}\beta_{k-1} \rangle_{\mathbf{B}} \Big) / \gamma_k. \qquad (2.22\text{b})\end{aligned}$$

If $m \le k - 2$, all terms on the right-hand side of (2.22b) vanish. For $m = k - 1$, we use (2.20) and (2.2) to obtain on the right-hand side of (2.22a) $(\beta_{k-1}\delta_{k-1} - 0 - \beta_{k-1}\delta_{k-1})/\gamma_k = 0$. Next, using (2.21a) and (2.2) we get for $m = k$ in (2.22a) $(\alpha_k\delta_k - \alpha_k\delta_k)/\gamma_k = 0$. Analogous arguments yield $\langle \widetilde{\mathbf{y}}_{k+1}, \mathbf{y}_n \rangle_{\mathbf{B}} = 0$ for $n = 0, \ldots, k$. Finally, by (2.21g)–(2.21j), $\langle \widetilde{\mathbf{y}}_{k+1}, \mathbf{y}_{k+1} \rangle_{\mathbf{B}} = \delta_k$. This completes the induction. □

In summary, the Lanczos process generates a pair of biorthogonal bases of a pair of Krylov spaces and does this with a pair of three-term recurrences that are closely linked to each other. In each step only two pairs of orthogonality conditions need to be enforced, which, due to the linkage, reduce to just one pair and thus require only two inner products in total. All the other orthogonality conditions are inherited: they are satisfied automatically, at least in theory. Of course, if we want normalized basis vectors, we need to invest another two inner products per step. Clearly, we also need two matrix-vector products per step to expand the two Krylov spaces.

2.2. Matrix relations

The matrix relations (2.9)–(2.11) can be considered as a shorthand notation for the BIO algorithm: the relations (2.10) describe the recurrences for $\mathbf{y}_n$

and $\widetilde{\mathbf{y}}_n$, while (2.9) and (2.11) summarize the formulae for determining the elements of $\mathbf{T}_\nu$, $\widetilde{\mathbf{T}}_\nu$, and $\mathbf{D}_{\delta;\nu}$. These matrix relations hold as well after $n < \nu$ steps, and in fact are obtained for such an intermediate stage by considering submatrices of the appropriate size.

Equation (2.10) can be simplified by introducing the $(n+1) \times n$ leading principal submatrices of $\mathbf{T}_\nu$ and $\widetilde{\mathbf{T}}_\nu$, the *extended* tridiagonal matrices[5]

$$\underline{\mathbf{T}}_n := \left[\begin{array}{c} \mathbf{T}_n \\ \hline \gamma_{n-1}\mathbf{l}_n^{\mathsf{T}} \end{array}\right] = \begin{bmatrix} \alpha_0 & \beta_0 & & & \\ \gamma_0 & \alpha_1 & \beta_1 & & \\ & \gamma_1 & \alpha_2 & \ddots & \\ & & \ddots & \ddots & \beta_{n-2} \\ & & & \gamma_{n-2} & \alpha_{n-1} \\ & & & & \gamma_{n-1} \end{bmatrix},$$

and the analogous matrices with tildes. Then we have altogether

$$\mathbf{A}\mathbf{Y}_n = \mathbf{Y}_{n+1}\underline{\mathbf{T}}_n, \quad \mathbf{A}^\star\widetilde{\mathbf{Y}}_n = \widetilde{\mathbf{Y}}_{n+1}\underline{\widetilde{\mathbf{T}}}_n \quad (n \le \nu), \tag{2.23}$$

$$\widetilde{\mathbf{Y}}_n^\star\mathbf{B}\mathbf{Y}_n = \mathbf{D}_{\delta;n}, \quad \widetilde{\mathbf{Y}}_n^\star\mathbf{B}\mathbf{A}\mathbf{Y}_n = \mathbf{D}_{\delta;n}\mathbf{T}_n \quad (n \le \nu), \tag{2.24}$$

$$\mathbf{D}_{\delta;n}\mathbf{T}_n = \widetilde{\mathbf{T}}_n^\star\mathbf{D}_{\delta;n} \quad (n \le \nu). \tag{2.25}$$

The first relation in (2.23) means that $\underline{\mathbf{T}}_n$ is the representation in the basis $\{\mathbf{y}_0, \ldots, \mathbf{y}_n\}$ of the restriction of $\mathbf{A}$ to the subspace $\mathcal{K}_n$, which is mapped into $\mathcal{K}_{n+1}$. The subspace $\mathcal{K}_n$ is 'nearly invariant' as its image requires only one additional space dimension. When it turns out that the component of $\mathbf{A}\mathbf{y}_{n-1}$ in this direction is relatively short, then we can expect that the eigenvalues of $\mathbf{T}_n$ are close approximations of eigenvalues of $\mathbf{A}$. This is vaguely the reasoning for applying the BIO algorithm to eigenvalue computations. There are several reasons that make this argument dangerous. First, the basis is not orthogonal, and thus $\mathbf{T}_n$ is linked to $\mathbf{A}$ by an oblique projection only: while in the Hermitian case the eigenvalues of $\mathbf{T}_n$ are *Ritz values*, that is, *Galerkin* or *Rayleigh–Ritz approximations*, they are in the non-Hermitian case only *Petrov–Galerkin* approximations, which are sometimes referred to as *Petrov values*. Second, since neither $\mathbf{A}$ nor $\mathbf{T}_n$ are Hermitian in the case considered here, eigenvalues need not behave nicely under perturbation.

A notorious problem with the Lanczos process is that in finite precision arithmetic round-off affects it strongly. In particular, since only two orthogonality conditions are enforced at every step, while orthogonality with respect to earlier vectors is inherited, a loss of (bi)orthogonality is noticed

[5] By underlining $\mathbf{T}_n$ we want to indicate that we augment this matrix by an additional row. We suggest reading $\underline{\mathbf{T}}_n$ as 'T sub n extended'. The same notation will be used on other occasions.

in practice, which means that $\mathbf{D}_{\delta;n}$ is not diagonal. This is also a serious problem in the Hermitian case; we will return to it briefly in Section 18.

2.3. Normalization; simplification due to symmetry

As discussed before, in the BIO algorithm (Algorithm 1) the coefficients α_n, $\widetilde{\alpha}_n$, β_{n-1}, $\widetilde{\beta}_{n-1}$, γ_n, $\widetilde{\gamma}_n$, and δ_{n+1} are defined uniquely except that in (2.21i) we are free to choose two of the three quantities γ_n, $\widetilde{\gamma}_n$, and δ_{n+1}. Special versions of the algorithm are found by making a particular choice for two of the coefficients γ_n, $\widetilde{\gamma}_n$, and δ_{n+1}, and capitalizing upon the particular choice. The classical choices for theoretical work are $\gamma_n := \widetilde{\gamma}_n := 1$ (Lanczos 1950, Lanczos 1952, Rutishauser 1953, Householder 1964) or $\widetilde{\gamma}_n = \overline{\gamma_n}$ and $\delta_{n+1} := 1$. The latter makes the two vector sequences biorthonormal and yields $\beta_{n-1} = \gamma_{n-1}$; that is, $\mathbf{T}_\nu = \mathbf{T}_\nu^\top$ is real or complex symmetric, *cf.* (2.18). (Consequently, (2.16) and (2.17) then coincide.) For numerical computations the former choice is risky with respect to overflow, and the latter is inappropriate for real nonsymmetric matrices since it may lead to some complex γ_n. Therefore, in the real nonsymmetric case, the two choices

$$\gamma_n := \widetilde{\gamma}_n := \sqrt{|\delta_{\text{temp}}|}\,, \quad \delta_{n+1} := \delta_{\text{temp}}/(\gamma_n\widetilde{\gamma}_n) = \delta_{\text{temp}}/|\delta_{\text{temp}}|, \tag{2.26}$$

$$\gamma_n := \|\mathbf{y}_{\text{temp}}\|\,, \quad \widetilde{\gamma}_n := \|\widetilde{\mathbf{y}}_{\text{temp}}\|\,, \quad \delta_{n+1} := \delta_{\text{temp}}/(\gamma_n\widetilde{\gamma}_n), \tag{2.27}$$

are normally suggested, but there may be a special reason for yet another one. Note that (2.27) requires two additional inner products. These are often justified anyway by the necessity of stability and round-off error control; see Section 18.

Replacing, say, $\gamma_n = 1$ (for all n) by some other choice means replacing $\mathbf{y}_n$ by

$$\widehat{\mathbf{y}}_n := \mathbf{y}_n/\Gamma_n, \quad \text{where } \Gamma_n := \gamma_0\gamma_1\cdots\gamma_{n-1}, \tag{2.28}$$

which in view of (2.23) amounts to replacing $\mathbf{T}_n$ by

$$\widehat{\mathbf{T}}_n := \mathbf{D}_{\Gamma;n}^{-1}\mathbf{T}_n\mathbf{D}_{\Gamma;n}, \quad \text{where } \mathbf{D}_{\Gamma;n} := \operatorname{diag}(\Gamma_0, \Gamma_1, \ldots, \Gamma_{n-1}), \tag{2.29}$$

If γ_n and $\widetilde{\gamma}_n$ are chosen independently of $\mathbf{y}_{\text{temp}}$ and $\widetilde{\mathbf{y}}_{\text{temp}}$, the formulae (2.21e)–(2.21j) can be simplified; *cf.* Algorithm 2 below.

If $\mathbf{A}$ and $\mathbf{B}$ are Hermitian and $\mathbf{B}$ is positive definite, starting with $\widetilde{\mathbf{y}}_0 = \mathbf{y}_0$ and making the natural choice $\widetilde{\gamma}_n := \overline{\gamma_n}$ and $\delta_n > 0$ leads to $\widetilde{\mathbf{T}}_\nu = \overline{\mathbf{T}_\nu} = \mathbf{T}_\nu$ (that is, $\widetilde{\mathbf{T}}_\nu = \mathbf{T}_\nu$ is real) and $\widetilde{\mathbf{y}}_n = \mathbf{y}_n$ (for all n). Thus, the recursion (2.21f) for $\widetilde{\mathbf{y}}_n$ is redundant, the costs are reduced to roughly half, and the Lanczos vectors are orthogonal to each other. Moreover, one can choose $\gamma_n > 0$ (for all n), which then implies that $\beta_{n-1} > 0$ also. Finally, choosing $\delta_n := \delta_0$ (for all n) makes $\mathbf{T}_\nu$ real symmetric. Then the BIO algorithm becomes the symmetric Lanczos algorithm, which is often just called the *Lanczos algorithm.*

If $\mathbf{A}$ is complex symmetric and $\widetilde{\mathbf{y}}_0 = \overline{\mathbf{y}_0}$, then $\widetilde{\mathbf{y}}_n = \overline{\mathbf{y}_n}$ (for all n). Again, the costs reduce to about half. Now, setting $\delta_n := 1$ (for all n) makes $\mathbf{T}_\nu$ complex symmetric, but this can be achieved in general and has nothing to do with $\mathbf{A}$ being complex symmetric. See Freund (1992) for further details on this case.

In Section 6 we will discuss yet other cases where the BIO algorithm simplifies.

3. Termination, breakdowns and convergence

The BIO algorithm *stops* with $\nu := n+1$ when $\delta_{\text{temp}} = 0$, since α_{n+1} would become infinite or indefinite. We call ν here the *index of first breakdown or termination.* Of course, ν is bounded by the maximum dimension of the Krylov spaces, but ν may be smaller for several reasons. The maximum dimension of the subspaces $\mathcal{K}_n$ defined by (2.1) depends not only on $\mathbf{A}$ but also on $\mathbf{y}_0$; it is called the *grade of* $\mathbf{y}_0$ *with respect to* $\mathbf{A}$ and is here denoted by $\bar{\nu}(\mathbf{y}_0, \mathbf{A})$. As is easy to prove, it satisfies

$$\bar{\nu}(\mathbf{y}_0, \mathbf{A}) = \min \left\{ n : \; \dim \mathcal{K}_n = \dim \mathcal{K}_{n+1} \right\},$$

and it is at most equal to the degree $\bar{\nu}(\mathbf{A})$ of the minimum polynomial of $\mathbf{A}$. Clearly, the Lanczos process stops with $\mathbf{y}_{\text{temp}} = \widetilde{\mathbf{y}}_{\text{temp}} = \mathbf{o}$ when $\nu = \bar{\nu}(\mathbf{A})$. If this *full termination* due to $\mathbf{y}_{\text{temp}} = \widetilde{\mathbf{y}}_{\text{temp}} = \mathbf{o}$ happens before the degree of the minimum polynomial is reached, that is, if $\nu < \bar{\nu}(\mathbf{A})$, we call it an *early full termination.* However, the BIO algorithm can also stop with either $\mathbf{y}_{\text{temp}} = \mathbf{o}$ or $\widetilde{\mathbf{y}}_{\text{temp}} = \mathbf{o}$, and even with $\langle \widetilde{\mathbf{y}}_{\text{temp}}, \mathbf{y}_{\text{temp}} \rangle_{\mathbf{B}} = 0$ when $\mathbf{y}_{\text{temp}} \neq \mathbf{o}$ and $\widetilde{\mathbf{y}}_{\text{temp}} \neq \mathbf{o}$. Then we say that it *breaks down*, or, more exactly, that we have a *one-sided termination* or a *serious breakdown*, respectively[6].

Lanczos (1950, 1952) was already aware of these various breakdowns. They have since been discussed by many authors; see, in particular, Faddeev and Faddeeva (1964), Gutknecht (1990), Householder (1964), Joubert (1992), Parlett (1992), Parlett et al. (1985), Rutishauser (1953), Saad (1982), and Taylor (1982).

Of course, in floating-point arithmetic, a *near-breakdown* passed without taking special measures may lead to stability problems. Therefore, in practice, breakdown conditions '$= \mathbf{o}$' and '$= 0$' have to be replaced by other conditions that should not depend on scaling and should prevent us from numerical instability. We will return to this question in Section 19.

[6] Parlett (1992) refers to a *benign breakdown* when we have either an early full termination or a one-sided termination.

3.1. Full termination

In the case of *full termination*, that is, when the BIO algorithm terminates with

$$\mathbf{y}_{\text{temp}} = \widetilde{\mathbf{y}}_{\text{temp}} = \mathbf{o}, \tag{3.1}$$

we can conclude from (2.10) that

$$\mathbf{A}\mathbf{Y}_\nu = \mathbf{Y}_\nu \mathbf{T}_\nu, \quad \mathbf{A}^\star \widetilde{\mathbf{Y}}_\nu = \widetilde{\mathbf{Y}}_\nu \widetilde{\mathbf{T}}_\nu. \tag{3.2}$$

This means that $\mathcal{K}_\nu$ and $\widetilde{\mathcal{K}}_\nu$ are invariant subspaces of dimension ν of $\mathbf{A}$ and $\mathbf{A}^\star$, respectively. The set of eigenvalues of $\mathbf{T}_\nu$ is then a subset of the spectrum of $\mathbf{A}$.

The formula $\mathbf{A}\mathbf{Y}_\nu = \mathbf{Y}_\nu \mathbf{T}_\nu$ points to an often cited objective of the BIO algorithm: the similarity reduction of a given matrix $\mathbf{A}$ to a tridiagonal matrix $\mathbf{T}_\nu$. For the latter the computation of the eigenvalues is much less costly, in particular in the Hermitian case. However, unless $\nu = N$ or the Lanczos process is restarted (possibly several times) with some new pair $\widetilde{\mathbf{y}}_\nu$, $\mathbf{y}_\nu$ satisfying (2.4), it can never determine the geometric multiplicity of an eigenvalue, since it can at best find the factors of the minimal polynomial of $\mathbf{A}$. In theory, to find the whole spectrum, we could continue the algorithm after a full termination with $\nu < N$ by constructing first a new pair $(\mathbf{y}_\nu, \widetilde{\mathbf{y}}_\nu)$ of nonzero vectors that are biorthogonal to the pair of vector sequences constructed so far, *i.e.*, satisfy (2.4); see, for instance, Householder (1964), Lanczos (1950), Lanczos (1952), and Rutishauser (1953). Starting from a trial pair $(\mathbf{y}, \widetilde{\mathbf{y}})$ one would have to construct

$$\mathbf{y}_\nu := \mathbf{y} - \sum_{k=0}^{\nu-1} \mathbf{y}_k \langle \widetilde{\mathbf{y}}_k, \mathbf{y} \rangle_{\mathbf{B}} / \delta_k, \tag{3.3a}$$

$$\widetilde{\mathbf{y}}_\nu := \widetilde{\mathbf{y}} - \sum_{k=0}^{\nu-1} \widetilde{\mathbf{y}}_k \langle \mathbf{y}_k, \widetilde{\mathbf{y}} \rangle_{\mathbf{B}^\star} / \overline{\delta_k}, \tag{3.3b}$$

and hope that the two resulting vectors are nonzero. Then, one can set $\gamma_{\nu-1} := \beta_{\nu-1} := 0$, so that after the restart the relations (2.10) hold even beyond this ν. If no breakdown or one-sided termination later occurs, and if any further early full termination is also followed by such a restart, then in theory the algorithm must terminate with $\mathbf{y}_{\text{temp}} = \widetilde{\mathbf{y}}_{\text{temp}} = \mathbf{o}$ and $\nu = N$. The relations (3.2) then hold with all the matrices being square of order N. The tridiagonal matrix $\mathbf{T}_N$ may have some elements $\gamma_k = \beta_k = 0$ due to the restarts, and the same will then happen to $\widetilde{\mathbf{T}}_N$. Since $\mathbf{Y}_N$ and $\widetilde{\mathbf{Y}}_N$ are nonsingular, $\mathbf{T}_N$ is similar to $\mathbf{A}$, and $\widetilde{\mathbf{T}}_N$ is similar to $\mathbf{A}^\star$. Unfortunately, in practice this all works only for very small N: first, as is seen from (3.3a)–(3.3b), the restart with persisting biorthogonality (2.2) requires all previously computed vectors of the two sequences, but these are normally not stored; second, completely avoiding loss of (bi)orthogonality during the iteration would require full reorthogonalization with respect to previous Lanczos vectors, which means giving up the greatest advantages of the Lanczos process, the short recurrences.

For the same reasons, finding all the roots of the minimal polynomial with the Lanczos process is in practice normally beyond reach: due to loss of (bi)orthogonality or because $\bar{\nu}(\mathbf{A})$ is too large and the process has to be stopped early, only few of the eigenvalues are found with acceptable accuracy; fortunately, these are often those that are of prime interest.

3.2. One-sided termination

When the BIO algorithm stops due to

$$\text{either } \mathbf{y}_{\text{temp}} = \mathbf{o} \text{ or } \widetilde{\mathbf{y}}_{\text{temp}} = \mathbf{o}$$

(but not $\mathbf{y}_{\text{temp}} = \widetilde{\mathbf{y}}_{\text{temp}} = \mathbf{o}$), we call this a *one-sided termination.* In some applications this is welcome, but in others it may still be a serious difficulty. In view of (2.10), it means that either $\mathcal{K}_\nu$ or $\widetilde{\mathcal{K}}_\nu$ is an invariant subspace of dimension ν of $\mathbf{A}$ or $\mathbf{A}^\star$, respectively. For eigenvalue computations, this is very useful information, although sometimes one may need to continue the algorithm with a pair $(\mathbf{y}_\nu, \widetilde{\mathbf{y}}_\nu)$ that is biorthogonal to the pairs constructed so far, in order to find further eigenvalues. Determining the missing vector of this pair will again require the expensive orthogonalization of a trial vector with respect to a ν-dimensional Krylov subspace, that is, either (3.3a) or (3.3b).

In contrast, when we have to solve a linear system, then $\mathbf{y}_{\text{temp}} = \mathbf{o}$ is all we aim at, as we will see in the next section. Unfortunately, when $\widetilde{\mathbf{y}}_{\text{temp}} = \mathbf{o}$ but $\mathbf{y}_{\text{temp}} \neq \mathbf{o}$, we have a nasty situation where we have to find a replacement for $\widetilde{\mathbf{y}}_{\text{temp}}$ that is orthogonal to $\mathcal{K}_\nu$. In practice, codes either just use some $\widetilde{\mathbf{y}}_{\text{temp}}$ that consists of scaled up round-off errors or restart the BIO algorithm. In either case the convergence slows down. The best precaution against this type of breakdown seems to be choosing as left initial vector $\widetilde{\mathbf{y}}_0$ a random one.

3.3. Serious breakdowns

Let us now discuss the *serious breakdowns* of the BIO algorithm, which we also call *Lanczos breakdowns* to distinguish them from a second type of serious breakdown that can occur additionally in the standard form of the biconjugate gradient method. Hence, assume that the BIO algorithm stops due to

$$\delta_{\text{temp}} = 0, \quad \text{but neither } \mathbf{y}_{\text{temp}} = \mathbf{o} \text{ nor } \widetilde{\mathbf{y}}_{\text{temp}} = \mathbf{o}.$$

In the past, the recommendation was to restart the algorithm from scratch. Nowadays one should implement what is called *look-ahead.* It is the curing of these breakdowns and the corresponding near-breakdowns that is addressed by the *look-ahead Lanczos algorithms* which have attracted so much attention recently. We will discuss the look-ahead BIO algorithm and some of the related theory in Section 19, where we will also give detailed references. In most cases, look-ahead is successful. Oversimplifying matters, we can say that curing a serious breakdown with look-ahead requires that $\langle \widetilde{\mathbf{y}}_\nu, \mathbf{A}^k \mathbf{y}_\nu \rangle_\mathbf{B} \neq 0$ for some k, while the breakdown is *incurable* when

$$\langle \widetilde{\mathbf{y}}_\nu, \mathbf{A}^k \mathbf{y}_\nu \rangle_\mathbf{B} = 0 \quad (\text{for all } k \geq 0).$$

In theory, the condition $\langle \widetilde{\mathbf{y}}_\nu, \mathbf{A}^k \mathbf{y}_\nu \rangle_\mathbf{B} \neq 0$ has to be replaced by a positive lower bound for the smallest singular value of a $k \times k$ matrix, but in practice even this condition is not safe.

The serious breakdown does not occur if $\mathbf{A}$ and $\mathbf{B}$ are Hermitian, $\mathbf{B}$ is positive definite, $\widetilde{\mathbf{y}}_0 = \mathbf{y}_0$, $\widetilde{\gamma}_n = \overline{\gamma_n}$, and $\delta_n > 0$ (for all n), since then $\widetilde{\mathbf{y}}_n = \mathbf{y}_n$ (for all n), and $\langle ., . \rangle_\mathbf{B}$ is an inner product. Also, choosing γ_n, $\widetilde{\gamma}_n$, or δ_n differently will not destroy this property. On the other hand, serious breakdowns can still occur for a real symmetric matrix $\mathbf{A}$ if $\widetilde{\mathbf{y}}_0 \neq \mathbf{y}_0$.

Under the standard assumption $\mathbf{B} = \mathbf{I}$ it was shown by Rutishauser (1953) (for another proof see Householder (1964)) that there exist $\mathbf{y}_0$ and $\widetilde{\mathbf{y}}_0$ such that neither a serious breakdown nor a premature termination occurs; that is, such that the process does not end before the degree of the minimal polynomial is attained. Unfortunately, such a pair $(\mathbf{y}_0, \widetilde{\mathbf{y}}_0)$ is in general not known. Joubert (1992) even showed that, in a probabilistic sense, nearly all pairs have this property; that is, the assumption of having neither a premature termination nor a breakdown is a generic property. This is no longer true if the matrix is real and one restricts the initial vectors by requiring $\widetilde{\mathbf{y}}_0 = \mathbf{y}_0 \in \mathbb{R}^N$. Joubert gives an example where serious breakdowns then occur for almost all $\mathbf{y}_0$. Of course, the set of pairs $(\mathbf{y}_0, \widetilde{\mathbf{y}}_0)$ that lead to a near-breakdown never has measure zero. But practice shows that near-breakdowns that have a devastating effect on the process are fairly rare.

3.4. Convergence

Although in theory the BIO algorithm either terminates or breaks down in at most $\min\{\bar{\nu}(\mathbf{y}_0, \mathbf{A}), \bar{\nu}(\widetilde{\mathbf{y}}_0, \mathbf{A}^\star)\}$ steps, this rarely happens in practice, and the process can be continued far beyond N. For small matrices this is sometimes necessary when one wants to find all eigenvalues or to solve a linear system. But the BIO algorithm is usually applied to very large matrices and stopped at some $n \ll N$, so that only (2.10) and (2.11) hold, but not (3.2). The n eigenvalues of $\mathbf{T}_n$ are then considered as approximations of n eigenvalues of $\mathbf{A}$, and typically they tend to approximate eigenvalues that lie near the border of the spectrum. For eigenvalues of small absolute value, the absolute error is comparable to the one for large eigenvalues; hence, the relative error tends to be large. Therefore, the method is best at finding the dominant eigenvalues. Lanczos (1950, p. 270) found a heuristic explanation for this important phenomenon. Some remarks and references on the convergence of Lanczos-type solvers will be made in the next section when we discuss the basic properties of the BiCG method.

An additional difficulty is that due to the loss of biorthogonality, eigenvalues of $\mathbf{A}$ may reappear in $\mathbf{T}_n$ with too high multiplicity. One refers to these extra eigenvalues as *ghost eigenvalues*. We will come back to this problem in Section 18.

4. The BiORes form of the BiCG method

Let us now turn to applying the Lanczos BiO algorithm to the problem of solving linear systems of equations $\mathbf{Ax} = \mathbf{b}$. We first review the basic properties of the conjugate gradient (CG) and the conjugate residual (CR) methods and then describe a first version, BiORes, of the biconjugate gradient (BiCG) method. In the next section we will further cover the MinRes algorithm for the CR method and a first version of the QMR method.

We assume that $\mathbf{A}$ is nonsingular and denote the solution of $\mathbf{Ax} = \mathbf{b}$ by $\mathbf{x}_{\mathsf{ex}}$, its initial approximation by $\mathbf{x}_0$, the nth approximation (or *iterate*) by $\mathbf{x}_n$, and the corresponding *residual* by $\mathbf{r}_n := \mathbf{b} - \mathbf{A}\mathbf{x}_n$. Additionally, we let $\mathbf{y}_0 := \mathbf{r}_0$, or $\mathbf{y}_0 := \mathbf{r}_0/\|\mathbf{r}_0\|$ if we aim for normalized Lanczos vectors. As in (2.1), $\mathcal{K}_n$ is the nth Krylov space generated by $\mathbf{A}$ from $\mathbf{y}_0$, which now has the direction of the initial residual. From time to time we also refer to the nth *error*, $\mathbf{x}_{\mathsf{ex}} - \mathbf{x}_n$. Note that $\mathbf{r}_n = \mathbf{A}(\mathbf{x}_{\mathsf{ex}} - \mathbf{x}_n)$.

4.1. The conjugate gradient and conjugate residual methods

We first assume that $\mathbf{A}$ and $\mathbf{B}$ are commuting Hermitian positive definite matrices and recall some facts about the *conjugate gradient* (CG) *method* of Hestenes and Stiefel (1952). It is characterized by the property that the nth iterate $\mathbf{x}_n$ minimizes the quadratic function

$$\mathbf{x} \mapsto \langle(\mathbf{x}_{\mathsf{ex}} - \mathbf{x}), \mathbf{A}(\mathbf{x}_{\mathsf{ex}} - \mathbf{x})\rangle_{\mathbf{B}}$$

among all $\mathbf{x} \in \mathbf{x}_0 + \mathcal{K}_n$. The standard case is again $\mathbf{B} = \mathbf{I}$. By differentiation one readily verifies that the minimization problem is equivalent to the *Galerkin condition*

$$\langle \mathbf{y}, \mathbf{r}_n\rangle_{\mathbf{B}} = 0 \ \ (\text{for all } \mathbf{y} \in \mathcal{K}_n), \quad \textit{i.e.}, \quad \mathcal{K}_n \perp_{\mathbf{B}} \mathbf{r}_n. \tag{4.1}$$

Note that

$$\mathbf{r}_n = \mathbf{b} - \mathbf{A}\mathbf{x}_0 + \mathbf{A}(\mathbf{x}_0 - \mathbf{x}_n) \in \mathbf{r}_0 + \mathbf{A}\mathcal{K}_n \subset \mathcal{K}_{n+1}. \tag{4.2}$$

In view of (4.1) and (4.2), $\mathbf{r}_n$ spans the orthogonal complement of $\mathcal{K}_n$ in $\mathcal{K}_{n+1}$. Hence, if we let $\mathbf{y}_0 := \mathbf{r}_0$ and generate the basis $\{\mathbf{y}_k\}_{k=0}^n$ of $\mathcal{K}_{n+1}$ by recursively orthogonalizing $\mathbf{A}\mathbf{y}_m$ with respect to $\mathcal{K}_{m+1}$ $(m = 0, \ldots, n-1)$, then $\mathbf{y}_n$ is proportional to $\mathbf{r}_n$, and by suitable normalization we can achieve $\mathbf{y}_n = \mathbf{r}_n$. This recursive orthogonalization procedure is none other than the symmetric Lanczos process, that is, the BIO algorithm with $\widetilde{\mathbf{y}}_0 = \mathbf{y}_0$ and commuting Hermitian matrices $\mathbf{A}$ and $\mathbf{B}$. Note also that by (4.2), $\mathbf{r}_n = p_n(\mathbf{A})\mathbf{y}_0$, where p_n is a polynomial of exact degree n satisfying $p_n(0) = 1$. This property, which is equivalent to $\mathbf{x}_n \in \mathbf{x}_0 + \mathcal{K}_n$, means that CG is a Krylov space solver, as defined by (1.1). Of course, up to normalization, the *residual polynomial* p_n is here the Lanczos polynomial of (2.3).

The directions $\mathbf{x}_n - \mathbf{x}_{n-1}$ can be seen to be conjugate to each other (*i.e.*, $\mathbf{A}$-orthogonal with respect to the $\mathbf{B}$-inner product, or, simply, $\mathbf{AB}$-orthogonal), whence CG is a special *conjugate direction* method. In their classical CG method Hestenes and Stiefel (1952) chose $\mathbf{B} = \mathbf{I}$ and thus minimized $\mathbf{x} \mapsto \langle(\mathbf{x}_{\text{ex}} - \mathbf{x}), \mathbf{A}(\mathbf{x}_{\text{ex}} - \mathbf{x})\rangle$, which is the square of the $\mathbf{A}$-norm of the error. (This is a norm since $\mathbf{A}$ is Hermitian positive definite.) With respect to the inner product induced by $\mathbf{A}$, we then have from (4.1) and (4.2)

$$\mathcal{K}_n \perp_{\mathbf{A}} (\mathbf{x}_{\text{ex}} - \mathbf{x}_n), \qquad (\mathbf{x}_{\text{ex}} - \mathbf{x}_n) - (\mathbf{x}_{\text{ex}} - \mathbf{x}_0) \in \mathcal{K}_n.$$

This means that $\mathbf{x}_0 - \mathbf{x}_n = (\mathbf{x}_{\text{ex}} - \mathbf{x}_n) - (\mathbf{x}_{\text{ex}} - \mathbf{x}_0)$ is the $\mathbf{A}$-orthogonal projection of the initial error $\mathbf{x}_{\text{ex}} - \mathbf{x}_0$ into $\mathcal{K}_n$, and the error $\mathbf{x}_{\text{ex}} - \mathbf{x}_n$ is the difference between the initial error and its projection. Therefore, the CG method can also be viewed as an *orthogonal projection method* in the error space endowed with the $\mathbf{A}$-norm. Moreover, it can be understood as an orthogonal projection method in the residual space endowed with the $\mathbf{A}^{-1}$-norm.

If $\mathbf{B} = \mathbf{A} = \mathbf{A}^\star$ instead, we have

$$\langle(\mathbf{x}_{\text{ex}} - \mathbf{x}), \mathbf{A}(\mathbf{x}_{\text{ex}} - \mathbf{x})\rangle_{\mathbf{B}} = \|\mathbf{A}(\mathbf{x}_{\text{ex}} - \mathbf{x})\|^2 = \|\mathbf{b} - \mathbf{A}\mathbf{x}\|^2, \tag{4.3}$$

which shows that the residual norm is now minimized. The $\mathbf{B}$-orthogonality of the residuals means here that they are conjugate. The method is therefore called *conjugate residual* (CR) or *minimum residual* method. Normally, the abbreviation MINRES stands for a particular algorithm due to Paige and Saunders (1975) for this method. We will come back to it in Section 5. Like some other versions of the CR method, MINRES is also applicable to Hermitian indefinite systems; see Ashby, Manteuffel and Saylor (1990), Fletcher (1976).

From (4.1) and (4.2) we can conclude here that $\mathbf{r}_n$ is chosen so that it is orthogonal to $\mathbf{A}\mathcal{K}_n$ and $\mathbf{r}_0 - \mathbf{r}_n$ lies in $\mathbf{A}\mathcal{K}_n$. In other words, with respect to the standard inner product in Euclidean space, $\mathbf{r}_0 - \mathbf{r}_n$ is the orthogonal projection of $\mathbf{r}_0$ onto $\mathbf{A}\mathcal{K}_n$. Therefore, the CR method is an orthogonal projection method in the residual space.

From the fact that the CG-residuals are mutually orthogonal and the CR-residuals are mutually conjugate, it follows in particular that in both cases $\mathbf{r}_\nu = \mathbf{o}$ for some $\nu \leq N$, and thus $\mathbf{x}_\nu = \mathbf{x}_{\text{ex}}$. However, in practice this *finite termination property* is fairly irrelevant, as it is severely spoiled by round-off.

So far we only know how to construct the residuals, but we still need another recurrence for the iterates $\mathbf{x}_n$ themselves. As we will see in a moment, such a recurrence is found by multiplying by $\mathbf{A}^{-1}$ the one for the residuals, that is, the one for the appropriately scaled right Lanczos vectors. This then leads to the three-term version[7], ORES, of the conjugate gradient method (Hestenes 1951). The standard version of the CG method, OMIN, instead uses coupled two-term recurrences also involving the direction vectors, which are multiples of the corrections $\mathbf{x}_{n+1} - \mathbf{x}_n$. In the rest of this section and in Section 8 we want to describe generalizations to the nonsymmetric case.

4.2. Basic properties of the BICG *method*

If $\mathbf{A}$ is non-Hermitian, the construction of an orthogonal basis $\{\mathbf{y}_n\}$ of the Krylov space becomes expensive and memory-intensive, since the recurrences for $\mathbf{y}_n$ generally involve all previous vectors (as first assumed in (2.7)). Therefore, the resulting *Arnoldi* or *full orthogonalization method* (FOM) (Arnoldi 1951, Saad 1981) has to be either restarted periodically or truncated, which means that some of the information that was built up is lost. The same applies to the *generalized conjugate residual* (GCR) *method* (Eisenstat, Elman and Schultz 1983) and its special form, the GMRES algorithm of Saad and Schultz (1986), which extends the MINRES algorithm to the nonsymmetric case.

However, we know how to construct efficiently a pair of biorthogonal sequences $\{\mathbf{y}_n\}$, $\{\widetilde{\mathbf{y}}_n\}$, namely by the BIO algorithm of Section 2. By requiring that iterates $\mathbf{x}_n \in \mathbf{x}_0 + \mathcal{K}_n$ satisfy the *Petrov–Galerkin condition*

$$\langle \widetilde{\mathbf{y}}, \mathbf{r}_n \rangle_{\mathbf{B}} = 0 \quad (\text{for all } \widetilde{\mathbf{y}} \in \widetilde{\mathcal{K}}_n), \qquad \textit{i.e.}, \quad \widetilde{\mathcal{K}}_n \perp_{\mathbf{B}} \mathbf{r}_n, \tag{4.4}$$

we find the *biconjugate gradient* (BICG) *method*. In contrast to the Galerkin condition (4.1) of the CG method, this one does not belong to a minimization problem in a fixed norm[8].

[7] The acronyms ORES, OMIN, and ODIR were introduced in Ashby et al. (1990) as abbreviations for ORTHORES, ORTHOMIN, and ORTHODIR. We suggest using the short form whenever the basic recurrences of the method are short. Note that our acronyms BIORES, BIOMIN, and BIODIR for the various forms of the BICG method fit into this pattern, as these algorithms also feature short recurrences.

[8] Barth and Manteuffel (1994) showed that BICG and QMR fit into the framework of *variable metric methods*: in exact arithmetic, if the methods do not break down or terminate with $\nu < N$, then the iterates that have been created minimize the error in a

Now $\mathbf{r}_n$ is chosen so that it is orthogonal to $\widetilde{\mathcal{K}}_n$ and so that $\mathbf{r}_0 - \mathbf{r}_n$ lies in $\mathbf{A}\mathcal{K}_n$. This means that $\mathbf{r}_0 - \mathbf{r}_n$ is the projection of $\mathbf{r}_0$ onto $\mathbf{A}\mathcal{K}_n$ along a direction that is orthogonal to another space, namely $\widetilde{\mathcal{K}}_n$, and, hence, is in general oblique with respect to the projection space $\mathbf{A}\mathcal{K}_n$. Therefore, the BiCG method is said to be an *oblique projection method* (Saad 1982, Saad 1996).

Since both the residual $\mathbf{r}_n$ and the right Lanczos vector $\mathbf{y}_n$ satisfy (4.4) and both lie in $\mathcal{K}_{n+1}$, and since we have seen in Section 2 that $\mathbf{y}_n$ is determined up to a scalar factor by these conditions, we can again conclude that the residual must be a scalar multiple of the Lanczos vector and that by appropriate normalization of the latter we could attain $\mathbf{r}_n = \mathbf{y}_n$.

The most straightforward way of taking into account the two conditions $\mathbf{x}_n \in \mathbf{x}_0 + \mathcal{K}_n$ and $\widetilde{\mathcal{K}}_n \perp_{\mathbf{B}} \mathbf{r}_n$ is the following one. Representing $\mathbf{x}_n - \mathbf{x}_0$ in terms of the Lanczos vectors we can write

$$\mathbf{x}_n = \mathbf{x}_0 + \mathbf{Y}_n\mathbf{k}_n, \qquad \mathbf{r}_n = \mathbf{r}_0 - \mathbf{A}\mathbf{Y}_n\mathbf{k}_n, \tag{4.5}$$

with some coordinate vector $\mathbf{k}_n$. Using $\mathbf{A}\mathbf{Y}_n = \mathbf{Y}_{n+1}\underline{\mathbf{T}}_n$, see (2.23), and

$$\mathbf{r}_0 = \mathbf{y}_0\rho_0 = \mathbf{Y}_{n+1}\underline{\mathbf{e}}_1\rho_0,$$

with $\underline{\mathbf{e}}_1 := [\begin{array}{ccccc} 1 & 0 & 0 & \cdots \end{array}]^\top \in \mathbb{R}^{n+1}$ and $\rho_0 := \|\mathbf{r}_0\|$ (assuming $\|\mathbf{y}_0\| = 1$ here), we find that

$$\mathbf{r}_n = \mathbf{Y}_{n+1}\left(\underline{\mathbf{e}}_1\rho_0 - \underline{\mathbf{T}}_n\mathbf{k}_n\right). \tag{4.6}$$

In view of $\widetilde{\mathbf{Y}}_n^\star\mathbf{B}\mathbf{Y}_{n+1} = [\begin{array}{c|c} \mathbf{D}_{\delta;n} & \mathbf{o} \end{array}]$, the Petrov–Galerkin condition (4.4), which may be written as $\widetilde{\mathbf{Y}}_n^\star\mathbf{B}\mathbf{r}_n = \mathbf{o}$, finally yields the square tridiagonal linear system

$$\mathbf{T}_n\mathbf{k}_n = \mathbf{e}_1\rho_0, \tag{4.7}$$

where now $\mathbf{e}_1 \in \mathbb{R}^n$. By solving it for $\mathbf{k}_n$ and inserting the solution into (4.5) we could compute $\mathbf{x}_n$. However, this approach, which is sometimes called the Lanczos method for solving linear systems, is very memory-intensive, as one has to store all right Lanczos vectors for evaluating (4.5). Fortunately, there are more efficient versions of the BiCG method that generate not only the residuals (essentially the right Lanczos vectors) but also the iterates with short recurrences. We could try to find such recurrences from the above relations, but we will derive them in a more general and more elegant way.

Unless one encounters a serious breakdown, the BiCG method terminates theoretically with $\mathbf{r}_\nu = \mathbf{o}$ or $\widetilde{\mathbf{y}}_\nu = \mathbf{o}$ for some ν. Therefore, the BiCG method also has the finite termination property, except that it is spoiled not only by round-off but also by the possibility of a breakdown (a serious

norm that depends on the created basis, that is, on $\mathbf{A}$ and $\mathbf{y}_0$. This result also follows easily from one of Hochbruck and Lubich (1997*a*) one or a left-sided termination). We must emphasize again, however, that it is misleading to motivate the CG method or the BiCG method (in any of their forms) by this *finite termination property*, because this property is irrelevant when large linear systems are solved. What really counts are certain approximation properties that make the residuals (and errors) of the iterates $\mathbf{x}_n$ decrease rapidly. There is a simple, standard error bound that implies at least linear convergence for the CG and CR methods (see, for instance, Kaniel (1966), Saad (1980, 1994, 1996)), but in practice superlinear convergence is observed; there are indeed more sophisticated estimates that explain the superlinearity under certain assumptions on the spectrum (van der Sluis and van der Vorst 1986, 1987, Strakoš 1991, van der Sluis 1992, Hanke 1997). These bounds are no longer valid in the nonsymmetric case, but some of the considerations can be extended to it (van der Vorst and Vuik 1993, Ye 1991). The true mechanism of convergence lies deeper, and seems to remain the same in the nonsymmetric case. For the CR method and its generalization to nonsymmetric systems it has been analysed by Nevanlinna (1993). For the BiCG method convergence seems harder to analyse, however. Recently, a unified approach to error bounds for BiCG, QMR, FOM, and GMRES, as well as comparisons among their residual norms, have been established in Hochbruck and Lubich (1997*a*).

The BiCG method is based on Lanczos (1952) and Fletcher (1976), but, as we will see, there are various algorithms that realize it.

4.3. Recurrences for the BiCG *iterates; the consistency condition*

The recurrence for the iterates $\mathbf{x}_n$ is obtained from the one for the residuals by following a general rule that we will use over and over again. By definition, $\mathbf{x}_n - \mathbf{x}_0 \in \mathcal{K}_n$ for any Krylov space solver, and thus (4.2) holds; here $\mathcal{K}_n$ is still the nth Krylov space generated by $\mathbf{A}$ from $\mathbf{y}_0 = \mathbf{r}_0$. Since $\mathbf{r}_n = p_n(\mathbf{A})\mathbf{y}_0$ with a polynomial p_n of exact degree n, the vectors $\mathbf{r}_0, \ldots, \mathbf{r}_{n-1}$ span $\mathcal{K}_n$ (even when they are linearly dependent, in which case $\mathcal{K}_n = \mathcal{K}_{n-1}$). Therefore, if we let

$$\mathbf{R}_n := [\ \mathbf{r}_0 \quad \mathbf{r}_1 \quad \cdots \quad \mathbf{r}_{n-1}\ ], \qquad \mathbf{X}_n := [\ \mathbf{x}_0 \quad \mathbf{x}_1 \quad \cdots \quad \mathbf{x}_{n-1}\ ],$$

and define the extended $(n+1) \times n$ Frobenius (or companion) matrix

$$\underline{\mathbf{F}}_n := \begin{bmatrix} -1 & -1 & \cdots & -1 \\ 1 & & & \\ & 1 & & \\ & & \ddots & \\ & & & 1 \end{bmatrix},$$

then we have, in view of $\mathbf{x}_n - \mathbf{x}_0 \in \mathcal{K}_n$,

$$\mathbf{X}_{n+1}\underline{\mathbf{F}}_n = -\mathbf{R}_n\mathbf{U}_n \tag{4.8}$$

with some upper triangular $n \times n$ matrix $\mathbf{U}_n$ and an extra minus sign. Each column sum in $\underline{\mathbf{F}}_n$ is zero, that is, $[\ 1\ \ 1\ \ \cdots\ \ 1\ ]\ \underline{\mathbf{F}}_n = \mathbf{o}^\top$, and therefore, for an arbitrary $\mathbf{b} \in \mathbb{C}^N$, multiplication of $\underline{\mathbf{F}}_n$ from the left by the $N \times (n+1)$ matrix $[\ \mathbf{b}\ \ \mathbf{b}\ \ \cdots\ \ \mathbf{b}\ ]$ yields an $N \times n$ zero matrix. Therefore,

$$\mathbf{R}_{n+1}\underline{\mathbf{F}}_n = ([\ \mathbf{b}\ \ \cdots\ \ \mathbf{b}\ ] - \mathbf{A}\mathbf{X}_{n+1})\,\underline{\mathbf{F}}_n = -\mathbf{A}\mathbf{X}_{n+1}\underline{\mathbf{F}}_n = \mathbf{A}\mathbf{R}_n\mathbf{U}_n. \quad (4.9)$$

Since $\mathbf{r}_m$ and $\mathbf{A}\mathbf{r}_{m-1}$ are both represented by polynomials of exact degree m, the diagonal elements of $\mathbf{U}_n$ cannot vanish. Hence, if we let

$$\underline{\mathbf{H}}_n := \underline{\mathbf{F}}_n\mathbf{U}_n^{-1},$$

we can write (4.8) and (4.9) as

$$\mathbf{R}_n = -\mathbf{X}_{n+1}\underline{\mathbf{H}}_n, \qquad \mathbf{A}\mathbf{R}_n = \mathbf{R}_{n+1}\underline{\mathbf{H}}_n, \quad (4.10)$$

where $\underline{\mathbf{H}}_n$ is an $(n+1) \times n$ upper Hessenberg matrix that satisfies[9]

$$\mathbf{e}^\top\underline{\mathbf{H}}_n = \mathbf{o}^\top, \qquad \text{where } \mathbf{e}^\top := [\ 1\ \ 1\ \ \cdots\ \ 1\ ], \quad (4.11)$$

as a consequence of $\mathbf{e}^\top\underline{\mathbf{F}}_n = \mathbf{o}^\top$. This is the matrix form of the *consistency condition* for Krylov space solvers. It means that *in each column of* $\underline{\mathbf{H}}_n$ *the elements must sum up to* 0; see, for instance, Gutknecht (1989*b*). This property is inherited from $\underline{\mathbf{F}}_n$. The relations in (4.10) are the matrix representations of the recurrences for computing the iterates and the residuals: $\mathbf{x}_n$ is a linear combination of $\mathbf{r}_{n-1}$ and $\mathbf{x}_0, \ldots, \mathbf{x}_{n-1}$, and $\mathbf{r}_n$ is a linear combination of $\mathbf{A}\mathbf{r}_{n-1}$ and $\mathbf{r}_0, \ldots, \mathbf{r}_{n-1}$. Note that the recurrence coefficients, which are stored in $\underline{\mathbf{H}}_n$, are the same in both formulae.

Another, equivalent form of the consistency condition is the property $p_n(0) = 1$ of the residual polynomials.

We call a Krylov space solver *consistent* if it generates a basis consisting of the residuals (and not of some multiples of them).

4.4. The BIORES *algorithm*

In the usual, consistent forms of the BICG method, the Lanczos vectors $\mathbf{y}_n$ are equal to the residuals $\mathbf{r}_n$ and thus the Lanczos polynomials satisfy the consistency condition $p_n(0) = 1$. To apply the above approach, we have to set $\underline{\mathbf{H}}_n := \underline{\mathbf{T}}_n$ and $\mathbf{R}_n := \mathbf{Y}_n$. Therefore, the zero column sum condition requires us to choose $\gamma_n := -\alpha_n - \beta_{n-1}$. However, this can lead to yet another type of breakdown, namely when $\alpha_n + \beta_{n-1} = 0$. Following Bank and Chan (1993) we call this a *pivot breakdown* (for reasons we will describe later, in Section 9), while a breakdown due to $\delta_{\text{temp}} = 0$ in the BIO algorithm is referred to as a *Lanczos breakdown*[10], as before.

[9] The dimension of the vectors $\mathbf{o}$ and $\mathbf{e}$ is always defined by the context.

[10] In Gutknecht (1990) we suggested calling a pivot breakdown a *normalization breakdown*, which is an appropriate name in view of its analogous occurrence in other Krylov space

Recalling that the formulae for the BIO algorithm can be simplified if γ_n is independent of δ_{temp}, and adding the appropriate recurrence for the approximants $\mathbf{x}_n$, we find the following BIORES version of the BICG method.

ALGORITHM 2. (BIORES FORM OF THE BICG METHOD)
For solving $\mathbf{Ax} = \mathbf{b}$, choose an initial approximation $\mathbf{x}_0$, set $\mathbf{y}_0 := \mathbf{b} - \mathbf{Ax}_0$, and choose $\widetilde{\mathbf{y}}_0$ such that $\delta_0 := \langle \widetilde{\mathbf{y}}_0, \mathbf{y}_0 \rangle_{\mathbf{B}} \neq 0$. Then apply Algorithm 1 (BIO) with

$$\gamma_n := -\alpha_n - \beta_{n-1} \tag{4.12}$$

and some $\widetilde{\gamma}_n \neq 0$, so that (2.21e)–(2.21j) simplify to

$$\begin{aligned} \mathbf{y}_{n+1} &:= (\mathbf{A}\mathbf{y}_n - \mathbf{y}_n\alpha_n - \mathbf{y}_{n-1}\beta_{n-1})/\gamma_n, && (4.13a) \\ \widetilde{\mathbf{y}}_{n+1} &:= (\mathbf{A}^\star\widetilde{\mathbf{y}}_n - \widetilde{\mathbf{y}}_n\widetilde{\alpha}_n - \widetilde{\mathbf{y}}_{n-1}\widetilde{\beta}_{n-1})/\widetilde{\gamma}_n, && (4.13b) \\ \delta_{n+1} &:= \langle \widetilde{\mathbf{y}}_{n+1}, \mathbf{y}_{n+1} \rangle_{\mathbf{B}}\,. && (4.13c) \end{aligned}$$

Additionally, compute the vectors

$$\mathbf{x}_{n+1} := -(\mathbf{y}_n + \mathbf{x}_n\alpha_n + \mathbf{x}_{n-1}\beta_{n-1})/\gamma_n. \tag{4.13d}$$

If $\gamma_n = 0$, the algorithm breaks down ('pivot breakdown'), and we set $\dot{\nu} := n$. If $\mathbf{y}_{n+1} = \mathbf{o}$, it terminates and $\mathbf{x}_{n+1}$ is the solution; if $\mathbf{y}_{n+1} \neq \mathbf{o}$, but $\delta_{n+1} = 0$, the algorithm also breaks down ('Lanczos breakdown' if $\widetilde{\mathbf{y}}_{n+1} \neq \mathbf{o}$, 'left termination' if $\widetilde{\mathbf{y}}_{n+1} = \mathbf{o}$). In these two cases we set $\dot{\nu} := n + 1$.

First we verify the relation between residuals and iterates.

Lemma 4.1 In Algorithm 2 (BIORES) the vector $\mathbf{y}_n$ is the residual of the nth iterate $\mathbf{x}_n$; that is, $\mathbf{b} - \mathbf{Ax}_n = \mathbf{y}_n$ $(n = 0, 1, \ldots, \dot{\nu})$.

Proof. First, $\mathbf{b} - \mathbf{Ax}_0 = \mathbf{y}_0$ by definition of $\mathbf{y}_0$. Assuming $n \geq 1$, $\mathbf{b} - \mathbf{Ax}_n = \mathbf{y}_n$, and $\mathbf{b} - \mathbf{Ax}_{n-1} = \mathbf{y}_{n-1}$, and using (4.13a)–(4.13d), (2.21e), (2.21j), and (4.12), we get

$$\begin{aligned} \mathbf{b} - \mathbf{Ax}_{n+1} &= \mathbf{b} + (\mathbf{Ay}_n + \mathbf{Ax}_n\alpha_n + \mathbf{Ax}_{n-1}\beta_{n-1})/\gamma_n \\ &= \mathbf{b} + (\mathbf{Ay}_n - \mathbf{y}_n\alpha_n - \mathbf{y}_{n-1}\beta_{n-1} + \mathbf{b}(\alpha_n + \beta_{n-1}))/\gamma_n \\ &= \mathbf{y}_{n+1}, \end{aligned}$$

which is what is needed for the induction. When $n = 0$, the same relations hold without the terms involving β_{-1}. □

solvers. Joubert (1992) calls the pivot breakdown a *hard breakdown* since it causes all three standard versions of the BICG method discussed in Jea and Young (1983) to break down, as we will see in Section 9. In his terminology the Lanczos breakdown is a *soft breakdown*. Brezinski, Redivo Zaglia and Sadok (1993) use the terms *true breakdown* and *ghost breakdown*, respectively, while Freund and Nachtigal (1991) refer to *breakdowns of the second kind* and *breakdowns of the first kind*. However, we will see that in the algorithms most often used in practice, it is easier to circumvent a pivot (or hard, or true, or second kind) breakdown than a Lanczos breakdown.

The shorthand notation (2.23)–(2.24) for the BIO algorithm can easily be extended to the BIORES algorithm. Due to the additional possibility of a pivot breakdown, the index of first breakdown or termination is now $\dot{\nu}$ $(\leq \nu)$. We also have to add the matrix representation of the recurrence for the iterates, (4.13d),

$$\mathbf{Y}_n = -\mathbf{X}_{n+1}\underline{\mathbf{T}}_n \quad (n \leq \dot{\nu}). \tag{4.14}$$

Analogously to (4.11), the column sum condition (4.12) can be expressed as

$$\mathbf{e}^\top \underline{\mathbf{T}}_n = \mathbf{o}^\top \quad (n \leq \dot{\nu}). \tag{4.15}$$

4.5. The inconsistent BIORES *algorithm*

We claim that by a small modification introduced in Gutknecht (1990) it is possible to avoid the pivot breakdown that may occur in Algorithm 2 (BIORES).

ALGORITHM 3. (INCONSISTENT BIORES ALGORITHM)
Initially, let $\mathbf{y}_0 := (\mathbf{b} - \mathbf{A}\mathbf{x}_0)/\gamma_{-1}$ with some $\gamma_{-1} \neq 0$, and redefine $\mathbf{x}_0 := \mathbf{x}_0/\gamma_{-1}$. (For example, choose $\gamma_{-1} := \|\mathbf{b} - \mathbf{A}\mathbf{x}_0\|$ or $\gamma_{-1} := 1$.) Modify Algorithm 2 (BIORES) by always choosing $\gamma_n \neq 0$ (instead of setting $\gamma_n := -\alpha_n - \beta_{n-1}$). Compute additionally the sequence $\{\dot{\pi}_n\}$ that is defined recursively by

$$\dot{\pi}_0 := 1/\gamma_{-1}, \quad \dot{\pi}_{n+1} := -(\alpha_n \dot{\pi}_n + \beta_{n-1}\dot{\pi}_{n-1})/\gamma_n, \quad n = 0, 1, \ldots, \nu - 1. \tag{4.16}$$

We will see later in Theorem 12.1 that $\dot{\pi}_n$ is the value at 0 of the Lanczos polynomial p_n of (2.3), which up to normalization is also the residual polynomial of $\mathbf{x}_n$. We will also see that p_n, if normalized to be monic, is the characteristic polynomial of the $n \times n$ leading principal submatrix of $\mathbf{T}_\nu$. The problem with BIORES is that this value may become zero, and hence there may not exist a residual polynomial normalized to be 1 at $\zeta = 0$. In other words, inconsistent BIORES works with 'unnormalized residual polynomials' not satisfying the consistency condition. The same idea can be applied to other Krylov space solvers that break down for the same reason. It follows immediately that the pivot breakdown is avoided.

Lemma 4.2 The index of first breakdown or termination ν of the BIO algorithm and the one of the inconsistent BIORES algorithm are identical; the index of first breakdown or termination $\dot{\nu}$ of the consistent BIORES algorithm can, but need not, be smaller.

Moreover, in view of the following result, inconsistent BIORES delivers the solution of $\mathbf{A}\mathbf{x} = \mathbf{b}$ whenever it does not break down.

Lemma 4.3 In the inconsistent BIORES algorithm, $\mathbf{y}_n$ and $\mathbf{x}_n$ are related by

$$\mathbf{y}_n = \mathbf{b}\dot{\pi}_n - \mathbf{A}\mathbf{x}_n. \tag{4.17}$$

If $\mathbf{y}_\nu \gamma_\nu = 0$, then $\dot{\pi}_\nu \neq 0$, and $\mathbf{x}_{\text{ex}} = \mathbf{x}_\nu / \dot{\pi}_\nu$ is the solution of $\mathbf{A}\mathbf{x} = \mathbf{b}$.

Proof. For $n = 0$, (4.17) is correct. Assume it is correct up to the index n. Then by (4.13a)–(4.13d) and (4.16)

$$\begin{aligned}
&(\mathbf{b}\dot{\pi}_{n+1} - \mathbf{A}\mathbf{x}_{n+1})\gamma_n \\
&\quad = -\mathbf{b}(\alpha_n \dot{\pi}_n + \beta_{n-1}\dot{\pi}_{n-1}) + \mathbf{A}\mathbf{y}_n + \mathbf{A}\mathbf{x}_n \alpha_n + \mathbf{A}\mathbf{x}_{n-1}\beta_{n-1} \\
&\quad = \mathbf{A}\mathbf{y}_n - \mathbf{y}_n \alpha_n - \mathbf{y}_{n-1}\beta_{n-1} = \mathbf{y}_{n+1}\gamma_n.
\end{aligned}$$

As mentioned above, $\dot{\pi}_n$ is, up to normalization, the value at 0 of the characteristic polynomial p_n of the $n \times n$ leading principal submatrix of $\mathbf{T}_\nu$. In particular, $\dot{\pi}_\nu$ is a nonzero multiple of the value at 0 of the characteristic polynomial p_ν of $\mathbf{T}_\nu$. We know that when the algorithm terminates due to $\mathbf{y}_{\text{temp}} = \mathbf{y}_\nu \gamma_\nu = \mathbf{o}$, then the eigenvalues of $\mathbf{T}_\nu$ are also eigenvalues of $\mathbf{A}$. Hence, $\dot{\pi}_\nu = 0$ would imply that $\mathbf{A}$ is singular, contrary to our assumption in this chapter. □

Lemma 4.3 indicates that in practice termination should be based on $||\mathbf{y}_n||/|\dot{\pi}_n|$ being small. If we let

$$\dot{\mathbf{p}}_n := [\ \dot{\pi}_0 \quad \cdots \quad \dot{\pi}_{n-1}\ ]^\top,$$

we can formulate the extra recurrence (4.16) of inconsistent BIORES and its residual relation (4.17) as

$$\dot{\mathbf{p}}_{n+1}^\top \underline{\mathbf{T}}_n = \mathbf{o}^\top \quad (n \leq \nu),$$

$$\mathbf{Y}_n = [\ \mathbf{b} \quad \cdots \quad \mathbf{b}\ ]\,\text{diag}(\dot{\pi}_0, \cdots, \dot{\pi}_{n-1}) - \mathbf{A}\mathbf{X}_n \quad (n \leq \nu + 1).$$

From (4.17) we conclude that $\dot{\pi}_n$ and the choice of γ_m $(m \leq n)$ only affect the scaling of $\mathbf{y}_n$ and $\mathbf{x}_n$. It is clear from this formula that whenever $\dot{\pi}_n \neq 0$ for all $n < \nu$, one can rescale $\mathbf{y}_n$, $\mathbf{x}_n$ $(n \leq \nu)$ to get the corresponding vectors of (consistent) BIORES. But once $\dot{\pi}_n = 0$ for some $n < \nu$, this is impossible and BIORES breaks down, that is, $\dot{\nu} < \nu$. In contrast, here one can still go on, and if $\mathbf{y}_\nu = \mathbf{o}$ one finds a solution that is not accessible through Algorithm 2 (using the same initial data). In practice, where vanishing of $\dot{\pi}_n$ is unlikely, but near-vanishing matters, inconsistent BIORES must be considered as a slightly stabilized version of BIORES that eliminates the possibility of overflow or division by zero. In floating-point arithmetic, however, there is no other stability pitfall caused by the particular scaling of (consistent) BIORES or by any other scaling.

5. The QMR solution of a linear system

While the BiCG method yields a Petrov–Galerkin approximation of the solution of a linear system, the *Quasi-Minimal Residual* (QMR) *method* of Freund and Nachtigal (Freund 1992, Freund and Nachtigal 1991, Freund and Nachtigal 1994) produces a solution whose residual has a coordinate vector of minimum length. However, since the basis of the space is – for economy reasons – the one generated by the Lanczos process, and thus is not orthonormal, in general, the residual vector itself is not of minimum length.

Basically, the QMR method takes the right Krylov space basis generated by the Lanczos process and solves a least squares problem in coordinate space in the same way as the MinRes algorithm of Paige and Saunders (1975) and the GMRes algorithm of Saad and Schultz (1986). However, the QMR algorithm in (Freund and Nachtigal 1991) has additional features: its Lanczos part includes an implementation of look-ahead from Freund et al. (1993), and its least squares part allows for weights, which, however, are rarely used and therefore dropped in our presentation[11].

In principle, the QMR philosophy has a wide scope of applications, which goes beyond what has been treated in the literature. In particular, we can apply it to any Krylov space generation procedure producing a relation of the form $\mathbf{A}\mathbf{Y}_n = \mathbf{Y}_{n+1}\underline{\mathbf{H}}_n$ (preferably with column vectors of norm 1) or of certain equivalent forms. We will return to this in Sections 10, 15 and 17.

Since MinRes plays an essential role in QMR, we need to look at it first. We are going to discuss a variation of it that is suitable for QMR, since it is easily adapted to allow for look-ahead.

5.1. *The* MinRes *algorithm*

The MinRes algorithm of Paige and Saunders (1975), as well as QMR and GMRes, start from the representation (4.6) of the residual. MinRes is a particular algorithm for the CR method for Hermitian systems, and thus the aim is to minimize the residual norm. The method makes use of the isometry induced by the coordinate mapping of an inner product space with an orthonormal basis. This isometry is also manifested by the well-known Parseval relation. It implies that instead of minimizing the residual we can minimize its coordinate vector. In fact, by running the symmetric Lanczos algorithm with $\mathbf{B} = \mathbf{I}$ (despite the fact that the inner product matrix $\mathbf{B} = \mathbf{A}$ is used in the minimization problem (4.3) of the CR method) and normalizing the resulting orthogonal basis $\{\mathbf{y}_m\}$ of the Krylov space,

[11] In Tong (1994) the diagonal weight matrix is replaced by a block diagonal one with 2×2 or 3×3 blocks that are chosen suitably. However, the numerical results show little gain in efficiency, if any at all.

we have $\mathbf{Y}_{n+1}^{\star}\mathbf{Y}_{n+1} = \mathbf{I}_{n+1}$, and therefore from (4.6)

$$\|\mathbf{r}_n\|^2 = \|\underline{\mathbf{e}}_1\rho_0 - \underline{\mathbf{T}}_n\mathbf{k}_n\|^2. \tag{5.1}$$

This is a least squares problem in coordinate space: $\underline{\mathbf{e}}_1\rho_0$ has to be approximated by a linear combination of the columns of the tridiagonal $(n+1)\times n$ matrix $\underline{\mathbf{T}}_n$.

For example, this problem can be solved using the QR or the LQ decomposition of the matrix and, due to the tridiagonality, these decompositions only require n or $n-1$ Givens rotations, respectively[12]. The QR decomposition of an upper Hessenberg matrix of the same size still only requires $n+1$ rotations, and that is why both the GMRES and the QMR algorithms apply QR, while Paige and Saunders used an LQ decomposition. In GMRES $\underline{\mathbf{T}}_n$ is replaced by a Hessenberg matrix, and in the QMR algorithm $\underline{\mathbf{T}}_n$ is tridiagonal except for a few extra nonzero elements above the upper codiagonal if look-ahead is needed; hence, it is a nearly tridiagonal Hessenberg matrix. Although we assume in our presentation of the QMR method in this section that look-ahead does not occur, we choose to work with the QR decomposition, and we modify the original MINRES algorithm accordingly. Our treatment is adapted from Freund and Nachtigal (1991).

Let $\underline{\mathbf{T}}_n = \mathbf{Q}_n\underline{\mathbf{R}}_n^{\mathrm{MR}}$ be a QR decomposition of $\underline{\mathbf{T}}_n$. The last row of the upper triangular $(n+1)\times n$ matrix $\underline{\mathbf{R}}_n^{\mathrm{MR}}$ is zero. If we denote its upper square $n \times n$ submatrix by $\mathbf{R}_n^{\mathrm{MR}}$ (not to be confused with the matrix $\mathbf{R}_n$ of residual vectors) and let

$$\underline{\mathbf{h}}_n := \left[\begin{array}{c} \mathbf{h}_n \\ \hline \widetilde{\eta}_{n+1} \end{array}\right] := \mathbf{Q}_n^{\star}\underline{\mathbf{e}}_1\rho_0, \tag{5.2}$$

we see that

$$\mathbf{k}_n := (\mathbf{R}_n^{\mathrm{MR}})^{-1}\mathbf{h}_n \tag{5.3}$$

is the solution of our least squares problem since

$$\begin{aligned} \|\underline{\mathbf{e}}_1\rho_0 - \underline{\mathbf{T}}_n\mathbf{k}_n\|^2 &= \|\mathbf{Q}_n^{\star}\underline{\mathbf{e}}_1\rho_0 - \underline{\mathbf{R}}_n^{\mathrm{MR}}\mathbf{k}_n\|^2 \\ &= \|\underline{\mathbf{h}}_n - \underline{\mathbf{R}}_n^{\mathrm{MR}}\mathbf{k}_n\|^2 \qquad (5.4) \\ &= \|\mathbf{h}_n - \mathbf{R}_n^{\mathrm{MR}}\mathbf{k}_n\|^2 + |\widetilde{\eta}_{n+1}|^2 \\ &= |\widetilde{\eta}_{n+1}|^2. \qquad (5.5) \end{aligned}$$

In fact, multiplying the least squares problem (5.1) by the unitary matrix $\mathbf{Q}_n^{\star}$ turns it into one with an upper triangular matrix, see (5.4), where the choice of $\mathbf{k}_n$ no longer influences the defect of the last equation, and thus the problem is solved by choosing $\mathbf{k}_n$ such that the first n equations are fulfilled.

[12] An alternative is to apply Householder transformations; see Walker (1988).

From (5.1) and (5.5) we see in particular that the minimum residual norm is equal to $|\widetilde{\eta}_{n+1}|$ and hence can be found without computing $\mathbf{k}_n$ or the residual. The unitary matrix $\mathbf{Q}_n$ is only determined in its factored form, as the product of n Givens rotations that are chosen to annihilate the subdiagonal elements of the tridiagonal (or Hessenberg) matrix

$$\mathbf{Q}_n := \left[\begin{array}{c|c} \mathbf{Q}_{n-1} & \mathbf{o} \\ \hline \mathbf{o}^\top & 1 \end{array}\right] \mathbf{G}_n \quad \text{with} \quad \mathbf{G}_n := \left[\begin{array}{c|cc} \mathbf{I}_{n-1} & \mathbf{o} & \mathbf{o} \\ \hline \mathbf{o}^\top & c_n & -s_n \\ \mathbf{o}^\top & \overline{s_n} & c_n \end{array}\right], \tag{5.6}$$

where $c_n \geq 0$ and $s_n \in \mathbb{C}^n$ satisfying $c_n^2 + |s_n^2| = 1$ are chosen such that

$$\mathbf{G}_n^\star \begin{bmatrix} \star \\ \vdots \\ \star \\ \mu_n \\ \nu_n \end{bmatrix} = \begin{bmatrix} \star \\ \vdots \\ \star \\ c_n\mu_n + s_n\nu_n \\ 0 \end{bmatrix} \quad \text{with} \quad \begin{bmatrix} \star \\ \vdots \\ \star \\ \mu_n \\ \nu_n \end{bmatrix} := \left[\begin{array}{c|c} \mathbf{Q}_{n-1}^\star & \mathbf{o} \\ \hline \mathbf{o}^\top & 1 \end{array}\right] \underline{\mathbf{T}}_n \begin{bmatrix} 0 \\ \vdots \\ 0 \\ 1 \end{bmatrix},$$

which means that

$$\begin{aligned} c_n &:= \frac{|\mu_n|}{\sqrt{|\mu_n|^2 + |\nu_n|^2}}, & s_n &:= c_n \overline{\frac{\nu_n}{\mu_n}}, & &\text{if } \mu_n \neq 0, \\ c_n &:= 0, & s_n &:= 1, & &\text{if } \mu_n = 0. \end{aligned} \tag{5.7}$$

If $\underline{\mathbf{T}}_n$ is real, c_n and s_n are the cosine and sine of the rotation angle.

The formula for updating $\underline{\mathbf{h}}_n$ is therefore very simple:

$$\left[\begin{array}{c} \mathbf{h}_n \\ \hline \widetilde{\eta}_{n+1} \end{array}\right] = \underline{\mathbf{h}}_n = \mathbf{G}_n^\star \left[\begin{array}{c} \underline{\mathbf{h}}_{n-1} \\ \hline 0 \end{array}\right] = \mathbf{G}_n^\star \left[\begin{array}{c} \mathbf{h}_{n-1} \\ \hline \widetilde{\eta}_n \\ 0 \end{array}\right] = \left[\begin{array}{c} \mathbf{h}_{n-1} \\ \hline c_n\,\widetilde{\eta}_n \\ -\overline{s_n}\,\widetilde{\eta}_n \end{array}\right]. \tag{5.8}$$

In particular, it follows that

$$\|\underline{\mathbf{e}}_1\rho_0 - \underline{\mathbf{T}}_n\mathbf{k}_n\| = |\widetilde{\eta}_{n+1}| = |s_n\widetilde{\eta}_n| = |s_1\, s_2 \cdots s_n|\, \|\mathbf{r}_0\|, \tag{5.9}$$

since $\widetilde{\eta}_1 = \|\mathbf{r}_0\|$. Even more important is the fact that $\mathbf{h}_n \in \mathbb{C}^n$ emerges from $\mathbf{h}_{n-1} \in \mathbb{C}^{n-1}$ by just appending an additional component $c_n\widetilde{\eta}_n$. By rewriting the first equation in (4.5) using (5.3) as

$$\mathbf{x}_n = \mathbf{x}_0 + \mathbf{Z}_n\mathbf{h}_n, \quad \text{where } \mathbf{Z}_n := [\, \mathbf{z}_0 \;\; \ldots \;\; \mathbf{z}_{n-1} \,] := \mathbf{Y}_n(\mathbf{R}_n^{\mathrm{MR}})^{-1}$$

contains the QMR direction vectors, we can conclude that

$$\mathbf{x}_n = \mathbf{x}_{n-1} + \mathbf{z}_{n-1}c_n\widetilde{\eta}_n. \tag{5.10}$$

Finally, since $\mathbf{R}_n^{\mathrm{MR}}$ is a banded upper tridiagonal matrix with bandwidth three, the relation

$$\mathbf{Y}_n = \mathbf{R}_n^{\mathrm{MR}} \mathbf{Z}_n \tag{5.11}$$

can be viewed as the matrix representation of a three-term recurrence for generating the vectors $\{\mathbf{z}_k\}_{k=0}^{n-1}$. (In contrast, in GMRES $\mathbf{R}_n^{\mathrm{MR}}$ is no longer banded, and therefore this recurrence is not short.)

Multiplying (5.10) by $\mathbf{A}$ we could find an analogous recurrence for the residuals, but since it would require an extra matrix-vector product, it is of no interest. There is another, cheaper way of updating the residual. First, inserting $\underline{\mathbf{T}}_n = \mathbf{Q}_n \underline{\mathbf{R}}_n^{\mathrm{MR}}$ and (5.3) into (4.6) and taking (5.2) into account we get

$$\begin{aligned}
\mathbf{r}_n &= \mathbf{Y}_{n+1}\left(\underline{\mathbf{e}}_1\rho_0 - \mathbf{Q}_n\underline{\mathbf{R}}_n^{\mathrm{MR}}(\mathbf{R}_n^{\mathrm{MR}})^{-1}\mathbf{h}_n\right) = \mathbf{Y}_{n+1}\left(\underline{\mathbf{e}}_1\rho_0 - \mathbf{Q}_n\left[\begin{array}{c}\mathbf{h}_n\\ \hline 0\end{array}\right]\right)\\
&= \mathbf{Y}_{n+1}\mathbf{Q}_n\,\mathbf{l}_{n+1}\widetilde{\eta}_{n+1}, \quad \text{where } \mathbf{l}_{n+1} = [\,0 \ \ \ldots \ \ 0 \ \ 1\,]^{\top} \in \mathbb{R}^{n+1}
\end{aligned} \tag{5.12}$$

as before. Using (5.6) we conclude further that

$$\begin{aligned}
\mathbf{r}_n &= [\,\mathbf{Y}_n \mid \mathbf{y}_n\,]\left[\begin{array}{c|c}\mathbf{Q}_{n-1} & \mathbf{o}\\ \hline \mathbf{o}^{\top} & 1\end{array}\right]\mathbf{G}_n\left[\begin{array}{c}\mathbf{o}\\ \hline 1\end{array}\right]\widetilde{\eta}_{n+1}\\
&= -\mathbf{Y}_n\mathbf{Q}_{n-1}\mathbf{l}_n s_n\widetilde{\eta}_{n+1} + \mathbf{y}_n c_n\widetilde{\eta}_{n+1}.
\end{aligned}$$

Finally, using (5.12) and $\widetilde{\eta}_{n+1} = -\overline{s_n}\,\widetilde{\eta}_n$ (see (5.8)) to simplify the first term on the right-hand side, we get the recursion

$$\mathbf{r}_n = \mathbf{r}_{n-1}|s_n|^2 + \mathbf{y}_n c_n \widetilde{\eta}_{n+1}. \tag{5.13}$$

However, recall that updating the residual is unnecessary for MINRES since its norm is equal to $|\widetilde{\eta}_{n+1}|$. But (5.13) also holds for GMRES, and, since we have not used the fact that $\mathbf{Y}_n$ has orthogonal columns, it will become clear that it remains true for QMR.

5.2. *A first version of the* QMR *method:* BIOQMR

The basic version of the QMR method without look-ahead is now easily explained: the BIO algorithm with normalized Lanczos vectors (that is, with normalization (2.27)) is applied to build up bases of the growing Krylov spaces $\mathcal{K}_n$ and $\widetilde{\mathcal{K}}_n$. As in inconsistent BIORES, the right initial vector $\mathbf{y}_0 := \mathbf{r}_0/\|\mathbf{r}_0\|$ is the normalized initial residual, while the left one, $\widetilde{\mathbf{y}}_0$, can be chosen arbitrarily. The relations (4.5)–(4.6) remain valid, but (5.1) is no longer true, since the basis $\{\mathbf{y}_k\}_{k=0}^{n-1}$ is no longer orthonormal when $\mathbf{A}$ is not Hermitian or $\widetilde{\mathbf{y}}_0 \neq \mathbf{y}_0$.

Since finding the minimum residual becomes too expensive, Freund and Nachtigal (Freund 1992, Freund and Nachtigal 1991) instead promoted minimizing the coefficient vector of the residual with respect to that basis, the so-called *quasi-residual*

$$\mathbf{q}_n := \underline{\mathbf{e}}_1 \rho_0 - \underline{\mathbf{T}}_n \mathbf{k}_n, \quad \text{satisfying } \mathbf{r}_n = \mathbf{Y}_{n+1} \mathbf{q}_n, \tag{5.14}$$

see (4.6). Minimizing $||\mathbf{q}_n||$ is accomplished exactly as described in the previous subsection, and even the recurrences (5.10) and (5.13) for updating the iterates and residuals, respectively, remain valid.

What differs, however, is that $||\mathbf{r}_n|| = ||\mathbf{q}_n||$ no longer holds, in general. Instead we just have

$$||\mathbf{r}_n|| \leq ||\mathbf{Y}_{n+1}|| \, ||\mathbf{q}_n|| \leq \sqrt{n+1} \, |\widetilde{\eta}_{n+1}|,$$

since $\mathbf{Y}_{n+1}$ has columns of length 1, and $||\mathbf{q}_n|| = |\widetilde{\eta}_{n+1}|$ as before; see (5.5). The factor $\sqrt{n+1}$ normally leads to a large overestimate, so that the bound is of limited value. However, the relationship between the residual and the quasi-residual may suggest sparing the work for updating the residual and computing its norm until the norm $|\widetilde{\eta}_{n+1}|$ of the quasi-residual has dropped below a certain tolerance. In the following summary of a (simplified) version of the QMR method we nevertheless assume that the residual is updated. We choose to call it BIOQMR for distinction, to indicate that it is based on the BIO algorithm without look-ahead, whose results are then piped into the QMR least squares process.

ALGORITHM 4. (BIOQMR VERSION OF THE QMR METHOD)
For solving $\mathbf{Ax} = \mathbf{b}$, choose an initial approximation $\mathbf{x}_0 \in \mathbb{C}^N$, let $\mathbf{r}_0 := (\mathbf{b} - \mathbf{Ax}_0)$ and $\mathbf{y}_0 := \mathbf{r}_0/||\mathbf{r}_0||$, choose $\widetilde{\mathbf{y}}_0$ of unit length, and apply Algorithm 1 (BIO) with the option (2.27) producing normalized Lanczos vectors. Within step $n-1$ of the main loop, after generating $\mathbf{y}_n$ and $\widetilde{\mathbf{y}}_n$,

(1) update the QR factorization $\underline{\mathbf{T}}_n = \mathbf{Q}_n \underline{\mathbf{R}}_n^{\mathrm{MR}}$ according to (5.6)–(5.7)
(2) compute the coefficient vector $\mathbf{h}_n$ by appending the component $c_n \widetilde{\eta}_n$ to $\mathbf{h}_{n-1}$, and compute the new last component $\widetilde{\eta}_{n+1} := -\overline{s_n}\,\widetilde{\eta}_n$ of $\underline{\mathbf{h}}_n$
(3) compute $\mathbf{z}_{n-1}$ according to the three-term recurrence implied by (5.11)
(4) compute $\mathbf{x}_n$ and $\mathbf{r}_n$ according to (5.10) and (5.13), respectively
(5) stop if $||\mathbf{r}_n||/||\mathbf{r}_0||$ is sufficiently small.

Note that the extra cost (in excess of those for the BIO algorithm) is very small. On the other hand, the smoothing effect of the QMR method is often very striking: while the Petrov–Galerkin condition imposed in the BICG method sometimes leads to a rather erratic residual norm plot, the norms of the QMR residuals typically decrease nearly monotonically, though not necessarily completely monotonically; see, for instance, the examples in Freund and Nachtigal (1991).

5.3. The relation between (Petrov–)Galerkin and the (Quasi-)Minimal Residual solutions

Between the BiCG iterates and residuals and those of the QMR method exist relationships inherited from CG and CR. For the latter two methods, most of them were found by Paige and Saunders (1975) as a byproduct of their derivation of the MinRes algorithm from the symmetric Lanczos process, but more transparent derivations and some new results and interpretations have been found more recently. These relations between Galerkin-based and minimal-residual-based solutions carry over in a straightforward way to the corresponding orthogonalization methods for nonsymmetric systems, the Arnoldi method (or FOM) and the GMRes algorithm for the GCR method, as was shown by Brown (1991).

The transition from CG to CR and from FOM to GCR is also possible by applying to the CG residuals or the Arnoldi residuals, respectively, the minimal residual smoothing process that we will discuss in Section 17. In particular, we will see there why a peak in the FOM residual norm plot leads to a plateau in the one of GCR.

Here we follow first the treatment of Freund and Nachtigal (1991); see also Paige and Saunders (1975, pp. 625–626).

Recall that the Galerkin condition of CG and the Petrov–Galerkin condition of BiCG yield in coordinate space the linear system (4.7),

$$\mathbf{T}_n \mathbf{k}_n^{\mathrm{G}} = \mathbf{e}_1 \rho_0, \tag{5.15}$$

while MinRes and QMR require us to minimize the quasi-residual

$$\mathbf{q}_n := \underline{\mathbf{e}}_1 \rho_0 - \underline{\mathbf{T}}_n \mathbf{k}_n^{\mathrm{MR}} \tag{5.16}$$

of (5.14). Now we have to distinguish the two coordinate vectors, and we will likewise denote the respective iterates and residuals by $\mathbf{x}_n^{\mathrm{G}}$, $\mathbf{r}_n^{\mathrm{G}}$ and $\mathbf{x}_n^{\mathrm{MR}}$, $\mathbf{r}_n^{\mathrm{MR}}$. The results we are going to derive hold for any of the three relations CG–CR, BiCG–QMR, and FOM–GMRes, but some minor modifications in the derivation are needed for the last pair.

The minimization of $||\mathbf{q}_n||^2$ is the least squares problem that we solved in the first subsection by QR decomposition of the $(n+1) \times n$ tridiagonal matrix $\underline{\mathbf{T}}_n$. (The latter could be replaced by the $(n+1)\times n$ upper Hessenberg matrix produced by GMRes or by QMR with look-ahead.) Inserting the update formula (5.6) for $\mathbf{Q}_n$ into $\underline{\mathbf{T}}_n = \mathbf{Q}_n \underline{\mathbf{R}}_n^{\mathrm{MR}}$ and moving the accumulated left factor to the left-hand side of this equation, we get

$$\left[\begin{array}{c|c} \mathbf{Q}_{n-1}^{\star} & \mathbf{o} \\ \hline \mathbf{o}^{\mathsf{T}} & 1 \end{array}\right] \underline{\mathbf{T}}_n = \mathbf{G}_n \underline{\mathbf{R}}_n^{\mathrm{MR}}.$$

Deleting the last row yields

$$\mathbf{Q}_{n-1}^{\star}\mathbf{T}_n = \left[\begin{array}{c|c} \mathbf{I}_{n-1} & \mathbf{o} \\ \hline \mathbf{o}^{\mathsf{T}} & c_n \end{array}\right] \mathbf{R}_n^{\mathrm{MR}} =: \mathbf{R}_n^{\mathrm{G}},$$

where $\mathbf{R}_n^{\mathrm{G}}$ is again upper triangular and, hence, $\mathbf{T}_n = \mathbf{Q}_{n-1}\mathbf{R}_n^{\mathrm{G}}$ is the QR decomposition of the square tridiagonal matrix $\mathbf{T}_n$. Of course, we can solve (5.15) using this decomposition, getting

$$\begin{aligned} \mathbf{k}_n^{\mathrm{G}} &= \left(\mathbf{R}_n^{\mathrm{G}}\right)^{-1} \mathbf{Q}_{n-1}^{\star}\mathbf{e}_1\rho_0 = (\mathbf{R}_n^{\mathrm{MR}})^{-1} \left[\begin{array}{c|c} \mathbf{I}_{n-1} & \mathbf{o} \\ \hline \mathbf{o}^{\mathsf{T}} & c_n^{-1} \end{array}\right] \mathbf{Q}_{n-1}^{\star}\mathbf{e}_1\rho_0 \\ &= (\mathbf{R}_n^{\mathrm{MR}})^{-1} \left(\left[\begin{array}{c|c} \mathbf{I}_{n-1} & \mathbf{o} \\ \hline \mathbf{o}^{\mathsf{T}} & c_n \end{array}\right] + \left[\begin{array}{c|c} \mathbf{O}_{n-1} & \mathbf{o} \\ \hline \mathbf{o}^{\mathsf{T}} & c_n^{-1} - c_n \end{array}\right] \right) \\ &\quad \times \mathbf{Q}_{n-1}^{\star}\mathbf{e}_1\rho_0. \end{aligned} \tag{5.17}$$

On the other hand, from (5.2) and (5.3) we conclude by inserting (5.6) that

$$\begin{aligned} \mathbf{k}_n^{\mathrm{MR}} &= (\mathbf{R}_n^{\mathrm{MR}})^{-1}\mathbf{h}_n = (\mathbf{R}_n^{\mathrm{MR}})^{-1} [\ \mathbf{I}_n \mid \mathbf{o}\ ]\ \mathbf{Q}_n^{\star}\underline{\mathbf{e}}_1\rho_0 \\ &= (\mathbf{R}_n^{\mathrm{MR}})^{-1} [\ \mathbf{I}_n \mid \mathbf{o}\ ] \left[\begin{array}{c|cc} \mathbf{I}_{n-1} & \mathbf{o} & \mathbf{o} \\ \hline \mathbf{o}^{\mathsf{T}} & c_n & s_n \\ \mathbf{o}^{\mathsf{T}} & -\overline{s_n} & c_n \end{array}\right] \left[\begin{array}{c|c} \mathbf{Q}_{n-1}^{\star} & \mathbf{o} \\ \hline \mathbf{o}^{\mathsf{T}} & 1 \end{array}\right] \underline{\mathbf{e}}_1\rho_0 \\ &= (\mathbf{R}_n^{\mathrm{MR}})^{-1} \left[\begin{array}{c|c} \mathbf{I}_{n-1} & \mathbf{o} \\ \hline \mathbf{o}^{\mathsf{T}} & c_n \end{array}\right] \mathbf{Q}_{n-1}^{\star}\mathbf{e}_1\rho_0, \end{aligned}$$

which shows that the first of the two terms in (5.17) is just $\mathbf{k}_n^{\mathrm{G}}$. To simplify the other we note that by (5.2) the last component of $\mathbf{Q}_{n-1}^{\star}\mathbf{e}_1\rho_0$ is $\widetilde{\eta}_n$, so that altogether:

$$\mathbf{k}_n^{\mathrm{G}} = \mathbf{k}_n^{\mathrm{MR}} + (\mathbf{R}_n^{\mathrm{MR}})^{-1} \left[\begin{array}{c} \mathbf{o} \\ \hline c_n^{-1} - c_n \end{array}\right] \widetilde{\eta}_n. \tag{5.18}$$

For the iterates, which are in both cases of the form $\mathbf{x}_n = \mathbf{x}_0 + \mathbf{Y}_n\mathbf{k}_n$, we find the relation

$$\mathbf{x}_n^{\mathrm{G}} = \mathbf{x}_n^{\mathrm{MR}} + \mathbf{Y}_n(\mathbf{R}_n^{\mathrm{MR}})^{-1} \left[\begin{array}{c} \mathbf{o} \\ \hline c_n^{-1} - c_n \end{array}\right] \widetilde{\eta}_n = \mathbf{x}_n^{\mathrm{MR}} + \mathbf{z}_{n-1}(c_n^{-1} - c_n)\widetilde{\eta}_n,$$

or, since $c_n^{-1} - c_n = |s_n|^2/c_n$, finally,

$$\mathbf{x}_n^{\mathrm{G}} = \mathbf{x}_n^{\mathrm{MR}} + \mathbf{z}_{n-1}\frac{|s_n|^2\widetilde{\eta}_n}{c_n}. \tag{5.19}$$

This formula allows us to compute the BICG iterates from quantities produced by the BIOQMR algorithm. But, of course, we could also generate these iterates recursively according to inconsistent BIORES, that is, from (4.13d), (4.16), and $\mathbf{x}_n^{\mathrm{G}} := \mathbf{x}_n/\dot{\pi}_n$; see Lemma 4.3.

Multiplication of (5.19) by $\mathbf{A}$ and subtraction from $\mathbf{b}$ yields an analogue relation for the residuals. However, its direct usage would cost an extra matrix-vector multiplication. Moreover, inserting $\mathbf{z}_{n-1}$ according to (5.10) and making use of (5.13) leads in a few lines to

$$\mathbf{r}_n^{\mathrm{G}} = \mathbf{y}_n \frac{\widetilde{\eta}_{n+1}}{c_n}. \tag{5.20}$$

This is no surprise since we know that the CG and BICG residuals are multiples of the Lanczos vectors. The analogue also holds for the FOM residuals. What we learn is that, using (5.9) and (5.16), we can express the residual norm as

$$\|\mathbf{r}_n^{\mathrm{G}}\| = \frac{|\widetilde{\eta}_{n+1}|}{c_n} = \frac{1}{c_n}\|\mathbf{q}_n\| = \frac{|s_1 s_2 \cdots s_n|}{c_n}\|\mathbf{r}_0\|. \tag{5.21}$$

As shown by Paige and Saunders (1975, p. 623), formula (5.20) is easily obtained directly: splitting $\mathbf{Y}_{n+1}$ up into its first n and its last column, we get from (4.6)

$$\mathbf{r}_n^{\mathrm{G}} = \mathbf{Y}_n(\mathbf{e}_1\rho_0 - \mathbf{T}_n\mathbf{k}_n^{\mathrm{G}}) - \mathbf{y}_n\gamma_n(\mathbf{l}^{\top}\mathbf{k}_n^{\mathrm{G}}).$$

Here, the first term vanishes, and for the second we see from (5.17) that the last component of $\mathbf{k}_n^{\mathrm{G}}$ is $\mathbf{l}^{\top}\mathbf{k}_n^{\mathrm{G}} = \widetilde{\eta}_n/(\rho_{n,n}c_n)$, where $\rho_{n,n}$ is the (n,n)-element of $\mathbf{R}_n^{\mathrm{MR}}$, which, in view of $\underline{\mathbf{T}}_n = \mathbf{Q}_n\underline{\mathbf{R}}_n^{\mathrm{MR}}$ and (5.6), is linked to γ_n by $\gamma_n = \overline{s_n}\rho_{n,n}$, so that

$$\mathbf{r}_n^{\mathrm{G}} = -\mathbf{y}_n \frac{\gamma_n\widetilde{\eta}_n}{\rho_{n,n}c_n} = -\mathbf{y}_n \frac{\overline{s_n}\,\widetilde{\eta}_n}{c_n} = \mathbf{y}_n \frac{\widetilde{\eta}_{n+1}}{c_n}.$$

This result also means that the quasi-residual $\mathbf{q}_n^{\mathrm{G}}$ that one can associate according to (5.14) with a Galerkin method is given by

$$\mathbf{q}_n^{\mathrm{G}} = \mathbf{l}_{n+1} \frac{\widetilde{\eta}_{n+1}}{c_n}.$$

From (5.20) we conclude further that (5.13) can be rewritten as

$$\mathbf{r}_n^{\mathrm{MR}} = \mathbf{r}_{n-1}^{\mathrm{MR}}|s_n|^2 + \mathbf{r}_n^{\mathrm{G}}c_n^2, \tag{5.22}$$

and, in view of $|s_n|^2 + c_n^2 = 1$, subtraction from $\mathbf{b}$ and premultiplication by $\mathbf{A}^{-1}$ yields

$$\mathbf{x}_n^{\mathrm{MR}} = \mathbf{x}_{n-1}^{\mathrm{MR}}|s_n|^2 + \mathbf{x}_n^{\mathrm{G}}c_n^2. \tag{5.23}$$

Finally, once again using $|s_n|^2 + c_n^2 = 1$ and $\widetilde{\eta}_{n+1} = -\overline{s_n}\,\widetilde{\eta}_n$ (see (5.8)), we see that

$$\frac{1}{|\tilde{\eta}_{n+1}|^2} = \frac{1}{|\tilde{\eta}_n|^2} + \frac{c_n^2}{|\tilde{\eta}_{n+1}|^2} = \frac{1}{|\tilde{\eta}_n|^2} + \frac{1}{||\mathbf{r}_n^G||^2}, \tag{5.24}$$

which allows us to find $|\tilde{\eta}_{n+1}|^2$ recursively without the QR decomposition, just from the residual norms of the Galerkin method. Then, weights in (5.22) and (5.23) are obtained from

$$c_n^2 = \frac{|\tilde{\eta}_{n+1}|^2}{||\mathbf{r}_n^G||^2}, \quad |s_n|^2 = 1 - c_n^2. \tag{5.25}$$

Recall that $|\tilde{\eta}_{n+1}| = ||\mathbf{r}_n^{\mathrm{MR}}||$ in the CR and FOM settings, while $|\tilde{\eta}_{n+1}| = ||\mathbf{q}_n||$ is the norm of the quasi-residual if we apply the above to the BiCG–QMR connection. The relations (5.22)–(5.25) open up an alternative way to compute the QMR (or CR or GCR) iterates and residuals from the corresponding Galerkin residuals, that is, the BiCG (or CG or FOM) residuals. Zhou and Walker (1994), who introduced this approach, call it QMR *smoothing*. We will return to it in Section 17.

The relation (5.22) has its root in the analogue one that holds for the coordinate vectors,

$$\mathbf{k}_n^{\mathrm{MR}} = \left[\begin{array}{c} \mathbf{k}_{n-1}^{\mathrm{MR}} \\ \hline 0 \end{array}\right] |s_n|^2 + \mathbf{k}_n^{\mathrm{G}} c_n^2,$$

which was given in Freund (1993, Lemma 4.1).

6. Variations of the Lanczos BiO algorithm

6.1. Further cases where the BiO *and* BiORes *algorithms simplify*

We have mentioned before that in the symmetric case the BiO algorithm simplifies: the left and the right Lanczos vectors coincide, and therefore only one matrix-vector product is needed per step. In fact, this simplification applies in a somewhat more general situation.

For every square matrix $\mathbf{A}$ there exists a nonsingular matrix $\mathbf{S}$ such that $\mathbf{A}^{\top} = \mathbf{SAS}^{-1}$, but, in general, the spectral decomposition of $\mathbf{A}$ is needed to construct $\mathbf{S}$, and thus $\mathbf{S}$ is normally not available. See, for instance, Horn and Johnson (1985, p. 134) for a proof of this result. Rutishauser (1953) and, later, Fletcher (1976) noticed that choosing $\tilde{\mathbf{y}}_0 = \overline{\mathbf{Sy}_0}$ in the BiO algorithm yields $\tilde{\mathbf{y}}_n = \overline{\mathbf{Sy}_n}$ $(n = 0, 1, \ldots, \nu - 1)$. (Rutishauser's and Fletcher's remarks are restricted to real matrices, but generalize in the way indicated above to the complex case.) Of course, it then suffices to generate $\{\mathbf{y}_n\}$ by the three-term recurrence, which means that the BiO algorithm becomes transpose-free and only one matrix-vector product involving $\mathbf{A}$ is needed per step. Hence, storage and work are then reduced to roughly half, except for the additional multiplication by $\mathbf{S}$ for temporarily creating $\widetilde{\mathbf{y}}_n$, which appears in the inner products for α_n and δ_{temp}. Moreover, under these assumptions one-sided termination cannot happen: $\mathbf{y}_n = \mathbf{o}$ if and only if $\widetilde{\mathbf{y}}_n = \mathbf{o}$. However, serious breakdowns are in general still possible.

Fortunately, there are several interesting situations where the matrix $\mathbf{S}$ is known and is simple to multiply with. A trivial case is when $\mathbf{A} = \mathbf{A}^\top$ is symmetric (real or complex), and thus $\mathbf{S} = \mathbf{I}$. Freund (1994) lists several classes of $\mathbf{S}$-*symmetric* and $\mathbf{S}$-*Hermitian* matrices satisfying by definition $\mathbf{A}^\top\mathbf{S} = \mathbf{S}\mathbf{A}$, $\mathbf{S} = \mathbf{S}^\top$ and $\mathbf{A}^\star\mathbf{S} = \mathbf{S}\mathbf{A}$, $\mathbf{S} = \mathbf{S}^\star$, respectively. In particular, every Toeplitz matrix is $\mathbf{S}$-symmetric with $\mathbf{S}$ the antidiagonal unit matrix. Real Hamiltonian matrices multiplied by $i := \sqrt{-1}$ are also $\mathbf{S}$-symmetric. However, note that the conditions $\mathbf{S} = \mathbf{S}^\top$ or $\mathbf{S} = \mathbf{S}^\star$ are not needed for the simplification. Also, the class of $\mathbf{S}$-Hermitian matrices is rather restricted since any such matrix has a real spectrum.

Incidentally, the transformation $\mathbf{S}$ is also crucial in a paper of Jea and Young (1983, Def. 1.1, Thm 4.1).

6.2. The one-sided Lanczos algorithm

It has been pointed out by Saad (see Algorithm 3 in (Saad 1982)), that one can exploit additional freedom in choosing the sequence $\{\widetilde{\mathbf{y}}_n\}$ of left Lanczos vectors without affecting the sequence $\{\mathbf{y}_n\}$ of right Lanczos vectors: we can use for the former any sequence that spans the nested Krylov spaces $\widetilde{\mathcal{K}}_n$ successively. In fact, a closer look at our derivation in Section 2 shows that up to a scalar factor the right Lanczos vectors are fully determined by the orthogonality condition $\widetilde{\mathcal{K}}_n \perp_{\mathbf{B}} \mathbf{y}_n$. Therefore, it does not matter which set of nested bases is used for the left Krylov spaces $\widetilde{\mathcal{K}}_n$. All we need is that

$$\widetilde{\mathbf{y}}_n = \overline{t_n}(A^*)\widetilde{\mathbf{y}}_0$$

with a polynomial t_n of exact degree n or, equivalently, that for $\widetilde{\mathbf{y}}_n$ a recurrence holds that is of type (2.7) with $\widetilde{\tau}_{n,n-1} \neq 0$. Since this means giving up the mutual biorthogonality of the two vector sequences, we have to rederive the formulae for α_n and β_{n-1}. Again taking inner products of the first Lanczos recurrence in (2.19) with $\widetilde{\mathbf{y}}_{n-1}$ and $\widetilde{\mathbf{y}}_n$, we see that the formula (2.20) for β_{n-1} does not change, but the one for α_n changes into one of the following two:

$$\begin{aligned}\alpha_n &:= \langle \widetilde{\mathbf{y}}_n, \mathbf{A}\mathbf{y}_n - \mathbf{y}_{n-1}\beta_{n-1}\rangle_{\mathbf{B}} / \delta_n \\ &= \Big(\langle \widetilde{\mathbf{y}}_n, \mathbf{A}\mathbf{y}_n\rangle_{\mathbf{B}}\, \widetilde{\gamma}_{n-1} - \langle \widetilde{\mathbf{y}}_{n-1}, \mathbf{A}\mathbf{y}_{n-1}\rangle_{\mathbf{B}}\, \beta_{n-1} \\ &\qquad + \beta_{n-1}\delta_{n-1}\overline{\widetilde{\tau}_{n-1,n-1}} \Big) / (\widetilde{\gamma}_{n-1}\delta_n).\end{aligned}$$

These formulae, together with the standard recurrence for the right Lanczos vectors and a nearly arbitrary recurrence for the left Lanczos vectors leads to Saad's variation of the BIO algorithm that we call here the *one-sided Lanczos algorithm*; in contrast to Saad we prefer the first formula for α_n, which is also used in Gutknecht and Ressel (1996).

ALGORITHM 5. (ONE-SIDED LANCZOS ALGORITHM)
Choose $\mathbf{y}_0, \widetilde{\mathbf{y}}_0 \in \mathbb{C}^N$ such that $\delta_0 := \langle \widetilde{\mathbf{y}}_0, \mathbf{y}_0 \rangle_{\mathbf{B}} \neq 0$, and set $\beta_{-1} := 0$. For $n = 0, 1, \ldots$ compute

$$\beta_{n-1} := \overline{\widetilde{\gamma}_{n-1}} \delta_n / \delta_{n-1}, \quad (\text{if } n > 0), \tag{6.1a}$$

$$\alpha_n := \langle \widetilde{\mathbf{y}}_n, \mathbf{A}\mathbf{y}_n - \mathbf{y}_{n-1}\beta_{n-1} \rangle_{\mathbf{B}} / \delta_n, \tag{6.1b}$$

$$\mathbf{y}_{\text{temp}} := \mathbf{A}\mathbf{y}_n - \mathbf{y}_n \alpha_n - \mathbf{y}_{n-1}\beta_{n-1}, \tag{6.1c}$$

$$\widetilde{\mathbf{y}}_{\text{temp}} := \mathbf{A}^\star \widetilde{\mathbf{y}}_n - \widetilde{\mathbf{y}}_n \widetilde{\tau}_{n,n} - \cdots - \widetilde{\mathbf{y}}_0 \widetilde{\tau}_{0,n}, \tag{6.1d}$$

$$\delta_{\text{temp}} := \langle \widetilde{\mathbf{y}}_{\text{temp}}, \mathbf{y}_{\text{temp}} \rangle_{\mathbf{B}}\,; \tag{6.1e}$$

if $\delta_{\text{temp}} = 0$, choose $\gamma_n \neq 0$ and $\widetilde{\gamma}_n \neq 0$, set

$$\nu := n+1, \quad \mathbf{y}_\nu := \mathbf{y}_{\text{temp}}/\gamma_n, \quad \widetilde{\mathbf{y}}_\nu := \widetilde{\mathbf{y}}_{\text{temp}}/\widetilde{\gamma}_n, \quad \delta_{n+1} := 0,$$

and stop; otherwise, choose $\gamma_n \neq 0$, $\widetilde{\gamma}_n \neq 0$, and δ_{n+1} such that

$$\gamma_n \overline{\widetilde{\gamma}_n} \delta_{n+1} = \delta_{\text{temp}},$$

set

$$\mathbf{y}_{n+1} := \mathbf{y}_{\text{temp}}/\gamma_n, \quad \widetilde{\mathbf{y}}_{n+1} := \widetilde{\mathbf{y}}_{\text{temp}}/\widetilde{\gamma}_n,$$

and proceed with the next step.

Here, the coefficients $\widetilde{\tau}_{k,n}$ have been assumed to be given, but there are situations where one might want to determine them from recently computed right Lanczos vectors. It is easy to adapt this algorithm to the problem of solving $\mathbf{A}\mathbf{x} = \mathbf{b}$ and to specify the resulting *one-sided consistent* BIORES *algorithm* with $\gamma_n := -\alpha_n - \beta_{n-1}$, or the *one-sided inconsistent* BIORES *algorithm*, or to combine it with the QMR approach.

Theoretically, one could use for the left sequence the Krylov vectors $\widetilde{\mathbf{y}}_n := (\mathbf{A}^\star)^n \widetilde{\mathbf{y}}_0$, thus simplifying (6.1d) to $\widetilde{\mathbf{y}}_{\text{temp}} := \mathbf{A}^\star \widetilde{\mathbf{y}}_n$, but in practice these soon become nearly multiples of each other (and of the eigenvector associated with the absolutely largest eigenvalue); therefore, even for moderate n, they are useless as a basis for $\widetilde{\mathcal{K}}_n$, and methods that rely on them do not work for most problems. On the other hand, to use a long recurrence for the left vectors would be a waste. Hence, only two-term or three-term recurrences are a serious option, and in most situations, the normal left Lanczos recurrences will be the best one.

Saad points out that this algorithm reminds us that the orthogonality of the left vectors, that is, $\widetilde{\mathbf{y}} \perp_{\mathbf{B}} \mathcal{K}_n$, is not essential, and that, therefore, in the BIO algorithm there is no reason to improve this orthogonality by reorthogonalization.

However, there are at least two situations where the one-sided Lanczos algorithm is valuable. First, a purely theoretical application is that it can serve as an intermediate step in the derivation of Lanczos-type product methods; see Section 16. Second, for the numerical stability of the Lanczos process it is most important that the space-expanding term in the recursion for the Lanczos vectors, that is, $\mathbf{Ay}$ or $\mathbf{A}^\star\widetilde{\mathbf{y}}$, respectively, is not too small compared to the two other terms. If this happens to the right sequence, we have to switch to a look-ahead step (see Section 19). However, if only the left sequence is affected, we can just switch to the one-sided Lanczos algorithm instead.

In fact, for numerical stability, the optimal choice for the left sequence would be the one generated by the Arnoldi process applied to $\mathbf{A}^\star$ with starting vector $\widetilde{\mathbf{y}}_0$. However, the cost of this process forbids this: recall that the main advantage of the Lanczos process over the Arnoldi process is the large reduction of memory and computational costs. As a cost-effective compromise we may choose the ORTHORES(2) process instead, which amounts to making $\widetilde{\mathbf{y}}_{\text{temp}}$ orthogonal to $\widetilde{\mathbf{y}}_n$ and $\widetilde{\mathbf{y}}_{n-1}$ (instead of $\mathbf{y}_n$ and $\mathbf{y}_{n-1}$):

$$\widetilde{\mathbf{y}}_{\text{temp}} := \mathbf{A}^\star\widetilde{\mathbf{y}}_n - \widetilde{\mathbf{y}}_n\langle\widetilde{\mathbf{y}}_n, \mathbf{A}^\star\widetilde{\mathbf{y}}_n\rangle_{\mathbf{B}} - \widetilde{\mathbf{y}}_{n-1}\langle\widetilde{\mathbf{y}}_{n-1}, \mathbf{A}^\star\widetilde{\mathbf{y}}_n\rangle_{\mathbf{B}},$$

except that the last term does not exist when $n = 1$. However, it is better to implement this according to the modified Gram–Schmidt process, and thus compute

$$\begin{aligned} \widetilde{\mathbf{y}}_{\text{temp}} &:= \mathbf{A}^\star\widetilde{\mathbf{y}}_n, \\ \widetilde{\mathbf{y}}_{\text{temp}} &:= \widetilde{\mathbf{y}}_{\text{temp}} - \widetilde{\mathbf{y}}_n\langle\widetilde{\mathbf{y}}_n, \widetilde{\mathbf{y}}_{\text{temp}}\rangle_{\mathbf{B}}, \\ \widetilde{\mathbf{y}}_{\text{temp}} &:= \widetilde{\mathbf{y}}_{\text{temp}} - \widetilde{\mathbf{y}}_{n-1}\langle\widetilde{\mathbf{y}}_{n-1}, \widetilde{\mathbf{y}}_{\text{temp}}\rangle_{\mathbf{B}}, \quad (\text{if } n > 0). \end{aligned}$$

6.3. An abstract setting for the Lanczos process

We have introduced the Lanczos process for a real or complex $N \times N$ matrix $\mathbf{A}$. Such a matrix can always be thought of as a linear operator $A : \mathbb{R}^N \to \mathbb{R}^N$ or $A : \mathbb{C}^N \to \mathbb{C}^N$, respectively. For generality, let us concentrate on the complex case: the Euclidean space $\mathbb{C}^N$ is a finite-dimensional inner product space, and we have made use of its inner product $\langle .,. \rangle$ when defining the formal inner product $\langle\widetilde{\mathbf{y}}, \mathbf{y}\rangle_{\mathbf{B}} := \langle\widetilde{\mathbf{y}}, \mathbf{By}\rangle = \widetilde{\mathbf{y}}^\star\mathbf{By}$, which involves a matrix $\mathbf{B}$ that commutes with $\mathbf{A}$ and is in most applications just the identity $\mathbf{I}$. The norm that comes with $\langle .,. \rangle$ was occasionally used to normalize vectors, but, as we have seen, the Lanczos process can work with unnormalized vectors, and thus the norm is only needed as soon as we want to measure convergence.

It is straightforward to reformulate the Lanczos process for an infinite-dimensional Hilbert space, and there are indeed applications for this setting; see, for instance, Hayes (1954), Kreuzer, Miller and Berger (1981) and Lanczos (1950). However, as pointed out by Parlett (1992) (and further developed in a private discussion), one can go a big step futher with respect to generality.

Parlett makes the point that the Lanczos algorithm can be defined in a plain vector space. There is no need to normalize the Lanczos vectors because they can be defined by the monic Lanczos polynomials. There is no need for $\mathbf{A}^\star$ or $\mathbf{A}^\top$ provided that one set of Lanczos vectors consists of row vectors and the other consists of column vectors. There is no need for a normed space, let alone an inner product space. The point is that there is no norm or inner product natural to the Lanczos algorithm. Different applications might require different norms. The choice of norm or inner product needs to be justified, not assumed.

Let $\mathcal{V}$ be a linear space over the field $\mathbb{C}$ (for simplicity), and $A: \mathcal{V} \to \mathcal{V}$ a linear operator. The linear functionals defined on $\mathcal{V}$ form another linear space, the *algebraic dual space* $\mathcal{V}^\times$ of $\mathcal{V}$; see, for instance, Kreyszig (1978, Section 2.8-8). A linear operator $A^\times : \mathcal{V}^\times \to \mathcal{V}^\times$ *adjoint* to A can be defined by

$$(A^\times \tilde{y})(y) := \tilde{y}(Ay) \quad (\text{for all } y \in \mathcal{V}, \tilde{y} \in \mathcal{V}^\times).$$

If $\mathcal{V}$ is in addition normed, one considers typically the *normed dual space* $\mathcal{V}'$ of $\mathcal{V}$ consisting only of the linear functionals that are bounded. Then, if A is bounded, the restriction of $A^\times$ to $\mathcal{V}$ becomes a bounded linear operator, the *adjoint* A' of A on $\mathcal{V}'$. It has the same (operator) norm as A; see, for instance, Kreyszig (1978, Section 4.5-2), Rudin (1973, pp. 92–93).

For either of these two situations we can define the Lanczos process if we replace '$\tilde{\mathbf{y}} \in \mathbb{C}^N$ is ($\mathbf{B}$-)orthogonal to $\mathbf{y} \in \mathbb{C}^N$' by '$y \in \mathcal{V}$ is a zero of $\tilde{y} \in \mathcal{V}^\times$ [or: $\mathcal{V}'$]':

$$\langle \tilde{\mathbf{y}}, \mathbf{y} \rangle_\mathbf{B} = 0 \quad \rightsquigarrow \quad \tilde{y}(y) = 0$$

In particular, instead of ($\mathbf{B}$-)biorthogonal bases satisfying $\langle \tilde{\mathbf{y}}_m, \mathbf{y}_n \rangle_\mathbf{B} = \delta_{m,n}\delta_n$, the Lanczos process then produces *dual bases* satisfying

$$\tilde{y}_m(y_n) = \delta_{m,n}\delta_n.$$

We need, however, to point out a difference between this setting and the one in a complex Hilbert space, in particular $\mathbb{C}^N$. The inner product $\langle \tilde{\mathbf{y}}, \mathbf{y} \rangle$ in $\mathbb{C}^N$ is sesquilinear, while $\tilde{y}(y)$ is bilinear. Therefore, the above defined adjoint operators $A^\times$ and A' are not identical with the Hermitian transpose $\mathbf{A}^\star$ if $\mathcal{V} = \mathbb{C}^N$ (or with the Hilbert space adjoint if $\mathcal{V}$ is a Hilbert space), but rather with the transpose $\mathbf{A}^\top$ of $\mathbf{A}$; see, for instance, Kreyszig (1978, Section 4.5-3), Rudin (1973, pp. 297–298). For this reason, the formulae of our algorithms require some small modifications if translated into the abstract setting of dual spaces.

7. Coupled recurrences: the BIOC algorithm

In his second paper on the subject, Lanczos (1952) suggested under the section heading 'The complete algorithm for minimized iterations' an alternative algorithm for computing the sequences $\{\mathbf{y}_n\}$, $\{\widetilde{\mathbf{y}}_n\}$ generated by the BIO algorithm. He also discussed in detail how to apply this algorithm for solving linear systems of equations. While the BIO algorithm for the nonsymmetric Lanczos process described in Section 2 is based on a three-term recurrence we turn now to another algorithm based on a coupled pair of two-term recurrences for the same process. The relationship between the two types of recurrence is the same as that between a linear second-order ordinary differential equation and an equivalent pair of coupled first-order equations: we just introduce an auxiliary quantity. In addition to the pair of biorthogonal (*i.e.*, $\mathbf{B}$-biorthogonal) Krylov space bases, a second pair of biconjugate (*i.e.*, $\mathbf{BA}$-biorthogonal) bases for the same Krylov space is now generated. That is why we introduce here the acronym BIOC for this algorithm.

While the BIO algorithm supplies us with a tridiagonal matrix $\mathbf{T}$ that represents a projection of $\mathbf{A}$, the new algorithm produces the two bidiagonal matrices that are the LU-factors of $\mathbf{T}$. If an LU-factorization (without pivoting) of $\mathbf{T}$ does not exist, the BIOC algorithm breaks down early. This seems to be a disadvantage of the Lanczos process based on coupled two-term recurrences (BIOC algorithm) compared to the one based on three-term recurrences (BIO algorithm). However, practice shows that nevertheless the BIOC algorithm is often numerically preferable, as round-off seems to have less impact.

The usual application of the BIOC algorithm is to solve a linear system of equations $\mathbf{Ax} = \mathbf{b}$, again either by additionally computing iterates that satisfy the Petrov–Galerkin condition of the BICG method, or by solving the least squares problem in coordinate space of the QMR method. Further investigations are necessary to find out if the BIOC algorithm is also advisable for eigenvalue computations, where it has hardly ever been applied until now. Parlett (1995) lists several advantages of the factored form. In particular, if we assume that $\mathbf{T}_n$ and its LU-factors are known to a certain precision, then the factors implicitly determine the entries of $\mathbf{T}_n$ to higher precision. The factors are also the input data of Rutishauser's differential QD algorithm of 1970 (see Rutishauser (1990)), which has recently been enhanced by Fernando and Parlett (1994). Enriched by a suitable shift strategy, it has become the method of choice for the bidiagonal singular value problem and the eigenvalue problem of a real symmetric positive definite tridiagonal matrix. By avoiding explicit shifts, it can be made competitive to the QR algorithm even for the general real symmetric tridiagonal eigenvalue problem. Making its nonsymmetric version sufficiently stable seems to be a long way ahead, however. It would be less expensive than the normally applied QR algorithm, as it works with bidiagonal matrices, while QR transforms the nonsymmetric tridiagonal into an upper Hessenberg matrix.

7.1. The BIOC *algorithm*

We start with the formulation of the BIOC algorithm and a discussion of its main properties.

ALGORITHM 6. (BIOC ALGORITHM)
Choose $\mathbf{y}_0, \widetilde{\mathbf{y}}_0 \in \mathbb{C}^N$ such that $\delta_0 := \langle \widetilde{\mathbf{y}}_0, \mathbf{y}_0 \rangle_{\mathbf{B}} \neq 0$ and $\delta_0' := \langle \widetilde{\mathbf{y}}_0, \mathbf{A}\mathbf{y}_0 \rangle_{\mathbf{B}} \neq 0$, and set $\mathbf{v}_0 := \mathbf{y}_0$, $\widetilde{\mathbf{v}}_0 := \widetilde{\mathbf{y}}_0$. For $n = 0, 1, \ldots$, choose $\gamma_n \neq 0$, $\widetilde{\gamma}_n \neq 0$ and compute

$$\varphi_n := \delta_n' / \delta_n, \tag{7.1a}$$

$$\widetilde{\varphi}_n := \overline{\varphi_n}, \tag{7.1b}$$

$$\mathbf{y}_{n+1} := (\mathbf{A}\mathbf{v}_n - \mathbf{y}_n \varphi_n)/\gamma_n, \tag{7.1c}$$

$$\widetilde{\mathbf{y}}_{n+1} := (\mathbf{A}^\star \widetilde{\mathbf{v}}_n - \widetilde{\mathbf{y}}_n \widetilde{\varphi}_n)/\widetilde{\gamma}_n, \tag{7.1d}$$

$$\delta_{n+1} := \langle \widetilde{\mathbf{y}}_{n+1}, \mathbf{y}_{n+1} \rangle_{\mathbf{B}}\,; \tag{7.1e}$$

$$\psi_n := \overline{\widetilde{\gamma}_n} \delta_{n+1} / \delta_n', \tag{7.1f}$$

$$\widetilde{\psi}_n := \overline{\gamma_n \delta_{n+1} / \delta_n'}, \tag{7.1g}$$

$$\mathbf{v}_{n+1} := \mathbf{y}_{n+1} - \mathbf{v}_n \psi_n, \tag{7.1h}$$

$$\widetilde{\mathbf{v}}_{n+1} := \widetilde{\mathbf{y}}_{n+1} - \widetilde{\mathbf{v}}_n \widetilde{\psi}_n, \tag{7.1i}$$

$$\delta_{n+1}' := \langle \widetilde{\mathbf{v}}_{n+1}, \mathbf{A}\mathbf{v}_{n+1} \rangle_{\mathbf{B}}\,. \tag{7.1j}$$

If $\delta_{n+1} = 0$ or $\delta_{n+1}' = 0$, set $\dot{\nu} := n+1$ and stop; otherwise proceed with the next step.

The formulae (7.1c)–(7.1d) and (7.1h)–(7.1i) are known as *coupled two-term recurrences.* We will see below that by eliminating $\mathbf{v}_n$ and $\widetilde{\mathbf{v}}_m$ from them we get back to the three-term recurrences (2.19) of the BIO algorithm.
The basic result for this BIOC algorithm is the following one.

Theorem 7.1 The sequences $\{\mathbf{y}_n\}_{n=0}^{\dot{\nu}}$, $\{\widetilde{\mathbf{y}}_n\}_{n=0}^{\dot{\nu}}$ generated by the BIOC algorithm are biorthogonal, and the sequences $\{\mathbf{v}_n\}_{n=0}^{\dot{\nu}}$ and $\{\widetilde{\mathbf{v}}_n\}_{n=0}^{\dot{\nu}}$ are biconjugate (with respect to $\mathbf{A}$) except that $\langle \widetilde{\mathbf{y}}_{\dot{\nu}}, \mathbf{y}_{\dot{\nu}} \rangle_{\mathbf{B}} = 0$ or $\langle \widetilde{\mathbf{v}}_{\dot{\nu}}, \mathbf{A}\mathbf{v}_{\dot{\nu}} \rangle_{\mathbf{B}} = 0$. That is, for $m, n = 0, 1, \ldots, \dot{\nu}$,

$$\langle \widetilde{\mathbf{y}}_m, \mathbf{y}_n \rangle_{\mathbf{B}} = \begin{cases} 0, & m \neq n, \\ \delta_n, & m = n, \end{cases} \tag{7.2}$$

$$\langle \widetilde{\mathbf{v}}_m, \mathbf{A}\mathbf{v}_n \rangle_{\mathbf{B}} = \begin{cases} 0\,, & m \neq n, \\ \delta_n' = \delta_n \varphi_n\,, & m = n, \end{cases} \tag{7.3}$$

where $\delta_n \neq 0$ and $\delta'_n \neq 0$ for $0 \leq n \leq \dot{\nu} - 1$, but $\delta_{\dot{\nu}} = 0$ or $\delta'_{\dot{\nu}} = 0$. Moreover, for $n = 1, \ldots, \dot{\nu} - 1$ holds in addition to (2.5)

$$\mathbf{v}_n \in \mathcal{K}_{n+1} \backslash \mathcal{K}_n \,, \quad \widetilde{\mathbf{v}}_n \in \widetilde{\mathcal{K}}_{n+1} \backslash \widetilde{\mathcal{K}}_n. \tag{7.4}$$

Proof. We provide an adaptation of Fletcher's proof (Fletcher 1976) to the complex case and to our adjustable normalization. For $m = n = 0$, (7.2)–(7.4) and (2.5) clearly hold. Assume that they hold for $m, n = 0, \ldots, k$ ($< \dot{\nu}$). From (7.1c) and (7.1i) we get

$$\begin{aligned} &\langle \widetilde{\mathbf{y}}_m, \mathbf{y}_{k+1} \rangle_{\mathbf{B}} && (7.5) \\ &\quad = \langle \widetilde{\mathbf{y}}_m, \mathbf{A}\mathbf{v}_k - \mathbf{y}_k \varphi_k \rangle_{\mathbf{B}} / \gamma_k \\ &\quad = \Big(\langle \widetilde{\mathbf{v}}_m, \mathbf{A}\mathbf{v}_k \rangle_{\mathbf{B}} + \overline{\widetilde{\psi}_{m-1}} \langle \widetilde{\mathbf{v}}_{m-1}, \mathbf{A}\mathbf{v}_k \rangle_{\mathbf{B}} - \langle \widetilde{\mathbf{y}}_m, \mathbf{y}_k \rangle_{\mathbf{B}} \, \varphi_k \Big) / \gamma_k. && (7.6) \end{aligned}$$

Here, for $m \leq k - 1$, all terms are zero by assumption. On the other hand, if $m = k$, (7.1a), (7.1e), and (7.1j) inserted into (7.6) yield $\langle \widetilde{\mathbf{y}}_k, \mathbf{y}_{k+1} \rangle_{\mathbf{B}} = (\varphi_k \delta_k + 0 - \varphi_k \delta_k)/\gamma_k = 0$; hence, $\langle \widetilde{\mathbf{y}}_m, \mathbf{y}_{k+1} \rangle_{\mathbf{B}} = 0$ for $m \leq k$. By symmetry, $\langle \widetilde{\mathbf{y}}_{k+1}, \mathbf{y}_m \rangle_{\mathbf{B}} = 0$ for $m \leq k$ too, and together with (7.1e) it follows that (7.2) holds up to $k + 1$.

Similarly, using (7.1h) and (7.1d) we get

$$\begin{aligned} &\langle \widetilde{\mathbf{v}}_m, \mathbf{A}\mathbf{v}_{k+1} \rangle_{\mathbf{B}} \\ &\quad = \langle \widetilde{\mathbf{v}}_m, \mathbf{A}(\mathbf{y}_{k+1} - \mathbf{v}_k \psi_k) \rangle_{\mathbf{B}} \\ &\quad = \overline{\widetilde{\gamma}_m} \langle \widetilde{\mathbf{y}}_{m+1}, \mathbf{y}_{k+1} \rangle_{\mathbf{B}} + \overline{\widetilde{\varphi}_m} \langle \widetilde{\mathbf{y}}_m, \mathbf{y}_{k+1} \rangle_{\mathbf{B}} - \langle \widetilde{\mathbf{v}}_m, \mathbf{A}\mathbf{v}_k \rangle_{\mathbf{B}} \, \psi_k. \end{aligned}$$

Here, too, for $m \leq k-1$, all terms are zero by assumption. If $m = k$, we see from (7.1h), (7.1d), (7.2), (7.3), and (7.1f) that $\langle \widetilde{\mathbf{v}}_k, \mathbf{A}\mathbf{v}_{k+1} \rangle_{\mathbf{B}} = \overline{\widetilde{\gamma}_k} \delta_{k+1} + 0 - \delta'_k \psi_k = 0$. Hence, $\langle \widetilde{\mathbf{v}}_m, \mathbf{A}\mathbf{v}_{k+1} \rangle_{\mathbf{B}} = 0$ for $m \leq k$, and by symmetry $\langle \widetilde{\mathbf{v}}_{k+1}, \mathbf{A}\mathbf{v}_m \rangle_{\mathbf{B}} = 0$ too. Finally, the equation in (7.3) for $m = n = k + 1$ results from (7.1a).

The formulae (7.1c), (7.1d), (7.1h), (7.1i), (2.5), and (7.4) show clearly that $\mathbf{y}_{k+1}, \mathbf{v}_{k+1} \in \mathcal{K}_{k+2}$ and $\widetilde{\mathbf{y}}_{k+1}, \widetilde{\mathbf{v}}_{k+1} \in \widetilde{\mathcal{K}}_{k+2}$. As in Section 2, $\delta_n \neq 0$ implies that $\mathbf{y}_{k+1} \notin \mathcal{K}_{k+1}$ and $\widetilde{\mathbf{y}}_{k+1} \notin \widetilde{\mathcal{K}}_{k+1}$. By (7.1h) and (7.1i) it thus follows that (7.4) holds for $n = k + 1$. This completes the induction. □

7.2. Matrix relations

The BIOC algorithm has a matrix interpretation, which quickly reveals the relation to the BIO algorithm. In addition to $\mathbf{Y}_n$ and $\widetilde{\mathbf{Y}}_n$ of (2.8) we need the $N \times n$ matrices

$$\mathbf{V}_n := [\; \mathbf{v}_0 \quad \mathbf{v}_1 \quad \cdots \quad \mathbf{v}_{n-1} \;], \quad \widetilde{\mathbf{V}}_n := [\; \widetilde{\mathbf{v}}_0 \quad \widetilde{\mathbf{v}}_1 \quad \cdots \quad \widetilde{\mathbf{v}}_{n-1} \;],$$

the $n \times n$ matrices

$$\mathbf{L}_n := \begin{bmatrix} \varphi_0 & & & & \\ \gamma_0 & \varphi_1 & & & \\ & \gamma_1 & \varphi_2 & & \\ & & \ddots & \ddots & \\ & & & \gamma_{n-2} & \varphi_{n-1} \end{bmatrix}, \quad \mathbf{U}_n := \begin{bmatrix} 1 & \psi_0 & & & \\ & 1 & \psi_1 & & \\ & & 1 & \ddots & \\ & & & \ddots & \psi_{n-2} \\ & & & & 1 \end{bmatrix},$$

which are lower and upper bidiagonal, respectively, and the *extended* bidiagonal matrices

$$\underline{\mathbf{L}}_n := \left[\begin{array}{c} \mathbf{L}_n \\ \hline \gamma_{n-1}\mathbf{l}_n^\top \end{array}\right] = \begin{bmatrix} \varphi_0 & & & & \\ \gamma_0 & \varphi_1 & & & \\ & \gamma_1 & \varphi_2 & & \\ & & \ddots & \ddots & \\ & & & \gamma_{n-2} & \varphi_{n-1} \\ & & & & \gamma_{n-1} \end{bmatrix}$$

and

$$\mathbf{U\!\vert}_n := \left[\begin{array}{c|c} \mathbf{U}_n & \mathbf{l}_n\psi_{n-1} \end{array}\right] = \begin{bmatrix} 1 & \psi_0 & & & & \\ & 1 & \psi_1 & & & \\ & & 1 & \ddots & & \\ & & & \ddots & \psi_{n-2} & \\ & & & & 1 & \psi_{n-1} \end{bmatrix},$$

with an additional row and column, respectively[13]. Analogously, we define $\widetilde{\mathbf{L}}_n$, $\widetilde{\mathbf{U}}_n$, $\widetilde{\underline{\mathbf{L}}}_n$, and $\widetilde{\mathbf{U\!\vert}}_n$ in the obvious way. Then, according to (7.1h), (7.1i), (7.1c), and (7.1d),

$$\mathbf{Y}_n = \mathbf{V}_n\mathbf{U}_n, \quad \widetilde{\mathbf{Y}}_n = \widetilde{\mathbf{V}}_n\widetilde{\mathbf{U}}_n \quad (n \leq \dot{\nu})$$

and

$$\mathbf{A}\mathbf{V}_n = \mathbf{Y}_{n+1}\underline{\mathbf{L}}_n, \quad \mathbf{A}^\star\widetilde{\mathbf{V}}_n = \widetilde{\mathbf{Y}}_{n+1}\widetilde{\underline{\mathbf{L}}}_n \quad (n \leq \dot{\nu}).$$

After setting[14]

$$\mathbf{T}_n := \mathbf{L}_n\mathbf{U}_n, \quad \mathbf{T}'_n := \mathbf{U\!\vert}_n\underline{\mathbf{L}}_n \tag{7.7}$$

and

$$\underline{\mathbf{T}}_n := \underline{\mathbf{L}}_n\mathbf{U}_n, \quad \underline{\mathbf{T}}'_n := \mathbf{U}_{n+1}\underline{\mathbf{L}}_n, \tag{7.8}$$

[13] Note that the additional column is indicated by a vertical line. We suggest reading this symbol as 'U sub n extended'.

[14] The prime does *not* mean transposition.

and likewise defining $\widetilde{\mathbf{T}}_n$, $\widetilde{\mathbf{T}}'_n$, $\widetilde{\underline{\mathbf{T}}}_n$, and $\widetilde{\underline{\mathbf{T}}}'_n$, we conclude by eliminating $\mathbf{V}_n$ and $\widetilde{\mathbf{V}}_n$ or $\mathbf{Y}_n$ and $\widetilde{\mathbf{Y}}_n$, respectively, that

$$\mathbf{A}\mathbf{Y}_n = \mathbf{Y}_{n+1}\underline{\mathbf{T}}_n, \quad \mathbf{A}^\star\widetilde{\mathbf{Y}}_n = \widetilde{\mathbf{Y}}_{n+1}\widetilde{\underline{\mathbf{T}}}_n \tag{7.9}$$

and

$$\mathbf{A}\mathbf{V}_n = \mathbf{V}_{n+1}\underline{\mathbf{T}}'_n, \quad \mathbf{A}^\star\widetilde{\mathbf{V}}_n = \widetilde{\mathbf{V}}_{n+1}\widetilde{\underline{\mathbf{T}}}'_n. \tag{7.10}$$

Note that $\mathbf{T}'_n = \mathbf{U\!l}_n\underline{\mathbf{L}}_n$ and $\mathbf{U}_n\mathbf{L}_n$ differ only by the rank-one matrix $\mathbf{l}_n\gamma_{n-1}\psi_{n-1}\mathbf{l}_n^\mathsf{T}$, which just modifies the element in the lower right corner of $\mathbf{U}_n\mathbf{L}_n$ by adding $\gamma_{n-1}\psi_{n-1}$ to it.

Since the matrix $\mathbf{T}_n$ with the recurrence coefficients α_m, β_m, γ_m of the BIO algorithm, and the matrices $\mathbf{L}_n$ and $\mathbf{U}_n$ with the recurrence coefficients φ_n and ψ_{n-1} of the BIOC algorithm, as well as the tridiagonal matrix $\mathbf{T}'_n$ with elements α'_m, β'_m, γ'_m are related by (7.7), these three sets of parameters are easily converted into each other. After setting $\psi_{-1} := 0$, we obtain

(i) from $\mathbf{T}_{\dot\nu} = \mathbf{L}_{\dot\nu}\mathbf{U}_{\dot\nu}$:

$$\alpha_n = \varphi_n + \gamma_{n-1}\psi_{n-1}, \quad \beta_n = \varphi_n\psi_n, \quad \gamma_n = \widetilde{\gamma}_n,$$

(ii) from $\mathbf{T}'_{\dot\nu} = \mathbf{U\!l}_{\dot\nu}\underline{\mathbf{L}}_{\dot\nu}$:

$$\alpha'_n = \varphi_n + \gamma_n\psi_n, \quad \beta'_n = \varphi_{n+1}\psi_n, \quad \widetilde{\gamma}_n = \gamma_n.$$

These are essentially the *rhombus rules* of the QD *algorithm* (Rutishauser 1957). Of course, the same formulae also hold with tildes.

Since (7.9) is identical to (2.23), except for the possibility that $\dot\nu < \nu$, we obtain the following result.

Theorem 7.2 If the same starting vectors $\mathbf{y}_0$ and $\widetilde{\mathbf{y}}_0$ and the same scale factors γ_n and $\widetilde{\gamma}_n$ are chosen in the BIO and the BIOC algorithms, then the same biorthogonal vector sequences $\{\mathbf{y}_n\}$ and $\{\widetilde{\mathbf{y}}_n\}$ are produced, except that the BIOC algorithm may break down earlier due to $\delta'_{\dot\nu} = 0$. The bidiagonal matrices $\mathbf{L}_{\dot\nu}$, $\widetilde{\mathbf{L}}_{\dot\nu}$, $\mathbf{U}_{\dot\nu}$, and $\widetilde{\mathbf{U}}_{\dot\nu}$ of the recurrence coefficients of the BIOC algorithm can be obtained by LU decomposition of the tridiagonal matrix $\mathbf{T}_{\dot\nu}$ with the recurrence coefficients of the first $\dot\nu$ steps of the BIO algorithm. The possible earlier breakdown of the BIOC algorithm is due to the possible nonexistence of the LU decomposition (without pivoting) of $\mathbf{T}_\nu$.

This result implies in particular that from the bidiagonal matrices constructed in the BIOC algorithm we can still compute the eigenvalues of $\mathbf{T}_n$, the so-called Petrov values (or, Ritz values in the Hermitian case), as approximations for eigenvalues of $\mathbf{A}$. As mentioned at the beginning of this section, there are a number of reasons why the bidiagonal matrices are preferable; see Parlett (1995).

Theorem 7.2 holds analogously for the ORTHORES and ORTHOMIN versions of the generalized CG and the GCR methods, except that $\mathbf{T}_{\tilde{\nu}}$ is then upper Hessenberg and $\mathbf{U}_{\tilde{\nu}}$ is upper triangular, while $\mathbf{L}_{\tilde{\nu}}$ is still lower bidiagonal; see Gutknecht (1993*a*). There is also a similar result that links ORTHOMIN with ORTHODIR. We will encounter its analogue in Section 9.

7.3. Normalization; simplification due to symmetry

For the BIO algorithm (Algorithm 1) we have chosen a somewhat complicated formulation to make explicit the freedom in choosing two of the three quantities γ_n, $\widetilde{\gamma}_n$, and δ_{n+1}. The same freedom exists in our formulation of BIOC, although we have not made that explicit. In particular, as normalization we can still enforce (2.26) or (2.27). In these two cases, it is necessary to define in a straightforward manner $\mathbf{y}_{\text{temp}}$, $\widetilde{\mathbf{y}}_{\text{temp}}$, and δ_{temp}, as in Algorithm 1. The second choice, (2.27), will lead to Lanczos vectors of length 1. However, if we also wanted to have normalized direction vectors, we would have to introduce additional scale factors in (7.1h) and (7.1i), and to compensate for them in some of the other formulae.

What has been said in Sections 2.1 and 2.6 regarding simplification due to symmetry also carries over to the BIOC algorithm. In particular, if $\mathbf{A}$ is Hermitian, complex symmetric, $\mathbf{S}$-Hermitian, or $\mathbf{S}$-symmetric, then the matrix-vector multiplication by $\mathbf{A}^{\star}$ can be replaced by multiplication by $\mathbf{S}$.

8. The BIOMIN form of the BICG method

A consistent version of the BICG method based on the BIOC algorithm was presented by Fletcher (1976). He referred to it as the *biconjugate gradient* (BICG) *algorithm*, while later, Jea and Young (1983) called it Lanczos/ORTHOMIN. Here we use the name BIOMIN in order to stress the analogy to the OMIN (Hestenes–Stiefel) version of the conjugate gradient (CG) method (Hestenes and Stiefel 1952) and the differently flavoured analogy to BIORES and BIODIR. The latter is discussed later. The BIOC algorithm is related to the BIOMIN version of BICG and to the OMIN version of CG in the same way as the Lanczos BIO algorithm is related to the BIORES version of BICG and to the ORES version of CG.

We will refer to BIOMIN also as the *standard version* of BICG. We keep using the abbreviation BICG (like CG) as the generic name for the various biconjugate gradient algorithms that are mathematically equivalent except for possible deviations in the breakdown conditions.

8.1. The BIOMIN *algorithm*

When applying the BIO algorithm to solving linear systems in such a way that the right Lanczos vectors became the residuals, we had to stick to a particular choice of γ_n, namely $\gamma_n := -\alpha_n - \beta_{n-1}$, in order to fulfil the consistency condition for Krylov space solvers. Likewise, γ_n is determined here by the latter condition. In fact, since $\underline{\mathbf{L}}_n = \underline{\mathbf{T}}_n \mathbf{U}_n^{-1}$, the bidiagonal matrix $\underline{\mathbf{L}}_n$ inherits from $\underline{\mathbf{T}}_n$ the property of zero column sums, which, as we recall, was inherited to $\underline{\mathbf{T}}_n$ from $\underline{\mathbf{F}}_n$; that is, we need

$$\mathbf{e}^\top \underline{\mathbf{L}}_n = \mathbf{o}^\top,$$

or, in terms of matrix elements,

$$\gamma_n := -\varphi_n. \tag{8.1}$$

Again, we can then define the iterates $\mathbf{x}_n$ in such a way that $\mathbf{y}_n$ is the nth residual: multiplicaton of (4.14) from the right by $\mathbf{U}_n^{-1}$ yields

$$\mathbf{V}_n = -\mathbf{X}_{n+1}\underline{\mathbf{L}}_n \quad (n \leq \dot{\nu}),$$

or,

$$\mathbf{x}_{n+1} := \mathbf{x}_n + \mathbf{v}_n \omega_n$$

if we set

$$\omega_n := \frac{1}{\varphi_n} = -\frac{1}{\gamma_n}. \tag{8.2}$$

Basically we are free to choose $\widetilde{\gamma}_n$, but the normal choice is $\widetilde{\gamma}_n := \overline{\gamma_n}$, which implies that $\widetilde{\psi}_n := \overline{\psi_n}$ and $\psi_n := -\delta_{n+1}/\delta_n$, see (7.1f)–(7.1g). This leads us to the standard BIOMIN form of the BICG method.

ALGORITHM 7. (BIOMIN FORM OF THE BICG METHOD)
For solving $\mathbf{Ax} = \mathbf{b}$ choose an initial approximation $\mathbf{x}_0$, set $\mathbf{v}_0 := \mathbf{y}_0 := \mathbf{b} - \mathbf{Ax}_0$, and choose $\widetilde{\mathbf{y}}_0$ such that $\delta_0 := \langle \widetilde{\mathbf{y}}_0, \mathbf{y}_0 \rangle_\mathbf{B} \neq 0$ and $\delta'_0 := \langle \widetilde{\mathbf{y}}_0, \mathbf{Av}_0 \rangle_\mathbf{B} \neq 0$. Then let $\widetilde{\mathbf{v}}_0 := \widetilde{\mathbf{y}}_0$, and apply Algorithm 6 (BIOC) with $\gamma_n := -\varphi_n$ and $\widetilde{\gamma}_n := -\overline{\varphi_n}$, so that after substituting $\omega_n := 1/\varphi_n$ the nth step consists of:

$$\begin{aligned}
\omega_n &:= \delta_n / \delta'_n, && (8.3a)\\
\mathbf{y}_{n+1} &:= \mathbf{y}_n - \mathbf{Av}_n\omega_n, && (8.3b)\\
\widetilde{\mathbf{y}}_{n+1} &:= \widetilde{\mathbf{y}}_n - \mathbf{A}^\star \widetilde{\mathbf{v}}_n \overline{\omega_n}, && (8.3c)\\
\mathbf{x}_{n+1} &:= \mathbf{x}_n + \mathbf{v}_n\omega_n, && (8.3d)\\
\delta_{n+1} &:= \langle \widetilde{\mathbf{y}}_{n+1}, \mathbf{y}_{n+1} \rangle_\mathbf{B}, && (8.3e)\\
\psi_n &:= -\delta_{n+1}/\delta_n, && (8.3f)\\
\mathbf{v}_{n+1} &:= \mathbf{y}_{n+1} - \mathbf{v}_n\psi_n, && (8.3g)\\
\widetilde{\mathbf{v}}_{n+1} &:= \widetilde{\mathbf{y}}_{n+1} - \widetilde{\mathbf{v}}_n\overline{\psi_n}, && (8.3h)\\
\delta'_{n+1} &:= \langle \widetilde{\mathbf{v}}_{n+1}, \mathbf{Av}_{n+1} \rangle_\mathbf{B}. && (8.3i)
\end{aligned}$$

If $\mathbf{y}_{n+1} = 0$ the process terminates and $\mathbf{x}_{n+1}$ is the solution; if $\delta_{n+1} = 0$ (and hence $\psi_n = 0$) or $\delta'_{n+1} = 0$, but $\mathbf{y}_{n+1} \neq 0$, the algorithm breaks down. In all cases we set $\dot{\nu} := n + 1$.

Assuming $\mathbf{b} - \mathbf{A}\mathbf{x}_n = \mathbf{y}_n$ and using (8.3d) and (8.3b) we in fact get

$$\mathbf{b} - \mathbf{A}\mathbf{x}_{n+1} = \mathbf{b} - \mathbf{A}\mathbf{x}_n - \mathbf{A}\mathbf{v}_n\omega_n = \mathbf{y}_n - \mathbf{A}\mathbf{v}_n\omega_n = \mathbf{y}_{n+1},$$

so that by induction $\mathbf{b} - \mathbf{A}\mathbf{x}_n = \mathbf{y}_n$ for $n = 0, 1, \ldots, \dot{\nu}$. Consequently, if $\mathbf{y}_{\dot{\nu}} = 0$, then $\mathbf{x}_{\dot{\nu}}$ is the solution of the system.

Note that by definition of $\dot{\nu}$ we have

$$\delta_n \neq 0, \quad \psi_{n-1} \neq 0, \quad \varphi_n \neq 0, \quad \omega_n \neq 0, \quad n = (0), 1, \ldots, \dot{\nu} - 1, \tag{8.4a}$$

and one of the following three cases:

$$\delta_{\dot{\nu}} \neq 0, \quad \delta'_{\dot{\nu}} = 0 \quad \Longrightarrow \quad \psi_{\dot{\nu}-1} \neq 0, \quad \varphi_{\dot{\nu}} = 0, \quad \omega_{\dot{\nu}} = \infty, \tag{8.4b}$$

$$\delta_{\dot{\nu}} = 0, \quad \delta'_{\dot{\nu}} \neq 0 \quad \Longrightarrow \quad \psi_{\dot{\nu}-1} = 0, \quad \varphi_{\dot{\nu}} = \infty, \quad \omega_{\dot{\nu}} = 0, \tag{8.4c}$$

$$\delta_{\dot{\nu}} = 0, \quad \delta'_{\dot{\nu}} = 0 \quad \Longrightarrow \quad \psi_{\dot{\nu}-1} = 0, \quad \varphi_{\dot{\nu}}, \ \omega_{\dot{\nu}} \text{ undefined.} \tag{8.4d}$$

In the first case, (8.4b), a pivot breakdown occurs. The second case, (8.4c), is a Lanczos breakdown. Here, we could still compute $\mathbf{y}_{\dot{\nu}+1} = \mathbf{y}_{\dot{\nu}}$, $\widetilde{\mathbf{y}}_{\dot{\nu}+1} = \widetilde{\mathbf{y}}_{\dot{\nu}}$, and $\delta_{\dot{\nu}+1} = 0$; but then $\psi_{\dot{\nu}}$ would be indefinite, and we would have $\mathbf{v}_{\dot{\nu}+1} \in \mathcal{K}_{\dot{\nu}+1}$, $\widetilde{\mathbf{v}}_{\dot{\nu}+1} \in \widetilde{\mathcal{K}}_{\dot{\nu}+1}$ for any value of $\psi_{\dot{\nu}}$. In other words, the Krylov space generation is stopped. Hence, the algorithm *stalls permanently*, with no chance to recover. In the next section we will derive yet another version, BIODIR, of the BICG method, which under a certain assumption can recover in this situation. The last case, (8.4d), is simultaneously a Lanczos and a pivot breakdown.

Formula (8.3d) shows clearly that the approximations $\mathbf{x}_n$ are modified by moving along the *direction vectors* $\mathbf{v}_n$, and it allows us to express $\mathbf{x}_n - \mathbf{x}_0$ as a sum of corrections:

$$\mathbf{x}_n = \mathbf{x}_0 + \sum_{j=0}^{n-1} \mathbf{v}_j\omega_j.$$

This formula was actually the starting point of Lanczos' application of the BIOC algorithm to linear systems (Lanczos 1952, p. 37). When $\mathbf{A}$ is Hermitian positive definite and $\mathbf{B} = \mathbf{I}$, that is, in the classical CG situation, one can say that for finding $\mathbf{x}_{n+1}$ one moves along the straight line determined by the approximation $\mathbf{x}_n$ and the direction $\mathbf{v}_n$ until one reaches the minimum of the quadratic function $\mathbf{x} \mapsto \frac{1}{2}\mathbf{x}^\star\mathbf{A}\mathbf{x} - \mathbf{b}^\star\mathbf{x}$ on this line, or equivalently, the minimum of $\mathbf{x} \mapsto (\mathbf{x}_{\text{ex}} - \mathbf{x})^\star\mathbf{A}(\mathbf{x}_{\text{ex}} - \mathbf{x})$. In fact, as we have mentioned in Section 4, this is then also the minimum among all $\mathbf{x} \in \mathbf{x}_0 + \mathcal{K}_n$. At $\mathbf{x}_n$ the gradient (direction of steepest ascent) of this function happens to be $-\mathbf{y}_n$. Thus the gradients (residuals) are orthogonal to each other. This geometric interpretation leads readily to a variety of generalizations of the conjugate gradient method to nonlinear minimization problems; see, for instance, Murray (1972, Chapter 5) and Hestenes (1980).

8.2. The inconsistent BIOMIN *algorithm*

After successfully eliminating the pivot breakdown of the BIORES algorithm by making a minor modification we may wonder whether it is possible to eliminate this type of breakdown here too by introducing an inconsistent version of BIOMIN. It is not difficult to define such a version, but it turns out that the goal is missed. Nevertheless, the inconsistent BIOMIN algorithm is of some interest due to its close relationship to the coupled two-term version of the QMR method.

ALGORITHM 8. (INCONSISTENT BIOMIN FORM OF THE BICG METHOD) Let $\mathbf{y}_0 := (\mathbf{b} - \mathbf{A}\mathbf{x}_0)/\gamma_{-1}$ with some $\gamma_{-1} \neq 0$, redefine $\mathbf{x}_0 := \mathbf{x}_0/\gamma_{-1}$, and set $\dot{\pi}_0 := 1/\gamma_{-1}$ as in Algorithm 3 (inconsistent BIORES). Modify Algorithm 6 (BIOC) by also computing, at the nth step,

$$\mathbf{x}_{n+1} := -(\mathbf{x}_n \varphi_n + \mathbf{v}_n)/\gamma_n$$

and

$$\dot{\pi}_{n+1} := -\dot{\pi}_n \varphi_n / \gamma_n. \tag{8.5}$$

In this algorithm the vectors $\mathbf{y}_n$ and $\mathbf{x}_n$ are again related by (4.17), with the same scale factors $\dot{\pi}_n$. The proof is once again by induction:

$$(\mathbf{b}\dot{\pi}_{n+1} - \mathbf{A}\mathbf{x}_{n+1})\gamma_n = -\mathbf{b}\dot{\pi}_n\varphi_n + \mathbf{A}\mathbf{x}_n\varphi_n + \mathbf{A}\mathbf{v}_n = -(\mathbf{y}_n\varphi_n - \mathbf{A}\mathbf{v}_n) = \mathbf{y}_{n+1}\gamma_n.$$

However, the BIOC algorithm, and thus also the inconsistent BIOMIN algorithm, in any case break down when $\delta'_{n+1} = 0$. Therefore, in exact arithmetic, the latter algorithm does not bring any substantial advantage. However, as we mentioned before, the BIOC algorithm seems to be less affected by round-off than the BIO algorithm. While we do not expect a big difference between consistent and inconsistent BIOMIN in this respect, round-off error control measures are somewhat easier to implement when the Lanczos vectors are normalized.

Concerning the relation to BIORES the following holds.

Lemma 8.1 If the same starting vectors are used, then the three algorithms BIORES (Algorithm 2), BIOMIN (Algorithm 7), and inconsistent BIOMIN (Algorithm 8) are mathematically equivalent, that is, they break down at the same time and they produce the same iterates $\mathbf{x}_n$, except that those of inconsistent BIOMIN are scaled by the factors $\dot{\pi}_n$.

Proof. The relation between the BIO and the BIOC algorithm was established in Theorem 7.2. Compared to the BIO algorithm, BIORES additionally requires that $\gamma_n := -\alpha_n - \beta_{n-1} \neq 0$. But since γ_n is the same for the BIO and the BIOC algorithm, and $\gamma_n := -\varphi_n$ in BIOMIN, this condition is equivalent to $\varphi_n \neq 0$, which has to be observed in the BIOC algorithm anyway, and which, conversely, is the only additional condition there compared to the BIO algorithm. □

9. The BIODIR form of the BICG method; comparison

9.1. *The* BIC *algorithm*

Formula (8.3d) suggests that the iterates $\mathbf{x}_n$ can be computed by building up the biconjugate sequences $\{\mathbf{v}_n\}$ and $\{\widetilde{\mathbf{v}}_n\}$ and using the vectors $\mathbf{y}_n$ and $\widetilde{\mathbf{y}}_n$ only for the determination of the step size $1/\varphi_n$, which depends on them via δ_n; see (7.1a). Since the biconjugate sequences can be thought of as being biorthogonal with respect to the formal inner product $\langle \widetilde{\mathbf{v}}, \mathbf{v} \rangle_{\mathbf{BA}} := \widetilde{\mathbf{v}}^\star \mathbf{BAv}$, we can construct them with the BIO algorithm if we substitute this inner product there. The sequences generated in this way can differ in scaling from those produced by BIOMIN, but we will see in a moment how to make them identical. We call the resulting algorithm the biconjugation or BIC algorithm, and although it is identical to the BIO algorithm except for the change in the formal inner product, we give it here in detail, because it will be a part of the BIODIR form of BICG, and because we want to fix the notation. We distinguish the new recurrence coefficients and the inner products by primes from those of the BIO algorithm.

ALGORITHM 9. (BIC ALGORITHM)
Choose $\mathbf{v}_0, \widetilde{\mathbf{v}}_0 \in \mathbb{C}^N$ such that $\delta'_0 := \langle \widetilde{\mathbf{v}}_0, \mathbf{Av}_0 \rangle_{\mathbf{B}} \neq 0$, and set $\beta'_{-1} := 0$. For $n = 0, 1, \ldots$ compute

$$\begin{aligned}
\alpha'_n &:= \langle \mathbf{A}^\star \widetilde{\mathbf{v}}_n, \mathbf{Av}_n \rangle_{\mathbf{B}} / \delta'_n, && (9.1a)\\
\widetilde{\alpha}'_n &:= \overline{\alpha'_n}, && (9.1b)\\
\beta'_{n-1} &:= \overline{\widetilde{\gamma}'_{n-1}} \delta'_n / \delta'_{n-1}, \quad (\text{if } n > 0), && (9.1c)\\
\widetilde{\beta}'_{n-1} &:= \overline{\gamma'_{n-1} \delta'_n / \delta'_{n-1}} = \overline{\beta'_{n-1} \gamma'_{n-1} / \widetilde{\gamma}'_{n-1}}, \quad (\text{if } n > 0), && (9.1d)\\
\mathbf{v}_{\text{temp}} &:= \mathbf{Av}_n - \mathbf{v}_n \alpha'_n - \mathbf{v}_{n-1} \beta'_{n-1}, && (9.1e)\\
\widetilde{\mathbf{v}}_{\text{temp}} &:= \mathbf{A}^\star \widetilde{\mathbf{v}}_n - \widetilde{\mathbf{v}}_n \widetilde{\alpha}'_n - \widetilde{\mathbf{v}}_{n-1} \widetilde{\beta}'_{n-1}, && (9.1f)\\
\delta'_{\text{temp}} &:= \langle \widetilde{\mathbf{v}}_{\text{temp}}, \mathbf{Av}_{\text{temp}} \rangle_{\mathbf{B}}\,; && (9.1g)
\end{aligned}$$

if $\delta'_{\text{temp}} = 0$, choose $\gamma'_n \neq 0$ and $\widetilde{\gamma}'_n \neq 0$, set

$$\nu' := n + 1, \quad \mathbf{v}_{\nu'} := \mathbf{v}_{\text{temp}} / \gamma'_n, \quad \widetilde{\mathbf{v}}_{\nu'} := \widetilde{\mathbf{v}}_{\text{temp}} / \widetilde{\gamma}'_n, \quad \delta'_{n+1} := 0,$$

and stop; otherwise, choose $\gamma'_n \neq 0$, $\gamma'_n \neq 0$, and δ'_{n+1} such that

$$\gamma'_n \overline{\widetilde{\gamma}'_n} \delta'_{n+1} = \delta'_{\text{temp}},$$

set

$$\mathbf{v}_{n+1} := \mathbf{v}_{\text{temp}} / \gamma'_n, \quad \widetilde{\mathbf{v}}_{n+1} := \widetilde{\mathbf{v}}_{\text{temp}} / \widetilde{\gamma}'_n, \qquad (9.1\text{h})$$

and proceed with the next step.

It is most natural to choose $\gamma'_n := \|\mathbf{y}_{\text{temp}}\|$, $\widetilde{\gamma}'_n := \|\widetilde{\mathbf{y}}_{\text{temp}}\|$, so that the $\|\mathbf{v}_{n+1}\| = \|\widetilde{\mathbf{v}}_{n+1}\| = 1$, in analogy to (2.27).

Theorem 2.1 now transforms into the following statement.

Corollary 9.1 The sequences $\{\mathbf{v}_n\}_{n=0}^{\nu'}$ and $\{\widetilde{\mathbf{v}}_n\}_{n=0}^{\nu'}$ generated by the BIC algorithm are biconjugate (with respect to $\mathbf{A}$), except that $\langle \widetilde{\mathbf{v}}_{\nu'}, \mathbf{A}\mathbf{v}_{\nu'} \rangle_{\mathbf{B}} = 0$. That is, for $m, n = 0, 1, \ldots, \nu'$,

$$\langle \widetilde{\mathbf{v}}_m, \mathbf{A}\mathbf{v}_n \rangle_{\mathbf{B}} = \begin{cases} 0, & m \neq n, \\ \delta'_n, & m = n, \end{cases} \tag{9.2}$$

where $\delta'_n \neq 0$ for $0 \leq n \leq \nu' - 1$, but $\delta'_{\nu} = 0$. Moreover, for $n = 1, \ldots, \nu' - 1$, (7.4) is valid for $\mathbf{v}_n$ and $\widetilde{\mathbf{v}}_n$. Conversely, the sequences $\{\mathbf{v}_n\}_{n=0}^{\nu'}$ and $\{\widetilde{\mathbf{v}}_n\}_{n=0}^{\nu'}$ are uniquely determined up to scaling by the condition (9.2) with $\delta'_n \neq 0$ $(n = 0, 1, \ldots, \nu' - 1)$, $\delta'_{\nu'} = 0$, and the assumption $\mathbf{v}_n \in \mathcal{K}_{n+1}$, $\widetilde{\mathbf{v}}_n \in \widetilde{\mathcal{K}}_{n+1}$ $(n = 0, 1, \ldots, \nu')$.

In view of (7.10) it should be clear that the recurrence coefficients α'_n, β'_n, γ'_n and $\widetilde{\alpha}'_n$, $\widetilde{\beta}'_n$, $\widetilde{\gamma}'_n$ in the BIC algorithm are the elements of the matrices $\underline{\mathbf{T}}'_n$ and $\underline{\widetilde{\mathbf{T}}}'_n$ that we introduced in Section 7. In particular, (7.7)–(7.8) hold, and the shorthand notation for recurrences is

$$\mathbf{A}\mathbf{V}_n = \mathbf{V}_{n+1}\underline{\mathbf{T}}'_n, \quad \mathbf{A}^\star\widetilde{\mathbf{V}}_n = \widetilde{\mathbf{V}}_{n+1}\underline{\widetilde{\mathbf{T}}}'_n.$$

9.2. The BIODIR *algorithm*

For the application of the BIC algorithm to linear systems of equations an additional recurrence for $\mathbf{x}_n$ is needed again. We keep the freedom of scaling the direction vectors $\mathbf{v}_n$ and $\widetilde{\mathbf{v}}_n$ arbitrarily, but then have to determine the step length appropriately.

ALGORITHM 10. (BIODIR FORM OF THE BICG METHOD)
For solving $\mathbf{A}\mathbf{x} = \mathbf{b}$ choose an initial approximation $\mathbf{x}_0$, set $\mathbf{v}_0 := \mathbf{y}_0 := \mathbf{b} - \mathbf{A}\mathbf{x}_0$, and apply Algorithm 9 (BIC), additionally computing

$$\omega'_n := \langle \widetilde{\mathbf{v}}_n, \mathbf{y}_n \rangle_{\mathbf{B}} / \delta'_n, \tag{9.3a}$$

$$\mathbf{x}_{n+1} := \mathbf{x}_n + \mathbf{v}_n\omega'_n, \tag{9.3b}$$

$$\mathbf{y}_{n+1} := \mathbf{y}_n - \mathbf{A}\mathbf{v}_n\omega'_n. \tag{9.3c}$$

If $\mathbf{y}_{n+1} = \mathbf{o}$, the process terminates and $\mathbf{x}_{n+1}$ is the solution; if $\delta'_{n+1} = 0$, but $\mathbf{y}_{n+1} \neq \mathbf{o}$, the algorithm breaks down. In both cases we set $\nu' := n + 1$.

Again, $\mathbf{y}_n$ is the nth residual. In fact, by assuming that $\mathbf{b} - \mathbf{A}\mathbf{x}_n = \mathbf{y}_n$ and using (9.3b) and (9.3c) we get

$$\mathbf{b} - \mathbf{A}\mathbf{x}_{n+1} = \mathbf{b} - \mathbf{A}\mathbf{x}_n - \mathbf{A}\mathbf{v}_n\omega'_n = \mathbf{y}_n - \mathbf{A}\mathbf{v}_n\omega'_n = \mathbf{y}_{n+1},$$

as required. The formula (9.3a) for ω'_n enforces $\langle \tilde{\mathbf{v}}_n, \mathbf{y}_{n+1} \rangle_{\mathbf{B}} = 0$. In Theorem 9.2 below, we will verify that BIODIR creates the same iterates as BIOMIN and BIORES, but has different breakdown conditions.

9.3. Comparison of breakdown conditions of BICG *algorithms*

The BIODIR algorithm can only break down due to $\delta'_{\text{temp}} = 0$. When $\omega'_n = 0$, the algorithm can recover: although it *stalls* for one step, that is, $\mathbf{x}_{n+1} = \mathbf{x}_n$ and $\mathbf{y}_{n+1} = \mathbf{y}_n$, a new direction $\mathbf{v}_{n+1}$ is created. Therefore, in general, the method cannot be equivalent to BIORES or BIOMIN since necessarily $\mathbf{y}_{n+1} \neq \mathbf{y}_n$ in both. But if $\omega'_n \neq 0$ $(n = 0, \ldots, \nu - 1)$, it is indeed equivalent to these two methods, as was noted in Jea and Young (1983, p. 411) and is proven next. In the following theorem the various conditions for a first breakdown, a stagnation point, or the termination are summarized and compared.

Theorem 9.2 Assume that the same initial approximation $\mathbf{x}_0$ and the same initial vectors $\mathbf{v}_0 := \mathbf{y}_0 := \mathbf{b} - \mathbf{A}\mathbf{x}_0$ and $\tilde{\mathbf{v}}_0 := \tilde{\mathbf{y}}_0$ are used in BIODIR, consistent or inconsistent BIOMIN, and consistent and inconsistent BIORES. Let ν', ν, and $\dot{\nu}$ be the indices of first breakdown or termination of BIODIR, inconsistent BIORES, and the other three algorithms, respectively.

Then $\dot{\nu} = \min\{\nu, \nu'\}$, and the following five conditions for a Lanczos breakdown are equivalent:

(i') $\omega'_m \neq 0$ (for all $m < n$) and $\langle \tilde{\mathbf{v}}_n, \mathbf{y}_n \rangle_{\mathbf{B}} = 0$ in BIODIR
(ii') $\delta_n = 0$ in BIOMIN
(iii') $\delta_n = 0$ in (consistent) BIORES
(iv') $\alpha_m + \beta_{m-1} \neq 0$ (for all $m < n$) and $\delta_n = 0$ in inconsistent BIORES
(v') $n = \nu = \dot{\nu}$.

Condition (i') implies that either $\omega'_n = 0$ or $\delta'_n = 0$; in the latter case, ω'_n is undefined and $n = \nu = \dot{\nu} = \nu'$. In the former, we have stagnation but no breakdown of BIODIR.

Likewise, the next four conditions for a pivot breakdown are equivalent:

(i") $\langle \tilde{\mathbf{v}}_m, \mathbf{y}_m \rangle_{\mathbf{B}} \neq 0$ (for all $m < n$) and $\delta'_n = 0$ in BIODIR
(ii") $\delta'_n = 0$ in BIOMIN
(iii") $\alpha_n + \beta_{n-1} = 0$ in (consistent) BIORES
(iv") $\dot{\pi}_n = 0$ in inconsistent BIORES
(v") $n = \dot{\nu} = \nu'$.

The conditions (ii')–(iv') and (i")–(iii") all cause either the termination or a breakdown of the respective algorithm. However, (iv") does not stop inconsistent BIORES.

The three consistent algorithms produce the same approximants $\mathbf{x}_1, \ldots, \mathbf{x}_{\dot\nu}$ and hence the same residuals $\mathbf{y}_1, \ldots, \mathbf{y}_{\dot\nu}$. If we choose

$$\gamma'_n := \gamma_n \; (= -\varphi_n = -1/\omega_n) \tag{9.4}$$

in BIODIR, then, for $n = 0, \ldots, \dot\nu$, the vectors $\mathbf{v}_n =: \mathbf{v}_n^{\mathrm{DIR}}$ and $\widetilde{\mathbf{v}}_n =: \widetilde{\mathbf{v}}_n^{\mathrm{DIR}}$ produced by BIODIR are the same as the biconjugate vectors $\mathbf{v}_n =: \mathbf{v}_n^{\mathrm{MIN}}$ and $\widetilde{\mathbf{v}}_n =: \widetilde{\mathbf{v}}_n^{\mathrm{MIN}}$ of BIOMIN. This implies that for $0 \le n \le \dot\nu - 1$ the parameters $\alpha'_n, \beta'_{n-1}, \gamma'_n$ of BIODIR are the elements of the matrix $\underline{\mathbf{T}}'_{\dot\nu}$ of (7.8).

In general, without the particular choice (9.4), we have

$$\mathbf{v}_n^{\mathrm{MIN}}\Gamma_n = \mathbf{v}_n^{\mathrm{DIR}}\Gamma'_n, \quad \widetilde{\mathbf{v}}_n^{\mathrm{MIN}}\widetilde{\Gamma}_n = \widetilde{\mathbf{v}}_n^{\mathrm{DIR}}\widetilde{\Gamma}'_n \quad (n \le \dot\nu), \tag{9.5}$$

where

$$\Gamma_n := \gamma_0\gamma_1\cdots\gamma_{n-1}, \quad \Gamma'_n := \gamma'_0\gamma'_1\cdots\gamma'_{n-1}. \tag{9.6}$$

The submatrix $\mathbf{T}'_{\dot\nu}$ of the tridiagonal matrix $\mathbf{T}'_{\nu'}$ with the parameters α'_n, β'_{n-1}, γ'_n of BIODIR and the bidiagonal matrices $\mathbf{L}_{\dot\nu}$ and $\mathbf{U}_{\dot\nu}$ of BIOMIN are then related by

$$\mathbf{D}_{\Gamma';\dot\nu}^{-1}\,\mathbf{T}'_{\dot\nu}\,\mathbf{D}_{\Gamma';\dot\nu} = \mathbf{D}_{\Gamma;\dot\nu}^{-1}\,\mathbf{U}\!\!\mid_{\dot\nu}\underline{\mathbf{L}}_{\dot\nu}\,\mathbf{D}_{\Gamma;\dot\nu}, \tag{9.7}$$

where

$$\mathbf{D}_{\Gamma;n} := \mathrm{diag}(1, \Gamma_1, \ldots, \Gamma_{n-1}), \quad \mathbf{D}_{\Gamma';n} := \mathrm{diag}(1, \Gamma'_1, \ldots, \Gamma'_{n-1}). \tag{9.8}$$

The step size ω'_n in BIODIR can be expressed as

$$\omega'_n = \omega_n\Gamma'_n/\Gamma_n = -\Gamma'_n/\Gamma_{n+1}, \quad n = 0, 1, \ldots, \dot\nu - 1.$$

Proof. It is clear that up to a possible earlier breakdown of BIOMIN the formulae (9.1a)–(9.1h) of the BIC algorithm produce the same biconjugate sequences $\{\mathbf{v}_n\}_{n=0}^{\dot\nu}$ and $\{\widetilde{\mathbf{v}}_n\}_{n=0}^{\dot\nu}$ as BIOMIN if the $(n+1, n)$-element γ'_n of the matrix $\mathbf{T}'_{\dot\nu}$ satisfying (7.10) is chosen appropriately, namely, so that $\mathbf{T}'_{\dot\nu}$ and $\mathbf{T}_{\dot\nu}$ are related by (7.7). (In fact, according to Theorem 2.1, applied with $\mathbf{B} := \mathbf{BA}$, these sequences are uniquely determined by the biconjugacy condition and the scaling.) Since $\mathbf{T}_{\dot\nu}$ and $\mathbf{T}'_{\dot\nu}$ then have the same subdiagonal elements, we need $\gamma'_n = \gamma_n$ for identical sequences. Choosing γ'_n differently just rescales the two vector sequences. From (7.1c), (7.1d), (7.1h), and (7.1i), it follows easily that (9.5) and (9.6) hold, which, in view of (7.7) and (7.10), lead readily to (9.7).

In the formulae (9.3a)–(9.3c), a scale factor for $\mathbf{v}_n$ inherited by δ'_n yields the inverse factor for ω'_n, which cancels when $\mathbf{y}_{n+1}$ and $\mathbf{x}_{n+1}$ are evaluated. Moreover, a scale factor for $\widetilde{\mathbf{v}}_n$ has no effect. Hence, to prove that the latter two vectors are the same as in consistent BIOMIN it suffices to verify this for a fixed scaling, say for the one induced by $\gamma'_n := \gamma_n = -\varphi_n$, when $\mathbf{v}_n$ and $\widetilde{\mathbf{v}}_n$ are the same as in BIOMIN. For the induction proof we assume that the pair $\mathbf{y}_n$, $\mathbf{x}_n$ is the same as in BIOMIN. Comparing (9.3b) and (9.3c) with (8.3b) and (8.3d) we see that the pair $\mathbf{y}_{n+1}$, $\mathbf{x}_{n+1}$ is again the same as in BIOMIN if and only if $\omega'_n = \omega_n$. By (7.2), (7.4), and (8.3h), $\langle \widetilde{\mathbf{y}}_n, \mathbf{y}_n \rangle_{\mathbf{B}} = \langle \widetilde{\mathbf{v}}_n, \mathbf{y}_n \rangle_{\mathbf{B}}$. Furthermore, if $\delta'_n \neq 0$, then by (7.2), (8.3a), and (9.3a) we can conclude that indeed

$$\omega_n = \frac{\delta_n}{\delta'_n} = \frac{\langle \widetilde{\mathbf{y}}_n, \mathbf{y}_n \rangle_{\mathbf{B}}}{\delta'_n} = \frac{\langle \widetilde{\mathbf{v}}_n, \mathbf{y}_n \rangle_{\mathbf{B}}}{\delta'_n} = \omega'_n .$$

This line also exhibits that $\omega'_n \neq 0, n = 0, \ldots, \dot{\nu} - 1$. Since $\langle \widetilde{\mathbf{y}}_n, \mathbf{y}_n \rangle_{\mathbf{B}} = \langle \widetilde{\mathbf{v}}_n, \mathbf{y}_n \rangle_{\mathbf{B}}$ is still valid for $n = \dot{\nu}$, the equivalence of (i')–(iii') holds, and since $\delta_{\dot{\nu}} = 0$ is the common breakdown condition of consistent and inconsistent BIORES, (iv) and (v') are equivalent too.

In view of (8.2) and (8.3a) the condition $\varphi_{\dot{\nu}} = 0$ (which by definition of $\dot{\nu}$ implies $\delta_{\dot{\nu}} \neq 0$; see (8.4a)–(8.4d)) is clearly equivalent to $\delta'_{\dot{\nu}} = 0$. Furthermore, since the consistent versions of BIOMIN and BIORES are related by $\varphi_n = -\gamma_n = \alpha_n + \beta_{n-1}$, the equivalence of (i")–(iii") and (v") follows. The fact that, in inconsistent BIORES, $\dot{\pi}_n = 0$ signals a pivot breakdown will follow in Section 12 (Theorem 12.1), where we will prove that $\dot{\pi}_n$ is the value at 0 of the nth Lanczos polynomial. It is also indicated by the infinity of the nth approximant $\mathbf{x}_n/\dot{\pi}_n$ of the solution $\mathbf{x}_{\text{ex}}$. Finally, since all conditions except (i') for BIODIR and (iv") for inconsistent BIORES imply a breakdown or the termination of the respective algorithm, and since no other types of breakdown exist, it follows that $\dot{\nu} \leq \nu'$ and $\dot{\nu} \leq \nu$, and that always at least one equality sign holds. □

What more can we say about the case where $\dot{\nu} < \nu'$, that is, $\omega'_n = 0$ for some $n < \nu'$? As we have seen, BIODIR stalls but does not break down. The sequences $\{\mathbf{v}_m\}_{m=0}^{\nu'-1}$ and $\{\widetilde{\mathbf{v}}_m\}_{m=0}^{\nu'-1}$ are still biconjugate (since they are generated by the BIC algorithm), and $\mathbf{y}_m$ is still the residual of $\mathbf{x}_m$. Using the connection to Padé approximation one can easily show that in this situation $\omega'_n = 0$ can only occur for isolated values of n. The proof was given in Gutknecht (1990, p. 30); the facts it is based on, namely the Lanczos–Padé connection and the block structure theorem, were also presented in Gutknecht (1994*b*). This has the following immediate consequence.

Theorem 9.3 If in the BIODIR algorithm $\omega'_n = 0$ for some n, then $\omega'_{n-1} \neq 0$ (if $n > 0$) and $\omega'_{n+1} \neq 0$ (if $n < \nu' - 1$). The sequence $\{\mathbf{y}_n\}_{n=0}^{\nu'-1}$ of residuals generated by BIODIR satisfies

$$\widetilde{\mathcal{K}}_{n+1} \perp_{\mathbf{B}} \mathbf{y}_{n+1} \quad \text{if} \quad \omega'_n \neq 0,$$

$$\widetilde{\mathcal{K}}_n \perp_{\mathbf{B}} \mathbf{y}_{n+1} = \mathbf{y}_n \quad \text{if} \quad \omega'_n = 0.$$

Table 1. *Matrix relations that describe the recurrences of the* BIO, BIOC, *and* BIC *algorithms. Only those for the right Lanczos and direction vectors are shown. Additionally, on the left, the biorthogonality and biconjugacy conditions are shown, while, on the right, the relations between the matrices of the recurrence coefficients are listed. The lower relation,* $\underline{\mathbf{T}}_n' = \mathbf{U}_{n+1}\underline{\mathbf{L}}_n$, *assumes that* $\gamma_k' = \gamma_k$ *(for all* k*)*

Algorithm	Biorthogonality	Recurrences	Relationships
1 BIO	$\widetilde{\mathbf{Y}}_n^\star \mathbf{B}\mathbf{Y}_n = \mathbf{D}_{\delta;n}$	$\mathbf{A}\mathbf{Y}_n = \mathbf{Y}_{n+1}\underline{\mathbf{T}}_n$	$\underline{\mathbf{T}}_n = \underline{\mathbf{L}}_n\mathbf{U}_n$
6 BIOC		$\mathbf{Y}_n = \mathbf{V}_n\mathbf{U}_n \quad \mathbf{A}\mathbf{V}_n = \mathbf{Y}_{n+1}\underline{\mathbf{L}}_n$	
9 BIC	$\widetilde{\mathbf{V}}_n^\star \mathbf{B}\mathbf{A}\mathbf{V}_n = \mathbf{D}_{\delta';n}$	$\mathbf{A}\mathbf{V}_n = \mathbf{V}_{n+1}\underline{\mathbf{T}}_n'$	$\underline{\mathbf{T}}_n' = \mathbf{U}_{n+1}\underline{\mathbf{L}}_n$

9.4. An overview of BICG *algorithms*

In Table 1 we first summarize the principle matrix relations of the BIO, BIOC, and BIC algorithms. To get an overview of the five BICG algorithms that we have discussed, we list in Table 2 the various vectors and the corresponding polynomials that come up (except for the iterates $\mathbf{x}_n$, which appear everywhere, of course). Those vectors that are listed in several algorithms are, up to scaling, the same. The table does not give full information about the memory requirement, however, as sometimes previously computed vectors or results of matrix-vector products have to be stored. In the last two columns of the table it is indicated if a Lanczos breakdown ('L') or a pivot breakdown ('P') can occur in the respective algorithm.

In Table 3 we compile the names of the coefficients and the corresponding matrices that belong to each method. However, we do not include those with tildes, as they are closely related to those without tildes.

10. Alternative ways to apply the QMR approach to BICG

As an alternative to the BICG method that is based on a Petrov–Galerkin condition, we discussed in Section 5 the QMR approach for solving a non-Hermitian linear system of equations. Starting from the representations

$$\mathbf{x}_n = \mathbf{x}_0 + \mathbf{Y}_n\mathbf{k}_n, \quad \mathbf{r}_n = \mathbf{r}_0 - \mathbf{A}\mathbf{Y}_n\mathbf{k}_n, \tag{10.1}$$

for the nth approximate solution and its residual, we saw by inserting $\mathbf{A}\mathbf{Y}_n = \mathbf{Y}_{n+1}\underline{\mathbf{T}}_n$ and $\mathbf{r}_0 = \mathbf{y}_0\rho_0 = \mathbf{Y}_{n+1}\underline{\mathbf{e}}_1\rho_0$ that

$$\mathbf{r}_n = \mathbf{Y}_{n+1}\mathbf{q}_n, \quad \mathbf{q}_n := \underline{\mathbf{e}}_1\rho_0 - \underline{\mathbf{T}}_n\mathbf{k}_n. \tag{10.2}$$

Table 2. *Krylov space vectors and corresponding polynomials that appear in our five forms of the* BiCG *method. Different scaling is mirrored by upper indices in the notation for the polynomials, but not in the one for the vectors. The complex conjugate polynomials, which are associated with left vectors, can be multiplied by yet another scale factor, in general. In the columns 'L' and 'P' we indicate if an algorithm is susceptible to Lanczos breakdown or pivot breakdown, respectively*

Algorithm	Vectors	Polynomials	L	P
2 BiORes (consistent)	$\mathbf{y}_n$, $\widetilde{\mathbf{y}}_n$	p_n, $\overline{p_n}$	√	√
3 BiORes (inconsistent)	$\mathbf{y}_n$, $\widetilde{\mathbf{y}}_n$	p_n^{INC}, $\overline{p_n^{\mathrm{INC}}}$	√	
7 BiOMin (standard BiCG)	$\mathbf{y}_n$,$\widetilde{\mathbf{y}}_n$,$\mathbf{v}_n$,$\widetilde{\mathbf{v}}_n$	p_n, $\overline{p_n}$, $\widehat{p}_n$, $\overline{\widehat{p}_n}$	√	√
8 BiOMin (inconsistent)	$\mathbf{y}_n$, $\widetilde{\mathbf{y}}_n$, $\mathbf{v}_n$, $\widetilde{\mathbf{v}}_n$	p_n^{INC}, $\overline{p_n^{\mathrm{INC}}}$, $\widehat{p}_n^{\mathrm{INC}}$, $\overline{\widehat{p}_n^{\mathrm{INC}}}$	√	√
10 BiODir	$\mathbf{y}_n$, $\mathbf{v}_n$, $\widetilde{\mathbf{v}}_n$	p_n, $\widehat{p}_n^{\mathrm{DIR}}$, $\overline{\widehat{p}_n^{\mathrm{DIR}}}$		√

Table 3. *Coefficients, inner products, and corresponding matrices that appear in our five forms of the* BiCG *method*

Algorithm	Coefficients	Inner products	Matrices
2 BiORes (consistent)	α_n, β_{n-1}, γ_n	δ_{n+1}	$\underline{\mathbf{T}}_n$, $\mathbf{D}_{\delta;n}$
3 BiORes (inconsistent)	α_n, β_{n-1}, γ_n, $\dot{\pi}_{n+1}$	δ_n	$\mathbf{T}_n$, $\mathbf{D}_{\delta;n}$
7 BiOMin (stand. BiCG)	ω_n, ψ_n	δ_n, δ'_n	$\underline{\mathbf{L}}_n$, $\mathbf{U}_n$, $\mathbf{D}_{\delta;n}$, $\mathbf{D}_{\delta';n}$
8 BiOMin (inconsistent)	ω_n, ψ_n, γ_n,	δ_n, δ'_n	$\underline{\mathbf{L}}_n$, $\mathbf{U}_n$, $\mathbf{D}_{\delta;n}$, $\mathbf{D}_{\delta';n}$
10 BiODir	α'_n, β'_n, γ'_n, ω'_n	δ'_{n+1}	$\mathbf{T}'_n$, $\underline{\mathbf{L}}_n$, $\mathbf{D}_{\delta';n}$

Recall that here the columns of $\mathbf{Y}_{n+1}$ are expected to be normalized. The QMR method then minimizes the quasi-residual $\mathbf{q}_n$ (instead of the residual), and this amounts to solving the least squares problem

$$\|\underline{\mathbf{e}}_1\rho_0 - \underline{\mathbf{T}}_n\mathbf{k}_n\|^2 = \min! \tag{10.3}$$

with the $(n+1) \times n$ tridiagonal matrix $\underline{\mathbf{T}}_n$ (which will have some additional fill-in in the upper triangle if look-ahead is needed).

Instead of using the BiO algorithm, can we also apply the BiOC or even the BiC algorithm to find the QMR iterates? Replacing (10.1) by

$$\mathbf{x}_n = \mathbf{x}_0 + \mathbf{V}_n\widehat{\mathbf{k}}_n, \quad \mathbf{r}_n = \mathbf{r}_0 - \mathbf{A}\mathbf{V}_n\widehat{\mathbf{k}}_n,$$

and now substituting $\mathbf{AV}_n = \mathbf{Y}_{n+1}\underline{\mathbf{L}}_n$, we find

$$\mathbf{r}_n = \mathbf{Y}_{n+1}\mathbf{q}_n, \quad \mathbf{q}_n := \underline{\mathbf{e}}_1\rho_0 - \underline{\mathbf{L}}_n\widehat{\mathbf{k}}_n.$$

Since $\mathbf{r}_n$ is again written as a linear combination of the columns of $\mathbf{Y}_{n+1}$, the coefficient vector $\mathbf{q}_n$ is the same as in (10.2). The least squares problem

$$||\underline{\mathbf{e}}_1\rho_0 - \underline{\mathbf{L}}_n\widehat{\mathbf{k}}_n||^2 = \min!$$

that has now to be solved involves a lower bidiagonal matrix, but it is equivalent to the one of (10.3): it could have been obtained from the latter by inserting $\underline{\mathbf{T}}_n = \underline{\mathbf{L}}_n\mathbf{U}_n$ and $\mathbf{U}_n\mathbf{k}_n = \widehat{\mathbf{k}}_n$. Hence, all we have done is a linear change of variables: while the coefficient vector $\widehat{\mathbf{k}}_n$ is different from $\mathbf{k}_n$, the approximate solution $\mathbf{x}_n$ is the same as before. (Of course, exact arithmetic is assumed here.) We suggest calling this form of the QMR method BIOCQMR for distinction. A complete description of it and its implementation, including a version of the BIOC algorithm with look-ahead, was given by Freund and Nachtigal (1994, 1993).

We can also construct a BICQMR form of the QMR approach. Writing

$$\mathbf{x}_n = \mathbf{x}_0 + \mathbf{V}_n\mathbf{k}'_n, \quad \mathbf{r}_n = \mathbf{r}_0 - \mathbf{AV}_n\mathbf{k}'_n,$$

and inserting $\mathbf{AV}_n = \mathbf{V}_{n+1}\underline{\mathbf{T}}'_n$ and $\mathbf{r}_0 = \mathbf{y}_0\rho_0 = \mathbf{v}_0\rho_0 = \mathbf{V}_{n+1}\underline{\mathbf{e}}_1\rho_0$, we obtain

$$\mathbf{r}_n = \mathbf{V}_{n+1}\mathbf{q}'_n, \quad \mathbf{q}'_n := \underline{\mathbf{e}}_1\rho_0 - \underline{\mathbf{T}}'_n\mathbf{k}'_n.$$

However, the resulting least squares problem

$$||\underline{\mathbf{e}}_1\rho_0 - \underline{\mathbf{T}}'_n\mathbf{k}'_n||^2 = \min!$$

is no longer equivalent to the two previous ones, since $\mathbf{r}_n$ is here represented in a different basis, the columns of $\mathbf{V}_{n+1}$. One must expect that in typical examples these columns are even further away from orthogonal than those of $\mathbf{Y}_{n+1}$ (which are orthogonal in the symmetric case), and that therefore the norm of the true residual is larger than in BIOQMR and BIOCQMR.

11. Preconditioning

By suitable preconditioning the convergence of iterative linear equation solvers is often improved dramatically. In fact, in practice large linear systems of equations resulting from the discretization of partial differential equations are often so hard to treat that iterative methods do not converge at all unless the system is preconditioned, which means that the coefficient matrix is – implicitly or explicitly – replaced by another one with better convergence properties. However, the more effective preconditioners are, the more costly they tend to be, both regarding their one-time computation and the additional cost of evaluation (for instance matrix multiplication) per step.

In general, preconditioning can be viewed as replacing the given system, say,

$$\widehat{\mathbf{A}}\widehat{\mathbf{x}} = \widehat{\mathbf{b}} \quad \text{with initial approximation } \widehat{\mathbf{x}}_0, \tag{11.1}$$

by an equivalent system,

$$\mathbf{A}\mathbf{x} = \mathbf{b} \quad \text{with initial approximation } \mathbf{x}_0,$$

where either

$$\begin{aligned} \mathbf{A} &:= \mathbf{C}_L\widehat{\mathbf{A}}\mathbf{C}_R, & \mathbf{b} &:= \mathbf{C}_L\widehat{\mathbf{b}}, \\ \mathbf{x}_0 &:= \mathbf{C}_R^{-1}\widehat{\mathbf{x}}_0, & \widehat{\mathbf{x}} &:= \mathbf{C}_R\mathbf{x}, \end{aligned} \tag{11.2}$$

or

$$\begin{aligned} \mathbf{A} &:= \mathbf{C}_L\widehat{\mathbf{A}}\mathbf{C}_R, & \mathbf{b} &:= \mathbf{C}_L(\widehat{\mathbf{b}} - \widehat{\mathbf{A}}\widehat{\mathbf{x}}_0), \\ \mathbf{x}_0 &:= \mathbf{o}, & \widehat{\mathbf{x}} &:= \widehat{\mathbf{x}}_0 + \mathbf{C}_R\mathbf{x}. \end{aligned} \tag{11.3}$$

The second version combines the preconditioning with a shift of the origin at the beginning of the iteration, as suggested by Freund and Nachtigal (1991). Note that the same shift of origin is also applied in the general concept of iterative refinement, and in Section 18 we will suggest also applying it at later stages.

We call $\mathbf{C}_L$ and $\mathbf{C}_R$ the *left* and the *right preconditioner* (as, for instance, in Ashby et al. (1990)). Many other authors (for instance, Golub and van Loan (1989), Saad (1996), Barrett, Berry, Chan, Demmel, Donato, Dongarra, Eijkhout, Pozo, Romine and van der Vorst (1994)) refer to $\mathbf{M}_L := \mathbf{C}_L^{-1}$ and $\mathbf{M}_R := \mathbf{C}_R^{-1}$ as left and right preconditioners. In fact, some preconditioning techniques, such as the various forms of incomplete LU factorizations (Meijerink and van der Vorst 1977), primarily generate $\mathbf{M}_L$ and $\mathbf{M}_R$, and then evaluate $\widetilde{\mathbf{y}} = \mathbf{A}\mathbf{y}$ by solving the two linear systems $\mathbf{M}_R\mathbf{t} = \mathbf{y}$ and $\mathbf{M}_L\widetilde{\mathbf{y}} = \widehat{\mathbf{A}}\mathbf{t}$. Other preconditioning techniques emerge directly as procedures for computing $\mathbf{t} = \mathbf{C}_R\mathbf{y}$ and $\widetilde{\mathbf{y}} = \mathbf{C}_L\widehat{\mathbf{A}}\mathbf{t}$. Finally, many algorithms can be reformulated so that the left and the right preconditioner can be combined into one matrix multiplication by the product $\mathbf{C}_R\mathbf{C}_L$ or one linear system with matrix $\mathbf{M}_L\mathbf{M}_R$ to solve per step; see, for instance, Saad (1996).

Often, only either a left or a right preconditioner is applied, that is, either $\mathbf{C}_R := \mathbf{I}$ or $\mathbf{C}_L := \mathbf{I}$. In the first case, $\mathbf{x} = \widehat{\mathbf{x}}$, so that the errors of the preconditioned system are the same as those of the original system. In the second case, the residuals remain unchanged. In the general situation, if we set $\mathbf{x}_{\text{ex}} = \mathbf{A}^{-1}\mathbf{b}$ and $\widehat{\mathbf{x}}_{\text{ex}} = \widehat{\mathbf{A}}^{-1}\widehat{\mathbf{b}}$, both (11.2) and (11.3) imply the relations

$$\mathbf{r}_n = \mathbf{C}_L\widehat{\mathbf{r}}_n, \quad \mathbf{x}_{\text{ex}} - \mathbf{x}_n = \mathbf{C}_R^{-1}(\widehat{\mathbf{x}}_{\text{ex}} - \widehat{\mathbf{x}}_n)$$

between the *preconditioned residuals* $\mathbf{r}_n := \mathbf{b} - \mathbf{A}\mathbf{x}_n$ and the *preconditioned errors* $\mathbf{x}_{\text{ex}} - \mathbf{x}_n$ on the one hand and the residuals $\widehat{\mathbf{r}}_n := \widehat{\mathbf{b}} - \widehat{\mathbf{A}}\widehat{\mathbf{x}}_n$ and errors $\widehat{\mathbf{x}}_{\text{ex}} - \widehat{\mathbf{x}}_n$ of the original system (11.1) on the other hand.

12. Lanczos and direction polynomials

Any vector $\mathbf{y}$ in the Krylov space $\mathcal{K}_{n+1} :=$ span $(\mathbf{y}_0, \mathbf{A}\mathbf{y}_0, \ldots, \mathbf{A}^n\mathbf{y}_0)$ can be written in the form $\mathbf{y} = p(\mathbf{A})\mathbf{y}_0$, where p is a polynomial of degree at most n, in which the matrix $\mathbf{A}$ is substituted for the argument. The induced mapping is an isomorphism as long as $\mathcal{K}_{n+1} \neq \mathcal{K}_n$, that is, as long as n is smaller than the grade of $\mathbf{y}_0$ with respect to $\mathbf{A}$. The sequence $\{\mathbf{y}_n\}_{n=0}^{\nu-1}$ of right Lanczos vectors generated by the BIO algorithm is in this way associated with the sequence $\{p_n\}_{n=0}^{\nu-1}$ of *Lanczos polynomials*, and from the three-term recurrence (2.19) for the former we get immediately a three-term recurrence formula for the latter. Of course, the analogue is true for the sequence $\{\widetilde{\mathbf{y}}_n\}_{n=0}^{\nu-1}$ of left Lanczos vectors, the operator $\mathbf{A}^\star$, and the corresponding nested sequence of Krylov spaces $\widetilde{\mathcal{K}}_{n+1} :=$ span $(\widetilde{\mathbf{y}}_0, \mathbf{A}^\star\widetilde{\mathbf{y}}_0, \ldots, (\mathbf{A}^\star)^n\widetilde{\mathbf{y}}_0)$. But from the recurrences (2.19) and the relations (2.21b) and (2.21d) we see that the coefficients of this second set of polynomials, $\{\widetilde{p}_n\}_{n=0}^{\nu-1}$, are just complex conjugate to those of the Lanczos polynomials if we choose $\widetilde{\gamma}_n = \overline{\gamma_n}$. (Otherwise, in view of (2.28), $\widetilde{p}_n$ would be a scalar multiple of $\overline{p_n}$; for simplicity, we assume $\widetilde{\gamma}_n = \overline{\gamma_n}$ in this section.) Similarly, from the BIOC formulae (7.1a)–(7.1j) it is seen that the vectors $\mathbf{v}_n \in \mathcal{K}_{n+1}$ and $\widetilde{\mathbf{v}}_n \in \widetilde{\mathcal{K}}_{n+1}$ can then be represented by a polynomial $\widehat{p}_n$ of degree n and the one with complex conjugate coefficients, $\overline{\widehat{p}_n}$, respectively. Since $\mathbf{v}_n$ is a direction vector, we call $\widehat{p}_n$ a *direction polynomial.* The coupled two-term recurrences of the BIOC algorithm translate into coupled two-term recurrences for the two polynomial sequences $\{p_n\}$ and $\{\widehat{p}_n\}$. A three-term recurrence for the direction polynomials alone follows by eliminating p_n from the coupled recurrences. Recall that the analogous elimination brought us from the BIOC to the BIC algorithm. The recurrences remain valid up to $n = \nu$, but the correspondence between $\mathcal{P}_n$ and $\mathcal{K}_n$ may no longer be one-to-one for $n = \nu$. When taking the different indices of first breakdown or termination of the various algorithms into account, we obtain altogether the following result.

Theorem 12.1 Let $\{\mathbf{y}_n\}_{n=0}^{\nu}$ and $\{\widetilde{\mathbf{y}}_n\}_{n=0}^{\nu}$ be the biorthogonal vector sequences generated by the BIO algorithm with $\widetilde{\gamma}_n = \gamma_n$, and let $\{\mathbf{v}_n\}_{n=0}^{\nu'}$ and $\{\widetilde{\mathbf{v}}_n\}_{n=0}^{\nu'}$ be the biconjugate vector sequences generated by the BIC algorithm using $\gamma'_n = \gamma_n$ and $\widetilde{\gamma}'_n = \overline{\gamma_n}$. Then there is a pair of sequences of polynomials, $\{p_n\}_{n=0}^{\nu}$ and $\{\widehat{p}_n\}_{n=0}^{\nu'}$, such that

$$\mathbf{y}_n = p_n(\mathbf{A})\mathbf{y}_0, \quad \widetilde{\mathbf{y}}_n = \overline{p_n}(\mathbf{A}^\star)\widetilde{\mathbf{y}}_0, \quad n = 0, 1, \ldots, \nu,$$

and

$$\mathbf{v}_n = \widehat{p}_n(\mathbf{A})\mathbf{v}_0, \quad \widetilde{\mathbf{v}}_n = \overline{\widehat{p}_n}(\mathbf{A}^\star)\widetilde{\mathbf{v}}_0, \quad n = 0, 1, \ldots, \nu'.$$

For $n < \dot{\nu} = \min\{\nu, \nu'\}$ these polynomial sequences satisfy the coupled two-term recurrences

$$\begin{aligned} p_0(\zeta) &:\equiv 1, \\ \widehat{p}_0(\zeta) &:\equiv 1, \\ p_{n+1}(\zeta) &:= [\zeta\widehat{p}_n(\zeta) - \varphi_n p_n(\zeta)] / \gamma_n, \\ \widehat{p}_{n+1}(\zeta) &:= p_{n+1}(\zeta) - \psi_n \widehat{p}_n(\zeta), \quad n = 0, 1, \ldots, \dot{\nu} - 1. \end{aligned} \tag{12.1}$$

They also satisfy individual three-term recurrences, namely

$$\begin{aligned} p_0(\zeta) &:\equiv 1, \\ p_1(\zeta) &:= (\zeta - \alpha_0)p_0(\zeta)/\gamma_0, \\ p_{n+1}(\zeta) &:= [(\zeta - \alpha_n)p_n(\zeta) - \beta_{n-1}p_{n-1}(\zeta)]/\gamma_n, \quad n = 1, \ldots, \nu - 1, \end{aligned} \tag{12.2}$$

and

$$\begin{aligned} \widehat{p}_0(\zeta) &:\equiv 1, \\ \widehat{p}_1(\zeta) &:= (\zeta - \alpha'_0)\widehat{p}_0(\zeta)/\gamma_0, \\ \widehat{p}_{n+1}(\zeta) &:= [(\zeta - \alpha'_n)\widehat{p}_n(\zeta) - \beta'_{n-1}\widehat{p}_{n-1}(\zeta)]/\gamma_n, \quad n = 1, \ldots, \nu' - 1, \end{aligned} \tag{12.3}$$

respectively. Both p_n and $\widehat{p}_n$ have exact degree n, and both have the leading coefficient Γ_n^{-1}, where

$$\Gamma_n := \gamma_0\gamma_1 \cdots \gamma_{n-1}. \tag{12.4}$$

If $\gamma_n = -\varphi_n$ (for all n) as in (8.1) or, equivalently, if $\gamma_n = -\alpha_n - \beta_{n-1}$ (for all n) as in (4.12), then

$$p_n(0) = 1, \quad n = 0, 1, \ldots, \nu. \tag{12.5}$$

Otherwise, the values $\dot{\pi}_n := p_n(0)$ can be computed recursively according to (4.16) or (8.5).

Proof. The recurrences (4.16) and (8.5) follow from (12.2) and (12.1), respectively, by inserting $\zeta = 0$. It remains to verify the formulae (12.4) for the leading coefficients and (12.5) on the normalization at $\zeta = 0$. Both follow by induction from the recurrences. □

Since the Lanczos vector $\mathbf{y}_n$ is the nth residual of the three consistent versions of the BiCG method that we discussed in Sections 4, 8 and 9, p_n is for each of these algorithms the so-called *residual polynomial*. Property (12.5) is the standard consistency condition for residual polynomials of Krylov space solvers.

For the general Krylov space solver that we briefly considered in Section 4, the recurrence relations (4.10) for the residuals also imply recurrence relations for the residual polynomials, namely, in shorthand notation,

$$\zeta \,[\, p_0 \ \cdots \ p_{n-1} \,] = [\, p_0 \ \cdots \ p_{n-1} \ \ p_n \,]\, \underline{\mathbf{H}}_n. \tag{12.6}$$

From this formula it is easy to see that the condition $p_k(0) = 1$ $(k \le n)$ is equivalent to the zero column sum condition (4.11) for $\underline{\mathbf{H}}_n$ and the choice of $p_0(\zeta) \equiv 1$ as the constant polynomial. In analogy to (12.6) the recurrences (12.2) and (12.3) can be written as

$$\zeta\,[\, p_0 \;\; \cdots \;\; p_{n-1} \,] = [\, p_0 \;\; \cdots \;\; p_{n-1} \;\; p_n \,]\,\underline{\mathbf{T}}_n,$$
$$\zeta\,[\, \widehat{p}_0 \;\; \cdots \;\; \widehat{p}_{n-1} \,] = [\, \widehat{p}_0 \;\; \cdots \;\; \widehat{p}_{n-1} \;\; \widehat{p}_n \,]\,\underline{\mathbf{T}}'_n.$$

Let us now define on the linear space $\mathcal{P}$ of all polynomials two linear functionals Φ and Φ'. They are specified by the values they take on the monomial basis:

$$\Phi(\zeta^k) := \mu_k, \quad \Phi'(\zeta^k) := \mu_{k+1} \quad (k \in \mathbb{N}), \tag{12.7}$$

where

$$\mu_k := \langle \widetilde{\mathbf{y}}_0, \mathbf{A}^k \mathbf{y}_0 \rangle_{\mathbf{B}} = \widetilde{\mathbf{y}}_0^\star \mathbf{B} \mathbf{A}^k \mathbf{y}_0. \tag{12.8}$$

Here μ_k is called the kth *moment* associated with $\mathbf{A}$, $\mathbf{y}_0$, and $\widetilde{\mathbf{y}}_0$. In engineering, the set of moments is referred to as *impulse response* or as the set of *Markov parameters* of a system; in the older mathematical literature they are sometimes called the *Schwarz constants* (Rutishauser 1957).

Note that for arbitrary polynomials s and t and corresponding Krylov space vectors $s(\mathbf{A})\mathbf{y}_0$, $t(\mathbf{A}^\star)\widetilde{\mathbf{y}}_0$ we have

$$\begin{aligned} \langle t(\mathbf{A}^\star)\widetilde{\mathbf{y}}_0, s(\mathbf{A})\mathbf{y}_0 \rangle_{\mathbf{B}} &= \Phi(ts), \\ \langle t(\mathbf{A}^\star)\widetilde{\mathbf{y}}_0, \mathbf{A}s(\mathbf{A})\mathbf{y}_0 \rangle_{\mathbf{B}} &= \Phi'(ts) = \Phi(\zeta ts). \end{aligned} \tag{12.9}$$

When we represent s and t by their coefficients,

$$s(\zeta) =: \sum \sigma_k \zeta^k, \quad t(\zeta) =: \sum \tau_k \zeta^k, \tag{12.10}$$

and introduce the infinite coefficient vectors (extended with zeros),

$$\mathbf{s} := [\, \sigma_0 \;\; \sigma_2 \;\; \sigma_3 \;\; \cdots \,]^\top, \quad \mathbf{t} := [\, \tau_0 \;\; \tau_2 \;\; \tau_3 \;\; \cdots \,]^\top \tag{12.11}$$

as well as the infinite *moment matrices*

$$\mathbf{M} := \begin{bmatrix} \mu_0 & \mu_1 & \mu_2 & \cdots \\ \mu_1 & \mu_2 & & \\ \mu_2 & & & \\ \vdots & & & \end{bmatrix}, \quad \mathbf{M}' := \begin{bmatrix} \mu_1 & \mu_2 & \mu_3 & \cdots \\ \mu_2 & \mu_3 & & \\ \mu_3 & & & \\ \vdots & & & \end{bmatrix}$$

associated with the functionals Φ and Φ', we see from

$$\Phi(ts) = \sum_k \sum_l \tau_k \sigma_k \Phi(\zeta^{k+l})$$

and from the analogous formula for Φ' that

$$\Phi(ts) = \mathbf{t}^\top \mathbf{M} \mathbf{s}, \quad \Phi'(ts) = \mathbf{t}^\top \mathbf{M}' \mathbf{s}. \tag{12.12}$$

The matrices $\mathbf{M}$ and $\mathbf{M}'$ have Hankel structure: the (k,l)-element only depends on $k+l$. This is a consequence of the fact that the product ζ^{k+l} of ζ^k and ζ^l only depends on $k+l$.

What can we say about the relationship between the functional Φ and the original data of the problem: the matrices $\mathbf{A}$ and $\mathbf{B}$ and the initial vectors $\mathbf{y}_0$ and $\widetilde{\mathbf{y}}_0$? (Recall that we are mainly interested in the case $\mathbf{B} = \mathbf{I}$, where we can forget $\mathbf{B}$.)

To explore this relationship, let us for a few lines assume that the matrices $\mathbf{A}$ and $\mathbf{B}$ are diagonalizable. Then, since they commute by assumption, they have a common complete system of eigenvectors (Wilkinson 1965, p. 52): there is a nonsingular $N \times N$ matrix $\mathbf{W}$ of eigenvectors such that

$$\mathbf{AW} = \mathbf{WD}_\lambda, \quad \mathbf{BW} = \mathbf{WD}_\kappa$$

where $\mathbf{D}_\lambda$ and $\mathbf{D}_\kappa$ are diagonal matrices containing the eigenvalues $\lambda_1, \ldots, \lambda_N$ of $\mathbf{A}$ and the eigenvalues $\kappa_1, \ldots, \kappa_N$ of $\mathbf{B}$, respectively. From $\mathbf{A}^\star\mathbf{W}^{-\star} = \mathbf{W}^{-\star}\mathbf{D}_\lambda^\star$ it is clear that the columns of $\mathbf{W}^{-\star}$ are a set of eigenvectors of $\mathbf{A}^\star$. We let

$$\mathbf{W} =: [\mathbf{w}_1, \mathbf{w}_2, \ldots, \mathbf{w}_N], \quad \mathbf{W}^{-\star} =: [\widetilde{\mathbf{w}}_1, \widetilde{\mathbf{w}}_2, \ldots, \widetilde{\mathbf{w}}_N],$$

and represent $\mathbf{y}_0$ in the basis $\{\mathbf{w}_k\}$, $\widetilde{\mathbf{y}}_0$ in the basis $\{\widetilde{\mathbf{w}}_k\}$:

$$\mathbf{y}_0 =: \sum_{j=1}^N \mathbf{w}_j\eta_j = \mathbf{W}\begin{bmatrix}\eta_1\\ \vdots \\ \eta_N\end{bmatrix}, \quad \widetilde{\mathbf{y}}_0 =: \sum_{j=1}^N \widetilde{\mathbf{w}}_j\widetilde{\eta}_j = \mathbf{W}^{-\star}\begin{bmatrix}\widetilde{\eta}_1\\ \vdots \\ \widetilde{\eta}_N\end{bmatrix}.$$

Then

$$\mu_k := \langle \widetilde{\mathbf{y}}_0, \mathbf{A}^k\mathbf{y}_0\rangle_\mathbf{B} = [\ \overline{\widetilde{\eta}_1} \ \ \ldots \ \ \overline{\widetilde{\eta}_N}\ ]\mathbf{D}_\kappa\mathbf{D}_\lambda^k\begin{bmatrix}\eta_1\\ \vdots \\ \eta_N\end{bmatrix} = \sum_{j=1}^N \lambda_j^k\kappa_j\eta_j\overline{\widetilde{\eta}_j}.$$

Therefore, μ_k can formally be written as the kth moment of a *discrete measure* $d\mu(\lambda)$ with masses $\kappa_j\eta_j\overline{\widetilde{\eta}_j}$ at the points λ_j:

$$\mu_k = \int \lambda^k d\mu(\lambda), \quad \text{where } d\mu(\lambda) := \sum_{j=1}^N \kappa_j\eta_j\overline{\widetilde{\eta}_j}\,\delta(\lambda - \lambda_j)d\lambda.$$

(In the last formula, δ is the Dirac function. Of course, if some of the eigenvalues coincide, the corresponding masses have to be added.) More generally,

$$\Phi(s) = \langle \widetilde{\mathbf{y}}_0, s(\mathbf{A})\mathbf{y}_0\rangle_\mathbf{B} = \int s(\lambda)d\mu(\lambda).$$

In the symmetric case ($\mathbf{A} = \mathbf{A}^\star$, $\widetilde{\mathbf{y}}_0 = \mathbf{y}_0$), where $\lambda_j \in \mathbb{R}$ and $\widetilde{\eta}_j = \eta_j$, $d\mu(\lambda)$ is indeed a discrete positive measure whose support consists of the eigenvalues λ_j that are represented in $\mathbf{y}_0$ (*i.e.*, for which $\eta_j \neq 0$).

By the induced mapping from the Krylov space to polynomials, the biorthogonality (2.2) of the Lanczos vectors and the biconjugacy (7.3) of the sequences $\{\mathbf{v}_n\}$ and $\{\widetilde{\mathbf{v}}_n\}$ now take the form

$$\widetilde{\mathbf{y}}_0^\star \mathbf{B} p_m(\mathbf{A}) p_n(\mathbf{A}) \mathbf{y}_0 = \delta_{mn}\delta_n, \quad n = 0, 1, \ldots, \nu,$$

and

$$\widetilde{\mathbf{y}}_0^\star \mathbf{B} \widehat{p}_m(\mathbf{A}) \mathbf{A} \widehat{p}_n(\mathbf{A}) \mathbf{y}_0 = \delta_{mn}\delta'_n, \quad n = 0, 1, \ldots, \nu',$$

respectively, in view of $\mathbf{AB} = \mathbf{BA}$. With the above definitions, we can translate these conditions into

$$\Phi(p_m p_n) = \delta_{mn}\delta_n, \quad m, n = 0, 1, \ldots, \nu,$$

$$\Phi'(\widehat{p}_m \widehat{p}_n) = \delta_{mn}\delta'_n, \quad m, n = 0, 1, \ldots, \nu'.$$

From the linearity of Φ and Φ' one concludes further that

$$\Phi(\zeta^m p_n) = \begin{cases} 0 & \text{if } 0 \le m < n, \\ \delta_n \Gamma_n & \text{if } m = n, \end{cases} \tag{12.13}$$

and

$$\Phi'(\zeta^m \widehat{p}_n) = \begin{cases} 0 & \text{if } 0 \le m < n, \\ \delta'_n \Gamma_n & \text{if } m = n, \end{cases} \tag{12.14}$$

which is another way to characterize these polynomial sequences. At this point we need to recall the following definition; see, for instance, Gutknecht (1992).

Definition 12.1 If the polynomial p_n of exact degree n satisfies (12.13) for some $\delta_n \neq 0$ and is uniquely determined by these conditions, it is called a *regular formal orthogonal polynomial (FOP)* of the functional Φ.

In other words, (12.13) and (12.14) mean that $\{p_n\}_{n=0}^{\nu}$ is a sequence of regular FOPs of the functional Φ, and that $\{\widehat{p}_n\}_{n=0}^{\nu'}$ is such a sequence for Φ'.

If p_n and $\widehat{p}_n$ are expressed in the monomial basis,

$$p_n(\zeta) =: \sum_{k=0}^{n} \pi_k^{(n)} \zeta^k,$$

and

$$\widehat{p}_n(\zeta) =: \sum_{k=0}^{n} \widehat{\pi}_k^{(n)} \zeta^k,$$

where $\pi_n^{(n)} = \widehat{\pi}_n^{(n)} = \Gamma_n^{-1}$, then the first n equations of (12.13) and (12.14) become homogeneous $n \times (n+1)$ linear systems for the coefficients $\pi_k^{(n)}$ and $\widehat{\pi}_k^{(n)}$, respectively. Since $\pi_n^{(n)}$ and $\widehat{\pi}_n^{(n)}$ are known, we can move them on the right-hand side:

$$\mathbf{M}_n \begin{bmatrix} \pi_0^{(n)} \\ \pi_1^{(n)} \\ \vdots \\ \pi_{n-1}^{(n)} \end{bmatrix} = - \begin{bmatrix} \mu_n \\ \mu_{n+1} \\ \vdots \\ \mu_{2n-1} \end{bmatrix} \pi_n^{(n)} \text{ with } \mathbf{M}_n := \begin{bmatrix} \mu_0 & \mu_1 & \cdots & \mu_{n-1} \\ \mu_1 & \mu_2 & & \mu_n \\ \vdots & & & \vdots \\ \mu_{n-1} & \mu_n & \cdots & \mu_{2n-2} \end{bmatrix}, \tag{12.15}$$

$$\mathbf{M}'_n \begin{bmatrix} \widehat{\pi}_0^{(n)} \\ \widehat{\pi}_1^{(n)} \\ \vdots \\ \widehat{\pi}_{n-1}^{(n)} \end{bmatrix} = - \begin{bmatrix} \mu_{n+1} \\ \mu_{n+2} \\ \vdots \\ \mu_{2n} \end{bmatrix} \widehat{\pi}_n^{(n)} \text{ with } \mathbf{M}'_n := \begin{bmatrix} \mu_1 & \mu_2 & \cdots & \mu_n \\ \mu_2 & \mu_3 & & \mu_{n+1} \\ \vdots & & & \vdots \\ \mu_n & \mu_{n+1} & \cdots & \mu_{2n-1} \end{bmatrix}.$$

In the engineering literature, linear systems of this form are called *Yule–Walker equations.* The matrices $\mathbf{M}_n$ and $\mathbf{M}'_n$ are the nth leading principal submatrices of the infinite moment matrices $\mathbf{M}$ and $\mathbf{M}'$. Since (12.13) and (12.14) are equivalent to (2.2) and (7.3), respectively, it follows from the uniqueness statements in Theorem 2.1 and Corollary 9.1 that for prescribed leading coefficients $\pi_n^{(n)}$ and $\widehat{\pi}_n^{(n)}$ the solutions of these linear systems are uniquely determined as long as $1 \le n \le \nu$ in the first and $1 \le n \le \nu'$ in the second system. Clearly, this existence and uniqueness statement is equivalent to the nonsingularity of the matrices $\mathbf{M}_n$ and $\mathbf{M}'_n$ in the respective range of indices. For future use we state part of this result so that it also applies for $n > \nu$ and $n > \nu'$, respectively.

Lemma 12.2 A Lanczos polynomial p_n of exact degree n that is a regular FOP of Φ (that is, which satisfies the orthogonality conditions $\Phi(\zeta^m p_n) = 0$ $(0 \le m < n)$ and is up to a scalar multiple uniquely determined by them) exists if and only if $\mathbf{M}_n$ is nonsingular.

Likewise, a direction polynomial $\widehat{p}_n$ of exact degree n that is a regular FOP of Φ' (that is, which satisfies the orthogonality conditions $\Phi'(\zeta^m \widehat{p}_n) = 0$ $(0 \le m < n)$ and is up to a scalar multiple uniquely determined by them) exists if and only if $\mathbf{M}'_n$ is nonsingular.

In particular, the indices of first breakdown or termination, ν and ν', of the BIO algorithm and the BIC algorithm (Algorithms 1 and 9 in Sections 2 and 9), respectively, satisfy

$$\begin{aligned} \nu &= \min \{n : \mathbf{M}_{n+1} \text{ singular}\}, \\ \nu' &= \min \{n : \mathbf{M}'_{n+1} \text{ singular}\}. \end{aligned}$$

Proof. It remains to show that $\mathbf{M}_{\nu+1}$ and $\mathbf{M}'_{\nu'+1}$ are singular. Set $n := \nu$. First, for given $\pi_n^{(n)} \ne 0$, the linear system (12.15) uniquely determines the coefficients of a polynomial p_n that corresponds to a vector $\mathbf{y}_n \in \mathcal{K}_{n+1}$

satisfying $\widetilde{\mathcal{K}}_n \perp_{\mathbf{B}} \mathbf{y}_n$. At the same time, $\overline{p_n}$ is mapped into $\widetilde{\mathbf{y}}_n \in \widetilde{\mathcal{K}}_{n+1}$ satisfying $\widetilde{\mathbf{y}}_n \perp_{\mathbf{B}} \mathcal{K}_n$. We want to prove that, in contrast to the definition of ν, the conditions (2.1)–(2.2) could be fulfilled up to $\nu+1$ if $\mathbf{M}_{\nu+1}$ were nonsingular. It remains to show that $\langle \widetilde{\mathbf{y}}_n, \mathbf{y}_n \rangle_{\mathbf{B}} = 0$ unless $\mathbf{M}_{\nu+1}$ is nonsingular. In fact, if this inner product is 0, then the additional equation $\mu_n \pi_0^{(n)} + \cdots + \mu_{2n-1}\pi_{n-1}^{(n)} + \mu_{2n}\pi_n^{(n)} = 0$ holds, which extends system (12.15) by an extra row at the bottom. Consequently, the coefficients of p_n satisfy $\mathbf{M}_{\nu+1}[\ \pi_0^{(n)}\ \cdots\ \pi_n^{(n)}\ ]^{\mathsf{T}} = \mathbf{o}$, which implies that $\mathbf{M}_{\nu+1}$ is singular.

Of course, the singularity of $\mathbf{M}'_{\nu'+1}$ is shown the same way. □

If the consistency condition (12.5) holds, then $\pi_0^{(n)} = 1$ and we can move the first column of $\mathbf{M}_n$ to the right-hand side of (12.15), in exchange for the current right-hand side that is moved back to the left. In this way, $\mathbf{M}_n$ is replaced by $\mathbf{M}'_n$, and the system becomes

$$\mathbf{M}'_n \begin{bmatrix} \pi_1^{(n)} \\ \pi_2^{(n)} \\ \vdots \\ \pi_n^{(n)} \end{bmatrix} = - \begin{bmatrix} \mu_0 \\ \mu_1 \\ \vdots \\ \mu_{n-1} \end{bmatrix}. \tag{12.16}$$

Lemma 12.3 A residual polynomial p_n of exact degree n that satisfies the consistency condition (12.5) and the orthogonality conditions $\Phi(\zeta^m p_n) = 0$ $(0 \le m < n)$ and is uniquely determined by the latter exists if and only if $\mathbf{M}_n$ and $\mathbf{M}'_n$ are nonsingular.

In particular, if $\acute{\nu}$ denotes the index of first breakdown or termination of the BIORES and the BIOMIN algorithm (Algorithms 2 and 7 in Sections 4 and 8, then

$$\acute{\nu} = \min \{ n :\ \mathbf{M}_{n+1} \text{ or } \mathbf{M}'_{n+1} \text{ singular} \}.$$

Proof. From the previous Lemma we know that an essentially unique p_n satisfying the orthogonality conditions exists if and only if $\mathbf{M}_n$ is nonsingular. In order that p_n can be normalized by $p_n(0) = 1$, we need $\pi_n^{(n)} \neq 0$; then the normalized coefficients satisfy (12.16). If this system had more than one solution, it would have infinitely many solutions with $\pi_n^{(n)} \neq 0$. Renormalizing them, we would find infinitely many solutions of (12.15) with $\pi_n^{(n)} = 1$, in contrast to the nonsingularity of $\mathbf{M}_n$.

Conversely, if both matrices are nonsingular, then the residual polynomial with the stated properties clearly exists and is uniquely determined. □

The above derivation of Lemma 12.2 relies on the existence and uniqueness statements in Theorem 2.1 and Corollary 9.1, which describe the construction of the Krylov space vectors $\mathbf{y}_n$ and $\mathbf{v}_n$ that are the images of the polynomials p_n and $\widehat{p}_n$. There is another, more direct approach to this lemma. Since $\mathbf{M}_n$ and $\mathbf{M}'_n$ are nonsingular for $1 \le n \le \nu$ and $1 \le n \le \nu'$, respectively, the matrices $\mathbf{M}_\nu$ and $\mathbf{M}'_{\nu'}$ have LDU decompositions (without pivoting). In view of the Hankel structure, $\mathbf{M}_\nu$ and $\mathbf{M}'_{\nu'}$ are real or complex symmetric, so that these LDU decompositions are symmetric ones. It turns out that the upper triangular factors contain in their columns the coefficients of the polynomials p_m and $\widehat{p}_m$, respectively. In fact, let

$$\mathbf{P}_n := \begin{bmatrix} \pi_0^{(0)} & \pi_0^{(1)} & \cdots & \pi_0^{(n)} \\ & \pi_1^{(1)} & & \pi_1^{(n)} \\ & & \ddots & \vdots \\ & & & \pi_n^{(n)} \end{bmatrix}, \quad \widehat{\mathbf{P}}_n := \begin{bmatrix} \widehat{\pi}_0^{(0)} & \widehat{\pi}_0^{(1)} & \cdots & \widehat{\pi}_0^{(n)} \\ & \widehat{\pi}_1^{(1)} & & \widehat{\pi}_1^{(n)} \\ & & \ddots & \vdots \\ & & & \widehat{\pi}_n^{(n)} \end{bmatrix},$$

and, as before, let $\mathbf{Y}_n$ and $\widetilde{\mathbf{Y}}_n$ be the $N \times n$ matrices with columns $\mathbf{y}_m$ and $\widetilde{\mathbf{y}}_m$, $m = 0, 1, \ldots, n-1$, respectively, so that

$$\begin{aligned} \mathbf{Y}_n &= [\ \mathbf{y}_0 \quad \mathbf{A}\mathbf{y}_0 \quad \ldots \quad \mathbf{A}^{n-1}\mathbf{y}_0\ ]\,\mathbf{P}_n, \\ \widetilde{\mathbf{Y}}_n &= [\ \widetilde{\mathbf{y}}_0 \quad \mathbf{A}^\star\widetilde{\mathbf{y}}_0 \quad \ldots \quad (\mathbf{A}^\star)^{n-1}\widetilde{\mathbf{y}}_0\ ]\,\overline{\mathbf{P}_n}. \end{aligned} \tag{12.17}$$

(We still assume that $\widetilde{\gamma}_n = \overline{\gamma_n}$ for all n.) By the definition of the moments,

$$\begin{bmatrix} \widetilde{\mathbf{y}}_0^\star \\ \widetilde{\mathbf{y}}_0^\star\mathbf{A} \\ \vdots \\ \widetilde{\mathbf{y}}_0^\star\mathbf{A}^{n-1} \end{bmatrix} \mathbf{B}\,[\ \mathbf{y}_0 \quad \mathbf{A}\mathbf{y}_0 \quad \ldots \quad \mathbf{A}^{n-1}\mathbf{y}_0\ ] = \mathbf{M}_n.$$

Inserting here (12.17) and the orthogonality property $\widetilde{\mathbf{Y}}_n^\star\mathbf{B}\mathbf{Y}_n = \mathbf{D}_{\delta;n}$ (see (2.24)) yields

$$\mathbf{P}_n^{-\top}\,\mathbf{D}_{\delta;n}\,\mathbf{P}_n^{-1} = \mathbf{M}_n \quad (n \le \nu), \tag{12.18}$$

and likewise we find

$$\widehat{\mathbf{P}}_n^{-\top}\,\mathbf{D}_{\delta';n}\,\widehat{\mathbf{P}}_n^{-1} = \mathbf{M}'_n \quad (n \le \nu). \tag{12.19}$$

These are the claimed (symmetric) LDU decompositions of the nth moment submatrices $\mathbf{M}_n$ and $\mathbf{M}'_n$ if the polynomials are monic, that is, $\gamma_n = \widetilde{\gamma}_n = 1$ (for all n). Otherwise we get a nonstandard symmetric LDU decompositions with prescribed diagonal elements of the triangular factors. $\mathbf{P}_n^{-1}$ and $\widehat{\mathbf{P}}_n^{-1}$ are the upper triangular matrices that contain the coefficients of the monomials when expressed as linear combinations of the Lanczos and the direction polynomials, respectively. Of course, if the decompositions (12.18) and (12.19) exist for $n = \nu$ and $n = \nu'$, respectively, and if the diagonal matrices $\mathbf{D}_{\delta;\nu}$ and $\mathbf{D}_{\delta';n}$ are nonsingular, then these decompositions exist for all n in the range specified. On the other hand, the latter holds if and only if the leading principal submatrices $\mathbf{M}_n$ and $\mathbf{M}'_n$ are all nonsingular. Moreover, we can conclude conversely that the inverses of the triangular factors contain the coefficients of monic FOPs p_m ($m = 0, 1, \ldots, \nu - 1$) associated with Φ and of monic FOPs $\widehat{p}_m$ ($m = 0, 1, \ldots, \nu' - 1$) associated with Φ', respectively. Hence, the singularity of $\mathbf{M}_{\nu+1}$ and $\mathbf{M}'_{\nu'+1}$ as well as the other statements of Lemma 12.2 follow.

13. The Lanczos process for polynomials: the Stieltjes procedure

So far we have considered the Lanczos process as a tool for generating biorthogonal bases for a pair of Krylov spaces of the linear operators $\mathbf{A}$ and $\mathbf{A}^\star$ defined on $\mathbb{C}^N$ (or, more generally, a linear space and its dual space). In the symmetric case (that is, when $\mathbf{A}^\star = \mathbf{A}$ and $\widetilde{\mathbf{y}}_0 = \mathbf{y}_0$) the two bases coincide and are orthogonal. We can apply the same process to the multiplication operator M defined on the space $\mathcal{P}$ of all polynomials with complex coefficients:

$$M : s(\zeta) \mapsto \zeta\, s(\zeta) \quad (s \in \mathcal{P}).$$

We aim for a setting where the dual space is also $\mathcal{P}$. In the case of classical (real) orthogonal polynomials an inner product defined on $\mathcal{P} \times \mathcal{P}$ is given by some bilinear integral operator:

$$\langle t, s \rangle := \int t(\zeta)s(\zeta)d\mu(\zeta) =: \Phi(ts), \quad (t, s) \in \mathcal{P} \times \mathcal{P}. \tag{13.1}$$

Here $d\mu$ is a positive measure whose support is a subset of the real line. Note that the integral only depends on the product ts, and therefore the inner product is a linear functional Φ of this product.

For the Lanczos process applied to polynomials we want to preserve this property, but to relax the assumption on Φ. This means, however, that in the complex case our formal inner product is still based on a bilinear functional, and not on a sesquilinear[15] one. Therefore it is better to forget the integral and just let

$$\langle t, s \rangle := \Phi(ts) \tag{13.2}$$

with an arbitrary complex linear functional Φ defined on $\mathcal{P}$, which may, but need not be, defined by its moments μ_k, the values it takes on the monomials:

$$\Phi(\zeta^k) := \mu_k$$

This exhibits the connection to the Lanczos process for an operator $\mathbf{A}$, where the moments are given by (12.8). When we represent s and t by their coefficient vectors, as in (12.10)–(12.11), the inner product is still given by (12.12).

[15] A *sesquilinear* functional $\langle .,. \rangle$ on $\mathcal{P} \times \mathcal{P}$ is one for which $\langle \alpha t, \beta s \rangle = \overline{\alpha}\beta \langle t, s \rangle$.

Note that as a consequence of (13.2) we have

$$M^\star = M.$$

However, this does not mean that M is a selfadjoint operator, unless (13.2) is really an inner product, which is not true in general, even in the real case, as we do not require that $\langle s, s \rangle = \Phi(s^2) > 0$ if $s \neq 0$.

It is now an easy matter to reformulate the Lanczos process, be it in the BIO or BIOC form, as a recursive process for generating *formal orthogonal polynomials (FOPs)*. For the BIO form, we start with $p_0(\zeta) \equiv 1$; in the nth step, we apply the multiplication operator to p_{n-1} and then 'orthogonalize' the product $\zeta p(\zeta)$ with respect to p_{n-1} and p_{n-2} to get the new member p_n of the sequence. Here 'orthogonalize' refers to the formal inner product (13.2). Again it is seen that the orthogonality with respect to polynomials of lower degree follows automatically. This construction leads exactly to the three-term recurrence (12.2). A similar one that parallels the BIOC algorithm yields the coupled two-term recurrences (12.1).

Lanczos (1952) was well aware of this form of his process, but, at least for classical orthogonal polynomials, it was published long before by Stieltjes (1884). This polynomial version of the Lanczos process is therefore called the *Stieltjes procedure.*

It remains to give formulae for the recurrence coefficients in terms of values of Φ. Since Φ', which appears in the BIOC form, is linked to Φ by

$$\Phi'(s) = \Phi(\zeta s) \quad (s \in \mathcal{P})$$

(see (12.7)), it is easily substituted. As mentioned before, when the aim is the construction of orthogonal polynomials for a measure supported on a subset of the real line, Φ is the integral operator (13.1). In terms of Φ the BIO recurrence coefficients can be expressed as follows:

$$\alpha_n = \Phi(\zeta p_n^2)/\delta_n, \tag{13.3a}$$

$$\beta_{n-1} = \Phi(\zeta p_{n-1} p_n)/\delta_{n-1} = \gamma_{n-1}\delta_n/\delta_{n-1} \quad (\text{if } n > 0), \tag{13.3b}$$

$$\delta_{n+1} = \Phi(p_{n+1}^2). \tag{13.3c}$$

Of course, if the functional Φ is not definite, and thus can assume the value zero at p_{n+1}^2, the Stieltjes procedure can also break down.

For the BIOC version of the Stieltjes procedure we just need

$$\delta_{n+1} = \Phi(p_{n+1}^2), \tag{13.4a}$$

$$\delta'_{n+1} = \Phi(\zeta \widehat{p}_{n+1}^2), \tag{13.4b}$$

since ψ_n and φ_{n+1} are expressed in terms of these values.

14. The biconjugate gradient squared method

One of the disadvantages of the BiCG method is that it also requires matrix-vector multiplications with the transpose matrix $\mathbf{A}^\star$ or $\mathbf{A}^\top$, although the relevant Krylov space containing the residuals is generated only by $\mathbf{A}$. In practice, $\mathbf{A}$ is typically large and sparse, and providing an efficient subroutine for both these products can be a nontrivial task. Moreover, since two Krylov spaces are generated, two matrix-vector products are needed per dimension of the subspaces $\mathcal{K}_n$ that matter.

In 1984 Sonneveld (1989) proposed a new Lanczos-type method that circumvents these two disadvantages and proved to be very successful in practice. He called it the *conjugate gradient squared* (CGS) *algorithm*, although it is aimed at nonsymmetric problems and is not derived from a CG, but from a BiCG algorithm, namely BiOMin. Its nth residual polynomial is the square p_n^2 of the nth residual polynomial p_n of BiCG.

Here Sonneveld's version will be called BiOMinS, but we refer to the underlying approach as the *biconjugate gradient squared* (BiCGS) *method*. As for the BiCG method, there exist several different forms of the BiCGS method: in addition to BiOMinS there are consistent and inconsistent versions of BiOResS and two version of BiODirS. The latter two are not really competitive, and therefore they are not discussed here. Together with the two BiOResS versions they were presented in a separately distributed part (Section 7) of Gutknecht (1990).

All competitive versions of the BiCGS method require two applications of the operator $\mathbf{A}$ at each step; this is comparable to the two matrix-vector multiplications with $\mathbf{A}$ and $\mathbf{A}^\star$ in BiOMin, but in many applications it is an advantage that the multiplication with $\mathbf{A}^\star$ is replaced by one with $\mathbf{A}$. For example, this is true when certain preconditioning techniques are applied, when ordinary differential equations are solved with the help of the Lanczos process (see, for instance, Hochbruck and Lubich (1997*b*)), or, quite generally, when vector and parallel computers are used.

The BiCGS method typically converges nearly twice as fast as the BiCG method. However, the convergence is even less smooth, and in tough problems very erratic: the norm of the residual can suddenly increase again by several orders of magnitude and then drop to the former level after just one or a few steps. Such peaks in the residual norm plot indicate a reduction of the ultimate accuracy of the solution that can be attained, but in Section 18 we will describe a remedy for this loss.

14.1. The BIOMINS *form of the* BICGS *method*

For the derivation of BIOMINS we start from the recurrences (12.1), choosing $\gamma_n := -\varphi_n$ and substituting $\omega_n := 1/\varphi_n$, as in BIOMIN:

$$p_{n+1} := p_n - \omega_n \zeta \widehat{p}_n \,, \tag{14.1}$$

$$\widehat{p}_{n+1} := p_{n+1} - \psi_n \widehat{p}_n \,. \tag{14.2}$$

(In this section we use the sloppy notation $\zeta \widehat{p}_n$ instead of $\zeta \widehat{p}_n(\zeta)$.) Multiplying (14.1) by $\widehat{p}_n$ and (14.2) by p_{n+1} yields

$$p_{n+1}\widehat{p}_n = p_n \widehat{p}_n - \omega_n \zeta \widehat{p}_n^2, \tag{14.3}$$

$$p_{n+1}\widehat{p}_{n+1} = p_{n+1}^2 - \psi_n p_{n+1} \widehat{p}_n. \tag{14.4}$$

Next, squaring both sides of (14.1) and (14.2) and using (14.3) and (14.4), respectively, leads to

$$p_{n+1}^2 = p_n^2 - 2\omega_n \zeta p_n \widehat{p}_n + \omega_n^2 \zeta^2 \widehat{p}_n^2 \tag{14.5a}$$

$$= p_n^2 - \omega_n \zeta (p_n \widehat{p}_n + p_{n+1} \widehat{p}_n) \tag{14.5b}$$

and

$$\begin{aligned} \widehat{p}_{n+1}^2 &= p_{n+1}^2 - 2\psi_n p_{n+1} \widehat{p}_n + \psi_n^2 \widehat{p}_n^2 \\ &= p_{n+1}\widehat{p}_{n+1} - \psi_n p_{n+1} \widehat{p}_n + \psi_n^2 \widehat{p}_n^2. \end{aligned} \tag{14.6}$$

The point is that equations (14.3), (14.5b), (14.6), and (14.4) are, in this order, a system of recurrence relations for the four polynomial sequences $\{p_n \widehat{p}_{n-1}\}$, $\{p_n^2\}$, $\{\widehat{p}_n^2\}$, and $\{p_n \widehat{p}_n\}$. From (13.4a)–(13.4b) it follows that the recurrence coefficients can be computed from the values that the functional Φ defined by (12.7) takes at the polynomials $\zeta \widehat{p}_n^2$, p_{n+1}^2, and p_n^2. Here we need to express these values in terms of the new Krylov space vectors

$$\begin{aligned} \mathbf{r}_n &:= p_n^2(\mathbf{A})\mathbf{r}_0, & \widehat{\mathbf{r}}_n &:= \widehat{p}_n^2(\mathbf{A})\mathbf{r}_0, \\ \mathbf{s}_n &:= p_n(\mathbf{A})\widehat{p}_n(\mathbf{A})\mathbf{r}_0, & \mathbf{s}_n' &:= p_{n+1}(\mathbf{A})\widehat{p}_n(\mathbf{A})\mathbf{r}_0 \end{aligned} \tag{14.7}$$

and their inner products with an additional vector $\widetilde{\mathbf{y}}_0$, which is now only used for these inner products. In his seminal paper, Sonneveld (1989) chose $\widetilde{\mathbf{y}}_0 := \mathbf{r}_0$, but today it is known that a random vector is likely to yield better convergence; see, for instance, Joubert (1990, 1992).

Once we have written down the recurrences for the four vector sequences of (14.7), we have managed to 'square' a special case of the BIOC algorithm. Like BIOMIN, the BIOMINS algorithm is then based on the fact that one can additionally compute a vector sequence $\{\mathbf{x}_n\}$ with the property that $\mathbf{r}_n$ is the residual at $\mathbf{x}_n$. In summary, we obtain the following standard BIOMIN version of the BICGS method.

ALGORITHM 11. (BIOMINS FORM OF THE BICGS METHOD)
For solving $\mathbf{Ax} = \mathbf{b}$ choose an initial approximation $\mathbf{x}_0 \in \mathbb{C}^N$ and set $\mathbf{s}_0 := \mathbf{r}_0 := \widehat{\mathbf{r}}_0 := \mathbf{b} - \mathbf{A}\mathbf{x}_0$. Choose $\widetilde{\mathbf{y}}_0 \in \mathbb{C}^N$ such that $\delta_0 := \langle \widetilde{\mathbf{y}}_0, \mathbf{r}_0 \rangle_\mathbf{B} \neq 0$ and $\delta'_0 := \langle \widetilde{\mathbf{y}}_0, \mathbf{A}\widehat{\mathbf{r}}_0 \rangle_\mathbf{B} \neq 0$. Then compute for $n = 0, 1, \ldots$

$$
\begin{aligned}
\omega_n &:= \delta_n / \delta'_n, && (14.8a)\\
\mathbf{s}'_n &:= \mathbf{s}_n - \mathbf{A}\widehat{\mathbf{r}}_n \omega_n, && (14.8b)\\
\mathbf{r}_{n+1} &:= \mathbf{r}_n - \mathbf{A}(\mathbf{s}_n + \mathbf{s}'_n)\,\omega_n, && (14.8c)\\
\mathbf{x}_{n+1} &:= \mathbf{x}_n + (\mathbf{s}_n + \mathbf{s}'_n)\,\omega_n, && (14.8d)\\
\delta_{n+1} &:= \langle \widetilde{\mathbf{y}}_0, \mathbf{r}_{n+1} \rangle_\mathbf{B}\,, && (14.8e)\\
\psi_n &:= -\delta_{n+1}/\delta_n, && (14.8f)\\
\mathbf{s}_{n+1} &:= \mathbf{r}_{n+1} - \mathbf{s}'_n \psi_n, && (14.8g)\\
\widehat{\mathbf{r}}_{n+1} &:= \mathbf{s}_{n+1} - \mathbf{s}'_n \psi_n + \widehat{\mathbf{r}}_n \psi_n^2, && (14.8h)\\
\delta'_{n+1} &:= \langle \widetilde{\mathbf{y}}_0, \mathbf{A}\widehat{\mathbf{r}}_{n+1} \rangle_\mathbf{B}\,. && (14.8i)
\end{aligned}
$$

If $\mathbf{r}_{n+1} = \mathbf{o}$, the process terminates and $\mathbf{x}_{n+1}$ is the solution; if $\mathbf{r}_{n+1} \neq \mathbf{o}$ but $\delta_{n+1} = 0$ or $\delta'_{n+1} = 0$, the algorithm breaks down. In each case we set $\dot{\nu} := n + 1$.

The recurrences for the vectors $\mathbf{s}_n$, $\mathbf{s}'_n$, $\mathbf{r}_n$, and $\widehat{\mathbf{r}}_n$ are direct translations of equations (14.3)–(14.6), and the formulae for δ_{n+1} and δ'_{n+1} are in view of (12.9) and (14.7) equivalent to (13.4a)–(13.4b). Finally, the recurrence for $\mathbf{x}_n$ is chosen so that $\mathbf{r}_n = \mathbf{b} - \mathbf{A}\mathbf{x}_n$ (for all n): by (14.8d), (14.8c), and by induction we indeed get $\mathbf{r}_{n+1} = \mathbf{r}_n - \mathbf{A}(\mathbf{x}_{n+1} - \mathbf{x}_n) = \mathbf{b} - \mathbf{A}\mathbf{x}_{n+1}$.

From our derivation of this algorithm it is clear that the following holds.

Theorem 14.1 If BIOMIN and BIOMINS are started with the same $\mathbf{x}_0$ and $\widetilde{\mathbf{y}}_0$, the index of first breakdown or termination $\dot{\nu}$, the recurrence coefficients φ_n, ψ_{n-1}, and the inner products δ_n are for both methods the same. The residual polynomials are p_n and p_n^2, respectively.

Although, theoretically, if convergence were defined by $\mathbf{r}_n = \mathbf{o}$ exactly, the two algorithms would converge or break down at the same step, it is evident that in practice, where convergence is defined by a condition like $||\mathbf{r}_n|| \leq \epsilon ||\mathbf{r}_0||$, BIOMINS converges normally faster than BIOMIN, since $|p_n^2(\zeta)| = |p_n(\zeta)|^2 < |p_n(\zeta)|$ if the latter is smaller than 1.

Each step requires two applications of the operator $\mathbf{A}$, that is, two multiplications of $\mathbf{A}$ with a vector, namely $\mathbf{A}\widehat{\mathbf{r}}_n$ and $\mathbf{A}(\mathbf{s}_n + \mathbf{s}'_n)$.

Since the coefficients $\omega_n = 1/\varphi_n$ and ψ_{n-1} are the same as in BIOMIN, the bidiagonal matrices $\mathbf{L}_n$ and $\mathbf{U}_n$ are again available, and thus the tridiagonal matrix $\mathbf{T}_n = \mathbf{L}_n \mathbf{U}_n$ may still be used to obtain eigenvalue estimates for $\mathbf{A}$. In fact, deleting (14.8d) from Algorithm 11, we get what one might call a special squared BIOC algorithm, and we could easily modify it to allow for the freedom of choosing γ_n independently from φ_n, as in our formulation of BIOC.

14.2. BIOS*: the squared* BIO *algorithm*

At this point one may ask whether, in analogy to the other two main forms of the BICG method, BIORES and BIODIR, there are also other forms of the BICGS method. To derive the analogue of BIORES, we first need a squared BIO algorithm based on separate recurrences for the polynomials p_n^2 and $\widehat{p}_n^2$. We start with those for p_n^2: multiplying (12.2) by p_n and squaring (12.2) we obtain, respectively,

$$\gamma_n p_n p_{n+1} = (\zeta - \alpha_n) p_n^2 - \beta_{n-1} p_{n-1} p_n, \tag{14.9}$$

$$\begin{aligned} \gamma_n^2 p_{n+1}^2 &= (\zeta - \alpha_n)^2 p_n^2 - 2(\zeta - \alpha_n)\beta_{n-1} p_{n-1} p_n + \beta_{n-1}^2 p_{n-1}^2 \\ &= (\zeta - \alpha_n)(\gamma_n p_n p_{n+1} - \beta_{n-1} p_{n-1} p_n) + \beta_{n-1}^2 p_{n-1}^2 , \end{aligned} \tag{14.10}$$

where (14.9) has already been used to simplify (14.10). These two relations allow us to generate the two sequences $\{p_{n-1}p_n\}$ and $\{p_n^2\}$ recursively. The coefficients α_n, β_{n-1} are given by (13.3a)–(13.3c). So far, the parameters γ_n can be chosen freely ($\neq 0$), and this freedom persists if one aims at an inconsistent version of BIORESS. Later, we will want to choose $\gamma_n := -\alpha_n - \beta_{n-1}$ in the case of consistent BIORESS, since this condition is equivalent to $p_n(0) = 1$, which implies $p_n^2(0) = 1$, the consistency condition for the residual polynomial p_n^2. (The freedom of choosing the sign of $p_n(0)$ would not help to avoid any type of breakdown.)

In summary, we see that a method for generating

$$\mathbf{r}_n := p_n^2(\mathbf{A})\mathbf{r}_0, \quad \mathbf{r}'_n := p_n(\mathbf{A})p_{n+1}(\mathbf{A})\mathbf{r}_0 \tag{14.11}$$

can be based on (14.9), (14.10), and (13.3a)–(13.3c). This is the *squared biorthogonalization* or BIOS *algorithm.*

ALGORITHM 12. (BIOS ALGORITHM)
Choose $\mathbf{r}_0$, $\widetilde{\mathbf{y}}_0 \in \mathbb{C}^N$ such that $\delta_0 := \langle \widetilde{\mathbf{y}}_0, \mathbf{r}_0 \rangle_\mathbf{B} \neq 0$, and set $\mathbf{r}'_{-1} := \mathbf{o} \in \mathbb{C}^N$, $\beta_{-1} := 0$. Choosing arbitrary scale factors $\gamma_n \neq 0$, compute, for $n = 0, 1, \ldots$,

$$\alpha_n := \langle \widetilde{\mathbf{y}}_0, \mathbf{A}\mathbf{r}_n \rangle_\mathbf{B} / \delta_n, \tag{14.12a}$$

$$\beta_{n-1} := \gamma_{n-1}\delta_n / \delta_{n-1} \quad (\text{if } n > 0), \tag{14.12b}$$

$$\mathbf{r}'_n := [\mathbf{A}\mathbf{r}_n - \mathbf{r}_n\alpha_n - \mathbf{r}'_{n-1}\beta_{n-1}]/\gamma_n, \tag{14.12c}$$

$$\begin{aligned} \mathbf{r}_{n+1} := {} & [\mathbf{A}(\mathbf{r}'_n\gamma_n - \mathbf{r}'_{n-1}\beta_{n-1}) - (\mathbf{r}'_n\gamma_n - \mathbf{r}'_{n-1}\beta_{n-1})\alpha_n \\ & + \mathbf{r}_{n-1}\beta_{n-1}^2]/\gamma_n^2, \end{aligned} \tag{14.12d}$$

$$\delta_{n+1} := \langle \widetilde{\mathbf{y}}_0, \mathbf{r}_{n+1} \rangle_\mathbf{B} \tag{14.12e}$$

until $\delta_{n+1} = 0$, when we set $\nu := n + 1$.

Note that γ_n can be chosen to make $\mathbf{r}_n$ of unit length.

This algorithm was proposed independently both by Chronopoulos and Ma (1989) and by Gutknecht (1990); later it was rediscovered by Chan, de Pillis and van der Vorst (1991).

14.3. The BIORESS *forms of the* BICGS *method*

After we have 'squared' the BIO algorithm so that nested Krylov spaces of the double dimension are generated, it remains to find a way to compute the sequence $\{\mathbf{x}_n\}$ of approximants with the property that

$$\mathbf{r}_n = \mathbf{b} - \mathbf{A}\mathbf{x}_n$$

in the case of consistent BIORESS, or, more generally,

$$\mathbf{r}_n = \mathbf{b}\dot{\pi}_n^2 - \mathbf{A}\mathbf{x}_n, \tag{14.13}$$

where $\dot{\pi}_n := p_n(0)$ as before. This approach to solving a linear system of equations follows exactly the general scheme discussed in Section 4. Assuming that it works, we conclude from (14.12d) and (4.16) that

$$\begin{aligned} &\mathbf{A}\mathbf{x}_{n+1}\gamma_n^2 && (14.14)\\ &= (\mathbf{b}\dot{\pi}_{n+1}^2 - \mathbf{r}_{n+1})\gamma_n^2 \\ &= \mathbf{b}(\beta_{n-1}^2\dot{\pi}_{n-1}^2 - \alpha_n\gamma_n\dot{\pi}_n\dot{\pi}_{n+1} + \alpha_n\beta_{n-1}\dot{\pi}_n\dot{\pi}_{n-1}) \\ &\quad - \mathbf{A}(\mathbf{r}'_n\gamma_n - \mathbf{r}'_{n-1}\beta_{n-1}) + (\mathbf{r}'_n\gamma_n - \mathbf{r}'_{n-1}\beta_{n-1})\alpha_n - \mathbf{r}_{n-1}\beta_{n-1}^2 \\ &= \mathbf{A}\mathbf{x}_{n-1}\beta_{n-1}^2 - \mathbf{A}\mathbf{x}'_n\alpha_n\gamma_n + \mathbf{A}\mathbf{x}'_{n-1}\alpha_n\beta_{n-1} \\ &\quad - \mathbf{A}(\mathbf{r}'_n\gamma_n - \mathbf{r}'_{n-1}\beta_{n-1}), && (14.15) \end{aligned}$$

where

$$\mathbf{x}'_n := \mathbf{A}^{-1}(\mathbf{b}\dot{\pi}_n\dot{\pi}_{n+1} - \mathbf{r}'_n)\,,$$

that is,

$$\mathbf{r}'_{n-1} = \mathbf{b}\dot{\pi}_{n-1}\dot{\pi}_n - \mathbf{A}\mathbf{x}'_{n-1}\,. \tag{14.16}$$

Multiplying (14.15) by $\mathbf{A}^{-1}$ yields a recursive formula for $\mathbf{x}_{n+1}$. Similarly, using (14.12c) and (4.16) we get

$$\begin{aligned} \mathbf{A}\mathbf{x}'_n\gamma_n &= \mathbf{b}\dot{\pi}_n\dot{\pi}_{n+1}\gamma_n - \mathbf{r}'_n\gamma_n \\ &= -\,\mathbf{b}(\dot{\pi}_n^2\alpha_n + \dot{\pi}_{n-1}\dot{\pi}_n\beta_{n-1}) - \mathbf{A}\mathbf{r}_n + \mathbf{r}_n\alpha_n + \mathbf{r}'_{n-1}\beta_{n-1} \\ &= -\,\mathbf{A}\mathbf{x}_n\alpha_n - \mathbf{A}\mathbf{x}'_{n-1}\beta_{n-1} - \mathbf{A}\mathbf{r}_n. \end{aligned}$$

If we set $\dot{\pi}_{-1} := 0$, $\dot{\pi}_0 := 1$, then (14.13) and (14.16) hold for $\mathbf{r}_0 := \mathbf{b} - \mathbf{A}\mathbf{x}_0$, $\mathbf{r}'_{-1} := \mathbf{x}'_{-1} := \mathbf{o}$, and the recurrence can be started with these initial values.

ALGORITHM 13. (INCONSISTENT BIORESS FORM OF BICGS)
For solving $\mathbf{A}\mathbf{x} = \mathbf{b}$ choose an initial approximation $\mathbf{x}_0 \in \mathbb{C}^N$, set $\mathbf{r}_0 := (\mathbf{b} - \mathbf{A}\mathbf{x}_0)/\gamma_{-1}$ with some $\gamma_{-1} \neq 0$ (for instance, such that $||\mathbf{r}_0|| = 1$), and

redefine $\mathbf{x}_0 := \mathbf{x}_0/\gamma_{-1}$. Furthermore, let $\mathbf{r}'_{-1} := \mathbf{x}'_{-1} := \mathbf{o} \in \mathbb{C}^N$, $\beta_{-1} := 0$, $\dot{\pi}_{-1} := 0$, $\dot{\pi}_0 := 1/\gamma_{-1}$, and choose $\widetilde{\mathbf{y}}_0 \in \mathbb{C}^N$ such that $\delta_0 := \langle \widetilde{\mathbf{y}}_0, \mathbf{r}_0 \rangle_{\mathbf{B}} \neq 0$. Then apply Algorithm 12, additionally computing

$$\dot{\pi}_{n+1} := -(\alpha_n \dot{\pi}_n + \beta_{n-1}\dot{\pi}_{n-1})/\gamma_n, \tag{14.17a}$$

$$\mathbf{x}'_n := -(\mathbf{x}_n \alpha_n + \mathbf{x}'_{n-1}\beta_{n-1} + \mathbf{r}_n)/\gamma_n, \tag{14.17b}$$

$$\mathbf{x}_{n+1} := [\mathbf{x}_{n-1}\beta^2_{n-1} - \mathbf{x}'_n \alpha_n \gamma_n + \mathbf{x}'_{n-1}\alpha_n\beta_{n-1} - (\mathbf{r}'_n\gamma_n - \mathbf{r}'_{n-1}\beta_{n-1})]/\gamma_n^2. \tag{14.17c}$$

If $\mathbf{r}_{n+1}\gamma_n^2 = \mathbf{o}$ and $\dot{\pi}_{n+1} \neq 0$, the algorithm terminates and $\mathbf{x}_{\mathrm{ex}} := \mathbf{x}_{n+1}/\dot{\pi}^2_{n+1}$ is the solution of $\mathbf{Ax} = \mathbf{b}$; likewise, if $\mathbf{r}'_n\gamma_n = \mathbf{o}$, $\dot{\pi}_n \neq 0$, and $\dot{\pi}_{n+1} \neq 0$, the algorithm terminates and $\mathbf{x}_{\mathrm{ex}} := \mathbf{x}'_n/(\dot{\pi}_n\dot{\pi}_{n+1})$ is the solution; if $\mathbf{r}_{n+1} \neq \mathbf{o}$ but $\delta_{n+1} = 0$, the algorithm stops due to a Lanczos breakdown. In each case we set $\nu := n + 1$.

ALGORITHM 14. (CONSISTENT BIORESS FORM OF BICGS)
Modify Algorithm 13 by choosing $\gamma_{-1} := 1$ and $\gamma_n := -\alpha_n - \beta_{n-1}$ $(n \geq 0)$, so that $\dot{\pi}_n = 1$ $(n \geq 0)$. If $\gamma_n = 0$ for some n, the algorithm stops due to a pivot breakdown, and we set $\dot{\nu} := n$. Otherwise, $\dot{\nu} := \nu$.

In both versions of BIORESS each step again requires two applications of the operator $\mathbf{A}$, namely for $\mathbf{Ar}_n$ and $\mathbf{A}(\mathbf{r}'_n\gamma_n - \mathbf{r}'_{n-1}\beta_{n-1})$. But note that these algorithms also produce two iterates and the corresponding residuals per step. The 'normal' BICGS iterates are $\mathbf{x}_n$, but in practice the intermediate iterates $\mathbf{x}'_n$ are often better. In fact, in Fokkema, Sleijpen and van der Vorst (1996), a *shifted* CGS *algorithm* is proposed whose residual polynomials are $(1 - \mu\zeta)p_{n-1}(\zeta)p_n(\zeta)$, where μ is a preselected value. This algorithm converges somewhat more smoothly than BICGS. With BIORESS we actually get a similar kind of iterates in addition to the usual ones. However, the three-term recurrence may spoil the accuracy more than the two-term recurrence of shifted CGS.

According to the derivation of these two algorithms the recurrence coefficients and the breakdown conditions are the same as those of the respective version of BIORES. Hence, in view of Theorem 9.2, the following result is straightforward.

Theorem 14.2 If consistent BIORES and consistent BIORESS are started with the same $\mathbf{x}_0$ and $\widetilde{\mathbf{y}}_0$, then the index of first breakdown or termination $\dot{\nu}$, the recurrence coefficients α_n, β_{n-1} (and thus $\gamma_n := -\alpha_n - \beta_{n-1}$), and the inner products δ_n are the same for both algorithms. Moreover, consistent BIORESS and BIOMINS produce the same iterates $\mathbf{x}_n$ and thus also the same residuals $\mathbf{r}_n = \mathbf{b} - \mathbf{Ax}_n$ and the same residual polynomials p_n^2.

Likewise, if inconsistent BIORES and inconsistent BIORESS are started with the same $\mathbf{x}_0$ and $\widetilde{\mathbf{y}}_0$, and if the same constants γ_n are used in both

Table 4. *Krylov space vectors and corresponding polynomials that appear in our five forms of the* BiCGS *method. Different scaling is mirrored by upper indices in the notation for the polynomials, but not in the one for the vectors. In the columns 'L' and 'P' it is indicated if an algorithm is susceptible to Lanczos breakdown or pivot breakdown, respectively*

Algorithm	Vectors	Polynomials	L	P
11 BiOMinS (CGS)	$\mathbf{r}_n, \mathbf{s}_n, \mathbf{s}'_n, \widehat{\mathbf{r}}_n$	$p_n^2,\ p_n\widehat{p}_n,\ p_{n+1}\widehat{p}_n,\ (\widehat{p}_n)^2$	√	√
13 BiOResS (cons.)	$\mathbf{r}_n,\ \mathbf{r}'_n$	$p_n^2,\ p_{n+1}p_n$	√	√
14 BiOResS (incons.)	$\mathbf{r}_n,\ \mathbf{r}'_n$	$(p_n^{\mathrm{INC}})^2,\ p_{n+1}^{\mathrm{INC}}p_n^{\mathrm{INC}}$	√	
BiODirS$_1$	$\mathbf{r}_n,\ \widehat{\mathbf{r}}_n,\ \widehat{\mathbf{w}}_n$	$p_n^2,\ (\widehat{p}_n^{\mathrm{DIR}})^2,\ \widehat{p}_{n+1}^{\mathrm{DIR}}\widehat{p}_n^{\mathrm{DIR}}$	√	√
BiODirS$_2$	$\mathbf{r}_n,\ \mathbf{s}_n,\ \widehat{\mathbf{r}}_n,\ \widehat{\mathbf{w}}_n$	$p_n^2,\ p_n\widehat{p}_n^{\mathrm{DIR}},\ (\widehat{p}_n^{\mathrm{DIR}})^2,\ \widehat{p}_{n+1}^{\mathrm{DIR}}\widehat{p}_n^{\mathrm{DIR}}$		√

algorithms, then the index of first breakdown or termination ν, the recurrence coefficients α_n, β_{n-1}, and the inner products δ_n are the same for both algorithms. The iterates $\mathbf{x}_n$ and the vectors $\mathbf{r}_n$ are related by (4.17) and (14.13), respectively.

It follows in particular that BiOResS yields as a by-product the same tridiagonal submatrices $\mathbf{T}_n$, and thus, optionally, the same eigenvalue estimates as BiORes.

14.4. An overview of BiCGS *algorithms*

For the reader's convenience we list in Table 4 the various vectors and the corresponding polynomials that come up in the three BiCGS algorithms that we have discussed and the two, BiODirS$_1$ and BiODirS$_2$ from Section 7 of Gutknecht (1990), that we only alluded to. The vectors that are listed in several algorithms are, up to scaling, identical. However, neither the iterates $\mathbf{x}_n$ nor the auxiliary vectors $\mathbf{x}'_n$ that appear in BiOResS and in the two forms of BiODirS, respectively, are contained in the list. If we wanted to judge the memory requirements, we would also have to take into account the storage of the results of matrix-vector products and the sometimes required storage of previously computed vectors. Also indicated in Table 4 are the breakdown possibilities.

Let us repeat that since the coefficients computed in the squared methods are the same as those of the respective unsquared method, one still implicitly generates the matrices $\underline{\mathbf{T}}_n$, $\underline{\mathbf{L}}_n$ and $\mathbf{U}_n$, or $\underline{\mathbf{T}}'_n$, respectively. Therefore, theoretically, the squared methods can be used for eigenvalue computations. However, it does not seem so easy to mimic ideas like selective reorthogonalization or to find other ways to enhance the numerical stability. Therefore, in practice, it may be difficult to obtain reliable eigenvalue information from these methods.

15. The transpose-free QMR algorithm

In Section 14 we have seen that 'squaring' the BICG method leads to a very effective method, BICGS, which, however, typically exhibits a somewhat erratic convergence behaviour. An obvious remedy would be to 'square' the QMR method instead, since it converges as fast as BICG, but more smoothly. However, it is not so obvious how this can be achieved. There are various answers to this question, but only one turns out to be convincing: Freund (1993) found a way to apply the QMR approach to bases built up from Krylov space vectors that correspond to squares and products of Lanczos and direction polynomials. The details are given below. This *transpose-free* QMR (TFQMR) *algorithm*, as he called it, is roughly equally fast but much more smoothly converging than the BICGS method, and the cost per step is only slightly higher. Since smoother convergence often helps to reduce round-off, one may expect that there are examples where the algorithm outperforms BIOMINS in speed, but the examples in Freund's paper do not yet confirm this. However, Freund has examples where TFQMR clearly outperforms the BICGSTAB algorithm of Section 16.

In Freund and Szeto (1991) an alternative strategy was followed: the *quasi-minimal residual squared* (QMRS) *algorithm* generates residuals whose residual polynomials are the squares of the QMR residual polynomials. Consequently, the convergence is fast and smooth. However, this method requires three matrix-vector products per step in contrast to two in TFQMR and BIOMINS, thus increasing the work per step by roughly 50%.

While for both these methods the residual lies in $\mathcal{K}_{2n}$ after n iterations, this is not true for the *transpose-free implementation of the* QMR *method* proposed by Chan et al. (1991). Here the idea is simply to run the BIOS algorithm for determining the Lanczos coefficients α_n, β_{n-1}, and γ_n, and then to construct the QMR iterates (or alternatively, the BICG iterates, if a transpose-free BICG algorithm is sought) by additionally executing the recurrences of the QMR (or the BIORES) algorithm, except for those that generate $\widetilde{\mathcal{K}}_n$. Clearly, such an approach requires considerable extra work, in particular three matrix-vector products per step instead of two. Nevertheless, the convergence speed is at best equal to the one of the BICG method, not the one of the BICGS method. Moreover, the Lanczos coefficients found by the BIOS algorithm, although in theory identical to those of inconsistent BIORES, turn out to be more contaminated by round-off. Therefore, convergence is in practice often worse than for QMR and BICG, respectively.

Algorithms using this approach cannot compete with other transpose-free methods.

Let us now turn to the preferred approach, the TFQMR algorithm. Multiplying (14.1) by p_n and p_{n-1}, we get the two relations

$$\begin{aligned} p_n p_{n+1} &:= p_n^2 - \omega_n \zeta p_n \widehat{p}_n, \\ p_{n+1}^2 &:= p_n p_{n+1} - \omega_n \zeta p_{n+1} \widehat{p}_n. \end{aligned}$$

Recalling the definitions (see (14.7) and (14.11))

$$\begin{aligned} \mathbf{r}_n &:= p_n^2(\mathbf{A})\mathbf{r}_0, & \mathbf{r}'_n &:= p_{n+1}(\mathbf{A})p_n(\mathbf{A})\mathbf{r}_0, \\ \mathbf{s}_n &:= p_n(\mathbf{A})\widehat{p}_n(\mathbf{A})\mathbf{r}_0, & \mathbf{s}'_n &:= p_{n+1}(\mathbf{A})\widehat{p}_n(\mathbf{A})\mathbf{r}_0, \end{aligned}$$

we can translate this into the recurrences

$$\mathbf{r}'_n := \mathbf{r}_n - \mathbf{A}\mathbf{s}_n\omega_n, \tag{15.1}$$

$$\mathbf{r}_{n+1} := \mathbf{r}'_n - \mathbf{A}\mathbf{s}'_n\omega_n, \tag{15.2}$$

which, together with (14.8b), (14.8c), and

$$\mathbf{A}\widehat{\mathbf{r}}_{n+1} := \mathbf{A}\mathbf{s}_{n+1} - \mathbf{A}\mathbf{s}'_n\psi_n + \mathbf{A}\widehat{\mathbf{r}}_n\psi_n^2,$$

allow us to build up the Krylov space. Note that (14.8h) generates $\mathbf{A}\widehat{\mathbf{r}}_n$ recursively, so that there is no need to determine $\widehat{\mathbf{r}}_n$ itself. Only the two matrix-vector products $\mathbf{A}\mathbf{s}_n$ and $\mathbf{A}\mathbf{s}'_n$ are required per step, and they boost the dimension of the space by two.

Defining the matrices

$$\begin{aligned} \mathbf{R}_{2n} &:= [\ \mathbf{r}_0 \ \ \mathbf{r}'_0 \ \ \mathbf{r}_1 \ \ \cdots \ \ \mathbf{r}_{n-1} \ \ \mathbf{r}'_{n-1}\ ], \\ \mathbf{S}_{2n} &:= [\ \mathbf{s}_0 \ \ \mathbf{s}'_0 \ \ \mathbf{s}_1 \ \ \cdots \ \ \mathbf{s}_{n-1} \ \ \mathbf{s}'_{n-1}\ ], \end{aligned}$$

each containing a basis for $\mathcal{K}_{2n}$, and extending this definition to odd indices by dropping the last component, we can write (15.1)–(15.2) as

$$\mathbf{A}\mathbf{S}_m = \mathbf{R}_{m+1}\underline{\mathbf{L}}_m^{[1]}\mathbf{D}_{\omega|\omega;m}^{-1}, \tag{15.3}$$

where

$$\underline{\mathbf{L}}_m^{[1]} := \begin{bmatrix} 1 & & & & \\ -1 & 1 & & & \\ & -1 & 1 & & \\ & & \ddots & \ddots & \\ & & & -1 & 1 \\ & & & & -1 \end{bmatrix}$$

is $(m+1) \times m$ lower bidiagonal and $\mathbf{D}_{\omega|\omega;m}$ is the $m \times m$ diagonal matrix

$$\mathbf{D}_{\omega|\omega;m} := \mathrm{diag}(\omega_0, \omega_0, \omega_1, \omega_1, \omega_2, \ldots).$$

Note that m can either be even, $m = 2n$, or odd, $m = 2n+1$.

Once (15.3) is found, the usual QMR approach applies. We represent the mth iterate as

$$\mathbf{x}_m^{\mathrm{TFQMR}} = \mathbf{x}_0 + \mathbf{S}_m \mathbf{k}_m,$$

so that $\mathbf{r}_m^{\mathrm{TFQMR}} = \mathbf{r}_0 - \mathbf{A}\mathbf{S}_m\mathbf{k}_m$ holds for the residual. Inserting (15.3) yields

$$\mathbf{r}_m^{\mathrm{TFQMR}} = \mathbf{R}_{m+1}\left(\underline{\mathbf{e}}_1 - \underline{\mathbf{L}}_m^{[1]}\mathbf{D}_{\omega|\omega;m}^{-1}\mathbf{k}_m\right) = \mathbf{R}_{m+1}\mathbf{D}_{\mathbf{r}|\mathbf{r}';m+1}^{-1}\mathbf{q}_m,$$

with the diagonal matrix

$$\mathbf{D}_{\mathbf{r}|\mathbf{r}';m+1} := \mathrm{diag}(||\mathbf{r}_0||, ||\mathbf{r}_0'||, ||\mathbf{r}_1||, ||\mathbf{r}_1'||, \ldots)$$

used to normalize the columns of $\mathbf{R}_{m+1}$ and the quasi-residual

$$\mathbf{q}_m := \underline{\mathbf{e}}_1\rho_0 - \underline{\mathbf{L}}_m^{\mathrm{TFQMR}}\mathbf{k}_m \quad \text{with } \underline{\mathbf{L}}_m^{\mathrm{TFQMR}} := \mathbf{D}_{\mathbf{r}|\mathbf{r}';m+1}\underline{\mathbf{L}}_m^{[1]}\mathbf{D}_{\omega|\omega;m}^{-1}.$$

This $(m+1) \times m$ least squares problem is solved by a QR decomposition based on Givens rotations as before. Once again, the quasi-residual norms are found for free, and the iterates can be updated as in Section 5.

Writing $\mathbf{t}$ instead of $\mathbf{A}\widehat{\mathbf{r}}$ we can formulate the TFQMR algorithm as follows.

ALGORITHM 15. (TFQMR ALGORITHM)
For solving $\mathbf{A}\mathbf{x} = \mathbf{b}$ choose an initial approximation $\mathbf{x}_0 \in \mathbb{C}^N$, set $\mathbf{s}_0 := \mathbf{r}_0 := \mathbf{b} - \mathbf{A}\mathbf{x}_0$, $\mathbf{t}_0 := \mathbf{A}\mathbf{r}_0$, and choose $\widetilde{\mathbf{y}}_0 \in \mathbb{C}^N$ such that $\delta_0 := \langle \widetilde{\mathbf{y}}_0, \mathbf{r}_0 \rangle_{\mathbf{B}} \neq 0$ and $\delta_0' := \langle \widetilde{\mathbf{y}}_0, \mathbf{A}\widehat{\mathbf{r}}_0 \rangle_{\mathbf{B}} \neq 0$. Then compute for $n = 0, 1, \ldots$

$$\omega_n := \delta_n / \delta_n', \tag{15.4a}$$
$$\mathbf{s}_n' := \mathbf{s}_n - \mathbf{t}_n\omega_n, \tag{15.4b}$$
$$\mathbf{r}_n' := \mathbf{r}_n - \mathbf{A}\mathbf{s}_n\omega_n, \tag{15.4c}$$
$$\mathbf{r}_{n+1} := \mathbf{r}_n' - \mathbf{A}\mathbf{s}_n'\omega_n, \tag{15.4d}$$
$$\delta_{n+1} := \langle \widetilde{\mathbf{y}}_0, \mathbf{r}_{n+1} \rangle_{\mathbf{B}}, \tag{15.4e}$$
$$\psi_n := -\delta_{n+1}/\delta_n, \tag{15.4f}$$
$$\mathbf{s}_{n+1} := \mathbf{r}_{n+1} - \mathbf{s}_n'\psi_n, \tag{15.4g}$$
$$\mathbf{t}_{n+1} := \mathbf{A}\mathbf{s}_{n+1} - \mathbf{A}\mathbf{s}_n'\psi_n + \mathbf{t}_n\psi_n^2, \tag{15.4h}$$
$$\delta_{n+1}' := \langle \widetilde{\mathbf{y}}_0, \mathbf{t}_{n+1} \rangle_{\mathbf{B}}. \tag{15.4i}$$

Within this loop, for $m := 2n+1$ and $2n+2$, additionally

(1) update the QR factorization $\underline{\mathbf{L}}_m^{\mathrm{TFQMR}} = \mathbf{Q}_m\underline{\mathbf{R}}_n^{\mathrm{TFQMR}}$, analogous to (5.6)–(5.7) (with $\underline{\mathbf{T}}_n$ replaced by $\underline{\mathbf{L}}_m^{\mathrm{TFQMR}}$)
(2) according to (5.8), compute the coefficient vector $\mathbf{h}_m$ by appending the component $c_m\widetilde{\eta}_m$ to $\mathbf{h}_{m-1}$, and compute the new last component $\widetilde{\eta}_{m+1} := -\overline{s_m}\,\widetilde{\eta}_m$ of $\underline{\mathbf{h}}_m$
(3) compute $\mathbf{z}_{m-1}$ according to the two-term recurrence implied by $\mathbf{S}_m = \mathbf{R}_m^{\mathrm{TFQMR}}\mathbf{Z}_m$

(4) compute $\mathbf{x}'_n$ and $\mathbf{x}_{n+1}$ according to

$$\mathbf{x}'_n := \mathbf{x}_n + \mathbf{z}_{2n} c_{2n+1} \widetilde{\eta}_{2n+1}, \qquad \mathbf{x}_{n+1} := \mathbf{x}'_n + \mathbf{z}_{2n+1} c_{2n+2} \widetilde{\eta}_{2n+2},$$

respectively,

(5) if the norm of the quasi-residual, $\widetilde{\eta}_m$, is below a certain bound, check the norm of the true residual; stop if it is also small enough.

Like BIORESS, the TFQMR algorithm has the benefit that it delivers two iterates and corresponding residual norms per step. We will encounter the same property again in the following section on Lanczos-type product methods.

By comparison with Algorithm 11, note that by inserting the assignment

$$\mathbf{x}^{\mathrm{G}}_{n+1} := \mathbf{x}^{\mathrm{G}}_n + (\mathbf{s}_n + \mathbf{s}'_n)\, \omega_n$$

after (15.4d), we could additionally produce the BICGS iterates $\mathbf{x}^{\mathrm{G}}_n$ at almost no cost.

16. Lanczos-type product methods

As we mentioned, Sonneveld's BICGS method has the disadvantage that convergence is often interrupted by a sudden large increase of the residual norm, followed by an equally fast decrease to the previous order of magnitude. Although such spikes normally do not prevent convergence, they may reduce the speed of convergence and, in particular, the ultimate accuracy of the solution. Actually, it is sometimes rather the maximum norm of the iterates (not the residuals) that counts; see Section 18. Most users of iterative methods prefer a smoother, if not monotone, residual norm plot.

By improving an unpublished idea of Sonneveld, van der Vorst (1992) was the first to find a modification of the BICGS method with smoother convergence. In retrospect, his BICGSTAB algorithm can be understood as the application of the BICGS approach to a coupled two-term version of the one-sided Lanczos process of Section 6 instead of BIOMIN. In other words, we make use of the freedom to choose the left polynomials t_n different from the Lanczos polynomials. The residual polynomials of the resulting method are no longer the squares p_n^2 but the products $p_n t_n$, where t_n is an arbitrary polynomial of exact degree n satisfying $t_n(0) = 1$. We therefore call the class of such methods *Lanczos-type product methods* (*LTPMs*).

In BICGSTAB the polynomials t_n are determined indirectly by local one-dimensional residual minimization, and in BICGSTAB2 (Gutknecht 1993*c*) we have extended this approach to local two-dimensional minimization, which is more appropriate in view of the typically complex spectrum of non-Hermitian matrices. In the same paper we presented the formulae for using an arbitrary set of polynomials t_n satisfying a three-term recurrence, such as, for instance, suitably shifted and scaled Chebyshev polynomials as they are used in the Chebyshev method for solving linear systems. Other authors have also contributed to the class of LTPMs; see in particular Brezinski and Redivo Zaglia (1995), Fokkema et al. (1996), Zhang (1997). There are several algorithms that compete for being the most efficient one for a broad variety of examples: among these are BiCGStab2 (Gutknecht 1993*c*), BiCGStab(ℓ) (Sleijpen and Fokkema 1993), and BiCG×MR2 (Gutknecht 1994*c*), which are treated as examples below.

We concentrate here on consistent LTPMs based on the coupled two-term recurrences for the Lanczos polynomials and a three-term recurrence for the second set of polynomials, but we will indicate modifications needed for other classes. Consistent and inconsistent LTPMs based on three-term recurrences for both sets are treated in Gutknecht and Ressel (1996), where the application of look-ahead to these methods was also achieved. Eijkhout (1994) also derived a three-term version of BiCGStab, but his way of finding the Lanczos recurrence coefficients is unnecessarily complicated. Brezinski and Redivo Zaglia (1995) suggested a different way of doing look-ahead; see Section 19 for comments.

Algorithms that combine the BiCGS method or an LTPM with smoothing processes attain convergence speeds equal to the best LTPMs and an even smoother residual norm plot. In particular, the BiCGS method and LTPMs can be combined with quasi-residual minimization. In the former case one finds Freund's TFQMR algorithm (Freund 1993) of Section 15. A QMR-smoothed BiCGStab algorithm was introduced in Chan, Gallopoulos, Simoncini, Szeto and Tong (1994), while QMR minimization for general LTPMs is described by Ressel and Gutknecht (1996). A very effective alternative is the minimum residual (MR) smoothing process, which can be applied to any Krylov space solver and will be described in Section 17.

Most of the methods discussed in this section only make sense if $\mathbf{B}$ is Hermitian positive definite. Since $\mathbf{B}$ is also required to commute with $\mathbf{A}$, there are hardly any interesting examples other than $\mathbf{B} = \mathbf{I}$. We therefore assume here that this holds, that is, we drop $\mathbf{B}$.

16.1. LTPMs based on coupled two-term recurrences

When choosing the polynomials t_n of an LTPM we mainly aim at two properties: (i) fast and smooth convergence of the resulting solver; (ii) low memory requirements, which, in particular, means short recurrences. In BiCGStab the recurrence is two-term, but this is a strong restriction for a basis of polynomials of increasing degree. In contrast, three-term recurrences are satisfied by a broad class of such bases including all sets of classical orthogonal polynomials. We assume the recurrence to be of the form

$$t_{l+1}(\zeta) := (\xi_l + \eta_l\zeta)\, t_l(\zeta) + (1 - \xi_l)\, t_{l-1}(\zeta), \tag{16.1}$$

with $t_{-1}(\zeta) :\equiv 0$, $t_0(\zeta) :\equiv 1$ and $\xi_0 = 1$. Note that this form conserves the consistency condition: $t_l(0) = 1$ (for all n). The formulae for the resulting LTPM were given in Gutknecht (1993*c*), but we choose here a different exposition that is analogous to the one in Gutknecht and Ressel (1996) and Ressel and Gutknecht (1996) for LTPMs based on the Lanczos three-term recurrences.

We define two doubly indexed sequences of *product vectors*:

$$\begin{aligned}\mathbf{w}_n^l &:= t_l(\mathbf{A})\mathbf{y}_n = t_l(\mathbf{A})p_n(\mathbf{A})\mathbf{y}_0,\\ \widehat{\mathbf{w}}_n^l &:= t_l(\mathbf{A})\mathbf{v}_n = t_l(\mathbf{A})\widehat{p}_n(\mathbf{A})\mathbf{y}_0.\end{aligned}$$

Here, $\mathbf{y}_n$ and $\mathbf{v}_n$ are the right Lanczos and direction vectors of BIOMIN; they will not appear in the final algorithm. Additionally, we introduce a doubly indexed sequence of *product iterates* $\mathbf{x}_n^l$ with the properties

$$\mathbf{b} - \mathbf{A}\mathbf{x}_n^l = \mathbf{w}_n^l \quad \text{and} \quad \mathbf{x}_n^l \in \mathbf{x}_0 + \mathcal{K}_{n+l}. \tag{16.2}$$

The diagonal sequences $\{\mathbf{x}_n^n\}$ and $\{\mathbf{w}_n^n\}$ contain the iterates and residuals we really aim at. Some other product iterates and product vectors will appear in the recurrences and can occasionally also be considered as useful approximations to the solution vector $\mathbf{x}_{\text{ex}}$ of $\mathbf{Ax} = \mathbf{b}$ and the corresponding residual, respectively.

Let us arrange the product vectors into a $\mathbf{w}$-*table* and a $\widehat{\mathbf{w}}$-*table*. These two tables are very helpful for explaining the underlying mechanism of LTPMs. To fix matters, we let the n-axis point downwards and the l-axis point to the right. Then the iteration will essentially proceed from the upper left corner (with the initial vectors $\mathbf{w}_0^0 := \widehat{\mathbf{w}}_0^0 := \mathbf{r}_0 := \mathbf{b} - \mathbf{A}\mathbf{x}_0$) to the lower right. The entries that are actually needed in competitive algorithms lie on or close to a diagonal of the tables, that is, $|n - l|$ is small for them. For moving down the tables, we apply the consistent coupled two-term Lanczos recurrences (14.1)–(14.2). Multiplying them by $t_l(\zeta)$ and translating into Krylov space notation we obtain

$$\mathbf{w}_{n+1}^l := \mathbf{w}_n^l - \mathbf{A}\widehat{\mathbf{w}}_n^l\omega_n, \tag{16.3a}$$

$$\widehat{\mathbf{w}}_{n+1}^l := \mathbf{w}_{n+1}^l - \widehat{\mathbf{w}}_n^l\psi_n. \tag{16.3b}$$

On the other hand, for moving to the right we multiply (16.1) by p_n and $\widehat{p}_n$, respectively. Translating into Krylov space notation we then get

$$\mathbf{w}_n^{l+1} := \mathbf{A}\mathbf{w}_n^l\eta_l + \mathbf{w}_n^l\xi_l + \mathbf{w}_n^{l-1}(1 - \xi_l), \tag{16.3c}$$

$$\widehat{\mathbf{w}}_n^{l+1} := \mathbf{A}\widehat{\mathbf{w}}_n^l\eta_l + \widehat{\mathbf{w}}_n^l\xi_l + \widehat{\mathbf{w}}_n^{l-1}(1 - \xi_l). \tag{16.3d}$$

Of course, since the Lanczos two-term recurrences are coupled, elements of both tables appear in (16.3a) and (16.3b), while (16.3c) and (16.3d) each affect only one table. Constructing recursively the diagonals of the two tables requires us to apply these formulae cyclically for a certain sequence

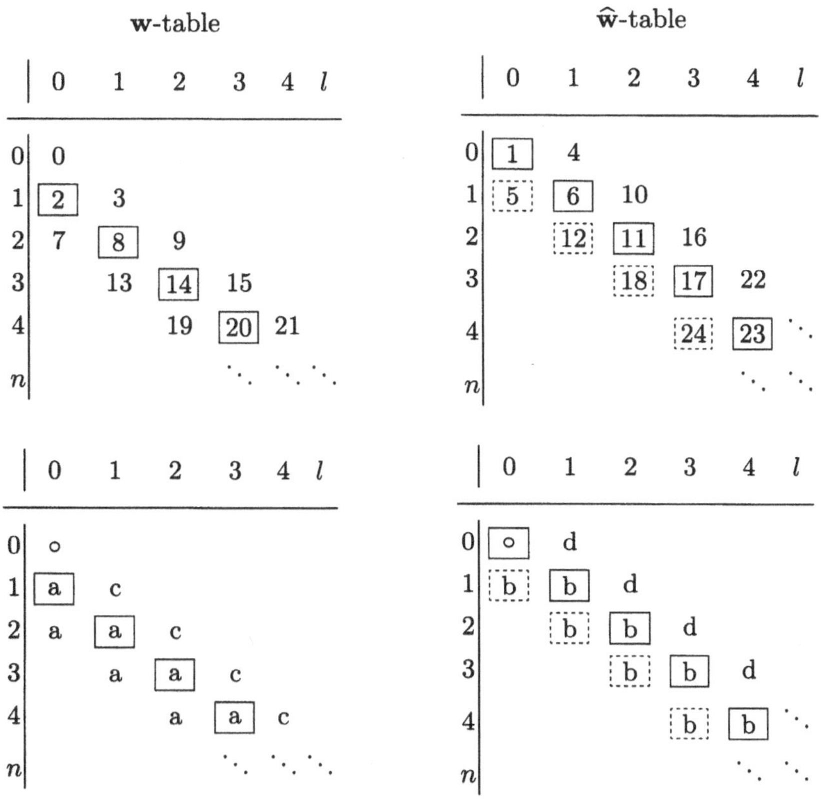

Fig. 1. The **w**-table and the $\widehat{\mathbf{w}}$-table of an LTPM based on coupled two-term Lanczos recurrences and a three-term recurrence for the polynomials t_l. The numbers in the upper tables specify the order in which the elements $\mathbf{w}_n^l$ and $\widehat{\mathbf{w}}_n^l$ of the two tables are computed. The letters in the lower tables indicate which formula among (16.3a)–(16.3d) to use for computing the corresponding entry. An entry in a solid box indicates that its product with **A** also has to be computed. For the entries in dashed boxes the product with **A** can be computed indirectly, without executing a matrix-vector product

of (l, n)-pairs. The order in which the formulae are applied is indicated by the numbers in the upper half of Figure 1, while the letters in the lower half specify the formula applied.

Inserting (16.2) into formulae (16.3a) and (16.3c), subtracting **b** on each side of them, and multiplying by the inverse of **A** we readily get corresponding recurrences for the product iterates:

$$\begin{aligned}
\mathbf{x}_{n+1}^l &:= \mathbf{x}_n^l + \widehat{\mathbf{w}}_n^l \omega_n, \\
\mathbf{x}_n^{l+1} &:= -\mathbf{w}_n^l \eta_l + \mathbf{x}_n^l \xi_l + \mathbf{x}_n^{l-1}(1 - \xi_l).
\end{aligned}$$

Of course, we could arrange them in an $\mathbf{x}$-table and display it along with the $\widehat{\mathbf{w}}$-table, very much like Figure 1. Note that $\mathbf{x}_0^0 := \mathbf{x}_0$.

It remains to specify formulae for determining the coefficients ω_n and ψ_n. Recall that in the Lanczos process we have $\widetilde{\mathcal{K}}_n \perp \mathbf{y}_n$ and $\widetilde{\mathcal{K}}_n \perp \mathbf{A}\mathbf{v}_n$. Because $\{t_l(\mathbf{A})\widetilde{\mathbf{y}}_0\}_{l=0}^{n-1}$ is a basis of $\widetilde{\mathcal{K}}_n$ (as long as n does not exceed the grade of $\widetilde{\mathbf{y}}_0$ with respect to $\mathbf{A}^\star$), we conclude that

$$\langle \widetilde{\mathbf{y}}_0, \mathbf{w}_n^l \rangle = 0, \quad \langle \widetilde{\mathbf{y}}_0, \mathbf{A}\widehat{\mathbf{w}}_n^l \rangle = 0, \quad \text{if } l < n. \tag{16.5}$$

In particular, if we set $l := n-1$ and then replace n by $n+1$, we have

$$\langle \widetilde{\mathbf{y}}_0, \mathbf{w}_{n+1}^n \rangle = 0, \quad \langle \widetilde{\mathbf{y}}_0, \mathbf{A}\widehat{\mathbf{w}}_{n+1}^n \rangle = 0.$$

Taking this orthogonality into account in the recursions (16.3a)–(16.3b) for $l = n$ and defining

$$\widetilde{\delta}_n := \langle \widetilde{\mathbf{y}}_0, \mathbf{w}_n^n \rangle, \quad \widetilde{\delta}'_n := \langle \widetilde{\mathbf{y}}_0, \mathbf{A}\widehat{\mathbf{w}}_n^n \rangle$$

leads us to

$$\omega_n := \frac{\langle \widetilde{\mathbf{y}}_0, \mathbf{w}_n^n \rangle}{\langle \widetilde{\mathbf{y}}_0, \mathbf{A}\widehat{\mathbf{w}}_n^n \rangle} = \frac{\widetilde{\delta}_n}{\widetilde{\delta}'_n},$$

and

$$\psi_n := \frac{\langle \widetilde{\mathbf{y}}_0, \mathbf{A}\mathbf{w}_{n+1}^n \rangle}{\langle \widetilde{\mathbf{y}}_0, \mathbf{A}\widehat{\mathbf{w}}_n^n \rangle} = \frac{\widetilde{\delta}_{n+1}}{\widetilde{\delta}'_n \eta_n}.$$

For the last equation we have expressed $\mathbf{A}\mathbf{w}_{n+1}^n$ according to (16.3c) and taken (16.5) into account to see that $\langle \widetilde{\mathbf{y}}_0, \mathbf{A}\mathbf{w}_{n+1}^n \rangle = \langle \widetilde{\mathbf{y}}_0, \mathbf{w}_{n+1}^{n+1} \rangle / \eta_n$.

Summarizing, we find the following *generic consistent* $(3, 2\times 2)$*-type LTPM*. Its type $(3, 2 \times 2)$ indicates that we apply a three-term recurrence for the polynomials t_l and two coupled two-term recurrences for the Lanczos and the direction polynomials. In Ressel and Gutknecht (1996) the analogous $(3,3)$-type LTPM is called BIO×THREE.

ALGORITHM 16. (GENERIC CONSISTENT $(3, 2 \times 2)$-TYPE LTPM)
For solving $\mathbf{A}\mathbf{x} = \mathbf{b}$ choose an initial approximation $\mathbf{x}_0^0 \in \mathbb{C}^N$ and set $\mathbf{w}_0^0 := \widehat{\mathbf{w}}_0^0 := \mathbf{b} - \mathbf{A}\mathbf{x}_0^0$. Choose $\widetilde{\mathbf{y}}_0 \in \mathbb{C}^N$ such that $\widetilde{\delta}_0 := \langle \widetilde{\mathbf{y}}_0, \mathbf{w}_0^0 \rangle \neq 0$ and $\widetilde{\delta}'_0 := \langle \widetilde{\mathbf{y}}_0, \mathbf{A}\widehat{\mathbf{w}}_0^0 \rangle \neq 0$. Then compute for $n = 0, 1, \ldots$

$$\omega_n := \widetilde{\delta}_n / \widetilde{\delta}'_n, \tag{16.6a}$$

$$\mathbf{w}_{n+1}^{n-1} := \mathbf{w}_n^{n-1} - \mathbf{A}\widehat{\mathbf{w}}_n^{n-1}\omega_n, \quad \text{if } n > 0, \tag{16.6b}$$

$$\mathbf{x}_{n+1}^{n-1} := \mathbf{x}_n^{n-1} + \widehat{\mathbf{w}}_n^{n-1}\omega_n, \quad \text{if } n > 0, \tag{16.6c}$$

$$\mathbf{w}_{n+1}^n := \mathbf{w}_n^n - \mathbf{A}\widehat{\mathbf{w}}_n^n \omega_n, \tag{16.6d}$$

$$\mathbf{x}_{n+1}^n := \mathbf{x}_n^n + \widehat{\mathbf{w}}_n^n \omega_n, \tag{16.6e}$$

$$\mathbf{w}_{n+1}^{n+1} := \mathbf{A}\mathbf{w}_{n+1}^n \eta_n + \mathbf{w}_{n+1}^n \xi_n + \mathbf{w}_{n+1}^{n-1}(1 - \xi_n), \tag{16.6f}$$

$$\mathbf{x}_{n+1}^{n+1} := -\mathbf{w}_{n+1}^{n}\eta_n + \mathbf{x}_{n+1}^{n}\xi_n + \mathbf{x}_{n+1}^{n-1}(1-\xi_n), \tag{16.6g}$$

$$\widetilde{\delta}_{n+1} := \langle \widetilde{\mathbf{y}}_0, \mathbf{w}_{n+1}^{n+1} \rangle, \tag{16.6h}$$

$$\widehat{\mathbf{w}}_{n}^{n+1} := \mathbf{A}\widehat{\mathbf{w}}_{n}^{n}\eta_n + \widehat{\mathbf{w}}_{n}^{n}\xi_n + \widehat{\mathbf{w}}_{n}^{n-1}(1-\xi_n), \tag{16.6i}$$

$$\psi_n := \widetilde{\delta}_{n+1}/(\widetilde{\delta}'_n\eta_n), \tag{16.6j}$$

$$\widehat{\mathbf{w}}_{n+1}^{n} := \mathbf{w}_{n+1}^{n} - \widehat{\mathbf{w}}_{n}^{n}\psi_n, \tag{16.6k}$$

$$\mathbf{A}\widehat{\mathbf{w}}_{n+1}^{n} := \mathbf{A}\mathbf{w}_{n+1}^{n} - \mathbf{A}\widehat{\mathbf{w}}_{n}^{n}\psi_n, \tag{16.6l}$$

$$\widehat{\mathbf{w}}_{n+1}^{n+1} := \mathbf{w}_{n+1}^{n+1} - \widehat{\mathbf{w}}_{n}^{n+1}\psi_n, \tag{16.6m}$$

$$\widetilde{\delta}'_{n+1} := \langle \widetilde{\mathbf{y}}_0, \mathbf{A}\widehat{\mathbf{w}}_{n+1}^{n+1} \rangle, \tag{16.6n}$$

If $\mathbf{w}_{n+1}^{l} = \mathbf{o}$ for $l = n-1$, n, or $n+1$, then the algorithm terminates and $\mathbf{x}_{n+1}^{l}$ is the solution of $\mathbf{Ax} = \mathbf{b}$. However, if none of these residuals vanishes, but $\widetilde{\delta}_{n+1} = 0$ or $\widetilde{\delta}'_{n+1} = 0$, the algorithm breaks down.

Note that $\mathbf{A}\widehat{\mathbf{w}}_{n+1}^{n}$ is obtained without using a matrix-vector product, hence only two of them are needed per step: $\mathbf{A}\mathbf{w}_{n+1}^{n}$ and $\mathbf{A}\widehat{\mathbf{w}}_{n+1}^{n+1}$. In the above form, the algorithm requires considerable storage (although of course we assume that entries on the same (co-)diagonal of any of our four tables are stored at the same memory location). However, some of the vectors can be spared, namely those that are used only once later and do not depend on a quantity that is overwritten before the vector is used. An extra benefit of the given algorithm is that we not only obtain one approximate solution per step but three, since each of the vectors $\mathbf{x}_n^l$ can be considered as one and its residual is available. However, checking the length of such a residual requires an inner product. On the other hand, in more sophisticated versions of the algorithm some of these inner products may get computed anyway in order to determine if it is worth recomputing the vector by applying reorthogonalization; see Section 18.

If the polynomials t_n satisfy a two-term recurrence instead of the three-term recurrence, that is, if $\xi_l = 1$ (for all l), then we do not need the quantities on the lowest codiagonal of each table. This is the most important advantage of BiCGStab over other LTPMs.

16.2. The BiCGStab *algorithm*

In van der Vorst's BiCGStab method (van der Vorst 1992), the polynomials t_l are built up in factored form,

$$t_l(\zeta) = (1-\chi_1\zeta)(1-\chi_2\zeta)\cdots(1-\chi_l\zeta),$$

and, hence, they satisfy a two-term recurrence:

$$t_{l+1}(\zeta) = (1-\chi_{l+1}\zeta)\, t_l(\zeta).$$

This means that in the three-term recurrence of (16.1) we have

$$\xi_l = 1, \quad \eta_l = -\chi_{l+1}.$$

Note that such a two-term recurrence necessarily means that the zeros remain fixed, except that an additional one is added at each step.

In the nth step, the reciprocal χ_{n+1} of the $(n+1)$th zero is chosen such that $\mathbf{w}_{n+1}^{n+1}$ has minimum length:

$$\|\mathbf{w}_{n+1}^{n+1}\| := \min_{\chi} \|\mathbf{w}_{n+1}^{n} - \mathbf{A}\mathbf{w}_{n+1}^{n}\chi\|. \tag{16.7}$$

This one-dimensional minimization problem is solved by making $\mathbf{w}_{n+1}^{n+1}$ orthogonal to $\mathbf{A}\mathbf{w}_{n+1}^{n}$, which means that

$$\chi_{n+1} := \frac{\langle \mathbf{A}\mathbf{w}_{n+1}^{n}, \mathbf{w}_{n+1}^{n}\rangle}{\|\mathbf{A}\mathbf{w}_{n+1}^{n}\|^2}. \tag{16.8}$$

This remains correct in the case of complex data; see Gutknecht (1993*c*).

Due to the two-term recurrence, there is no need to compute $\mathbf{w}_{n+1}^{n-1}$, $\widehat{\mathbf{w}}_{n+1}^{n}$, and $\mathbf{A}\widehat{\mathbf{w}}_{n+1}^{n}$, so that the corresponding lines in Algorithm 16 can be dropped. Moreover, in order to minimize the memory requirements we insert the formulae for $\mathbf{x}_{n+1}^{n}$ and $\widehat{\mathbf{w}}_{n}^{n+1}$ at the points where these vectors are actually used.

ALGORITHM 17. (BICGSTAB ALGORITHM)
For solving $\mathbf{A}\mathbf{x} = \mathbf{b}$ choose an initial approximation $\mathbf{x}_0^0 \in \mathbb{C}^N$ and set $\mathbf{w}_0^0 := \widehat{\mathbf{w}}_0^0 := \mathbf{b} - \mathbf{A}\mathbf{x}_0^0$. Choose $\widetilde{\mathbf{y}}_0 \in \mathbb{C}^N$ such that $\widetilde{\delta}_0 := \langle \widetilde{\mathbf{y}}_0, \mathbf{w}_0^0\rangle \neq 0$ and $\widetilde{\delta}_0' := \langle \widetilde{\mathbf{y}}_0, \mathbf{A}\widehat{\mathbf{w}}_0^0\rangle \neq 0$. Then compute for $n = 0, 1, \ldots$

$$\omega_n := \widetilde{\delta}_n / \widetilde{\delta}_n', \tag{16.9a}$$

$$\mathbf{w}_{n+1}^{n} := \mathbf{w}_n^n - \mathbf{A}\widehat{\mathbf{w}}_n^n\omega_n, \tag{16.9b}$$

$$\chi_{n+1} := \langle \mathbf{A}\mathbf{w}_{n+1}^{n}, \mathbf{w}_{n+1}^{n}\rangle / \|\mathbf{A}\mathbf{w}_{n+1}^{n}\|^2, \tag{16.9c}$$

$$\mathbf{w}_{n+1}^{n+1} := \mathbf{w}_{n+1}^{n} - \mathbf{A}\mathbf{w}_{n+1}^{n}\chi_{n+1}, \tag{16.9d}$$

$$\mathbf{x}_{n+1}^{n+1} := \mathbf{x}_n^n + \widehat{\mathbf{w}}_n^n\omega_n + \mathbf{w}_{n+1}^{n}\chi_{n+1}, \tag{16.9e}$$

$$\widetilde{\delta}_{n+1} := \langle \widetilde{\mathbf{y}}_0, \mathbf{w}_{n+1}^{n+1}\rangle, \tag{16.9f}$$

$$\psi_n := -\widetilde{\delta}_{n+1} / (\widetilde{\delta}_n'\chi_{n+1}), \tag{16.9g}$$

$$\widehat{\mathbf{w}}_{n+1}^{n+1} := \mathbf{w}_{n+1}^{n+1} - (\widehat{\mathbf{w}}_n^n - \mathbf{A}\widehat{\mathbf{w}}_n^n\chi_{n+1})\,\psi_n, \tag{16.9h}$$

$$\widetilde{\delta}_{n+1}' := \langle \widetilde{\mathbf{y}}_0, \mathbf{A}\widehat{\mathbf{w}}_{n+1}^{n+1}\rangle, \tag{16.9i}$$

If $\mathbf{w}_{n+1}^{n+1} = \mathbf{o}$ or $\mathbf{w}_{n+1}^{n} = \mathbf{o}$, the algorithm terminates and, respectively, $\mathbf{x}_{n+1}^{n+1}$ or $\mathbf{x}_{n+1}^{n}$ (defined by $\mathbf{x}_{n+1}^{n} := \mathbf{x}_n^n + \widehat{\mathbf{w}}_n^n\omega_n$) is the solution of $\mathbf{A}\mathbf{x} = \mathbf{b}$. Otherwise, if $\widetilde{\delta}_{n+1} = 0$ or $\widetilde{\delta}_{n+1}' = 0$, the algorithm breaks down.

Note that this algorithm, like TFQMR, provides two residuals and corresponding iterates per step, although, as suggested in our formulation, we need to compute only one iterate per step and can determine the other only once its residual satisfies the convergence criterion.

The breakdown conditions $\widetilde{\delta}_{n+1} = 0$ and $\widetilde{\delta}'_{n+1} = 0$ seem to indicate a Lanczos or pivot breakdown, respectively. However, BICGSTAB is also subject to a somewhat hidden danger of breakdown. In fact, due to the different leading coefficients of p_n and t_n, the inner products δ_{n+1}, δ'_{n+1} of the Lanczos process and $\widetilde{\delta}_{n+1}$, $\widetilde{\delta}'_{n+1}$ of BICGSTAB are related by

$$\widetilde{\delta}_{n+1} = \delta_{n+1} \frac{\chi_1 \chi_2 \cdots \chi_{n+1}}{\omega_0 \omega_1 \cdots \omega_n}, \qquad \widetilde{\delta}'_{n+1} = \delta'_{n+1} \frac{\chi_1 \chi_2 \cdots \chi_{n+1}}{\omega_0 \omega_1 \cdots \omega_n}.$$

Consequently, not only the Lanczos breakdown ($\delta_{n+1} = 0$) and the pivot breakdown ($\delta'_{n+1} = 0$) take their toll, but also the *minimization breakdown* $\chi_{n+1} = 0$. But the latter also surfaces as a vanishing of $\widetilde{\delta}_{n+1}$ (and $\widetilde{\delta}'_{n+1}$). It is no surprise that a vanishing χ_{n+1} causes a disaster, since the Krylov space is then no longer expanded by (16.9d).

Unfortunately, there are applications where χ_n tends to be small, namely if $\mathbf{A}$ has eigenvalues with small real part, but non-small imaginary part. An example are matrices resulting from the discretization of convection-dominated convection–diffusion equations.

BICGSTAB has the additional disadvantage that, for real matrices and right-hand sides (and real initial vectors), the zeros $1/\chi_k$ of t_l are necessarily always real, and therefore they cannot efficiently help to damp error components associated with eigenvalues of $\mathbf{A}$ with large imaginary part.

16.3. The BICGSTAB2 *algorithm*

Both above-mentioned disadvantages of BICGSTAB can be removed by replacing the local one-dimensional minimization by a two-dimensional one in every other step. In between, in the odd steps, we may still do a one-dimensional minimization, but it will have no effect on what follows, at least not in exact arithmetic, as long as we advance in the Krylov space. This is the idea behind BICGSTAB2 (Gutknecht 1993*c*). Here, if required to be consistent, t_l satisfies recurrences of the form

$$\begin{aligned} t_{l+1}(\zeta) &:= (1 - \chi_{l+1}\zeta)\, t_l(\zeta), \quad \text{if } l \text{ is even}, \\ t_{l+1}(\zeta) &:= (\xi_l + \eta_l\zeta)\, t_l(\zeta) + (1 - \xi_l)\, t_{l-1}(\zeta), \quad \text{if } l \text{ is odd}. \end{aligned}$$

The pair (ξ_n, η_n) is again chosen to minimize $\mathbf{w}_{n+1}^{n+1}$, and in view of the relevant recurrence (16.3c) this means that the pair is the minimizer of

$$\|\mathbf{w}_{n+1}^{n+1}\| = \min_{\xi,\eta} \|\mathbf{A}\mathbf{w}_{n+1}^{n}\eta + \mathbf{w}_{n+1}^{n}\xi + \mathbf{w}_{n+1}^{n-1}(1-\xi)\|. \qquad (16.11)$$

This is a standard least squares problem for two unknowns. It could be solved via the 2×2 system of normal equations, but since this system can be ill-conditioned, it may be preferable to use a QR decomposition. In theory, the system cannot be rank-deficient except when the Krylov space is exhausted: rank-deficiency means here that $\mathbf{A}\mathbf{w}^n_{n+1} \in \mathcal{K}_{2n+3}$ and $\mathbf{w}^n_{n+1} - \mathbf{w}^{n-1}_{n+1} \in \mathcal{K}_{2n+2}$ are linearly dependent.

In Gutknecht (1993*c*) we chose χ_n such that in the odd steps the one-dimensional minimization problem (16.7) is solved. At the cost of one additional inner product[16], this choice has the advantage of also producing at odd steps a new, normally better iterate that may satisfy the convergence criterion. On the other hand, the result of the next, even step does not depend on χ_n as long as $\chi_n \neq 0$, except for a possible effect on round-off.

Sleijpen and Fokkema (1993) pointed out that round-off may indeed be a problem if $|\chi_{n+1}|$ is small, which is likely to happen for some n if $\mathbf{A}$ has eigenvalues close to the imaginary axis. In contrast to BiCGStab, which will then have the same round-off problem, we can get around it here. In this case, for example, we could just redefine χ_{n+1} as 1 (or some other suitably chosen value). We could even go further, forget the one-dimensional minimization problem completely, and just let $\chi_{n+1} := 1$ for all n. This has the additional advantage that one inner product can be saved. The choice of 1 as value of χ_{n+1} (and, thus, of an additional zero at 1 of t_n if n is odd) can be justified by the fact that well preconditioned matrices often have a cluster of eigenvalues around 1. However, numerical experiments indicate that $\chi_{n+1} := 1$ is not always the best choice.

In fact, choosing χ_{n+1} requires us to compromise between several objectives (see Section 18 for more details):

(i) finding small residuals, or at least, avoiding large intermediate iterates $\mathbf{x}^n_n$ and $\mathbf{x}^{n-1}_n$

(ii) avoiding round-off errors in the computation of the inner products $\widetilde{\delta}_{n+1}$ and $\widetilde{\delta}'_{n+1}$

(iii) avoiding stagnation of the Krylov space generation.

An analysis (Sleijpen and van der Vorst 1995*a*) shows that the last two objectives go hand in hand and suggests choosing χ_{n+1} as minimizer of

$$\frac{||\mathbf{w}^{n+1}_{n+1}||}{|\chi|} = \frac{||\mathbf{w}^n_{n+1} - \mathbf{A}\mathbf{w}^n_{n+1}\chi||}{|\chi|}$$

instead of (16.7), which optimizes the first objective. This requires us to

[16] The other one is needed anyway for solving (16.11).

make $\mathbf{A}\mathbf{w}_{n+1}^n - \mathbf{w}_{n+1}^n/\chi_{n+1}$ orthogonal to $\mathbf{w}_{n+1}^n$, that is, to choose

$$\chi_{n+1} := \frac{\|\mathbf{w}_{n+1}^n\|^2}{\langle \mathbf{w}_{n+1}^n, \mathbf{A}\mathbf{w}_{n+1}^n \rangle}. \tag{16.12}$$

Compromising suitably between (16.8) and (16.12), taking the value of

$$\widehat{\delta}_{n+1} := \frac{\langle \mathbf{w}_{n+1}^n, \mathbf{A}\mathbf{w}_{n+1}^n \rangle}{\|\mathbf{w}_{n+1}^n\| \, \|\mathbf{A}\mathbf{w}_{n+1}^n\|}$$

into account, ultimately leads to our recommendation (Gutknecht and Ressel 1997) to let in BiCGStab2

$$\chi_{n+1} := \begin{cases} \dfrac{\|\mathbf{w}_{n+1}^n\|}{\|\mathbf{A}\mathbf{w}_{n+1}^n\|} & \text{if } \widehat{\delta}_{n+1} \le \frac{1}{2}\sqrt{2}, \\[2ex] \dfrac{\langle \mathbf{A}\mathbf{w}_{n+1}^n, \mathbf{w}_{n+1}^n \rangle}{\|\mathbf{A}\mathbf{w}_{n+1}^n\|^2} & \text{if } \widehat{\delta}_{n+1} > \frac{1}{2}\sqrt{2}. \end{cases} \tag{16.13}$$

This is a variation of a proposal of Sleijpen and van der Vorst (1995*a*) for choosing χ_{n+1} in BiCGStab such that the effect of rounding errors in case of stagnation is minimized (Sleijpen and van der Vorst 1995*a*, Section 3.4).

In their BiCGStab(2) algorithm, which is mathematically equivalent to BiCGStab2, Sleijpen and Fokkema (1993) gave up the consistency condition of t_l for odd l and simply based the iteration on

$$\begin{aligned} t_{l+1}(\zeta) &:= \zeta\, t_l(\zeta), \quad \text{if } l \text{ is even}, & (16.14\text{a}) \\ t_{l+1}(\zeta) &:= (1 - \chi_l\zeta - \chi_{l+1}\zeta^2) t_{l-1}(\zeta) \\ &= t_{l-1}(\zeta) - \chi_l t_l(\zeta) - \chi_{l+1}\zeta t_l(\zeta) \quad \text{if } l \text{ is odd}, & (16.14\text{b}) \end{aligned}$$

where χ_l and χ_{l+1} are now the parameters of the two-dimensional minimization problem for $\mathbf{w}_{n+1}^{n+1}$.

This saves a few operations, but has the disadvantage that the odd steps do not produce a residual and a corresponding iterate. Therefore, only every four matrix-vector products does one have a chance to realize that the algorithm has converged. On the other hand, once the problems caused by small values of $|\chi_{n+1}|$ and ill-conditioned least squares problems have been eliminated in BiCGStab2, we can expect that it is numerically as stable as BiCGStab(2), in particular since (16.14b) seems to be more susceptible to a growing gap between recursively computed and true residuals, a difference barely noticed in practice, however; see Section 18 for more details. Alternative versions of BiCGStab(2) introduced in Sleijpen, van der Vorst and Fokkema (1994) also avoid these two drawbacks.

BiCGStab2 is easy to generalize by allowing any sequence of one-, two-, or even higher-dimensional minimizations. Let us define $\mathcal{S}_\ell$ as the set of indices $n+1$ where we perform an ℓ-dimensional minimization, and let $\mathcal{S}_0$

denote the set of indices where no minimization is performed. Since $n+1$ starts at 1 in our algorithms, we then have, for example, $\mathcal{S}_1 = \{1,3,5,\ldots\}$ and $\mathcal{S}_2 = \{2,4,6,\ldots\}$ for standard BiCGStab2, but $\mathcal{S}_0 = \{1,3,5,\ldots\}$ and $\mathcal{S}_2 = \{2,4,6,\ldots\}$ for BiCGStab(2). In the variant of BiCGStab2 obtained by observing (16.13), which in contrast to BiCGStab(2) is still consistent at each n, it is determined on the fly whether an odd index belongs to $\mathcal{S}_0$ or $\mathcal{S}_1$. But even if it does not, the residual will not normally grow so much, so that the loss of ultimate accuracy is kept within limits.

With this notation, BiCGStab2 and its consistent variants can be formulated as follows.

ALGORITHM 18. (GENERALIZED BICGSTAB2 ALGORITHM)
For solving $\mathbf{Ax} = \mathbf{b}$ choose an initial approximation $\mathbf{x}_0^0 \in \mathbb{C}^N$ and set $\mathbf{w}_0^0 := \widehat{\mathbf{w}}_0^0 := \mathbf{b} - \mathbf{Ax}_0^0$. Choose $\widetilde{\mathbf{y}}_0 \in \mathbb{C}^N$ such that $\widetilde{\delta}_0 := \langle \widetilde{\mathbf{y}}_0, \mathbf{w}_0^0 \rangle \neq 0$ and $\widetilde{\delta}_0' := \langle \widetilde{\mathbf{y}}_0, \mathbf{A}\widehat{\mathbf{w}}_0^0 \rangle \neq 0$. Then, for $n = 0, 1, \ldots,$ compute

$$\begin{aligned}
\omega_n &:= \widetilde{\delta}_n / \widetilde{\delta}_n', && (16.15a)\\
\mathbf{w}_{n+1}^{n-1} &:= \mathbf{w}_n^{n-1} - \mathbf{A}\widehat{\mathbf{w}}_n^{n-1}\omega_n, \quad \text{if } n+1 \in \mathcal{S}_2, && (16.15b)\\
\mathbf{x}_{n+1}^{n-1} &:= \mathbf{x}_n^{n-1} + \widehat{\mathbf{w}}_n^{n-1}\omega_n, \quad \text{if } n+1 \in \mathcal{S}_2, && (16.15c)\\
\mathbf{w}_{n+1}^{n} &:= \mathbf{w}_n^{n} - \mathbf{A}\widehat{\mathbf{w}}_n^{n}\omega_n, && (16.15d)\\
\mathbf{x}_{n+1}^{n} &:= \mathbf{x}_n^{n} + \widehat{\mathbf{w}}_n^{n}\omega_n. && (16.15e)
\end{aligned}$$

If $n+1 \notin \mathcal{S}_2$, define χ_{n+1}, for example, by (16.13) and let

$$\begin{aligned}
\mathbf{w}_{n+1}^{n+1} &:= \mathbf{w}_{n+1}^{n} - \mathbf{A}\mathbf{w}_{n+1}^{n}\chi_{n+1}, && (16.15f)\\
\mathbf{x}_{n+1}^{n+1} &:= \mathbf{x}_{n+1}^{n} + \mathbf{w}_{n+1}^{n}\chi_{n+1}, && (16.15g)\\
\widetilde{\delta}_{n+1} &:= \langle \widetilde{\mathbf{y}}_0, \mathbf{w}_{n+1}^{n+1} \rangle, && (16.15h)\\
\widehat{\mathbf{w}}_n^{n+1} &:= \widehat{\mathbf{w}}_n^{n-1} - \mathbf{A}\widehat{\mathbf{w}}_n^{n}\chi_{n+1}, && (16.15i)\\
\psi_n &:= -\widetilde{\delta}_{n+1}/(\widetilde{\delta}_n'\chi_{n+1}), && (16.15j)\\
\widehat{\mathbf{w}}_{n+1}^{n} &:= \mathbf{w}_{n+1}^{n} - \widehat{\mathbf{w}}_n^{n}\psi_n, && (16.15k)\\
\mathbf{A}\widehat{\mathbf{w}}_{n+1}^{n} &:= \mathbf{A}\mathbf{w}_{n+1}^{n} - \mathbf{A}\widehat{\mathbf{w}}_n^{n}\psi_n, \quad \text{if } n+2 \in \mathcal{S}_2, && (16.15l)\\
\widehat{\mathbf{w}}_{n+1}^{n+1} &:= \mathbf{w}_{n+1}^{n+1} - \widehat{\mathbf{w}}_n^{n+1}\psi_n, && (16.15m)
\end{aligned}$$

else (that is, if $n+1 \in \mathcal{S}_2$) determine ξ_n and η_n as solutions of (16.11) and proceed with

$$\begin{aligned}
\mathbf{w}_{n+1}^{n+1} &:= \mathbf{A}\mathbf{w}_{n+1}^{n}\eta_n + \mathbf{w}_{n+1}^{n}\xi_n + \mathbf{w}_{n+1}^{n-1}(1-\xi_n), && (16.15n)\\
\mathbf{x}_{n+1}^{n+1} &:= -\mathbf{w}_{n+1}^{n}\eta_n + \mathbf{x}_{n+1}^{n}\xi_n + \mathbf{x}_{n+1}^{n-1}(1-\xi_n), && (16.15o)\\
\widehat{\mathbf{w}}_n^{n+1} &:= \mathbf{A}\widehat{\mathbf{w}}_n^{n}\eta_n + \widehat{\mathbf{w}}_n^{n}\xi_n + \widehat{\mathbf{w}}_n^{n-1}(1-\xi_n), && (16.15p)
\end{aligned}$$

$$\widetilde{\delta}_{n+1} := \langle \widetilde{\mathbf{y}}_0, \mathbf{w}_{n+1}^{n+1} \rangle, \tag{16.15q}$$

$$\psi_n := \widetilde{\delta}_{n+1} / (\widetilde{\delta}'_n \eta_n), \tag{16.15r}$$

$$\widehat{\mathbf{w}}_{n+1}^n := \mathbf{w}_{n+1}^n - \widehat{\mathbf{w}}_n^n \psi_n, \tag{16.15s}$$

$$\mathbf{A}\widehat{\mathbf{w}}_{n+1}^n := \mathbf{A}\mathbf{w}_{n+1}^n - \mathbf{A}\widehat{\mathbf{w}}_n^n \psi_n, \tag{16.15t}$$

$$\widehat{\mathbf{w}}_{n+1}^{n+1} := \mathbf{w}_{n+1}^{n+1} - \widehat{\mathbf{w}}_n^{n+1} \psi_n. \tag{16.15u}$$

Finally, set

$$\widetilde{\delta}'_{n+1} := \langle \widetilde{\mathbf{y}}_0, \mathbf{A}\widehat{\mathbf{w}}_{n+1}^{n+1} \rangle. \tag{16.15v}$$

If $\mathbf{w}_{n+1}^l = \mathbf{o}$ for $l = n-1$, n, or $n+1$, then the algorithm terminates and $\mathbf{x}_{n+1}^l$ is the solution of $\mathbf{Ax} = \mathbf{b}$. However, if none of these residuals vanishes, but $\widetilde{\delta}_{n+1} = 0$ or $\widetilde{\delta}'_{n+1} = 0$, the algorithm breaks down.

Note that as in BiCGStab storing $\mathbf{x}_{n+1}^n$ and $\mathbf{x}_{n+1}^{n-1}$ could be avoided, and that $\widehat{\mathbf{w}}_{n+1}^n$ could be stored over $\mathbf{w}_{n+1}^n$.

16.4. The BiCG×MR2 *and* BiCG×Cheby *methods*

In the framework of Algorithm 18 we may in particular let $\mathcal{S}_2 = \{2, 3, 4, \ldots\}$, which means that we do a two-dimensional minimization at every step except the first one (where $n+1 = 1$). This yields a method we call BiCG×MR2 or, if we want to indicate that we mean the version based on coupled two-term recurrences, the BiOC×MR2 algorithm. It was proposed in Gutknecht (1994*c*) and also independently by Cao (1997*b*) and by Zhang (1997); a look-ahead variant based on the three-term look-ahead Lanczos recurrences is given in Gutknecht and Ressel (1996), while the combination with the QMR approach is explored in Ressel and Gutknecht (1996). This method has the advantage of avoiding the sometimes doubtful one-dimensional minimization step and yet producing an iterate and its residual at every step. Experiments show that it typically converges slightly faster than BiCGStab2, which often converges markedly faster than BiCGStab. The latter can be considered as a BiCG×MR1 method.

Compared to BiCGStab2, the character of the polynomials t_n changes drastically in BiCG×MR2: while in the former method just two new zeros are added in every other step, but all the other zeros remain fixed, here all the zeros get modified in each step, as is the case with orthogonal polynomials, for which the zeros interlace.

A further alternative is to look at product methods where the second set of polynomials, $\{t_l\}$, consists of the residual polynomials of some other Krylov space solver. For example, one might choose them as suitably shifted and scaled Chebyshev polynomials, as mentioned as a possibility in van der Vorst (1992) and Gutknecht (1993*c*). This requires the construction of an ellipse containing the eigenvalues of $\mathbf{A}$. A technique for this task was devised by Manteuffel (1977). The recurrences of the resulting BICG×CHEBY *algorithm* are the same as those for BICG×MR2, but the coefficients η_n and ξ_n are known in advance; they are determined by the foci of the ellipse.

16.5. The BICGSTAB(ℓ) *algorithm*

Sleijpen and Fokkema (1993) went one step further in the generalization of BICGSTAB and BICGSTAB2: with BICGSTAB(ℓ) (where ℓ is a positive integer) they introduced a method that performs an ℓ-dimensional minimization every ℓ steps. In other words, the polynomials t_n are built up by appending polynomial factors of degree ℓ. The original paper suggested generating the Krylov space in between by simply multiplying $\mathbf{w}_{k\ell}^{k\ell}$ successively by $\mathbf{A}$ as in (16.14a). We have mentioned already that this may cause a departure of the recursively computed from the true residual. Therefore, two other realizations of this method were proposed and compared in Sleijpen et al. (1994): a 'stabilized matrix' version, which is, for $\ell > 4$, considerably more costly, and an 'orthogonal matrix' version, which sometimes seems to converge more slowly; see Tables 1 and 5 in Sleijpen et al. (1994). Further tests ultimately led to the enhanced implementation of BICGSTAB(ℓ) of Fokkema (1996*a*, 1996*b*).

There are examples where $\ell = 4$ is markedly superior to $\ell = 2$, and there are even cases where $\ell = 8$ works well, while neither $\ell < 8$ nor any other method tried converged. But in most examples the convergence rate seems to be about the same. There are two reasons to expect better convergence: first, the residual reduction due to ℓ-dimensional minimization is clearly stronger than for ℓ times of one-dimensional minimization; second, it can be seen that the Lanczos process is less affected by round-off if one works with larger ℓ.

16.6. Further LTPMs

As is clear from the definition of LTPMs, there exist infinitely many such methods, even when we allow at most three terms in the recursion for t_n, and quite a few of them may seem to make sense for one reason or another. But our interest is of course restricted to those that are competitive.

Moreover, as mentioned, LTPMs come in various versions depending on the recursions used. We have chosen here to describe what we call the $(3, 2 \times 2)$-type LTPM, which applies a three-term recurrence for the polynomials t_l and two coupled two-term recurrences for the Lanczos and the direction polynomials. We could equally well use the three-term recurrence for the Lanczos polynomials, as in Gutknecht and Ressel (1996) and Ressel and Gutknecht (1996), thus getting $(3, 3)$-type LTPMs. Then there is no $\widehat{\mathbf{w}}$-table, but the $\mathbf{w}$-table has in the generic case bandwidth 4. On the other hand, one can also turn to coupled two-term recurrences for both polynomial sequences, that is, $(2\times2, 2\times2)$-type LTPMs, as suggested by Fokkema et al. (1996) as well as Zhang (1997). This has the advantage that the standard BIOMINS version of the BICGS method fits into the pattern, but, on the other hand, methods like BICGSTAB2 and BICG×MR2 are less easy to reformulate. To display these versions in a way that is analogous to our Figure 1, we would have to introduce four tables for four different types of product vectors, and in each table the bandwidth would be 2.

Among the recently proposed LTPMs is the already mentioned *shifted* CGS *algorithm* of Fokkema et al. (1996) whose residual polynomials are $(1-\mu\zeta)p_{n-1}(\zeta)p_n(\zeta)$, where μ is a fixed chosen value, for instance the inverse of an estimate for the largest eigenvalue for $\mathbf{A}$. The authors' examples indicate that its residual norm histories have typically less dramatic peaks than BICGS. The same is true for the CGS2 *method* introduced in the same paper: there, both polynomials are Lanczos polynomials of the same degree, but they correspond to two different left initial vectors. A number of further LTPMs were suggested by Brezinski and Redivo Zaglia (1995), but examples for demonstrating their usefulness are missing.

17. Smoothing processes

The erratic convergence behaviour of the basic Lanczos-type solvers, the BICG and BICGS methods, often gives rise to criticism. Plots of the residual norm are the most often used tool when algorithms are compared and, therefore, a method sells well if it converges quickly and smoothly. Thus, it is not surprising that there is an interest in smoothing processes that modify the BICG or BICGS iterates so that the residual norm plot becomes smoother. However, we should say that smoothing is also dubious. In fact, what really counts in practice is that a method should find the solution (up to a certain error) as quickly as possible and, since the error cannot be checked, the residual is monitored instead. The smoothness of the convergence does not matter from that point of view. Nevertheless, smoothing processes are of some interest since they can speed up the convergence slightly. On the other hand, they hide some useful information: a peak in the residual norm plot indicates a temporary stagnation of convergence, while in the smoothed residual plot we cannot distinguish temporary from permanent stagnation.

A question that has been resolved concerns the relationship between the convergence behaviour of CG and CR, as well as FOM and GMRES. This relationship approximately carries over to BICG and QMR, and makes us understand why peaks in the residual norm plot of BICG are matched by plateaux in the one of QMR. This result is closely related to our previous discussion of the relationship between (Petrov–)Galerkin and (quasi-)minimal residual methods, in Section 5.

17.1. Trivial minimal residual smoothing

If a monotone residual norm plot is all we aim at, then there is a trivial recipe. We let[17]

$$\widetilde{\mathbf{x}}_n := \begin{cases} \widetilde{\mathbf{x}}_{n-1}, \\ \mathbf{x}_n, \end{cases} \quad \widetilde{\mathbf{r}}_n := \begin{cases} \widetilde{\mathbf{r}}_{n-1}, & \text{if } \|\widetilde{\mathbf{r}}_{n-1}\| < \|\mathbf{r}_n\|, \\ \mathbf{r}_n, & \text{if } \|\widetilde{\mathbf{r}}_{n-1}\| \geq \|\mathbf{r}_n\|. \end{cases}$$

This clearly implies that the residuals $\widetilde{\mathbf{r}}_n$ of the 'smoothed' iterates satisfy

$$\|\widetilde{\mathbf{r}}_n\| = \min\{\|\widetilde{\mathbf{r}}_{n-1}\|, \|\mathbf{r}_n\|\} = \min\{\|\mathbf{r}_0\|, \|\mathbf{r}_1\|, \ldots, \|\mathbf{r}_n\|\}.$$

We call this *trivial minimal residual* (TMR) *smoothing.* We do not consider this as a serious proposal, but mention it, because it is sometimes applied when numerical results are presented. It reflects the position mentioned above, that all that really matters is to fulfil a prescribed bound for the residual norm as quickly as possible. This is a legitimate reason for applying TMR smoothing when publishing results, but the code of conduct requires that authors declare it.

Note that TMR neither increases nor reduces round-off.

17.2. Minimal residual smoothing

Given any pair of sequences of iterates and residuals, for example those produced by the BiCG method, Schönauer (1987) proposed to replace them by the smoothed sequences

$$\widetilde{\mathbf{x}}_n := \widetilde{\mathbf{x}}_{n-1}(1-\theta_n) + \mathbf{x}_n\theta_n, \quad \widetilde{\mathbf{r}}_n := \widetilde{\mathbf{r}}_{n-1}(1-\theta_n) + \mathbf{r}_n\theta_n, \tag{17.1}$$

where θ_n is chosen to make the residual as small as possible, which requires that

$$\widetilde{\mathbf{r}}_{n-1} - \mathbf{r}_n \perp \widetilde{\mathbf{r}}_n, \tag{17.2}$$

or

$$\theta_n := \frac{\langle \widetilde{\mathbf{r}}_{n-1} - \mathbf{r}_n, \widetilde{\mathbf{r}}_{n-1} \rangle}{\|\widetilde{\mathbf{r}}_{n-1} - \mathbf{r}_n\|^2}. \tag{17.3}$$

From the relations (17.1) and (17.2) we conclude by Pythagoras' theorem that

$$\|\widetilde{\mathbf{r}}_n\|^2 = \|\widetilde{\mathbf{r}}_{n-1}\|^2 - \|\widetilde{\mathbf{r}}_{n-1} - \mathbf{r}_n\|^2 |\theta_n|^2. \tag{17.4}$$

The idea was further developed and investigated by Weiss (1990, 1994),

[17] Before, we used tildes for the left Lanczos vectors and the left direction vectors, but we did not define $\widetilde{\mathbf{x}}_n$ and $\widetilde{\mathbf{r}}_n$. These vectors are now the smoothed iterates and residuals.

and then taken up by Gutknecht (1993*a*) as well as Zhou and Walker (1994); see also Walker (1995). Following Zhou and Walker, we call it *minimal residual* (MR) *smoothing.* Note that we perform a *local*, one-dimensional minimization, as we did in BiCGStab. Of course, we could generalize this approach to a local MR(ℓ) smoothing, that is, an ℓ-dimensional minimization additionally involving $\widetilde{\mathbf{r}}_{n-2}, \ldots, \widetilde{\mathbf{r}}_{n-\ell}$. However, numerical tests show that this is hardly worth the extra work (Zhou and Walker 1994).

Obviously, MR smoothing is some kind of a recursive weighted mean process, but, in general, the weights need not be positive. From the definition it is clear, however, that the resulting residual norm plot is monotonically decreasing. Therefore, this is a very effective smoothing process. Note that the given sequence is piped through the process without generating any feedback.

The main theoretical result of Weiss' thesis is that applying MR smoothing to the FOM iterates yields the GCR (or, GMRes) iterates: the orthogonal residuals of the former become conjugate and minimal residuals of the latter method. For a short proof see Gutknecht (1993*a*, p. 49). *A fortiori*, applying the MR smoothing to the CG iterates yields the CR iterates. Hence, in these two cases the transformation must be identical to the relations (5.22) and (5.23), and we conclude that

$$\theta_n = c_n^2 \ \ (\geq 0), \quad 1 - \theta_n = |s_n|^2 \ \ (\geq 0). \tag{17.5}$$

Actually, in these cases of orthogonal residual methods, where $\mathbf{r}_n \perp \mathcal{K}_n$, (17.3) and (17.4) simplify since we have $\mathbf{r}_n \perp \widetilde{\mathbf{r}}_{n-1} \in \mathcal{K}_n$ and thus, again by Pythagoras, $\|\widetilde{\mathbf{r}}_{n-1} - \mathbf{r}_n\|^2 = \|\widetilde{\mathbf{r}}_{n-1}\|^2 + \|\mathbf{r}_n\|^2$, so that

$$\theta_n = \frac{\|\widetilde{\mathbf{r}}_{n-1}\|^2}{\|\widetilde{\mathbf{r}}_{n-1}\|^2 + \|\mathbf{r}_n\|^2} \quad \in \ (0, 1].$$

Inserting this into (17.4) and taking the reciprocal yields

$$\frac{1}{\|\widetilde{\mathbf{r}}_n\|^2} = \frac{\|\widetilde{\mathbf{r}}_{n-1}\|^2 + \|\mathbf{r}_n\|^2}{\|\widetilde{\mathbf{r}}_{n-1}\|^2 \|\mathbf{r}_n\|^2} = \frac{1}{\|\widetilde{\mathbf{r}}_{n-1}\|^2} + \frac{1}{\|\mathbf{r}_n\|^2} = \sum_{k=0}^{n} \frac{1}{\|\mathbf{r}_k\|^2}. \tag{17.6}$$

Here, for the last equality, we applied induction in order to represent the norm of the smoothed residual in terms of the original residual norms. Conversely, solving for $\|\mathbf{r}_n\|^2$ leads to

$$\|\mathbf{r}_n\|^2 = \frac{\|\widetilde{\mathbf{r}}_n\|^2}{1 - \|\widetilde{\mathbf{r}}_n\|^2 / \|\widetilde{\mathbf{r}}_{n-1}\|^2} \quad (n \geq 1). \tag{17.7}$$

We will comment on these formulae later, but want to remind the reader at this point that they only hold if the given residuals $\mathbf{r}_n$ are mutually orthogonal.

In Gutknecht (1993*a*) we pointed out that there is also an algorithm to do the inverse of MR smoothing. One motivation for using it can be that in

Galerkin methods the errors $\mathbf{x}_n - \mathbf{x}_{\text{ex}}$ have a different spectral decomposition and are often smaller than in methods that minimize the residual.

17.3. Quasi-minimal residual smoothing

When we apply MR smoothing to the BiCG iterates, the resulting smoothed iterates differ from those of the QMR method. But in Section 5 we have seen that nevertheless the BiCG iterates $\mathbf{x}_n := \mathbf{x}_n^{\mathrm{G}}$ and their residuals $\mathbf{r}_n := \mathbf{r}_n^{\mathrm{G}}$ are related to the QMR iterates $\widetilde{\mathbf{x}}_n := \mathbf{x}_n^{\mathrm{MR}}$ by the relations (5.22) and (5.23), which are of the form (17.1) with θ_n satisfying (17.5). Now θ_n is no longer determined by one-dimensional minimization in the residual space, which leads to the choice (17.3), but given by (17.5), where, as we have seen in (5.25), c_n^2 and $|s_n|^2$ can be expressed in terms of the norms of the BiCG residual, $||\mathbf{r}_n^G||$, and the quasi-residual, $||\mathbf{q}_n|| = |\widetilde{\eta}_{n+1}|$:

$$\theta_n := c_n^2 = \frac{||\mathbf{q}_n||^2}{||\mathbf{r}_n||^2}. \tag{17.8}$$

Moreover, $||\mathbf{q}_n||^2$ satisfies the recursion (5.24):

$$\frac{1}{||\mathbf{q}_n||^2} = \frac{1}{||\mathbf{q}_{n-1}||^2} + \frac{1}{||\mathbf{r}_n||^2}. \tag{17.9}$$

Of course, in practice, a substitution $v_n := ||\mathbf{q}_n||^2$ will be made, because we do not compute $\mathbf{q}_n$.

Zhou and Walker (1994) suggest applying the smoothing process (17.1) with this choice of weights to any kind of Krylov space solver. They call this QMR *smoothing*.

Comparing (17.9) with (17.6) we see that $||\mathbf{q}_n||$ here takes the place of $||\widetilde{\mathbf{r}}_n||$. In particular, as in (17.6) and (17.7), we now have

$$\frac{1}{||\mathbf{q}_n||^2} = \sum_{k=0}^{n} \frac{1}{||\mathbf{r}_k||^2}, \quad ||\mathbf{r}_n||^2 = \frac{||\mathbf{q}_n||^2}{1 - ||\mathbf{q}_n||^2/||\mathbf{q}_{n-1}||^2} \quad (n \geq 1). \tag{17.10}$$

There is an alternative to formulae (17.8) and (17.9) that leads to the same weights: according to (5.21) we have

$$||\mathbf{r}_n|| = \frac{1}{c_n}\,||\mathbf{q}_n|| = \frac{|s_1 \cdots s_n|}{c_n}\,||\mathbf{r}_0|| = \frac{|s_n|}{c_n}\,||\mathbf{q}_{n-1}||.$$

In view of the interpretation of s_n and c_n as sine and cosine, this suggests defining

$$\tau_n := \arctan \frac{||\mathbf{r}_n||}{||\mathbf{q}_{n-1}||} \in \left[0, \frac{\pi}{2}\right).$$

Then we let

$$c_n := \cos \tau_n = \frac{||\mathbf{q}_{n-1}||}{\sqrt{||\mathbf{r}_n||^2 + ||\mathbf{q}_{n-1}||^2}}, \qquad s_n := \sin \tau_n = \frac{||\mathbf{r}_n||}{\sqrt{||\mathbf{r}_n||^2 + ||\mathbf{q}_{n-1}||^2}}.$$

It is easy to verify that this construction is consistent with the former one of (17.8) and (17.9), and that, as before, $\|\mathbf{r}_n\|\, c_n = \|\mathbf{q}_n\| = \|\mathbf{q}_{n-1}\|\, s_n$. Note that here $s_n \in [0, 1)$, while in the QMR method s_n can take complex values.

17.4. An alternative smoothing algorithm using direction vectors

Zhou and Walker (1994) noticed that when MR or QMR smoothing is applied to BIOMIN or BIOMINS, better numerical results can be obtained with a reformulated algorithm that updates the smoothed iterates using direction vectors. Assume that our iterates and residuals are generated by a formula of the form

$$\mathbf{x}_{n+1} := \mathbf{x}_n + \mathbf{v}_n\omega_n, \qquad \mathbf{r}_{n+1} := \mathbf{r}_n - \mathbf{A}\mathbf{v}_n\omega_n. \tag{17.11}$$

Here, $\mathbf{v}_n$ and ω_n can, but need not, be the quantities from BIOMIN. In order to rewrite the smoothing formulae (17.1) we introduce the difference $\mathbf{u}_n := \mathbf{x}_n - \widetilde{\mathbf{x}}_{n-1}$, so that

$$\widetilde{\mathbf{x}}_n := \widetilde{\mathbf{x}}_{n-1} + \mathbf{u}_n\theta_n, \qquad \widetilde{\mathbf{r}}_n := \widetilde{\mathbf{r}}_{n-1} - \mathbf{A}\mathbf{u}_n\theta_n. \tag{17.12}$$

Next we need update formulae for $\mathbf{u}_n$ and $\mathbf{A}\mathbf{u}_n$; we do not want to spend an extra matrix-vector product on the latter and consider it as a single vector. Substituting the above update formulae into $\mathbf{u}_{n+1} = \mathbf{x}_{n+1} - \widetilde{\mathbf{x}}_n$ yields

$$\mathbf{u}_{n+1} := \mathbf{v}_n\omega_n + \mathbf{u}_n(1-\theta_n), \qquad \mathbf{A}\mathbf{u}_{n+1} := \mathbf{A}\mathbf{v}_n\omega_n + \mathbf{A}\mathbf{u}_n(1-\theta_n). \tag{17.13}$$

Consequently, if for a Krylov space solver (17.11) holds, then the formulae (17.12) and (17.13) are an alternative form of the smoothing process based on (17.1). For MR smoothing there is also a simplified formula for the weight θ_n:

$$\theta_n := \frac{\langle \mathbf{A}\mathbf{u}_n, \widetilde{\mathbf{r}}_{n-1}\rangle}{\|\mathbf{A}\mathbf{u}_n\|^2}. \tag{17.14}$$

17.5. The peak–plateau connection

Numerical experiments with the QMR methods and with MR smoothing show that peaks in the residual norm plot of BICG or BICGS are always matched by plateaux in the plot for the smoothed method, a *plateau* being a region where the residual norm stagnates or decreases only slowly. This phenomenon was studied in Brown (1991) and Cullum (1995), but it was finally Cullum and Greenbaum (1996) who came up with a simple explanation based on formulae we derived above, notably (17.7) and the analogue one in (17.10); see also Walker (1995). They also proved by example that the peaks in a FOM residual plot need not come from a near-singular $\mathbf{H}_n$.

Indeed, according to (17.7), if $\|\mathbf{r}_n\| \gg \|\widetilde{\mathbf{r}}_n\|$, then the denominator must be small, that is, $\|\widetilde{\mathbf{r}}_n\| \approx \|\widetilde{\mathbf{r}}_{n-1}\|$, and *vice versa*. On the other hand, if the norm of the smoothed residual decreases quickly, then the denominator will be close to 1, and thus the residual of the original method is nearly as small as the smoothed one. Consequently, in a very vague sense, either both the original method and the smoothed one do well, or both do badly. Equation (17.6) also makes clear that the smoothed residual cannot be much smaller than the original one; in fact, it follows that

$$||\tilde{\mathbf{r}}_n|| \geq \frac{1}{\sqrt{n+1}} \min_{0 \leq k \leq n} ||\mathbf{r}_k||,$$

and equality only holds if $||\mathbf{r}_k|| = ||\mathbf{r}_n||$ for $k \leq n$. However, we need to recall that (17.6) and (17.7) assume orthogonal original residuals. Therefore, the application of MR smoothing to BiCG residuals is not covered.

Regarding the QMR method (or the application of QMR smoothing to the BiCG residuals), we can draw the same conclusions on the plateau behaviour of the quasi-residuals $\mathbf{q}_n$, since these appear in (17.10). Moreover, we know that the QMR residual is at most $\sqrt{n+1}$ times larger than the quasi-residual, and in practice the factor is often closer to 1. Since the peaks in the BiCG residual norm plot can be several orders of magnitude high, this factor is rather unimportant in this discussion.

18. Accuracy considerations

Unfortunately, in finite precision arithmetic, inherent round-off problems jeopardize the use of the BiO and BiOC processes and related algorithms. There are at least four effects that can cause trouble:

(i) the loss of (bi)orthogonality
(ii) the low relative accuracy of certain inner products, notably δ_n and δ'_n
(iii) inaccurate Krylov space extension
(iv) the deviation between the recursively computed and the true residual.

There has been considerable work on analysing some of these effects, and also on finding ways to reduce them or compensate for them. Regarding the nonsymmetric Lanczos process, this is an area where results are very recent and investigations are still going on: one wants to know where and why Lanczos-type algorithms lose accuracy, and how one can avoid that at a reasonable price. The production of quality software relies heavily on such findings. But this also means that a good implementation of these algorithms is much more complicated than one would expect from the basic descriptions we have given here. We can only give a very brief and superficial overview of this area, however.

Considerable effort has also been invested into the backwards error analysis of the symmetric Lanczos and the standard CG algorithms (Greenbaum 1989, Greenbaum 1994*b*, Greenbaum and Strakos 1992). More recently, this effort has been extended to BiOMin (Tong and Ye 1995). But we cannot discuss this work here.

18.1. Loss of biorthogonality

Recall that, in the nth step of the BIO algorithm, the vectors $\mathbf{y}_{n+1}$ and $\widetilde{\mathbf{y}}_{n+1}$ are determined so that $\widetilde{\mathbf{y}}_n \perp_{\mathbf{B}} \mathbf{y}_{n+1}$, $\widetilde{\mathbf{y}}_{n-1} \perp_{\mathbf{B}} \mathbf{y}_{n+1}$ and $\widetilde{\mathbf{y}}_{n+1} \perp_{\mathbf{B}} \mathbf{y}_n$, $\widetilde{\mathbf{y}}_{n+1} \perp_{\mathbf{B}} \mathbf{y}_{n-1}$. Nevertheless, the biorthogonality $\widetilde{\mathbf{y}}_m \perp_{\mathbf{B}} \mathbf{y}_{n+1}$, $\widetilde{\mathbf{y}}_{n+1} \perp_{\mathbf{B}} \mathbf{y}_m$ theoretically holds for all $m \leq n$. But due to round-off one encounters a loss of this inherited biorthogonality when $n - m$ becomes large. This is no surprise: orthogonal projection necessarily reduces the size of a vector and thus its relative accuracy. If recursively generated projections of vectors are the relevant data used in the process of building up dual bases, and if one counts on inherited orthogonality, it is not surprising that working with finite precision may have a strong effect.

Lanczos (1952, pp. 39–40) was aware of this loss and suggested, on one hand, full reorthogonalization as a possible yet expensive remedy and, on the other hand, for the iterative solution of linear systems, the modification of the right-hand side by damping the components that correspond to the large eigenvalues. This second remedy, which he called 'purification' of the right-hand side, is similar in spirit to what we now call polynomial preconditioning, but he applied it only before the Lanczos process and not at every step. By suitable preconditioning the loss of (bi)orthogonality is reduced and, at the same time, becomes less relevant since the residual becomes sufficiently small long before n is comparable to N in size.

On the other hand, for the eigenvalue problem, where only few preconditioning techniques apply and the accuracy of the tridiagonal matrix is crucial, the loss of orthogonality is a serious problem even in the symmetric case, for which this numerical phenomenon was analysed by Paige (1971, 1976, 1980). His theory is also discussed in Cullum and Willoughby (1985) and Parlett (1980). It allows us to recognize when a critical step occurs that will induce loss of orthogonality. This loss is coupled with the occurrence of extra, so-called *spurious* copies of eigenvalues and eigenvectors. This is nicely demonstrated by the numerical experiments displayed in Parlett (1994). For Hermitian $\mathbf{A}$, Parlett and his coworkers (Parlett 1980, Parlett and Nour-Omid 1989, Parlett and Reid 1981, Parlett and Scott 1979, Parlett et al. 1985, Simon 1984*a*, Simon 1984*b*) as well as Cullum and Willoughby (Cullum and Willoughby 1985, Cullum 1994), and others have explored various ways to get around this problem in practice. While Cullum and Willoughby developed a method to distinguish spurious eigenvalues from true ones, Parlett's group chose to avoid the spurious ones from the beginning by *partial* or *selective reorthogonalization*, that is, by orthogonalizing $\mathbf{y}_{\text{temp}}$ additionally with respect to those basis vectors that were constructed earlier in certain critical steps. All this earlier work is on the symmetric Lanczos process based on three-term recurrences.

Recently, Bai (1994) generalized Paige's theory to the nonsymmetric case and Parlett's student Day (Day, III 1993, Day 1997) adopted the partial reorthogonalization technique; he refers to it in this case as *maintaining duality*. Again, certain earlier computed Lanczos vectors are included in the now two-sided Gram–Schmidt process. This is quite an effective remedy for the loss of biorthogonality, but of course, it causes considerable overhead in memory requirement, computational cost, and program complexity. For eigenvalue computations this extra effort may well be worthwhile, but for linear solvers it seems too costly. Moreover, it is impossible to extend this technique to squared and product methods.

Day also aims at reducing the local error as much as possible. First, since the BIO algorithm involves two-sided Gram–Schmidt orthogonalization, we can implement it in the modified form, that is, replace (2.21e) and (2.21f) by

$$\begin{aligned} \mathbf{y}_{\text{temp}} &:= \mathbf{A}\mathbf{y}_n - \mathbf{y}_{n-1}\beta_{n-1}, & \widetilde{\mathbf{y}}_{\text{temp}} &:= \underline{\mathbf{A}^\star\widetilde{\mathbf{y}}_n} - \widetilde{\mathbf{y}}_{n-1}\widetilde{\beta}_{n-1}, \\ \alpha_n &:= \langle \widetilde{\mathbf{y}}_n, \mathbf{y}_{\text{temp}}\rangle_{\mathbf{B}} / \delta_n, & \widetilde{\alpha}_n &:= \overline{\alpha_n}, \\ \mathbf{y}_{\text{temp}} &:= \mathbf{y}_{\text{temp}} - \mathbf{y}_n\alpha_n, & \widetilde{\mathbf{y}}_{\text{temp}} &:= \widetilde{\mathbf{y}}_{\text{temp}} - \widetilde{\mathbf{y}}_n\widetilde{\alpha}_n. \end{aligned}$$

Then we can make a tiny correction in order to reinforce the orthogonality that we just took into account:

$$\begin{aligned} \partial\alpha_n &:= \langle \widetilde{\mathbf{y}}_n, \mathbf{y}_{\text{temp}}\rangle_{\mathbf{B}} / \delta_n, & \partial\widetilde{\alpha}_n &:= \langle \mathbf{y}_n, \widetilde{\mathbf{y}}_{\text{temp}}\rangle_{\mathbf{B}} / \overline{\delta_n}, \\ \alpha_n &:= \alpha_n + \partial\alpha_n, & \widetilde{\alpha}_n &:= \widetilde{\alpha}_n + \partial\widetilde{\alpha}_n, \\ \mathbf{y}_{\text{temp}} &:= \mathbf{y}_{\text{temp}} - \mathbf{y}_n\partial\alpha_n, & \widetilde{\mathbf{y}}_{\text{temp}} &:= \widetilde{\mathbf{y}}_{\text{temp}} - \widetilde{\mathbf{y}}_n\partial\widetilde{\alpha}_n. \end{aligned}$$

While Day proposes to apply this local reorthogonalization at every step, one can choose to include it only when $||\mathbf{y}_{\text{temp}}||$ or $||\widetilde{\mathbf{y}}_{\text{temp}}||$ is much shorter than $||\mathbf{A}\mathbf{y}_n||$ or $||\mathbf{A}^\star\widetilde{\mathbf{y}}_n||$, respectively. We can also implement this enhancement in LTPMs like BICGSTAB2.

18.2. Low relative accuracy of certain inner products

The inner products $\delta_n := \langle \widetilde{\mathbf{y}}_n, \mathbf{y}_n\rangle_{\mathbf{B}}$ and $\delta'_n := \langle \widetilde{\mathbf{v}}_n, \mathbf{A}\mathbf{v}_n\rangle_{\mathbf{B}}$ are crucial for determining the recurrence coefficients of the BIO and BIOMIN algorithms:

$$\alpha_n := \frac{\langle \widetilde{\mathbf{y}}_n, \mathbf{A}\mathbf{y}_n\rangle_{\mathbf{B}}}{\delta_n}, \quad \beta_{n-1} := \frac{\gamma_n\delta_n}{\delta_{n-1}}, \quad \omega_n := \frac{\delta_n}{\delta'_n}, \quad \psi_n := -\frac{\delta_{n+1}}{\delta_n}.$$

Most of the algorithms we discussed break down if one of these inner products is used but vanishes. Then one has to resort to look-ahead, which is discussed in the next section. However, typically, the look-ahead tolerance is chosen to be rather small. It is therefore quite normal to proceed without look-ahead, although

$$\frac{\delta_n}{||\widetilde{\mathbf{y}}_n||\,||\mathbf{B}\mathbf{y}_n||} \ll 1 \quad \text{or} \quad \frac{\delta'_n}{||\widetilde{\mathbf{v}}_n||\,||\mathbf{B}\mathbf{A}\mathbf{v}_n||} \ll 1,$$

and this means that δ_n and δ'_n suffer from low relative accuracy. It does not help to compute them in double precision.

Note that this is a difficulty that occurs only in the nonsymmetric case and, regarding δ'_n, the symmetric indefinite case. A possible remedy consists of switching to a suitable one-sided Lanczos process (see Section 6), but so far tests have not been so successful.

The same issue comes up in LTPMs, but there the inner products for δ_n and δ'_n are different. Taking appropriate measures leads to improved versions of algorithms from the BiCGStab family; see Sleijpen and van der Vorst (1995*a*, 1995*b*) and Gutknecht and Ressel (1997).

18.3. Inaccurate Krylov space extension

The generation of well-conditioned Krylov space bases is the prime aim of the Lanczos process. In this regard it is important that in the recursions the term that increases the dimension is not small compared to those that lie in the current subspace. For example, in the BiO algorithm it is dangerous if $||\mathbf{A}\mathbf{y}_n||$ is considerably smaller than $||\mathbf{y}_n\alpha_n||$ or $||\mathbf{y}_{n-1}\beta_{n-1}||$; in the BiOC algorithm it is dangerous if $||\mathbf{A}\mathbf{v}_n||$ is much smaller than $||\mathbf{y}_n\varphi_n||$; and in the BiCGStab algorithm $||\mathbf{A}\widehat{\mathbf{w}}_n^n\omega_n||$ should not be small compared to $||\mathbf{w}_n^n||$, nor should $||\mathbf{A}\mathbf{w}_{n+1}^n\chi_{n+1}||$ be small compared to $||\mathbf{w}_{n+1}^n||$. While one has a choice to modify χ_{n+1} in BiCGStab2, the other cases call for the application of look-ahead; see Section 19.

18.4. Deviation between recursive and true residual

Krylov space solvers normally update the residuals recursively, since the computation of the true residual $\mathbf{b} - \mathbf{A}\mathbf{x}_n$ costs an extra matrix-vector multiplication and, according to folklore, convergence is slower if the true residuals are used for further computation; see, for instance, van der Vorst (1992) and Greenbaum (1997). However, while the size of the true residuals is bounded below by the round-off errors that occur in its evaluation, recursively computed residuals keep getting smaller and smaller if a method converges. This is easy to understand for most Krylov space solvers: they are scale-invariant with respect to the size of the residuals, as long as we neglect the iterates. Therefore, at some point the recursively computed and the true residuals necessarily start to deviate from each other completely. Normally, from then on the true residual remains on roughly the same level, while the recursive one continues to decrease.

Numerical experiments readily show that this branch point is not just determined by the round-off in the evaluation of the true residual, but that it may be reached much earlier. In particular, examples with BiCG or BiCGS with peaks in the residual norm plot that are much higher than $||\mathbf{r}_0||$ seem to indicate that the tallest peak, $\max ||\mathbf{r}_n||$, determines the ultimate level of the true residual. Sleijpen et al. (1994) provide a theory to support this observation. However, a more careful analysis of Greenbaum (1994*a*, 1997) shows that under the assumption that the Krylov space solver uses direction vectors, as in (17.11), it is the maximum norm of the iterates (or the corrections) that matters. Of course, very high peaks in the residual norm plot normally go along with iterates whose norm exceeds by far the one of $\mathbf{x}_{\text{ex}}$. Therefore such peaks normally imply a loss of accuracy.

By estimating the round-off in the evaluation of $\mathbf{x}_{n+1}$ and $\mathbf{r}_{n+1}$ according to (17.11), Greenbaum (1997) finds the following result.

Theorem 18.1 If iterates and residuals are updated according to (17.11), the difference between the true residual $\mathbf{b} - \mathbf{A}\mathbf{x}_n$ and the recursively computed residual $\mathbf{r}_n$ satisfies

$$\frac{||\mathbf{b} - \mathbf{A}\mathbf{x}_n - \mathbf{r}_n||}{||\mathbf{A}||\;||\mathbf{x}||} \leq (\epsilon + \mathcal{O}(\epsilon^2))\,[n + 2 + (1 + \gamma + (n+1)(10 + 2\gamma))\Theta_n]\,,$$

where ϵ denotes the machine-epsilon, γ is a constant that is needed to estimate the round-off in the matrix-vector product according to

$$||\mathbf{A}\mathbf{v}_n - \mathrm{fl}(\mathbf{A}\mathbf{v}_n)|| \leq \gamma\epsilon||\mathbf{A}||\;||\mathbf{v}_n||,$$

and

$$\Theta_n := \max_{k \leq n} \frac{||\mathbf{x}_k||}{||\mathbf{x}_{\text{ex}}||}.$$

This estimate is only based on (17.11), and it does not matter where the direction vectors $\mathbf{v}_n$ and the step sizes ω_n come from. Therefore, errors in these quantities do not influence the gap between true and recursive residuals. Clearly, the theorem not only applies to BIOMIN, but, for instance, also to the smoothing process (17.12) and to BICGSTAB if we force (16.9e) to be executed in the given order. It does not directly apply to our general $(3, 2 \times 2)$-type LTPM algorithm (Algorithm 16), since the latter also involves three-term recurrences. For these the maximum residual plays an essential role too. Applying these considerations to the BICGSTAB(2) recursions (16.14a)–(16.14b), we conclude that if the Krylov space vectors associated with $\zeta t_{l-1}(\zeta)$ and $\zeta^2 t_{l-1}(\zeta)$ are close to being linearly dependent, the corresponding terms in

$$\mathbf{w}_{n+1}^{n+1} := \mathbf{w}_{n+1}^{n-1} + \mathbf{A}\mathbf{w}_{n+1}^{n-1}\chi_n + \mathbf{A}^2\mathbf{w}_{n+1}^{n-1}\chi_{n+1}$$

could be large compared to $\mathbf{w}_{n+1}^{n-1}$ and $\mathbf{w}_{n+1}^{n+1}$, thus causing a large deviation between $\mathbf{x}_{n+1}^{n+1}$ and its recursively computed residual $\mathbf{w}_{n+1}^{n+1}$.

It is often possible to overcome the loss of accuracy discussed above by a modification that was first proposed for BIOMINS by Neumaier (1994), but is equally applicable to other methods, although for some it will cost an additional matrix-vector multiplication per step; see Sleijpen and van der

Vorst (1996). One can think of it as a repeated shift of the origin or an *implicit iterative refinement.* First, we let

$$\mathbf{b}' := \mathbf{b} - \mathbf{A}\mathbf{x}_0, \quad \mathbf{x}' := \mathbf{x}_0, \quad \mathbf{x}_0 := \mathbf{o},$$

so that $\mathbf{b} - \mathbf{A}\mathbf{x} = \mathbf{b}' - \mathbf{A}\mathbf{h}$, where $\mathbf{h} := \mathbf{x} - \mathbf{x}'$. We then apply our algorithm of choice to $\mathbf{A}\mathbf{h} = \mathbf{b}'$. At step n, if the *update condition*

$$\|\mathbf{r}_n\| < \|\mathbf{b}'\|\,\gamma' \quad (\text{where } \gamma' \in (0,1] \text{ is given}) \tag{18.1}$$

is satisfied, we include the reassignments

$$\mathbf{b}' := \mathbf{b}' - \mathbf{A}\mathbf{x}_n, \quad \mathbf{x}' := \mathbf{x}' + \mathbf{x}_n, \quad \mathbf{x}_n := \mathbf{o}. \tag{18.2}$$

Note that at every step, we then have

$$\mathbf{r}_n = \mathbf{b}' - \mathbf{A}\mathbf{x}_n = \mathbf{b} - \mathbf{A}(\mathbf{x}' + \mathbf{x}_n).$$

Neumaier actually computed the true residual at every step and chose $\gamma' = 1$, which means that the update is performed at every step where the residual decreases, hence, nearly always. Sleijpen and van der Vorst (1996) followed up on this idea, provided an analysis, and suggested several alternatives to the update condition (18.1). According to our numerical tests, the best strategy depends on the particular example.

In general, each update (18.2) requires an extra matrix-vector product. However, Neumaier found a way to use it in BIOMINS for replacing one of the two other such products, and Sleijpen and van der Vorst achieved the same for BICGSTAB.

19. Look-ahead Lanczos algorithms

Look-ahead Lanczos algorithms are extensions of Lanczos-type algorithms, in particular the BIO and BIOC algorithms, that circumvent the breakdowns one might encounter, or at least most of them. The look-ahead BIO algorithm was first thought of by Gragg (1974) in a paper on matrix interpretations of recursions for continued fractions and Padé fractions; however, he only followed up on his idea in the context of the partial realization problem of system theory (Gragg and Lindquist 1983). Years later it materialized in Taylor's thesis (Taylor 1982) and the paper by Parlett et al. (1985), who, however, concentrated on steps of length two only and used a very different approach, namely by thinking of the BIO algorithm as a two-sided Gram–Schmidt process. Look-ahead was then rediscovered by Gutknecht (1992, 1994*a*) in connection with work on continued fractions associated with rational interpolation (Gutknecht 1989*a*) and joint work with Gene Golub on the modified Chebyshev algorithm (Golub and Gutknecht 1990). At the same time[18], the subject was also taken up by Joubert (1990) and Parlett (1992), and latterly also by Boley, Elhay, Golub and Gutknecht (1991), Freund et al. (1993), Nachtigal (1991), Brezinski, Redivo Zaglia and Sadok (1992*b*), Hochbruck (1992), and others. Much of the earlier work (including Gragg's note) was on exact breakdowns, and the idea was to treat near-breakdowns as if they were exact ones. But Gutknecht (1994*a*, Sections 9 and 10), Freund et al. (1993), Hochbruck (1992), Nachtigal (1991), and Parlett (1992) addressed explicitly the general near-breakdown case. Moreover, in contrast to the continued fraction approach, the two-sided Gram–Schmidt approach requires no modification for near-breakdowns.

We first derive a *look-ahead* BIO *algorithm*, or LABIO for short, and then treat an analogue *look-ahead* BIOC *algorithm* with mixed recurrences. In both cases we first describe the standard versions as given in Gutknecht (1994*a*), Freund et al. (1993), and Freund and Nachtigal (1994) and then describe recently found simplifications. We do not state recursions for the iterates, because these two look-ahead algorithms are normally coupled either with QMR or with 'inconsistent' update formulae for the Galerkin iterates, as in our inconsistent BIORES and BIOMIN algorithms. Finally we refer to some other look-ahead algorithms, including the *composite-step* BICG *algorithm* of Bank and Chan (1993) and Bank and Chan (1994) and the GMRZ and other algorithms of Brezinski, Redivo Zaglia and Sadok (1991).

In retrospect, the LABIO algorithm can be understood as follows: if a breakdown or near-breakdown of the BIO algorithm occurs, we avoid it by temporarily reducing the number of orthogonality conditions that have to be fulfilled, in fact dropping the biorthogonality to the most recent basis vectors. It turns out that in all but certain very exceptional situations (which are referred to as *incurable breakdowns*) one can return to the full set of conditions after just one or a few steps. The resulting pairs of Lanczos vectors are then in a certain way block biorthogonal.

One could also consider making use of the freedom capitalized upon in the one-sided Lanczos algorithm and apply look-ahead only to the right Lanczos vectors.

As before, we could allow for a formal inner product matrix $\mathbf{B}$ that commutes with $\mathbf{A}$, but for simplicity we assume $\mathbf{B} = \mathbf{I}$.

19.1. LABIO*: the look-ahead* BIO *algorithm*

We know from Lemma 12.2 that a regular Lanczos polynomial, and, thus, a pair of *regular* Lanczos vectors $\widetilde{\mathbf{y}}_n$, $\mathbf{y}_n$ ($\neq \mathbf{o}$) satisfying (2.4), exists if and only if the nth leading principal submatrix $\mathbf{M}_n$ of the moment matrix

[18] The publication dates are misleading; one has to look at the dates where the papers were submitted or preprints were made available.

is nonsingular. In finite precision arithmetic we also need to avoid near-singular sections; hence, we want to compute a pair of regular Lanczos vectors only if $\mathbf{M}_n$ is in a certain sense well-conditioned. However, we cannot enforce this requirement directly: first, $\mathbf{M}_n$ itself is not available and even if it were, we would not want to have to compute its condition number; second, leading principal submatrices of a Hankel matrix are notorious for their bad condition; and third, the formulae below will show that, primarily, certain other matrices need to be well-conditioned.

We let $0 = n_0 < n_1 < n_2 < \ldots$ be a set of indices, where we can and want to enforce all orthogonality conditions, that is, where

$$\widetilde{\mathcal{K}}_{n_j} \perp \mathbf{y}_{n_j}, \quad \widetilde{\mathbf{y}}_{n_j} \perp \mathcal{K}_{n_j}. \tag{19.1}$$

We will call these indices *well-conditioned*, and likewise the corresponding Lanczos polynomial and vectors will be referred to as well-conditioned. Clearly, if a Lanczos polynomial is well-conditioned, then it is also regular. A step for constructing well-conditioned Lanczos vectors is therefore sometimes called a *regular step*. When n is not a well-conditioned index, we refer to $\mathbf{y}_n$ and $\widetilde{\mathbf{y}}_n$ as *inner vectors*; and we call a step for computing these vectors an *inner step*. Of course, the well-conditioned indices will be chosen such that the problem of finding the Lanczos polynomial and the Lanczos vectors is well-conditioned in the usual sense. We aim at generating sequences of Lanczos vectors that satisfy (2.5) and (2.6) as before, but where (2.4) is relaxed to

$$\widetilde{\mathcal{K}}_{n_l} \perp \mathbf{y}_n, \quad \widetilde{\mathbf{y}}_n \perp \mathcal{K}_{n_l} \quad \text{if } n_l \le n < n_{l+1}. \tag{19.2}$$

This implies that these sequences are block-biorthogonal: if we define matrix blocks containing groups of Lanczos vectors,

$$Y_l := [\ \mathbf{y}_{n_l} \quad \mathbf{y}_{n_l+1} \quad \cdots \quad \mathbf{y}_{n_{l+1}-1}\ ], \quad \widetilde{Y}_l := [\ \widetilde{\mathbf{y}}_{n_l} \quad \widetilde{\mathbf{y}}_{n_l+1} \quad \cdots \quad \widetilde{\mathbf{y}}_{n_{l+1}-1}\ ], \tag{19.3}$$

and matrix blocks containing the inner products of these vectors,

$$D_l := \widetilde{Y}_l^\star Y_l = \left(\langle \widetilde{\mathbf{y}}_i, \mathbf{y}_k \rangle\right)_{i,k=n_l}^{n_{l+1}-1}, \tag{19.4}$$

then we have

$$\widetilde{Y}_j^\star Y_l = \begin{cases} D_l & \text{if } j = l, \\ O & \text{if } j \neq l. \end{cases} \tag{19.5}$$

For the derivation of the recurrences we assume that every n is associated with the l for which (19.2) holds. We let

$$h_l := n_{l+1} - n_l,$$

and denote the possibly incompleted last blocks by

$$Y_{l;n} := [\ \mathbf{y}_{n_l} \quad \mathbf{y}_{n_l+1} \quad \cdots \quad \mathbf{y}_n\ ], \quad \widetilde{Y}_{l;n} := [\ \widetilde{\mathbf{y}}_{n_l} \quad \widetilde{\mathbf{y}}_{n_l+1} \quad \cdots \quad \widetilde{\mathbf{y}}_n\ ], \tag{19.6}$$

and

$$D_{l;n} := (\langle \widetilde{\mathbf{y}}_i, \mathbf{y}_k \rangle)_{i,k=n_l}^{n} . \tag{19.7}$$

Then we have, in consistency with our earlier notation,

$$\mathbf{Y}_{n+1} := [\; Y_0 \quad \cdots \quad Y_{l-1} \quad Y_{l;n} \;], \quad \widetilde{\mathbf{Y}}_{n+1} := [\; \widetilde{Y}_0 \quad \cdots \quad \widetilde{Y}_{l-1} \quad \widetilde{Y}_{l;n} \;], \tag{19.8}$$

and (19.5) leads to

$$\widetilde{\mathbf{Y}}_{n+1}^{\star} \mathbf{Y}_{n+1} = \mathbf{D}_{n+1} := \text{block diag}\,(D_0, \ldots, D_{l-1}, D_{l;n}). \tag{19.9}$$

We still have to show that such sequences of Lanczos vectors exist, and how they can be constructed. Let us for the moment assume that they exist. Clearly, like any Krylov space basis, they could be constructed according to (2.7), or, in shorthand notation, by

$$\mathbf{A}\mathbf{Y}_n = \mathbf{Y}_{n+1}\underline{\mathbf{T}}_n, \quad \mathbf{A}^{\star}\widetilde{\mathbf{Y}}_n = \widetilde{\mathbf{Y}}_{n+1}\underline{\widetilde{\mathbf{T}}}_n. \tag{19.10}$$

Assuming that the last block is completed, we conclude from (19.1) or (19.5) that $\widetilde{\mathbf{Y}}_n^{\star}\mathbf{Y}_{n+1} = [\; \mathbf{D}_n \mid \mathbf{o} \;]$ and $\widetilde{\mathbf{Y}}_{n+1}^{\star}\mathbf{Y}_n = [\; \mathbf{D}_n^{\top} \mid \mathbf{o} \;]^{\top}$, so that, as in (2.12),

$$\mathbf{D}_n\mathbf{T}_n = \widetilde{\mathbf{Y}}_n^{\star}\mathbf{A}\mathbf{Y}_n = \widetilde{\mathbf{T}}_n^{\star}\mathbf{D}_n \quad \text{if } n = n_{l+1} - 1.$$

Here, the product on the left is a block upper Hessenberg matrix, while the one on the right is a block lower Hessenberg matrix. Consequently, it must be block tridiagonal, and thus $\mathbf{T}_n$ and $\widetilde{\mathbf{T}}_n$ are themselves block tridiagonal, in addition to being Hessenberg matrices. When all this is formulated in terms of the Lanczos polynomials instead of the Lanczos vectors, as in Gutknecht (1994*a*), it is obvious that we can still choose $\widetilde{\mathbf{T}}_n = \overline{\mathbf{T}_n}$. If we want to allow for independent scale factors $\widetilde{\gamma}_n$ and γ_n for the left and right Lanczos vectors, we can achieve this by diagonal scaling as in (2.29). Therefore, in the following we only derive the formulae for the elements of $\mathbf{T}_n$; we know that those for $\widetilde{\mathbf{T}}_n$ look analogous and do not require computing additional inner products.

So, when the lth block is just completed, that is, when $n = n_{l+1} - 1$, then $\mathbf{T}_n$ is of the form

$$\mathbf{T}_n =: \begin{bmatrix} A_0 & B_0 & & & \\ C_0 & A_1 & B_1 & & \\ & C_1 & A_2 & \ddots & \\ & & \ddots & \ddots & B_{l-1} \\ & & & C_{l-1} & A_l \end{bmatrix},$$

where

$$B_{l-1} =: [\; b_{n_l} \quad \cdots \quad b_n \;]$$

is a block of size $h_{l-1} \times h_l$ that is in general full, C_{l-1} is a $h_{l-1} \times h_l$ block that is zero except for the element γ_{n_l-1} in the upper right corner, and A_l is a $h_l \times h_l$ block of Hessenberg form that we write as

$$A_l =: \begin{bmatrix} \alpha_{n_l} & & & & \\ \gamma_{n_l} & a_{n_l+1} & \cdots & a_{n-1} & a_n \\ & \gamma_{n_l+1} & & & \\ & & \ddots & & \\ & & & \gamma_{n-1} & \end{bmatrix}$$

The extended matrix $\underline{\mathbf{T}}_n$ has at the bottom the additional row $[\ \mathbf{o}^\top \mid \gamma_n\ ]$. In this notation the recurrences for the Lanczos vectors can be written as

$$\left.\begin{aligned} \mathbf{y}_{n+1} &:= (\mathbf{A}\mathbf{y}_n - Y_{l;n}a_n - Y_{l-1}b_n)/\gamma_n \\ \widetilde{\mathbf{y}}_{n+1} &:= (\mathbf{A}^\star\widetilde{\mathbf{y}}_n - \widetilde{Y}_{l;n}\widetilde{a}_n - \widetilde{Y}_{l-1}\widetilde{b}_n)/\widetilde{\gamma}_n \end{aligned}\right\} \quad \text{if } n_l < n+1 \le n_{l+1}. \quad (19.11)$$

Here, γ_n and $\widetilde{\gamma}_n$ are again used to normalize the Lanczos vectors.

By now we know that if there exist Lanczos vectors that fulfil (19.2), then they satisfy the above recurrence. It is straightforward to see by induction that, conversely, these recurrences produce such vectors if

(i) we choose the well-conditioned indices such that the diagonal blocks D_l are well-conditioned

(ii) we determine b_n such that the biorthogonality to the previous block is enforced

(iii) we determine a_n in a regular step (that is, when $n+1 = n_{l+1}$) such that the biorthogonality to the just completed block is enforced.

In the inner steps, a_n can be chosen arbitrarily. In particular, to reduce the computational work, we can let it be the zero vector, although for stability we might want to choose differently. For example, one can use these parameters to make the right Lanczos vectors within a block orthogonal to each other, as proposed in Boley et al. (1991).

Enforcing the mentioned conditions readily yields

$$\begin{aligned} b_n &:= D_{l-1}^{-1}\widetilde{Y}_{l-1}^\star\mathbf{A}\mathbf{y}_n, & \widetilde{b}_n &:= D_{l-1}^{\star-1}Y_{l-1}^\star\mathbf{A}^\star\widetilde{\mathbf{y}}_n, & &\text{if } n_l < n+1 \le n_{l+1}, \\ a_n &:= D_l^{-1}\widetilde{Y}_l^\star\mathbf{A}\mathbf{y}_n, & \widetilde{a}_n &:= D_l^{\star-1}Y_l^\star\mathbf{A}^\star\widetilde{\mathbf{y}}_n, & &\text{if } n+1 = n_{l+1}. \end{aligned} \quad (19.12)$$

Formulae (19.11) and (19.12) are the standard ones for a look-ahead step, as given in Gutknecht (1994*a*, §9) and Freund et al. (1993). In the latter paper it is pointed out that by making use of recursions among the inner products the large number of inner products that seem to be needed for evaluating (19.4) and (19.12) can be reduced to just $2h_l$, the same number as for h_l normal steps. Normalizing the Lanczos vectors costs another $2h_l$ inner products. Moreover, one has to store the current and the previous pair of blocks of Lanczos vectors.

However, there is a way to simplify recurrence (19.11). In the formula for b_n in (19.12), we note that due to (19.2) only the last column of $\widetilde{Y}_{l-1}$ contributes to $\widetilde{Y}_{l-1}^{\star}\mathbf{A}\mathbf{y}_n$. Thus, we can replace $\widetilde{Y}_{l-1}^{\star}$ by $\mathbf{l}\,\widetilde{\mathbf{y}}_{n_l-1}^{\star}$, where $\mathbf{l} := \mathbf{l}_{h_{l-1}} := [\ 0 \ \cdots \ 0 \ 1\ ]^{\top} \in \mathbb{R}^{h_{l-1}}$. Consequently, if we let

$$\mathbf{y}'_{l-1} := Y_{l-1}D_{l-1}^{-1}\mathbf{l}, \quad \widetilde{\mathbf{y}}'_{l-1} := \widetilde{Y}_{l-1}D_{l-1}^{\star-1}\mathbf{l},$$

and

$$\beta'_{n-1} := \widetilde{\mathbf{y}}_{n_l-1}^{\star}\mathbf{A}\mathbf{y}_n, \quad \widetilde{\beta}'_{n-1} := \mathbf{y}_{n_l-1}^{\star}\mathbf{A}^{\star}\widetilde{\mathbf{y}}_n \quad (n_l \le n < n_{l+1}),$$

then

$$Y_{l-1}b_n = (Y_{l-1}D_{l-1}^{-1}\mathbf{l})(\widetilde{\mathbf{y}}_{n_l-1}^{\star}\mathbf{A}\mathbf{y}_n) = \mathbf{y}'_{l-1}\beta'_{n-1},$$

and likewise $\widetilde{Y}_{l-1}\widetilde{b}_n = \widetilde{\mathbf{y}}'_{l-1}\widetilde{\beta}'_{n-1}$. Moreover, using the same argument again,

$$(\widetilde{\mathbf{y}}'_{l-1})^{\star}\mathbf{A}\mathbf{y}_n = \mathbf{l}^{\top}D_{l-1}^{-1}\widetilde{Y}_{l-1}^{\star}\mathbf{A}\mathbf{y}_n = (\mathbf{l}^{\top}D_{l-1}^{-1}\mathbf{l})\widetilde{\mathbf{y}}_{n_l-1}^{\star}\mathbf{A}\mathbf{y}_n = (\mathbf{l}^{\top}D_{l-1}^{-1}\mathbf{l})\beta'_{n-1}.$$

Thus we can redefine β'_{n-1} in terms of the new vector $\widetilde{\mathbf{y}}'_{l-1}$ instead of $\widetilde{\mathbf{y}}_{n_l-1}$, and the analogue holds for $\widetilde{\beta}'_{n-1}$:

$$\left.\begin{aligned} \beta'_{n-1} &:= (\mathbf{l}^{\top}D_{l-1}^{-1}\mathbf{l})^{-1}(\widetilde{\mathbf{y}}'_{l-1})^{\star}\mathbf{A}\mathbf{y}_n \\ \widetilde{\beta}'_{n-1} &:= (\mathbf{l}^{\top}D_{l-1}^{-1}\mathbf{l})^{-1}(\mathbf{y}'_{l-1})^{\star}\mathbf{A}^{\star}\widetilde{\mathbf{y}}_n \end{aligned}\right\} \quad (n_l < n+1 \le n_{l+1}). \tag{19.13}$$

The parenthesis contains only the bottom right element of D_{l-1}^{-1}. Putting the pieces together we see that the recursions (19.11) simplify to

$$\left.\begin{aligned} \mathbf{y}_{n+1} &:= (\mathbf{A}\mathbf{y}_n - Y_{l;n}a_n - \mathbf{y}'_{l-1}\beta'_{n-1})/\gamma_n \\ \widetilde{\mathbf{y}}_{n+1} &:= (\mathbf{A}^{\star}\widetilde{\mathbf{y}}_n - \widetilde{Y}_{l;n}\widetilde{a}_n - \widetilde{\mathbf{y}}'_{l-1}\widetilde{\beta}'_{n-1})/\widetilde{\gamma}_n \end{aligned}\right\} \quad \text{if } n_l < n+1 \le n_{l+1}. \tag{19.14}$$

In other words, the previous blocks are replaced by single vectors $\mathbf{y}'_{l-1}$ and $\widetilde{\mathbf{y}}'_{l-1}$, respectively. It can be shown (Hochbruck 1996) that

$$\widetilde{\mathcal{K}}_{n_l-1} \perp \mathbf{y}'_{l-1}, \quad \widetilde{\mathbf{y}}'_{l-1} \perp \mathcal{K}_{n_l-1}. \tag{19.15}$$

At the root of this simplification is the fact that the blocks

$$B_{l-1} = D_{l-1}^{-1}\mathbf{l}\,\widetilde{\mathbf{y}}_{n_l-1}^{\star}\mathbf{A}Y_l$$

are of rank one, as was pointed out in Gutknecht (1992). Using a similar argument Freund and Zha (1993) implicitly capitalized upon this in their Hankel solver, but only recently it was pointed out by Hochbruck (1996), who used yet another approach, that this leads to the above simplification of the look-ahead Lanczos process. In the polynomial formulation the vectors $\mathbf{y}'_{l-1}$ and $\mathbf{y}_{n_l}$ correspond to polynomials that are the denominators of a regular (or, well-conditioned) pair of Padé approximants; see Gutknecht (1993*b*), Gutknecht and Gragg (1994). In the case of an exact breakdown, the above simplification is irrelevant since the block B_{l-1} then has only a single nonzero element, the one in the upper right corner, and $\mathbf{y}'_{l-1}$ is then a multiple of $\mathbf{y}_{n_{l-1}}$; see Gutknecht (1992).

A look-ahead algorithm always requires a *look-ahead strategy*, a recipe as to when to start a look-ahead step and when to terminate it. The above formulae clearly indicate that we need to avoid singular or near-singular blocks D_j, and this suggests requiring that the smallest singular value, $\sigma_{\min}(D_j)$, be larger than a certain tolerance. Indeed, Parlett (1992) showed that the following quantitative result holds: if the Lanczos vectors are normalized, then

$$\min \left\{\sigma_{\min}(\widetilde{\mathbf{Y}}_{n_l}),\ \sigma_{\min}(\mathbf{Y}_{n_l})\right\} \geq \frac{1}{\sqrt{n_l+1}} \min_{0\leq j\leq l} \sigma_{\min}(D_j),$$

However, numerical examples reported in Freund et al. (1993) showed that, in finite precision arithmetic, we should not rely on this result and not monitor $\sigma_{\min}(D_{l;n})$ alone. Instead, it is suggested that the 1-norms of the coefficient vectors in (19.11) be kept below a certain bound when $n+1 = n_{l+1}$, that is, when the new index $n+1$ is declared as well-conditioned:

$$||a_n||_1 \leq \Upsilon,\ \ ||\widetilde{a}_n||_1 \leq \Upsilon,\ \ ||b_n||_1 \leq \Upsilon,\ \ ||\widetilde{b}_n||_1 \leq \Upsilon \implies n_{l+1} := n+1.$$

Here, the bound Υ depends on $\mathbf{A}$ and is typically of the order of $||\mathbf{A}||$. For example, one can start with $\Upsilon := \max\ \{||\mathbf{A}\mathbf{y}_0||, ||\mathbf{A}^\star\widetilde{\mathbf{y}}_0||\}$ and then adjust this bound dynamically during the algorithm. See Freund et al. (1993) for more details. If the simplified formulae (19.14) are used, the bound for b_n and $\widetilde{b}_n$ can be replaced by one for $\mathbf{y}'_{l-1}\beta'_{n-1}$ and $\widetilde{\mathbf{y}}'_{l-1}\widetilde{\beta}'_{n-1}$. There is, unfortunately, the possibility that $D_{l;n}$ remains singular or ill-conditioned for all $n > n_l$. This is what is called an *incurable breakdown*. Then there is no way to escape a restart of the algorithm, but when solving a linear system one can of course restart from the most recent approximation.

Clearly, the same approach can be used to specify an LABIC algorithm. The look-ahead approach for Lanczos-type product methods (LTPMs) introduced in Gutknecht and Ressel (1996) is based on the standard look-ahead procedure, but can also accommodate the above simplification.

19.2. LABIOC: *the look-ahead* BIOC *algorithm*

The BIOC algorithm is susceptible to both Lanczos and pivot breakdowns, and thus a look-ahead generalization of it should be able to cope with both. In addition to the sequence of regular indices n_l for the Lanczos vectors, we therefore have a second sequence of regular indices m_k for the direction vectors. From Lemma 12.2 we know that $\det \mathbf{M}_{n_l} \neq 0$ and $\det \mathbf{M}'_{m_k} \neq 0$ but, as in the last subsection, we only consider here those regular indices that are in some sense well-conditioned. In addition to (19.2) we now want to enforce

$$\widetilde{\mathcal{K}}_{m_k} \perp \mathbf{A}\mathbf{v}_n, \quad \mathbf{A}^\star\widetilde{\mathbf{v}}_n \perp \mathcal{K}_{m_k}, \quad \text{if } m_k \le n < m_{k+1}. \tag{19.16}$$

In the following, we associate with every n a pair (k, l) such that

$$m_k < n \le m_{k+1} \quad \text{and} \quad n_l \le n < n_{l+1} \tag{19.17}$$

hold, and we let $h'_k := m_{k+1} - m_k$. Defining, in analogy to (19.3)–(19.4) and (19.6)–(19.8), blocks V_k, $\widetilde{V}_k$, D'_k, $V_{k;n}$, $\widetilde{V}_{k;n}$, and $D'_{k;n}$, we then have, as in (19.9),

$$\widetilde{\mathbf{V}}^\star_{n+1}\mathbf{A}\mathbf{V}_{n+1} = \mathbf{D}'_{n+1} := \text{block diag}\,(D'_0, \dots, D'_{k-1}, D'_{k;n}). \tag{19.18}$$

Since we want each of the sets $\{\mathbf{y}_i\}$, $\{\widetilde{\mathbf{y}}_i\}$, $\{\mathbf{v}_i\}$, $\{\widetilde{\mathbf{v}}_i\}$ to be a nested basis of the respective Krylov space, it is clear that there should exist recurrences with a matrix representation

$$\begin{aligned} \mathbf{Y}_n = \mathbf{V}_n\mathbf{U}_n, \qquad \mathbf{A}\mathbf{V}_n &= \mathbf{Y}_{n+1}\underline{\mathbf{L}}_n, \\ \widetilde{\mathbf{Y}}_n = \widetilde{\mathbf{V}}_n\widetilde{\mathbf{U}}_n, \qquad \mathbf{A}^\star\widetilde{\mathbf{V}}_n &= \widetilde{\mathbf{Y}}_{n+1}\underline{\widetilde{\mathbf{L}}}_n, \end{aligned} \tag{19.19}$$

where $\underline{\mathbf{L}}_n$ and $\underline{\widetilde{\mathbf{L}}}_n$ are of upper Hessenberg form and $\mathbf{U}_n$ and $\widetilde{\mathbf{U}}_n$ are unit upper triangular. From the polynomial formulation of the BIO algorithm we could again readily conclude that we can assume $\underline{\widetilde{\mathbf{L}}}_n$ and $\widetilde{\mathbf{U}}_n$ are diagonally scaled versions of $\overline{\underline{\mathbf{L}}_n}$ and $\overline{\mathbf{U}_n}$, respectively. Moreover, as in Section 7, we can return to (19.10) by eliminating $\mathbf{V}_n$ and $\widetilde{\mathbf{V}}_n$ from (19.19), while by eliminating $\mathbf{Y}_n$ and $\widetilde{\mathbf{Y}}_n$ we find a block version of (7.10), if we assume the definitions $\underline{\mathbf{T}}_n := \underline{\mathbf{L}}_n\mathbf{U}_n$ and $\underline{\mathbf{T}}'_n := \mathbf{U}_{n+1}\underline{\mathbf{L}}_n$ from (7.8). But we cannot conclude that these are block LU and block UL factorizations of the block triangular matrices $\underline{\mathbf{T}}_n$ and $\underline{\mathbf{T}}'_n$, respectively. The block triangularity of $\underline{\mathbf{T}}'_n$ can be verified as above for $\underline{\mathbf{T}}_n$, but we must keep in mind that the blocks can be of different sizes to those of $\underline{\mathbf{T}}_n$.

If only exact breakdowns are considered, it has been shown that we indeed have block factorizations and that the block sizes are linked to each other in a well-defined way; see Gutknecht (1994*a*). The link between the block sizes follows directly from the block structure theorem for the Padé table. If near-breakdowns are included, there are still arguments to have the breakdowns linked, but not in the same rigid way: typically, for every n_l there is an m_k such that $n_l = m_k$ or $n_l = m_k + 1$. In the first case, $(\widehat{p}_{n_l-1}, p_{n_l})$ is a row-regular (or a row-well-conditioned) pair; in the second case, $(p_{n_l-1}, \widehat{p}_{n_l-1})$ is a column-regular (or a column-well-conditioned) pair; see Hochbruck (1996). The notions of row-regularity and column-regularity, which play a crucial role here, were introduced in Gutknecht (1993*b*). They mean that the respective polynomials (which are FOPS for the two closely related functionals Φ and Φ' of Section 12) are not scalar multiples of each other; and they imply that these polynomials are even relatively prime; see also Gutknecht and Hochbruck (1995), Gutknecht and Gragg (1994).

But let us first consider the general case: after replacing n by $n+1$ in the first line of (19.19), we can always write the last column of these matrix equations as

$$\begin{aligned} \mathbf{y}_n &= \mathbf{V}_{m_k} g_n + V_{k;n-1} g_{n;k} + \mathbf{v}_n, \\ \mathbf{A}\mathbf{v}_n &= \mathbf{Y}_{n_l} f_n + Y_{l;n} f_{n;l} + \mathbf{y}_{n+1}\gamma_n. \end{aligned} \tag{19.20}$$

The conditions (19.2) and (19.16) that led to (19.9) and (19.18), respectively, yield under assumption (19.17)

$$g_n := \mathbf{D}'^{-1}_{m_k} \widetilde{\mathbf{V}}^\star_{m_k} \mathbf{A}\mathbf{y}_n, \quad f_n := \mathbf{D}^{-1}_{n_l} \widetilde{\mathbf{Y}}^\star_{n_l} \mathbf{A}\mathbf{v}_n. \tag{19.21}$$

These formulae suggest that the recurrences (19.20) are long. However, in (19.21) only few blocks of the block diagonal matrices are multiplied by nonzero blocks of the vectors that follow. Recall that we have

$$\widetilde{\mathbf{v}}_i^\star \mathbf{A}\mathbf{y}_n = 0 \quad \text{if } i < n_l - 1, \quad \widetilde{\mathbf{y}}_i^\star \mathbf{A}\mathbf{v}_n = 0 \quad \text{if } i < m_k. \tag{19.22}$$

Therefore, if we define

$$\begin{aligned} k^\star &:= \max\,\{j : j \le k,\ m_j \le \max\{n_l - 1, 0\}\}, \\ l^\star &:= \max\,\{j : j \le l,\ n_j \le m_k\}, \end{aligned}$$

then in (19.21) only the blocks $D'_{k^\star}, \dots, D'_{k-1}$ of $\mathbf{D}'_{m_k}$ and $D_{l^\star}, \dots, D_{l-1}$ of $\mathbf{D}_{n_l}$ matter. Therefore, (19.20) becomes

$$\begin{aligned} \mathbf{v}_n &:= \mathbf{y}_n - \textstyle\sum_{j=k^\star}^{k-1} V_j g_{n;j} - V_{k;n-1} g_{n;k}, \\ \mathbf{y}_{n+1} &:= \left(\mathbf{A}\mathbf{v}_n - \textstyle\sum_{j=l^\star}^{l-1} Y_j f_{n;j} - Y_{l;n} f_{n;l}\right) \frac{1}{\gamma_n}, \end{aligned} \tag{19.23}$$

where

$$\begin{aligned} g_{n;j} &:= D'^{-1}_j \widetilde{V}^\star_j \mathbf{A}\mathbf{y}_n \quad && (j = k^\star, \dots, k-1 \text{ if } n < m_{k+1}, \\ & && \ j = k^\star, \dots, k \text{ if } n = m_{k+1}), \\ f_{n;j} &:= D^{-1}_j \widetilde{Y}^\star_j \mathbf{A}\mathbf{v}_n \quad && (j = l^\star, \dots, l-1 \text{ if } n < n_{l+1} - 1, \\ & && \ j = l^\star, \dots, l \text{ if } n = n_{l+1} - 1), \end{aligned} \tag{19.24}$$

while $g_{n;k}$ and $f_{n;l}$ are arbitrary if $n < m_{k+1}$ or $n < n_{l+1} - 1$, respectively. This means that if we compute inner vectors, then these two coefficients can be chosen as zero vectors, so that the corresponding terms in (19.23) can be dropped. The recurrences (19.23)–(19.24) are due to Freund and Nachtigal (1994); see also Freund and Nachtigal (1993). Of course, analogous formulae exist for the left-hand side vectors. This look-ahead version of the BIOC algorithm is more general than the one sketched in §10 of Gutknecht (1994*a*), because the two sequences of well-conditioned indices, $\{n_l\}$ and $\{m_k\}$ are not assumed to be linked in a certain way.

As mentioned above, typically either $n_l = m_k$ or $n_l = m_k + 1$, and then we can draw further conclusions:

$$\begin{array}{llll} n_l = m_k & \Longrightarrow & k^\star = k-1, & g_{n;k-1} = D'^{-1}_{k-1}\,\mathbf{l}\,\widetilde{\mathbf{v}}^\star_{m_k-1}\mathbf{A}\mathbf{y}_n, \\ & & l^\star = l & (i.e.,\ f_n = \mathbf{o}), \\ n_l = m_k + 1 & \Longrightarrow & k^\star = k & (i.e.,\ g_n = \mathbf{o}), \\ & & l^\star = l-1, & f_{n;l-1} = D^{-1}_{l-1}\,\mathbf{l}\,\widetilde{\mathbf{y}}^\star_{n_l-1}\mathbf{A}\mathbf{v}_n, \end{array} \tag{19.25}$$

Here, we have already taken into account that, in view of (19.22), in these situations only the last columns of $\widetilde{V}_{k-1}$ and $\widetilde{Y}_{l-1}$, respectively, yield a nonzero contribution to $\widetilde{V}^\star_{k-1}\mathbf{A}\mathbf{y}_n$ and $\widetilde{Y}^\star_{l-1}\mathbf{A}\mathbf{v}_n$. This gives rise to a simplification analogous to the one that led from (19.12) to (19.13)–(19.14). Letting

$$\begin{array}{rclrcl} \mathbf{v}'_{k-1} &:=& V_{k-1}D'^{-1}_{k-1}\mathbf{l}, & \widetilde{\mathbf{v}}'_{k-1} &:=& \widetilde{V}_{k-1}D'^{\star-1}_{k-1}\mathbf{l}, \\ \psi'_n &:=& (\mathbf{l}^\top D'^{-1}_{k-1}\mathbf{l})^{-1}(\widetilde{\mathbf{v}}'_{k-1})^\star\mathbf{A}\mathbf{y}_n, & \widetilde{\psi}'_n &:=& (\mathbf{l}^\top D'^{-1}_{k-1}\mathbf{l})^{-1}(\mathbf{v}'_{k-1})^\star\mathbf{A}^\star\widetilde{\mathbf{y}}_n, \\ \mathbf{y}'_{l-1} &:=& Y_{l-1}D^{-1}_{l-1}\mathbf{l}, & \widetilde{\mathbf{y}}'_{l-1} &:=& \widetilde{Y}_{l-1}D^{\star-1}_{l-1}\mathbf{l}, \\ \varphi'_n &:=& (\mathbf{l}^\top D^{-1}_{l-1}\mathbf{l})^{-1}(\widetilde{\mathbf{y}}'_{l-1})^\star\mathbf{A}\mathbf{v}_n, & \widetilde{\varphi}'_n &:=& (\mathbf{l}^\top D^{-1}_{l-1}\mathbf{l})^{-1}(\mathbf{y}_{l-1})^\star\mathbf{A}^\star\widetilde{\mathbf{v}}_n, \end{array} \tag{19.26}$$

we finally obtain the following simplified recurrences for the right-hand side vectors: if $n_l = m_k$, then

$$\begin{array}{rcll} \mathbf{v}_n &=& \mathbf{y}_n - \mathbf{v}'_{k-1}\psi'_n & \text{if } m_k < n < m_{k+1}, \\ \mathbf{v}_n &=& \mathbf{y}_n - V_{k;n-1}g_{n;k} - \mathbf{v}'_{k-1}\psi'_n & \text{if } n = m_{k+1}, \\ \mathbf{y}_{n+1} &=& \mathbf{A}\mathbf{v}_n/\gamma_n & \text{if } n_l \le n < n_{l+1}-1, \\ \mathbf{y}_{n+1} &=& (\mathbf{A}\mathbf{v}_n - Y_{l;n}f_{n;l})/\gamma_n & \text{if } n = n_{l+1}-1, \end{array} \tag{19.27}$$

while if $n_l = m_k + 1$, then

$$\begin{array}{rcll} \mathbf{v}_n &=& \mathbf{y}_n & \text{if } m_k < n < m_{k+1}, \\ \mathbf{v}_n &=& \mathbf{y}_n - V_{k;n-1}g_{n;k} & \text{if } n = m_{k+1}, \\ \mathbf{y}_{n+1} &=& (\mathbf{A}\mathbf{v}_n - \mathbf{y}'_{l-1}\varphi'_n)/\gamma_n & \text{if } n_l \le n < n_{l+1}-1, \\ \mathbf{y}_{n+1} &=& (\mathbf{A}\mathbf{v}_n - Y_{l;n}f_{n;l} - \mathbf{y}'_{l-1}\varphi'_n)/\gamma_n & \text{if } n = n_{l+1}-1. \end{array} \tag{19.28}$$

Again, analogous recurrences exist for the left-hand side vectors. The above formulae appear in polynomial formulation based on a different derivation in Hochbruck (1996). Again, the previous blocks are replaced by a single vector, and, actually, only one of the Lanczos vectors or one of the direction vectors is needed. Now, these two auxiliary vectors satisfy, respectively, (19.15) and

$$\widetilde{\mathcal{K}}_{m_k-1} \perp \mathbf{A}\mathbf{v}'_{k-1}, \quad \mathbf{A}^\star\widetilde{\mathbf{v}}'_{l-1} \perp \mathcal{K}_{m_k-1}. \tag{19.29}$$

If we also know that $m_{k+1} = n_{l+1}$ or $m_{k+1} + 1 = n_{l+1}$ holds at the end of the look-ahead step, then we can further capitalize upon this: only one of the two recurrences for the regular step in (19.27) or (19.28), respectively, is needed, but which of the two recurrences can be dropped depends on the situation at the end of the step; see Hochbruck (1996).

19.3. Other look-ahead Lanczos algorithms

A simple, but also limited, look-ahead approach is the *composite step* BiCG *algorithm* of Bank and Chan (1993, 1994): it is the poor man's look-ahead Lanczos solver. It requires that no Lanczos breakdowns occur, which implies that $|m_{k+1} - m_k| \leq 2$, as one can show by arguments involving the moment matrices or the block structure theorem of the Padé table; see, for instance, Gutknecht (1990, Theorem 3.6). The algorithm is a variation of BiOMin in which such pivot breakdowns are cured by a special double step: an undefined Galerkin iterate $\mathbf{x}_{n+1}$ is skipped, and $\mathbf{x}_{n+2}$ and its residual are then constructed according to

$$\mathbf{x}_{n+2} := \mathbf{x}_n + \mathbf{v}_n\omega_n' + \mathbf{z}_{n+1}\omega_n'', \qquad \mathbf{r}_{n+2} := \mathbf{r}_n - \mathbf{A}\mathbf{v}_n\omega_n' - \mathbf{A}\mathbf{z}_{n+1}\omega_n'',$$

where $\mathbf{z}_{n+1}$ is an auxiliary vector that is itself a linear combination of $\mathbf{r}_n$ and $\mathbf{A}\mathbf{v}_n$. As we know, there are other algorithms that are not susceptible to pivot breakdowns, in particular, BiOQMR and inconsistent BiORes. The composite step BiCG algorithm has the merit that it uses the standard two-term BiOMin version of the BiCG method as default.

The composite step approach has been extended to both BiOMinS (Chan and Szeto 1994) and to LTPMs based on the coupled two-term Lanczos recurrences (Chan and Szeto 1996). Incidentally, the idea can be traced back to Luenberger (1979) and Fletcher (1976), who designed for symmetric indefinite systems CG algorithms that, in case of a breakdown, make use of such double steps by exploiting the concept of *hyperbolic pairs*.

We mentioned earlier that by choosing the inner vectors appropriately in a look-ahead algorithm we can further improve its numerical stability. For the algorithm of Parlett et al. (1985), which was also restricted to steps of length at most two, Khelifi (1991) investigated this freedom and specified an optimal choice.

Recurrences for formal orthogonal polynomials (FOPs) and the closely related Padé approximants, which are so-called convergents ('partial sums') of certain continued fractions, can serve as the basis for look-ahead algorithms restricted to exact breakdowns; see Gutknecht (1992) and (1994*a*), where these algorithms are called *nongeneric*. The relevant continued fractions, the so-called *q-fractions*, can be traced back at least to Chebyshev; see Gutknecht and Gragg (1994) for some historical remarks. Brezinski, Redivo Zaglia, and Sadok have taken up this approach in a series of papers. In the first one, Brezinski et al. (1992*b*), they introduce the MRZ algorithm, which is basically the same as the nongeneric BIODIR algorithm of Gutknecht (1994a). (One difference is that Brezinski et al. (1992b) suggest using $(\mathbf{A}^\star)^i\widetilde{\mathbf{y}}_0$ as the left vectors, which does not work in finite precision arithmetic due to the extremely bad condition of this basis.) In Brezinski and Sadok (1991) the same recurrences are used to define a nongeneric version of the BIODIRS$_2$ algorithm from Section 7 of Gutknecht (1990), and in Cao (1997a) they are applied to the BICGSTAB family.

Brezinski et al. (1991) first suggest two methods (SMRZ and BMRZ) that apply mixed recurrences, but can handle only exact pivot breakdowns, in contrast to other nongeneric algorithms that can cure exact breakdowns of both types, such as, for instance, nongeneric BIOMIN from (Gutknecht 1994a). The authors then turn to the treatment of near-breakdowns in the BIODIR algorithm. Their GMRZ algorithm is based on polynomial recurrences of the form

$$\begin{aligned} \widehat{p}_{m_{k+1}}(\zeta) &:= c'_k(\zeta)\widehat{p}_{m_k}(\zeta) + d'_k(\zeta)\widehat{p}_{m_{k-1}}(\zeta), \\ \widehat{p}_{m_{k+2}}(\zeta) &:= c''_k(\zeta)\widehat{p}_{m_k}(\zeta) + d''_k(\zeta)\widehat{p}_{m_{k-1}}(\zeta), \end{aligned} \tag{19.30}$$

for the direction polynomials, where it is assumed that $\widehat{p}_{m_{k-1}}$ and $\widehat{p}_{m_k}$ as well as $\widehat{p}_{m_{k+1}}$ and $\widehat{p}_{m_{k+2}}$ are pairs of successive regular FOPs (with no ill-conditioned, but in exact arithmetic regular FOPs between them). If, for simplicity, we exclude the possibility of exact breakdowns, this means that we require that

$$m_k = m_{k-1} + 1 \quad \text{and} \quad m_{k+2} = m_{k+1} + 1. \tag{19.31}$$

In (19.30), c'_k, d'_k, c''_k, and d''_k are polynomials of the appropriate degrees, namely h'_k, $h'_k - 1$, $h'_k + 1$, and h'_k, respectively, whose coefficients are determined by enforcing the biconjugacy conditions. Translation into Krylov space notation yields recurrences for the direction vectors. Note that even when these are only applied to the right-hand side vectors, this costs $2h_k + 1$ matrix-vector multiplications with $\mathbf{A}$ (not including those that might be needed for inner products), as opposed to the $h_k + 1$ required by LABIC. Of course, the left Krylov space needs to be generated too, and the authors again suggest using $\{(\mathbf{A}^\star)^n\widetilde{\mathbf{v}}_0\}$ as basis.

Additional recurrences are needed to update the iterates and the residuals of BIODIR. They are based on the following recursion for the residual polynomials:

$$p_{m_{k+1}}(\zeta) := c_k(\zeta)\widehat{p}_{m_k}(\zeta) + (1 + \zeta e_k(\zeta))p_{m_k}(\zeta), \tag{19.32}$$

where c_k and e_k have degrees $h'_k - 1$ and $h'_k - 2$ at most. Since the residual polynomials are normalized at $\zeta = 0$ and need not have full degree, one can easily verify that they are regular (in a suitably adapted sense) for the same indices m_k as the direction polynomials. However, if $\widehat{p}_{m_k}$ is a multiple of p_{m_k} (and $m_k > 0$), the above formula cannot be true since it would imply that $p_{m_{k+1}}$ is also a multiple of p_{m_k}, while successive regular polynomials are known to be relatively prime.

The authors indeed realized later that an extra condition has to be observed; see Brezinski et al. (1993), a paper that gives an overview of the methods proposed by this group. Incidentally, the missing condition is equivalent to $(\widehat{p}_{m_k}, p_{m_k})$ being a *column-regular pair.* As mentioned, this means that the two polynomials are not scalar multiples of each other; even more, they are then automatically relatively prime. Likewise, the restriction $m_k = m_{k-1} + 1$ of (19.31) means that $\widehat{p}_{m_{k-1}}$ and $\widehat{p}_{m_k}$ are a *regular pair*, which also implies that they are relatively prime; see Gutknecht and Gragg (1994). Fortunately, column-regularity implies regularity. Hence, if we do not consider the possibility of exact breakdowns, then the GMRZ algorithm requires us to treat successive look-ahead blocks as a single large one, until we find a pair $(\widehat{p}_{m_k}, p_{m_k})$ of well-conditioned column-regular (*i.e.*, column-well-conditioned) polynomials. This means more overhead and less numerical stability than when blocks of minimum length can be used.

In Brezinski et al. (1991) (see also Brezinski, Redivo Zaglia and Sadok (1992*a*)), a BSMRZ algorithm is introduced additionally. It is supposed to cure near-breakdowns of BIOMIN and is based on (19.32) and an analogue recurrence for generating $\widehat{p}_{m_{k+1}}$. Hence, it proceeds from the pair $(p_{m_k}, \widehat{p}_{m_k})$ to the pair $(p_{m_{k+1}}, \widehat{p}_{m_{k+1}})$, and thus these pairs are again required to be column-regular. Consequently, in this algorithm too, the steps are in general longer than in our LABIOC algorithm, discussed above, even if we only allowed steps that start and end with a column-regular or row-regular pair and thus always applied either (19.27) or (19.28). Moreover, the overhead is again higher, since two matrix-vector products are needed to expand the right Krylov space.

We emphasize that this approach, which was later also applied to the BICGS method (Brezinski and Redivo Zaglia 1994) and LTPMs (Brezinski and Redivo Zaglia 1995), differs considerably from ours, described in detail above, not only because of the preference for BIODIR and BIOMIN, but because of the different type of recurrence. The connection between the two types has been clarified by Hochbruck (1996). The recursions (19.30) with the restriction (19.31) represent a special case of those of Cabay and Meleshko (1993); see also Gutknecht and Gragg (1994). To attain the generality of the Cabay–Meleshko recurrences we would have to replace $\widehat{p}_{m_{k-1}}$ and $\widehat{p}_{m_{k+1}}$ by differently defined polynomials of maximum degree m_{k-1} and m_{k+1}, respectively.

As look-ahead strategy, Brezinski et al. suggest choosing the step size h'_k such that, for a suitably chosen $\varepsilon_1 > 0$,

$$\langle (\mathbf{A}^\star)^{m_k}\widetilde{\mathbf{v}}_0, \mathbf{A}^{i+1}\mathbf{v}_{m_k}\rangle \begin{cases} \leq \varepsilon_1 & \text{if } 0 \leq i < h'_k - 1, \\ > \varepsilon_1 & \text{if } i = h'_k - 1. \end{cases} \tag{19.33}$$

Note that this condition does not guarantee the nonsingularity of the block D'_k that one would use in LABIC, and thus it cannot guarantee that $\mathbf{v}_{m_{k+1}}$ is regular, let alone well-conditioned. (As an example, consider the situation where all the off-diagonal inner products in (19.33) are ε_1 and those on the diagonal are slightly larger.) The authors did not in fact prove that the resulting linear system for the coefficients is nonsingular, but suggest prescribing in addition to (19.33) a threshold for the size of the pivots in the Gaussian elimination; see Brezinski et al.(1992*a*, 1993).

Yet another proposal for curing Lanczos breakdowns has been made by Ye (1994). However, it requires storing all the Lanczos vectors so that when a breakdown occurs, $\tilde{\mathbf{y}}_{n+1}$ can be replaced by a 'newstart vector' that is orthogonal to all previously computed right Lanczos vectors, as in (3.3b). A corresponding linear solver with QMR smoothing is described in Tong and Ye (1996).

20. Outlook

We hope to have convinced the reader that despite some obvious difficulties (such as breakdowns and loss of biorthogonality due to round-off) the unsymmetric Lanczos process is the basis of a series of very effective and reliable algorithms. We do not expect that yet another new algorithm of this type will markedly surpass all those that we know already, but nevertheless there is still research to be done in this area: convergence is not yet well understood, further investigations on how to improve the accuracy and stability of these algorithms are worthwhile, and there is still a shortage of quality software both for conventional and parallel computer architectures.

Acknowledgments
The author is first of all indebted to Klaus Ressel for carefully reading several versions of this paper and for performing some of the recent research that is described. The section on look-ahead profited crucially from hints and comments provided by Marlis Hochbruck. Further important input came from Bill Gragg, Anne Greenbaum, Beresford Parlett, Gerard Sleijpen and Eric de Sturler. Finally, I have to thank Arieh Iserles for his patience and flexibility.

REFERENCES

J. I. Aliaga, V. Hernandez and D. L. Boley (1994), Using the block clustered nonsymmetric Lanczos algorithm to solve control problems for MIMO linear systems, in *Proceedings of the Cornelius Lanczos International Centenary Conference* (J. D. Brown, M. T. Chu, D. C. Ellison and R. J. Plemmons, eds), SIAM, Philadelphia, pp. 387–389.

W. E. Arnoldi (1951), 'The principle of minimized iterations in the solution of the matrix eigenvalue problem', *Quart. Appl. Math.* **9**, 17–29.

S. F. Ashby, T. A. Manteuffel and P. E. Saylor (1990), 'A taxonomy for conjugate gradient methods', *SIAM J. Numer. Anal.* **27**, 1542–1568.

Z. Bai (1994), 'Error analysis of the Lanczos algorithm for the nonsymmetric eigenvalue problem', *Math. Comp.* **62**, 209–226.

R. E. Bank and T. F. Chan (1993), 'An analysis of the composite step biconjugate gradient method', *Numer. Math.* **66**, 295–319.

R. E. Bank and T. F. Chan (1994), 'A composite step bi-conjugate gradient algorithm for nonsymmetric linear systems', *Numerical Algorithms* **7**, 1–16.

R. Barrett, M. Berry, T. F. Chan, J. Demmel, J. Donato, J. Dongarra, V. Eijkhout, R. Pozo, C. Romine and H. van der Vorst (1994), *Templates for the Solution of Linear Systems: Building Blocks for Iterative Methods*, SIAM, Philadelphia.

T. Barth and T. Manteuffel (1994), Variable metric conjugate gradient methods, in *Advances in Numerical Methods for Large Sparse Sets of Linear Systems* (M. Natori and T. Nodera, eds), number 10 *in* 'Parallel Processing for Scientific Computing', Keio University, Yokahama, Japan, pp. 165–188.

D. Boley and G. H. Golub (1991), 'The nonsymmetric Lanczos algorithm and controllability', *Systems Control Lett.* **16**, 97–105.

D. L. Boley (1994), Krylov space methods in linear control and model reduction: A survey, in *Proceedings of the Cornelius Lanczos International Centenary Conference* (J. D. Brown, M. T. Chu, D. C. Ellison and R. J. Plemmons, eds), SIAM, Philadelphia, PA, pp. 377–379.

D. L. Boley, S. Elhay, G. H. Golub and M. H. Gutknecht (1991), 'Nonsymmetric Lanczos and finding orthogonal polynomials associated with indefinite weights', *Numerical Algorithms* **1**, 21–43.

C. Brezinski and M. Redivo Zaglia (1994), 'Treatment of near-breakdown in the CGS algorithms', *Numerical Algorithms* **7**, 33–73.

C. Brezinski and M. Redivo Zaglia (1995), 'Look-ahead in Bi-CGSTAB and other product methods for linear systems', *BIT* **35**, 169–201.

C. Brezinski and H. Sadok (1991), 'Avoiding breakdown in the CGS algorithm', *Numerical Algorithms* **1**, 199–206.

C. Brezinski, M. Redivo Zaglia and H. Sadok (1991), 'Avoiding breakdown and near-breakdown in Lanczos type algorithms', *Numerical Algorithms* **1**, 261–284.

C. Brezinski, M. Redivo Zaglia and H. Sadok (1992*a*), 'Addendum to "Avoiding breakdown and near-breakdown in Lanczos type algorithms"', *Numerical Algorithms* **2**, 133–136.

C. Brezinski, M. Redivo Zaglia and H. Sadok (1992*b*), 'A breakdown-free Lanczos' type algorithm for solving linear systems', *Numer. Math.* **63**, 29–38.

C. Brezinski, M. Redivo Zaglia and H. Sadok (1993), Breakdowns in the implementation of the Lánczos method for solving linear systems, Technical Report ANO-320, Université Lille Flandres Artois.

P. N. Brown (1991), 'A theoretical comparison of the Arnoldi and GMRES algorithms', *SIAM J. Sci. Statist. Comput.* **12**, 58–78.

S. Cabay and R. Meleshko (1993), 'A weakly stable algorithm for Padé approximants and the inversion of Hankel matrices', *SIAM J. Matrix Anal. Appl.* **14**, 735–765.

Z.-H. Cao (1997*a*), 'Avoiding breakdown in variants of the BI-CGSTAB algorithm', *Linear Algebra Appl.* To appear.

Z.-H. Cao (1997*b*), 'On the QMR approach for iterative methods including coupled three-term recurrences for solving nonsymmetric linear systems', *Int. J. Num. Math. Engin.* To appear.

T. F. Chan and T. Szeto (1994), 'A composite step conjugate gradient squared algorithm for solving nonsymmetric linear systems', *Numerical Algorithms* **7**, 17–32.

T. F. Chan and T. Szeto (1996), 'Composite step product methods for solving nonsymmetric linear systems', *SIAM J. Sci. Comput.* **17**, 1491–1508.

T. F. Chan, L. de Pillis and H. van der Vorst (1991), A transpose-free squared Lanczos algorithm and application to solving nonsymmetric linear systems, Technical Report CAM 91-17, Dept. of Mathematics, University of California, Los Angeles.

T. F. Chan, E. Gallopoulos, V. Simoncini, T. Szeto and C. H. Tong (1994), 'A quasi-minimal residual variant of the Bi-CGSTAB algorithm for nonsymmetric systems', *SIAM J. Sci. Comput.* **15**, 338–347.

A. T. Chronopoulos and S. Ma (1989), On squaring Krylov subspace iterative methods for nonsymmetric linear systems, Technical Report 89-67, Computer Science Department, University of Minnesota.

J. Cullum (1995), 'Peaks, plateaus, numerical instabilities in a Galerkin/minimal residual pair of methods for solving $Ax = b$', *Appl. Numer. Math.* **19**, 255–278.

J. Cullum and A. Greenbaum (1996), 'Relations between Galerkin and norm-minimizing iterative methods for solving linear systems', *SIAM J. Matrix Anal. Appl.* **17**, 223–247.

J. K. Cullum (1994), Lanczos algorithms for large scale symmetric and nonsymmetric matrix eigenvalue problems, in *Proceedings of the Cornelius Lanczos International Centenary Conference* (J. D. Brown, M. T. Chu, D. C. Ellison and R. J. Plemmons, eds), SIAM, Philadelphia, PA, pp. 11–31.

J. K. Cullum and R. A. Willoughby (1985), *Lanczos Algorithms for Large Symmetric Eigenvalue Computations (2 Vols.)*, Birkhäuser, Boston-Basel-Stuttgart.

D. Day (1997), 'An efficient implementation of the non-symmetric Lanczos algorithm', *SIAM J. Matrix Anal. Appl.* To appear.

D. M. Day, III (1993), Semi-duality in the two-sided Lanczos algorithm, PhD thesis, University of California at Berkeley.

V. Eijkhout (1994), LAPACK Working Note 78: Computational variants of the CGS and BiCGstab methods, Technical Report UT-CS-94-241, Computer Science Department, University of Tennessee.

S. C. Eisenstat, H. C. Elman and M. H. Schultz (1983), 'Variational iterative methods for nonsymmetric systems of linear equations', *SIAM J. Numer. Anal.* **20**, 345–357.

D. K. Faddeev and V. N. Faddeeva (1964), *Numerische Verfahren der linearen Algebra*, Oldenbourg, München. This is not the same book as *Computational Methods of Linear Algebra.*

K. V. Fernando and B. N. Parlett (1994), 'Accurate singular values and differential qd algorithms', *Numer. Math.* **67**, 191–229.

R. Fletcher (1976), Conjugate gradient methods for indefinite systems, in *Numerical Analysis, Dundee, 1975* (G. A. Watson, ed.), Vol. 506 of *Lecture Notes in Mathematics*, Springer, Berlin, pp. 73–89.

D. R. Fokkema (1996*a*), Enhanced implementation of BiCGstab(ℓ) for solving linear systems of equations, Preprint 976, Department of Mathematics, Utrecht University.

D. R. Fokkema (1996*b*), Subspace Methods for Linear, Nonlinear, and Eigen Problems, PhD thesis, Utrecht University.

D. R. Fokkema, G. L. G. Sleijpen and H. A. van der Vorst (1996), 'Generalized conjugate gradient squared', *J. Comput. Appl. Math.* **71**, 125–146.

R. W. Freund (1992), 'Conjugate gradient-type methods for linear systems with complex symmetric coefficient matrices', *SIAM J. Sci. Statist. Comput.* **13**, 425–448.

R. W. Freund (1993), 'A transpose-free quasi-minimal residual algorithm for non-Hermitian linear systems', *SIAM J. Sci. Comput.* **14**, 470–482.

R. W. Freund (1994), Lanczos-type algorithms for structured non-Hermitian eigenvalue problems, in *Proceedings of the Cornelius Lanczos International Centenary Conference* (J. D. Brown, M. T. Chu, D. C. Ellison and R. J. Plemmons, eds), SIAM, Philadelphia, PA, pp. 243–245.

R. W. Freund and M. Malhotra (1997), 'A block-QMR algorithm for non-Hermitian linear systems with multiple right-hand sides', *Linear Algebra Appl.* To appear.

R. W. Freund and N. M. Nachtigal (1991), 'QMR: a quasi-minimal residual method for non-Hermitian linear systems', *Numer. Math.* **60**, 315–339.

R. W. Freund and N. M. Nachtigal (1993), Implementation details of the coupled QMR algorithm, in *Numerical Linear Algebra* (L. Reichel, A. Ruttan and R. S. Varga, eds), W. de Gruyter, pp. 123–140.

R. W. Freund and N. M. Nachtigal (1994), 'An implementation of the QMR method based on coupled two-term recurrences', *SIAM J. Sci. Comput.* **15**, 313–337.

R. W. Freund and N. M. Nachtigal (1996), 'QMRPACK: a package of QMR algorithms', *ACM Trans. Math. Software* **22**, 46–77.

R. W. Freund and T. Szeto (1991), A quasi-minimal residual squared algorithm for non-Hermitian linear systems, Technical Report 91.26, RIACS, NASA Ames Research Center, Moffett Field, CA.

R. W. Freund and H. Zha (1993), 'A look-ahead algorithm for the solution of general Hankel systems', *Numer. Math.* **64**, 295–321.

R. W. Freund, M. H. Gutknecht and N. M. Nachtigal (1993), 'An implementation of the look-ahead Lanczos algorithm for non-Hermitian matrices', *SIAM J. Sci. Comput.* **14**, 137–158.

G. Golub and C. van Loan (1989), *Matrix Computations*, 2nd edn, Johns Hopkins University Press, Baltimore, MD.

G. H. Golub and M. H. Gutknecht (1990), 'Modified moments for indefinite weight functions', *Numer. Math.* **57**, 607–624.

G. H. Golub and D. P. O'Leary (1989), 'Some history of the conjugate gradient and Lanczos algorithms: 1949–1976', *SIAM Rev.* **31**, 50–102.

G. Golub, B. Kågström and P. Van Dooren (1992), 'Direct block tridiagonalization of single-input single-output systems', *Systems Control Lett.* **18**, 109–120.

W. B. Gragg (1974), 'Matrix interpretations and applications of the continued fraction algorithm', *Rocky Mountain J. Math.* **4**, 213–225.

W. B. Gragg and A. Lindquist (1983), 'On the partial realization problem', *Linear Algebra Appl.* **50**, 277–319.

A. Greenbaum (1989), 'Predicting the behavior of finite precision Lanczos and conjugate gradient computations', *Linear Algebra Appl.* **113**, 7–63.

A. Greenbaum (1994*a*), Accuracy of computed solutions from conjugate-gradient-like methods, in *Advances in Numerical Methods for Large Sparse Sets of Linear Systems* (M. Natori and T. Nodera, eds), number 10 *in* 'Parallel Processing for Scientific Computing', Keio University, Yokahama, Japan, pp. 126–138.

A. Greenbaum (1994*b*), The Lanczos and conjugate gradient algorithms in finite precision arithmetic, in *Proceedings of the Cornelius Lanczos International Centenary Conference* (J. D. Brown, M. T. Chu, D. C. Ellison and R. J. Plemmons, eds), SIAM, Philadelphia, PA, pp. 49–60.

A. Greenbaum (1997), 'Estimating the attainable accuracy of recursively computed residual methods', *SIAM J. Matrix Anal. Appl.* To appear.

A. Greenbaum and Z. Strakos (1992), 'Predicting the behavior of finite precision Lanczos and conjugate gradient computations', *SIAM J. Matrix Anal. Appl.* **13**, 121–137.

M. H. Gutknecht (1989*a*), 'Continued fractions associated with the Newton–Padé table', *Numer. Math.* **56**, 547–589.

M. H. Gutknecht (1989*b*), 'Stationary and almost stationary iterative (k, l)-step methods for linear and nonlinear systems of equations', *Numer. Math.* **56**, 179–213.

M. H. Gutknecht (1990), 'The unsymmetric Lanczos algorithms and their relations to Padé approximation, continued fractions, and the qd algorithm', in *Preliminary Proceedings of the Copper Mountain Conference on Iterative Methods, April 1–5, 1990*, `http://www.scsc.ethz.ch/~mgh/pub/CopperMtn90.ps.Z` and `CopperMtn90-7.ps.Z`.

M. H. Gutknecht (1992), 'A completed theory of the unsymmetric Lanczos process and related algorithms, Part I', *SIAM J. Matrix Anal. Appl.* **13**, 594–639.

M. H. Gutknecht (1993*a*), 'Changing the norm in conjugate gradient type algorithms', *SIAM J. Numer. Anal.* **30**, 40–56.

M. H. Gutknecht (1993*b*), 'Stable row recurrences in the Padé table and generically superfast lookahead solvers for non-Hermitian Toeplitz systems', *Linear Algebra Appl.* **188/189**, 351–421.

M. H. Gutknecht (1993*c*), 'Variants of BiCGStab for matrices with complex spectrum', *SIAM J. Sci. Comput.* **14**, 1020–1033.

M. H. Gutknecht (1994*a*), 'A completed theory of the unsymmetric Lanczos process and related algorithms, Part II', *SIAM J. Matrix Anal. Appl.* **15**, 15–58.

M. H. Gutknecht (1994*b*), The Lanczos process and Padé approximation, in *Proceedings of the Cornelius Lanczos International Centenary Conference* (J. D. Brown, M. T. Chu, D. C. Ellison and R. J. Plemmons, eds), SIAM, Philadelphia, PA, pp. 61–75.

M. H. Gutknecht (1994*c*), 'Local minimum residual smoothing', Talk at Oberwolfach, Germany.

M. H. Gutknecht and W. B. Gragg (1994), Stable look-ahead versions of the Euclidean and Chebyshev algorithms, IPS Research Report 94-04, IPS, ETH Zürich.

M. H. Gutknecht and M. Hochbruck (1995), 'Look-ahead Levinson and Schur algorithms for non-Hermitian Toeplitz systems', *Numer. Math.* **70**, 181–227.

M. H. Gutknecht and K. J. Ressel (1996), Look-ahead procedures for Lanczos-type product methods based on three-term recurrences, Tech. Report TR-96-19, Swiss Center for Scientific Computing.

M. H. Gutknecht and K. J. Ressel (1997), Attempts to enhance the BiCGStab family, Technical report, Swiss Center for Scientific Computing. In preparation.

M. Hanke (1997), 'Superlinear convergence rates for the Lanczos method applied to elliptic operators', *Numer. Math.* To appear.

R. M. Hayes (1954), Iterative methods of solving linear problems on Hilbert space, in *Contributions to the Solution of Simultaneous Linear Equations and the Determination of Eigenvalues* (O. Taussky, ed.), Vol. 49 of *Applied Mathematics Series*, National Bureau of Standards, pp. 71–103.

P. Henrici (1974), *Applied and Computational Complex Analysis, Vol. 1*, Wiley, New York.

M. R. Hestenes (1951), Iterative methods for solving linear equations, NAML Report 52-9, National Bureau of Standards, Los Angeles, CA. Reprinted in *J. Optim. Theory Appl.* 11, 323–334 (1973).

M. R. Hestenes (1980), *Conjugate Direction Methods in Optimization*, Springer, Berlin.

M. R. Hestenes and E. Stiefel (1952), 'Methods of conjugate gradients for solving linear systems', *J. Res. Nat. Bur. Standards* **49**, 409–435.

M. Hochbruck (1992), Lanczos- und Krylov-Verfahren für nicht-Hermitesche lineare Systeme, PhD thesis, Fakultät für Mathematik, Universität Karlsruhe.

M. Hochbruck (1996), 'The Padé table and its relation to certain numerical algorithms', Habilitationsschrift, Universität Tübingen, Germany.

M. Hochbruck and C. Lubich (1997*a*), 'Error analysis of Krylov methods in a nutshell', *SIAM J. Sci. Comput.* To appear.

M. Hochbruck and C. Lubich (1997*b*), 'On Krylov subspace approximations to the matrix exponential operator', *SIAM J. Numer. Anal.* To appear.

R. A. Horn and C. R. Johnson (1985), *Matrix Analysis*, Cambridge University Press.

A. S. Householder (1964), *The Theory of Matrices in Numerical Analysis*, Dover, New York.

K. C. Jea and D. M. Young (1983), 'On the simplification of generalized conjugate-gradient methods for nonsymmetrizable linear systems', *Linear Algebra Appl.* **52**, 399–417.

W. D. Joubert (1990), Generalized conjugate gradient and Lanczos methods for the solution of nonsymmetric systems of linear equations, PhD thesis, Center for Numerical Analysis, University of Texas at Austin. Tech. Rep. CNA-238.

W. D. Joubert (1992), 'Lanczos methods for the solution of nonsymmetric systems of linear equations', *SIAM J. Matrix Anal. Appl.* **13**, 926–943.

S. Kaniel (1966), 'Estimates for some computational techniques in linear algebra', *Math. Comp.* **20**, 369–378.

M. Khelifi (1991), 'Lanczos maximal algorithm for unsymmetric eigenvalue problems', *Appl. Numer. Math.* **7**, 179–193.

K. Kreuzer, H. Miller and W. Berger (1981), 'The Lanczos algorithm for self-adjoint operators', *Physics Letters* **81A**, 429–432.

E. Kreyszig (1978), *Introductory Functional Analysis with Applications*, Wiley.

C. Lanczos (1950), 'An iteration method for the solution of the eigenvalue problem of linear differential and integral operators', *J. Res. Nat. Bur. Standards* **45**, 255–281.

C. Lanczos (1952), 'Solution of systems of linear equations by minimized iterations', *J. Res. Nat. Bur. Standards* **49**, 33–53.

D. G. Luenberger (1979), 'Hyperbolic pairs in the method of conjugate gradients', *SIAM J. Appl. Math.* **17**, 1263–1267.

T. A. Manteuffel (1977), 'The Tchebyshev iteration for nonsymmetric linear systems', *Numer. Math.* **28**, 307–327.

J. A. Meijerink and H. A. van der Vorst (1977), 'An iterative solution method for linear equations systems of which the coefficient matrix is a symmetric M-matrix', *Math. Comp.* **31**, 148–162.

J. A. Meijerink and H. A. van der Vorst (1981), 'Guidelines for the usage of incomplete decompositions in solving sets of linear equations as they occur in practical problems', *J. Comput. Phys.* **44**, 134–155.

W. Murray, ed. (1972), *Numerical Methods for Unconstrained Optimization*, Academic, London.

N. M. Nachtigal (1991), A look-ahead variant of the Lanczos algorithm and its application to the quasi-minimal residual method for non-Hermitian linear systems, PhD thesis, Department of Mathematics, MIT.

A. Neumaier (1994), 'Iterative regularization for large-scale ill-conditioned linear systems', Talk at Oberwolfach.

O. Nevanlinna (1993), *Convergence of Iterations for Linear Equations*, Birkhäuser, Basel.

C. C. Paige (1971), The computations of eigenvalues and eigenvectors of very large sparse matrices, PhD thesis, University of London.

C. C. Paige (1976), 'Error analysis of the Lanczos algorithms for tridiagonalizing a symmetric matrix', *J. Inst. Math. Appl.* **18**, 341–349.

C. C. Paige (1980), 'Accuracy and effectiveness of the Lanczos algorithm for the symmetric eigenproblem', *Linear Algebra Appl.* **34**, 235–258.

C. C. Paige and M. A. Saunders (1975), 'Solution of sparse indefinite systems of linear equations', *SIAM J. Numer. Anal.* **12**, 617–629.

B. N. Parlett (1980), *The Symmetric Eigenvalue Problem*, Prentice-Hall, Englewood Cliffs, NJ.

B. N. Parlett (1992), 'Reduction to tridiagonal form and minimal realizations', *SIAM J. Matrix Anal. Appl.* **13**, 567–593.

B. N. Parlett (1994), Do we fully understand the symmetric Lanczos algorithm yet?, in *Proceedings of the Cornelius Lanczos International Centenary Conference* (J. D. Brown, M. T. Chu, D. C. Ellison and R. J. Plemmons, eds), SIAM, Philadelphia, PA, pp. 93–107.

B. N. Parlett (1995), 'The new qd algorithms', in *Acta Numerica*, Vol. 4, Cambridge University Press, pp. 459–491.

B. N. Parlett and B. Nour-Omid (1989), 'Towards a black box Lanczos program', *Computer Physics Comm.* **53**, 169–179.

B. N. Parlett and J. K. Reid (1981), 'Tracking the progress of the Lanczos algorithm for large symmetric matrices', *IMA J. Numer. Anal.* **1**, 135–155.

B. N. Parlett and D. S. Scott (1979), 'The Lanczos algorithm with selective reorthogonalization', *Math. Comp.* **33**, 217–238.

B. N. Parlett, D. R. Taylor and Z. A. Liu (1985), 'A look-ahead Lanczos algorithm for unsymmetric matrices', *Math. Comp.* **44**, 105–124.

K. J. Ressel and M. H. Gutknecht (1996), QMR-smoothing for Lanczos-type product methods based on three-term recurrences, Tech. Report TR-96-18, Swiss Center for Scientific Computing.

W. Rudin (1973), *Functional Analysis*, McGraw-Hill, New York.

A. Ruhe (1979), 'Implementation aspects of band Lanczos algorithms for computation of eigenvalues of large sparse symmetric matrices', *Math. Comp.* **33**, 680–687.

H. Rutishauser (1953), 'Beiträge zur Kenntnis des Biorthogonalisierungs-Algorithmus von Lanczos', *Z. Angew. Math. Phys.* **4**, 35–56.

H. Rutishauser (1957), *Der Quotienten-Differenzen-Algorithmus*, Mitt. Inst. angew. Math. ETH, Nr. 7, Birkhäuser, Basel.

H. Rutishauser (1990), *Lectures on Numerical Mathematics*, Birkhäuser, Boston.

Y. Saad (1980), 'Variations on Arnoldi's method for computing eigenelements of large unsymmetric systems', *Linear Algebra Appl.* **34**, 269–295.

Y. Saad (1981), 'Krylov subspace methods for solving large unsymmetric systems', *Math. Comp.* **37**, 105–126.

Y. Saad (1982), 'The Lanczos biorthogonalization algorithm and other oblique projection methods for solving large unsymmetric systems', *SIAM J. Numer. Anal.* **2**, 485–506.

Y. Saad (1994), Theoretical error bounds and general analysis of a few Lanczos-type algorithms, in *Proceedings of the Cornelius Lanczos International Centenary Conference* (J. D. Brown, M. T. Chu, D. C. Ellison and R. J. Plemmons, eds), SIAM, Philadelphia, PA, pp. 123–134.

Y. Saad (1996), *Iterative Methods for Sparse Linear Systems*, PWS Publishing, Boston.

Y. Saad and M. H. Schultz (1986), 'GMRES: a generalized minimal residual algorithm for solving nonsymmetric linear systems', *SIAM J. Sci. Statist. Comput.* **7**, 856–869.

W. Schönauer (1987), *Scientific Computing on Vector Computers*, Elsevier, Amsterdam.

H. D. Simon (1984*a*), 'Analysis of the symmetric Lanczos algorithm with reorthogonalization methods', *Linear Algebra Appl.* **61**, 101–131.

H. D. Simon (1984*b*), 'The Lanczos algorithm with partial reorthogonalization', *Math. Comp.* **42**, 115–142.

G. L. G. Sleijpen and D. R. Fokkema (1993), 'BiCGstab(l) for linear equations involving unsymmetric matrices with complex spectrum', *Electronic Trans. Numer. Anal.* **1**, 11–32.

G. L. G. Sleijpen and H. A. van der Vorst (1995*a*), 'Maintaining convergence properties of bicgstab methods in finite precision arithmetic', *Numerical Algorithms* **10**, 203–223.

G. L. G. Sleijpen and H. A. van der Vorst (1995*b*), 'An overview of approaches for the stable computation of hybrid BiCG methods', *Appl. Numer. Math.* **19**, 235–254.

G. L. G. Sleijpen and H. A. van der Vorst (1996), 'Reliable updated residuals in hybrid Bi-CG methods', *Computing* **56**, 141–163.

G. L. G. Sleijpen, H. A. van der Vorst and D. R. Fokkema (1994), 'BiCGstab(l) and other hybrid Bi-CG methods', *Numerical Algorithms* **7**, 75–109.

P. Sonneveld (1989), 'CGS, a fast Lanczos-type solver for nonsymmetric linear systems', *SIAM J. Sci. Statist. Comput.* **10**, 36–52.

G. W. Stewart (1994), Lanczos and linear systems, in *Proceedings of the Cornelius Lanczos International Centenary Conference* (J. D. Brown, M. T. Chu, D. C. Ellison and R. J. Plemmons, eds), SIAM, Philadelphia, PA, pp. 135–139.

T. J. Stieltjes (1884), 'Quelques recherches sur la théorie des quadratures dites mécaniques', *Ann. Sci. École Norm. Paris Sér. 3* **1**, 409–426. [*Oeuvres*, vol. 1, pp. 377–396].

Z. Strakoš (1991), 'On the real convergence rate of the conjugate gradient method', *Linear Algebra Appl.* **154–156**, 535–549.

D. R. Taylor (1982), Analysis of the Look Ahead Lanczos Algorithm, PhD thesis, Dept. of Mathematics, University of California, Berkeley.

C. H. Tong (1994), 'A family of quasi-minimal residual methods for nonsymmetric linear systems', *SIAM J. Sci. Comput.* **15**, 89–105.

C. H. Tong and Q. Ye (1995), 'Analysis of the finite precision bi-conjugate gradient algorithm for nonsymmetric linear systems'. Preprint.

C. H. Tong and Q. Ye (1996), 'A linear system solver based on a modified Krylov subspace method for breakdown recovery', *Numerical Algorithms* **12**, 233–251.

A. van der Sluis (1992), The convergence behaviour of conjugate gradients and Ritz values in various circumstances, in *Iterative Methods in Linear Algebra* (R. Beauwens and P. de Groen, eds), Elsevier (North-Holland), Proceedings IMACS Symposium, Brussels, 1991, pp. 49–66.

A. van der Sluis and H. A. van der Vorst (1986), 'The rate of convergence of conjugate gradients', *Numer. Math.* **48**, 543–560.

A. van der Sluis and H. A. van der Vorst (1987), 'The convergence behavior of Ritz values in the presence of close eigenvalues', *Linear Algebra Appl.* **88/89**, 651–694.

H. A. van der Vorst (1992), 'Bi-CGSTAB: a fast and smoothly converging variant of Bi-CG for the solution of nonsymmetric linear systems', *SIAM J. Sci. Statist. Comput.* **13**, 631–644.

H. A. van der Vorst and C. Vuik (1993), 'The superlinear convergence behaviour of GMRES', *J. Comput. Appl. Math.* **48**, 327–341.

H. F. Walker (1988), 'Implementation of the GMRES method using Householder transformations', *SIAM J. Sci. Statist. Comput.* **9**, 152–163.

H. F. Walker (1995), 'Residual smoothing and peak/plateau behavior in Krylov subspace methods', *Appl. Numer. Math.* **19**, 279–286.

R. Weiss (1990), Convergence behavior of generalized conjugate gradient methods, PhD thesis, University of Karlsruhe.

R. Weiss (1994), 'Properties of generalized conjugate gradient methods', *J. Numer. Linear Algebra Appl.* **1**, 45–63.

J. H. Wilkinson (1965), *The Algebraic Eigenvalue Problem*, Clarendon Press, Oxford.

Q. Ye (1991), 'A convergence analysis for nonsymmetric Lanczos algorithms', *Math. Comp.* **56**, 677–691.

Q. Ye (1994), 'A breakdown-free variation of the nonsymmetric Lanczos algorithms', *Math. Comp.* **62**, 179–207.

S.-L. Zhang (1997), 'GPBI-CG: generalized product-type methods based on Bi-CG for solving nonsymmetric linear systems', *SIAM J. Sci. Comput.* **18**, 537–551.

L. Zhou and H. F. Walker (1994), 'Residual smoothing techniques for iterative methods', *SIAM J. Sci. Comput.* **15**, 297–312.

Acta Numerica (1997), *pp.* 399–436

Numerical solution of multivariate polynomial systems by homotopy continuation methods

T. Y. Li *

Department of Mathematics
Michigan State University
East Lansing, MI 48824–1027
USA
E-mail: li@math.msu.edu

CONTENTS

1. Introduction

Let $P(x) = 0$ be a system of n polynomial equations in n unknowns. Denoting $P = (p_1, \ldots, p_n)$, we want to find all isolated solutions of

$$\begin{aligned} p_1(x_1, \ldots, x_n) &= 0, \\ &\vdots \\ p_n(x_1, \ldots, x_n) &= 0 \end{aligned} \tag{1.1}$$

for $x = (x_1, \ldots, x_n)$. This problem is very common in many fields of science and engineering, such as formula construction, geometric intersection problems, inverse kinematics, power flow problems with PQ-specified bases, computation of equilibrium states, *etc.* Elimination theory-based methods, most notably the Buchberger algorithm (Buchberger 1985) for constructing Gröbner bases, are the classical approach to solving (1.1), but their reliance on symbolic manipulation makes those methods seem somewhat unsuitable for all but small problems.

* This research was supported in part by the NSF under Grant DMS-9504953 and a Guggenheim Fellowship.

In 1977, Garcia and Zangwill (1979) and Drexler (1977) independently presented theorems suggesting that homotopy continuation could be used to find numerically the full set of isolated solutions of (1.1). During the last two decades, this method has been developed into a reliable and efficient numerical algorithm for approximating all isolated zeros of polynomial systems. Modern scientific computing is marked by the advent of vector and parallel computers and the search for algorithms that are to a large extent naturally parallel. A great advantage of the homotopy continuation algorithm for solving polynomial systems is that it is to a large degree parallel, in the sense that each isolated zero can be computed independently. This natural parallelism makes the method an excellent candidate for a variety of architectures. In this respect, it stands in contrast to the highly serial Gröbner bases method.

The homotopy continuation method for solving (1.1) is to define a trivial system $Q(x) = (q_1(x), \dots, q_n(x)) = 0$ and then follow the curves in the real variable t which make up the solution set of

$$0 = H(x,t) = (1-t)Q(x) + tP(x). \tag{1.2}$$

More precisely, if $Q(x) = 0$ is chosen correctly, the following three properties hold:

Property 0 (*Triviality*). The solutions of $Q(x) = 0$ are known.

Property 1 (*Smoothness*). The solution set of $H(x,t) = 0$ for $0 \le t < 1$ consists of a finite number of smooth paths, each parametrized by t in $[0, 1)$.

Property 2 (*Accessibility*). Every isolated solution of $H(x, 1) = P(x) = 0$ can be reached by some path originating at $t = 0$. It follows that this path starts at a solution of $H(x, 0) = Q(x) = 0$.

When the three properties hold, the solution paths can be followed from the initial points (known because of Property 0) at $t = 0$ to all solutions of the original problem $P(x) = 0$ at $t = 1$ using standard numerical techniques; see Allgower and Georg (1990, 1993).

Several authors have suggested choices of Q that satisfy the three properties: *cf.* Chow, Mallet-Paret and Yorke (1979), Li (1983), Morgan (1986), Wright (1985) and Zulener (1988) for a partial list. A typical suggestion is

$$\begin{aligned} q_1(x) &= a_1 x_1^{d_1} - b_1, \\ &\vdots \\ q_n(x) &= a_n x_n^{d_n} - b_n, \end{aligned} \tag{1.3}$$

where $d_1, \dots, d_n$ are the degrees of $p_1(x), \dots, p_n(x)$ respectively, and a_i, b_i are random complex numbers (and therefore nonzero, with probability one).

So in one sense, the original problem posed is solved. All solutions of $P(x) = 0$ are found at the end of the $d_1 \cdots d_n$ paths that make up the solution set of $H(x,t) = 0, 0 \leq t < 1$.

In this article, we report on some recent developments that make this method more convenient to apply.

The reason the problem is not satisfactorily solved by the above considerations is the existence of *extraneous paths.* Although the above method produces $d = d_1 \cdots d_n$ paths, the system $P(x) = 0$ may have fewer than d solutions. We call such a system *deficient.* In this case, some of the paths produced by the above method will be extraneous paths.

More precisely, even though Properties 0–2 imply that each solution of $P(x) = 0$ will lie at the end of a solution path, it is also consistent with these properties that some of the paths may diverge to infinity as the parameter t approaches 1 (the smoothness property rules this out for $t \to t_0 < 1$). In other words, it is quite possible for $Q(x) = 0$ to have more solutions than $P(x) = 0$. In this case, some of the paths leading from roots of $Q(x) = 0$ are extraneous, and diverge to infinity when $t \to 1$ (see Figure 1).

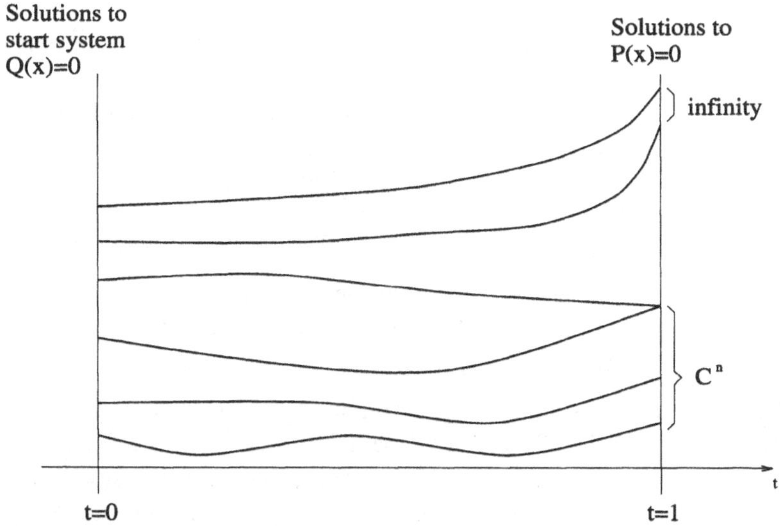

Fig. 1.

Empirically, we find that most systems arising in applications are deficient. A great majority of the systems have fewer than, and in some cases only a small fraction of, the 'expected number' of solutions. For a typical example of this sort, let us look at the following Cassou–Nogues system

$$\begin{aligned}
p_1 &= 15b^4cd^2 + 6b^4c^3 + 21b^4c^2d - 144b^2c - 8b^2c^2e \\
&\quad -28b^2cde - 648b^2d + 36b^2d^2e + 9b^4d^3 - 120, \\
p_2 &= 30b^4c^3d - 32cde^2 - 720b^2cd - 24b^2c^3e - 432b^2c^2 + 576ce \\
&\quad -576de + 16b^2cd^2e + 16d^2e^2 + 16c^2e^2 + 9b^4c^4 + 39b^4c^2d^2 \\
&\quad +18b^4cd^3 - 432b^2d^2 + 24b^2d^3e - 16b^2c^2de - 240c + 5184, \qquad (1.4) \\
p_3 &= 216b^2cd - 162b^2d^2 - 81b^2c^2 + 1008ce - 1008de + 15b^2c^2de \\
&\quad -15b^2c^3e - 80cde^2 + 40d^2e^2 + 40c^2e^2 + 5184, \\
p_4 &= 4b^2cd - 3b^2d^2 - 4b^2c^2 + 22ce - 22de + 261.
\end{aligned}$$

Since $d_1 = 7, d_2 = 8, d_3 = 6$ and $d_4 = 4$ for this system, the system $Q(x)$ in (1.3) will produce $d_1 \times d_2 \times d_3 \times d_4 = 7 \times 8 \times 6 \times 4 = 1344$ paths for the homotopy in (1.2). However, the system (1.4) has only 16 isolated zeros. Consequently, most of the paths are extraneous. Sending out 1344 paths in search of 16 solutions is a highly wasteful computation.

The choice of $Q(x)$ in (1.3) to solve the system $P(x) = 0$ requires an amount of computational effort proportional to $d_1 \cdots d_n$ and, roughly, proportional to the size of the system. We would like to derive methods for solving deficient systems for which the computational effort is instead proportional to the actual number of solutions.

To organize our discussion, we will at times use a notation that makes the coefficients and variables in $P(x) = 0$ explicit. Thus, when the dependence on coefficients is important, we will consider the system $P(c, x) = 0$ of n polynomial equations in n unknowns, where $c = (c_1, \ldots, c_M)$ are coefficients and $x = (x_1, \ldots, x_n)$ are unknowns. Two different problems can be posed:

Problem A Solve the system of equations $P(x) = 0$.

Problem B For each of several different choices of coefficients c, solve the system of equations $P(c, x) = 0$.

We divide our discussion on dealing with and eliminating extraneous paths for Problem A in Section 2, and for Problem B in Section 3. In Section 4, an algorithm is presented which, in some sense, uses the method for Problem B to treat Problem A. Some numerical considerations, the use of projective coordinates and real homotopies, are given in Section 5.

2. Methods for Problem A

Progress on Problem A has been the least satisfactory among the areas we discuss. For deficient systems, there are some partial results that use algebraic geometry to reduce the number of extraneous paths, with various degrees of success.

2.1. Random product homotopy

For a specific example that is quite simple, consider the system

$$\begin{aligned} p_1(x) &= x_1(a_{11}x_1 + \cdots + a_{1n}x_n) + b_{11}x_1 + \cdots + b_{1n}x_n + c_1 = 0, \\ &\vdots \\ p_n(x) &= x_1(a_{n1}x_1 + \cdots + a_{nn}x_n) + b_{n1}x_1 + \cdots + b_{nn}x_n + c_n = 0. \end{aligned} \tag{2.1}$$

This system has total degree $d = d_1 \cdots d_n = 2^n$. Thus the 'expected number' of solutions is 2^n, and the classical homotopy continuation method using the start system $Q(x) = 0$ in (1.3) sends out 2^n paths from 2^n trivial starting points. However, the system $P(x) = 0$ has only $n+1$ isolated solutions (even fewer for special choices of coefficients). This is a deficient system; at least $2^n - n - 1$ paths will be extraneous. It is never known from the start which of the paths will end up being extraneous, so they must all be followed to the end: wasteful computation.

The random product homotopy was developed in Li, Sauer and Yorke (1987*a*, 1987*b*) to alleviate this problem. According to that technique, a more efficient choice for the trivial system $Q(x) = 0$ is

$$\begin{aligned} q_1(x) &= (x_1 + e_{11})(x_1 + x_2 + \cdots + x_n + e_{12}), \\ q_2(x) &= (x_1 + e_{21})(x_2 + e_{22}), \\ &\vdots \\ q_n(x) &= (x_1 + e_{n1})(x_n + e_{n2}). \end{aligned} \tag{2.2}$$

Set

$$H(x,t) = (1-t)cQ(x) + tP(x).$$

It is clear by inspection that for a generic choice of the complex numbers e_{ij}, $Q(x) = 0$ has exactly $n+1$ roots. Thus there are only $n+1$ paths starting from $n+1$ starting points for this choice of homotopy. It is proved in Li, Sauer and Yorke (1987*b*) that Properties 0–2 hold for this choice of $H(x,t)$ for almost all complex numbers e_{ij} and c. Thus all solutions of $P(x) = 0$ are found at the end of the $n+1$ paths. The result of Li et al. (1987*b*) is then both a mathematical result (that there can be at most $n+1$ solutions to (2.1)) and the basis of a numerical procedure for approximating the solutions.

The reason this works is quite simple. The solution paths of (1.2) which do not proceed to a solution of $P(x) = 0$ in $\mathbb{C}^n$ diverge to infinity. If the system (1.2) is viewed in projective space

$$\mathbb{P}^n = \mathbb{C}^{n+1} \backslash \{(0,\ldots,0)\} / \sim,$$

where the equivalent relation '$\sim$' is given by $x \sim y$ if $x = cy$ for some nonzero $c \in \mathbb{C}$, the diverging paths simply proceed to a 'point at infinity' in $\mathbb{P}^n$.

For a polynomial $f(x_1, \dots, x_n)$ of degree d, denote the associated homogeneous polynomial by

$$\tilde{f}(x_0, x_1, \dots, x_n) = x_0^d f(\frac{x_1}{x_0}, \dots, \frac{x_n}{x_0}).$$

The solutions of $f(x) = 0$ at infinity are those zeros of $\tilde{f}$ in $\mathbb{P}^n$ with $x_0 = 0$, and the remaining zeros of $\tilde{f}$ with $x_0 \neq 0$ are the solutions of $f(x) = 0$ in $\mathbb{C}^n$ when x_0 is set to be 1.

Viewed in projective space $\mathbb{P}^n$ the system $P(x) = 0$ in (2.1) has some roots at infinity. The roots at infinity make up a nonsingular variety, specifically the linear space $\mathbb{P}^{n-2}$ defined by $x_0 = x_1 = 0$. A Chern class formula from intersection theory (Fulton 1984, 9.1.1, 9.1.2) shows that the contribution of a linear variety of solutions of dimension e to the 'total degree' $(d_1 \times \cdots \times d_n)$, or the total expected number of solutions, of the system is at least s, where s is the coefficient of t^e in the Maclaurin series expansion of

$$(1+t)^{e-n} \prod_{i=1}^{n} (1 + d_i t).$$

In our case, $d_1 = \cdots = d_n = 2$, and $e = n - 2$, hence,

$$\frac{(1+2t)^n}{(1+t)^2} = \frac{\sum_{i=0}^{n} (1+t)^{n-i} t^i \binom{n}{i}}{(1+t)^2} = \sum_{i=0}^{n} (1+t)^{n-i-2} t^i \binom{n}{i}$$

and $s = \sum_{i=0}^{n-2} \binom{n}{i}$, meaning there are at least $\sum_{i=0}^{n-2} \binom{n}{i}$ solutions of $P(x) = 0$ at infinity. Thus there are at most

$$2^n - s = (1+1)^n - \sum_{i=0}^{n-2} \binom{n}{i} = n + 1$$

solutions of $P(x) = 0$ in $\mathbb{C}^n$. The system $Q(x) = 0$ is chosen to have the same nonsingular variety at infinity, and this variety stays at infinity as the homotopy progresses from $t = 0$ to $t = 1$. As a result, the infinity solutions stay infinite, the finite solution paths stay finite, and no extraneous paths exist.

This turns out to be a fairly typical situation. Even though the system $P(x) = 0$ to be solved has isolated solutions, when viewed in projective space there may be large number of roots at infinity, and quite often high-dimensional manifolds of roots at infinity. Extraneous paths are those that are drawn to the manifolds lying at infinity. If $Q(x) = 0$ can be chosen correctly, extraneous paths can be eliminated.

As another example, consider the algebraic eigenvalue problem

$$Ax = \lambda x,$$

where

$$A = \begin{bmatrix} a_{11} & \cdots & a_{1n} \\ \vdots & & \\ a_{n1} & \cdots & a_{nn} \end{bmatrix}$$

is an $n \times n$ matrix. This problem is actually one of n polynomial equations in the $n+1$ variables $\lambda, x_1, \ldots, x_n$:

$$\begin{gathered} \lambda x_1 - (a_{11}x_1 + \cdots + a_{1n}x_n) = 0, \\ \vdots \\ \lambda x_n - (a_{n1}x_1 + \cdots + a_{nn}x_n) = 0. \end{gathered}$$

Augmenting the system with a linear equation

$$c_1x_1 + \cdots + c_nx_n + c_{n+1} = 0,$$

where $c_1, \ldots, c_{n+1}$ are chosen at random, we have a polynomial system of $n+1$ equations in $n+1$ variables. This system has total degree 2^n. However, it can have at most n isolated solutions. So, the system is deficient. But the system $Q(x)$ in random product form:

$$\begin{aligned} q_1 &= (\lambda + e_{11})(x_1 + e_{12}), \\ q_2 &= (\lambda + e_{21})(x_2 + e_{22}), \\ &\vdots \\ q_n &= (\lambda + e_{n1})(x_n + e_{n2}), \\ q_{n+1} &= c_1x_1 + \cdots + c_nx_n + c_{n+1} \end{aligned}$$

has n isolated zeros for randomly chosen e_{ij}s. This $Q(x)$ will produce n curves for the homotopy in (1.3) that proceed to all solutions of the eigenvalue problem. Implicit in this is the fact that the algebraic eigenvalue problem has at most n solutions. Moreover, the generic eigenvalue problem has exactly n solutions.

To be more precise, we state the main random product homotopy result, Theorem 2.2 of Li et al. (1987*b*). Let $V_\infty(Q)$ and $V_\infty(P)$ denote the variety of roots at infinity of $Q(x) = 0$ and $P(x) = 0$ respectively.

Theorem 2.1 *If $V_\infty(Q)$ is nonsingular and contained in $V_\infty(P)$, then Properties 1 and 2 hold.*

Of course, Properties 1 and 2 are not enough. Without starting points, the path-following method cannot begin. Thus $Q(x) = 0$ should also be chosen to be of random product form, as in (2.2), these being trivial to solve.

This result was superseded by the result in Li and Sauer (1989). The complex numbers e_{ij} are chosen at random in Li et al. (1987*b*) to ensure Properties 1 and 2. In Li and Sauer (1989), it was proved that e_{ij} can be any fixed numbers; as long as the complex number c is chosen at random,

Properties 1 and 2 still hold. In fact, the result in Li and Sauer (1989) implies that the start system $Q(x) = 0$ in Theorem 2.1 need not be in product form. It can be any chosen polynomial system as long as its zeros in $\mathbb{C}^n$ are known or easy to obtain and its variety of roots at infinity $V_\infty(Q)$ is nonsingular and contained in $V_\infty(P)$.

Theorem 2.1 in Li and Wang (1991) goes one step further. Even when the set $V_\infty(Q)$ of roots at infinity of $Q(x) = 0$ has singularities, if the set is contained in $V_\infty(P)$ counting multiplicities, that is, containment in the sense of *scheme* theory of algebraic geometry, then Properties 1 and 2 still hold. To be more precise, let $I =< \tilde{q}_1, \ldots, \tilde{q}_n >$ and $J =< \tilde{p}_1, \ldots, \tilde{p}_n >$ be the homogeneous ideals spanned by homogenizations of q_is and p_is respectively. For a point p at infinity, if the *local rings* I_p and J_p satisfy

$$I_p \subset J_p,$$

then Properties 1 and 2 hold. However, this hypothesis can be much more difficult to verify than the singularity of the set. This limits the usefulness of this approach for practical examples.

2.2. m-homogeneous structure

In Morgan and Sommese (1987*b*), another interesting approach to Problem A is developed, using the concept of m-homogeneous structure.

The complex n-space $\mathbb{C}^n$ can be naturally embedded in $\mathbb{P}^n$. Similarly, the space $\mathbb{C}^{k_1} \times \cdots \times \mathbb{C}^{k_m}$ can be naturally embedded in $\mathbb{P}^{k_1} \times \cdots \times \mathbb{P}^{k_m}$. A point $(y_1, \ldots, y_m)$ in $\mathbb{C}^{k_1} \times \cdots \times \mathbb{C}^{k_m}$ with $y_i = (y_1^{(i)}, \ldots, y_{k_i}^{(i)})$, $i = 1, \ldots, m$, corresponds to a point $(z_1, \ldots, z_m)$ in $\mathbb{P}^{k_1} \times \cdots \times \mathbb{P}^{k_m}$ with $z_i = (z_0^{(i)}, \ldots, z_{k_i}^{(i)})$ and $z_0^{(i)} = 1$, $i = 1, \ldots, m$. The set of such points in $\mathbb{P}^{k_1} \times \cdots \times \mathbb{P}^{k_m}$ is usually called the *affine space* in this setting. The points in $\mathbb{P}^{k_1} \times \cdots \times \mathbb{P}^{k_m}$ with at least one $z_0^{(i)} = 0$ are called the *points at infinity.*

Let f be a polynomial in the n variables $x_1, \ldots, x_n$. If we partition the variables into m groups $y_1 = (x_1^{(1)}, \ldots, x_{k_1}^{(1)}), y_2 = (x_1^{(2)}, \ldots, x_{k_2}^{(2)}), \ldots, y_m = (x_1^{(m)}, \ldots, x_{k_m}^{(m)})$ with $k_1 + \cdots + k_m = n$ and let d_i be the degree of f with respect to y_i (more precisely, to the variables in y_i), then we can define its m-homogenization as

$$\tilde{f}(z_1, \ldots, z_m) = (z_0^{(1)})^{d_1} \times \cdots \times (z_0^{(m)})^{d_m} f(y_1/z_0^{(1)}, \ldots, y_m/z_0^{(m)}).$$

This polynomial is homogeneous with respect to each $z_i = (z_0^{(i)}, \ldots, z_{k_i}^{(i)})$, $i = 1, \ldots, m$. Here $z_j^{(i)} = x_j^{(i)}$, for $j \neq 0$. Such a polynomial is said to be *m-homogeneous*, and $(d_1, \ldots, d_m)$ is the m-homogeneous degree of f. To illustrate this definition, let us consider the polynomial $p_i(x)$ in (2.1):

$$\begin{aligned} p_i(x) &= x_1(a_{i1}x_1 + \cdots + a_{in}x_n) + b_{i1}x_1 + \cdots + b_{in}x_n + c_i \\ &= a_{i1}x_1^2 + x_1(a_{i2}x_2 + \cdots + a_{in}x_n + b_{i1}) + b_{i2}x_2 + \cdots + b_{in}x_n + c_i. \end{aligned}$$

It is sufficient to set $y_1 = (x_1), y_2 = (x_2, \ldots, x_n)$ and $z_1 = (x_0^{(1)}, x_1)$, $z_2 = (x_0^{(2)}, x_2, \ldots, x_n)$. The degree of $p_i(x)$ is two with respect to y_1 and is one with respect to y_2. Hence, its 2-homogenization is

$$\begin{aligned} \widetilde{p}_i(z_1, z_2) &= a_{i1}x_1^2 x_0^{(2)} + x_1 x_0^{(1)}(a_{i2}x_2 + \cdots + a_{in}x_n + b_{i1}x_0^{(2)}) \\ &+ (x_0^{(1)})^2(b_{i2}x_2 + \cdots + b_{in}x_n + c_i x_0^{(2)}). \end{aligned}$$

which is homogeneous with respect to both z_1 and z_2. When the system (2.1) is viewed in $\mathbb{P}^n = \{(x_0, x_1, \ldots, x_n)\}$ with the homogenization

$$\begin{aligned} \widetilde{p}_1(x_0, x_1, \ldots, x_n) &= x_1(a_{11}x_1 + \cdots + a_{1n}x_n) \\ &\quad +(b_{11}x_1 + \cdots + b_{1n}x_n)x_0 + c_1 x_0^2 = 0, \\ &\vdots \\ \widetilde{p}_n(x_0, x_1, \ldots, x_n) &= x_1(a_{n1}x_1 + \cdots + a_{nn}x_n) \\ &\quad +(b_{n1}x_1 + \cdots + b_{nn}x_n)x_0 + c_n x_0^2 = 0, \end{aligned}$$

its total degree, or Bézout number, is $d = d_1 \cdots d_n = 2^n$. However, when (2.1) is viewed in $\mathbb{P}^1 \times \mathbb{P}^{n-1} = \{(z_1, z_2) = ((x_0^{(1)}, x_1), (x_0^{(2)}, x_2, \ldots, x_n))\}$ with the 2-homogenization

$$\begin{aligned} \widetilde{p}_1(z_1, z_2) &= a_{11}x_1^2 x_0^{(2)} + x_1 x_0^{(1)}(a_{12}x_2 + \cdots + a_{in}x_n + b_{11}x_0^{(2)}) \\ &\quad +(x_0^{(1)})^2(b_{12}x_2 + \cdots + b_{1n}x_n + c_1 x_0^{(2)}), \\ &\vdots \\ \widetilde{p}_n(z_1, z_2) &= a_{n1}x_1^2 x_0^{(2)} + x_1 x_0^{(1)}(a_{n2}x_2 + \cdots + a_{nn}x_n + b_{n1}x_0^{(2)}) \\ &\quad +(x_0^{(1)})^2(b_{n2}x_2 + \cdots + b_{nn}x_n + c_n x_0^{(2)}), \end{aligned} \tag{2.3}$$

the Bézout number d is different, and equals the coefficient of $\alpha_1^1\alpha_2^{n-1}$ in the product $(2\alpha_1 + \alpha_2)^n$. Thus, $d = 2n$. In general, for an m-homogeneous system

$$\begin{aligned} \widetilde{p}_1(z_1, \ldots, z_m) &= 0, \\ &\vdots \\ \widetilde{p}_n(z_1, \ldots, z_m) &= 0, \end{aligned} \tag{2.4}$$

in $\mathbb{P}^{k_1} \times \cdots \times \mathbb{P}^{k_m}$ with $\widetilde{p}_i$ having m-homogeneous degree $(d_1^{(i)}, \ldots, d_m^{(i)})$, $i = 1, \ldots, n$, with respect to $(z_1, \ldots, z_m)$, then the m-homogeneous Bézout number d of the system with respect to $(z_1, \ldots, z_m)$ is the coefficient of $\alpha_1^{k_1} \times \cdots \times \alpha_m^{k_m}$ in the product

$$(d_1^{(1)}\alpha_1 + \cdots + d_m^{(1)}\alpha_m)(d_1^{(2)}\alpha_1 + \cdots + d_m^{(2)}\alpha_m) \cdots (d_1^{(n)}\alpha_1 + \cdots + d_m^{(n)}\alpha_m)$$

(Shafarevich 1977). The classical Bézout Theorem says the system (2.4) has no more than d isolated solutions, counting multiplicities, in $\mathbb{P}^{k_1} \times \cdots \times \mathbb{P}^{k_m}$. Applying this to our example in (2.3), the upper bound on the number of isolated solutions of (2.3), in affine space and at infinity, is $2n$. When solving the original system in (2.1), we may choose the start system $Q(x) = 0$ in the homotopy

$$H(x,t) = (1-t)cQ(x) + tP(x)$$

in random product form to respect the 2-homogeneous structure of $P(x)$. For instance, we may choose $Q(x) = 0$ to be

$$\begin{aligned} q_1(x) &= (x_1 + e_{11})(x_1 + e_{12})(x_2 + \cdots + x_n + e_{13}), \\ q_2(x) &= (x_1 + e_{21})(x_1 + e_{22})(x_2 + e_{23}), \\ &\vdots \\ q_n(x) &= (x_1 + e_{n1})(x_1 + e_{n2})(x_n + e_{n3}), \end{aligned} \tag{2.5}$$

which has the same 2-homogeneous structure as $P(x)$ with $y_1 = (x_1)$ and $y_2 = (x_2, \ldots, x_n)$. Namely, each $q_i(x)$ has degree two with respect to y_1 and degree one with respect to y_2. It is easy to see that for randomly chosen complex numbers e_{ij}, $Q(x) = 0$ has $2n$ solutions in $\mathbb{C}^n (= \mathbb{C}^1 \times \mathbb{C}^{n-1})$ (thus, no solutions at infinity when viewed in $\mathbb{P}^1 \times \mathbb{P}^{n-1}$). Hence there are $2n$ paths starting from $2n$ starting points for this choice of homotopy. It is shown in Morgan and Sommese (1987*b*) that Properties 1 and 2 hold for all complex numbers c, except those lying on a finite number of rays starting at the origin. Thus, all solutions of $P(x) = 0$ are found at the end of $n+1$ paths. The number of extraneous paths, $2n - (n+1) = n - 1$, is far less than the corresponding number, namely $2^n - n - 1$, arising via classical homotopy with $Q(x) = 0$ in (1.3).

More precisely, we state the main theorem in Morgan and Sommese (1987*b*).

Theorem 2.2 Let $Q(x)$ be a system of equations chosen to have the same m-homogeneous form as $P(x)$ with respect to a certain partition of the variables $(x_1, \ldots, x_n)$. Assume that $Q(x) = 0$ has exactly the Bézout number of nonsingular solutions with respect to this partition, and define

$$H(x,t) = (1-t)cQ(x) + tP(x),$$

where $t \in [0,1]$ and $c \in \mathbb{C}$. If $c = re^{i\theta}$ for some positive r, then, for all but finitely many θ, Properties 1 and 2 hold.

In general, if $x = (x_1, \ldots, x_n)$ is partitioned into $x = (y_1, \ldots, y_m)$ where

$$y_1 = (x_1^{(1)}, \ldots, x_{k_1}^{(1)}),\ y_2 = (x_1^{(2)}, \ldots, x_{k_2}^{(2)}),\ \ldots,\ y_m = (x_1^{(m)}, \ldots, x_{k_m}^{(m)}),$$

with $k_1 + \cdots + k_m = n$, and $p_i(x)$ has degree $(d_1^{(i)}, \ldots, d_m^{(i)})$ with respect to $(y_1, \ldots, y_m)$, $i = 1, \ldots, n$, then we may choose the start system $Q(x) = (q_1(x), \ldots, q_n(x))$,

$$q_i(x) = \prod_{j=1}^{m} \prod_{\ell=1}^{d_j^{(i)}} (c_{\ell 1}^{(j)} x_1^{(j)} + \cdots + c_{\ell k_j}^{(j)} x_{k_j}^{(j)} + c_{\ell 0}^{(j)}), \quad i = 1, \ldots, n. \quad (2.6)$$

Clearly, $q_i(x)$ has degree $(d_1^{(i)}, \ldots, d_m^{(i)})$ with respect to $(y_1, \ldots, y_m)$, the same degree structure of $p_i(x)$. Further, it is not hard to see that, for random coefficients, $Q(x)$ has exactly an m-homogeneous Bézout number, with respect to this particular partition $x = (y_1, \ldots, y_m)$, of nonsingular isolated solutions in $\mathbb{C}^n$. Those solutions are easy to obtain: the system $Q(x)$ in (2.5) is constructed according to this principle. In Wampler (1994), the product in (2.6) is modified along the same principle to be more efficient to evaluate.

In the example above, there are still $n-1$ extraneous paths. This is because, even when it is viewed in $\mathbb{P}^1 \times \mathbb{P}^{n-1}$, $P(x)$ has zeros at infinity. One can see in (2.3) that

$$S = \{((x_0^{(1)}, x_1), (x_0^{(2)}, x_2, \ldots, x_n)) \in \mathbb{P}^1 \times \mathbb{P}^{n-1} : x_0^{(1)} = 0, x_0^{(2)} = 0\}$$

is a set of zeros of $P(x)$ at infinity. So, to lower the number of those extraneous paths further, we may choose the start system to have the same nonsingular variety of roots as $P(x) = 0$ at infinity, in addition to sharing the same 2-homogeneous structure of $P(x)$. For instance, the system $Q(x) = (q_1(x), \ldots, q_n(x))$ where

$$\begin{aligned} q_1(x) &= (x_1 + e_{11})(x_1 + x_2 + \cdots + x_n + e_{12}), \\ q_2(x) &= (x_1 + e_{21})(x_1 + x_2 + e_{22}), \\ &\vdots \\ q_n(x) &= (x_1 + e_{n1})(x_1 + x_n + e_{n2}) \end{aligned}$$

shares the same 2-homogeneous structure of $P(x)$ with $y_1 = (x_1)$ and $y_2 = (x_2, \ldots, x_n)$, namely, each $q_i(x)$ has degree two with respect to y_1 and degree one with respect to y_2. On the other hand, when viewed in $(z_1, z_2) \in \mathbb{P}^1 \times \mathbb{P}^{n-1}$ with $z_1 = (x_0^{(1)}, x_1)$ and $z_2 = (x_0^{(2)}, x_2, \ldots, x_n)$, this system has the same nonsingular variety S at infinity as $P(x)$. The system $Q(x) = 0$ also has $n+1$ solutions in $\mathbb{C}^n$ for generic e_{ij}s, and there are no extraneous paths. It can be shown (Li and Wang 1991, Morgan and Sommese 1987*a*) that if $Q(x) = 0$ in

$$H(x, t) = (1-t)cQ(x) + tP(x)$$

is chosen to have the same m-homogeneous form as $P(x)$, and the set of zeros $V_\infty(Q)$ of $Q(x)$ at infinity is nonsingular and contained in $V_\infty(P)$, then Properties 1 and 2 hold for $c = re^{i\theta}$, $r > 0$, and for all but finitely many θ.

The zeros of an m-homogeneous polynomial system $\widetilde{P}(z_1, \ldots, z_m)$ at infinity in $\mathbb{P}^{k_1} \times \cdots \times \mathbb{P}^{k_m}$ may sometimes be difficult to obtain. Nevertheless, the choice of $Q(x) = 0$ in Theorem 2.2, assuming no zeros at infinity regardless of the structure of the zeros at infinity of $P(x)$, can still reduce the number of extraneous paths dramatically simply by sharing the same m-homogeneous structure of $P(x)$.

Let us consider the system

$$\begin{aligned} p_1(x) &= x_1(a_{11}x_1 + \cdots + a_{1n}x_n) + b_{11}x_1 + \cdots + b_{1n}x_n + c_1 = 0, \\ &\vdots \\ p_n(x) &= x_1(a_{n1}x_1 + \cdots + a_{nn}x_n) + b_{n1}x_1 + \cdots + b_{nn}x_n + c_n = 0 \end{aligned}$$

in (2.1) again. This time we partition the variables $x_1, \ldots, x_n$ into $y_1 = (x_1, x_2)$ and $y_2 = (x_3, \ldots, x_n)$. For this partition, the 2-homogeneous degree structure of $p_i(x)$ stays the same; namely, the degree of $p_i(x)$ is two with respect to y_1 and is one with respect to y_2. However, the Bézout number with respect to this partition becomes the coefficient of $\alpha_1^2\alpha_2^{n-2}$ in the product $(2\alpha_1 + \alpha_2)^n$. This number is

$$\binom{n}{2} \times 2^2 = 2n(n-1),$$

which is greater than the original Bézout number $2n$ with respect to the partition $y_1 = (x_1)$ and $y_2 = (x_2, \ldots, x_n)$ when $n > 2$. Apparently, the Bézout number is highly sensitive to the chosen partition: different ways of partitioning the variables produce different Bézout numbers. By using Theorem 2.2, we follow the Bézout number (with respect to the chosen partition) of curves to obtain all the isolated zeros of $P(x)$. To minimize the number of extraneous paths, it is certainly desirable to find a partition which provides the lowest Bézout number possible. In Wampler (1992), an algorithm to this end was given. By using this algorithm, one can determine, for example, the partition $\mathcal{P} = \{(b), (c, d, e)\}$ which gives the lowest possible Bézout number 368 for the Cassou–Nogues system in (1.4). Consequently, we may construct a random product start system $Q(x)$ to respect the degree structure of the system with respect to this partition. The start system $Q(x)$ will have 368 isolated zeros in $\mathbb{C}^n$, and, according to Theorem 2.2, only 368 homotopy curves need to be followed to obtain all 16 isolated zeros of the Cassou–Nogues system, in contrast to following the 1344 curves, 1344 being the total degree of the system.

The usefulness of the methods yet developed for Problem A is restricted to application on an *ad hoc* basis. The challenge is, in a specific case, to find a $Q(x)$ that is simple to solve (Property 0) and also produces minimal extraneous paths.

3. Methods for Problem B

The situation for Problem B is different. A method called the 'cheater's homotopy' has been developed, which is, in some sense, an optimum solution procedure; see Li, Sauer and Yorke (1988) and Li, Sauer and Yorke (1989) (a similar procedure can be found in Morgan and Sommese (1989)). Problem B asks that the system $P(c, x) = 0$ be solved for several different values of the coefficients c. In other words, we think of $P(c, x) = 0$ as a system with the same structure or sparsity.

The idea of the method is to establish Properties 1 and 2 theoretically by deforming a sufficiently generic system (in a precise sense to be given later) and then to 'cheat' on Property 0 by using a preprocessing step. The amount of computation per preprocessing step may be large, but is shared among the several solving characteristics of Problem B.

We begin with an example. Let $P(x)$ be the system

$$\begin{aligned} p_1(x) &= x_1^3 x_2^2 + c_1 x_1^3 x_2 + x_2^2 + c_2 x_1 + c_3 = 0, \\ p_2(x) &= c_4 x_1^4 x_2^2 - x_1^2 x_2 + x_2 + c_5 = 0. \end{aligned} \tag{3.1}$$

This is a system of two polynomial equations in two unknowns x_1 and x_2. We want to solve Problem B, that is, we want to solve the system of equations several times, for various specific choices of $c = (c_1, \ldots, c_5)$.

It turns out that, for any choice of coefficients c, system (3.1) has at most 10 isolated solutions. More precisely, there is an open dense subset S of $\mathbb{C}^5$ such that, for $c \in S$, there are 10 solutions of (3.1). Moreover, 10 is an upper bound for the number of isolated solutions for all c in $\mathbb{C}^5$. The total degree of the system is $6 \times 5 = 30$, meaning that if we had taken a generic system of two polynomials in two variables of degree 5 and 6, there would be 30 solutions. Thus (3.1), with any choice of c, is a deficient system.

Classical homotopy using the start system $Q(x) = 0$ in (1.3) produces $d = 30$ paths, beginning at 30 trivial starting points. Thus there are (at least) 20 extraneous paths.

The cheater's homotopy continuation approach begins by solving (3.1) with *randomly chosen* complex coefficients $c^* = (c_1^*, \ldots, c_5^*)$; let X^* be the set of 10 solutions. No work is saved, since 30 paths need to be followed and 20 paths are wasted. However, the 10 elements of the set X^* are the seeds for the remainder of the process. Subsequently, for each choice of coefficients $c = (c_1, \ldots, c_5)$ for which the system (3.1) needs to be solved, we use the homotopy continuation method to follow a straight-line homotopy from the system with coefficient c^* to the system with coefficient c, and we follow the 10 paths beginning at the 10 elements of X^*. Thus Property 0, the existence of trivial starting points, is satisfied. The fact that Properties 1 and 2 are also satisfied is the content of Theorem 3.1 below. Thus for each fixed c, all 10 (or fewer) isolated solutions of (3.1) lie at the end of 10 smooth homotopy paths beginning at the seeds in X^*. After the initial step of finding the seeds, the complexity of all further solvings of (3.1) is proportional to the number of solutions 10, rather than the total degree 30.

Furthermore, this method, unlike the method for Problem A, requires no *a priori* analysis of the system. The first preprocessing step of finding the seeds establishes a sharp upper bound on the number of isolated solutions as a by-product of the computation; further solving of the system uses the optimal number of paths to be followed.

We earlier characterized a successful homotopy continuation method as having three properties: triviality, smoothness, and accessibility (Properties 0, 1 and 2, respectively). Given an arbitrary system of polynomial equations, such as (3.1), it is not too hard (through generic perturbations) to find a family of systems with the last two properties. The problem is that one member of the family must be trivial to solve, or the path-following cannot begin. The idea of the cheater's homotopy is simply to 'cheat' on this part of the problem, and run a preprocessing step (the computation of the seeds X^*) which gives us Property 0 (triviality) in a roundabout way. Hence the name, the 'cheater's homotopy'.

A statement of the theoretical result we need follows. Let

$$\begin{aligned} p_1(c_1,\dots,c_M,x_1,\dots,x_n) &= 0,\\ &\vdots\\ p_n(c_1,\dots,c_M,x_1,\dots,x_n) &= 0, \end{aligned} \tag{3.2}$$

be a system of polynomial equations in the variables $c_1,\dots,c_M,x_1,\dots,x_n$. For each choice of $c=(c_1,\dots,c_M)$ in $\mathbb{C}^M$, this is a system of polynomial equations in the variables $x_1,\dots,x_n$. Let d be the total degree of the system for a generic choice of c.

Theorem 3.1 Let c belong to $\mathbb{C}^M$. There exists an open, dense, full-measure subset U of $\mathbb{C}^{n+M}$ such that for $(b_1^*,\dots,b_n^*,c_1^*,\dots,c_M^*)\in U$, the following holds.

(a) The set X^* of solutions $x=(x_1,\dots,x_n)$ of

$$\begin{aligned} q_1(x_1,\dots,x_n) &= p_1(c_1^*,\dots,c_M^*,x_1,\dots,x_n)+b_1^*=0,\\ &\vdots\\ q_n(x_1,\dots,x_n) &= p_n(c_1^*,\dots,c_M^*,x_1,\dots,x_n)+b_n^*=0, \end{aligned} \tag{3.3}$$

consists of d_0 isolated points, for some $d_0\le d$.

(b) Properties 1 and 2 (smoothness and accessibility) hold for the homotopy

$$H(x,t) = P((1-t)c_1^* + tc_1, \ldots, (1-t)c_M^* + tc_M, x_1, \ldots, x_n) + (1-t)b^* \quad (3.4)$$

where $b^* = (b_1^*, \ldots, b_n^*)$. It follows that every solution of $P(x) = 0$ is reached by a path beginning at a point of X^*.

A proof of Theorem 3.1 can be found in Li et al. (1989). The theorem is used as part of the following procedure. Let $P(c,x) = 0$ as in (3.2) denote the system to be solved for various values of the coefficients c.

Cheater's homotopy procedure

(1) Choose complex numbers $(b_1^*, \ldots, b_n^*, c_1^*, \ldots, c_M^*)$ at random, and use the classical homotopy continuation method to solve $Q(x) = 0$ in (3.3). Let d_0 denote the number of solutions found (this number is bounded above by the total degree d). Let X^* denote the set of d_0 solutions.
(2) For each new choice of coefficients $c = (c_1, \ldots, c_M)$, follow the d_0 paths defined by $H(x,t) = 0$ in (3.4), beginning at the points in X^*, to find all solutions of $P(c,x) = 0$.

In step (1) above, for random complex numbers $(c_1^*, \ldots, c_M^*)$, using classical homotopy continuation methods to solve $Q(x) = 0$ in (3.3) may itself sometimes be computationally expensive. It is desirable that those numbers do not have to be random. For illustration, consider the linear system

$$\begin{aligned} c_{11}x_1 + \cdots + c_{1n}x_n &= b_1, \\ &\vdots \\ c_{n1}x_1 + \cdots + c_{nn}x_n &= b_n, \end{aligned} \quad (3.5)$$

which may be considered as a polynomial system with each equation having degree one. For generic c_{ij}s, (3.5) has a unique solution which is not available right away. However, if we choose $c_{ij} = \delta_{ij}$ (the Kronecker delta), the solution is obvious.

For this purpose, an alternative is suggested in Li and Wang (1992). When a system $P(c,x) = 0$ with a particular parameter c^0 is solved, this c^0 may be chosen arbitrarily instead of being chosen randomly; then for any parameter $c \in \mathbb{C}^M$ consider the nonlinear homotopy

$$H(a,x,t) = P((1 - [t - t(1-t)a])c^0 + (t - t(1-t)a)c, x) = 0. \quad (3.6)$$

It is shown in Li and Wang (1992) that for a randomly chosen complex a the solution paths of (3.6) emanating from the solutions of $P(c^0, x) = 0$ will reach the isolated solutions of $P(c,x) = 0$ under the natural assumption that, for generic c, $P(c,x)$ has the same number of isolated zeros in $\mathbb{C}^n$.

The most important advantage of the homotopy in (3.6) is that the parameter c^0 of the start system $P(c^0, x) = 0$ is arbitrary so long as $P(c^0, x) = 0$ has the same number of solutions as $P(c, x) = 0$ for generic c. Therefore, in some situations, when the solutions of $P(c, x) = 0$ are easily available for certain c^0, the system $P(c^0, x) = 0$ may be used as the start system (3.6) and the extra effort of solving $P(c, x) = 0$ for a randomly chosen c would be saved.

To finish, we give a more non-trivial example of the procedure described in this section.

Consider the indirect position problem for revolute-joint kinematic manipulators. Each joint is associated with a one-dimensional parametrization, namely the angular position of the joint. If all angular positions are known, then of course the position and orientation of the end of the manipulator (the hand) are determined. The indirect position problem is the inverse problem: given the desired position and orientation of the hand, find a set of angular parameters for the (controllable) joints which will place the hand in the desired state.

The indirect position problem for six joints is reduced to a system of eight nonlinear equations in eight unknowns in Tsai and Morgan (1985). The coefficients of the equations depend on the desired position and orientation, and a solution of the system (an eight-vector) represents the sines and cosines of the angular parameters. Whenever the manipulator's position is changed, the system needs to be resolved with new coefficients. The equations are too long to repeat here; see the appendix of Tsai and Morgan (1985). Suffice it to say that it is a system of eight degree-two polynomial equations in eight unknowns which is rather deficient. The total degree of the system is $2^8 = 256$, but there are at most 32 isolated solutions.

The nonlinear homotopy of (3.6) requires only 32 paths to solve the system with different sets of parameters (Li and Wang 1990, 1992). The system contains 26 coefficients, and a specific set of coefficients is chosen for which the system has 32 solutions. For subsequent solving of the system, for any choice of the coefficients $c_1, \ldots, c_{26}$, all solutions can be found at the end of exactly 32 paths, by using nonlinear homotopy in (3.6) with randomly chosen complex a.

4. Polyhedral homotopy

In the last few years, a major computational breakthrough has occurred in the solution of polynomial systems by the homotopy continuation method. The new method takes great advantage of the Bernshteín theory, which gives a much tighter bound, in general, for the number of isolated zeros of a polynomial system in the algebraic tori $(\mathbb{C}^*)^n$, where $\mathbb{C}^* = \mathbb{C}\backslash\{0\}$. In Huber and Sturmfels (1995), this root count was used to actually find all the isolated zeros of the polynomial system by establishing polyhedral homotopies. For a given polynomial system, the new method solves a new polynomial system with the same monomials, but with randomly chosen coefficients. The new system is then used as the start system in the cheater's homotopy described in Section 3 to solve the original polynomial system. In a way, the new method uses the method for Problem B to solve Problem A. The new algorithm is very promising. In particular, for polynomial systems without special structure, the new algorithm substantially outperformed other methods.

We take the following example (Huber and Sturmfels 1995) as our point of departure. Setting $x = (x_1, x_2)$, consider the system $P(x) = (p_1(x), p_2(x))$, where

$$\begin{aligned} p_1 &= c_{11}x_1x_2 + c_{12}x_1 + c_{13}x_2 + c_{14} &&= 0, \quad \text{and} \\ p_2 &= c_{21}x_1x_2^2 + c_{22}x_1^2x_2 + c_{23} &&= 0. \end{aligned} \tag{4.1}$$

Here, $c_{ij} \in \mathbb{C}^* = \mathbb{C}\backslash\{0\}$. The monomials $\{1, x_1x_2, x_1, x_2\}$ in p_1 can be written as $x_1x_2 = x_1^1x_2^1$, $x_1 = x_1^1x_2^0$, $x_2 = x_1^0x_2^1$ and $1 = x_1^0x_2^0$. The set of their exponents

$$S_1 = \{(0,0), (1,0), (1,1), (0,1)\}$$

is called the *support* of p_1, and its convex hull $Q_1 = \text{conv}(S_1)$ is called the *Newton polytope* of p_1. Similarly, p_2 has support $S_2 = \{(0,0), (2,1), (1,2)\}$ with Newton polytope $Q_2 = \text{conv}(S_2)$. Using multi-index notation $x^q = x_1^{q_1}x_2^{q_2}$ where $q = (q_1, q_2)$, we may rewrite (4.1) as

$$p_1(x) = \sum_{q \in S_1} c_q x^q \quad \text{and} \quad p_2(x) = \sum_{q \in S_2} c_q x^q.$$

The *Minkowski sum* $R_1 + R_2$ of polytopes R_1 and R_2 is defined as

$$R_1 + R_2 = \{r_1 + r_2 : r_1 \in R_1 \text{ and } r_2 \in R_2\}$$

(polytopes Q_1, Q_2 and $Q_1 + Q_2$ for (4.1) are shown in Figure 2). Now, let us consider the area of the convex polytope $\lambda_1 Q_1 + \lambda_2 Q_2$ with non-negative variables λ_1 and λ_2 for the system (4.1). From elementary geometry, the area of a triangle on the plane with vertices u, v and w equals

$$\frac{1}{2}\left|\det\begin{pmatrix} u - v \\ w - v \end{pmatrix}\right|. \tag{4.2}$$

Thus, to compute the area $f(\lambda_1, \lambda_2)$ of $\lambda_1 Q_1 + \lambda_2 Q_2$, one may partition the polytope into a collection of triangles, $A_1, A_2, \ldots, A_k$. These triangles are mutually disjoint, and the vertices take the form $\lambda_1 q_1 + \lambda_2 q_2$, with $q_1 \in Q_1$ and $q_2 \in Q_2$. In other words, the vertices of these triangles coincide with the vertices of the polytope $\lambda_1 Q_1 + \lambda_2 Q_2$. It follows from (4.2) that the area of each A_i is a second-degree homogeneous polynomial in λ_1 and λ_2.

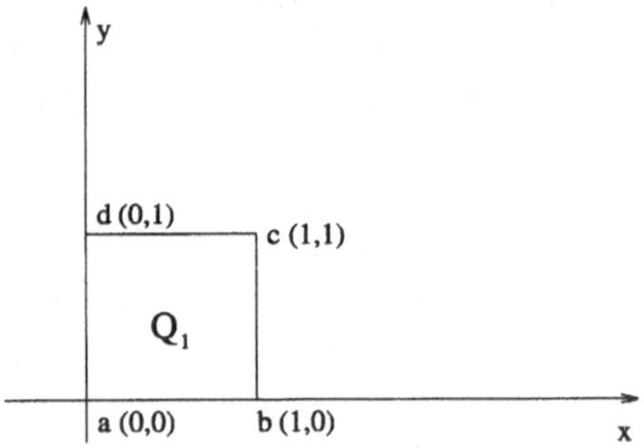

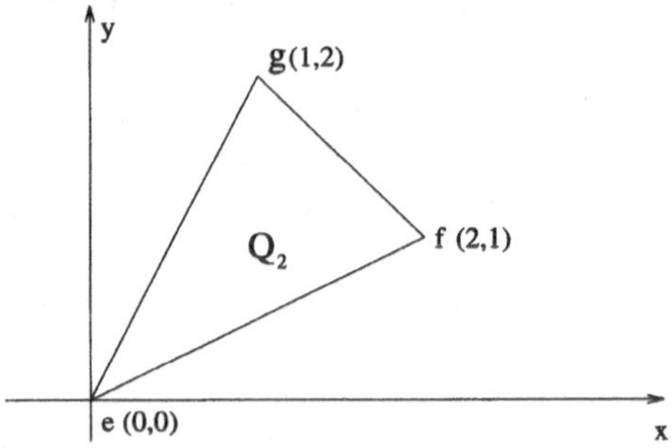

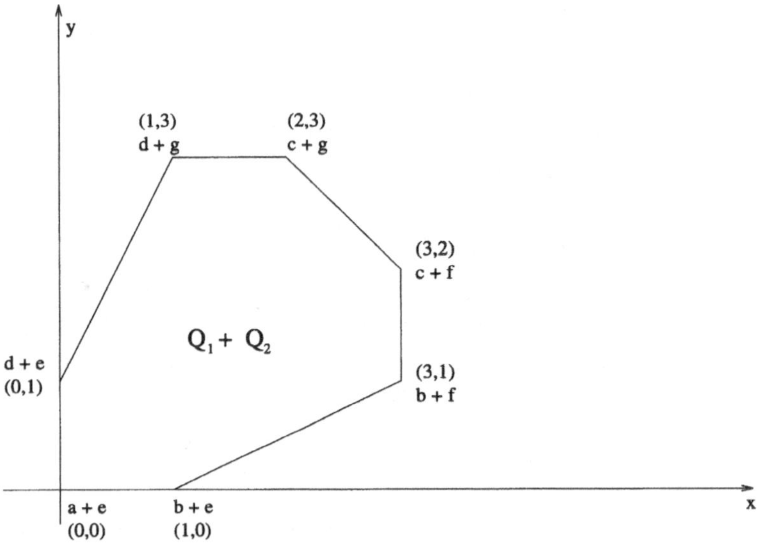

Fig. 2.

Therefore, $f(\lambda_1, \lambda_2)$, as a sum of the areas of $A_1, ..., A_k$, is also a second-degree homogeneous polynomial in λ_1 and λ_2. Writing

$$f(\lambda_1, \lambda_2) = a_1\lambda_1^2 + a_2\lambda_2^2 + a_{12}\lambda_1\lambda_2,$$

the coefficient a_{12} of $\lambda_1\lambda_2$ in f is called the *mixed volume* of the polytopes Q_1 and Q_2. We denote it by $\mathcal{M}(Q_1, Q_2)$, or $\mathcal{M}(S_1, S_2)$ when no ambiguity exists.

Clearly,

$$\begin{aligned} a_{12} &= f(1,1) - f(1,0) - f(0,1) \\ &= \text{ area of } (Q_1 + Q_2) - \text{ area of } (Q_1) - \text{ area of } (Q_2). \end{aligned}$$

For (4.1), it is easy to see that the areas of Q_1+Q_2, Q_1 and Q_2 are 6.5, 1 and 3.5 respectively. Therefore, $a_{12}=6.5-1-1.5=4$. On the other hand, one can also easily see that system (4.1) has two zeros $(0,0,1)$ and $(0,1,0)$ at infinity in $\mathbb{P}^2$; hence it can have at most 4 isolated zeros in $\mathbb{C}^2$, or in $(\mathbb{C}^*)^2$ in particular. According to the Bernshteín theory, this is not a coincidence: the number of isolated zeros of (4.1) in $(\mathbb{C}^*)^2$, counting multiplicities, is bounded above by the mixed volume of its Newton polytopes. Further, when the coefficients in (4.1) are chosen generically, then these two numbers are exactly the same.

To state the Bernshteín theory in a more general form, we first allow monomials $x_1^{a_1}\cdots x_n^{a_n}$ to have negative exponents; such a polynomial is called a *Laurent polynomial.* With $x=(x_1,\ldots,x_n)$, let $p(x)=(p_1(x),\ldots,p_n(x))$ be a system of n Laurent polynomials with supports $S_1,\ldots,S_n$ respectively in $\mathbb{Z}^n$. The corresponding Newton polytopes are $Q_1,\ldots,Q_n$. Following reasoning similar to that described above, the n-dimensional volume of the polytope $\lambda_1Q_1+\cdots+\lambda_nQ_n$, with non-negative variables $\lambda_1,\ldots,\lambda_n$, is a homogeneous polynomial in $\lambda_1,\ldots,\lambda_n$ of degree n. The coefficient of $\lambda_1\times\lambda_2\times\cdots\times\lambda_n$ in this polynomial is defined as the *mixed volume* of $Q_1,\ldots,Q_n$, denoted by $\mathcal{M}(Q_1,\ldots,Q_n)$ or $\mathcal{M}(S_1,\ldots,S_n)$.

Theorem 4.1 (Bernshteín 1975) The number of isolated zeros, counting multiplicities, of $P(x)=(p_1(x),\ldots,p_n(x))$ in $(\mathbb{C}^*)^n$ is bounded above by the mixed volume $\mathcal{M}(S_1,\ \ldots,\ S_n)$. For generically chosen coefficients, the system $P(x)=0$ has exactly $\mathcal{M}(S_1,\ \ldots,\ S_n)$ roots in $(\mathbb{C}^*)^n$.

In Canny and Rojas (1991), this bound was nicknamed the *BKK bound* after its inventors, Bernshteín (1975), Khovanskií (1978) and Kushnirenko (1976). It turns out that this root count is very helpful in using the polyhedral homotopy to solve sparse polynomial systems, sparse in the sense that each polynomial in the system contains few terms. This sparseness is by no means a big restriction. After all, almost all the polynomial systems we encountered in application belong to this category.

An apparent limitation of the above theorem is that it counts only the roots of a polynomial system in $(\mathbb{C}^*)^n$, but not necessarily all roots in affine space $\mathbb{C}^n$. This problem was first attempted in Canny and Rojas (1991) and Rojas (1994) by introducing the notion of the *shadowed* sets, and a bound in $\mathbb{C}^n$ was obtained. Later, a significantly tighter bound was discovered in the following theorem.

Theorem 4.2 (Li and Wang 1996) The number of isolated zeros in $\mathbb{C}^n$, counting multiplicities, of a polynomial system $P(x)=(p_1(x),\ldots,p_n(x))$ with supports $S_1,\ldots,S_n$ is bounded above by the mixed volume

$$\mathcal{M}(S_1\bigcup\{0\},\ldots,S_n\bigcup\{0\}).$$

This theorem was further extended in several ways by Huber and Sturmfels (1997) and Rojas and Wang (1996). When $0 \in S_i$ for all $i = 1, \ldots, n$, so that each p_i has a nontrivial constant term, then Theorem 4.2 implies that the BKK bound of Theorem 4.1 gives the number of zeros of the polynomial system in $\mathbb{C}^n$. In fact, the proof of Theorem 4.2 uses the important fact that generic constant perturbations of a polynomial system can only have isolated zeros in $(\mathbb{C}^*)^n$, and all isolated zeros become nonsingular.

Now consider the system (4.1) again. To compute the area of $Q_1 + Q_2$, we can certainly subdivide $Q_1 + Q_2$ as we wish. The subdivision may not consist of all triangles as before. However, the subdivision shown in Figure 3 – call it subdivision B – is of particular interest. By a *cell* of a subdivision we mean any member of the subdivision. It can be easily verified that all the cells in subdivision B have the following special properties.

Proposition 4.1

(a) Each one is a Minkowski sum of the convex hull of a subset C_1 in S_1 and the convex hull of a subset C_2 in S_2.
(b) For $i = 1, 2, \operatorname{conv}(C_i)$ is a simplex of dimension $\#(C_i) - 1$, where $\#(C_i)$ is the number of points in C_i.
(c) Simplices $\operatorname{conv}(C_1)$ and $\operatorname{conv}(C_2)$ are complementary to each other in the sense that $\dim(\operatorname{conv}(C_1)) + \dim(\operatorname{conv}(C_2)) = \dim(\operatorname{conv}(C_1) + \operatorname{conv}(C_2))$.

In light of properties (a) and (b), each cell $C = \operatorname{conv}(C_1) + \operatorname{conv}(C_2)$ in B can be identified as a cell of type (l_1, l_2), where $l_1 = \dim(\operatorname{conv}(C_1))$ and $l_2 = \dim(\operatorname{conv}(C_2))$. Property (c) mainly says that simplices $\operatorname{conv}(C_1)$ and $\operatorname{conv}(C_2)$ are 'linearly independent', for otherwise their Minkowski sum would be lower dimensional.

In $\mathbb{R}^n$, consider the n-dimensional volume of the Minkowski sum of simplices $A_1, \ldots, A_n$ with dimensions $k_1, \ldots, k_n$, respectively, where $k_i \geq 0$ for $1 \leq i \leq n$ and $k_1 + k_2 + \cdots + k_n = n$. For $i = 1, \ldots, n$, let $A_i = \operatorname{conv}\{q_0^{(i)}, \ldots, q_{k_i}^{(i)}\}$ and let V be the $n \times n$ matrix whose rows are $q_j^{(i)} - q_0^{(i)}$ for $1 \leq i \leq n$ and $1 \leq j \leq k_i$. Notice that any 0-dimensional simplex consists of only one point, and therefore contributes no rows to V. It can be shown that

$$\operatorname{Vol}_n(A_1 + \cdots + A_n) = \frac{1}{k_1! \cdots k_n!} |\det V|. \tag{4.3}$$

Here, we use Vol_n to denote the n-dimensional volume; of course, $\operatorname{Vol}_2(C)$ represents the area of C. Applying (4.3) to cell ①$= \operatorname{conv}\{a, d\} + \operatorname{conv}\{e, g\}$ in subdivision B, we have

$$\operatorname{Vol}_2(\text{cell ①}) = \left| \det \begin{pmatrix} d - a \\ g - e \end{pmatrix} \right|.$$

Now, when Q_1 and Q_2 are scaled by λ_1 and λ_2, respectively, cell ① becomes $\text{conv}\{\lambda_1 a, \lambda_2 d\} + \text{conv}\{\lambda_2 e, \lambda_2 g\}$ and its volume becomes

$$\begin{aligned} \left|\det\begin{pmatrix}\lambda_1 d - \lambda_1 a\\ \lambda_2 g - \lambda_2 e\end{pmatrix}\right| &= \left|\det\begin{pmatrix}d-a\\ g-e\end{pmatrix}\right| \times \lambda_1\lambda_2 \\ &= (\text{volume of cell ① before scaling}) \times \lambda_1\lambda_2. \end{aligned}$$

From the definition of the mixed volume, it follows that the volume of the original cell ① constitutes part of the mixed volume of Q_1 and Q_2. On the other hand, after scaling, cell ② in subdivision B becomes $\text{conv}\{\lambda_1 a, \lambda_1 c, \lambda_1 d\} + \{\lambda_2 g\}$ and its volume becomes, according to (4.3),

$$\begin{aligned} \frac{1}{2}\left|\det\begin{pmatrix}\lambda_1 c - \lambda_1 a\\ \lambda_1 d - \lambda_a a\end{pmatrix}\right| &= \frac{1}{2}\left|\det\begin{pmatrix}c-a\\ d-a\end{pmatrix}\right| \times \lambda_1^2 \\ &= (\text{volume of cell ② before scaling}) \times \lambda_1^2. \end{aligned}$$

Apparently, the volume of the original cell ② has no contribution to the mixed volume of Q_1 and Q_2.

In summary, only cells of type (1,1) contribute to the mixed volume $\mathcal{M}(Q_1, Q_2)$ of Q_1 and Q_2 and, therefore,

$$\begin{aligned} \mathcal{M}(Q_1, Q_2) &= \text{the sum of the volumes of cells of type } (1,1) \\ &= \text{volume of cell ① + volume of cell ③ + volume of cell ⑤} \\ &= 1 + 2 + 1 = 4. \end{aligned}$$

The type of subdivisions of $Q_1 + Q_2$ that share the same special properties in Proposition 4.1 as subdivision B is called the *fine mixed subdivision.* To state a formal definition with less notation, we omit '+' and 'conv', except where absolutely necessary. For instance, $(S_1, \ldots, S_n)$ will replace $Q_1 + \cdots + Q_n (= \text{conv}(S_1) + \cdots + \text{conv}(S_n))$ as the key object.

Let $S = (S_1, \ldots, S_n)$ be a sequence of finite subsets of $\mathbb{Z}^n$, whose union affinely spans $\mathbb{R}^n$. By a *cell* of S we mean an n-tuple $C = (C_1, \ldots, C_n)$ of subsets $C_i \subset S_i$, for $i = 1, \ldots, n$. Define

$$\begin{aligned} \text{type}(C) &:= (\dim(\text{conv}(C_1)), \ldots, \dim(\text{conv}(C_n))), \\ \text{conv}(C) &:= \text{conv}(C_1) + \cdots + \text{conv}(C_n), \end{aligned}$$

and Vol(C):=Vol(conv(C)). A *face* of C is a subcell $F = (F_1, \ldots, F_n)$ of C where $F_i \subset C_i$ and some linear functional $\alpha \in (\mathbb{R}^n)^\vee$ attains its minimum over C_i at F_i, for $i = 1, \ldots, n$. We call such an α an *inner normal* of F. If F is a face of C then conv(F_i) is a face of the polytope conv(C_i) for $i = 1, \ldots, n$.

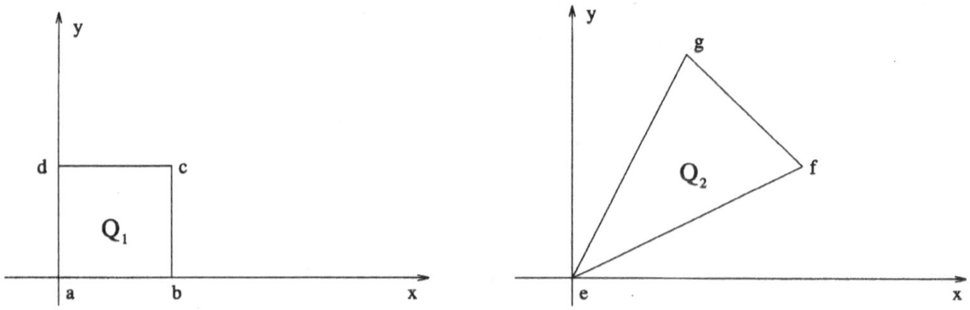

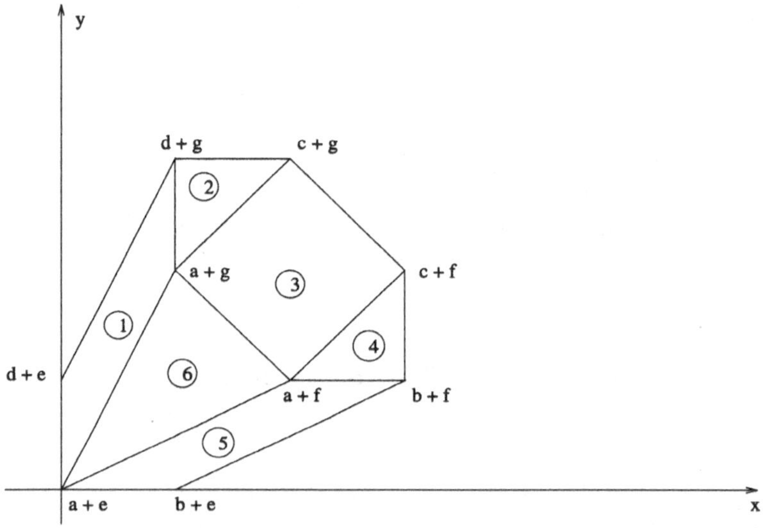

Fig. 3. Subdivision B for $Q_1 + Q_2$

Definition 4.1 A *fine mixed subdivision* of S is a set $\{C^{(1)}, \ldots, C^{(m)}\}$ of cells such that:

(a) for all $j = 1, \ldots, m$, $\dim(\text{conv}(C^{(j)})) = n$
(b) $\text{conv}(C^{(j)}) \cap \text{conv}(C^{(k)})$ is a proper common face of $\text{conv}(C^{(j)})$ and $\text{conv}(C^{(k)})$ when it is nonempty for $j \neq k$
(c) $\bigcup_{j=1}^{m} \text{conv}(C^{(j)}) = \text{conv}(S)$
(d) for $j = 1, \ldots, m$, write $C^{(j)} = (C_1^{(j)}, \ldots, C_n^{(j)})$. Then, each $\text{conv}(C_i^{(j)})$ is a simplex of dimension $\#C_i^{(j)} - 1$, and for each j,

$$\dim(\text{conv}(C_1^{(j)})) + \cdots + \dim(\text{conv}(C_n^{(j)})) = n.$$

As we have discussed for the special system (4.1), when a polynomial system $P(x) = (p_1(x), \ldots, p_n(x))$ in $\mathbb{C}[x_1, \ldots, x_n]$ is given with support $S = (S_1, ..., S_n)$, where S_i is the support of p_i, and if we can find a fine mixed subdivision for S, then the mixed volume $\mathcal{M}(S_1, \ldots, S_n)$ will be the sum of the volumes of cells of type (1,...,1). Thus formula (4.3), together with condition (d) above, makes the volume computation of this type of cell quite easy.

A fine mixed subdivision for $S = (S_1, \ldots, S_n)$ can be found by the following standard process: choose real-valued functions $\omega^{(i)} : S_i \to \mathbb{R}$, for $i = 1, \ldots, n$; call the n-tuple $\omega = (\omega^{(i)}, \ldots, \omega^{(i)})$ a *lifting function* on S, and say that ω lifts S_i to its graph $\hat{S}_i = \{(q, \omega^{(i)}(q)) : q \in S_i\} \subset \mathbb{R}^{n+1}$. This notation is extended in the obvious way: $\hat{S} = (\hat{S}_1, \ldots, \hat{S}_n)$, $\hat{Q}_i = \text{conv}(\hat{S}_i)$, $\hat{Q} = \hat{Q}_1 + \cdots + \hat{Q}_n$, *etc.* Let S_ω be the set of cells $\{C\}$ of S which satisfy

(a) $\dim(\text{conv}(\hat{C})) = n$,
(b) $\hat{C}$ is a facet (an n-dimensional face) of $\hat{S}$ whose inner normal $\alpha \in (\mathbb{R}^{n+1})^\vee$ has positive last coordinate.

In other words, $\text{conv}(\hat{C})$ is a facet of the lower hull of $\hat{Q}$. The fact is that when the lifting function ω is chosen generically, S_ω always gives a fine mixed subdivision for S (Gel'fand, Kapranov and Zelevinskiĭ 1994, Lee 1991). The subdivision B in Figure 3 for system (4.1) is, in fact, induced by the lifting $\omega = ((0, 1, 1, 1), (0, 0, 0))$, that is,

$$\hat{S} = (\{(a, 0), (b, 1), (c, 1), (d, 1)\}, \{(e, 0), (f, 0), (g, 0)\})$$

(see Figure 4). While this lifting does not seem so generic, it is sufficient to give a fine mixed subdivision.

Let us return to our main issue: how can this Bernshteĭn theory help us to solve polynomial systems by homotopy continuation methods? Actually, the lifting function introduced above has already provided a nonlinear homotopy. This ingenious idea is due to Huber and Sturmfels (1995).

For a given polynomial system $P(x) = (p_1(x), \ldots, p_n(x))$ in $\mathbb{C}[x_1, \ldots, x_n]$, to find all isolated zeros of $P(x)$ in $\mathbb{C}^n$ instead of $(\mathbb{C}^*)^n$, we first, according to Theorem 4.2, augment the monomial $x^0 (= 1)$ to those p_is which do not have constant terms. We then choose the coefficients of all the monomials in $P(x)$ at random. For simplicity, we abuse notation and retain the name $P(x) = (p_1(x), \ldots, p_n(x))$ for this system. We wish to solve this system first, and then, by using the cheater's homotopy introduced in Section 3, it can be used as the start system for solution of the original system by linear homotopy.

Let S_i be the support of p_i, so that

$$p_i(x) = \sum_{q \in S_i} c_q x^q, \qquad i = 1, \ldots, n,$$

where $q = (q_1, \ldots, q_n)$ and $x^q = x_1^{q_1} \cdots x_n^{q_n}$. Let t denote a new complex

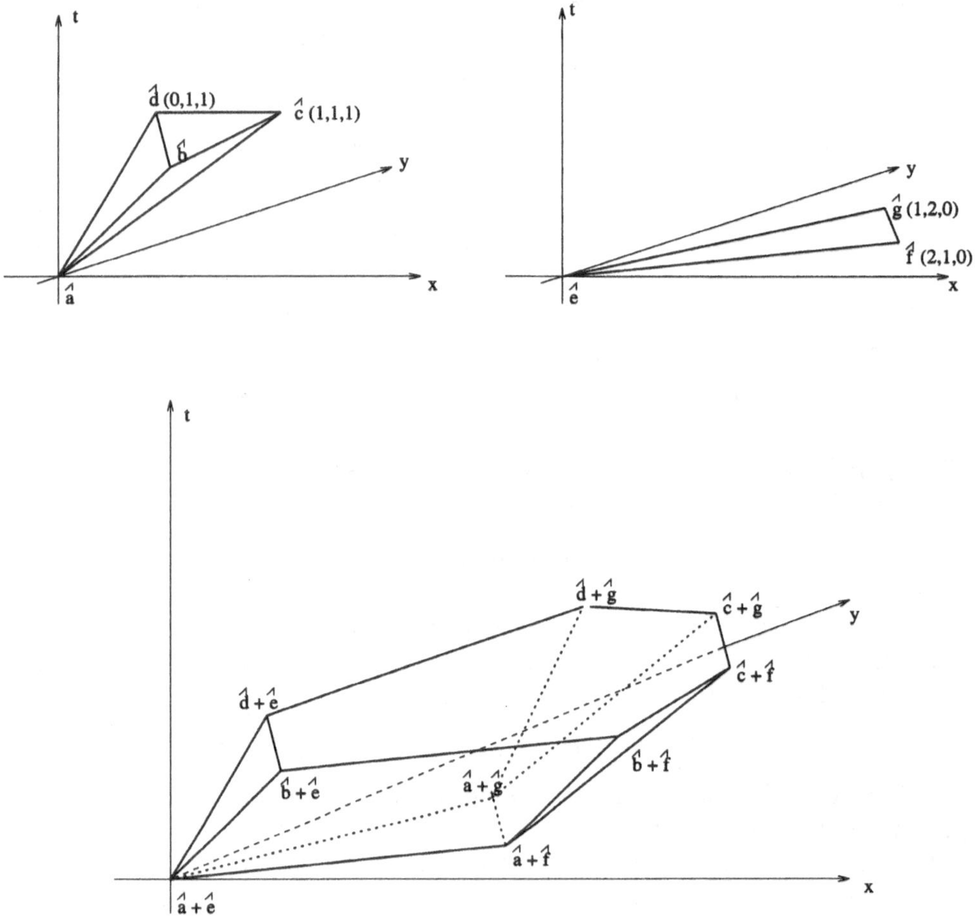

Fig. 4.

variable and consider the polynomials in $n+1$ variables given by

$$\hat{p}_i(x,t) = \sum_{q \in S_i} c_q x^q t^{\omega_i(q)}, \qquad i = 1, \ldots, n, \tag{4.4}$$

where each $\omega_i : S_i \to \mathbb{R}$ for $i = 1, \ldots, n$ is chosen generically. The support of $\hat{p}_i$ is now $\hat{S}_i = \{\hat{q} = (q, \omega_i(q)) : q \in S_i\}$ with Newton polytope $\hat{Q}_i = \operatorname{conv}(\hat{S}_i)$. The function $\omega = (\omega_1, \ldots, \omega_n)$ can be viewed as a lifting function on $S = (S_1, \ldots, S_n)$ which lifts S_i to $\hat{S}_i$. The induced subdivision S_ω on S is then a fine mixed subdivision and the mixed volume $\mathcal{M}(S_1, \ldots, S_n)$ equals the sum of the volumes of cells of type $(1,\ldots,1)$ in S_ω. Recall that, for each $t \in (0, 1]$, the isolated zeros of the system

$$\hat{P}(x,t) = (\hat{p}_1(x,t), \ldots, \hat{p}_n(x,t))$$

are all nonsingular and, by the Bernshteín theory, the total number of those zeros is equal to $\mathcal{M}(S_1,\ldots,S_n)$. We may write these zeros as $x^1(t),\ldots,x^k(t)$ where $k=\mathcal{M}(S_1,\ldots,S_n)$, so $\hat{P}(x^j(t),t)=0$ for each $t\in(0,1]$ and $j=1,\ldots,k$.

Let $C=(C_1,\ldots,C_n)$ be a cell of type $(1,\ldots,1)$ in S_ω. For $i=1,\ldots,n$, let $C_i=\{q_i^{(0)},q_i^{(1)}\}\subset S_i$ and $v_i=q_i^{(1)}-q_i^{(0)}$. Since S_ω is a fine mixed subdivision, $\{v_1,\ldots,v_n\}$ is linearly independent; otherwise, $\dim(\mathrm{conv}(C_1))+\cdots+\dim(\mathrm{conv}(C_n))<n$. So,

$$\mathrm{Vol}_n(C)=\left|\det\begin{bmatrix}v_1\\ \vdots\\ v_n\end{bmatrix}\right|.$$

On the other hand, $\hat{C}=(\hat{C}_1,\ldots,\hat{C}_n)$ is a facet of $\hat{S}=(\hat{S}_1,\ldots,\hat{S}_n)$ whose inner normal $\hat{\alpha}\in(\mathbb{R}^{n+1})^\vee$ has positive last coordinate. Let $\hat{\alpha}=(\alpha_1,\ldots,\alpha_n,1)$ and $\alpha=(\alpha_1,\ldots,\alpha_n)$, so $\hat{\alpha}=(\alpha,1)$. Let $x(t)$ represent general solution curves $x^1(t),\ldots,x^k(t)$ of $\hat{P}(x,t)=0$. Setting $x(t)=(x_1(t),\ldots,x_n(t))$, let

$$\begin{aligned}t^{\alpha_1}y_1(t)&=x_1(t),\\ &\vdots\\ t^{\alpha_n}y_n(t)&=x_n(t).\end{aligned}$$

Or, simply, $t^\alpha y(t)=x(t)$. Substituting this into (4.4) yields

$$\begin{aligned}\hat{p}_i&=\sum_{q\in S_i}c_q y^q t^{\alpha q}t^{\omega_i(q)}\\ &=\sum_{q\in S_i}c_q y^q t^{\langle(\alpha,1),(q,\omega_i(q))\rangle}\\ &=\sum_{q\in S_i}c_q y^q t^{\langle\hat{\alpha},\hat{q}\rangle},\qquad i=1,\ldots,n.\end{aligned}\tag{4.5}$$

Let $\beta_i=\min_{q\in S_i}\langle\hat{\alpha},\hat{q}\rangle$. Since $\hat{C}$ is a facet of $\hat{S}$, $\hat{C}_i=\{\hat{q}_i^{(0)},\hat{q}_i^{(1)}\}$ is a face of $\hat{S}_i$ and $\hat{\alpha}=(\alpha,1)$ also serves as an inner normal of $\hat{C}_i$. It follows that $\langle\hat{\alpha},\hat{q}_i^{(0)}\rangle=\langle\hat{\alpha},\hat{q}_i^{(1)}\rangle=\beta_i$ and $\langle\hat{\alpha},\hat{q}\rangle>\beta_i$ for $\hat{q}\in\hat{S}_i\backslash\hat{C}_i$. Hence, factoring out t^{β_i} in (4.5), we have

$$\hat{p}_i=t^{\beta_i}(c_{i0}y^{q_i^{(0)}}+c_{i1}y^{q_i^{(1)}}+R_i(y,t)),\qquad i=1,\ldots,n,$$

where $c_{i0}=c_{q_i^{(0)}}, c_{i1}=c_{q_i^{(1)}}$ and

$$R_i(y,t)=\sum_{q\in S_i\backslash C_i}c_q y^q t^{\langle\hat{\alpha},\hat{q}\rangle-\beta_i}.$$

Evidently, $R_i(y,0)=0$ for each i, since $\langle\hat{\alpha},\hat{q}\rangle-\beta_i>0$ for $q\in S_i\backslash C_i$. Now, consider the homotopy $H(y,t) = (h_1(y,t),\ldots,h_n(y,t)) = 0$ where

$$h_i(y,t) = c_{i0}y^{q_i^{(0)}} + c_{i1}y^{q_i^{(1)}} + R_i(y,t), \qquad i = 1,\ldots,n. \tag{4.6}$$

The solutions $(y(t),t)$ of this homotopy satisfy

$$c_{j0}y^{q_j^{(0)}} + c_{j1}y^{q_j^{(1)}} = 0, \qquad j = 1,\ldots,n, \tag{4.7}$$

at $t = 0$. For $t \neq 0$, they agree with the zeros of (4.5) and, since $t^{\alpha}y(t) = x(t)$ for $y(t)$ in (4.5), they also agree with the zeros of (4.4) at $t = 1$. In other words, $y(1)$ of (4.6) are solutions of $P(x) = 0$. So, by following the solution curves $(y(t),t)$ of the homotopy $H(y,t) = 0$ defined by (4.6), we may reach the solutions of $P(x) = 0$, at $t = 1$. Of course, we need to solve the system (4.7) at $t = 0$ to begin with. It can be shown that for randomly chosen c_{ij}, for $i = 1,\ldots,n$ and $j = 0,1$, system (4.7) has

$$\left|\det\begin{bmatrix} v_1 \\ \vdots \\ v_n \end{bmatrix}\right| = \text{the volume of } C$$

solutions in $(\mathbb{C}^*)^n$; recall that $v_i = q_i^{(1)} - q_i^{(0)}$ for $i = 1,\ldots,n$. To see how to solve (4.7) in $(\mathbb{C}^*)^n$, we rewrite (4.7) as

$$\begin{aligned} y^{v_1} &= b_1, \\ &\ \vdots \\ y^{v_n} &= b_n, \end{aligned} \tag{4.8}$$

where $b_1 \cdot b_2 \cdot \ldots \cdot b_n \neq 0$, and let

$$V = \begin{bmatrix} v_1 \\ \vdots \\ v_n \end{bmatrix}.$$

For brevity, write $y^V = (y^{v_1},\ldots,y^{v_n})$ and $b = (b_1,\ldots,b_n)$. Then (4.8) becomes

$$y^V = b. \tag{4.9}$$

With this notation, one can easily check that for an $n \times n$ integer matrix U, the following holds:

$$(y^U)^V = y^{(VU)}.$$

When V is a lower nonsingular triangular integer matrix

$$V = \begin{bmatrix} v_{11} & & & \\ v_{21} & v_{22} & & 0 \\ \vdots & \vdots & \ddots & \\ v_{n1} & v_{n2} & \cdots & v_{nn} \end{bmatrix},$$

(4.8) becomes

$$\begin{aligned} y_1^{v_{11}} &= b_1, \\ y_1^{v_{21}} y_2^{v_{22}} &= b_2, \\ &\ \vdots \\ y_1^{v_{n1}} y_2^{v_{n2}} \cdots y_n^{v_{nn}} &= b_n. \end{aligned} \tag{4.10}$$

Obviously, by forward substitution, (4.10) has $|v_{11}| \times \cdots \times |v_{nn}| = |\det V|$ solutions. In general, we may lower triangularize V by multiplying on the right by an integer matrix U with $|\det U| = 1$, which can be found by the following procedure. Firstly, the greatest common divisor d of two integers a and b is

$$d = \gcd(a, b) = ka + lb, \qquad \text{for certain } k, l \in Z.$$

Let

$$M = \begin{bmatrix} k & l \\ -\frac{b}{d} & \frac{a}{d} \end{bmatrix}.$$

Then $\det(M) = 1$ and

$$M \begin{bmatrix} a \\ b \end{bmatrix} = \begin{bmatrix} k & l \\ -\frac{b}{d} & \frac{a}{d} \end{bmatrix} \begin{bmatrix} a \\ b \end{bmatrix} = \begin{bmatrix} d \\ 0 \end{bmatrix}.$$

In view of this, a series of $n \times n$ matrices like M can be used to produce zeros in matrices in a similar way to the use of Givens rotations for the QR factorization. For instance, if a and b are the ith and jth components of an n-dimensional vector v, that is,

$$v = \begin{bmatrix} \vdots \\ a \\ \vdots \\ b \\ \vdots \end{bmatrix} \begin{matrix} \\ \to i\text{th} \\ \\ \to j\text{th}, \\ \\ \end{matrix}$$

then we set

$$U(i,j,a,b) = \begin{array}{c} \begin{array}{ccccccccccc} & & & i & & & & j & & & \end{array} \\ \left[\begin{array}{ccccccccccc} 1 &&&&&&&&&& \\ & 1 &&&&&&&&& \\ && \ddots &&&&&&&& \\ &&& k &&&& l &&& \\ &&&& 1 &&&&&& \\ &&&&& \ddots &&&&& \\ &&&&&& 1 &&&& \\ &&& -\frac{b}{d} &&&& \frac{a}{d} &&& \\ &&&&&&&& 1 && \\ &&&&&&&&& \ddots & \\ &&&&&&&&&& 1 \end{array}\right] \end{array} \begin{array}{c} \\ \\ \\ \\ i \\ \\ \\ \\ j \\ \\ \\ \\ \end{array},$$

so that $\det(U(i,j,a,b)) = 1$ and the jth component of $U(i,j,a,b)v$ will vanish. Thus, a product of a series of matrices in the form of $U(i,j,a,b)$ can be chosen to upper triangularize a matrix from the left. To lower triangularize V, let U be an integer matrix with $|\det U| = 1$ such that $U^T V^T$ is upper triangular; hence, VU is lower triangular.

Now, let $z^U = y$ and substitute it into (4.9); we have

$$y^V = (z^U)^V = z^{VU} = b. \tag{4.11}$$

Since VU is lower triangular, $z = (z_1, \ldots, z_n)$ in (4.11) can be solved and the number of solutions is equal to $|\det(VU)| = |\det(V)| \cdot |\det(U)| = |\det(V)|$. Consequently, we have as many solutions of $y = (y_1, \ldots, y_n)$ in (4.9).

In summary, to find all the isolated zeros of a polynomial system $P(x) = (p_1(x), \ldots, p_n(x))$ in $\mathbb{C}[x_1, \ldots, x_n]$, we augment x^0 to those p_is without constant terms first, then equip all the monomials in $P(x)$ with generic coefficients. In the same notation, we construct $\hat{P}(x,t) = (\hat{p}_1(x,t), \ldots, \hat{p}_n(x,t))$, where

$$\hat{p}_i(x,t) = \sum_{q \in S_i} c_q x^q t^{\omega_i(q)}, \qquad i = 1, \ldots, n.$$

Here S_i is the support of p_i and the lifting function $\omega = (\omega_1, \ldots, \omega_n)$ is chosen at random. Then each cell C of type (1,...,1) in the induced fine mixed subdivision S_ω provides a set of k starting points for the homotopy $H(y,t) = 0$ defined by (4.6), where k denotes the volume of C. Following the solution curves of this homotopy with those k starting points from 0 to 1, we reach k of the solutions of $P(x) = 0$. By Bernshteín's theory, the total number of isolated zeros of $P(x)$ equals the sum of the volumes of all cells of this type. We are thus able to find all the isolated zeros of $P(x)$, and this modified system can then be used as a start system of the linear homotopy to find all the isolated zeros of the original system.

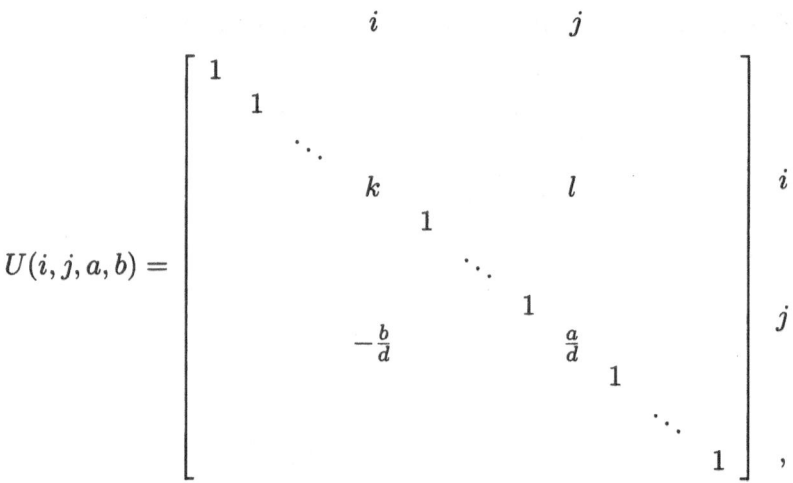

What seems to be missing in the process described above is a constructive way of finding cells of type (1,...,1) in the induced fine mixed subdivision S_ω corresponding to the lifting ω. This issue was discussed in Emiris (1994), Verschelde (1996) and Verschelde, Gatermann and Cools (1996), papers which provided different ways to deal with this problem. At present, the most efficient technique for finding those cells is still undetermined.

The algorithm has been implemented with remarkable success. Recall that the Cassou–Nogues system in (1.4) has total degree 1344 and optimal m-homogeneous Bézout number 368. This system has 16 isolated zeros and its mixed volume equals 24. So, by using polyhedral nonlinear homotopies, one need only follow 24 paths to reach all isolated zeros of the system.

Originally, a more general version of the above process was presented in Huber and Sturmfels (1995). If some of the p_is have the same supports, then cells of the 'appropriate' types, instead of cells of type $(1,\ldots,1)$, can serve the same purpose. The method can be made much more efficient by taking this special structure into consideration. For simplicity, we describe here only the special, and more common, case where the supports of the p_is are all different.

Polyhedral homotopies have been applied to solve symmetric polynomial systems by means of constructing symmetric polyhedral homotopies (Verschelde and Cools 1994, Verschelde and Gatermann 1995). On the other hand, the Bernshteĭn theory is also used for constructing random product start systems for linear homotopies with various degrees of success (Li, Wang and Wang 1996, Li and Wang 1994).

5. Numerical considerations

5.1. Projective coordinates

As described in Section 1, solution paths of (1.2) that do not proceed to a solution of $P(x) = 0$ in $\mathbb{C}^n$ diverge to infinity: a very poor state of affairs for numerical methods. However, there is a simple idea from classical mathematics which improves the situation. If the system (1.2) is viewed in $\mathbb{P}^n$, the diverging paths are simply proceeding to a 'point at infinity' in projective space. Since projective space is compact, we can force all paths, including the extraneous ones, to have finite length by using projective coordinates.

For $P(x) = (p_1(x_1,\ldots,x_n),\ldots,p_n(x_1,\ldots,x_n)) = 0$, consider the system of $n+1$ equations in $n+1$ unknowns after homogenization,

$$\widetilde{P}: \quad \left\{ \begin{array}{lcl} \widetilde{p}_1(x_0,\ldots,x_n) & = & 0, \\ & \vdots & \\ \widetilde{p}_n(x_0,\ldots,x_n) & = & 0, \\ a_0x_0 + \cdots + a_nx_n - 1 & = & 0, \end{array} \right.$$

where $a_0, \ldots, a_n$ are complex numbers. When a start system

$$Q(x) = (q_1(x_1, \ldots, x_n), \ldots, q_n(x_1, \ldots, x_n)) = 0$$

is chosen, we also homogenize $Q(x)$ and consider the system

$$\tilde{Q}: \quad \begin{cases} \tilde{q}_1(x_0, \ldots, x_n) & = & 0, \\ & \vdots & \\ \tilde{q}_n(x_0, \ldots, x_n) & = & 0, \\ a_0 x_0 + \cdots + a_n x_n - 1 & = & 0. \end{cases}$$

We then use the classical homotopy continuation procedure to follow all the solution paths of the homotopy

$$\widetilde{H}(x_0, x_1, \ldots, x_n, t) = (1-t)c\widetilde{Q}(x_0, \ldots, x_n) + t\widetilde{P}(x_0, \ldots, x_n).$$

For almost every choice of $a_0, \ldots, a_n$, the paths stay in $\mathbb{C}^{n+1}$. It only remains to ignore solutions with $x_0 = 0$. Of the remaining solutions with $x_0 \neq 0$, it is easy to see that $x = (x_1/x_0, \ldots, x_n/x_0)$ is the corresponding solution of $P(x) = 0$.

A similar technique is described in Morgan and Sommese (1987*a*), where it is called a 'projective transformation'. It differs from the above in the following way. Instead of increasing the size of the problem from $n \times n$ to $(n+1) \times (n+1)$, they implicitly consider solving the last equation for x_0 and substituting in the other equations, essentially retaining n equations in n unknowns. Then the chain rule is used for the Jacobian calculations needed for path following. In many cases, it seems that this may create extra work. Suppose, for example, that the tenth equation in the system is $p_{10}(x) = x_7^3 - x_1 x_2$; its homogeneous version is $\widetilde{p}_{10}(x) = x_7^3 - x_0 x_1 x_2$. Since x_0 is now considered as a function of all other variables, the partial derivative of p_{10} with respect to every variable is suddenly nonzero. This results in added computation for each Jacobian evaluation, and is particularly problematic if the original problem is large and/or sparse.

A more advanced technique, the *projected Newton method*, was suggested in Shub and Smale (1993). A typical step to follow a solution curve in $\mathbb{C}^n$ of homotopy $H(x,t) = 0$, a system of n equations in $n+1$ variables, consists of two major steps: *prediction* and *correction*. The prediction step locates a point $(x^{(0)}, t_0)$. For fixed t_0, $H(x, t_0) = 0$ is a system of n equations in n unknowns. With starting points $x^{(0)}$, Newton's iteration,

$$x^{(m+1)} = x^{(m)} - [H_x(x^{(m)}, t_0)]^{-1} H(x^{(m)}, t_0), \qquad m = 0, 1, \ldots$$

can be applied to find the solution of $H(x, t_0) = 0$. If $x^{(0)}$ is suitably chosen by the prediction step, the iteration will converge to a solution of $H(x, t_0) = 0$ close to $x^{(0)}$. This is called the correction step. To follow the solution curve in projective space $\mathbb{P}^n$ after homogenization, $H(x,t) = 0$ becomes $\widetilde{H}(\widetilde{x}, t) = 0$ and, for fixed t_0, $\widetilde{H}(\widetilde{x}, t_0) = 0$ is now a system of n

equations in $n+1$ variables: $x_0, \ldots, x_n$. It is, therefore, unsuitable for the classical Newton iteration at the correction step. However, for any nonzero constant $c \in \mathbb{C}$, $\tilde{x}$ and $c\tilde{x}$ in $\mathbb{C}^{n+1}$ are considered to be equal in $\mathbb{P}^n$, whence the magnitude of $\tilde{x}$ in $\mathbb{C}^{n+1}$ is no longer significant in $\mathbb{P}^n$. Therefore it is reasonable to project every step of the Newton iteration onto the hyperplane perpendicular to the current point in $\mathbb{C}^{n+1}$. At $\tilde{x}^{(m)} \in \mathbb{C}^{n+1}$, we now have $n+1$ equations in $n+1$ unknowns, namely

$$\bar{H}(\tilde{x}, t_0) = \begin{cases} \tilde{H}(\tilde{x}, t_0) = 0 \\ (\tilde{x} - \tilde{x}^{(m)}) \cdot \tilde{x}^{(m)} = 0 \end{cases}$$

and one step of Newton's iteration for this system can be used to obtain

$$\tilde{x}^{(m+1)} = \tilde{x}^{(m)} - [\bar{H}_{\tilde{x}}(\tilde{x}^{(m)}, t_0)]^{-1} \bar{H}(\tilde{x}^{(m)}, t_0).$$

The efficiency of this strategy, known as the projected Newton iteration, when applied to following the homotopy curve in $\mathbb{P}^n$, is intuitively clear. See Figure 5. It frequently allows a bigger step size at the prediction stage.

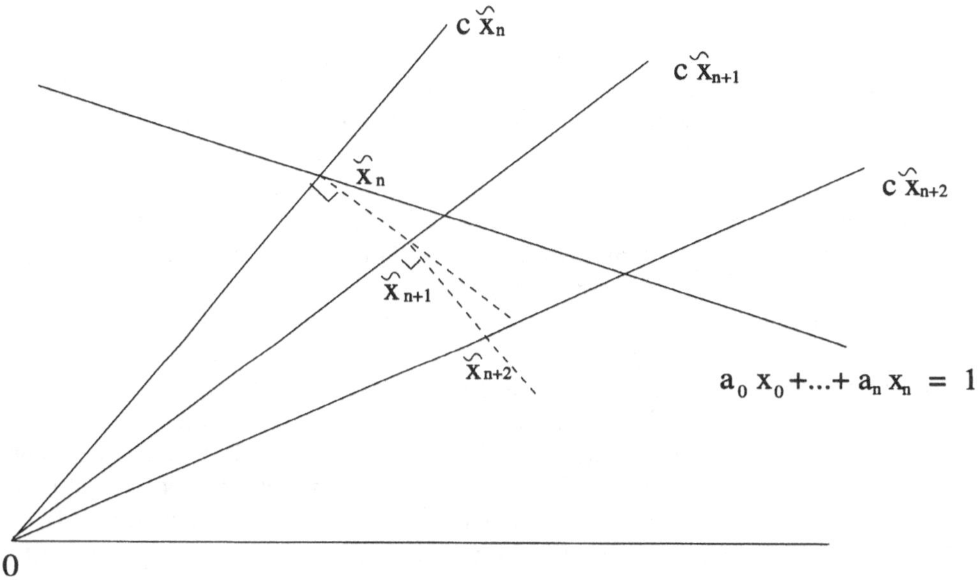

Fig. 5.

For practical considerations, we revise the above procedure as follows. At $t = t_1$, let

$$\tilde{x}(t_1) = (x_0(t_1), x_1(t_1), \ldots, x_n(t_n)) \in \mathbb{P}^n$$

be the corresponding point on a homotopy curve $(\tilde{x}(t), t)$ of $\tilde{H}(\tilde{x}(t), t) = 0$.

Let

$$|x_i(t_1)| = \max(|x_0(t_1)|, \ldots, |x_n(t_1)|).$$

We then fix the variable x_i in $\widetilde{H}(\widetilde{x}, t) = 0$ by the number $x_i(t_1)$, and thereafter, $\widetilde{H}(x_0, \ldots, x_{i-1}, x_{i+1}, \ldots, x_n, t) = 0$ becomes a system of n equations in $n+1$ variables. A standard prediction–correction procedure can now be applied to arrive at a new point $(x_0(t_2), \ldots, x_{i-1}(t_2), x_{i+1}(t_2), \ldots, x_n(t_2))$, which satisfies

$$\widetilde{H}(x_0, \ldots, x_{i-1}, x_{i+1}, \ldots, x_n, t_2) = 0.$$

Letting $\widetilde{x}(t_2) = (x_0(t_2), \ldots, x_{i-1}(t_2), x_i(t_1), x_{i+1}(t_2), \ldots, x_n(t_2))$, the point on the curve $(\widetilde{x}(t), t)$ for $t = t_2$ is obtained. A major advantage of this revision is that the size of the problem remains $n \times n$ throughout the procedure.

5.2. Real homotopy

Most polynomial systems arising in applications consist of polynomials with real coefficients, and most often the only desired solutions are real solutions. This suggests the use of real homotopies. That is, when the coefficients of the target polynomial system $P(x) = 0$ we want to solve are all real, we may choose a start system $Q(x) = 0$ with real coefficients, ensuring that the homotopy $H(x, t) = 0$ has real coefficients for all t. Thus, for fixed t, if x is a solution of $H(x, t) = 0$, so is its conjugate $\bar{x}$. Accordingly, a major advantage of real homotopy is that following a complex homotopy path $(x(s), t(s))$ provides its conjugate homotopy path $(\bar{x}(s), t(s))$ as a by-product without any further computation. On the other hand, although the homotopy $H(x, t)$ is still a map from $\mathbb{C}^n \times [0, 1]$ to $\mathbb{C}^n$, when a real homotopy path is traced, we may consider $H(x, t)$ as a map from $\mathbb{R}^n \times [0, 1]$ to $\mathbb{R}^n$, and hence the computation can be achieved in real arithmetic. In this way, a considerable reduction in computation is achieved.

There are numerous computational problems associated with the path following algorithms of real homotopies. In particular, when real homotopies are used, in contrast to the complex homotopy, bifurcation of some of the homotopy paths is inevitable. Hence, efficient algorithms must be developed to identify the bifurcation points and to follow the path after bifurcation. We can no longer parametrize the homotopy path of $H(x, t) = 0$ by t conventionally. Instead, the arclength s can be used as a parameter, and both x and t are considered to be independent variables. We now have

$$H(x(s), t(s)) = 0$$

and

$$H_x \dot{x} + H_t \dot{t} = 0,$$

where $\dot{x} \equiv \dfrac{dx}{ds}$, $\dot{t} \equiv \dfrac{dt}{ds}$ and $\|\dot{x}\|^2 + |\dot{t}|^2 = 1$. It is easy to see that bifurcations can only occur at *turning points*, points (x^*, t^*) for which $\dot{t} = 0$ and $H_x(x^*, t^*) = 0$ is singular. To identify the bifurcation point, let $a_0 = (x^{(0)}, t_0)$ be a point on the homotopy path Γ with $\dot{t}(a_0) > 0$. After a standard Euler prediction with step size h_0 and Newton corrections (Allgower and Georg 1990, 1993), we obtain a point $a_1 = (x^{(1)}, t_1)$ on Γ. When the tangent vector $(\dot{x}, \dot{t})$ is calculated at a_1 with $\dot{t}(a_1) < 0$, a turning point $a^* = (x^*, t^*)$ apparently exists in this situation (see Figure 6).

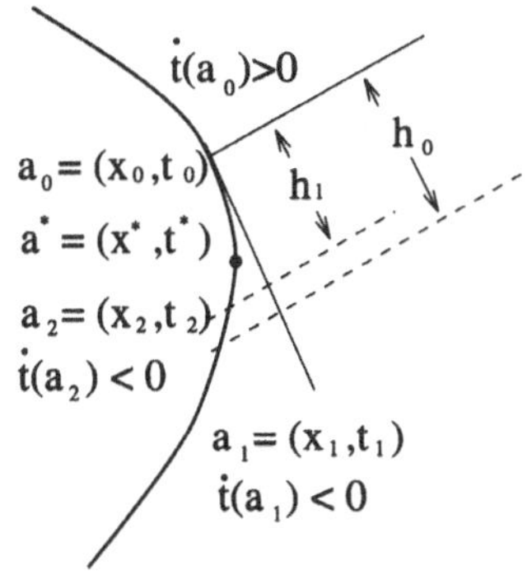

Fig. 6.

To identify a^*, we take the following procedure.

(1) Let h_1 be the solution of the equation ...

$$\frac{h}{h_0}\dot{t}(a_0) + \frac{h_0 - h}{h_0}\dot{t}(a_1) = 0.$$

Taking the Euler prediction at a_0 with step size h_1 followed by Newton corrections, we obtain a new point a_2 on Γ.

(2) If $\dot{t}(a_2) > 0$, we replace a_0 by a_2 and replace h_0 by the real part of the inner product of $(a_1 - a_2)$ and the unit tangent vector at a_2. If $\dot{t}(a_2) < 0$, we replace a_1 by a_2 and h_0 by h_1.

(3) (3) Repeat step 1 until $\dot{t}(a_2)$ is sufficiently small. Then, a_2 will be taken as a bifurcation point $a^* = (x^*, t^*)$.

When the bifurcation point a^* is identified, in order to follow the bifurcation branches, tangent vectors of the branches need to be characterized. It turns out that for the following special kind of turning point the bifurcation phenomenon is rather simple.

Definition 5.1 A singular point $(x^*, t^*) \in \mathbb{C}^n \times [0, 1]$ is said to be a quadratic turning point of $H(x, t) = 0$ if

(1) $\text{Rank}_R H_x(x^*, t^*) = 2n - 2$
(2) $\text{Rank}_R[H_x(x^*, t^*), H_t(x^*, t^*)] = 2n - 1$

(3) For $y \in \mathbb{C}^n \setminus \{0\}$ satisfying $H_x(x^*, t^*)y = 0$, we have

$$\text{Rank}_R[H_x(x^*, t^*), H_{xx}(x^*, t^*)yy) = 2n.$$

Here, Rank_R denotes the real rank.

Proposition 5.1 (Li and Wang 1994) Let (x^*, t^*) be a quadratic turning point. Then, there are only two branches of solution paths Γ and Γ' passing through (x^*, t^*). If ϕ is the tangent vector of the path Γ at (x^*, t^*), then the tangent vector of Γ' is the direction of $i\phi$ (see Figure 7).

When Γ is a real path, the assertion of this proposition can be considered as a special case of Allgower (1984) and Henderson and Keller (1990). The most general version, where Γ and x^* may both be complex, was proved in Li and Wang (1993).

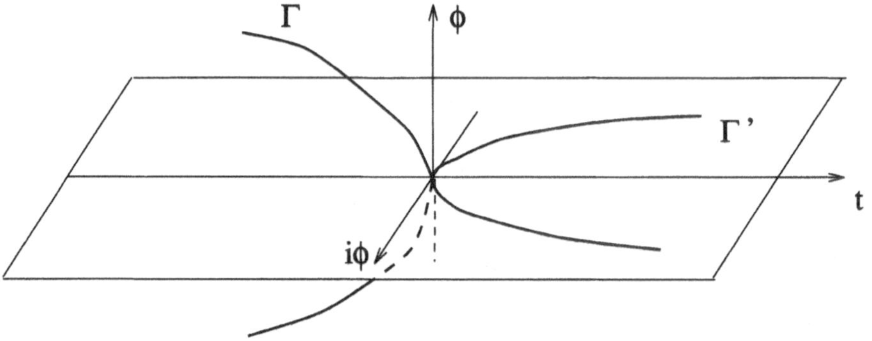

Fig. 7.

To follow the bifurcation branch Γ' at a quadratic turning point, we consider the following three situations.

(1) *Γ is a real path.*
Then, ϕ is real and $i\phi$ is pure imaginary. Apparently, the bifurcation branch Γ' consists of a complex path and its complex conjugacy. We need only to follow one of them with tangent vector $i\phi$ or $-i\phi$.

(2) *Γ is a complex path and (x^*, t^*) is real.*
Then, Γ consists of complex conjugate pairs (x, t) and $(\bar{x}, t)$ for each $t < t^*$. The tangent vector at (x^*, t^*) is

$$\phi = \lim_{s_1 - s_2) \to 0} \frac{(x(s_1), t(s_1)) - (x(s_2), t(s_2))}{s_1 - s_2},$$

where $x(s_2) = \bar{x}(s_1), t(s_1) = t(s_2)$ is clearly pure imaginary. Hence, $i\phi$ is real. Consequently, the bifurcation branch Γ' consists of two real paths. We may follow them in real space $\mathbb{R}^n \times [0, 1]$ with real tangent vectors $i\phi$ and $-i\phi$ respectively.

(3) Γ *is a complex path and* x^* *is complex.*
The bifurcation branch Γ', in this case, consists of two complex solution paths. They are not conjugate to each other. We may follow them with tangent vector $i\phi$ and $-i\phi$ respectively.

It was conjectured in Brunovský and Meravý (1984) and proved in Li and Wang (1993) that, generically, real homotopies contain no singular points other than a finite number of quadratic turning points.

5.3. Software

Several software packages dedicated to solving polynomial systems by homotopy continuation are publicly available. HOMPACK (Morgan, Sommese and Watson 1989) and CONSOL (Morgan 1987) are written in FORTRAN 77. HOMPACK is a general package for homotopy continuation with a polynomial driver. It has been parallelized to various architectures (Allison, Chakraborty and Watson 1989, Harimoto and Watson 1989). The code for CONSOL is contained in Morgan (1987), Appendix 6. The programs pss (Malajovich, software) and Pelican (Huber, software) are written in C. The pss contains facilities for parallel continuation and Pelican provides the polyhedral methods. The package PHC and MVC (Verschelde 1995) is written in Ada and compiled on three different hardware platforms, for which executables are available on request. Two main features of this package are the wide variety of homotopy methods and the powerful facilities for mixed volume computation.

Nonetheless, a more efficient and user-friendly code including all the features described in this article is still under development. In particular, a better understanding of the convex geometry with a clever use of linear programming techniques will make the polyhedra homotopy method described in Section 3 much more powerful.

REFERENCES

E. L. Allgower (1984), Bifurcation arising in the calculation of critical points via homotopy methods, in *Numerical Methods for Bifurcation Problems* (T. Kupper, H. D. Mittelman and H. Weber, eds), Birkhäuser, Basel, pp. 15–28.

E. L. Allgower and K. Georg (1990), *Numerical Continuation Methods, an Introduction*, Springer, Berlin. Springer Series in Computational Mathematics, Vol. 13.

E. L. Allgower and K. Georg (1993), Continuation and path following, in *Acta Numerica*, Vol. 2, Cambridge University Press, pp. 1–64.

D. C. S. Allison, A. Chakraborty and L. T. Watson (1989), 'Granularity issues for solving polynomial systems via globally convergent algorithms on a hypercube', *J. Supercomputing* **3**, 5–20.

D. N. Bernshteín (1975), 'The number of roots of a system of equations', *Functional Anal. Appl.* **9**, 183–185. Translated from *Funktsional. Anal. i Prilozhen.*, **9**, 1–4.

P. Brunovský and P. Meravý (1984), 'Solving systems of polynomial equations by bounded and real homotopy', *Numer. Math.* **43**, 397–418.

B. Buchberger (1985), Gröbner basis: An algorithmic method in polynomial ideal theory, in *Multidimensional System Theory* (N. Bose, ed.), D. Reidel, Dordrecht, pp. 184–232.

J. Canny and J. M. Rojas (1991), An optimal condition for determining the exact number of roots of a polynomial system, in *Proceedings of the 1991 International Symposium on Symbolic and Algebraic Computation*, ACM, pp. 96–101.

S. N. Chow, J. Mallet-Paret and J. A. Yorke (1979), Homotopy method for locating all zeros of a system of polynomials, in *Functional Differential Equations and Approximation of Fixed Points* (H. O. Peitgen and H. O. Walther, eds), Lecture Notes in Mathematics, Vol. 730, Springer, Berlin, pp. 77–88.

F. J. Drexler (1977), 'Eine Methode zur Berechnung sämtlicher Lösungen von Polynomgleichungssystemen', *Numer. Math.* **29**, 45–58.

I. Emiris (1994), Sparse Elimination and Applications in Kinematics, PhD thesis, University of California at Berkeley.

I. Emiris and J. Canny (1995), 'Efficient incremental algorithms for the sparse resultant and the mixed volume', *J. Symb. Computation* **20**, 117–149.

W. Fulton (1984), *Intersection Theory*, Springer, Berlin.

C. B. Garcia and W. I. Zangwill (1979), 'Finding all solutions to polynomial systems and other systems of equations', *Mathematical Programming* **16**, 159–176.

I. M. Gel'fand, M. M. Kapranov and A. V. Zelevinskií (1994), *Discriminants, Resultants and Multidimensional Determinants*, Birkhäuser, Boston.

S. Harimoto and L. T. Watson (1989), The granularity of homotopy algorithms for polynomial systems of equations, in *Parallel Processing for Scientific Computing* (G. Rodrigue, ed.), SIAM, Philadelphia.

M. E. Henderson and H. B. Keller (1990), 'Complex bifurcation from real paths', *SIAM J. Appl. Math.* **50**, 460–482.

B. Huber (software), *Pelican Manual.* Available via the author's web page, `http://math.cornell.edu/~birk`.

B. Huber and B. Sturmfels (1995), 'A polyhedral method for solving sparse polynomial systems', *Math. Comp.* **64**, 1541–1555.

B. Huber and B. Sturmfels (1997), 'Bernstein's theorem in affine space', *Discrete Comput. Geom.* To appear.

A. G. Khovanskií (1978), 'Newton polyhedra and the genus of complete intersections', *Functional Anal. Appl.* **12**, 38–46. Translated from *Funktsional. Anal. i Prilozhen.*, **12**, 51–61.

A. G. Kushnirenko (1976), 'Newton polytopes and the Bézout theorem', *Functional Anal. Appl.* **10**, 233–235. Translated from *Funktsional. Anal. i Prilozhen.*, **10**, 82–83.

C. W. Lee (1991), Regular triangulations of convex polytopes, in *Applied Geometry and Discrete Mathematics – The Victor Klee Festschrift, DIMACS Series Vol. 4* (P. Gritzmann and B. Sturmfels, eds), American Mathematical Society, Providence, RI, pp. 443–456.

T. Y. Li (1983), 'On Chow, Mallet-Paret and Yorke homotopy for solving systems of polynomials', *Bulletin of the Institute of Mathematics, Acad. Sin.* **11**, 433–437.

T. Y. Li and T. Sauer (1989), 'A simple homotopy for solving deficient polynomial systems', *Japan J. Appl. Math.* **6**, 409–419.

T. Y. Li and X. Wang (1990), A homotopy for solving the kinematics of the most general six- and-five-degree of freedom manipulators, in *Proc. of ASME Conference on Mechanisms, D1-Vol. 25*, pp. 249–252.

T. Y. Li and X. Wang (1991), 'Solving deficient polynomial systems with homotopies which keep the subschemes at infinity invariant', *Math. Comp.* **56**, 693–710.

T. Y. Li and X. Wang (1992), 'Nonlinear homotopies for solving deficient polynomial systems with parameters', *SIAM J. Numer. Anal.* **29**, 1104–1118.

T. Y. Li and X. Wang (1993), 'Solving real polynomial systems with real homotopies', *Math. Comp.* **60**, 669–680.

T. Y. Li and X. Wang (1994), 'Higher order turning points', *Appl. Math. Comput.* **64**, 155–166.

T. Y. Li and X. Wang (1996), 'The BKK root count in $\mathbb{C}^n$', *Math. Comp.* **65**, 1477–1484.

T. Y. Li, T. Sauer and J. A. Yorke (1987*a*), 'Numerical solution of a class of deficient polynomial systems', *SIAM J. Numer. Anal.* **24**, 435–451.

T. Y. Li, T. Sauer and J. A. Yorke (1987*b*), 'The random product homotopy and deficient polynomial systems', *Numer. Math.* **51**, 481–500.

T. Y. Li, T. Sauer and J. A. Yorke (1988), 'Numerically determining solutions of systems of polynomial equations', *Bull. Amer. Math. Soc.* **18**, 173–177.

T. Y. Li, T. Sauer and J. A. Yorke (1989), 'The cheater's homotopy: an efficient procedure for solving systems of polynomial equations', *SIAM J. Numer. Anal.* **26**, 1241–1251.

T. Y. Li, T. Wang and X. Wang (1996), Random product homotopy with minimal BKK bound, in *Proceedings of the AMS-SIAM Summer Seminar in Applied Mathematics on Mathematics of Numerical Analysis: Real Number Algorithms, Park City, Utah*, pp. 503–512.

G. Malajovich (software), *pss 2.alpha, polynomial system solver, version 2.alpha, README file.* Distributed by the author through gopher, `http://www.labma.ufrj.br/~gregorio`.

A. P. Morgan (1986), 'A homotopy for solving polynomial systems', *Appl. Math. Comput.* **18**, 173–177.

A. P. Morgan (1987), *Solving Polynomial Systems using Continuation for Engineering and Scientific Problems*, Prentice Hall, Englewood Cliffs, NJ.

A. P. Morgan and A. J. Sommese (1987*a*), 'Computing all solutions to polynomial systems using homotopy continuation', *Appl. Math. Comput.* **24**, 115–138.

A. P. Morgan and A. J. Sommese (1987*b*), 'A homotopy for solving general polynomial systems that respect m-homogeneous structures', *Appl. Math. Comput.* **24**, 101–113.

A. P. Morgan and A. J. Sommese (1989), 'Coefficient-parameter polynomial continuation', *Appl. Math. Comput.* **29**, 123–160. Errata: *Appl. Math. Comput.* **51**, 207 (1992).

A. P. Morgan, A. J. Sommese and L. T. Watson (1989), 'Finding all isolated solutions to polynomial systems using HOMPACK', *ACM Trans. Math. Software* **15**, 93–122.

J. M. Rojas (1994), 'A convex geometric approach to counting the roots of a polynomial system', *Theoret. Comput. Sci.* **133**, 105–140.

J. M. Rojas and X. Wang (1996), 'Counting affine roots of polynomial systems via pointed Newton polytopes', *J. Complexity* **12**, 116–133.

I. R. Shafarevich (1977), *Basic Algebraic Geometry*, Springer, New York.

M. Shub and S. Smale (1993), 'Complexity of Bézout's theorem I: Geometric aspects', *J. Amer. Math. Soc.* **6**, 459–501.

L. W. Tsai and A. P. Morgan (1985), 'Solving the kinematics of the most general six- and five-degree-of-freedom manipulators by continuation methods', *ASME Journal of Mechanics, Transmissions, and Automation in Design* **107**, 189–200.

J. Verschelde (1995), PHC and MVC: two programs for solving polynomial systems by homotopy continuation, Technical report. Presented at the PoSSo workshop on software, Paris. Available by anonymous ftp to `ftp.cs.kuleuven.ac.be` in the directory /pub/NumAnal-ApplMath/PHC.

J. Verschelde (1996), Homotopy continuation methods for solving polynomial systems, PhD thesis, Katholieke Universiteit Leuven, Belgium.

J. Verschelde and R. Cools (1993), 'Symbolic homotopy construction', *Applicable Algebra in Engineering, Communication and Computing* **4**, 169–183.

J. Verschelde and R. Cools (1994), 'Symmetric homotopy construction', *J. Comput. Appl. Math.* **50**, 575–592.

J. Verschelde and K. Gatermann (1995), 'Symmetric Newton polytopes for solving sparse polynomial systems', *Adv. Appl. Math.* **16**, 95–127.

J. Verschelde, K. Gatermann and R. Cools (1996), 'Mixed volume computation by dynamic lifting applied to polynomial system solving', *Discrete Comput. Geom.* **16**, 69–112.

C. W. Wampler (1992), 'Bézout number calculations for multi-homogeneous polynomial systems', *Appl. Math. Comput.* **51**, 143–157.

C. W. Wampler (1994), 'An efficient start system for multi-homogeneous polynomial continuation', *Numer. Math.* **66**, 517–523.

A. H. Wright (1985), 'Finding all solutions to a system of polynomial equations', *Math. Comp.* **44**, 125–133.

W. Zulener (1988), 'A simple homotopy method for determining all isolated solutions to polynomial systems', *Math. Comp.* **50**, 167–177.

Acta Numerica (1997), *pp.* 437–483

Numerical solution of highly oscillatory ordinary differential equations

Linda R. Petzold
Department of Computer Science,
University of Minnesota,
4-192 EE/CS Bldg, 200 Union Street S.E.,
Minneapolis, MN 55455-0159, USA
E-mail: petzold@cs.umn.edu

Laurent O. Jay
Department of Computer Science,
University of Minnesota,
4-192 EE/CS Bldg, 200 Union Street S.E.,
Minneapolis, MN 55455-0159, USA
E-mail: na.ljay@na-net.ornl.gov

Jeng Yen
Army High Performance Computing Research Center,
University of Minnesota,
1100 Washington Ave. S.,
Minneapolis, MN 55415, USA
E-mail: yen@ahpcrc.umn.edu

CONTENTS

1. Introduction

One of the most difficult problems in the numerical solution of ordinary differential equations (ODEs) and in differential-algebraic equations (DAEs) is the development of methods for dealing with highly oscillatory systems. These types of systems arise, for example, in vehicle simulation when modelling the suspension system or tyres, in models for contact and impact, in flexible body simulation from vibrations in the structural model, in molecular dynamics, in orbital mechanics, and in circuit simulation. Standard numerical methods can require a huge number of time-steps to track the oscillations, and even with small stepsizes they can alter the dynamics, unless the method is chosen very carefully.

What is a highly oscillatory system, and what constitutes a solution of such a system? As we will see, this question is somewhat application-dependent, to the extent that it does not seem possible to give a precise mathematical definition which would include most of the problems that scientists, engineers and numerical analysts have described as highly oscillatory. *Webster's Ninth New Collegiate Dictionary* (1985) includes the following definitions for *oscillate*: 'to swing backward and forward like a pendulum; to move or travel back and forth between two points; to vary above and below a mean value.' Here we are mainly interested in systems whose solutions may be oscillatory in the sense that there is a *fast* solution which varies regularly about a *slow* solution. The problem will be referred to as *highly oscillatory* if the timescale of the fast solution is much shorter than the interval of integration.

We will begin with a simple example of an oscillating problem from multibody dynamics. In *Cartesian coordinates*, a simple stiff spring pendulum model with unit mass, length, and gravity, can be expressed as

$$0 = x' - u, \tag{1.1a}$$
$$0 = y' - v, \tag{1.1b}$$
$$0 = u' + x\lambda, \tag{1.1c}$$
$$0 = v' + y\lambda - 1, \tag{1.1d}$$
$$\epsilon^2\lambda = \frac{\sqrt{x^2+y^2}-1}{\sqrt{x^2+y^2}}, \tag{1.1e}$$

where $1/\epsilon^2 \gg 1$ is the spring constant. Preloading the spring by using $\epsilon = \sqrt{10^{-3}}$, the initial position $(x_0, y_0) = (0.9, 0.1)$, and the zero initial velocity $(u_0, v_0) = (0, 0)$, the results of the states (x, y, u, v) in the 0 to 10[s] simulation are shown in Fig. 1.

The solution to this problem consists of a low-amplitude, high-frequency oscillation superimposed on a slow solution. It is not immediately clear that a slow solution appears in the above problem. In fact, we can only identify

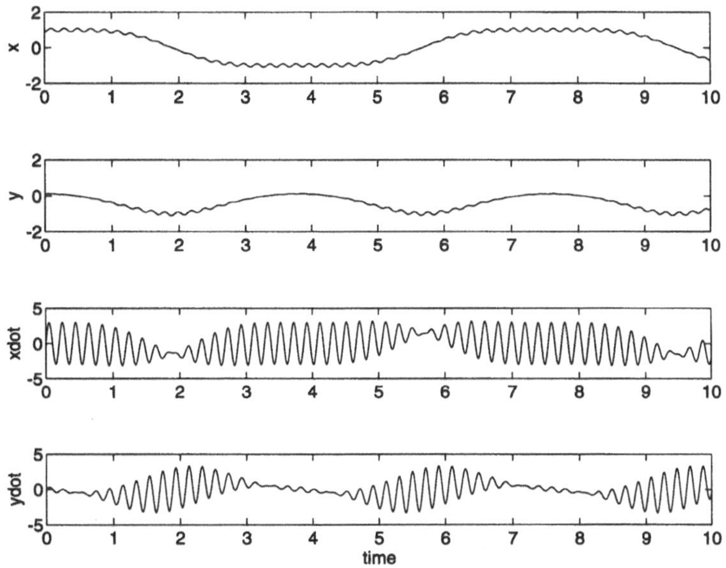

Fig. 1. Stiff spring pendulum in Cartesian coordinates

the slow solution of (1.1) using a proper nonlinear coordinate transformation $(x, y) = (r\cos(\theta), r\sin(\theta))$. In *polar coordinates* (r, θ), we obtain the equations of motion of (1.1):

$$0 = r' - z, \tag{1.2a}$$

$$0 = \theta' - \omega, \tag{1.2b}$$

$$0 = z' + r\omega^2 + \frac{1}{\epsilon^2}(r-1) - \sin\theta, \tag{1.2c}$$

$$0 = \omega' - \frac{1}{r}(2z\omega - \cos\theta), \tag{1.2d}$$

where (z, ω) is the velocity. In the 0 to 10[s] second simulation, using the same initial conditions, we obtain the solution in Fig. 2. It is clear that the length r represents the fast motion and the angle θ the slow motion.

One of the questions one must answer in selecting an appropriate mathematical or numerical method is: 'What do we mean by a solution?' For example, one might be interested only in finding the slow solution. On the other hand, in some situations it may be important to recover more information about the high-frequency oscillation, such as its amplitude, its energy or its envelope. The most detailed information about the high-frequency oscillation also includes its phase; this information is usually very difficult to recover, particularly over intervals which are long in comparison to the period of the oscillation. Efficiency is often an important consideration; one might be willing to give up on tracking some of the detailed information of

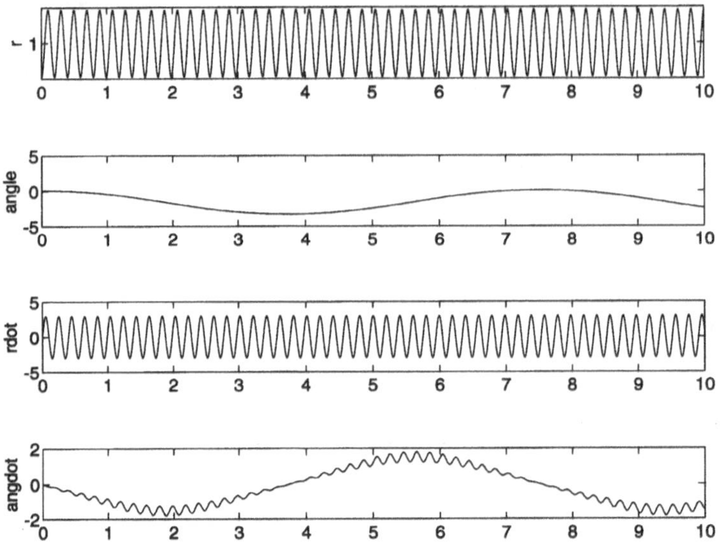

Fig. 2. Stiff spring pendulum in polar coordinates

the high-frequency oscillation in order to take much larger stepsizes. This is the case in real-time simulation of mechanical systems. For other problems, maintaining physical and mathematical properties in the numerical solution can be critical, particularly over long-time intervals. For example, in molecular dynamics it may be important to maintain invariants like the energy or the symplectic structure of the problem (Arnold 1989). What is meant by a solution is determined not only by the physical properties of the system and its mathematical structure, but also by how the information from the simulation is to be used.

The form and structure of the oscillating problem is highly application-dependent. Some problems are posed as a first-order ODE system, others as a second-order ODE system. Other problems include constraints, and hence are formulated as a DAE system. Often these ODE and DAE systems have a special mathematical structure. Some applications yield problems which are linear or nearly linear, while other applications require the solution of highly nonlinear oscillating systems. Some problems may have a single high frequency and be nearly periodic, whereas other problems may have multiple high-frequency components. Some oscillating problems, for example in ocean dynamics (Garrett and Munk 1979, Gjaja and Holm 1996), corrosion modelling (Tidblad and Graedel 1996), atmospheric modelling (Kopell 1985), nonlinear optics (Agrawal 1989), *ab initio* molecular dynamics (Tuckerman and Parrinello 1994) yield partial differential equation (PDE) systems; these problems are beyond the scope of the present paper although many of the same considerations and types of methods apply for the time integration.

This paper will deal mainly with *numerical methods* for oscillating systems. There is an extensive literature in applied mathematics (Bogoliubov and Mitropolski 1961, Minorsky 1974, Fenichel 1979, Kevorkian and Cole 1981) including the method of averaging, the method of multiple scales, and the stroboscopic method, on approximating the solution to oscillating problems. Some of these techniques are related to the numerical methods described here. Methods from applied mathematics can sometimes be combined advantageously with numerical methods (Kirchgraber 1982) for the solution of oscillating problems. Most of the mathematical techniques require a nearly linear structure of the problem. For some applications, the equations naturally occur in this structure or can be easily reformulated; for others, casting the problem in this form is difficult or impossible. There is also a tradition of physically motivated mathematical or numerical methods that reformulate the system prior to numerical solution, using approximations that the scientist or engineer deems to be valid. These methods can be quite powerful when used carefully. The LIN method (see Subsection 4.4) for molecular dynamics, and modal analysis methods (see Subsection 3.4) for structural analysis are examples of these kinds of methods.

A wide variety of numerical methods has been developed for highly oscillatory problems. The best method to use is strongly dependent on the application. Small-amplitude oscillations in linear or nearly linear systems can often be damped via highly stable implicit numerical methods. We will see that it is also feasible to damp the oscillation in certain structured, highly nonlinear oscillating problems from mechanical systems. Even with numerical methods based simply on damping the oscillation, there can be unforeseen difficulties due to the nonlinear oscillation, for example in automatic stepsize control and in obtaining convergence of the Newton iteration of implicit numerical methods. In other applications, damping the oscillation can destroy important properties of the solution. For these problems, much attention has been focused on preserving important physical and mathematical properties like the energy or the symplectic structure of the system. Many of the numerical methods that can do this require relatively small stepsizes. Efficiency is also an important consideration, making these problems quite challenging. Still other problems yield systems with a single high-frequency oscillation. Methods based on envelope following can yield the smooth solution in this case.

It is important to recognize that, in general, one should not expect to be able to numerically solve nonlinear highly oscillatory problems using stepsizes which are large relative to the timescale of the fast solution. Standard numerical ODE methods make use only of local information about the problem, obtained from evaluating the right-hand side of the differential equation. The methods which for some applications are able to take large stepsizes are able to do this by implicitly or explicitly making use of global information about the problem and/or its mathematical structure. For example, it is feasible to damp the oscillation for certain mechanical systems only because of a very specific mathematical structure. The LIN method of molecular dynamics makes use of both the mathematics and the physics of the problem. Envelope-following methods make use of the fact that for some problems it is known *a priori* that the fast solution has a single high-frequency component.

Unlike most stiff problems, which can be solved by strongly damped implicit numerical methods, effective solution of nonlinear highly oscillatory problems generally requires exploitation of the problem structure and a careful examination of the objectives and intended use of the computation. Therefore we have based the organization of this paper on classes of application problems. Section 2 covers linear problems and basic concepts that are fundamental to understanding numerical methods for highly oscillatory problems. Section 3 deals with highly oscillatory rigid and flexible mechanical systems, describing the nonlinear structure of these systems and implications for numerical methods, when and how the oscillation can be safely and efficiently damped, modal analysis techniques from structural analysis, and the problems and considerations in extending these techniques to flexible multibody systems. Section 4 briefly describes problems and numerical methods for molecular dynamics. Section 5 describes problems from circuit analysis and orbital mechanics for which envelope-following techniques are applicable, and describes those numerical methods.

2. Basic concepts and methods for linear oscillatory systems

Numerical methods used to treat oscillatory problems differ, depending on the formulation of the problem, the knowledge of certain characteristics of the solution, and the objectives of the computation (Gear 1984). However, certain concepts are common to most classes of methods. Since it is not possible to give a uniform presentation of these concepts, as an illustration we will consider the class of *partitioned Runge–Kutta* (PRK) methods which includes standard Runge–Kutta (RK) methods and other schemes of interest, such as the Verlet algorithm (4.2). For other classes of methods, the definitions are analogous.

To investigate the stability properties of numerical methods applied to oscillatory systems, the scalar *harmonic oscillator* equation

$$y'' = -\omega^2 y, \qquad (\omega > 0) \tag{2.1}$$

is chosen as a standard test equation. This is the analogue of Dahlquist's test equation $y' = \lambda y$ for first-order ODEs, although the situation is not totally parallel to problems with large negative eigenvalues of the Jacobian matrix. The solutions to (2.1) are given by the family of *sine curves* $y(t) = A\sin(\omega t + \phi)$, where the expression $\omega t + \phi$ is called the *phase*. The real parameters $A \geq 0$, ω, and ϕ are called, respectively, the *amplitude*, the *pulse*, and the *phase-lag*. The *period* of the solution is $T := 2\pi/\omega$ and its *frequency* is $f := 1/T = \omega/2\pi$. The parameters A and ϕ are determined from the initial conditions. Sine curves are archetypal oscillatory functions and they form the basis of Fourier analysis.

To apply PRK methods, we must first rewrite (2.1) as a first-order system by introducing a new variable $z := y'$, yielding

$$y' = z, \qquad z' = -\omega^2 y. \tag{2.2}$$

This is one of the simplest systems for which the eigenvalues $(\pm i\omega)$ of the Jacobian matrix of the system are purely imaginary. This is also a linear Hamiltonian system with Hamiltonian $H(y,z) = \left(\omega^2 y^2 + z^2\right)/2$. PRK methods take advantage of the intrinsic partitioning of the equations by making use of the conjunction of two sets of RK coefficients. One step of an *s-stage PRK method* applied to partitioned systems of the form

$$y' = f(t,y,z), \qquad z' = g(t,y,z),$$

with initial values (y_0, z_0) at t_0 and stepsize h, is defined by

$$y_1 = y_0 + h\sum_{i=1}^{s} b_i f(t_i, Y_i, Z_i), \qquad z_1 = z_0 + h\sum_{i=1}^{s} b_i g(t_i, Y_i, Z_i),$$

$$Y_i = y_0 + h\sum_{j=1}^{s} a_{ij} f(t_j, Y_j, Z_j), \qquad Z_i = z_0 + h\sum_{j=1}^{s} \widehat{a}_{ij} g(t_j, Y_j, Z_j),$$

where $t_i := t_0 + c_i h$. The coefficients (b_i, a_{ij}, c_i) and $(b_i, \widehat{a}_{ij}, c_i)$ are the coefficients of two RK methods based on the same quadrature formula (b_i, c_i). Applying the PRK method to (2.2) we get

$$\begin{pmatrix} y_1 \\ z_1 \end{pmatrix} = D_\omega M(\mu) D_\omega^{-1} \begin{pmatrix} y_0 \\ z_0 \end{pmatrix},$$

where $\mu := h\omega$, $D_\omega := \operatorname{diag}(1, \omega)$, and $M(\mu)$ is the 2×2 *stability matrix* of the PRK method. This matrix is given by

$$M(\mu) = I_2 + \mu \begin{pmatrix} O & b^T \\ b^T & O \end{pmatrix} \begin{pmatrix} I_s & -\mu A \\ \mu\widehat{A} & I_s \end{pmatrix}^{-1} \begin{pmatrix} \mathbb{1}_s & O \\ O & \mathbb{1}_s \end{pmatrix}, \tag{2.3}$$

where we have used $\mathbb{1}_s$ to denote the s-dimensional vector $(1, \ldots, 1)^T$, I_n for the identity matrix in $\mathbb{R}^{n \times n}$, $\{A, \widehat{A}\}$ for the matrices of the RK coefficients, and b for the vector of the RK weights. Other expressions for the stability matrix can be derived with the help of Van Der Houwen and Sommeijer

(1989, Lemma 2.1). The exact solution to (2.2) at $t_0 + h$ can be expressed by

$$\begin{pmatrix} y(t_0+h) \\ z(t_0+h) \end{pmatrix} = D_\omega \Theta(\mu) D_\omega^{-1} \begin{pmatrix} y_0 \\ z_0 \end{pmatrix}, \quad \Theta(\mu) = \begin{pmatrix} \cos(\mu) & \sin(\mu) \\ -\sin(\mu) & \cos(\mu) \end{pmatrix}.$$

The eigenvalues of the rotation matrix $\Theta(\mu)$ are of modulus one. This motivates the following definition.

Definition 2.1 For a PRK method, an interval I with $\{0\} \subset I \subset \mathbb{R}$ is an *interval of periodicity* if for all $\mu \in I$ the eigenvalues $\lambda_i(\mu)$ $(i = 1, 2)$ of the stability matrix $M(\mu)$ (2.3) satisfy $|\lambda_i(\mu)| = 1$ $(i = 1, 2)$ and, if $\lambda_1(\mu) = \lambda_2(\mu)$, then this eigenvalue must possess two distinct eigenvectors. A method is said to be *P-stable* if $\mathbb{R}$ is an interval of periodicity.

If the interval of periodicity is not reduced to $\{0\}$, the method is usually called *nondissipative*. These concepts are due to Lambert and Watson (1976) and were originally introduced for linear multistep methods applied to $y'' = g(t, y)$. They proved that nondissipative linear multistep methods must be symmetric. They also stated that P-stable linear multistep methods cannot have order greater than two. A proof of this result in a more general setting was given by Hairer (1979). This is a result similar to the famous Dahlquist's second barrier (Dahlquist 1963). To overcome this order barrier, several hybrid multistep methods have been derived (Cash 1981, Chawla and Rao 1985, Hairer 1979), for instance, a P-stable modification to the fourth-order Numerov method (Hairer 1979). For standard RK methods ($A = \widehat{A}$) the eigenvalues of the stability matrix are simply given by $\lambda_1(\mu) = R(\mathrm{i}\mu)$ and $\lambda_2(\mu) = R(-\mathrm{i}\mu) = \overline{\lambda_1(\mu)}$, where $R(z) = 1 + zb^T(I_s - zA)^{-1}\mathbb{1}_s$ is the usual stability function of the RK method. Hence, we have the following well-known result.

Theorem 2.1 Symmetric RK methods are P-stable.

For example, the implicit midpoint rule is P-stable. However, a similar theorem does not hold for PRK methods. For example, we can consider the coefficients of the two-stage Lobatto IIIA–IIIB method (Jay 1996)

$$A = \begin{pmatrix} 0 & 0 \\ 1/2 & 1/2 \end{pmatrix}, \qquad \widehat{A} = \begin{pmatrix} 1/2 & 0 \\ 1/2 & 0 \end{pmatrix}, \qquad b^T = \begin{pmatrix} 1/2 & 1/2 \end{pmatrix}.$$

This symmetric and symplectic method is equivalent to the famous leapfrog/Störmer/Encke/Verlet method (4.2) used for second-order ODEs. Necessary conditions on the coefficients of the stability matrix $M(\mu)$ to satisfy the conditions of Definition 2.1 are given by

$$\det(M(\mu)) = 1, \qquad |\operatorname{tr}(M(\mu))| \leq 2. \tag{2.4}$$

The stability matrix of the two-stage Lobatto IIIA–IIIB method is

$$M(\mu) = \begin{pmatrix} 1-\mu^2/2 & \mu \\ -\mu+\mu^3/4 & 1-\mu^2/2 \end{pmatrix}.$$

It satisfies $\det(M(\mu)) = 1$, but only $|\operatorname{tr}(M(\mu))| = |2-\mu^2|$. Thus, according to (2.4) this method is not P-stable; its interval of periodicity (and of absolute stability, see below) is $(-2, 2)$. As pointed out in Lambert and Watson (1976), the relevance of the property of P-stability seems restricted to situations exhibiting *periodic stiffness*, that is, where the oscillatory solution is of negligible amplitude. The reason is that the stepsize of a method is not only limited by stability requirements but is also dictated by accuracy requirements. A stepsize of the same magnitude as the period of oscillation with highest frequency is required even for P-stable methods to follow this oscillation in order to preserve the accuracy of the method, unless its amplitude is sufficiently small.

The weaker property of nondissipativity is of primary interest in celestial mechanics for orbital computation, where it is desired that the numerically computed orbits do not spiral inwards or outwards. In this context, a related notion is the property of *orbital stability* of Cooper (1987), that is, the preservation of quadratic invariants by the numerical method. The construction of nondissipative explicit Runge–Kutta–Nyström (RKN) methods of order two to five with a minimal number of function evaluations per step and possessing relatively large intervals of periodicity is given in Chawla and Sharma (1981*a*) and (1981*b*). In Portillo and Sanz-Serna (1995), it is shown with an example that, for Hamiltonian systems, nondissipative methods do not in general share the advantageous error propagation mechanism possessed by symplectic methods (Sanz-Serna and Calvo 1994). In this framework, an explicit symplectic method of effective order four with three function evaluations per step and with a maximal interval of periodicity is presented in López-Marcos, Sanz-Serna and Skeel (1995*b*).

In certain applications, it can be desirable to leave the fast oscillation modes unresolved. For example, in many structural dynamics applications (see Subsection 3.5), high-frequency oscillations are spurious and should be damped out. Hence, we can consider a less stringent notion of stability.

Definition 2.2 Replacing the condition $|\lambda_i(\mu)| = 1$ by $|\lambda_i(\mu)| \leq 1$ in Definition 2.1 we define the notions of an *interval of absolute stability* and of *I-stability*.

For standard RK methods we recover the usual definition of *I-stability* (Hairer and Wanner 1996). *L-stable* RK methods, *i.e.*, RK methods satisfying $R(\infty) = 0$ and $|R(z)| \leq 1$ when $\operatorname{Re}(z) \leq 0$, such as the implicit Euler method, may be appropriate to damp out highly oscillatory components (corresponding to large μ) since they are I-stable and satisfy $\lim_{\mu\to\infty} |\lambda_i(\mu)|$

$= 0$ $(i = 1, 2)$. When the eigenvalues of the stability matrix (2.3) are conjugate, we can write

$$\lambda_1(\mu) = \overline{\lambda_2(\mu)} = \rho(\mu)e^{i\theta(\mu)},$$

where $\rho(\mu)$ and $\theta(\mu)$ are real-valued functions. Notice that the exact solution of (2.2) in $\mathbb{C}$ is reproduced if $\rho(\mu) \equiv 1$ and $\theta(\mu) \equiv \mu$. Following Brusa and Nigro (1980), we can define the functions $a(\mu)$ and $b(\mu)$ by the relations $\rho(\mu) = e^{-\mu a(\mu)}$ and $\theta(\mu) = \mu b(\mu)$. The function $a(\mu)$ is called the *factor of numerical (or algorithmic) damping.* Owren and Simonsen (1995) have constructed families of L-stable *singly diagonally implicit Runge–Kutta* (SDIRK) methods with controllable numerical damping. The expression $|b(\mu) - 1|$ is called the *frequency distortion.* In the *phase-lag expansion* of the *relative period error* $b(\mu) - 1 = b_r(\mu)\mu^r + \mathcal{O}(\mu^{r+1})$ with $b_r(\mu) \not\equiv 0$, the exponent r is called the *dispersion order.* In Van Der Houwen and Sommeijer (1987) and (1989), several nondissipative RKN methods and diagonally implicit RK (DIRK) methods with high order of dispersion are derived. On certain test problems with oscillatory solutions, they show that the accuracy of the method is mostly determined by its dispersion rather than by its usual local truncation error. Other related error measures are often used in the literature to compare the merits of different methods, such as the *(relative) amplitude (or amplification) error* for $1 - \rho(\mu)$ and the *phase (or period) error (or dispersion or phase-lag)* for $\mu - \theta(\mu)$.

In addition to the natural free modes of oscillation of a system, modelled by the harmonic oscillator equation (2.2), the presence of forcing terms of oscillation may be considered. A simple inhomogeneous test equation in $\mathbb{C}$ is given by

$$y'' = -\omega^2 y + \delta e^{i\omega_f t}, \qquad (\omega \neq \omega_f, \omega > 0, \omega_f > 0),$$

where $\omega_f/2\pi$ represents the frequency of the forcing term. The exact solution is

$$y(t) = Ae^{i(\omega t + \phi)} + \frac{\delta}{\omega^2 - \omega_f^2} e^{i\omega_f t}.$$

It may be of interest to know how well a numerical method approximates the second term of the solution, corresponding to the forcing term. Nevertheless, it must be emphasized that the *inhomogeneous phase error* introduced by the forcing term remains constant, whereas the *homogeneous phase error* due to the free oscillation accumulates with time and is therefore the main source of errors (Van Der Houwen and Sommeijer 1987). Methods with no inhomogeneous phase error are said to have *in-phase forced oscillations* (Gladwell and Thomas 1983).

Several different methods have been proposed for problems whose solutions are known to be periodic and such that the period can be estimated *a priori.* In Section 5 we will treat in detail the envelope-following techniques. Another category of methods which can be interpreted as *exponentially fitted* methods (Liniger and Willoughby 1970) is based on the exact integration of the trigonometric polynomials $\cos(\ell\omega t), \sin(\ell\omega t)$ $(\ell = 1, \ldots, r)$ with ω fixed. Such methods depend on a parameter $\widetilde{\omega}$ approximating ω. They are exact when $\widetilde{\omega} = \omega$, but they may be sensitive to an inaccurate estimate of ω. Gautschi (1961) was the first to develop a basic theory for linear multistep methods with modified coefficients depending on $\widetilde{\mu} := \widetilde{\omega}h$. In the limit as $\widetilde{\mu} \to 0$, those methods reduce to the classical Adams and Störmer methods. As an example, the modified two-step explicit Störmer method of classical order $p = 2$ and of trigonometric order $r = 1$ applied to $y'' = g(t, y)$ is given by

$$y_{n+1} - 2y_n + y_{n-1} = h^2 \left(\frac{2\sin(\widetilde{\mu}/2)}{\widetilde{\mu}} \right)^2 g(t_n, y_n).$$

Methods of Nyström and Milne–Simpson type, less sensitive to inaccuracy in estimating ω, can be found in Neta and Ford (1984). Using different techniques from mixed interpolation (De Meyer, Vanthournout and Vanden Berghe 1990), Vanthournout, Vanden Berghe and De Meyer (1990) have constructed methods of Adams, Nyström, and Milne–Simpson type with an elegant derivation of their local truncation error. More general, and requiring more parameters, is the exact integration of products of ordinary polynomials and trigonometric functions with multiple frequencies given in Stiefel and Bettis (1969) and Bettis (1970), where methods of Störmer type are constructed. This was motivated from applications in celestial mechanics to take into account secular effects of orbit motion. Still in the same framework, the *minimax* methods of multistep type proposed by Van Der Houwen and Sommeijer (1984) attempt to minimize the local truncation error over a given interval of frequencies $[\omega_{\min}, \omega_{\max}]$. Such methods are less sensitive to inaccurate prediction of the frequencies. However, as for all methods mentioned in this paragraph, the presence of perturbations superimposed on the oscillations generally decreases dramatically the performance of these methods. Using an approach based on the 'principle of coherence' of Hersch (1958), numerical methods of multistep type for nearly linear ODEs are proposed in Denk (1993) and (1994), but they require the exact computation of the matrix exponential.

For certain problems with slow and fast components, *multirate methods* may be applied to reduce the total computational effort. A first method is used with one *macrostep* H to integrate the slow components and a second method is applied N times with a *microstep* h ($H = Nh$ to ensure synchronization) to integrate the fast components. The main difficulty with multirate methods is the assumption, before performing a macrostep, that the splitting between slow and fast components is known. In a counterintuitive but justified 'slowest first strategy' (Gear and Wells 1984), the slow components are integrated first using extrapolated values for the fast components and then the fast components are integrated using interpolated values for the slow components. Multirate Rosenbrock–Wanner (MROW) methods are analysed in detail in Günther and Rentrop (1993*a*) and (1993*b*). They have constructed a four-stage A-stable method of order three with a second-order embedded formula for error estimation. Their partitioning strategy is based on the stepsizes predicted for each component. In highly integrated electrical circuits applications, where most of the elements at any given time are inactive, they also make use of some information about the neighbourhood of the active elements to improve the performance of the partitioning strategy. A multirate extrapolation method based on the explicit Euler method has been developed by Engstler and Lubich (1995). An inexpensive partitioning strategy is implemented, which consists of stopping to build the extrapolation tableau for the components recognized as sufficiently accurate. Closely related to multirate methods are *multiple time-stepping* (MTS) methods. The right-hand side of the ODE is split as a sum of fast and slowly varying functions which are evaluated at different rates (see Subsection 4.5).

In the next sections we will deal with different classes of problems exhibiting oscillatory behaviour. For each class of problems we will discuss the structure of the equations, the objectives of the numerical simulation, the computational challenges, and some numerical methods that may be appropriate.

3. Mechanical systems

3.1. Multibody systems

The governing equations of motion of a mechanical system of stiff or highly oscillatory force devices may be written as a system of DAEs (Brenan, Campbell and Petzold 1995)

$$M(q)q'' + G^T(q)\lambda - (f^s(q', q, t) + f^n(q', q, t)) = 0, \qquad (3.1a)$$
$$g(q) = 0, \qquad (3.1b)$$

where $q = (q_1, \ldots, q_n)^T$ are the generalized coordinates, $q' = dq/dt$ the generalized velocities, $q'' = d^2q/dt^2$ the generalized accelerations, $\lambda = (\lambda_1, \ldots, \lambda_m)^T$ the Lagrange multipliers, M is the mass-inertia matrix, $g = (g_1, \ldots, g_m)^T$ the holonomic constraints, and $G = \partial g/\partial q$. The *stiff* or *oscillatory force* is $f^s = \sum_i^{n_f} f_i^s$, and f^n includes all the field forces and the external forces which are nonstiff compared to the stiff components, that is,

$$\left\| \frac{\partial f^s}{\partial(q, q')} \right\| \gg \left\| \frac{\partial f^n}{\partial(q, q')} \right\|.$$

The stiff force components in (3.1a) can often be written in the form

$$f_i^s = -B_i(q)\left(K_i\eta_i(q) + C_i\frac{d\eta_i}{dt}\right), \tag{3.2}$$

where η_i is smooth, $i \in \{1,\ldots,n_f\}$, $B_i = (\partial\eta_i/\partial q)^T$, and K_i, C_i are the associated stiffness and damping matrices. For some generalized coordinate sets, the functions η_i may be linear, or even the identity. When the components of the coefficient matrices K_i and C_i are large, these force components may cause rapid decay or high frequency oscillation in the solution of (3.1). It is well known that the characteristics of the fast or slow solution are determined not only by the modelling aspects, for example the coefficients of the stiffness and damping matrices, but also by the initial conditions and events that may excite stiff components in the system during the simulation.

To demonstrate some of the potential difficulties caused by highly oscillatory forces in mechanical systems, we consider two common oscillatory forces: a spring force (which is exemplified by the stiff spring pendulum of Section 1), and a 2D *bushing* force. The former is a very simple example of a type of system often seen in molecular dynamics (see Section 4), and the latter is a general form of modelling force devices in multibody mechanical systems.

Spring force

The stiff spring pendulum of Section 1 is an example of a point-mass connected to a stiff spring force. The equations of motion of the particle in Cartesian coordinates are given by (1.1), where the spring force is given by $(x\lambda, y\lambda)^T$. This problem is highly nonlinear, due to (1.1e). Since most of the mathematical methods for oscillatory problems assume a nearly linear form of the problem, and many numerical techniques are implicitly based on linearization, we will begin by examining the structure of the local linearized system. The eigenvalues of the *underlying ODE* of (1.1), that is, substituting (1.1e) into (1.1c, 1.1d), are illustrated for $\epsilon = \sqrt{10^{-3}}$ in Fig. 3. The dominant eigenvalues are $\pm\,\mathrm{i}/\epsilon$. As $\epsilon \to 0$, the dominant pair of eigenvalues approaches $\pm\infty$ along the imaginary axis. The other pair of eigenvalues oscillates on the complex plane, with amplitude and frequency approaching $\pm\infty$. The amplitude of the oscillations in the eigenvalues depends on the initial conditions for the problem. If the initial conditions are on the slow solution, then the amplitude is zero. In Fig. 3, we have chosen the initial conditions to be slightly off the slow solution, which is the situation for most numerical methods. From this we can see that methods based on linearization are likely to fail for this problem unless the stepsizes are very small or the linearization is performed exactly on the slow solution.

In Section 1, we showed that a slow solution for this problem could be identified by shifting to polar coordinates. One might guess that perhaps

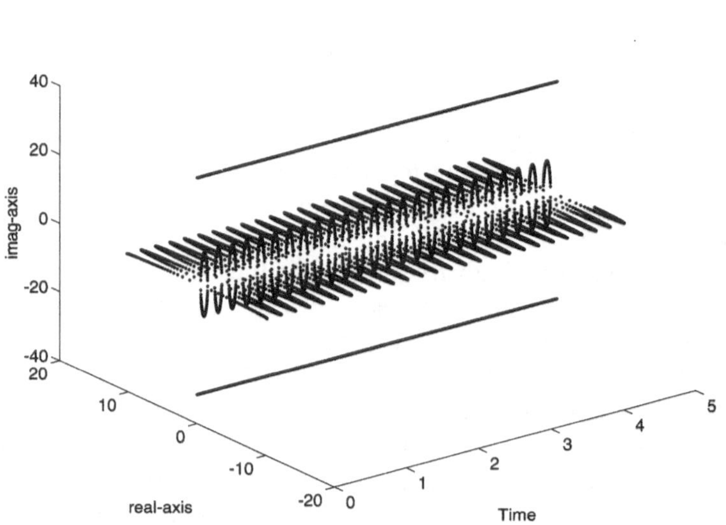

Fig. 3. Eigenvalues of stiff spring pendulum in Cartesian coordinates, $\epsilon = 10^{-1.5}$

the oscillation in the eigenvalues described above is due to the choice of the Cartesian coordinate system, which is unnatural for this problem. This is true, but only partly so. The eigenvalues along the solution trajectory in polar coordinates are shown in Fig. 4. The dominant eigenvalues are of the same magnitude as those in (1.1); see Fig. 3. This is because the coordinate transformation is linear with respect to the fast moving r. The oscillation of the other pair of eigenvalues along the real axis persists.

Bushing force

Nonlinear oscillations in general multibody systems are often generated by forces from components such as bushings. This type of component is used in modelling vehicle suspension systems. Unlike the spring, this element is usually an *anisotropic* force, that is, it has different spring coefficients along the principle axes of the bushing local coordinate frame. The bushing force between *body-i* and *body-j* may be defined using the relative displacement d_{ij}, its time derivative d'_{ij}, and the relative angle θ_{ij} and its time derivative θ'_{ij} of two body-fixed local coordinate frames at the bushing location on two bodies. Using the vectors s_i and s_j representing the bushing location in *body-i*'s and *body-j*'s centroid local coordinate systems, respectively, we have

$$d_{ij} = \begin{pmatrix} x_i \\ y_i \end{pmatrix} - \begin{pmatrix} x_j \\ y_j \end{pmatrix} + A_i s_i - A_j s_j,$$

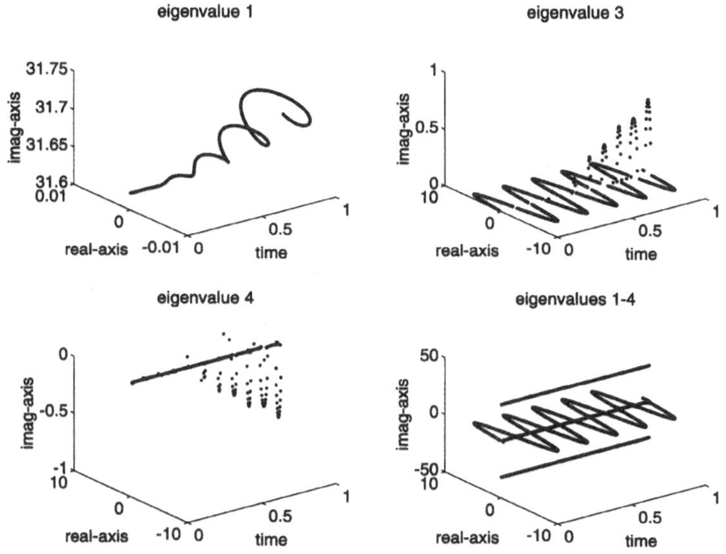

Fig. 4. Eigenvalues of stiff spring pendulum in polar coordinates, $\epsilon = 10^{-1.5}$

where the orientation transformation matrices A_k are

$$A_k = A(\theta_k) = \begin{pmatrix} \cos(\theta_k) & -\sin(\theta_k) \\ \sin(\theta_k) & \cos(\theta_k) \end{pmatrix},$$

and (x_k, y_k, θ_k) are coordinates at body-fixed frames. The bushing force f_b can then be written as

$$f_b = \begin{pmatrix} f_b^x \\ f_b^y \end{pmatrix} = A_i \begin{pmatrix} k^x & 0 \\ 0 & k^y \end{pmatrix} A_i^T d_{ij} + A_i \begin{pmatrix} c^x & 0 \\ 0 & c^y \end{pmatrix} A_i^T d'_{ij},$$

and the applied torque is

$$\tau_b = k^\theta \theta_{ij} + c^\theta \frac{d\theta_{ij}}{dt},$$

where k^x, k^y, and k^θ are the spring coefficients associated with the x, y, and θ coordinates, and c^x, c^y, and c^θ are the corresponding damping coefficients.

An example of a simple mechanical system incorporating this force may be obtained from this model using unit mass-inertia and gravity, and setting the bushing location on the body to $s = (-1/2, 0)$. A bushing element with no damping, attached at the global position of $(1/2, 0)$, yields

$$0 = x'' - k^x \left(\frac{1}{2} - x + \frac{\cos(\theta)}{2} \right), \tag{3.3a}$$

$$0 = y'' + k^y \left(y - \frac{\sin(\theta)}{2} \right) + 1, \tag{3.3b}$$

$$0 = \theta'' + k^\theta \theta - \frac{\sin(\theta)}{2} k^x \left(\frac{1}{2} - x + \frac{\cos(\theta)}{2}\right) - \frac{\cos(\theta)}{2} k^y \left(y - \frac{\sin(\theta)}{2}\right). \tag{3.3c}$$

It can be seen from (3.3) that the local eigenstructure of the system may change rapidly, depending on the size of the stiffness coefficients. Using the initial values $(x, y, \theta) = (1.1, 0.1, 0.0)$ with $(k^x, k^y, k^\theta) = (10^4, 10^4, 10^3)$, the solution of (3.3) exhibits high-frequency oscillations in all variables, as shown in Fig. 5. Solving the eigenvalue problem of (3.3) at each time-step yields three pairs of eigenvalues as illustrated in Fig. 6.

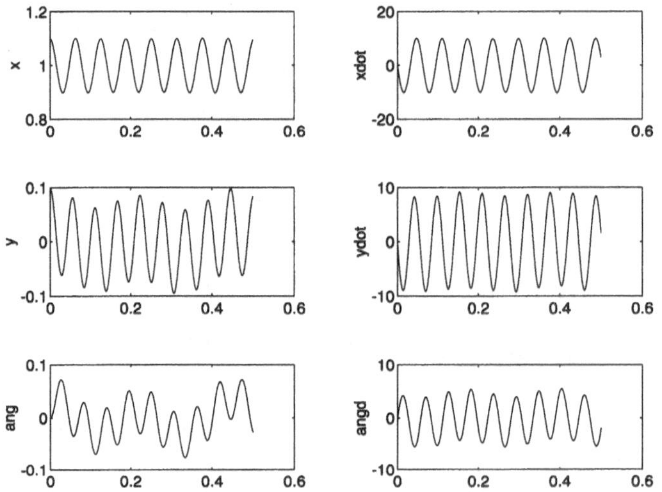

Fig. 5. Bushing problem in Cartesian coordinates

Structure of limiting DAE

Another source of difficulties for the numerical solution arises from the structure of stiff multibody systems. These systems are *singular singular perturbation problems* (O'Malley 1991). In the limit as the fast timescale tends to infinity, the system becomes a *high-index DAE*. For example, as $\epsilon \to 0$ in the stiff spring pendulum problem (1.1) or (1.2), the equations become those of a rigid pendulum. This index-3 DAE has a *Hessenberg structure*. Numerical solution of high-index DAE systems of Hessenberg structure has been extensively studied (Brenan et al. 1995, Hairer, Lubich and Roche 1989, Hairer and Wanner 1996). There are well-known difficulties with numerical accuracy, matrix conditioning, error control, and stepsize selection. Roughly speaking, the higher the index of a DAE, the more difficulties for its numerical solution. Hence, it is not surprising that there would be difficulties for the numerical solution of highly oscillatory mechanical systems.

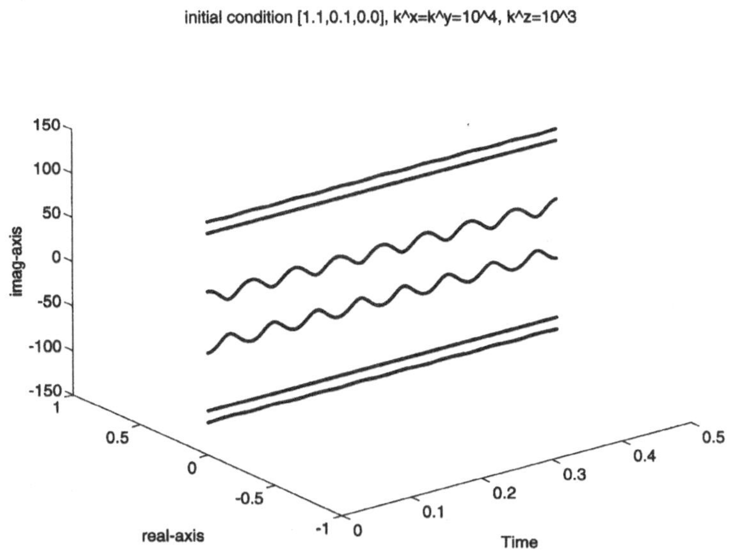

Fig. 6. Eigenvalues of bushing problem

3.2. Finding the slow solution

Given the situation of a rapidly changing local eigenstructure, perhaps the simplest strategy for numerical solution is to consider damping the oscillation, when it is of sufficiently small amplitude, via highly stable implicit numerical methods. First, we want to emphasize that damping the oscillation for general nonlinear systems is not safe and can easily lead to an erroneous solution! However, the system may have a very special structure such that this approach is appropriate.

Lubich (1993) has shown that the numerical solution of stiff spring mechanical systems of a strong potential energy (for instance, a stiff spring force such as in (1.1)) by a class of implicit Runge–Kutta methods with stepsize independent of the parameter ϵ, converges to the slowly varying part of the solution. These results have been extended to a class of multistep Runge–Kutta methods (Schneider 1995). Unfortunately, it is not clear that these results may apply directly to all the types of oscillatory components in multibody systems. As indicated in Lubich (1993), the representation of stiff or oscillatory components in an appropriate coordinate system is not always possible, that is, the constraints associated with the stiff or oscillatory potential force can be difficult to obtain in general. Nevertheless, for (3.1), an approximation of the dynamics of such local coordinates can be obtained for oscillatory force components of the form (3.2).

The amount of damping in a highly stable implicit method is controlled by the stepsize; for the types of methods that one would consider using for this purpose, the damping increases with the stepsize. One might hope that the automatic stepsize selection mechanisms used in variable-stepsize ODE/DAE codes would increase the stepsize whenever the magnitude of the oscillation is small compared with the local error tolerances. This works well, but only if the usual error control strategy is changed to one that is appropriate for the limiting high-index DAE.

There are also difficulties with Newton convergence for implicit numerical methods applied to highly oscillatory nonlinear mechanical systems (Lubich 1993, Yen and Petzold 1997). The Newton iteration at each time-step does not converge for large (relative to the period of the high-frequency oscillation) stepsizes. The problem is due to the linearization on which Newton's method is based. With the eigenstructure of the local Jacobian matrix changing so rapidly, Newton's method does not yield good directions to the slow solution unless the initial guess (prediction) is extremely accurate; such an accurate initial guess can only be attained by using very small stepsizes. Some variables can be predicted more accurately than others. Variables that play the role of Lagrange multipliers in (3.1) are not predicted well by polynomial extrapolation, which is used in many ODE/DAE codes. This is not surprising: these variables depend directly on the second derivatives of the highly oscillatory position variables.

Yen and Petzold (1997) have recently proposed a *coordinate-split* formulation of the equations of motion which eliminates difficulties due to obtaining an accurate predictor for the Lagrange multiplier variables, because these variables are no longer present in the computation. These methods are particularly effective for oscillatory multibody systems with components such as the stiff bushing. The coordinate-split formulation is described as follows. Direct numerical integration of the index-3 DAE (3.1) suffers from the well-known difficulties inherent in the solution of high-index DAEs. One way to lower the index involves introducing derivatives of the constraint $g(q)$, along with additional Lagrange multipliers μ. This yields the *stabilized index-2* or *GGL* formulation of the constrained equations of motion (Gear, Gupta and Leimkuhler 1985)

$$\begin{aligned} q' - v + G^T(q)\mu &= 0, & (3.4a)\\ M(q)v' + G^T(q)\lambda - f(v,q,t) &= 0, & (3.4b)\\ G(q)v &= 0, & (3.4c)\\ g(q) &= 0, & (3.4d) \end{aligned}$$

where $v = q'$ and $f = f^s + f^n$, which has been used widely in simulation. The Lagrange multiplier variables λ and μ fulfil the role of projecting the solution onto the *position* (3.4d) and the *velocity* (3.4c) constraints, respectively. Many of the numerical methods for multibody systems solve the system (3.4) directly. It is also possible to eliminate the Lagrange multipliers and reduce the size of the system to the number of degrees of freedom. One way to accomplish this begins with the stabilized index-2 system (3.4). Suppose that $G(p)$ is full-rank on the constraint manifold $\mathcal{M} = \{q \in \mathbb{R}^n : g(q) = 0\}$. Then one can find an annihilation matrix $P(q) \in \mathbb{R}^{(n-m)\times n}$ such that $P(q)G^T(q) = 0$, for all $q \in \mathcal{M}$. Premultiplying (3.4a) and (3.4b) by $P(q)$ yields an index-1 DAE

$$\begin{aligned} P(q)\,(q' - v) &= 0, && (3.5a)\\ P(q)\,(M(q)v' - f(v,q,t)) &= 0, && (3.5b)\\ G(q)v &= 0, && (3.5c)\\ g(q) &= 0. && (3.5d)\end{aligned}$$

An important practical consequence of (3.5) is that (μ, λ) have been eliminated from the DAE, via multiplication of (3.4a, 3.4b) by the nonlinear $P(q)$. Thus, the error test and the Newton iteration convergence test in a numerical implementation of (3.5) no longer need to include the problematic Lagrange multipliers (μ, λ).

The coordinate-split method gives an inexpensive way to find $P(q)$ via a splitting of the Cartesian basis (Yen and Petzold 1996). Discretizing the coordinate-split formulation by an implicit method like BDF or an implicit Runge–Kutta method, it seems at first glance that the local Jacobian matrix might be difficult to compute, because it involves derivatives of $P(q)$. However, this is easily overcome by using the formulæ for the derivative of a projector given by Golub and Pereyra (1973) and the resulting method lends itself to efficient implementation.

The performance of damped numerical methods for highly oscillatory mechanical systems is improved by using the coordinate-split formulation. However, for problems with very high-frequency oscillations, there are still difficulties for Newton convergence. To obtain rates of convergence which are independent of the frequency of the oscillation, Yen and Petzold (1997) have introduced a modification to the Newton iteration, that is, the *modified coordinate-split (CM)-iteration.* The basic idea of the CM-iteration is that there are terms in the Jacobian which involve derivatives of the projection onto the constraint manifold. These terms are large and complicated to compute, but small on the slow solution. For example, applying a (low-order) BDF formula to (3.5) yields the nonlinear system

$$\begin{aligned} P(q_n)h(\rho_h q_n - v_n) &= 0, && (3.6a)\\ P(q_n)h(M(q_n)\rho_h v_n - f(v_n, q_n, t_n)) &= 0, && (3.6b)\\ G(q_n)v_n &= 0, && (3.6c)\\ g(q_n) &= 0, && (3.6d)\end{aligned}$$

where ρ_h is the discretization operator, and h is the stepsize of the time discretization. Given an initial prediction $(q_n^{(0)}, v_n^{(0)})$, applying Newton-type methods to (3.6) requires the solution of a linear system

$$J(q_n, v_n)(\Delta q_n, \Delta v_n) = -r(q_n, v_n)$$

such that Δq_n and Δv_n are the increments of q_n and v_n,

$$J(q_n, v_n) = \begin{pmatrix} P(q_n)[(\frac{dG^T(q_n)}{dq_n})s_1 + h\frac{\partial \rho_h q_n}{\partial q_n}] & -hP(q_n) \\ P(q_n)[(\frac{dG^T(q_n)}{dq_n})s_2 + \frac{dr_2(q_n,v_n)}{dq_n}] & P(q_n)\frac{\partial r_2(q_n,v_n)}{\partial v_n} \\ \frac{\partial(G(q_n)v_n)}{\partial q_n} & G(q_n) \\ G(q_n) & 0 \end{pmatrix},$$

and

$$r(q_n, v_n) = (P(q_n)r_1(q_n, v_n), P(q_n)r_2(q_n, v_n), G(q_n)v_n, g(q_n)),$$

where $s_1 = -(GY)^{-T}Y^T r_1$, $s_2 = -(GY)^{-T}Y^T r_2$, $r_1 = h(\rho_h q_n - v_n)$, and $r_2 = h(M(q_n)\rho_h v_n - f(v_n, q_n, t_n))$. The terms which cause the Newton convergence problem are those involving s_1 and s_2. Away from the slow solution, small perturbations in the positions can result in large changes in these terms, leading to convergence difficulties for the Newton iteration. The CM-iteration sets these terms to zero, yielding a reliable direction towards the slow solution for the Newton-type iteration. Convergence results for the CM-iteration are given in Yen and Petzold (1997). For nonoscillatory mechanical systems, the convergence behaviour of the CM-modification is similar to that of standard Newton.

For nonlinear mechanical systems with small-amplitude, high-frequency oscillations, the CS formulation combined with a highly stable implicit method to damp the oscillations and the CM-modification to the Newton iteration can be highly effective. A two-body pendulum problem in 2D Cartesian coordinates, with a bushing force, as given in Subsection 3.1, which is the source of the high-frequency oscillation, is described in Yen and Petzold (1997). In experiments at very high frequencies using the BDF code DASSL (Brenan et al. 1995) with the method order restricted to two, the CS formulation is solved twice as efficiently as the GGL formulation (3.4). The CM modification to the Newton iteration further improves the efficiency by a factor of more than a hundred. At lower frequencies, of course, the comparison is less dramatic.

An alternative to numerical methods that use damping to find the slow solution is to approximate the slow solution directly. Reich (1995) has extended the *principle of slow manifold*, which has been widely used in the approximation of multiple timescale systems (Fenichel 1979, Kopell 1985), to the DAEs of multibody systems with highly oscillatory force terms. Algebraic constraints corresponding to the slow motion were introduced with a relaxation parameter to preserve the slow solution, while adding flexibility to it in the slow manifold approach.

3.3. Flexible multibody mechanical systems

The numerical solution of flexible multibody systems is required for nonlinear dynamic analysis of articulated structures. The need for modelling deformable bodies has been kindled by the dynamic simulation of physically large and massive mechanisms, such as aeroplanes, industrial robots and automobiles. These are structures in which kinematic connections permit large relative motion between components that undergo small elastic deformation. A source of difficulty in the solution of flexible multibody equations of motion is the coupling between the elastodynamic equations and the gross motion. The methods for analysing flexible mechanisms can generally be divided into two categories:

(i) methods that focus on the structure, while using the gross multibody motion as a source of dynamic loading
(ii) methods that incorporate flexibility effects into the multibody dynamic analysis.

Simulation of flexible multibody systems has been an active research topic for the last two decades. Many of the methods for flexible multibody systems have been implemented in multibody dynamic analysis codes (Haug 1989, Nikravesh 1988, Pereira and Ambrósio 1993). For such systems, an important feature of the solution is the nonlinear oscillations induced by the elastodynamics equations. Moreover, since the governing equations of flexible multibody systems are often modelled using algebraic constraints, the numerical solution of DAEs is required. As discussed in the two previous subsections, the numerical solution of the resulting highly oscillatory DAEs presents many challenging problems.

Modelling of flexibility effects in multibody systems can significantly alter the dimension and solutions of the governing equations of motion. It is well documented that adding flexible components to rigid body models can drastically increase the computational complexity. For instance, a typical rigid-body model of a ground vehicle, such as a passenger car, may consist of several rigid bodies of 10–100 coordinates. Replacing the chassis of the car with its flexible model can increase the number of coordinates to millions. Compounding the difficulty of an increased dimension are the high-frequency oscillations that arise from the modal stiffness and damping coefficients. They represent both the physical and geometrical approximations of the elastic, plastic, and viscoelastic effects of the flexible bodies, and their eigenvalues are usually of magnitudes greater than those of the gross motion. It has been shown for some flexible multibody systems that the coupling of unresolved high frequencies to the rigid motion may result in a nonlinear instability. In Simo, Tarnow and Doblare (1993), the numerical simulation of a flexible rod illustrates the nonlinear instability in Hamiltonian systems. For nonconservative flexible mechanisms, such problems can be found in the approximation of the deformation of elastic bodies in constrained multibody systems (Yoo and Haug 1986, Yen, Petzold and Raha 1996). Special care must be taken to maintain the stability of the oscillatory components in the solution. In the following, we give an overview of the numerical techniques used for handling oscillations in flexible multibody systems. We begin with a summary of computational methods used in structural dynamics.

3.4. Modal analysis of structures

Applying spatial discretization to the elastomechanical PDE, the dynamic equations of the response of a discrete structural model are given by

$$M^e u'' + C^e u' + K^e u = f(t), \tag{3.7}$$

where u is the nodal displacement, $f(t)$ is the load, and M^e, C^e, and K^e are constant mass, damping, and stiffness matrices of the node coordinates, respectively. Numerical methods have been developed based on spectral decomposition of this linear ODE system. Rewriting (3.7) as a first-order ODE, we obtain

$$\begin{pmatrix} 0 & M^e \\ M^e & C^e \end{pmatrix} \frac{\mathrm{d}}{\mathrm{d}t} \begin{pmatrix} u' \\ u \end{pmatrix} + \begin{pmatrix} -M^e & 0 \\ 0 & K^e \end{pmatrix} \begin{pmatrix} u' \\ u \end{pmatrix} = \begin{pmatrix} 0 \\ f(t) \end{pmatrix}. \tag{3.8}$$

Denoting $z = (u', u)^T$, the solution of (3.8) can be written explicitly,

$$z = e^{-tA} z_0 + F(t), \tag{3.9}$$

where z_0 and $F(t)$ are two vectors that depend on the initial values and loading function, and

$$A = \begin{pmatrix} 0 & M^e \\ M^e & C^e \end{pmatrix}^{-1} \begin{pmatrix} -M^e & 0 \\ 0 & K^e \end{pmatrix}.$$

Using (3.9), numerical solution techniques for (3.7) can be unified in the framework of approximating the exponential of the matrix A times a vector.

A straightforward approach for dealing with the high frequencies in (3.7) is to truncate the higher *eigenmodes*, which were obtained from the *generalized eigenvalue problem* of the matrix $K^e - \omega^2 M^e$ (Bathé and Wilson 1976, Craig and Bampton 1968). For most structures the eigenvectors, or *normal modes* ϕ_i, span the nodal coordinate space, and form the coordinate transformation, for instance the *modal matrix*, from the nodal coordinates u to the modal coordinates η,

$$u = \sum_{i=1}^{N} \phi_i \eta_i = \Phi \eta. \tag{3.10}$$

Note that the nodal and modal coordinates in (3.10) are those corresponding to the undamped system of (3.7).

To approximate the harmonic frequencies of (3.7) with fewer modes, one can apply the *Rayleigh–Ritz method* to the undamped system, employing the *Rayleigh quotient* and *Ritz vectors* (Bathé and Wilson 1976). In contrast to the *mode-superposition method*, which requires all the natural frequencies (eigenvalues) and modes (eigenvectors) to satisfy

$$(K^e - \omega_i^2 M^e)\phi_i = 0, \quad for\ i = 1, 2, \ldots, N,$$

the Rayleigh–Ritz method allows the use of a few *shape vectors* (Ritz vectors) to approximate the solution of (3.7). For some classes of problems, the Rayleigh–Ritz method is more efficient than eigenvector mode superposition methods for computing the dynamic response of (3.7). Efficient numerical procedures have been developed for determining a set of lowest orthonormalized Ritz vectors (Chen and Taylor 1989, Wilson, Yuan and Dickens 1982). For flexible multibody simulation, the oscillatory solution can be eliminated by removing the high modes, provided that the reduced structural model is consistent. More precisely, the deformations at the locations of kinematic joint and force attachment nodes must be taken into account for some proper mode shapes, for instance *constraint* or *attachment* modes (Craig 1981).

For damped systems, the aforementioned methods assume proportional damping or, more generally, modal damping of (3.7) (*i.e.*, that the damping matrix satisfies $\phi_i^T C^e \phi_j = 0, i \neq j$) for lack of a more realistic representation in many of the structural models. This approach may be too simplistic to be effective in some applications. A general approach to the numerical solution of (3.7) solves the generalized *unsymmetric eigenvalue problem* (Lanczos 1950), where the equations of dynamic equilibrium are first transformed into a first-order system (3.8) (Nour-Omid and Clough 1984). The development of numerical solution techniques for this problem has been one of the most active research topics in iterative solution of linear systems (Freund, Golub and Nachtigal 1992). Some efficient numerical methods developed in recent years are based on *Krylov subspace* approximations to (3.9) (Friesner, Tuckerman, Dornblaser and Russo 1989, Gallopoulos and Saad 1992). Such Krylov subspace approximations have been used in structural dynamics (Nour-Omid and Clough 1984) and chemical physics (Park and Light 1986). A recent study (Hochbruck, Lubich and Selhofer 1995) indicated that a class of *exponential integrators* has favourable properties in the numerical integration of large oscillatory systems. What remains to be seen is an effective application of these exponential integrators for simulating large *flexible* mechanisms.

Another approach to incorporating flexible components in multibody dynamics is to use *nonlinear beam theory*, which applies finite element approximation to the forces resulting from body deformation (Hughes 1987, Simo and Vu-Quoc 1986, Cardona and Géradin 1993). An appropriate nonlinear beam formalism requires in many cases incorporating geometric nonlinear effects such as *geometric stiffening*, which contribute inertia forces to the global motion. The approximation of the inertia force due to geometric nonlinearity usually depends on the nodal position and velocity, for instance, the damping and stiffness matrices of (3.7) become nonconstant. In some cases, these nonlinear forces introduce additional oscillations, which can hinder efficient numerical solution of flexible multibody systems (Simeon 1996).

3.5. Numerical integration methods

Time integration algorithms for solving structural dynamics problems have been developed since the late 1950s (Newmark 1959). General requirements and the foundations of these methods have been well documented (Bathé and Wilson 1976, Chung and Hulbert 1993, Hilber, Hughes and Taylor 1977, Hoff and Pahl 1988, Wood, Bossak and Zienkiewicz 1980). Although their main application area is to linear structural dynamics, these methods can be directly applied to initial value problems of nonlinear second-order ODEs

$$q'' = f(q', q, t). \tag{3.11}$$

Accuracy and stability analysis hold for the numerical methods, provided the discretized nonlinear equations have been solved accurately, that is, within a small enough tolerance. For example, the *HHT*-α method (Hilber et al. 1977) for (3.11) is given by

$$\begin{aligned} a_{n+1} &= (1+\alpha)f_{n+1} - \alpha f_n, && (3.12a)\\ q_{n+1} &= q_n + hv_n + h^2((\frac{1}{2} - \beta)a_n + \beta a_{n+1}), && (3.12b)\\ v_{n+1} &= v_n + h((1-\gamma)a_n + \gamma a_{n+1}), && (3.12c) \end{aligned}$$

where h is the stepsize, $\alpha \in [-1/3, 0]$, $\beta = (1-\alpha)^2/4$, and $\gamma = 1/2 - \alpha$. It is well known that the *HHT*-α family is second-order accurate and A-stable. Numerical damping is maximum for $\alpha = -0.3$, and zero for $\alpha = 0$. Controllable numerical damping and unconditional stability are needed to deal with the high-frequency modes which often result from standard finite element spatial discretization. For nonlinear oscillations, these properties are also required in the solution of flexible multibody systems. Rather than using *ad hoc* mode-selection processes, this approach is desirable because the elimination of higher frequencies is controlled by selection of the method parameters.

Recent work has dealt with extending these types of methods to treat flexible multibody systems (Cardona and Géradin 1989, Yen et al. 1996). The basic form of constrained multibody equations of motion is given by (3.1), which is a DAE of index-3. Due to the problems of numerical instability in solving index-3 DAEs, most of the solution techniques for (3.1) have been developed using differentiation of the constraints (3.1b). Assuming that M is invertible, direct application of (3.12) to the *underlying ODE* of (3.1) (Führer and Leimkuhler 1991), for instance,

$$q'' = \phi(q', q, t) = M^{-1}(q)(f(q', q, t) - G^T(q)\lambda), \tag{3.13}$$

where

$$\lambda = (GM^{-1}G^T)^{-1}\left(GM^{-1}f + \frac{\partial Gq'}{\partial q}q'\right),$$

can be carried out. However, the numerical solution will not generally preserve the constraint (3.1b) and its derivative. To enforce the constraints, the numerical solution should be projected onto the constraint manifold. Applying the method of Lagrange multipliers to combine the projection with the solution of (3.13), which has been discretized using (3.12), leads to the *DAE α-method* (Yen et al. 1996)

$$\begin{aligned}
M_{n+1}(q_{n+1} - \hat{q}_n) - \hat{\beta}h^2 f_{n+1} + G_{n+1}^T \nu_{n+1} &= 0, & (3.14a)\\
M_{n+1}(v_{n+1} - \hat{v}_n) - \hat{\gamma}h f_{n+1} + G_{n+1}^T \mu_{n+1} &= 0, & (3.14b)\\
G_{n+1}v_{n+1} &= 0, & (3.14c)\\
g(q_{n+1}) &= 0, & (3.14d)
\end{aligned}$$

where $\hat{\beta} = \beta(1+\alpha)$, $\hat{\gamma} = \gamma(1+\alpha)$,

$$\begin{aligned}
\hat{q}_n &= q_n + hv_n + h^2\left(\left(\frac{1}{2} - \beta\right)a_n - \beta\alpha\phi_n\right),\\
\hat{v}_n &= v_n + h\left((1-\gamma)a_n - \gamma\alpha\phi_n\right),
\end{aligned}$$

$\phi_n = M_n^{-1}(f_n - G_n^T\lambda_n)$, $a_0 = \phi_0$ and $a_n = (1+\alpha)\phi_n + \alpha\phi_{n-1}$ for $n \geq 1$. The *algebraic* variables ν_{n+1} and μ_{n+1} in (3.14) comprise $h^2\hat{\beta}\lambda_{n+1}$ and the corresponding *correction* terms, which project the position and velocity variables onto the constraint manifold. A convergence analysis of (3.14) was given by Yen et al. (1996).

The DAE α-methods are most effective when combined with the CS formulation and CM iteration described earlier. In the Lagrange multipliers formulation there may be convergence difficulties with the Newton iteration. Premultiplying (3.14a) and (3.14b) by the CS matrix $P(q)$ yields

$$P(q_{n+1})(M_{n+1}(q_{n+1} - \hat{q}_n) - \hat{\beta}h^2 f_{n+1}) = 0, \tag{3.15a}$$

$$
\begin{aligned}
P(q_{n+1})(M_{n+1}(v_{n+1}-\hat{v}_n)-\hat{\gamma}hf_{n+1}) &= 0, &(3.15b)\\
G_{n+1}v_{n+1} &= 0, &(3.15c)\\
g(q_{n+1}) &= 0. &(3.15d)
\end{aligned}
$$

Accuracy and stability of the α-methods for ODEs are preserved. More importantly, the high-index variables (ν_{n+1}, μ_{n+1}), which exhibit high-frequency oscillations of large amplitudes, are not present in (3.15). Compared to the Lagrangian form (3.14), much improved Newton convergence was observed in a number of flexible multibody simulations (Yen et al. 1996). When applying strong numerical damping to the higher modes, the CM iteration illustrated even better convergence in these examples.

4. Classical molecular dynamics

Classical molecular dynamics (MD) has become an important tool in the study of (bio)molecules, such as nucleic acids, polymers, and proteins (Allen and Tildesley 1987, Board Jr., Kalé, Schulten, Skeel and Schlick 1994, Gerschel 1995). In classical MD, quantum effects are neglected and the motion of the atoms is often described by *Newton's equations*

$$q' = v, \qquad Mv' = -\nabla U(q), \tag{4.1}$$

where the vector q contains the Cartesian coordinates of the atoms, the vector v contains their velocities, M is the diagonal matrix of atomic masses, and $U(q)$ is a semi-empirical potential energy function. Defining the momenta $p := Mv$, these equations form a Hamiltonian system with Hamiltonian $H(q,p) := \frac{1}{2}p^T M^{-1} p + U(q)$. Therefore, the Hamiltonian (the energy) and the symplectic form $dq \wedge dp$ are invariant under the action of the flow (Arnold 1989). More sophisticated dynamics are also often considered in MD simulation. In Langevin dynamics (4.5), stochastic and friction forces are introduced to model additional aspects (see Subsection 4.4). In Nosé dynamics, temperature and pressure constraints are included to treat nonequilibrium situations (Nosé 1984, Hoover 1991).

The potential energy function $U(q)$ is generally given by a repeated sum over the atoms of pairwise potentials modelling interactions of diverse type (Gerschel 1995): electrostatic, dipolar, polar, dispersive, repulsive, etc. These interactions vary with the interatomic distance and have different ranges of influence: localized for the covalent bondings, short-range for the Van der Waals forces, and long-range for the electrostatic forces. They also differ in their strength and timescale, making the dynamics of (bio)molecules very complex, even chaotic. The equations of MD are highly nonlinear and extremely sensitive to perturbations. A perturbation grows roughly by a factor of 10 every picosecond ($= 10^{-12}$[s]). Therefore, due to various sources of approximation and error in MD simulation, it is not reasonable from the viewpoint of forward error analysis to ask for an accurate representation of the molecular configuration after several picoseconds. The framework of MD is actually statistical mechanics. To emphasize this point, let us mention that the initial velocities of the atoms of a (bio)molecule are usually chosen randomly to follow a Boltzmann–Maxwell distribution. What is actually desired in MD is to generate a statistically acceptable motion or to obtain a good sampling of phase space over sufficiently long periods of time to provide spatial and temporal information; it is not usually necessary to follow an exact trajectory. Monte Carlo simulation, by generating random configurations, is another technique used in the study of molecular systems based on their statistical properties, but this falls outside the scope of this article. For large (bio)molecules, MD simulation is usually preferred.

Conformational changes of a (bio)molecule arise on a continuum from $1[\mathrm{ps}]$ to $10^2[\mathrm{s}]$. In MD simulation the main difficulty in the integration of the equations is the presence of a spectrum of very high-frequency oscillations of Brownian character. The fastest vibrations are the bond stretchings and the bond-angle bendings which are orders of magnitude stronger than the other interactions. For example a C–H stretch has an oscillation around an equilibrium position of approximate frequency $0.9 \cdot 10^{14}[\mathrm{Hz}]$ (Streitwieser Jr. and Heathcock 1985). This imposes a severe limit on the stepsize used by standard integration schemes in order to resolve these high-frequency oscillations; for example, a stepsize around $1[\mathrm{fs}]$ $(= 10^{-15}[\mathrm{s}])$ is necessary for the widely used Verlet algorithm (4.2). Computing the forces for a large system at each step is computationally expensive. Therefore, with today's computer technology this stepsize constraint limits the horizon of integration to the order of a nanosecond $(= 10^{-9}[\mathrm{s}])$, several orders of magnitude less than the biological timescale for which phenomena like protein folding $(\approx 10^{-1}[\mathrm{s}])$ take place. Decreasing the ratio of force evaluation per step is therefore a major goal to speed up the integration.

There are three ways of handling the high-frequency components in MD: resolve them, model their effects, or suppress them. Methods combining these different approaches are of course possible. The desire is that the dynamics should be correctly reproduced from the point of view of statistical mechanics. A recent detailed survey on MD integration methods is Schlick, Barth and Mandziuk (1997); other references are Skeel, Biesiadecki and Okunbor (1993) and Leimkuhler, Reich and Skeel (1995). In this section we will briefly present different approaches, stressing some of their strengths and weaknesses.

4.1. The Verlet algorithm

The most commonly used method in MD is the *Verlet algorithm* (Verlet 1967). Using the momenta $p = Mv$, this explicit second-order method applied to (4.1) can be expressed as follows:

$$\begin{aligned} p_{n+1/2} &= p_n - \frac{h}{2}\nabla U(q_n), \\ q_{n+1} &= q_n + hM^{-1}p_{n+1/2}, \\ p_{n+1} &= p_{n+1/2} - \frac{h}{2}\nabla U(q_{n+1}). \end{aligned} \tag{4.2}$$

Given the inaccuracy of the governing force field, such a low-order integration method is adequate in MD. Besides being relatively easy to program, this method possesses several attractive features. It preserves two important geometric properties of the flow: symplecticness and reversibility under the involution $p \mapsto -p$. For more details about symplectic discretization we refer the reader to Sanz-Serna (1992) and Sanz-Serna and Calvo (1994). The main interest in preserving the symplectic structure of the flow lies in the following result of mixed backward–forward error analysis: for constant stepsizes the numerical solution of a symplectic method can be interpreted over long-time intervals as being exponentially close to the exact solution of a perturbed Hamiltonian system (Hairer 1994, Hairer and Lubich 1997, Reich 1996*a*). This long-time stability property is the main distinction of the Verlet algorithm, compared to nonsymplectic methods used in MD for short-time integration, such as the Beeman algorithm (Beeman 1976). When applied to the harmonic oscillator (2.2), the Verlet algorithm also possesses the largest relative interval of periodicity among explicit RKN methods (Chawla 1985). However, its use with variable stepsizes destroys not only the aforementioned backward–forward error result but also the existence of an interval of periodicity (Skeel 1993). Nevertheless, a strategy has been discovered recently by Hairer (1996) and Reich (1996*a*) combining variable stepsizes with symplectic integration: the symplectic method is simply applied with constant stepsizes to a modified Hamiltonian function $s(q,p)\,(H(q,p) - H(q_0,p_0))$ where the scaling function $s(q,p)$ corresponds to a time-reparametrization of the original Hamiltonian system.

Since the Verlet method is explicit, the stepsize is usually limited to approximately 1[fs], to resolve the high-frequency vibrations. As for other symplectic integrators, resonance phenomena at certain stepsizes have also been observed (Mandziuk and Schlick 1995). At those given stepsizes, large fluctuations of energy or even instability may occur due to repeated sampling of a component at certain points.

4.2. Implicit symplectic methods

To overcome the stability barrier of the explicit Verlet algorithm while preserving its favourable long-time stability property, it is tempting to consider the application of implicit symplectic methods, for instance the implicit midpoint (IM) rule. Applied to (4.1), one step of IM is given by the solution of a nonlinear system

$$\begin{aligned} q_{n+1} &= q_n + \frac{h}{2}\left(v_n + v_{n+1}\right), \\ v_{n+1} &= v_n - hM^{-1}\nabla U\left(\frac{q_n + q_{n+1}}{2}\right). \end{aligned}$$

Having to solve a nonlinear system is the major drawback of implicit methods. Here, the solution can also be seen as a minimum of an optimization problem, for instance, for IM q_{n+1} is a minimum of a 'dynamics function'

$$\Phi(q) := \frac{1}{2}(q - q_n - hv_n)^T M(q - q_n - hv_n) + h^2 U\left(\frac{q + q_n}{2}\right). \tag{4.3}$$

Therefore, optimization techniques can be applied (Schlick and Fogelson 1992).

The IM method is known to be P-stable. However, in the limit of large stepsizes, the high-frequency oscillations are misrepresented by being aliased to one lower frequency. Moreover, as for the Verlet algorithm, instability at certain stepsizes may occur due to numerical resonance. Recently, Ascher and Reich (1997) have shown that, for implicit symmetric schemes applied to highly oscillatory Hamiltonian systems, unless the stepsize is restricted to the order of the square root of the period of the high-frequency oscillation, then even the errors in slowly varying quantities, like energy, can grow undesirably. This error growth is due to the fact that, at large stepsizes, the numerical method fails to accurately represent the time-dependent transformation that decouples the system into a slowly varying and highly oscillatory part (for example, the transformation from Cartesian to polar coordinates in the stiff spring pendulum). In Skeel, Zhang and Schlick (1997), a general one-parameter family of symplectic integrators has been studied in detail, including the explicit Verlet method and several implicit methods: IM, the trapezoidal rule, the Numerov method, and the scheme LIM2 of Zhang and Schlick (1995). Although the interval of periodicity of implicit symplectic methods is larger than that of the Verlet algorithm, implicit methods do not seem competitive in MD. Even when solving the nonlinear equations in parallel by functional iterations, the two-stage implicit Gauss RK method has been found on a standard test problem involving long-range forces to be less efficient than the Verlet algorithm (López-Marcos, Sanz-Serna and Díaz 1995*a*). Implicit symplectic methods are not recommended for resolving high-frequency oscillations efficiently, because of their large overhead for only a modest increase in the allowable stepsize.

4.3. Constrained dynamics and the Rattle algorithm

As mentioned previously, the highest frequencies in MD are due to the bond stretchings and to the bond-angle bendings. The potential due to these bonds can be expressed as follows,

$$U_{\text{bond}}(q) = \frac{1}{2} g^T(q) K g(q), \tag{4.4}$$

where K is a diagonal matrix of large force constants and the vector $g(q)$ contains the stretches $r(q) - \overline{r}$ and the angle bends $\phi(q) - \overline{\phi}$, where $\overline{r}$ and $\overline{\phi}$ are equilibrium values. The potential $U(q)$ can thus be decomposed as $U(q) = V(q) + U_{\text{bond}}(q)$. Introducing the new variable $\lambda := Kg(q)$, we can rewrite the corresponding Hamilton's equations as follows:

$$q' = M^{-1}p, \qquad p' = -\nabla V(q) - G^T(q)\lambda, \qquad K^{-1}\lambda = g(q),$$

where $G(q) := g_q(q)$. If the elements of K are all of the same size and are very large compared to $\|V_{qq}\|$, the last equation can be replaced by holonomic constraints

$$0 = g(q).$$

Mathematical conditions under which this approach is legitimate have been analysed in detail by Bornemann and Schütte (1995*b*). Constraining the bond interactions has the effect of suppressing the presence of the high-frequency oscillations associated with them, hence of allowing an increase in the stepsize at the cost of some added complexity per integration step. We have obtained a system of DAEs of index 3 where λ plays the role of a Lagrange multiplier (Brenan et al. 1995, Hairer and Wanner 1996, Jay 1996). Differentiating the constraint equations twice, we get two additional constraints:

$$\begin{aligned} 0 &= G(q)M^{-1}p, \\ 0 &= G_q(q)\left(M^{-1}p, M^{-1}p\right) - G(q)M^{-1}\left(\nabla V(q) + G^T(q)\lambda\right). \end{aligned}$$

To integrate the above DAE system numerically, a generalization of the Verlet algorithm is given by the *Rattle algorithm* (Andersen 1983)

$$\begin{aligned} p_{n+1/2} &= p_n - \frac{h}{2}\left(\nabla V(q_n) + G^T(q_n)\Lambda_n\right), \\ q_{n+1} &= q_n + hM^{-1}p_{n+1/2}, \\ 0 &= g(q_{n+1}), \\ p_{n+1} &= p_{n+1/2} - \frac{h}{2}\left(\nabla V(q_{n+1}) + G^T(q_{n+1})\lambda_{n+1}\right), \\ 0 &= G(q_{n+1})M^{-1}p_{n+1}. \end{aligned}$$

The computation of the projected value p_{n+1} can actually be avoided by using the relation

$$p_{n+1/2} = p_{n-1/2} - h\left(\nabla V(q_n) + G^T(q_n)\Lambda_n^*\right)$$

which is the basis of the *Shake algorithm* (Ryckaert, Ciccotti and Berendsen 1977). The method is semi-explicit in the sense that it requires only one evaluation of $\nabla V(q)$ per step. The above equations form a nonlinear system for the Lagrange multiplier Λ_n^* which can be solved iteratively. The Shake iterations consist of a combination of Newton and Gauss–Seidel iterations. An overrelaxation procedure that may improve the performance of the Shake iterations by up to a factor two has been advocated in Barth, Kuczera, Leimkuhler and Skeel (1995).

From a physical point of view, constraining a bond corresponds to freezing the interaction. The dynamics of the constrained system is called the *slow dynamics* (Reich 1994). Whereas this approach seems appropriate for bond stretchings, it is inappropriate for bond-angle bendings since the original dynamics is altered (Van Gunsteren and Karplus 1982). The justification in MD for a constrained dynamics borrows arguments from mathematics and statistical mechanics. From a mathematical point of view, one is interested in a running average of the solution

$$\begin{pmatrix} q_\alpha(t) \\ p_\alpha(t) \end{pmatrix} = \frac{1}{\alpha}\int_{-\infty}^{+\infty} \rho\left(\frac{t-s}{\alpha}\right) \cdot \begin{pmatrix} q(t) \\ p(t) \end{pmatrix} \mathrm{d}s$$

for $0 < \alpha \ll 1$ with an appropriate weight function ρ, for example,

$$\rho(x) = \begin{cases} 1 & \text{if } -1/2 \le x \le 1/2, \\ 0 & \text{otherwise.} \end{cases}$$

The goal is to find the dynamics of $(q_{\sqrt{\varepsilon}}(t), p_{\sqrt{\varepsilon}}(t))$ for $\varepsilon \approx \sqrt{\|V_{qq}\| \cdot \|K\|^{-1}}$ called the *smoothed dynamics* (Reich 1995, Schütte 1995) and which generally differs from the slow dynamics. By introduction of an additional soft potential $W(q)$ aimed at correcting the dynamics of the constrained system, the smoothed dynamics can be reestablished. The establishing of the correcting potential has been the subject of recent controversy (Bornemann and Schütte 1995*b*, Bornemann and Schütte 1995*a*, Reich 1996*b*). A standard correction is given by the *Fixman potential* (Fixman 1974)

$$W_F(q) = \frac{k_B T}{2} \log\left(\det\left(G(q) M^{-1} G^T(q)\right)\right),$$

where k_B is the Boltzmann constant and T is the temperature. The computation of the Fixman potential is rather expensive in practice, but it can be simplified by approximating the matrix $G(q)M^{-1}G^T(q)$ by block-diagonal parts (Reich 1997). To improve the correction it has also been proposed to replace the *hard* constraints $0 = g(q)$ by *soft* constraints (Reich 1995)

$$0 = \widetilde{g}(q) = g(q) + K^{-1}\left(G(q)M^{-1}G^T(q)\right)^{-1} G(q)M^{-1}\nabla V(q),$$

restoring some flexibility in the dynamics (Reich 1997). The controversy about the correctness of the Fixman potential seems to be due to the intrusion of physical arguments in its derivation. The principle of *equipartition of energy* of statistical mechanics (Diu, Guthmann, Lederer and Roulet 1989) used to derive the Fixman potential (Fixman 1974, Reich 1995, Reich 1997) is the likely source of the controversy, because of the hypothesis of ergodicity postulated in statistical mechanics.

4.4. Normal-mode techniques in Langevin dynamics

To take into account the effects of a heat bath, of the constant energy transfer between the slow and fast degrees of freedom due to molecular collisions, and of various simplifications in the model, a more realistic dynamics in MD is reflected by the *Langevin dynamics.* The Langevin equations are given by

$$q' = v, \qquad Mv' = -\nabla U(q) - \gamma M v + \zeta(t), \tag{4.5}$$

where γ is a collision frequency (friction) parameter and $\zeta(t)$ is a random force chosen to counterbalance the frictional damping to establish temperature equilibrium. One of the main motivations for the *Langevin/implicit-Euler/normal-mode* (LIN) method of Zhang and Schlick (1993) is to mitigate the undesirable severe high-frequency damping of the implicit Euler method, which may alter the dynamics, while maintaining its ability to take large stepsizes (Peskin and Schlick 1989). In LIN the solution is decomposed into *fast* and *slow* components, that is, $q = q_f + q_s$. A linear approximation to the Langevin equations is used for the fast components

$$q_f' = v_f, \qquad Mv_f' = -\nabla U(q_r) - \overline{H}(q_r)(q_f - q_r) - \gamma M v_f + \zeta(t), \tag{4.6}$$

where q_r is a reference point and $\overline{H}(q_r)$ is a sparse (usually block-diagonal) approximation to $U_{qq}(q_r)$. These equations are solved over a relatively large stepsize, for instance, by using standard *normal-mode techniques*: by diagonalizing the matrix $M^{-1/2}\overline{H}(q_r)M^{-1/2}$, the system (4.6) is rewritten as a set of decoupled equations

$$\widetilde{q}' = \widetilde{v}, \qquad \widetilde{v}' = -D\widetilde{q} - \gamma\widetilde{v} + \theta(t),$$

where $\widetilde{q} = TM^{1/2}(q_f - q_r)$, $\widetilde{v} = TM^{1/2}v_f$, $D = TM^{-1/2}\overline{H}(q_r)M^{-1/2}T^{-1}$ is diagonal, and $\theta(t) = TM^{-1/2}(\zeta(t) - \nabla U(q_r))$; these equations can then be solved analytically (Zhang and Schlick 1994). Nevertheless, it has been observed recently in Barth, Mandziuk and Schlick (1997) and Schlick et al. (1997) that the direct numerical integration of (4.6) can in fact be much faster than computing the normal modes, for instance, by application of the second-order Lobatto IIIA–IIIB PRK method with small inner stepsizes. The above procedure, consisting of computing the fast components, turns out to be a very competitive method in itself, and constitutes the *Langevin/normal* (LN) method. In LIN there is an additional correction step for the slowly varying anharmonic part of the solution

$$q_s' = v_s, \qquad Mv_s' = -\nabla U(q_f + q_s) + \nabla U(q_r) + \overline{H}(q_r)(q_f - q_r) - \gamma M v_s.$$

This system is integrated by the implicit Euler method with one large stepsize h or, equivalently, by minimizing a dynamics function similar to (4.3).

4.5. Multiple time-stepping methods in MD

Since the forces in MD can be decomposed as a sum of *hard* short-range interactions and *soft* long-range interactions on a different time-scale, it is natural to consider the application of *multiple time-stepping* (MTS) methods. The idea is to reduce the overall computational work by evaluating the soft forces less often than the hard forces (Streett, Tildesley and Saville 1978). In (4.1) the potential $U(q)$ is decomposed into hard and soft parts

$$U(q) = U^{\mathrm{hard}}(q) + U^{\mathrm{soft}}(q).$$

The bonded interactions (4.4) enter into the hard part. Moreover, an artificial partitioning of a long-range interaction into one hard and one soft part has been proposed in Skeel and Biesiadecki (1994), for example, an electrostatic interaction $V(r) = C/r$ can be decomposed as

$$V^{\mathrm{soft}}(r) = \begin{cases} C(3r_{\mathrm{cut}}^2 - r^2)/2r_{\mathrm{cut}}^3 & \text{if } r < r_{\mathrm{cut}}, \\ C/r & \text{if } r \geq r_{\mathrm{cut}}, \end{cases}$$

and $V^{\mathrm{hard}}(r) = V(r) - V^{\mathrm{soft}}(r)$ where r_{cut} is a cut-off distance. MTS methods do not generally preserve the symplectic and reversible character of the flow of (4.1). However, the *Verlet-I* algorithm (Grubmüller, Heller, Windemuth and Schulten 1991, Biesiadecki and Skeel 1993) or, equivalently, the *r-RESPA* method (Tuckerman, Berne and Martyna 1992), is an MTS method retaining these properties. One macrostep H of this method can be seen as a composition method: the Verlet algorithm is first applied with stepsize $H/2$ to

$$q' = 0, \qquad Mv' = -\nabla U^{\mathrm{soft}}(q); \tag{4.7}$$

then it is applied N times with a microstep $h = H/N$ to

$$q' = v, \qquad Mv' = -\nabla U^{\mathrm{hard}}(q);$$

finally it is again applied with stepsize $H/2$ to (4.7). Basically, the soft forces are evaluated every macrostep H while the hard forces are evaluated every microstep $h = H/N$. It must be mentioned that resonance and other problems have been reported (Grubmüller et al. 1991, Biesiadecki and Skeel 1993).

5. Circuit simulation

5.1. Introduction

There are a number of applications where the solution has the property that the fast solution is composed of an oscillation with a single high frequency. In this section we will explore some problems from circuit simulation, and a class of methods based on envelope-following ideas, which exploit this property. These methods are often able to take stepsizes that are much larger than the period of the oscillation. The problem of transient simulation in this case is very closely related to the problem of finding a periodic steady state; we will also discuss how similar ideas have been employed in numerical methods for this problem.

Circuit simulation programs like SPICE (Nagel 1975) often need to employ hundreds of thousands of time-steps to simulate the transient behaviour of clocked analog circuits like switching power converters and phase-locked loops. This is because in circuit simulation the stepsizes must be chosen (for accuracy) to be much smaller than a clock period, but the time interval of interest to a designer can be thousands of clock periods. Circuit designers are typically not interested in the details of the node voltage behaviour in every clock cycle, but instead are interested in the *envelope* of that behaviour. With that in mind, the *quasi-envelope* is defined to be a continuous function derived by the following process. Starting at the initial value or at some other point on the solution, define a discrete sequence of points by sampling the state of the system after every clock period T (see Fig. 7). The quasi-envelope is derived by interpolating that sequence to form a smooth curve. We note that the quasi-envelope is different from the more standard notion of envelope because the quasi-envelope is not unique but instead depends on the initial time used to generate the sequence.

Envelope-following methods are based on the idea that if the sequence of points formed by sampling the state at the beginning of each cycle changes slowly as a function of the cycle number, then the quasi-envelope will vary relatively slowly and we will be able to approximate it using stepsizes which are large relative to the length of a cycle.

Envelope-following methods are closely related to the *stroboscopic method* proposed in 1951 by Minorsky (1974) for the study of differential equations. Numerical methods using the envelope-following idea were first introduced by astronomers in 1957 for calculating the orbits of artificial satellites (Mace and Thomas 1960, Taratynova 1960) and were called *multirevolution methods.* Unlike circuit designers, who are interested in the envelope of the oscillations but not in the details, the astronomers were concerned with com-

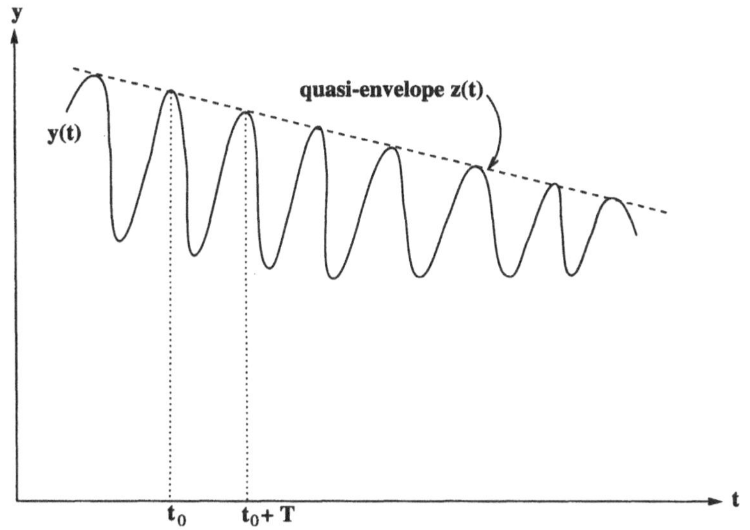

Fig. 7. ODE solution and quasi-envelope

puting future orbits accurately. The multirevolution methods developed in Mace and Thomas (1960), Taratynova (1960) and Graff and Bettis (1975) were generalizations of explicit multistep and Runge–Kutta methods for ODEs to approximately solve the difference equation which generates the sequence of points defining the quasi-envelope. Thus they would always take stepsizes which are multiples of the period of the oscillation. Rather than using an arbitrary starting point to define the quasi-envelope, as above, they used a physical reference point (for example, node, apogee, or perigee). Petzold (1981) extended these methods to more general systems by defining the smooth quasi-envelope as above (independent of any physical reference points), by providing a separate algorithm for finding the period of the oscillation in the fast solution, and by showing how to handle the case of a slowing changing period of the oscillation. Convergence results for envelope-following numerical methods were also given in Petzold (1981). Gear and Gallivan (Gallivan 1980, Gallivan 1983, Gear 1984) explored the design of general ODE codes which incorporate multirevolution techniques and attempt to detect the onset of oscillations. Kirchgraber (1982, 1983) proposed a novel class of methods which synthesize ideas from the method of averaging with envelope-following techniques. White et al. (White and Leeb 1991, Kundert, White and Sangiovanni-Vincentelli 1988*b*, Kundert, White and Sangiovanni-Vincentelli 1988*a*, Telichevesky, Kundert and White 1995, Telichevesky, Kundert and White 1996) applied envelope-following methods to circuit simulation, developing implicit methods which are quite efficient for this application and for finding the periodic steady state.

5.2. Envelope-following methods

Given the initial value problem

$$y' = f(y,t), \quad y(0) = y_0, \quad 0 \le t \le L, \tag{5.1}$$

where $y(t)$ is periodic or nearly periodic with period T, the quasi-envelope $z(t)$ is defined more precisely by

$$z(t+T) = z(t) + Tg(z(t),t), \quad 0 \le t \le L - T, \tag{5.2}$$

where

$$g(z,t) = \frac{1}{T}\Big(\tilde{y}(t+T,t) - \tilde{y}(t,t)\Big)$$

and

$$\frac{d}{ds}\tilde{y}(t+s,t) = f(\tilde{y}(t+s,t), t+s), \quad \tilde{y}(t,t) = z.$$

It is easy to see that if $z(0) = y(0)$ then $z(KT) = y(KT)$, $0 \le KT \le L$, so that z agrees with y at multiples of the period. Since y is nearly periodic, the values of z at points $\{KT\}$, K an integer, should change slowly. Solving (5.2) exactly amounts to solving the differential equation (5.1) over the entire interval $[0, L]$, because $g(z,t)$ is determined by integrating the differential equation over one period of the oscillation. The basis of envelope-following methods is to compute an approximation to z, that is, to solve the difference equation (5.2) approximately with stepsizes H much larger than T. For some applications, like circuit simulation, it is possible to define a smooth z over the entire interval $[0, L]$. For other applications like orbit calculations, it is best to consider z as a discrete function and to take stepsizes in the approximation method which are multiples of T. We note that the solution to the differential equation can be recovered at any time from the (discrete) quasi-envelope, by solving the original ODE with initial condition on the quasi-envelope for no more than one cycle.

Envelope-following methods that are generalizations of linear multistep methods or Runge–Kutta methods have been derived (Gallivan 1983, Graff and Bettis 1975, Petzold 1981, Taratynova 1960). For example, the 'trapezoidal' envelope-following method is given by

$$z_{n+1} = z_n + \left(\frac{H-T}{2}\right) g(z_{n+1}, t_{n+1}) + \left(\frac{H+T}{2}\right) g(z_n, t_n).$$

We note that the coefficients of these methods reduce to those of standard ODE methods as $T \to 0$, and that the methods are 'exact' (up to errors in solving the original ODE numerically over each individual cycle) when $T = H$. For efficiency, the objective is to be able to take $H \gg T$.

In some applications, the period (cycle length) of the oscillation might also be slowly varying. This is handled in envelope-following methods by means of a change of independent variable t so that in the new variable $\hat{t}$ the period τ of the oscillation is a constant, that is,

$$t(\hat{t}+\tau) - t(\hat{t}) = T(t(\hat{t})), \quad t(0) = 0.$$

Defining $\hat{y}(\hat{t}) := y(t(\hat{t}))$ and $\hat{z}(\hat{t}) := z(t(\hat{t}))$, the difference equations that define the quasi-envelope in the case of a slowly varying period of the oscillation are given by

$$\begin{aligned} \hat{z}(\hat{t}+\tau) &= \hat{z}(\hat{t}) + \tau\hat{g}(\hat{z}(\hat{t}),\hat{t}). \quad \hat{z}(0) = z(0), \\ t(\hat{t}+\tau) &= t(\hat{t}) + \tau\left(\frac{T(t(\hat{t}))}{\tau}\right), \quad t(0) = 0, \end{aligned}$$

where

$$\hat{g}(\hat{z},\hat{t}) := \left(\frac{T(t(\hat{t}))}{\tau}\right) g(\hat{z}, t(\hat{t})).$$

The accuracy and stability of these formulæ have been analysed by Petzold (1981).

5.3. Finding the period

In some applications, such as circuit simulation, the period of the oscillation is known *a priori*. In other applications, and in finding a periodic steady state, finding the period of the high-frequency oscillation is an important part of the method.

Several algorithms have been proposed for finding the period. Noting that if y were periodic, $\|y(t) - y(t+T)\| = 0$, Petzold (1981) proposed finding the period T by miminizing $\|y(t) - y(t+T)\|$ over one approximate period. More precisely, T_{m+1} is defined as the value of T^* that satisfies

$$\min_{0<\epsilon\leq T^*\leq I} \int_0^{T_m} (y(t) - y(t+T^*))^2 \, \mathrm{d}t.$$

In practice, in order to better model problems whose solutions are given by a fast oscillation superimposed on a slowly varying solution, the period is found by solving

$$\min_{0<\epsilon\leq T^*\leq I} \int_0^{T_m} (y - p_{m+1} - (y(t+T_{m+1}) - p_{m+1}(t+T_{m+1})))^2 \, \mathrm{d}t,$$

where p_{m+1} is a polynomial which approximates the slow solution that is found at each iteration for T_{m+1} via another minimization. It is shown in Petzold (1978) that this algorithm converges, given a sufficiently smooth initial guess. A similar approach has been used in finding the periodic steady state, as we will discuss below.

The above method of comparing the solution over two periods to find the period of the oscillation is quite general; however, it suffers from the drawback that it is somewhat inefficient. In particular, each time the period needs to be found, the original problem must be solved over two cycles, whereas the envelope-following formulæ require the solution over only one cycle. For problems with a slowly varying period (where the period needs to be recomputed often) this is a relatively large expense. To remedy this problem, another algorithm was proposed by Gallivan and Gear (Gallivan 1980, Gallivan 1983, Gear 1984). This algorithm is based on the idea of defining the period by identifying certain points on the solution at which a simple characterization is repeated, such as zero crossing. Astronomers did this for the multirevolution methods by identifying points of physical interest, such as node, apogee or perigee. For a general problem, the solution itself may have no zero crossing, and there may be difficulty in choosing any value which is crossed periodically. However, the derivative will have periodic sign changes, so the method examines the zero crossings of $c^T y'$, where c is a vector of constant weights. Since there may be more than one zero crossing in a single period, $\|y'(t_1) - y'(t_2)\|$ is also examined, where t_1 and t_2 are the times of zero crossings. If the norm is small, the possibility of a period is considered. For some problems, the solution may not start out oscillatory. This type of algorithm can be used to detect the onset of oscillations, by monitoring the sequence of periods T which are computed. In the event that highly oscillatory behaviour is detected, the software can switch to envelope-following methods.

5.4. Stiffness and implicit methods

As we have noted, the objectives of circuit designers for simulation differ from those of astronomers because circuit designers are not usually interested in the fine details of the oscillation. There are also significant differences in the properties of the ODE systems which influence the choice of numerical methods. In particular, circuit simulation problems are usually quite stiff. Thus they require the use of implicit versions of the envelope-following methods.

White et al. (White and Leeb 1991, Kundert et al. 1988*b*, Kundert et al. 1988*a*, Telichevesky et al. 1995, Telichevesky et al. 1996) have developed efficient implicit algorithms based on the envelope-following idea, and applied them to circuit simulation. The simplest implicit envelope-following method is based on the implicit Euler method and is given by

$$z_{n+1} = z_n + Hg(z_{n+1}, t_{n+1}). \tag{5.3}$$

Solving the nonlinear system (5.3) for z_{n+1}, which is accomplished in stiff ODE codes by a modified Newton iteration, requires an approximate Jacobian matrix $\partial g/\partial z$. Gallivan (1983) has considered the implementation of implicit envelope-following methods and suggests computing the Jacobian by finite difference approximation. This can work well for small problems, but for large systems it is prohibitively expensive because each evaluation of g requires solving the original problem over one cycle.

A more efficient and accurate approach (White and Leeb 1991) is to view this computation as finding the sensitivities of g with respect to perturbations in z. The sensitivity problem is solved concurrently with the original system over one cycle. This can be implemented very efficiently by noting that the Jacobian matrix at every time-step for the sensitivities with respect to each parameter is the same as the Jacobian matrix of the original system. Hence this matrix, if it is dense, can be formed and decomposed once, then used in the Newton iteration for each sensitivity. Differencing in this way is also more accurate than directly differencing the numerical solution over each cycle, because the original system and the sensitivity equations use the same sequence of stepsizes and orders (Hairer, Nørsett and Wanner 1993). White and Leeb (1991) have further noted that if the implicit Euler envelope-following method is used unmodified, the stepsize H will be constrained by the component of y with the fastest-changing envelope. This can be unnecessarily conservative; components of y which have rapidly changing envelopes in stiff problems are likely to be 'nearly algebraic' functions of other, more slowly changing components, over the timescale of one period. These nearly algebraic components of y are computed in White and Leeb (1991) directly from the other components via a DC (steady-state) analysis, and hence are not computed by the envelope-following method. A component y_i of y is considered *quasi-algebraic* if the ith column of the sensitivity matrix is nearly zero.

For large-scale systems, approximating the Jacobian matrix directly is too expensive, because there are so many sensitivities to be computed. Telichevesky et al. (1996) have applied preconditioned iterative methods to solve the linear system at each Newton iteration. These Krylov subspace methods have the property that the Jacobian is never needed directly. Instead, the iterative method needs the product of the Jacobian matrix times a given vector. This can be approximated by a directional difference (a sensitivity in the direction of the given vector). Further efficiency is attained by exploiting the structure of the system in a 'recycled' version of the Krylov algorithm. Telichevesky et al. (1996) found that this method can be as much as forty times faster than direct factorization, for large circuits.

Finally, we note that the solution of stiff oscillatory systems by implicit envelope-following methods has much in common with finding the periodic steady state. That problem can be described as finding y and T such that

$$y(T) - y(0) = 0,$$

which can also be written as

$$g(z(0), T) = 0.$$

Aprille Jr. and Trick (1972) proposed a Newton-type algorithm for solving the steady-state problem in circuit analysis. Telichevesky et al. (1995) have efficiently performed steady-state analysis for large-scale circuits making use of the Krylov subspace approach described above. Lust, Roose, Spence and Champneys (1997) have proposed an algorithm for computing periodic steady states of general ODE systems which combines the recursive projection method of Shroff and Keller (1993), which separates the slow from the fast components, with a Krylov method.

REFERENCES

G. P. Agrawal (1989), *Nonlinear Fiber Optics*, Academic Press.

M. P. Allen and D. J. Tildesley (1987), *Computer Simulation of Liquids*, Clarendon Press, Oxford.

H. C. Andersen (1983), 'Rattle: a velocity version of the Shake algorithm for molecular dynamics calculations', *J. Comput. Phys.* **52**, 24–34.

T. J. Aprille Jr. and T. N. Trick (1972), 'Steady-state analysis of nonlinear circuits with periodic inputs', *Proc. IEEE* **60**, 108–114.

V. I. Arnold (1989), *Mathematical Methods of Classical Mechanics*, Vol. 60 of *Graduate Texts in Mathematics*, 2nd edn, Springer, New York.

U. M. Ascher and S. Reich (1997), 'The midpoint scheme and variants for Hamiltonian systems: advantages and pitfalls'. Department of Computer Science, University of British Columbia, Canada. Preprint.

E. Barth, K. Kuczera, B. Leimkuhler and R. D. Skeel (1995), 'Algorithms for constrained molecular dynamics', *J. Comp. Chem.* **16**, 1192–1209.

E. Barth, M. Mandziuk and T. Schlick (1997), A separating framework for increasing the timestep in molecular dynamics, in *Computer Simulation of Biomolecular Systems: Theoretical and Experimental Applications* (W. F. van Gunsteren, P. K. Weiner and A. J. Wilkinson, eds), Vol. 3, ESCOM, Leiden, The Netherlands. To appear.

K.-J. Bathé and E. L. Wilson (1976), *Numerical Methods in Finite Element Analysis*, Prentice-Hall, Englewood Cliffs, NJ.

D. Beeman (1976), 'Some multistep methods for use in molecular dynamics calculations', *J. Comput. Phys.* **20**, 130–139.

D. G. Bettis (1970), 'Numerical integration of products of Fourier and ordinary polynomials', *Numer. Math.* **14**, 421–434.

J. J. Biesiadecki and R. D. Skeel (1993), 'Dangers of multiple-time-step methods', *J. Comput. Phys.* **109**, 318–328.

J. A. Board Jr., L. V. Kalé, K. Schulten, R. D. Skeel and T. Schlick (1994), 'Modeling biomolecules: larger scales, longer durations', *IEEE Comp. Sci. Eng.* **1**, 19–30.

N. N. Bogoliubov and Y. A. Mitropolski (1961), *Asymptotic Methods in the Theory of Nonlinear Oscillations*, Hindustan Publishing Corp., Delhi, India.

F. A. Bornemann and C. Schütte (1995*a*), Homogenization of highly oscillatory Hamiltonian systems, Technical Report SC 95-39, Konrad-Zuse-Zentrum, Berlin, Germany.

F. A. Bornemann and C. Schütte (1995*b*), A mathematical approach to smoothed molecular dynamics: correcting potentials for freezing bond angles, Technical Report SC 95-30, Konrad-Zuse-Zentrum, Berlin, Germany.

K. E. Brenan, S. L. Campbell and L. R. Petzold (1995), *Numerical Solution of Initial-Value Problems in Differential-Algebraic Equations*, 2nd edn, SIAM.

L. Brusa and L. Nigro (1980), 'A one-step method for direct integration of structural dynamics equations', *Internat. J. Numer. Methods Engrg.* **15**, 685–699.

A. Cardona and M. Géradin (1989), 'Time integration of the equations of motion in mechanism analysis', *Comput. & Structures* **33**, 801–820.

A. Cardona and M. Géradin (1993), Finite element modeling concepts in multibody dynamics, in *Computer Aided Analysis of Rigid and Flexible Mechanical Systems* (M. S. Pereira and J. A. C. Ambrósio, eds), Vol. 1 of *NATO ASI Series*, pp. 325–375.

J. R. Cash (1981), 'High order P-stable formulæ for the numerical integration of periodic initial value problems', *Numer. Math.* **37**, 355–370.

M. M. Chawla (1985), 'On the order and attainable intervals of periodicity of explicit Nyström methods for $y'' = f(x,y)$', *SIAM J. Numer. Anal.* **22**, 127–131.

M. M. Chawla and P. S. Rao (1985), 'High-accuracy P-stable methods for $y'' = f(x,y)$', *IMA J. Numer. Anal.* **5**, 215–220.

M. M. Chawla and S. R. Sharma (1981*a*), 'Families of fifth order Nyström methods for $y'' = f(x,y)$ and intervals of periodicity', *Computing* **26**, 247–256.

M. M. Chawla and S. R. Sharma (1981*b*), 'Intervals of periodicity and absolute stability of explicit Nyström methods for $y'' = f(x,y)$', *BIT* **21**, 455–464.

H. C. Chen and R. L. Taylor (1989), 'Using Lanczos vectors and Ritz vectors for computing dynamic responses', *Eng. Comput.* **6**, 151–157.

J. Chung and G. M. Hulbert (1993), 'A time integration algorithm for structural dynamics with improved numerical dissipation: the generalized-α method', *ASME J. Appl. Mech.* **93-APM-20**.

G. J. Cooper (1987), 'Stability of Runge–Kutta methods for trajectory problems', *IMA J. Numer. Anal.* **7**, 1–13.

R. Craig and M. Bampton (1968), 'Coupling of substructures for dynamic analysis', *AIAA J.* **6**, 1313–1319.

R. R. Craig (1981), *Structural Dynamics, an Introduction to Computer Methods*, Wiley, New York.

G. Dahlquist (1963), 'A special stability problem for linear multistep methods', *BIT* **3**, 27–43.

H. De Meyer, J. Vanthournout and G. Vanden Berghe (1990), 'On a new type of mixed interpolation', *J. Comput. Appl. Math.* **30**, 55–69.

G. Denk (1993), 'A new numerical method for the integration of highly oscillatory second-order ordinary differential equations', *APNUM* **13**, 57–67.

G. Denk (1994), The simulation of highly oscillatory circuits: an effective integration scheme, Technical Report TUM-M9413, Techn. Univ. München, Germany.

B. Diu, C. Guthmann, D. Lederer and B. Roulet (1989), *Physique Statistique*, Hermann, Paris.

C. Engstler and C. Lubich (1995), Multirate extrapolation methods for differential equations with different time scales, Technical Report 29, Math. Inst., Univ. Tübingen, Germany.

N. Fenichel (1979), 'Geometric singular perturbation theory for ordinary differential equations', *J. Diff. Eq.* **31**, 53–98.

M. Fixman (1974), 'Classical statistical mechanics of constraints: a theorem and application to polymers', *Proc. Nat. Acad. Sci.* **71**, 3050–5053.

R. W. Freund, G. H. Golub and N. M. Nachtigal (1992), Iterative solution of linear systems, in *Acta Numerica*, Vol. 1, Cambridge University Press, pp. 57–100.

R. A. Friesner, L. Tuckerman, B. Dornblaser and T. Russo (1989), 'A method for exponential propagation of large systems of stiff nonlinear differential equations', *J. Sci. Comp.* **4**, 327–254.

C. Führer and B. J. Leimkuhler (1991), 'Numerical solution of differential-algebraic equations for constrained mechanical motion', *Numer. Math.* **59**, 55–69.

K. A. Gallivan (1980), Detection and integration of oscillatory differential equations with initial stepsize, order and method selection, Technical report, Dept of Comput. Sci., Univ. of Illinois.

K. A. Gallivan (1983), An algorithm for the detection and integration of highly oscillatory ordinary differential equations using a generalized unified modified divided difference representation, PhD thesis, Dept of Comput. Sci., Univ. of Illinois.

E. Gallopoulos and Y. Saad (1992), 'Efficient solution of parabolic equations by Krylov approximation methods', *SIAM J. Sci. Statist. Comput.* **13**, 1236–1264.

C. Garrett and W. Munk (1979), 'Internal waves in the ocean', *Ann. Rev. Fluid Mech.* **14**, 339–369.

W. Gautschi (1961), 'Numerical integration of ordinary differential equations based on trigonometric polynomials', *Numer. Math.* **3**, 381–397.

C. W. Gear (1984), The numerical solution of problems which may have high frequency components, in *Computer Aided Analysis and Optimization of Mechanical System Dynamics* (E. J. Haug, ed.), Vol. F9 of *NATO ASI Series*, pp. 335–349.

C. W. Gear and D. R. Wells (1984), 'Multirate linear multistep methods', *BIT* **24**, 484–502.

C. W. Gear, G. K. Gupta and B. J. Leimkuhler (1985), 'Automatic integration of the Euler–Lagrange equations with constraints', *J. Comput. Appl. Math.* **12**, 77–90.

A. Gerschel (1995), *Liaisons Intermoléculaires*, Savoirs actuels, InterEditions/CNRS Editions.

I. Gjaja and D. D. Holm (1996), 'Self-consistent wave-mean flow interaction dynamics and its Hamiltonian formulation for a rotating stratified incompressible fluid', *Physica D*. To appear.

I. Gladwell and R. M. Thomas (1983), 'Damping and phase analysis for some methods for solving second-order ordinary differential equations', *Int. J. Numer. Meth. Eng.* **19**, 495–503.

G. H. Golub and V. Pereyra (1973), 'The differentiation of pseudo-inverses and nonlinear least squares problems whose variables separate', *SIAM J. Numer. Anal.* **10**, 413–432.

O. F. Graff and D. G. Bettis (1975), 'Methods of orbit computation with multirevolution steps', *Celestial Mechanics* **11**, 443–448.

H. Grubmüller, H. Heller, A. Windemuth and K. Schulten (1991), 'Generalized Verlet algorithm for efficient dynamics simulations with long-range interactions', *Mol. Simul.* **6**, 121–142.

M. Günther and P. Rentrop (1993*a*), 'Multirate ROW-methods and latency of electric circuits', *Appl. Numer. Math.* **13**, 83–102.

M. Günther and P. Rentrop (1993*b*), Partitioning and multirate strategies in latent electric circuits, Technical Report TUM-M9301, Technische Univ. München, Germany.

E. Hairer (1979), 'Unconditionally stable methods for second order differential equations', *Numer. Math.* **32**, 373–379.

E. Hairer (1994), 'Backward analysis of numerical integrators and symplectic methods', *Ann. Numer. Math.* **1**, 107–132.

E. Hairer (1996), Variable time step integration with symplectic methods, Technical report, Dept. of Math., Univ. of Geneva, Switzerland.

E. Hairer and C. Lubich (1997), 'The life-span of backward error analysis for numerical integrators', *Numer. Math.* To appear.

E. Hairer and G. Wanner (1996), *Solving Ordinary Differential Equations II. Stiff and Differential-Algebraic Problems*, Vol. 14 of *Comput. Math.*, 2nd revised edn, Springer, Berlin.

E. Hairer, C. Lubich and M. Roche (1989), *The Numerical Solution of Differential-Algebraic Systems by Runge–Kutta Methods*, Vol. 1409 of *Lect. Notes in Math.*, Springer, Berlin.

E. Hairer, S. P. Nørsett and G. Wanner (1993), *Solving Ordinary Differential Equations I. Nonstiff Problems*, Vol. 18 of *Comput. Math.*, 2nd revised edn, Springer, Berlin.

E. J. Haug (1989), *Computer Aided Kinematics and Dynamics of Mechanical systems. Volume I: basic methods*, Allyn-Bacon, MA.

J. Hersch (1958), 'Contribution à la méthode aux différences', *ZAMP* **9a(2)**, 129–180.

H. H. Hilber, T. J. R. Hughes and R. L. Taylor (1977), 'Improved numerical dissipation for time integration algorithms in structural dynamics', *Earthquake Engineering and Structural Dynamics* **5**, 283–292.

M. Hochbruck, C. Lubich and H. Selhofer (1995), Exponential integrators for large systems of differential equations, Technical report, Math. Inst., Univ. Tübingen, Germany.

C. Hoff and P. J. Pahl (1988), 'Development of an implicit method with numerical dissipation from a generalized single-step algorithm for structural dynamics', *Comput. Meth. Appl. Mech. Eng.* **67**, 367–385.

W. G. Hoover (1991), *Computational Statistical Mechanics*, Vol. 11 of *Studies in modern thermodynamics*, Elsevier, Amsterdam.

T. J. R. Hughes (1987), *The Finite Element Method*, Prentice-Hall, Englewood Cliffs, NJ.

L. O. Jay (1996), 'Symplectic partitioned Runge–Kutta methods for constrained Hamiltonian systems', *SIAM J. Numer. Anal.* **33**, 368–387.

J. Kevorkian and J. D. Cole (1981), *Perturbation Methods in Applied Mathematics*, Springer, New York.

U. Kirchgraber (1982), 'A numerical scheme for problems in nonlinear oscillations', *Mech. Res. Comm.* **9**, 411–417.

U. Kirchgraber (1983), Dynamical system methods in numerical analysis. Part I: An ODE-solver based on the method of averaging, Technical report, Seminar für Angew. Math., ETH Zürich, Switzerland.

N. Kopell (1985), 'Invariant manifolds and the initialization problem for some atmospheric equations', *Physica D* **14**, 203–215.

K. Kundert, J. White and A. Sangiovanni-Vincentelli (1988*a*), An envelope-following method for the efficient transient simulation of switching power and filter circuits, in *Proc. of IEEE International Conf. on Computer-Aided Design.*

K. Kundert, J. White and A. Sangiovanni-Vincentelli (1988*b*), A mixed frequency-time approach for finding the steady-state solution of clocked analog circuits, in *Proc. of IEEE 1988 Custom Integrated Circuits Conf.*

J. D. Lambert and I. A. Watson (1976), 'Symmetric multistep methods for periodic initial value problems', *J. Inst. Math. Appl.* **18**, 189–202.

C. Lanczos (1950), 'An iteration method for the solution of the eigenvalue problem of linear differential and integral operators', *J. Res. Nat. Bur. Standards* **45**, 255–281.

B. Leimkuhler, S. Reich and R. D. Skeel (1995), Integration methods for molecular dynamics, in *Mathematical Approaches to Biomolecular Structure and Dynamics* (J. Mesirov, K. Schulten and D. W. Sumners, eds), Vol. 82 of *IMA Volumes in Mathematics and its Applications*, Springer, New York, pp. 161–187.

W. Liniger and R. A. Willoughby (1970), 'Efficient integration methods for stiff systems of ordinary differential equations', *SIAM J. Numer. Anal.* **7**, 47–66.

M. A. López-Marcos, J. M. Sanz-Serna and J. C. Díaz (1995*a*), Are Gauss–Legendre methods useful in molecular dynamics?, Technical Report 4, Dept. of Math., Univ. of Valladolid, Spain.

M. A. López-Marcos, J. M. Sanz-Serna and R. D. Skeel (1995*b*), An explicit symplectic integrator with maximal stability interval, Technical report, Dept. of Math., Univ. of Valladolid, Spain.

C. Lubich (1993), 'Integration of stiff mechanical systems by Runge–Kutta methods', *ZAMP* **44**, 1022–1053.

K. Lust, D. Roose, A. Spence and A. Champneys (1997), 'An adaptive Newton–Picard algorithm with subspace iteration for computing periodic solutions', *SIAM J. Sci. Comput.* To appear.

D. Mace and L. H. Thomas (1960), 'An extrapolation method for stepping the calculations of the orbit of an artificial satellite several revolutions ahead at a time', *Astronomical Journal*.

M. Mandziuk and T. Schlick (1995), 'Resonance in the dynamics of chemical systems simulated by the implicit-midpoint scheme', *Chem. Phys. Lett.* **237**, 525–535.

N. Minorsky (1974), *Nonlinear Oscillations*, Robert E. Krieger Publ. Comp., Huntinton, NY.

L. W. Nagel (1975), SPICE2: A computer program to simulate semiconductor circuits, Technical report, Electronics Research Laboratory, Univ. of California at Berkeley.

B. Neta and C. H. Ford (1984), 'Families of methods for ordinary differential equations based on trigonometric polynomials', *J. Comput. Appl. Math.* **10**, 33–38.

N. M. Newmark (1959), 'A method of computation for structural dynamics', *ASCE J. Eng. Mech. Div.* **85**, 67–94.

P. E. Nikravesh (1988), *Computer-Aided Analysis of Mechanical Systems*, Prentice-Hall, Englewood Cliffs, NJ.

S. Nosé (1984), 'A unified formulation of the constant temperature molecular dynamics methods', *J. Chem. Phys.* **81**, 511–519.

B. Nour-Omid and R. W. Clough (1984), 'Dynamic analysis of structures using Lanczos coordinates', *Earthquake Engineering and Structural Dynamics.*

R. E. O'Malley (1991), *Singular Perturbation Methods for Ordinary Differential Equations*, Springer, New York.

B. Owren and H. H. Simonsen (1995), 'Alternative integration methods for problems in structural mechanics', *Comput. Meth. Appl. Mech. Eng.* **122**, 1–10.

T. J. Park and J. C. Light (1986), 'Unitary quantum time evolution by iterative Lanczos reduction', *J. Chem. Phys.* **85**, 5870–5876.

M. S. Pereira and J. A. C. Ambrósio, eds (1993), *Computer aided analysis of rigid and flexible mechanical systems*, Vol. 1 & 2, Tróia, Portugal.

C. S. Peskin and T. Schlick (1989), 'Molecular dynamics by the backward Euler's method', *Comm. Pure Appl. Math.* **42**, 1001–1031.

L. R. Petzold (1978), An efficient numerical method for highly oscillatory ordinary differential equations, PhD thesis, Dept. of Comput. Sci., Univ. of Illinois.

L. R. Petzold (1981), 'An efficient numerical method for highly oscillatory ordinary differential equations', *SIAM J. Numer. Anal.* **18**, 455–479.

A. Portillo and J. M. Sanz-Serna (1995), 'Lack of dissipativity is not symplecticness', *BIT* **35**, 269–276.

S. Reich (1994), Numerical integration of highly oscillatory Hamiltonian systems using slow manifolds, Technical Report UIUC-BI-TB-94-06, The Beckman Institute, Univ. of Illinois, USA.

S. Reich (1995), 'Smoothed dynamics of highly oscillatory Hamiltonian systems', *Physica D* **89**, 28–42.

S. Reich (1996*a*), Backward error analysis for numerical integrators, Technical report, Konrad-Zuse-Zentrum, Berlin, Germany.

S. Reich (1996*b*), Smoothed Langevin dynamics of highly oscillatory systems, Technical Report SC 96-04, Konrad-Zuse-Zentrum, Berlin, Germany.

S. Reich (1997), 'A free energy approach to the torsion dynamics of macromolecules', *Phys. Rev. E.* To appear.

J. P. Ryckaert, G. Ciccotti and H. J. C. Berendsen (1977), 'Numerical integration of the Cartesian equations of motion of a system with constraints: molecular dynamics of *n*-alkanes', *J. Comput. Phys.* **23**, 327–341.

J. M. Sanz-Serna (1992), Symplectic integrators for Hamiltonian problems: an overview, in *Acta Numerica*, Vol. 1, Cambridge University Press, pp. 243–286.

J. M. Sanz-Serna and M. P. Calvo (1994), *Numerical Hamiltonian Problems*, Chapman and Hall, London.

T. Schlick and A. Fogelson (1992), 'TNPACK-A truncated Newton minimization package for large-scale problems: I. Algorithm and usage', *ACM Trans. Math. Software* **18**, 46–70.

T. Schlick, E. Barth and M. Mandziuk (1997), 'Biomolecular dynamics at long timesteps: bridging the timescale gap between simulation and experimentation', *Ann. Rev. Biophys. Biomol. Struct.* To appear.

S. Schneider (1995), 'Convergence results for multistep Runge–Kutta on stiff mechanical systems', *Numer. Math.* **69**, 495–508.

C. Schütte (1995), Smoothed molecular dynamics for thermally embedded systems, Technical Report SC 95-15, Konrad-Zuse-Zentrum, Berlin, Germany.

G. Shroff and H. Keller (1993), 'Stabilization of unstable procedures: the recursive projection method', *SIAM J. Numer. Anal.* **30**, 1099–1120.

B. Simeon (1996), 'Modelling of a flexible slider crank mechanism by a mixed system of DAEs and PDEs', *Math. Model. of Systems* **2**, 1–18.

J. C. Simo and L. Vu-Quoc (1986), 'A three-dimensional finite strain rod model. Part II: computational aspects', *Comput. Meth. Appl. Mech. Eng.* **58**, 79–116.

J. C. Simo, N. Tarnow and M. Doblare (1993), Nonlinear dynamics of three-dimensional rods: exact energy and momentum conserving algorithms, Technical report, Div. of Appl. Mech., Dept. of Mech. Eng., Stanford Univ.

R. D. Skeel (1993), 'Variable step size destabilizes the Störmer/leapfrog/Verlet method', *BIT* **33**, 172–175.

R. D. Skeel and J. J. Biesiadecki (1994), 'Symplectic integration with variable stepsize', *Ann. Numer. Math.* **1**, 191–198.

R. D. Skeel, J. J. Biesiadecki and D. Okunbor (1993), Symplectic integration for macromolecular dynamics, in *Scientific Computation and Differential Equations*, World Scientific, pp. 49–61.

R. D. Skeel, G. Zhang and T. Schlick (1997), 'A family of symplectic integrators: stability, accuracy, and molecular dynamics applications', *SIAM J. Sci. Comput.* **18**, 203–222.

E. Stiefel and D. G. Bettis (1969), 'Stabilization of Cowell's method', *Numer. Math.* **13**, 154–175.

W. B. Streett, D. J. Tildesley and G. Saville (1978), 'Multiple time step methods in molecular dynamics', *Mol. Phys.* **35**, 639–648.

A. Streitwieser Jr. and C. H. Heathcock (1985), *Introduction to Organic Chemistry*, 3rd edn, Macmillan, New York.

G. P. Taratynova (1960), 'Numerical solution of equations of finite differences and their application to the calculation of orbits of artificial earth satellites', *AES J. Supplement* **4**, 56–85. Translated from *Artificial Earth Satellites.*

R. Telichevesky, K. Kundert and J. White (1995), Steady-state analysis based on matrix-free Krylov subspace methods, in *Proc. of Design Automation Conf., San Francisco.*

R. Telichevesky, K. Kundert and J. White (1996), Efficient AC and noise analysis of two-tone RF circuits, in *Proc. of Design Automation Conf., Las Vegas.*

J. Tidblad and T. E. Graedel (1996), 'GILDES model studies of aqueous chemistry. Initial SO_2-induced atmospheric corrosion of copper', *Corrosion Science*. To appear.

M. Tuckerman, B. J. Berne and G. J. Martyna (1992), 'Reversible multiple time scale molecular dynamics', *J. Chem. Phys.* **97**, 1990–2001.

M. E. Tuckerman and M. Parrinello (1994), 'Integrating the Car–Parrinello equations. I: Basic integration techniques', *J. Chem. Phys.* **101**, 1302–1315.

P. J. Van Der Houwen and B. P. Sommeijer (1984), 'Linear multistep methods with reduced truncation error for periodic initial-value problems', *IMA J. Numer. Anal.* **4**, 479–489.

P. J. Van Der Houwen and B. P. Sommeijer (1987), 'Explicit Runge–Kutta (–Nyström) methods with reduced phase errors for computing oscillating solutions', *SIAM J. Numer. Anal.* **24**, 595–617.

P. J. Van Der Houwen and B. P. Sommeijer (1989), 'Phase-lag analysis of implicit Runge–Kutta methods', *SIAM J. Numer. Anal.* **26**, 214–229.

W. F. Van Gunsteren and M. Karplus (1982), 'Effects of constraints on the dynamics of macromolecules', *Macromolecules* **15**, 1528–1544.

J. Vanthournout, G. Vanden Berghe and H. De Meyer (1990), 'Families of backward differentiation methods based on a new type of mixed interpolation', *Comput. Math. Appl.* **20**, 19–30.

L. Verlet (1967), 'Computer experiments on classical fluids. I: thermodynamical properties of Lennard–Jones molecules', *Phys. Rev.* **159**, 98–103.

Webster's Ninth New Collegiate Dictionary (1985), Merriam-Webster, Springfield, MA.

J. White and S. B. Leeb (1991), 'An envelope-following approach to switching power converter simulation', *IEEE Trans. Power Electronics* **6**, 303–307.

E. L. Wilson, M. Yuan and J. M. Dickens (1982), 'Dynamic analysis by direct superposition of Ritz vectors', *Earthquake Engineering and Structural Dynamics* **10**, 813–821.

W. L. Wood, M. Bossak and O. C. Zienkiewicz (1980), 'An alpha modification of Newmark's method', *Internat. J. Numer. Methods Engrg.* **15**, 1562–1566.

J. Yen and L. R. Petzold (1996), Numerical solution of nonlinear oscillatory multibody systems, in *Numerical Analysis 95* (D. F. Griffiths and G. A. Watson, eds), Vol. 344 of *Pitman Research Notes in Mathematics*, pp. 209–224.

J. Yen and L. R. Petzold (1997), 'An efficient Newton-type iteration for the numerical solution of highly oscillatory constrained multibody dynamic systems', *SIAM J. Sci. Comput.* To appear.

J. Yen, L. Petzold and S. Raha (1996), A time integration algorithm for flexible mechanism dynamics: the DAE α-method, Technical Report TR 96-024, Dept. of Comput. Sci., Univ. of Minnesota.

W. S. Yoo and E. J. Haug (1986), 'Dynamics of articulated structures. Part I: Theory', *J. Mech. Struct. Mach.* **14**, 105–126.

G. Zhang and T. Schlick (1993), 'LIN: a new algorithm to simulate the dynamics of biomolecules by combining implicit-integration and normal mode techniques', *J. Comput. Chem. Phys.* **14**, 1212–1233.

G. Zhang and T. Schlick (1994), 'The Langevin/implicit-Euler/normal-mode scheme for molecular dynamics at large time steps', *J. Chem. Phys.* **101**, 4995–5012.

G. Zhang and T. Schlick (1995), 'Implicit discretization schemes for Langevin dynamics', *Mol. Phys.* **84**, 1077–1098.

Acta Numerica (1997), *pp.* 485–521

Computational methods for semiclassical and quantum transport in semiconductor devices

Christian Ringhofer
Department of Mathematics, Arizona State University
Tempe, AZ 85287-1804, USA
E-mail: ringhofer@asu.edu

The progressive miniaturization of semiconductor devices, and the use of bulk materials other than silicon, necessitates the use of a wide variety of models in semiconductor device simulation. These include classical and semiclassical models, such as the Boltzmann equation and the hydrodynamic system, as well as quantum transport models such as the quantum Boltzmann equation and the quantum hydrodynamic system. This paper gives an overview of recently developed numerical methods for these systems. The focus is on Galerkin methods for the semiclassical and quantum kinetic systems and on difference methods for the classical and quantum hydrodynamic systems. The stability and convergence properties of these methods and their relation to the analytical properties of the continuous systems are discussed.

CONTENTS

1. Introduction

The goal of numerical semiconductor device simulation is to model the flow of electrons in a crystal in order to predict macroscopically measurable quantities, such as currents and heat fluxes, in given operating and environmental conditions, such as the bias applied to a given device and ambient temperature. Other than in process simulation, it is always the same physical process that is considered, namely the transport of charged particles in a solid state medium. Different mathematical models are used only because of the wide range of device dimensions and operating conditions. Since one and the same set of equations can be used to model a wide variety of devices, it is reasonable to develop customized numerical methods for the governing equations.

The key parameters influencing the choice of model equations are the mean free path of electrons (the average length of free flight before the electron undergoes a scattering event), the number of free electrons in a given device, the size of the Planck constant in relation to the dimensions of the simulation domain, and the ambient temperature. These parameters determine whether the electrons can be modelled as a continuum, as classical particles or via quantum mechanical descriptions.

The resulting model equations range from the Schrödinger equation for the evolution of the electron wave function to the drift–diffusion system for the evolution of an 'electron gas' which is close to a Maxwellian equilibrium. Because of the progressive miniaturization of semiconductor devices and the use of materials whose mean free path is considerably longer than that of silicon, the trend in device simulation is certainly towards a more and more microscopic description. Since the field is now so wide, one necessarily has to limit the scope of an overview of numerical techniques in device simulation. There are two types of models and simulation techniques which are extremely well developed and documented at this point. One consists of finite difference and finite element techniques for the drift–diffusion system, and the other of Monte Carlo methods for the Boltzmann equation. Since it would be impossible to do all this work justice in the space provided, we have instead decided to focus on more recent developments, and refer the reader to excellent reference works such as Kersch and Morokoff (1995) and Selberherr (1981) for these topics. The first category of methods presented in this paper deals with the intermediate regime between the Boltzmann equation and the drift–diffusion system. This category comprises methods based on series expansion of the Boltzmann equation and various forms of moment closure hierarchies, including the so-called hydrodynamic models. The second category includes methods for quantum kinetic equations and their moment closure hierarchies, such as the so-called quantum hydrodynamic model.

This paper is organized as follows. Section 2 presents a brief overview of the various models, pointing out some of the features relevant to numerical simulations. Section 3 deals with methods for semiclassical transport descriptions, based on the semiclassical Boltzmann equation. Series expansion methods around a Maxwell distribution are discussed in Section 3.1, numerical methods for the hydrodynamic model are discussed in Section 3.2, and extensions of hydrodynamic models are presented in Section 3.3. Section 4 is devoted to numerical methods for quantum transport models. In Section 4.1 numerical methods for the quantum Boltzmann equation are discussed.

As is the case for the classical Boltzmann equation, many of the interesting effects can be studied using much simpler macroscopic models based on moment hierarchies, leading to the quantum hydrodynamic model. Section 4.2 deals with numerical methods for the quantum hydrodynamic system.

2. Model equations

In this section a brief overview of the underlying model equation is presented. Models for semiconductor device simulations generally fall into two categories, namely semiclassical models, based on the semiclassical Boltzmann equation, and quantum mechanical models, derived from the Schrödinger equation. In Section 2.1 we discuss the semiclassical Boltzmann equation together with some of its features, such as conservation properties. In Section 2.2 its quantum mechanical equivalent, namely the quantum Boltzmann equation, is presented.

2.1. The semiclassical Boltzmann equation

The basis for the semiclassical description of electron transport is the Boltzmann equation in the form

$$\partial_t f + \operatorname{div}_x(v(k)f) - \frac{1}{\hbar}\operatorname{div}_k(E(x,t)f) = Q(f). \tag{2.1}$$

Here $f(x,k,t)$ denotes the density of electrons, x stands for position, k denotes the three-dimensional wave vector, and t time. If we let $\varepsilon(k)$ denote the energy of an electron with wave vector k in a certain band, the corresponding velocity in (2.1) is given by $v(k) = \hbar^{-1}\nabla_k\varepsilon$. In a vacuum, the classical Hamiltonian yields the energy-wave vector relationship

$$\varepsilon(k) = \frac{\hbar^2}{2m}|k|^2, \quad v(k) = \frac{\hbar}{m}k. \tag{2.2}$$

Thus the velocity v and the wave vector k are identical up to the constant $\hbar/m$ and the classical Boltzmann transport equation is obtained. In a crystal, the relationship between the wave vector and the energy is given by the parametrization of the eigenfunctions of the Schrödinger equation with a potential that is periodic on the crystal lattice, and the energy band function $\varepsilon(k)$ has to be computed. However, for small wave vectors, and consequently for small velocities, the energy band function is often approximated locally by a parabolic function via the effective mass approximation, for analytical purposes. The collision integral $Q(f)$ on the right-hand side of (2.1) is given by

$$Q(f)(x,k,t) = \int S(k,k')f'(1-f) - S(k',k)f(1-f')\,\mathrm{d}k', \tag{2.3}$$

where the notation

$$f = f(x,k,t), \quad f' = f(x,k',t) \tag{2.4}$$

is used. The collision integral $Q(f)$ models the interaction of electrons with the crystal lattice. These interactions include scattering with crystal impurities and acoustic and polar optical phonons (the vibrations of the lattice). A more complicated collision operator Q_{ee} is used to model the interaction of electrons with each other. The electron–electron collision operator is of the form

$$\begin{aligned} Q_{ee}(f)(x,k,t) = \\ &\int S_{ee}(x,k,k_1,k',k_1')f'f_1'(1-f)(1-f_1) - \\ &S_{ee}(x,k',k_1',k,k_1)ff_1(1-f')(1-f_1')\,\mathrm{d}k_1\,\mathrm{d}k'\,\mathrm{d}k_1', \end{aligned} \tag{2.5}$$

where $f = f(x,k,t)$, $f' = f(x,k',t)$, $f_1 = f(x,k_1,t)$, and $f_1' = f(x,k_1',t)$. However, other than in gas dynamics, particle–particle scattering is a rather rare event in most semiconductor devices and the operator Q_{ee} is rarely used.

Conservation and equilibrium

There are two important features of the collision operator Q that need to be reflected by any numerical method, namely conservation and the existence of a thermal equilibrium. A quantity $g(k)$ is said to be conserved if

$$\int g(k)Q(f)(x,k,t)\,\mathrm{d}k = 0 \tag{2.6}$$

holds for any density function f. For reasons of symmetry $g(k) = 1$, the number of particles, is obviously conserved by all collision operators. The second property is the existence of a thermal equilibrium, namely a density function f_e such that, because of the principle of detailed balance (Markowich, Ringhofer and Schmeiser 1990),

$$S(k,k')f_e'(1-f_e) = S(k',k)f_e(1-f_e') \tag{2.7}$$

holds. The thermal equilibrium density f_e is given by the Fermi–Dirac density function

$$f_e = F_D\left(\frac{\varepsilon(k)-\varepsilon_F}{k_BT}\right), \quad F_D(z) = \frac{1}{1+e^z}, \tag{2.8}$$

where ε_F is the Fermi energy, k_B is the Boltzmann constant, and T denotes the lattice temperature of the crystal. The principle of detailed balance (2.7) implies the relation

$$S(k,k') = M(k)s(k,k'), \quad M(k) = \exp\left(\frac{-\varepsilon(k)}{k_BT}\right) \tag{2.9}$$

for the scattering rate S, where M is called the Maxwellian distribution and s is symmetric in the variables k and k', so $s(k,k') = s(k',k)$ holds.

Low density and relaxation time approximations

In order to derive simplified models from the Boltzmann equation (2.1), it is often necessary to make simplifying approximations to the collision operator Q. The first approximation is to assume the density function f to be small, and therefore to drop the quadratic terms in the collision operator Q in (2.3), giving the linear operator

$$Q(f)(x,k,t) = \int s(k,k')f' - s(k',k)f \, dk'. \tag{2.10}$$

Next, it is assumed that the density function f is close to a Maxwellian distribution of the form

$$f(x,k,t) \approx \frac{n(x,t)}{n_0} \exp\left(\frac{-\varepsilon(k)}{k_B T}\right), \quad n_0 = \int \exp\left(\frac{-\varepsilon(k)}{k_B T}\right) dk. \tag{2.11}$$

Replacing f' in the linear collision operator (2.10) by the expression (2.11) gives

$$Q(f)(x,k,t) = \frac{1}{\tau(x,k)}\Big(n(x,t)M(k) - f\Big), \tag{2.12}$$

with

$$\frac{1}{\tau(x,k)} = \int s(k',k)\, dk', \quad n(x,t) = \int f(x,k,t)\, dk, \tag{2.13}$$

$$\text{and} \quad M(k) = \frac{1}{n_0} \exp\left(\frac{-\varepsilon(k)}{k_B T}\right).$$

The term $\tau(x,k)$ is called the relaxation time.

Collision frequency, mean free path and scaling

One of the most important quantitative parameters determining which model to choose is the mean free path. The mean free path is given by the shape of the scattering rate $s(k,k')$ and the the energy band function ε. If we define the collision frequency ω by

$$\omega(k) = \int M(k')s(k',k)\, dk', \tag{2.14}$$

then ω^{-1} is the average time an electron travels freely before undergoing a collision event. Scaling the velocity wave vector relationship

$$v(k) = \frac{1}{\hbar}\nabla_k \varepsilon(k) = v_0 \tilde{v}\left(\frac{k}{k_0}\right), \tag{2.15}$$

where v_0 and k_0 are chosen such that $\tilde{v}(k)$ is an $O(1)$ function, the expression

$$\lambda_0 = \frac{v_0}{\omega} \tag{2.16}$$

gives the average distance an electron travels between collision events. If we now scale the position variable x by the device length L, the wave vector k by k_0 and the time by $\gamma = \omega_0 L^2/v_0^2$, we obtain the scaled Boltzmann equation

$$\lambda^2 \partial_t f_s + \lambda \operatorname{div}_x\left(\tilde{v}(k) f_s\right) - \lambda \operatorname{div}_k\left(E_s f_s\right) = Q(f_s) \tag{2.17a}$$

$$Q(f_s) = \int_{\mathbb{R}^3} dk\ s_s\left(M_s f_s'(1-f_s) - M_s' f_s(1-f_s')\right), \tag{2.17b}$$

where x, k and t are now dimensionless and the scaled field and scattering rate E_s and s_s are given by

$$\begin{aligned} E(x,t) &= \frac{v_0 \hbar k_0}{L} E_s\left(\frac{x}{L}, \frac{t}{\gamma}\right), \\ s(k,k') &= \frac{\omega_0}{k_0^3} s_s\left(\frac{k}{k_0}, \frac{k}{k_0}\right), \\ M(k) &= M_s\left(\frac{k}{k_0}\right), \end{aligned} \tag{2.18}$$

and $\lambda = \lambda_0/L$ is the Knudsen number, the ratio of the mean free path and the size of the simulation domain. To what extent macroscopic models provide an accurate transport picture depends mainly on the size of the Knudsen number λ. In the limit for $\lambda \to 0$ one obtains from the Hilbert expansion (see Markowich et al. (1990)) that $f_s = n(x,t)M_s(k) + O(\lambda)$ holds, where the macroscopic electron density n satisfies the drift–diffusion equation

$$\partial_t n - \operatorname{div}_x (D\nabla_x n - mE_s n) = 0. \tag{2.19}$$

(We will drop the subscript s from now on.) Through miniaturization, the device length L decreases, and through the use of materials such as gallium arsenide, the mean free path λ_0 increases, making the drift–diffusion system less and less valid. Current state-of-the-art technology for devices like MOSFETs works with values of $\lambda = O(0.1)$, which makes simulations based on alternative models necessary.

Boundary conditions for the semiclassical Boltzmann equation

For the simulation of an actual device, the simulation domain will consist of a bounded region Ω, whose boundary $\partial\Omega$ is made up of segments $\partial\Omega_c, \partial\Omega_i, \partial\Omega_a$, corresponding to contacts, insulating surfaces and artificial boundaries, introduced to limit the size of the simulation domain. At contacts, the inflow of electrons according to a certain given distribution f_c is prescribed. At insulating surfaces we usually prescribe a specular reflection condition, and at artificial surface segments zero influx is required. The situation will be somewhat more complicated in the quantum case (see Section 4.1). So altogether we have

$$\begin{aligned}
\partial\Omega &= \partial\Omega_c \cup \partial\Omega_a \cup \partial\Omega_i && && \text{(2.20a)}\\
f(x,k,t) &= b(x,t) f_c(k) \quad \text{for} \quad x \in \partial\Omega_c, && k\cdot\nu < 0 && \text{(2.20b)}\\
f(x,k,t) &= 0 \quad \text{for} \quad x \in \partial\Omega_a, && k\cdot\nu < 0 && \text{(2.20c)}\\
f(x,k,t) &= f(x,-k,t) \quad \text{for} \quad x \in \partial\Omega_i, && k\cdot\nu < 0, && \text{(2.20d)}
\end{aligned}$$

where ν denotes the outward normal vector on the boundary $\partial\Omega$. The function $b(x,t)$ in (2.20b) gives the amount of electrons injected. Assuming that the device is part of a circuit, b is given by Ohm's law.

The Poisson equation

The electric field E is, in general, derived from Maxwell's equations. However, since we operate in a regime where the speed of light can safely be set to infinity, Maxwell's equations become

$$\operatorname{div}_x(\varepsilon_D E) = \rho, \qquad \nabla_x \times E = 0, \tag{2.21}$$

where ε_D denotes the dielectric constant of the material. The charge density ρ is given by $\rho = e(N_D - N_A - n)$, where e is the unit charge and N_D and N_A are the pre-concentrations of donor and acceptor atoms in the crystal, due to doping. Here n is the spatial density of free electrons, given by the zeroth order moment of the density function f in the Boltzmann equation. Introducing the potential V by $E = -\nabla_x V$, we obtain the Poisson equation

$$\operatorname{div}_x(\varepsilon_D E) = e(N_D - N_A - n), \quad E = -\nabla_x V, \quad n = \int_{\mathbb{R}^3} \mathrm{d}k\; n. \tag{2.22}$$

The Poisson equation (2.22) is coupled to the Boltzmann equation (2.1) via the charge density n in (2.12), and therefore the two equations have to be solved simultaneously. A bias is applied to the contacts of the device by prescribing a potential difference between the contacts, that is, by setting V at the boundary segments corresponding to contacts.

2.2. Quantum transport models

As device dimensions decrease, quantum mechanical transport phenomena play an increasing role in the function of devices. It is therefore necessary to develop simulation models that are capable of describing these effects. These models are a generalization of the classical models in the sense that they reduce to the Boltzmann transport picture in the classical limit, that is, when an appropriately scaled form of the Planck constant tends to zero. The quantum mechanical description of the motion of an electron in a vacuum under the influence of a potential field is given by the Schrödinger equation

$$i\hbar\partial_t\psi = H\psi, \quad H\psi = -\frac{\hbar^2}{2m}\Delta\psi - eV\psi, \tag{2.23}$$

where ψ denotes the wave function and V denotes the potential. The operator H is called the quantum Hamiltonian. The electron density n and the electron current density J are then given by

$$n = |\psi|^2 \quad \text{and} \quad eJ = \frac{\hbar}{m}Im(\psi\nabla\psi). \tag{2.24}$$

In order to describe transport in an actual device, several features have to be added to the above transport picture.

- An ensemble of Schrödinger equations must be considered in order to model the mixed state of an electron.
- The electron is moving in a crystal and not in a vacuum.
- Collisions, representing the interaction of the electron with the crystal lattice, have to be modelled.
- The system has to interact with the outside world via boundary conditions at device contacts and insulating surfaces.

Several steps are taken to achieve these goals. Some are mathematically precise, whereas some are purely phenomenological. First, the density matrix for mixed states of the form

$$\rho(r,s,t) = \sum_j \sigma(\omega_j)\psi(r,t)^*\psi(s,t) \tag{2.25}$$

is introduced, where each of the wave functions ψ_j satisfies the Schrödinger equation (2.23). The density matrix ρ then satisfies the quantum Liouville equation

$$i\hbar\partial_t\rho = (H_s - H_r)\rho = \frac{\hbar^2}{2m}(\Delta_r - \Delta_s)\rho + e\Big(V(r) - V(s)\Big)\rho, \tag{2.26}$$

and the electron and current densities n and J are given by

$$n(x,t) = \rho(x,x,t), \quad eJ(x,t) = \frac{i\hbar}{2m}(\nabla_r\rho - \nabla_s\rho)(r=x, s=x, t). \tag{2.27}$$

In order to relate the quantum picture to the classical picture it is convenient to introduce the Wigner function (Wigner 1932)

$$w(x,k,t) = (2\pi)^{-3}\int_{\mathbb{R}^3} d\eta\, \rho\left(x + \frac{1}{2}\eta, x - \frac{1}{2}\eta, t\right)e^{i\eta\cdot k}, \tag{2.28}$$

which then satisfies the Fourier transformed version of the quantum Liouville equation, often referred to as the Wigner equation

$$\partial_t w + \frac{\hbar}{m}k\cdot\nabla_x w + \frac{ie}{\hbar}\delta V\left(x, \frac{1}{2i}\nabla_k\right)w = 0, \tag{2.29}$$

and the electron and current densities are given by

$$n(x,t) = \int_{\mathbb{R}^3} \mathrm{d}k\; w, \quad J(x,t) = \frac{\hbar}{m}\int_{\mathbb{R}^3} \mathrm{d}k\; kw, \tag{2.30}$$

and the operator $\delta V\,(x, (1/2i)\nabla_k)$ in (2.29) is defined in the sense of pseudo-differential operators (Taylor 1981) as

$$\delta V\left(x, \frac{1}{2i}\nabla_k\right) w(x,k,t) =$$

$$(2\pi)^{-3}\int_{\mathbb{R}^3} \mathrm{d}\eta \int_{\mathbb{R}^3} \mathrm{d}k'\; \delta V\left(x, \frac{1}{2}\eta\right) w(x,k',t)e^{i\eta\cdot(k-k')},$$

$$\text{where} \quad \delta V\left(x, \frac{1}{2}\eta\right) = V\left(x+\frac{1}{2}\eta\right) - V\left(x+\frac{1}{2}\eta\right). \tag{2.31}$$

The advantage of the Wigner formulation lies in the fact that it relates the quantum mechanical picture to the classical picture. For quadratic potentials V, the Wigner equation (2.29) reduces to the Boltzmann equation without collision terms. It can be shown (Markowich and Ringhofer 1989) that the Wigner function converges to the solution of the collisionless Boltzmann equation in the limit of large time and spatial scales. However, from the point of view of device simulation, we are interested in quantum transport equations in regimes which are quite far away from the classical picture. Here, the advantage of the Wigner equation lies in the fact that it allows for a more phenomenological treatment of collision terms and boundary conditions. Clearly the Wigner equation (2.29) is the quantum equivalent of the Boltzmann equation with a parabolic band structure (2.2), since the starting point was the quantum Hamiltonian for a vacuum. In order to describe the motion of the electron in a crystal, a modified Hamiltonian of the form

$$H = -\frac{\hbar^2}{2m}\Delta_x - e(V_L + V) \tag{2.32}$$

has to be considered, where V_L denotes the potential due to a periodic crystal lattice (Ashcroft and Mermin 1976). So

$$V_L(x + \gamma z_j) = V_L(x), \quad j = 1,2,3 \tag{2.33}$$

holds, where the z_j are the lattice directions and γ is the length scale of the lattice, chosen such that $\det(Z) = 1$, where $Z = (z_1, z_2, z_3)$. It can be shown by using a Bloch wave decomposition (Arnold, Degond, Markowich and Steinrück 1989, Poupaud and Ringhofer 1995, Markowich, Mauser and Poupaud 1994) that the projection of the wave function onto the mth energy band satisfies the Schrödinger equation

$$i\hbar\partial_t\psi = \left(\frac{\gamma}{2\pi}\right)^3 \int_{\mathbf{B}} \mathrm{d}k \sum_m \psi(x+\gamma Zm)\varepsilon_m(k)e^{-i\gamma k^T Zm} - eV\psi, \tag{2.34}$$

where $\varepsilon_m(k)$ is the mth eigenvalue of the Hamiltonian $H_L = -(\hbar^2/2m)\Delta + V_L$ together with quasi-periodic boundary conditions. Here $\mathbf{B}$ denotes the Brillouin zone, the unit cell of the dual crystal lattice defined as $\mathbf{B} = Z^{-T}[-\pi/\gamma, \pi/\gamma]^3$. Performing the Wigner transformation for the mixed state, as before, now gives the Wigner equation in a crystal of the form

$$\partial_t w + \operatorname{div}_x \left(\int_{-1/2}^{1/2} \mathrm{d}s \sum_m \hat{v}(m) w(x + \gamma s Z m, k, t) \exp(i\gamma k^T Z m) \right) + \frac{ie}{\hbar} \delta V \left(x, \frac{1}{2i} \nabla_k \right) w = 0, \tag{2.35}$$

where all functions are now periodic in the wave vector k on the Brillouin zone $\mathbf{B}$ and Fourier transforms and pseudo-differential operators are appropriately reformulated as

$$\begin{aligned} v(k) &= \sum_m \hat{v}(m) \exp(i\gamma k^T Z m), \\ \delta V\left(x, \frac{1}{2i}\nabla_k\right) w(x,k,t) &= \\ \left(\frac{\gamma}{2\pi}\right)^3 \sum_n \int_{\mathbf{B}} \mathrm{d}k' \; &\delta V\left(x, \frac{\gamma}{2} Z n\right) w(x,k',t) \exp(i\gamma (k-k')^T Z n). \end{aligned} \tag{2.36}$$

Since the length scale γ of the crystal lattice will be small even for quantum mechanical simulations, the formal limit $\gamma \to 0$ is used in equation (2.35) for actual simulations giving

$$\partial_t w + \operatorname{div}_x (v(k) w) + \frac{ie}{\hbar} \delta V \left(x, \frac{1}{2i} \nabla_k \right) w = 0. \tag{2.37}$$

In the limit $\gamma \to 0$, the Brillouin zone becomes infinite and the definition of Fourier transforms and pseudo-differential operators reverts to (2.31).

Modelling the scattering processes of electrons with phonons quantum mechanically is a much more complicated task. Most models lead to equations which are too high-dimensional to be actually used in the simulation of devices (Ferry and Grubin 1995). Therefore, we are in practice reduced to two approaches to formulating the quantum Boltzmann equation

$$\partial_t w + \operatorname{div}_x (v(k) w) + \theta[V] w = Q(w), \quad \theta[V] w = \frac{ie}{\hbar} \delta V \left(x, \frac{1}{2i} \nabla_k \right) w, \tag{2.38}$$

namely the relaxation time model and the Fokker–Planck term. The relaxation time model is, as in the classical case, given by

$$\begin{aligned} Q(w) &= \frac{1}{\tau} \left(\frac{n}{n_0} w_0 - w \right), \\ n(x,t) &= \int_{\mathbb{R}^3} \mathrm{d}k \; w(x,k,t), \end{aligned}$$

$$n_0(x) \quad = \quad \int_{\mathbb{R}^3} \mathrm{d}k \; w_0(x,k), \tag{2.39}$$

where $w_0(x,k)$ denotes the quantum mechanical thermal equilibrium density, the quantum equivalent to the Maxwellian. The Fokker–Planck term model is given by

$$Q(w) = \frac{1}{\tau} \operatorname{div}_k \left(\frac{mT_0}{\hbar^2} \nabla_k w + kw \right), \tag{2.40}$$

where T_0 denotes the lattice temperature.

Thermal equilibrium

To carry out actual simulations it is necessary to compute a quantum mechanical thermal equilibrium solution. This is necessary for two reasons. First, the thermal equilibrium solution w_0 is used in the relaxation time approximation (2.39). Second, transient simulations are started by using the thermal equilibrium as initial datum. For a mixed state, the thermal equilibrium density matrix is defined by

$$\rho_{TE}(r,s) = \sum_j \sigma(\omega_j)\psi_j(r)\psi_j(s), \tag{2.41}$$

where the ψ_j and l_j are the eigenfunctions and eigenvalues of the quantum Hamiltonian and $\sigma(\omega)$ is the statistical distribution. For Boltzmann statistics, σ is of the form $\sigma(\omega) = \exp(-\omega/T_0)$.

3. Numerical methods for semiclassical and classical transport

In this section we describe two types of approaches to simulating semiconductor devices based on the semiclassical Boltzmann equation (2.17). Both approaches are more or less restricted to the case of parabolic band structures, so (2.2) is assumed. A relatively easy generalization is to assume a more general quadratic band structure of the form $\varepsilon(k) = (\hbar^2/2m)k^T Zk$ with a positive definite symmetric matrix Z. This corresponds to a Taylor expansion of the band energy ε for small wave vectors k and leads to the so-called effective mass approximation. Since generalizing the presented numerical methods to this case is straightforward, it will not be considered separately here. The discussed methods cannot be expected to represent the physical transport picture as accurately as a complete Monte Carlo simulation in all possible cases. They have, however, the great advantage of dealing with deterministic computational models that possess a well defined steady state. At the same time, they seem to give a reasonably accurate transport description for current device dimensions, as verified by comparisons with experiments and Monte Carlo calculations.

3.1. Series expansion methods

Series expansion methods for the Boltzmann transport equation have the advantage that they give, in some sense, a direct extension of macroscopic models such as the drift–diffusion system and the hydrodynamic system, to be discussed in the next section. Usually a spectral Galerkin approach is used in the wave vector direction, while some other standard finite difference or finite element discretization is employed in the spatial and time directions. Most series expansion methods assume a parabolic band structure (2.2). So, after an appropriate scaling, the wave vector can be identified with the velocity vector. We will discuss series expansion methods on the scaled Boltzmann transport equation

$$\lambda^2 \partial_t f + \lambda v \cdot \nabla_x f - \lambda E \cdot \nabla_v f = Q(f), \tag{3.1}$$

where λ denotes the Knudsen number, the ratio of the mean free path to the length scale of the simulation domain. Most expansion methods for the Boltzmann transport equation use the assumption of isotropic scattering, that is, that the scattering rates $s(v, v')$ in (2.3) as well as the Maxwellian depend only on the energy ε. So, after scaling, and assuming a parabolic band structure, the collision integral $Q(f)$ in (3.1) and the Maxwellian are of the form

$$\begin{aligned} Q(f)(x,v,t) &= \int_{\mathbb{R}^3} \mathrm{d}v \ \left\{ s(|v|,|v'|)\Big(Mf'(1-f) - M'f(1-f')\Big)\right\}, \\ M(v) &= \exp\left(\frac{-|v^2|}{2}\right). \end{aligned} \tag{3.2}$$

One of the drawbacks of series expansion methods is that the evaluation of the collision integral is quite complicated and expensive if no Monte Carlo approach is used. The dependence of the integral kernel on the energy alone, reduces the complexity of this problem considerably once polar coordinates

$$\begin{aligned} v &= (r\cos\theta, r\sin\theta\cos\phi, r\sin\theta\sin\phi)^T, \\ &\quad \text{where} \quad \theta \in [0,\pi], \quad \phi \in [-\pi,\pi], \quad r \in [0,\infty), \end{aligned} \tag{3.3}$$

are used. In polar coordinates, the collision integral Q then becomes

$$\begin{aligned} &Q(f)(x,r,\theta,\phi,t) = \\ &\int_0^\infty \mathrm{d}r' \int_0^\pi \mathrm{d}\theta' \int_{-\pi}^{\pi} \mathrm{d}\phi' \ r'^2 \sin(\theta) s(r,r')\Big(Mf'(1-f) - M'f(1-f')\Big), \end{aligned} \tag{3.4}$$

and the integration over the angular variables can be carried out explicitly, giving

$$Q(f)(x,r,\theta,\phi,t) = \int_0^\infty \mathrm{d}r' \ r'^2 s(r,r')\Big(MF'(1-f) - M'f(4\pi - F')\Big), \tag{3.5}$$

where $F(x,r,t)$ denotes the average of the density function over spheres of equal energy

$$F(x,r,t) = \int_0^{\pi} \mathrm{d}\theta \int_{-\pi}^{\pi} \mathrm{d}\phi \ \sin(\theta) f(x,r,\theta,\phi,t), \tag{3.6}$$

and again the terms f' and F' mean that the corresponding functions are evaluated at (r', θ', ϕ') rather than at (r, θ, ϕ). The advantage of the use of polar coordinates lies in the fact that the collision integral is now one-dimensional and the collision terms for scattering of acoustic and polar optical phonons, whose scattering rates are of the form

$$s(r,r') = \sum_{j=-1}^{1} \gamma(|j|)\delta(r^2 - r'^2 - j\omega_{ph}), \tag{3.7}$$

with ω_{ph} the energy of emission/absorption of a polar optical phonon, amount to pointwise evaluation of F. The Boltzmann transport equation in polar coordinates takes the form

$$\lambda^2 \partial_t f + \lambda r a \cdot \nabla_x f - \lambda E \cdot (a\partial_r f + b\partial_\theta f + c\partial_\phi f) = Q(f), \tag{3.8}$$

where

$$a = \frac{1}{r} v = \begin{pmatrix} \cos\theta \\ \sin\theta\cos\phi \\ \sin\theta\sin\phi \end{pmatrix}, \quad b = \frac{1}{r}\begin{pmatrix} -\sin\theta \\ \cos\theta\cos\phi \\ \cos\theta\sin\phi \end{pmatrix}, \quad c = \frac{1}{r\sin\theta}\begin{pmatrix} 0 \\ -\sin\phi \\ \cos\phi \end{pmatrix}.$$

Starting with Odeh, Gnudi, Baccarani, Rudan and Ventura (Ventura, Gnudi and Baccarani 1991, Ventura, Gnudi, Baccarani and Odeh 1992), and continuing with the work of Goldsman and Frey (Goldsman, Wu and Frey 1990, Goldsman, Henrickson and Frey 1991), spherical harmonic expansions of the Boltzmann transport equation in polar coordinates have been used with great success, meaning that good agreement with Monte Carlo simulations has been achieved for realistic devices using only a relatively small number of terms. We recall that spherical harmonic functions take the form

$$S_n(\theta,\phi) = L_n(\cos\theta)(\sin\theta)^{n_2} \exp(in_2\phi), \quad n = (n_1, n_2), \tag{3.9}$$

where L_n is the associated Legendre polynomial of degree (n_1, n_2). Thus L_n is a polynomial of degree n_1 satisfying the orthogonality relation

$$\int_{-1}^{1} L_{n_1,n_2}(y) L_{\nu_1,n_2}(y)(1-y^2)^{n_2}\,\mathrm{d}y = \frac{1}{2\pi}\delta(n_1 - \nu_1), \tag{3.10}$$

and consequently the spherical harmonics satisfy

$$\int_{-\pi}^{\pi} \mathrm{d}\phi \int_0^{\pi} \mathrm{d}\theta \sin(\theta) S_n^*(\theta,\phi) S_\nu(\theta,\phi) = \delta(n-\nu). \tag{3.11}$$

Expanding the density function f in spherical harmonics gives

$$f \approx f^N = \sum_{n \in N} f_n(x,r,t) S_n(\theta,\phi), \tag{3.12}$$

where N denotes some suitable index set, and using the standard Galerkin approach, we find

$$\lambda^2 \partial_t f_n + \lambda r A_{inm} \partial_{x_i} f_m - E_i \left(A_{inm} \partial_r f_m + B_{inm} f_m \right) = q_n, \tag{3.13}$$

where the summation convention is used in (3.13) and the coefficients A_{inm} and $B_{inm}(r)$ are given by

$$A_{inm} = \int_0^\pi d\theta \int_{-\pi}^\pi d\phi \ \sin(\theta) S_n a_i S_m, \tag{3.14a}$$

$$B_{inm} = \int_0^\pi d\theta \int_{-\pi}^\pi d\phi \ \sin(\theta) S_n (b_i \partial_\theta S_m + c_i \partial_\phi S_m) \tag{3.14b}$$

$$q_n = \int_0^\infty dr' \ r'^2 s(r, r') \left(M f_0' \left(4\pi\delta(n) - \sqrt{4\pi} f_n \right) - M' f_n \left(4\pi - \sqrt{4\pi} f_0' \right) \right). \tag{3.14c}$$

Stability and discretization

Equation (3.13) represents a hyperbolic first-order system in the spatial and time variables (x, t) and the energy variable $r = |v|$. In principle, the system (3.13) could be discretized by any number of methods suitable for hyperbolic systems. In Ventura et al. (1991) and (1992), standard finite differences are used in all variables. However, in addition to exhibiting the usual stiffness of PDE discretizations, (3.13) is extremely stiff in time close to the drift–diffusion regime, for small values of the Knudsen number λ, and in space for large values of the electric field E. It therefore pays to investigate the stability properties of the system (3.13) before writing down any approximation scheme. We will now give a simple linear stability estimate, first derived by Poupaud (1991), which indicates how to discretize the system in the spatial, time and energy variables. The Galerkin approach implies the equation

$$2 \int_0^\pi d\theta \int_{-\pi}^\pi d\phi \ \sin(\theta) f^N \left(\lambda^2 \partial_t f^N + \lambda v \cdot \nabla_x f^N - \lambda E \cdot \left(a \partial_r f^N + b \partial_\theta f^N + c \partial_\phi f^N \right) - Q(f^N) \right) = 0. \tag{3.15}$$

Integrating (3.15) by parts yields

$$\lambda^2 \partial_t G + \int_0^\pi d\theta \int_{-\pi}^\pi d\phi \ \sin(\theta) \lambda \left(\mathrm{div}_x \left(v f^{N2} \right) - E \cdot a \partial_r \left(f^{N2} \right) \right) + f^{N2} E \cdot \left(\partial_\theta (b \sin\theta) + \partial_\phi (\sin\theta c) \right) - 2 f^N Q \left(f^N \right) = 0, \tag{3.16}$$

where

$$G(x,r,t) = \int_0^\pi d\theta \int_{-\pi}^\pi d\phi \ \sin(\theta) f^{N2} = \sum_{n\in N} f_n(x,r,t)^2 \tag{3.17}$$

is the norm of the coefficient vector (f_n).

Because of the properties of polar coordinates, we have

$$\partial_\theta(b\sin\theta) + \partial_\phi(\sin\theta c) = -2\frac{\sin(\theta)}{r}a, \tag{3.18}$$

and (3.16) can be rewritten as

$$\begin{aligned}&\lambda^2\partial_t G\\ &\quad + \int_0^\pi d\theta \int_{-\pi}^\pi d\phi \ \sin(\theta)\lambda\left(\operatorname{div}_x\left(v f^{N2}\right) - E\cdot a\frac{1}{r^2}\partial_r\left(r^2 f^{N2}\right)\right)\\ &\quad -2f^N Q\left(f^N\right) = 0.\end{aligned} \tag{3.19}$$

Using the scaled Maxwellian $M(r) = \exp(-r^2/2)$, the term $r^{-2}\partial_r(r^2 f^{N2})$ can be rewritten as $r^{-2}\partial_r(r^2 f^{N2}) = (M/r^2)\partial_r((r^2/M)f^{N2}) - rf^{N2}$ and, since $v = ar$ and $E = -\nabla_x V$ holds, we have

$$E\cdot a\frac{1}{r^2}\partial_r\left(r^2 f^{N2}\right) = E\cdot a\frac{M}{r^2}\partial_r\left(\frac{r^2}{M}f^{N2}\right) + \nabla_x V\cdot v f^{N2}. \tag{3.20}$$

Thus equation (3.16) can be written in conservation form as

$$\begin{aligned}&\lambda^2\partial_t G\\ &\quad + \int_0^\pi d\theta \int_{-\pi}^\pi d\phi \ \sin(\theta)\lambda\left(e^{-V}\operatorname{div}_x\left(e^V v f^{N2}\right) - E\cdot a\frac{M}{r^2}\partial_r\left(\frac{r^2}{M}f^{N2}\right)\right)\\ &\quad -2f^N Q(f^N) = 0.\end{aligned} \tag{3.21}$$

Next, we note the following basic identity for the linearized collision operator $Q(f) = \int_{\mathbb{R}^3} dvs[Mf' - M'f]$:

$$\begin{aligned}&2\int_0^\pi d\theta \int_{-\pi}^\pi d\phi \ \sin(\theta) r^2 f Q(f)\\ &= -\int_0^\pi d\theta \int_{-\pi}^\pi d\phi \int_0^\pi d\theta' \int_{-\pi}^\pi d\phi' \ \sin(\theta) r^2 \sin(\theta') r'^2 s M M'\left(\frac{f}{M} - \frac{f'}{M'}\right)^2\\ &< 0,\end{aligned} \tag{3.22}$$

which can be verified by direct calculation. Therefore, multiplying (3.21) by $e^V r^2/M$ and integrating with respect to x and r gives

$$\begin{aligned}&\partial_t \int_\Omega dx \int_0^\infty dr\ e^{-V}\frac{r^2}{M}G(x,r,t) < \\ &\quad -\int_\Omega dx \int_0^\infty dr\ (\partial_t V)e^{-V}\frac{r^2}{M}G(x,r,t) - \frac{1}{\lambda}\int_{\partial\Omega}\int_0^\infty dr\frac{r^2}{M}e^{-V}(\nu\cdot v)f^{N2},\end{aligned} \tag{3.23}$$

where ν denotes the unit outward normal vector of the domain Ω. The second term on the right-hand side of (3.23) has to be controlled by the boundary fluxes and (3.23) yields a Gronwall inequality for the term

$$\int_\Omega \mathrm{d}x \int_0^\infty \mathrm{d}r\ \frac{r^2}{M} e^{-V} G(x,r,t) = \sum_{n\in N} \int_\Omega \mathrm{d}x \int_0^\infty \mathrm{d}r\ \frac{r^2}{M} e^{-V} f_n(x,r,t)^2. \tag{3.24}$$

The significance of the estimate (3.24) lies in the fact that it is independent of the Knudsen number λ, and therefore is a valid stability result even in the stiff case close to the drift–diffusion regime. Therefore, the discretization of the hyperbolic system (3.13) in the x, t and r variables should reflect this estimate. Without going into the explicit details, we will now show how to construct the discretization for system (3.13). Because of the equality (3.18) the matrices A_i and B_i, made up of the coefficients A_{inm} and B_{inm} in (3.13), satisfy

$$B_i + B_i^T = \frac{2}{r} A_i, \quad i = 1, 2, 3, \tag{3.25}$$

and, consequently, the matrices $B_i - (1/r)A_i$ are skew symmetric. In the linear case the terms q_n on the right-hand side of (3.13) are given by a matrix integral operator of the form

$$q_n = \sum_{n\in N} C_{nm} f_m, \tag{3.26}$$

$$\text{where} \quad C_{nm} g(x,r,t) = 4\pi \int_0^\infty \mathrm{d}r'\ r'^2 s\Big(\delta(n)\delta(m) M g' - \delta(n-m) M' g\Big),$$

and, because of (3.22), the matrix operator C satisfies

$$\int_0^\infty \mathrm{d}r\ \frac{r^2}{M} \mathbf{f}^T C \mathbf{f} < 0, \tag{3.27}$$

where $\mathbf{f}$ denotes the vector of coefficients (f_n). The stability estimate for the semi-discretized Boltzmann transport equation suggests rewriting the system as

$$\begin{aligned} &\lambda^2 \left(\partial_t \left(e^{-V/2}\mathbf{f}\right) + \frac{1}{2}(\partial_t V) e^{-V/2}\mathbf{f}\right) + \lambda r A_i \partial_{x_i} \left(e^{-V/2}\mathbf{f}\right) + \\ &\quad \frac{\lambda}{r}\left(M e^{-V}\right)^{1/2} E_i A_i \partial_r \left(r M^{-1/2}\mathbf{f}\right) + \lambda e^{-V/2} E_i \left(B_i - \frac{1}{r} A_i\right) \mathbf{f} = e^{-V/2} C \mathbf{f}. \end{aligned} \tag{3.28}$$

Multiplying (3.28) from the left with $(r^2/M)e^{-V/2}\mathbf{f}^T$ and integrating with respect to x and r gives the stability estimate

$$\partial_t \int_\Omega \mathrm{d}x \int_0^\infty \mathrm{d}r\ \frac{r^2}{M} e^{-V} |\mathbf{f}|^2 = -\int_\Omega \mathrm{d}x \int_0^\infty \mathrm{d}r\ \frac{r^2}{M} (\partial_t V) e^{-V} |\mathbf{f}|^2. \tag{3.29}$$

Therefore, any difference discretization of the system (3.13) should build on differencing the terms $e^{-V/2}\mathbf{f}$ in the spatial and $rM^{-1/2}\mathbf{f}$ in the energy direction. A completely time-implicit scheme for (3.13) would immediately reproduce the stability estimate (3.29). Unfortunately, completely time-implicit schemes are usually too computationally expensive. From (3.28) it is clear, however, that at least the collision term on the right-hand side should be discretized implicitly in time in order to reduce the $O(\lambda^2)$ stiffness of the system to $O(\lambda)$. So, system (3.13) should be discretized as

$$\lambda^2 \partial_t f_n - q_n = \lambda r \exp\left(\frac{V}{2}\right) A_{inm} \partial_{x_i} \left(\exp\left(-\frac{V}{2}\right) f_m\right) - \quad (3.30a)$$

$$\lambda E_i \left(\frac{1}{r} M^{1/2} A_{inm} \partial_r \left(r M^{-1/2} f_m\right) + \left(B_{inm} - \frac{1}{r} A_{inm}\right) f_m\right)$$

$$q_n = \sum_{j=-1}^{1} \gamma(|j|) \int_0^\infty dr' \; r'^2 \delta(r^2 - r'^2 - j\omega_{ph}) \quad (3.30b)$$

$$\left(M f_0' \left(4\pi\delta(n) - \sqrt{4\pi} f_n\right) - M' f_n \left(4\pi - \sqrt{4\pi} f_0'\right)\right),$$

where the terms on the left-hand side are taken implicitly, and a standard hyperbolic scheme, such as Lax–Wendroff, is used on the right-hand side, yielding a Courant–Friedrich–Lewy (CFL) condition of the form $\Delta t/(\lambda \Delta x) <$ const. In the case of a linear collision operator ($1 - f \approx 1$ in (3.30b)), if the mesh size in the energy direction is taken as an integer fraction of the energy ω_{ph} of the emission of a polar optical phonon, the collision operator C becomes a sparse matrix whose LU decomposition can be computed once and reused for every grid point in the x-direction (Ventura et al. 1991, 1992). An alternative to this approach is to use a spectral discretization in the energy direction as well. In order to preserve the stability estimate (3.23), it is necessary to use the function r^2/M as a weight function for the scalar product. This approach has, in principle, been used in Schmeiser and Zwirchmayr (1995) and (1997), although there, Cartesian coordinates in v and Laguerre polynomial basis functions are used. Consequently, the matrix collision operator matrix C becomes a full matrix and the scheme is restricted to relatively few terms in the expansion.

3.2. The hydrodynamic model

The series expansion methods described in the previous section are centred around an almost spherically symmetric density function. Although they present a non-perturbative theory and are always convergent, we can expect slow convergence far from equilibrium, that is, in the case of large group velocities. As device dimensions decrease, the value of the Knudsen number λ increases, and the transport picture is not dominated by the collision term any longer. In order to study ballistic transport in short channels of a transistor, an alternative model is used. This model corresponds to the compressible Euler equations for a gas driven by an external force. This model, not very aptly named the 'hydrodynamic model for semiconductors', is obtained by taking moments of the Boltzmann equation with respect to the wave vector k, corresponding to the macroscopic particle density, the group velocity and the energy. The hydrodynamic model usually assumes a parabolic band structure as given in (2.2). Multiplying the Boltzmann transport equation (2.17) by $1, k$ and $|k|^2$ and integrating with respect to the wave vector k gives the moment equations

$$\lambda^2\partial_t\langle 1\rangle + \lambda\partial_{x_l}\langle k_l\rangle = 0, \quad (3.31a)$$

$$\lambda^2\partial_t\langle k_j\rangle + \lambda\partial_{x_l}\langle k_l k_j\rangle + \lambda E_j\langle 1\rangle = \left\langle \frac{k_j}{f}Q(f)\right\rangle, \quad (3.31b)$$

$$\lambda^2\partial_t\left\langle \frac{1}{2}|k|^2\right\rangle + \lambda\partial_{x_l}\left\langle \frac{1}{2}k_l|k|^2\right\rangle + \lambda E_l\langle k_l\rangle = \left\langle \frac{|k|^2}{2f}Q(f)\right\rangle, \quad (3.31c)$$

where the symbol $\langle\cdot\rangle$ denotes the expectation of a quantity with respect to the density f. So

$$< g >= \int_{\mathbb{R}^3} \mathrm{d}k\ (gf) \quad (3.32)$$

holds and the summation convention is again used in (3.31). (3.31) is regarded as a system for the particle density $n = \langle 1\rangle$, the moment $\langle k\rangle = nv$ and the total energy $W = \langle(1/2)|k|^2\rangle$. The so-called closure problem is then given by expressing the higher order moments $k_j k_l$, $k_l|k|^2$ and the moments of the collision operator Q in terms of the primary variables n, v and W. This is achieved by the assumption that the density function f is approximately equal to a displaced Maxwellian of the form

$$f(x,k,t) \approx \mu(x,t)\exp\left(-\frac{|k-\lambda v(x,t)|^2}{2T(x,t)}\right), \quad (3.33)$$

where v denotes the macroscopic velocity and T the electron temperature. Under assumption (3.33), the higher order moments are of the form

$$\langle k\rangle = \lambda n v, \quad \langle k_j k_l\rangle = n(T\delta_{jl} + \lambda^2 v_j v_l), \quad (3.34)$$

$$W = \frac{1}{2}\left\langle |k|^2\right\rangle = \frac{n}{2}(3T + \lambda^2|v|^2), \quad \left\langle k_l|k|^2\right\rangle = \lambda v_l n\frac{1}{2}(5T + \lambda^2|v|^2),$$

and (3.31) can be written as

$$\partial_t\langle n\rangle + \partial_{x_l}\langle v_l n\rangle = 0 \quad (3.35a)$$

$$\lambda^2\partial_t(nv_j) + \lambda^2\partial_{x_l}(nv_l v_j) + \partial_{x_j}(nT) + E_j n = \left\langle \frac{k_j}{\lambda f}Q(f)\right\rangle \quad (3.35b)$$

$$\partial_t W + \partial_{x_l}(v_l W) + \partial_{x_l}(n v_l T) \qquad (3.35c)$$
$$+\lambda E_l v_l n - \operatorname{div}_x(\kappa \nabla_x T) = \left\langle \frac{|k|^2}{2f\lambda^2} Q(f) \right\rangle .$$

Here the term $\operatorname{div}_x(\kappa\nabla_x T)$ denotes the heat flux which is derived from a higher order perturbation theory, and κ denotes the heat conductivity. The moments of the collision terms are modelled phenomenologically as

$$\left\langle \frac{k_j}{f} Q(f) \right\rangle = -\frac{\lambda}{\tau_p} n v_j, \quad \left\langle \frac{|k|^2}{2f} Q(f) \right\rangle = -\frac{1}{2\tau_W}\left(3(T-T_0)+\lambda^2|v|^2\right) \qquad (3.36)$$

(see Baccarani and Wordeman (1985)).

Note that, for $\lambda \to 0$, the transport terms in (3.35) vanish and $T = T_0$ holds. Thus the drift–diffusion system is recovered in the limit. However, the hydrodynamic model is used in regimes where the active region of the device is of the same order as the mean free path and $\lambda = O(1)$ usually holds. For $\kappa = 0$, neglecting the heat flux, the hydrodynamic model represents a nonlinear hyperbolic system with sound speed $c = (1/\lambda)\sqrt{5T/3}$ which will usually exhibit shocks for $\lambda = O(1)$. For finite values of the heat conductivity κ, these shocks will have a finite width of order $O(\kappa)$ (Gardner 1991*b*). For short and relatively small active regions (of the order of 0.1–0.5 μm) the hydrodynamic model equations (3.35) usually lead to a sufficiently good agreement with Monte Carlo simulations (Gardner 1993*b*).

Steady state calculations

The most common approach to discretizing the hydrodynamic model equations in steady state is upwind box integration (Gardner 1992, Gardner 1991*a*). To this end, the steady state version of the model equations (3.35) is put in conservation form as

$$\partial_{x_l}(v_l G_{lj}) + H_j = 0, \quad j = 0, \ldots, 5, \qquad (3.37)$$

where the G_{lj} and H_j are given by

$$G_{l0} = n \quad \text{and} \quad G_{lj} = \lambda^2 n v_j, \quad j = 1,2,3, \qquad (3.38)$$

$$G_{l4} = (n/2)(5T + \lambda^2|v|^2) - n\partial_{x_l} V, \quad G_{l5} = 0,$$

$$H_0 = 0, \quad H_j = \partial_{x_j}(nT) - n\partial_{x_j} V - \frac{1}{\tau_p} n v_j, \quad j = 1,2,3,$$

$$H_4 = -\nabla_x \cdot (\kappa \nabla_x T) + \frac{1}{2\lambda^2 \tau_W}\left(3(T-T_0)+\lambda^2|v|^2\right),$$

$$\text{and} \quad H_5 = -\operatorname{div}_x(e\nabla_x V) = n - D,$$

where we have already included the Poisson equation for the potential V. Using the upwind box integration method, the terms $\partial_{x_l}(v_l G_{lj})$ in (3.37) are discretized by

$$\partial_{x_l}(v_l G_{lj})(x) = \frac{1}{(\mu_l \Delta x_l)} \delta_l \left((\mu_l v_l)(\mu_l G_{lj}) - \frac{1}{2} |\mu_l v_l| (\delta_l G_{lj}) \right), \tag{3.39}$$

where the discrete difference and averaging operators δ_l and μ_l are defined by

$$\begin{aligned} \delta_l z(x) &= z\left(x + \frac{1}{2}\Delta x_l e_l\right) - z\left(x - \frac{1}{2}\Delta x_l e_l\right), \\ \mu_l z(x) &= \frac{1}{2}\left(z\left(x + \frac{1}{2}\Delta x_l e_l\right) + z\left(x - \frac{1}{2}\Delta x_l e_l\right)\right), \end{aligned} \tag{3.40}$$

where Δx_l denotes the (variable) stepsize and e_l denotes the unit vector in the lth coordinate direction. The derivatives in the terms H_j in (3.37) are usually discretized by standard centred finite differences (Gardner 1991*b*, Gardner 1992). In order to deal with locally large electric fields, a modification of the Scharfetter–Gummel scheme (Selberherr 1981) may be used for $H_j, j = 1, 2, 3$. In this modification the derivative

$$H_j = \partial_{x_j}(nT) - n\partial_{x_j} V - \frac{n}{\tau_p} v_j \tag{3.41}$$

is written as

$$H_j = -\partial_{x_j}\left(\frac{V}{T}\right) \frac{\partial_{x_j}\left(nT \exp(-\frac{V}{T})\right)}{\partial_{x_j}\left(\exp(-\frac{V}{T})\right)} - nV\partial_{x_j}\left(\log(|T|)\right) - \frac{n}{\tau_p} v_j. \tag{3.42}$$

The derivatives in (3.42) are then discretized by using standard differences. For constant temperature T the discretization of (3.42) then reduces to the classical Scharfetter–Gummel scheme, which has the advantage of correctly performing the right upwinding in the direction of the electric field $E = -\nabla_x V$.

After carrying out the discretization, a large sparse nonlinear system of algebraic equations has to be solved. After linearization this leads to the solution of the linear system

$$J \Delta F = -F(z), \quad z = (n, v, T, V)^T, \tag{3.43}$$

at each step. Here the vector F denotes the discretization of (3.37) on the mesh and the Jacobian J has the block structure

$$J = \begin{pmatrix} \frac{\partial F_0}{\partial n} & \frac{\partial F_0}{\partial v} & 0 & 0 \\ \frac{\partial F_j}{\partial n} & \frac{\partial F_j}{\partial v} & \frac{\partial F_j}{\partial T} & \frac{\partial F_j}{\partial V} \\ \frac{\partial F_4}{\partial n} & \frac{\partial F_4}{\partial v} & \frac{\partial F_4}{\partial T} & \frac{\partial F_4}{\partial V} \\ \frac{\partial F_5}{\partial n} & 0 & 0 & \frac{\partial F_0}{\partial V} \end{pmatrix}. \tag{3.44}$$

In Lanzkron, Gardner and Rose (1991) and Gardner, Jerome and Rose (1989), detailed investigations of the convergence of block iterative methods for the solution of the Newton equations (3.43) have been carried out. The basic result is that block under-relaxation methods, in conjunction with conjugate gradient methods for the individual blocks, perform well as long as the equations for the density n and the velocities v are treated as one block. In this case chaotic under-relaxation methods on parallel architectures also give good results.

3.3. Generalizations of the hydrodynamic model – Grad systems

The relative simplicity of the hydrodynamic model equations and their certain shortcomings, such as the overestimation of velocities close to P-N junctions (see Ringhofer (1997)), suggests a generalization of the underlying principle to higher order moment methods. In the ballistic regime, that is, in the presence of large electron velocities, series expansion methods based on a perturbation of a spherical symmetric density function will in general not perform well. This suggests the introduction of a modified series expansion approach based not on a centred Maxwellian distribution function but on a wave vector displaced Maxwellian instead. This idea was first introduced by Grad (1949, 1958) for the study of the fine structure of shock waves in fluid dynamics. Since the assumption underlying the hydrodynamic model is that the density function is approximately of the form (3.33), it is natural to expand the Boltzmann equation around a Maxwellian distribution function in a stretched variable coordinate system in wave vector space with the macroscopic velocity u as the origin and the square root of the macroscopic temperature T as the stretching factor. Introducing the coordinate transformation

$$(x, v, t) \to (x, w, t), \quad v = \alpha(x,t)w + u(x,t) \tag{3.45}$$

gives the transformed Boltzmann equation

$$\lambda^2 \partial_t f + \lambda v \cdot \nabla_x f - \frac{\lambda}{\alpha} H \cdot \nabla_w f = Q(f) \tag{3.46}$$

$$H = E + \lambda((\partial_t \alpha)w + \partial_t u) + \left(w(\nabla_x \alpha)^T + \frac{\partial u}{\partial x}\right) v, \quad v = \alpha w + u.$$

Equation (3.46) is now approximated by a Galerkin method in the microscopic velocity variable v where α and u are still kept as free parameters. There are two conditions which we have to place on the choice of basis functions and the scalar product in order to obtain a generalization of the hydrodynamic model.

- The lowest order basis function is the displaced Maxwellian $e^{-|w|^2/2}$.
- The Galerkin approach should correspond to taking the moments of the Boltzmann equation (3.46).

This is achieved by choosing basis functions of the form

$$\psi_m = M(w)p_m(w), \quad M(w) = \exp\left(-\frac{|w|^2}{2}\right), \tag{3.47}$$

where the p_m in (3.47) are vector basis polynomials containing the polynomials $1, w$ and $|w|^2$. So

$$\{1, w, |w|^2\} \subseteq \operatorname{span}\{p_0, \dots, p_N\} \tag{3.48}$$

holds. Secondly, the scalar product is taken to be of the form

$$< f, g >= \int_{\mathbb{R}^3} dw \, \frac{1}{M(w)} f^T g. \tag{3.49}$$

Thus, taking the scalar product of the Boltzmann equation with the basis function ψ_m corresponds to integrating against the polynomial p_m, and the moments leading to the hydrodynamic model are reproduced. Expanding the density function

$$f(x, w, t) \approx \alpha^{-3} \sum_n f_n(x, t)\psi_n(w) \tag{3.50}$$

and using the Galerkin procedure yields the system

$$\lambda^2 \partial_t f_m + \lambda \partial_{x_l}[A_{lmn} f_n] + \frac{\lambda}{\alpha} B_{mn} f_n = C_{mn} f_n, \tag{3.51}$$

with

$$\begin{aligned} A_{lmn} &= \langle \psi_m, v_l \psi_n \rangle, \\ B_{mn} &= -\langle \psi_m, \operatorname{div}_w(H\psi_n) \rangle = \int_{\mathbb{R}^3} dw \, \psi_n H \cdot \nabla_w p_m, \\ C_{mn} &= \langle \psi_m, Q\psi_n \rangle. \end{aligned} \tag{3.52}$$

It remains to choose the parameters $\alpha(x, t)$ and $u(x, t)$. In the original Grad system (Grad 1949), they are chosen to correspond to the square root of the temperature and the group velocity:

$$u = \frac{\int_{\mathbb{R}^3} dv \, vf}{\int_{\mathbb{R}^3} dv \, f}, \quad |u|^2 + 3\alpha^2 = \frac{\int_{\mathbb{R}^3} dv \, |v|^2 f}{\int_{\mathbb{R}^3} dv \, f}, \tag{3.53}$$

which implies

$$\int_{\mathbb{R}^3} \mathrm{d}w \; [wf] = 0, \quad \int_{\mathbb{R}^3} \mathrm{d}w \; [|w|^2 - 3]f = 0. \tag{3.54}$$

If we denote the basis functions $M, w_1 M, w_2 M, w_3 M, |w|^2 M$ by $\psi_0, \ldots, \psi_4$, this becomes

$$f_j = 0, \quad j = 1, 2, 3 \quad \text{and} \quad f_4 = \text{const} f_0. \tag{3.55}$$

Thus $f_1, \ldots, f_4$ can be eliminated from the system and the corresponding equations determine the free parameters α and u. By virtue of construction the system reduces to the hydrodynamic model (without heat conduction) if only the basis functions $\psi_0, \ldots, \psi_4$ are used. One of the major problem of Grad systems is that they are not necessarily well-posed. Equation (3.51), together with the constraints (3.55), represents a hyperbolic differential algebraic system. If the coefficient functions $f_1, \ldots, f_4$ are eliminated from the system the resulting equations form a first-order system for the variables $(f_0, \alpha, u_1, u_2, u_3, f_5 \ldots)$ whose linearization can have complex eigenvalues, and therefore modes can grow proportionally to their frequencies. This ill-posedness occurs at quite moderate Mach numbers and has been analysed by Cordier (1994*a*, 1994*b*) for some special sets of basis functions. Several approaches to remedy this problem have been given by Ringhofer (1994*b*, 1994*a*, 1997). They involve relaxing the conditions (3.55) in some way or another, and lead to well-posed problems.

4. Numerical methods for quantum transport

The numerical methods discussed in this section are essentially mirror images of the methods for semiclassical transport from Section 3. The quantum Boltzmann equation (2.38) replaces the semiclassical Boltzmann equation (2.1) and its moment expansion gives the quantum hydrodynamic model. There are, however, several important differences which do not allow us to treat quantum transport phenomena as just a perturbation of semiclassical transport. First, the discretization of the pseudo-differential operator θ in (2.38) is not trivial. The transport term on the left-hand side of (2.38) does not possess classical characteristics. (They would be replaced by the paths in the Feynman path integrals.) Second, because of the nonlocality of the transport operator, the formulation of proper boundary conditions is more complicated than in the classical case. Finally, because of the dispersive nature of the underlying Schrödinger equation, moment models, such as the quantum hydrodynamic equations, will also be dispersive, that is, waves will be able to travel at all speeds. Therefore, the artificial diffusion introduced by a discretization scheme will play a crucial role in its accuracy.

4.1. Discretization of the quantum Boltzmann equation

We now turn to the design of numerical methods for the quantum Boltzmann equation (2.38). There is a variety of possible choices for discretization schemes in the spatial and temporal directions, which will be discussed in more detail later. The more fundamental problem is posed by the discretization of the wave vector k, in particular by the approximation of the pseudo-differential operator θ, the quantum equivalent of the operator $-\nabla_x V \cdot \nabla_k$ in the classical Boltzmann equation. Since the quantum Boltzmann equation does not possess characteristics in the classical sense, and the Wigner function w does not necessarily remain nonnegative (see Tatarski (1983)), a Monte Carlo type approach becomes too complicated to be practically feasible. On the other hand, since the operator θ is defined in terms of Fourier transforms, a spectral discretization using trigonometric basis functions seems natural.

First we note that the quantum Boltzmann equation (2.38) allows for a reduction in dimension. If the potential V is only dependent on the variables $x_1, \ldots, x_d$ with $d = 1$ or 2 (so there is no field pointing in the direction $x_{d+1}, \ldots, x_3$) the quantum Boltzmann equation with both collision terms (2.39),(2.40) allows for solutions of the form

$$w(x, k, t) = \exp\left(-\frac{1}{2}\left(k_{d+1}^2 + \cdots + k_3^2\right)\right) \tilde{w}(x_1, \ldots, x_d, k_1, \ldots, k_d, t). \quad (4.1)$$

The dimensionally reduced quantum Boltzmann equation for $\tilde{w}$ is then of the form

$$\partial_t \tilde{w} + \tilde{k} \cdot \nabla_{\tilde{x}} \tilde{w} + \theta[V]\tilde{w} = Q(\tilde{w}) \quad (4.2a)$$

$$\theta[V]w(\tilde{x}, \tilde{k}, t) = (2\pi)^{-d} \quad (4.2b)$$
$$\int_{\mathbb{R}^d} d\tilde{k}' \int_{\mathbb{R}^d} d\eta \, \frac{i}{h}\delta V\left(\tilde{x}, \frac{h}{2}\eta, t\right) w(\tilde{x}, \tilde{k}, t) \exp\bigl(i\eta \cdot (\tilde{k} - \tilde{k}')\bigr),$$

where $\tilde{x} = (x_1, \ldots, x_d)^T$, $\tilde{k} = (k_1, \ldots, k_d)^T$.

Here we have already used the quantum Boltzmann equation in a scaled and dimensionless form, where h denotes the scaled Planck constant $\hbar$. Of course, the Poisson equation has to be modified accordingly to take into account the integral of the Maxwellian in the directions $k_{d+1}, \ldots, k_3$. Since the reduced quantum Boltzmann equation has the same form as the three-dimensional equation, we will from now on drop the tilde symbol. Following Ringhofer (1990) and (1992), we approximate the Wigner function w by a trigonometric polynomial of the form

$$w \approx w_N(x, k, t) = \sum_{n \in \mathbf{N}} c(x, n, t)\phi_n(k), \quad \mathbf{N} = \{-N+1, \ldots, N\}^d, \quad (4.3)$$

$$\phi_n(k) = \left(\frac{\alpha}{2\pi}\right)^{d/2} \exp(i\alpha n \cdot k).$$

Thus we approximate the L^2 function w by the $(2\pi/\alpha)$-periodic function w_N and, consequently, α will have to go to zero in the limit to achieve convergence. The quantum Boltzmann equation (4.2) is simply approximated by collocation at the appropriate equally spaced nodes. So

$$\partial_t w_N + k \cdot \nabla_x w_N + \theta[V] w_N = Q_N(w_N) \quad \text{at} \quad k = k_j, \tag{4.4}$$

$$k_j = \beta j, \quad j \in \mathbf{N}, \quad \beta = \frac{\pi}{N\alpha},$$

holds. If the relaxation time approximation is used for the collision term Q then the corresponding approximation Q_N has to be modified accordingly to account for the fact that densities are now computed from periodic basis functions. So, in this case,

$$Q_N(w)(x,k,t) = \frac{1}{\tau}\left(\frac{n}{n_0} w_{0N} - w\right), \tag{4.5}$$

$$n = \int_{[-\pi/\alpha,\pi/\alpha]^d} w, \quad n_0 = \int_{[-\pi/\alpha,\pi/\alpha]^d} w_{0N}, \quad M_N = \sum_{n\in\mathbf{N}} c_0(x,n)\phi_n,$$

holds. Here w_{0N} denotes a suitable approximation of the thermal equilibrium density w_0. The advantage of this approach lies in the fact that the highly oscillatory integrals in the definition (4.2b) of the pseudo-differential operator θ can be evaluated exactly. The basis functions ϕ_n satisfy the orthogonality relations

$$\int_{[-\pi/\alpha,\pi/\alpha]^d} \phi_m^* \phi_n = \delta(m-n), \quad \beta^d \sum_{\nu\in\mathbf{N}} \phi_m^*(k_\nu)\phi_n(k_\nu) = \delta_N(m-n), \tag{4.6}$$

where δ_N denotes the Kronecker δ on $\mathbf{N}$ periodically extended over all integers. Using these orthogonality relations, a direct calculation yields

$$\begin{aligned} \theta w_N(x,k,t) &= \sum_{n\in\mathbf{N}} \frac{i}{h} \delta V(x, \frac{\alpha h}{2} n, t) c(x,n,t) \phi_n(k), \\ c(x,n,t) &= \beta^d \sum_{\nu\in\mathbf{N}} \phi_n^*(k_\nu) w_N(x,k_\nu,t). \end{aligned} \tag{4.7}$$

Collecting the function values of the trigonometric polynomial w_N at the collocation points k_ν into a vector W, one obtains the hyperbolic system

$$\partial_t W + \Lambda_j \partial_{x_j} W + B(V) W = QW, \tag{4.8}$$

where the Λ_j are diagonal matrices made up of the jth component of the collocation points k_ν, and the tensor B is given by

$$B(\mu,\nu) = \beta^d \sum_{n\in \mathbf{N}} \frac{i}{h}\delta V\left(x, \frac{\alpha h}{2}n, t\right) \phi_N^*(\beta\mu)\phi_n(\beta\nu). \tag{4.9}$$

Multiplication with the tensor B can now be carried out using FFTs, a significant advantage in higher dimensions.

Spectral accuracy

A complete convergence proof for the semi-discretized scheme can be found in Ringhofer (1990) and (1992), and turns out to be quite tedious. We will here only sketch the consistency of the discretization in order to indicate under what conditions and with what order the scheme is convergent. Note that the discretization scheme (4.8) is somewhat nonstandard. The basis functions ϕ_n are not elements of the same space as the exact solution, since we have approximated the L^2 solution by periodic functions. Let the interpolation operator P be defined by

$$Pw(x,k,t) = \sum_{n\in \mathbf{N}} c(x,n,t)\phi_n(k), \quad Pw(x,k_\nu,t) = w(x,k_\nu,t), \quad \nu \in \mathbf{N}. \tag{4.10}$$

Then the scheme can be formally written as

$$P\Big(\partial_t w_N + k\cdot\nabla_x w_N + \theta[V]w_N - Q_N w_N\Big) = 0. \tag{4.11}$$

Defining the global discretization error as $e = w_N - Pw$ we obtain

$$P\Big(\partial_t e + k\cdot\nabla_x e + \theta[V]e - Q_N e\Big) = L, \tag{4.12}$$

where the local discretization error L is given by

$$\begin{aligned} L &= -P\Big(\partial_t Pw + k\cdot\nabla_x Pw + \theta[V]Pw - Q_N Pw\Big) \\ &= (P\theta - \theta P)w + (Q_N P - PQ)w \end{aligned} \tag{4.13}$$

The interpolation operator P has the representation

$$Pf(k) = \beta^d \sum_{n\in \mathbf{N}} \sum_{\nu\in \mathbf{N}} f(k_\nu)\phi_\nu^*(k_\nu)\phi_\nu(k) \tag{4.14}$$

and, consequently, the interpolant of any $(2\pi/\alpha)$-periodic function f is given by

$$Pf = \sum_{n\in \mathbf{N}} \sum_{s\in Z^d} \hat{f}(n+2Ns)\phi_n, \quad f = \sum_{n\in Z^d} \hat{f}(n)\phi_n. \tag{4.15}$$

The formula (4.15) represents the usual aliasing error. The exact solution w is now smoothly decomposed into a part which vanishes identically outside the interval $[-\pi/\alpha, \pi/\alpha]^d$ and a part which vanishes identically inside a subinterval of $[-\pi/\alpha, \pi/\alpha]^d$, that is,

$$w = w_i + w_o, \quad w_i = 0 \quad \text{for} \quad k \notin [-\frac{\pi}{\alpha+\varepsilon}, \frac{\pi}{\alpha+\varepsilon}]^d,$$
$$w_o = 0 \quad \text{for} \quad k \in [-\frac{\pi}{\alpha+2\varepsilon}, \frac{\pi}{\alpha+2\varepsilon}]^d. \qquad (4.16)$$

So w_o denotes the tail of the distribution and w_i equals w in the smaller domain. Clearly w_i is $(2\pi/\alpha)$-periodic and, therefore, using (4.15),

$$\begin{aligned}(\theta P - P\theta)w_i &= \sum_{n\in\mathbf{N}} \sum_{s\in Z^d-\{0\}} \\ &\quad \hat{w}_i(n+2Ns)\Big(\delta V(x,\alpha n,t) - \delta V(x,\alpha n + 2Ns,t)\Big)\phi_n, \\ &\quad \text{where} \quad w_i = \sum_{n\in Z^d} \hat{w}_i(n)\phi_n, \end{aligned} \qquad (4.17)$$

holds. Note that the sum in equation (4.17) only contains Fourier coefficients with indices larger than N, and therefore

$$\|(\theta P - P\theta)w_i\|_{L^2([-\frac{\pi}{\alpha},\frac{\pi}{\alpha}]^d)} < c_q \|\delta V\|_\infty \|w_i\|_{H_p([-\frac{\pi}{\alpha},\frac{\pi}{\alpha}]^d)} \qquad (4.18)$$

holds, which gives the usual estimate for spectral accuracy of the discretization scheme.

Time discretization
After employing the spectral collocation scheme in the wave vector direction, it remains to discretize the first-order hyperbolic system (4.8) in space and time. Of course, every method for hyperbolic systems would do this job. However, the use of a standard hyperbolic scheme for (4.8) will result in a CFL condition of the form $\Delta t/(\alpha \Delta x) <$ const, which will be prohibitive in practice since $\alpha \to 0$ has to hold for the spectral discretization to be convergent. The best alternative, given in Arnold and Ringhofer (1995*b*) is to employ operator splitting to the semi discretizeed equation (4.8). In the operator splitting approach, one time step of length Δt for the equation (4.8), starting from $W(x, t_n)$ is performed by

$$\partial_t W_1 + \Lambda_j \partial_{x_j} W_1 = 0, \quad W_1(x,t) = W(x,t_n) \qquad (4.19\text{a})$$
$$\partial_t W_2 + B(V) W_2 = Q(W_2), \quad W_2(x,t_n) = W_1(x,t_n+\Delta t) \qquad (4.19\text{b})$$
$$W(x,t_{n+1}) = W_2(x,t_{n+1}), \quad t_{n+1} = t_n + \Delta t. \qquad (4.19\text{c})$$

This discretization is first-order accurate in time. A second-order accurate discretization can be achieved with a slight modification using so-called Strang splitting (Arnold and Ringhofer 1995*a*) . The step (4.19b) represents the solution of a system of ordinary differential equations. This can be achieved using any ODE integrator. (Actually, in the absence of the collision term Q, this step can be carried out exactly.) In theory, the first step

(4.19a) could be carried out exactly as well, since it is given by the shift

$$W_1(x, k_j, t_n + \Delta t) = W(x - k_j\Delta t, k_j, t_n), \tag{4.20}$$

eliminating any type of CFL condition. In practice, since the vector W is given on a fixed mesh on the x-axis, the term $W(x - k_j\Delta t, k_j, t_n)$ will have to be interpolated between the nearest gridpoints. Using second-order interpolation between nearest neighbours in the x-mesh and a first-order ODE integrator for step (4.19b) gives a first-order accurate scheme (Arnold and Ringhofer 1995*b*).

Boundary conditions

One of the major problems in the application of quantum kinetic models to the simulation of actual devices is the appropriate formulation of boundary conditions. In a device, the simulation region will be a bounded domain whose boundaries will consist of contacts, insulating surfaces or artificial boundaries, which are introduced to limit the size of the simulation domain. The quantum Boltzmann equation is nonlocal in the wave vector k and the transport term on the left-hand side of (2.38) does not possess classical characteristics. Nevertheless, the quantum Boltzmann equation allows for wave solutions since, at least in the collisionless case, it is equivalent to the Schrödinger equation. Thus, if care is not taken in the formulation of boundary conditions, artificial reflections of waves at the boundaries will occur, and these spurious waves will propagate back in the interior of the simulation domain. We will first treat the case of an artificial boundary, where the boundary conditions should be such that reflection of waves at the boundary is kept to a minimum. For simplicity, let us consider a one-dimensional model $x \in \mathbb{R}^1, k \in \mathbb{R}^1$ which is obtained from the Schrödinger equation in one spatial dimension. The presented methodology is given in detail in Ringhofer, Ferry and Kluksdahl (1989) and represents a generalization of the approach of Engquist and Majda (1977) for hyperbolic systems to the infinite-dimensional case. In the one-dimensional collisionless case the Wigner equation becomes

$$\partial_t w + k\partial_x w + \theta[V]w = 0, x, k \in R^1. \tag{4.21}$$

We will assume the boundary to be located at $x = 0$ and the simulation domain to be given by the half plane $x > 0$. Generalizations to more than one boundary are straightforward. In the absence of the pseudo-differential operator θ, the absorbing boundary condition would trivially be given by

$$w(x = 0, k, t) = 0, \quad \text{for} \quad k > 0, \tag{4.22}$$

since we assume that no waves enter the domain from outside the region. The same would be true for an equation of the form

$$\partial_t w + k\partial_x w + \Gamma w = 0, \quad \Gamma w = \int_{\mathbb{R}} \mathrm{d}k' \; G(x,t,k,k')w(x,k',t), \tag{4.23}$$

if the operator Γ is block diagonal in the sense that $G(x,t,k,k') = 0$ when $kk' < 0$, since the solution $w(x,k,t)$ for $k < 0$ is completely decoupled from the solution $w(x,k,t)$ for $k > 0$. The goal of the presented approach is to achieve such a decoupling asymptotically for large wave speeds. If we assume a plane wave solution in the x, t plane of the form $w(x,k,t) = g(k)\exp[i\xi(x-\omega t)]$ with a velocity ω and a frequency ξ, $\partial_t w = -i\xi\omega w$ holds and the time derivative will be proportional to the velocity ω. Thus, we formally decouple $k < 0$ from $k > 0$ by expanding the operator in powers of ∂_t^{-1} for 'large ∂_t'. This will be made more precise later. Setting formally

$$u = w - \partial_t^{-1}A[kw], \quad A[f](x,k,t) = \int_{\mathbb{R}} \mathrm{d}k' \; a(x,t,k,k')f(x,k',t), \tag{4.24}$$

we obtain

$$k\partial_x u = (1 - \partial_t^{-1}kA)[k\partial_x w] - k\partial_t^{-1}A_x[kw], \tag{4.25}$$

where the operator A_x arises from differentiating the product, so

$$A_x[f](x,k,t) = \int_{\mathbb{R}} \mathrm{d}k' \; d_x a(x,t,k,k')f(x,k',t) \tag{4.26}$$

holds. Using the differential equation (4.21) yields

$$k\partial_x u = (1 - \partial_t^{-1}kA)(-\partial_t - \theta[V])[w - k\partial_t^{-1}A_x(kw)]. \tag{4.27}$$

Asymptotically, the inverse of the operator $1 - \partial_t^{-1}Ak$ will be given by $1 + \partial_t^{-1}Ak$ and $w = u + \partial_t^{-1}A(ku) + O(\partial_t^{-2})$ holds. So, formally, up to terms of order $O(\partial_t^{-2})$ we obtain

$$\begin{aligned} k\partial_x u &= -\partial_t u + (\partial_t^{-1}kA\partial_t - Ak - \theta)u + O(\partial_t^{-2}u) \\ &= -\partial_t u + (kA - Ak - \theta)u + O(\partial_t^{-2}u). \end{aligned} \tag{4.28}$$

Therefore, we choose the operator A such that it diagonalizes the equation (4.28). If we write the pseudo-differential operator θ in terms of its kernel

$$\begin{aligned} \theta(u)(x,k,t) &= \int_{\mathbb{R}} \mathrm{d}k' \; D(x,t,k-k')u(x,k',t), \\ D(x,t,r) &= \int_{\mathbb{R}} \mathrm{d}\eta \; \frac{i}{\hbar}\delta V\left(x, \frac{h}{2}\eta\right) e^{i\eta r}, \end{aligned} \tag{4.29}$$

then $(k-k')a(x,t,k,k') - D(x,t,k-k') = 0$ has to hold for $kk' < 0$. So, we set

$$a(x,t,k,k') = \begin{cases} \frac{D(x,t,k-k')}{k-k'} & \text{for} \quad kk' < 0, \\ 0 & \text{otherwise,} \end{cases} \tag{4.30}$$

and the absorbing boundary condition reads

$$u = w - \partial_t^{-1} \int_{\mathbb{R}} \mathrm{d}k' \ a(x,t,k,k')k'w(x,k',t) = 0 \quad \text{for} \quad x = 0, \quad k > 0. \tag{4.31}$$

Differentiating (4.31) once with respect to time gives an implementable boundary condition. The above formal manipulations can be made precise to make the whole approach more plausible. Given a solution w of the Wigner equation (4.21), we define u by

$$\partial_t u = \partial_t w - \int_{\mathbb{R}} \mathrm{d}k' \ a(x,t,k,k')k'w(x,k',t), \quad u(x,k) = w(x,k). \tag{4.32}$$

A direct calculation gives the residuals for the inverse transformations

$$R = w - u, \quad S = \partial_t(w-u) - \int_{\mathbb{R}} \mathrm{d}k' \ a(x,t,k,k')k'u(x,k',t) \tag{4.33}$$

$$\partial_t R = A(kw), \quad S = A(kR).$$

Inserting the new variable u into the transport operator and differentiation with respect to time gives

$$\begin{aligned} &\partial_t\Big(\partial_t u + k\partial_x u\Big) \\ &= \partial_t\Big(\partial_t + k\partial_x\Big)w - \Big(\partial_t + k\partial_x\Big)A(kw) \\ &= -\partial_t\theta(w) - \Big(\partial_t + k\partial_x\Big)A(kw) \\ &= -\theta(\partial_t w) - A(k\partial_t w) - kA(k\partial_x w) - \Big(\theta_t w + A_t(kw) - kA_x(kw)\Big) \\ &= -\theta(\partial_t w) - A(k\partial_t w) - kA(\partial_t w + \theta(w)) - \Big(\theta_t w + A_t(kw) - kA_x(kw)\Big) \\ &= -\Gamma(\partial_t w) - L(w), \end{aligned} \tag{4.34}$$

where the operators θ_t, A_t and A_x are the ones obtained from differentiating the kernels with respect to x and t, the block diagonal operator Γ is given by $\Gamma(f) = \theta(f) + A(kf) - kA(f)$ and L is given by $L(f) = \theta_t(f) + A_t(kf) - kA_x(f) - kA(\theta(f))$. Setting $w = u + R$ and $\partial_t w = \partial_t u + A(ku) + S$, and integrating with respect to time gives

$$\begin{aligned} \partial_t u + k\partial_x u + \Gamma(u) &= H, \\ \partial_t H &= \Gamma_t(u) - \Gamma(A(ku)) - \Gamma(A(kR)) - L(u) - L(R), \\ \partial_t R &= A(k(u+R)). \end{aligned} \tag{4.35}$$

The above system is decoupled up to the lower order term H. Imposing the boundary condition $u(x=0,k>0,t)=0$ and inserting a plane wave of the form

$$\begin{aligned} u(x,k,t) &= g_u(k,\omega,\xi)\exp\Big(i\xi(x-\omega t)\Big), \\ H(x,k,t) &= g_H(k,\omega,\xi)\exp\Big(i\xi(x-\omega t)\Big), \\ R(x,k,t) &= g_R(k,\omega,\xi)\exp\Big(i\xi(x-\omega t)\Big) \end{aligned} \tag{4.36}$$

immediately gives that the waves travelling to the right (for $k>0$) have amplitudes of order $O(\omega^{-2})$, that is, $g_u(k,\omega,\xi)=O(\omega^{-2})$ for $k>0$ holds. So, the second-order absorbing boundary condition is of the form

$$\partial_t u(0,k>0,t) = \Big(\partial_t w - A(kw)\Big)(0,k>0,t) = 0. \tag{4.37}$$

In the case of an insulating surface, perfect reflection is imposed instead. So, in this case, the boundary condition $u(0,k,t)=u(0,-k,t)$, or

$$\Big(\partial_t w - A(kw)\Big)(0,k,t) = \Big(\partial_t w - A(kw)\Big)(0,-k,t), \quad \text{for} \quad k>0, \tag{4.38}$$

holds. Finally, in the case of a contact, we will impose a boundary condition modelling the injection of electrons according to a certain distribution. The corresponding boundary condition is then given such that nothing but the injected part of the distribution is reflected. Therefore, if we denote the injection distribution by $f(k)$, the absorbing boundary condition in (4.31) acts on $w-\rho(t)f$, giving

$$\begin{aligned} &\partial_t w - \int_{\mathbb{R}} \mathrm{d}k'\ a(x,t,k,k')k'w(x,k',t) = \qquad\qquad (4.39)\\ &\qquad f(k)\partial_t\rho(t) - \rho(t)\int_{\mathbb{R}} \mathrm{d}k'\ a(x,t,k,k')k'f(k') \quad \text{for} \quad x=0, \quad k>0, \end{aligned}$$

where the function $\rho(t)$ is chosen such that the total charge in the device is conserved.

4.2. Quantum hydrodynamic models

The calculation of quantum transport phenomena via the quantum Boltzmann equation becomes prohibitively expensive in more than two dimensions. However, certain essential effects, such as non-monotone voltage current characteristics or negative differential resistance (Gardner 1993*a*), which are characteristic of the behaviour of quantum devices, can be simulated using much simpler macroscopic models. Like their classical counterpart, these model equations, the so-called quantum hydrodynamic equations (Gardner 1994), are derived from a moment expansion of the underlying kinetic equation. So, in the classical limit for $\hbar \to 0$, they reduce to the hydrodynamic model equations treated previously. Denoting the momentum

$\hbar k$ by p and the corresponding moments by

$$\langle 1\rangle = \int_{\mathbb{R}^3} \mathrm{d}k\ w, \quad \langle p_j\rangle = \int_{\mathbb{R}^3} \mathrm{d}k\ \hbar k_j w,$$

$$\langle p_j p_l\rangle = \int_{\mathbb{R}^3} \mathrm{d}k\ \hbar^2 k_j k_l w, \quad \left\langle |p|^2\right\rangle = \int_{\mathbb{R}^3} \mathrm{d}k\ \hbar^2 |k|^2 w,$$

$$\left\langle p_j |p|^2\right\rangle = \int_{\mathbb{R}^3} \mathrm{d}k\ \hbar^3 k_j |k|^2 w, \tag{4.40}$$

and taking the first three moments of the quantum Boltzmann equation (2.38) gives

$$\partial_t \langle 1\rangle + \frac{1}{m}\partial_{x_l}\langle p_l\rangle = 0 \tag{4.41a}$$

$$\partial_t \langle p_j\rangle + \frac{1}{m}\partial_{x_l}\langle p_l p_j\rangle - e\partial_{x_j}V\langle 1\rangle = \langle p_j Q\rangle \tag{4.41b}$$

$$\partial_t \left\langle |p|^2\right\rangle + \frac{1}{m}\partial_{x_l}\left\langle p_l |p|^2\right\rangle - 2e\partial_{x_l}V\langle p_l\rangle = \left\langle |p|^2 Q\right\rangle. \tag{4.41c}$$

(In (4.41) the summation convention is used again.) The system has to be closed again by expressing the pseudo-expectations $\langle p_l p_j\rangle, \langle p_l |p|^2\rangle, \langle p_j Q\rangle$ and $\langle |p|^2 Q\rangle$ in terms of the primary variables $\langle 1\rangle, \langle p_j\rangle$ and $\langle |p|^2\rangle$. If the Fokker–Planck term (2.40) is used as a collision operator, the moments on the right-hand side of (4.41) become

$$\langle p_j Q\rangle = -\frac{1}{\tau}\langle p_j\rangle, \quad \left\langle |p|^2 Q\right\rangle = \frac{2}{\tau}\left(3mT_0\langle 1\rangle - \left\langle |p|^2\right\rangle\right). \tag{4.42}$$

As in the classical case, closure is achieved by assuming that the Wigner function w is close to a wave vector displaced equilibrium density. Note that the first three moments of the quantum Boltzmann equation are the same as in the classical case. Therefore, quantum effects will enter solely through the closure conditions. If we assume the form of a wave vector displaced equilibrium density, so that

$$w(x,k,t) = w_e\left(x, k - \frac{m}{\hbar}u(x,t)\right) \tag{4.43}$$

holds with some group velocity vector u, we obtain

$$\langle 1\rangle = n, \quad \langle p_j\rangle = mnu_j, \quad \langle p_j p_l\rangle = m^2 n u_j u_l - mP_{jl}, \tag{4.44}$$
$$\left\langle |p|^2\right\rangle = m^2 n|u|^2 + mP =: 2mW, \quad \left\langle p_j |p|^2\right\rangle = 2m^2\left(u_j W - P_{jl}u_l\right),$$

where the P_{jl} and P denote the second moments of the equilibrium density, that is,

$$P_{jl} = -\frac{\hbar}{m}\int_{\mathbb{R}^3} \mathrm{d}k\ k_j k_l w_e, \quad P = -\frac{1}{2}\sum_j P_{jj} \tag{4.45}$$

holds. (It can always be assumed that the equilibrium density is symmetric, which implies that the odd order moments of w_e vanish.) Following Gardner (1994), the approximate equilibrium density is taken as

$$w_e(x,k) = A(x,t) f_c \cdot \left[1 + \hbar^2 \left(-\frac{1}{8mT^2}\Delta_x V + \frac{1}{24mT^3}|\nabla_x V|^2 + \frac{p_k p_l}{24m^2T^3}\partial^2_{x_k x_l} V\right) + O(\hbar^4)\right], \tag{4.46}$$

where $p = \hbar k$ holds and f_c denotes the classical equilibrium density

$$f_c = A(x,t)\exp\left(-\frac{|p|^2}{2mT} + \frac{V}{T}\right). \tag{4.47}$$

The form (4.46) is derived from an $O(\hbar^4)$ approximation of the thermal equilibrium density first given by Wigner (1932). With this form of the equilibrium density the moments P_{jl} and P in (4.44) become

$$P_{jl} = -nT\delta_{jl} - \frac{\hbar^2 n}{12m}\partial^2_{x_j x_l}\log(n) + O(\hbar^4), \quad P = \frac{3}{2}nT + \frac{\hbar^2 n}{24m}\Delta_x \log(n) \tag{4.48}$$

and the quantum hydrodynamic equations become

$$\partial_t n + \partial_{x_l}\langle n u_l\rangle = 0 \tag{4.49a}$$

$$\partial_t\langle m n u_j\rangle + \partial_{x_l}(m n u_l u_j - P_{jl}) - e\partial_{x_j} V n = -\frac{1}{\tau} m n u_j \tag{4.49b}$$

$$\partial_t\left(\frac{1}{2}mn|u|^2 + P\right) + \partial_{x_l}\left(u_l\left(\frac{1}{2}mn|u|^2 + P\right) - P_{lj}u_j + q_l\right) - en\,\partial_{x_l} V u_l = \frac{2}{\tau}\left(3T_0 n - m|u|^2\right). \tag{4.49c}$$

The structure of the quantum hydrodynamic equations is considerably more complex than that of the classical hydrodynamic model. Because of the presence of the term $\frac{\hbar^2 n}{12m}\partial^2_{x_j x_l} n$ in the correction to the stress tensor P_{jk}, the quantum hydrodynamic equations show the same dispersive behaviour as the underlying Wigner or Schrödinger equation. More precisely, an analysis of the linearized problem shows that the corresponding matrix has two hyperbolic (pure imaginary) eigenvalues, two dispersive modes (real eigenvalues which are proportional to the frequency) of order $\hbar^2$ and one parabolic eigenvalue, due to the presence of the heat conduction term $\operatorname{div}_x(\kappa\nabla_x T)$ (Gardner 1993*a*). At present, there are essentially two approaches to discretizing the quantum hydrodynamic system (4.49). The first treats the quantum hydrodynamic equations as a perturbation of the classical hydrodynamic system and uses a discretization appropriate for hyperbolic conservation laws. In this approach, the system is written as

$$\partial_t Z_j + \partial_{x_l} F_{lj}(Z) = R_j(Z), \quad Z = (n, mnu, W_c^T), \tag{4.50}$$

where W_c denotes the classical energy term $W_c = (1/2)(3nT + mn|u|^2)$, and the flux F and the right-hand side R are given by

$$F(Z) = \begin{pmatrix} nu \\ mnuu^T + nT\mathbf{I} \\ (W_c + nT)u \end{pmatrix},$$

$$R(Z) = \begin{pmatrix} 0 \\ n\nabla_x V - \frac{mnu}{\tau} - \frac{n}{3}\nabla_x Q \\ nu^T\nabla_x V - (W_c - \frac{3}{2}nT)/\tau + \mathrm{div}_x(\kappa\nabla_x T) + Q_w \end{pmatrix}, \quad (4.51a)$$

$$Q = \frac{\hbar^2}{2m}\frac{1}{\sqrt{n}}\Delta_x\sqrt{n},$$

$$Q_w = \frac{\hbar^2}{24m\tau}\Delta_x\Big(\log(n)\Big) - \frac{\hbar^2}{24m}\mathrm{div}_x\Big(n\Delta_x u\Big) - \frac{n}{3}u^T\nabla_x Q. \quad (4.51b)$$

The term Q in (4.51b) is referred to as the Bohm potential. Writing the quantum hydrodynamic system in the form (4.51), numerical methods suitable for hyperbolic conservation laws are used. In Chen, Cockburn, Gardner and Jerome (1995), a discontinuous Galerkin method is used in the spatial direction to simulate hysteresis effects in resonant tunnelling diodes. The time variable is discretized by a second-order explicit Runge–Kutta method, where each of the intermediate stages are projected orthogonally onto the manifold given by the Poisson equation.

A different approach to the discretization of the quantum hydrodynamic system is used in Gardner (1993*a*) for one-dimensional steady state simulations. Here, as in the classical case, the system is written in a form suitable for upwinding methods as

$$\partial_x(uG_j) + H_j + S_j = 0, \quad j = 0, \dots, 3, \quad (4.52)$$

with the G_j, h_j and S_j given by

$$G_0 = n, \qquad G_1 = mnu,$$

$$G_2 = \frac{5}{2}nT + \frac{1}{2}mnu^2 - \frac{\hbar^2 n}{m}\partial_x^2\log(n) - nV, \qquad G_3 = 0 \quad (4.53a)$$

$$H_0 = 0, \qquad H_1 = \partial_x(nT) - \partial_x\left(\frac{\hbar^2 n}{12m}\partial_x^2\log(n)\right) - n\partial_x V,$$

$$H_2 = -\partial_x(\kappa\partial_x T), \qquad H_3 = e\partial_x^2 V \quad (4.53b)$$

$$S_0 = 0, \qquad S_1 = \frac{mnu}{\tau},$$

$$S_2 = \frac{3}{2}nT + \frac{1}{2}mnu^2 - \frac{\hbar^2 n}{24m}\partial_x^2\log(n) - \frac{3}{2}nT_0,$$

$$S_3 = e^2(N_D - N_A - n). \quad (4.53c)$$

Here the Poisson equation has already been included in the system. In this form the one-dimensional quantum hydrodynamic equations are discretized by conservative upwinding similar to the classical case. Thus, the term $\partial_x(uG_j)$ is discretized as

$$\partial_x(uG_j) \approx \frac{1}{\mu_x \Delta x} \delta_x \left((\mu_x u)(\mu_x G_j) - \frac{1}{2} |\mu_x u| (\delta_x G_j) \right), \tag{4.54}$$

where the averaging operator μ_x and the difference operator δ_x are defined as in (3.40). Notice, that the philosophy behind the upwind discretization differs from the one presented above since the quantum correction term $(\hbar^2 n/m)\partial_x^2 \log(n)$ is included in the transport term G. This is only possible in the one-dimensional case and results in the dispersive modes of the quantum hydrodynamic system being heavily damped out through the artificial diffusion produced by upwinding method. However, this one-dimensional scheme has proven nevertheless to be quite successful in the simulation of quantum mechanical phenomena, such as negative differential resistance in actual devices.

REFERENCES

A. Arnold and C. Ringhofer (1995*a*), 'An operator splitting method for the Wigner–Poisson problem', *SIAM J. Numer. Anal.* **32**, 1895–1921.

A. Arnold and C. Ringhofer (1995*b*), 'Operator splitting methods applied to spectral discretizations of quantum transport equations', *SIAM J. Numer. Anal.* **32**, 1876–1894.

A. Arnold, P. Degond, P. Markowich and H. Steinrück (1989), 'The Wigner–Poisson equation in a crystal', *Appl. Math. Lett.* **2**, 187–191.

N. Ashcroft and M. Mermin (1976), *Solid State Physics*, Holt-Saunders, New York.

G. Baccarani and M. Wordeman (1985), 'An investigation of steady state velocity overshoot effects in Si and GaAs devices', *Solid State Electr.* **28**, 407–416.

Z. Chen, B. Cockburn, C. Gardner and J. Jerome (1995), 'Quantum hydrodynamic simulation of hysteresis in the resonant tunneling diode', *J. Comput. Phys.* **117**, 274–280.

S. Cordier (1994*a*), 'Hyperbolicity of Grad's extension of hydrodynamic models for ionospheric plasmas I: the single species case', *Math. Mod. Meth. Appl. Sci.* **4**, 625–645.

S. Cordier (1994*b*), 'Hyperbolicity of Grad's extension of hydrodynamic models for ionospheric plasmas II: the two species case', *Math. Mod. Meth. Appl. Sci.* **4**, 647–667.

B. Engquist and A. Majda (1977), 'Absorbing boundary conditions for the numerical simulation of waves', *Math. Comput.* **31**, 629–651.

D. Ferry and H. Grubin (1995), 'Modelling of quantum transport in semiconductor devices', *Solid State Phys.* **49**, 283–448.

C. Gardner (1991*a*), 'Numerical simulation of a steady state electron shock wave in a submicrometer semiconductor device', *IEEE Trans. Electr. Dev.* **38**, 392–398.

C. Gardner (1991*b*), Shock waves in the hydrodynamic model for semiconductor devices, in *IMA Volumes in Mathematics and its Applications*, Vol. 59, pp. 123–134.

C. Gardner (1992), Upwind simulation of a steady state electron shock wave in a semiconductor device, in *Viscous Profiles and Numerical Methods for Shock Waves* (M. Shearer, ed.), pp. 21–30.

C. Gardner (1993*a*), The classical and the quantum hydrodynamic models, in *Proc. Int. Workshop on Computational Electronics, Leeds 1993* (J. Snowden, ed.), pp. 25–36.

C. Gardner (1993*b*), 'Hydrodynamic and Monte Carlo simulations of an electron shock wave in a $1\mu m\ n^+ - n - n^-$ diode', *IEEE Trans. Electr. Dev.* **40**, 455–457.

C. Gardner (1994), 'The quantum hydrodynamic model for semiconductor devices', *SIAM J. Appl. Math.* **54**, 409–427.

C. Gardner, J. Jerome and D. Rose (1989), 'Numerical methods for the hydrodynamic device model', *IEEE Trans. CAD* **8**, 501–507.

N. Goldsman, L. Henrickson and J. Frey (1991), 'A physics based analytical–numerical solution to the Boltzmann equation for use in semiconductor device simulation', *Solid State Electr.* **34**, 389.

N. Goldsman, J. Wu and J. Frey (1990), 'Efficient calculation of ionization coefficients in silicon from the energy distribution function', *J. Appl. Phys.* **68**, 1075.

H. Grad (1949), 'On the kinetic theory of rarefied gases', *Comm. Pure Appl. Math.* **2**, 331–407.

H. Grad (1958), 'Principles of the kinetic theory of gases', *Handbooks Phys.* **12**, 205–294.

A. Kersch and W. Morokoff (1995), *Transport Simulation in Microelectronics*, Birkhäuser, Basel.

P. Lanzkron, C. Gardner and D. Rose (1991), 'A parallel block iterative method for the hydrodynamic device model', *IEEE Trans. CAD* **10**, 1187–1192.

P. Markowich and C. Ringhofer (1989), 'An analysis of the quantum Liouville equation', *ZAMM* **69**, 121–127.

P. Markowich, N. Mauser and F. Poupaud (1994), 'A Wigner function approach to semiclassical limits', *J. Math. Phys.* **35**, 1066–1094.

P. Markowich, C. Ringhofer and C. Schmeiser (1990), *Semiconductor Equations*, Springer.

F. Poupaud (1991), 'Diffusion approximation of the linear Boltzmann equation: analysis of boundary layers', *Asympt. Anal.* **4**, 293–317.

F. Poupaud and C. Ringhofer (1995), 'Quantum hydrodynamic models in semiconductor crystals', *Appl. Math. Lett.* **8**, 55–59.

C. Ringhofer (1990), 'A spectral method for the numerical solution of quantum tunneling phenomena', *SIAM J. Numer. Anal.* **27**, 32–50.

C. Ringhofer (1992), 'On the convergence of spectral methods for the Wigner–Poisson problem', *Math. Mod. Meth. Appl. Sci.* **2**, 91–111.

C. Ringhofer (1994*a*), Galerkin methods for kinetic equations in time variant coordinate systems, in *Proc. 'Mathematical Methods in Semiconductor Simulation'* (R. Natalini, ed.), pp. 32–49.

C. Ringhofer (1994*b*), A series expansion method for the Boltzmann transport equation using variable coordinate systems, in *Proc. Int. Wkshp. on Comp. Electr.* (S. Goodnick, ed.), Portland, pp. 128–132.

C. Ringhofer (1997), 'An adaptive Galerkin procedure for the Boltzmann transport equation', *Math. Mod. Meth. Appl. Sci.* To appear.

C. Ringhofer, D. Ferry and N. Kluksdahl (1989), 'Absorbing boundary condition for the simulation of quantum transport phenomena', *Transp. Theory and Stat. Phys.* **18**, 331–346.

C. Schmeiser and A. Zwirchmayr (1995), Galerkin methods for the semiconductor Boltzmann equation, in *Proc. ICIAM 95, Hamburg.*

C. Schmeiser and A. Zwirchmayr (1997), 'Convergence of moment methods for the semiconductor Boltzmann equation', *SIAM J. Numer. Anal.* To appear.

S. Selberherr (1981), *Analysis of Semiconductor Devices*, 2nd edn, Wiley, New York.

V. Tatarski (1983), 'The Wigner representation of quantum mechanics', *Soviet. Phys. Uspekhi* **26**, 311–372.

M. Taylor (1981), *Pseudodifferential Operators*, Princeton University Press, Princeton.

D. Ventura, A. Gnudi and G. Baccarani (1991), One dimensional simulation of a bipolar transistor by means of spherical harmonics expansions of the Boltzmann equation, in *Proc. SISDEP 91 Conference (Zürich)* (W. Fichtner, ed.), pp. 203–205.

D. Ventura, A. Gnudi, G. Baccarani and F. Odeh (1992), 'Multidimensional spherical harmonics expansions for the Boltzmann equation for transport in semiconductors', *Appl. Math. Lett.* **5**, 85–90.

E. Wigner (1932), 'On the quantum correction for thermodynamic equilibrium', *Phys. Rev.* **40**, 749–759.

Acta Numerica (1997), *pp.* 523–551

Complexity theory and numerical analysis

Steve Smale
Department of Mathematics
City University of Hong Kong
Hong Kong
E-mail: masmale@math.cityu.edu.hk

CONTENTS

Preface

Section 5 is written in collaboration with Ya Yan Lu of the Department of Mathematics, City University of Hong Kong.

1. Introduction

Complexity theory of numerical analysis is the study of the number of arithmetic operations required to pass from the input to the output of a numerical problem.

To a large extent this requires the (global) analysis of the basic algorithms of numerical analysis. This analysis is complicated by the existence of ill-posed problems, conditioning and round-off error.

A complementary aspect ('lower bounds') is the examination of efficiency for all algorithms solving a given problem. This study is difficult and needs a formal definition of algorithm.

Highly developed complexity theory of computer science provides some inspiration to the subject at hand. Yet the nature of theoretical computer science, with its foundations in discrete Turing machines, prevents a simple transfer to a subject where real number algorithms such as Newton's method dominate.

One can indeed be sceptical about a formal development of complexity into the domain of numerical analysis, where problems are solved only to a certain precision and round-off error is central.

Recall that, according to computer science, an algorithm defined by a Turing machine is *polynomial time* if the computing time (measured by the number of Turing machine operations) $T(y)$ on input y satisfies:

$$T(y) \le K(\text{size}(y))^c. \tag{1.1}$$

Here, size(y) is the number of bits of y. A problem is said to be in P (or tractable) if there is a polynomial time algorithm (*i.e.* machine) solving it.

The most natural replacement for a Turing machine operation in a numerical analysis context is an arithmetical operation, since that is the basic measure of cost in numerical analysis. Thus, one can say with little objection that the problem of solving a linear system $Ax = b$ is tractable because the number of required Gaussian pivots is bounded by cn and the input size of the matrix A and vector b is about n^2. (There remain some crucial questions of conditioning to be discussed later.) In this way complexity theory is part of the tradition of numerical analysis.

But this situation is no doubt exceptional in numerical analysis in that one obtains an exact answer, and most algorithms in numerical analysis solve problems only approximately with, say, accuracy $\varepsilon > 0$, or precision $\log \varepsilon^{-1}$. Moreover, the time required depends more typically on the condition of the problem. Therefore it is reasonable for 'polynomial time' to be recast in the form:

$$T(y, \varepsilon) \le K\Big(\mu(y) + \text{size}(y) - \log \varepsilon\Big)^c. \tag{1.2}$$

Here, $y = (y_1, \cdots, y_n)$, with $y_i \in \mathbb{R}$ is the input of a numerical problem, with size$(y) = n$. The accuracy required is $\varepsilon > 0$ and $\mu(y)$ is a number representing the condition of the particular problem represented by y ($\mu(y)$ could be a condition number). There are situations where one might replace μ by $\log \mu$ or $\log \varepsilon^{-1}$ by $\log \log \varepsilon^{-1}$, for example. Moreover, using the notion of approximate zero, described below, the ε might be eliminated.

I see much of the complexity theory ('upper bound' aspect) of numerical analysis conveniently represented by a two-part scheme. Part 1 is the estimate (1.2). Part 2 is an estimate of the probability distribution of μ, and takes the form

$$\text{prob}\left\{y : \mu(y) \geq K\right\} \leq \left(\frac{1}{K}\right)^c, \tag{1.3}$$

where a probability measure has been put on the space of inputs.

Then Parts 1 and 2 combine, eliminating the μ, to give a probability bound of the complexity of the algorithm. The following sections illustrate this theme. One needs to understand the condition number μ with great clarity for the procedure to succeed.

I hope this gives some immediate motivation for a complexity theory of numerical analysis and even to indicate that, all along, numerical analysts have often been thinking in complexity terms.

Now, complexity theory of computer science has also studied extensively the problem of finding lower bounds for certain basic problems. For this one needs a formal definition of algorithm, and the Turing machine begins to play a serious role. That makes little sense when the real numbers of numerical analysis dominate the mathematics. However without too much fuss we can extend the concept of a machine to deal with real numbers, and one can also start dealing with lower bounds of real number algorithms. This last is not so traditional for numerical analysis, yet the real number machine leads to exciting new perspectives and problems.

In computer science, consideration of polynomial time bounds led to the fundamentally important and notoriously difficult problem 'P = NP?'. There is a corresponding problem for real number machines, namely 'P = NP over $\mathbb{R}$?'.

The above is a highly simplified, idealized snapshot of a complexity theory of numerical analysis. Some details follow in the sections below. Also see Blum, Cucker, Shub and Smale (1996), referred to hereafter as the Manifesto, and its references for more background, history and examples.

2. Fundamental theorem of algebra

The fundamental theorem of algebra (FTA) deserves special attention. Its study in the past has been a decisive factor in the discovery of algebraic numbers, complex numbers, group theory and more recently in the development of the foundations of algorithms.

Gauss gave four proofs of this result. The first was in his thesis which, in spite of a gap (see Ostrowski in *Gauss*), anticipates some modern algorithms (see Smale 1981). Constructive proofs of the FTA were given in 1924 by Brouwer and Weyl.

Further, Peter Henrici and his co-workers have given a substantial development for analysing algorithms and a complexity theory for the FTA. See Dejon and Henrici (1969) and Henrici (1977). Also, Collins (1975) gave a contribution to the complexity of FTA. See especially Pan (1996) and McNamee (1993) for historical background and references.

In 1981–82, two articles appeared with approximately the same title, Schönhage (1982) and Smale (1981), which systematically pursued the issue of complexity for the FTA. Coincidentally, both authors gave main talks at the International Congress of Mathematicians, Berkeley 1986, on this subject; see Schönhage (1987) and Smale (1987*a*).

These articles fully illustrate two contrasting approaches.

Schönhage's algorithm is in the tradition of Weyl, with a number of added features which give very good polynomial time complexity bounds. The Schönhage analysis includes the worst case and the implicit model is the Turing machine. On the other hand, the methods have never extended to more than one variable, and the algorithm is complicated. Some subsequent developments in a similar spirit include Renegar (1987*b*), Bini and Pan (1987), Neff (1994), and Neff and Reif (1996). See Pan (1997) for an account of this approach to the FTA.

In contrast, in Smale (1981), the algorithm is based on continuation methods such as Kellog, Li, and Yorke (1976), Smale (1976), Keller (1978), and Hirsch and Smale (1979). See Allgower and Georg (1990, 1993) for a survey. The complexity analysis of the 1981 paper was a probabilistic polynomial time bound on the number of arithmetic operations, but much cruder than Schönhage's. The algorithm, based on Newton's method, was simple, robust, easy to program, and extended eventually to many variables. The implicit machine model was that of Blum, Shub and Smale (1989), hereafter referred to as BSS (1989). Subsequent developments along these lines include Shub and Smale (1985, 1986), Kim (1988), Renegar (1987*b*), Shub and Smale (1993*a*, 1993*b*, 1993*c*, 1996 and 1994), hereafter referred to as Bez I–V, respectively, and Blum, Cucker, Shub and Smale (1997), hereafter referred to as BCSS (1997).

Here is a brief account of some of the ideas of Smale (1981). A point z is called an *approximate zero* if Newton's method starting at z converges well in a certain precise sense; see Section 4 below. The main theorem of this paper asserts the following.

Theorem 2.1 A sufficient number of steps of a modified Newton's method to obtain an approximate zero of a polynomial f (starting at 0) is polynomially bounded by the degree of the polynomial and $1/\sigma$, where σ is the probability of failure.

For the proof, an invariant $\mu = \mu(f)$ of f is defined akin to a condition number of f. Then the proof is broken into two parts.

Part 1: A sufficient number of modified Newton steps to obtain an approximate zero of f is polynomially bounded by $\mu(f)$.

The proof of Part 1 relies on a Loewner estimate related to the Bieberbach conjecture.

Part 2: The probability that $\mu(f)$ is larger than k is less than k^{-c}, some constant c.

The proof of Part 2 uses elimination theory of algebraic geometry and geometric probability theory, Crofton's formula, as in Santaló (1976).

The crude bounds given in Smale (1981), and the mathematics too, were substantially developed in Shub and Smale (1985, 1986).

Here is a more detailed, more developed, complexity theoretic version of the FTA in the spirit of numerical analysis. See BCSS (1997) for the details.

Assume given (or input):

a complex polynomial $f(z) = \sum a_i z^i$ in one complex variable,
a complex number z_0, and an $\varepsilon > 0$.

Here is the algorithm to produce a solution (output) z^* satisfying

$$|f(z^*)| < \varepsilon. \tag{2.1}$$

Let $t_0 = 0$, $t_i = t_{i-1} + \Delta t$, where $\Delta t = 1/k$, for some positive integer k; thus $t_k = 1$, and we have a partition of $[0,1]$. For any polynomial g, we define Newton's method by

$$N_g(z) = z - \frac{g(z)}{g'(z)}, \qquad \text{for all } z \in \mathbb{C}, \text{ such that } g'(z) \neq 0.$$

Let $f_t(z) = f(z) - (1-t)f(z_0)$. Then, generally, there is a unique path ζ_t such that $f_t(\zeta_t) = 0$ all $t \in [0,1]$ and $\zeta_0 = z_0$. Define inductively

$$z_i = N_{f_{t_i}}(z_{i-1}), \qquad i = 1, \dots, k, \quad z^* = z_k. \tag{2.2}$$

It is easily shown that for almost all (f, z_0), z_i will be defined, $i = 1, \dots, k$, provided Δt is small enough. We may say that $k = 1/\Delta t$ is the 'complexity'. It is the main measure of complexity in any case: the problem at hand is, 'how big may we choose Δt and still have z^* satisfying (2.1) and (2.2)?' (*i.e.* so that the complexity is the lowest).

Next a 'condition number' $\mu(f, z_0)$ is defined which measures how close ζ_t is to being ill-defined. (More precisely $\mu(f, z_0) = \operatorname{cosec}\theta$ where θ is the supremum of the angles of sectors about $f(z_0)$ for which the inverse f^{-1} mapping $f(z_0)$ to z_0 is defined.)

Theorem 2.2 A sufficient number k of Newton steps defined in (2.2) to achieve (2.1) is given by

$$k < 26\mu(f, z_0)\left(\log\frac{|f(z_0)|}{\varepsilon} + 1\right).$$

Remark 2.1

(a) We are assuming $0 < \varepsilon < 1/2$.
(b) Note that the degree d of f plays no role, and the result holds for any (f, z_0, ε).
(c) The proof is based on 'point estimates' (α-theory) (see Section 4 below) and an estimate of Loewner from Schlicht function theory. Thus it doesn't quite extend to n variables. It remains a good problem to find the connection between Theorem 2.2 and Theorem 6.1.

For the next result suppose that f has the form

$$f(z) = \sum_{i=0}^{d} a_i z^i, \qquad a_d = 1, \quad |a_i| \le 1.$$

Theorem 2.3 The set of points $z_0 \in \mathbb{C}$, $|z_0| = R > 2$, such that $\mu(f, z_0) > b$, is contained in the union of $2(d-1)$ arcs of total angle

$$\frac{2}{d}\left(\frac{1}{b} + \sin^{-1}\frac{1}{R-1}\right).$$

This result is an estimate on how infrequently poorly conditioned pairs (f, z_0) occur.

It is straightforward to combine Theorems 2.2 and 2.3 to eliminate the μ and obtain both probabilistic and deterministic complexity bounds for approximating a zero of a polynomial. The probabilistic estimate improves the deterministic one by a factor of d. Theorem 2.3 and these results are in Shub and Smale (1985, 1986), but see also BCSS (1997), and Smale (1985).

Remark 2.2 The above-mentioned development might be improved in sharpness in two ways.

(A) Replace the hypothesis on the polynomial f by assuming as in Renegar (1987*b*) and Pan (1996) that all the roots of f are in the unit disk.
(B) Suppose that the input polynomial f is described not by its coefficients, but by a 'program' for f.

3. Condition numbers

The condition number as studied by Wilkinson (1963), important in its own right in numerical analysis, also plays a key role in complexity theory. We review it now, especially some recent developments. For linear systems, $Ax = b$, the condition number is defined in most basic numerical analysis texts.

The Eckart and Young (1936) theorem is central, and may be stated as

$$\|A^{-1}\|^{-1} = d_f(A, \Sigma_n),$$

where A is a non-singular $n \times n$ matrix, with the operator norm on the left and the Frobenius distance on the right. Moreover, Σ_n is the subspace of singular matrices.

The case of 1-variable polynomials was studied by Wilkinson (1963) and Demmel (1987), among others. Demmel gave estimates on the condition number and the reciprocal of the distance to the set of polynomials with multiple roots.

We now give a more general context for condition numbers and give exact formulae for the condition number as the reciprocal of a distance to the set of ill-posed problems following Bez I, II, IV, Dedieu (1997*a*, 1997*b*, 1997*c*) and BCSS (1997).

Consider first the context of the implicit function theorem:

$$F : \mathbb{R}^k \times \mathbb{R}^m \to \mathbb{R}^m, \quad C^1, \quad F(a_0, y_0) = 0,$$
$$\frac{\partial F}{\partial y}(a_0, y_0) : \mathbb{R}^m \to \mathbb{R}^m \quad \text{non-singular.}$$

Then there exists an open neighbourhood $\mathcal{U}$ of a_0 in $\mathbb{R}^k$ and a C^1 map $G : \mathcal{U} \to \mathbb{R}^m$ such that $G(a_0) = y_0$ and $F(a, G(a)) = 0$, for $a \in \mathcal{U}$.

Regard $F_a : \mathbb{R}^m \to \mathbb{R}^m$, $F_a(y) = F(a, y)$, as a system of equations parameterized by $a \in \mathbb{R}^k$. Then a might be the input of a problem $F_a(y) = 0$ with output y; G is the 'implicit function'.

Let us call the derivative $DG(a_0) : \mathbb{R}^k \to \mathbb{R}^m$ the *condition matrix* at (a_0, y_0). Then the condition number $\mu(a_0, y_0) = \mu$, as in Wilkinson (1963), Rice (1966), Wozniakowski (1977), Demmel (1987), Bez IV, and Dedieu (1997*a*), is defined by

$$\mu(a_0, y_0) = \|DG(a_0)\|,$$

the operator norm. Thus $\mu(a_0, y_0)$ is the bound on the infinitesimal output error of the system $F_a(y) = 0$ in terms of the infinitesimal input error.

It is important to note that, while the map G is given only implicitly, the condition matrix

$$DG(a_0) = \frac{\partial F}{\partial y}(a_0, y_0)^{-1} \frac{\partial F}{\partial a}(a_0, y_0)$$

is given explicitly, as is its norm, the condition number $\mu(a_0, y_0)$.

An example, given by Wilkinson, is the case where $\mathbb{R}^k$ is the space of real polynomials f in one variable of degree $\leq k - 1$, and $\mathbb{R}^m = \mathbb{R}$ the space of ζ, $F(f, \zeta) = f(\zeta)$. One may compute that in this case

$$\mu(f, \zeta) = \frac{\left(\sum_0^d |\zeta^i|^2\right)^{1/2}}{|f'(\zeta)|}.$$

For the discussion of several variable polynomial systems, it is convenient to use complex numbers and homogeneous polynomials.

If $f : \mathbb{C}^n \to \mathbb{C}$ is a polynomial of degree d, we may introduce a new variable, say z_0, and define $\hat{f} : \mathbb{C}^{n+1} \to \mathbb{C}$ by $\hat{f}(1, z_1, \ldots, z_n) = f(z_1, \ldots, z_n)$ and $\hat{f}(\lambda z_0, \lambda z_1, \ldots, \lambda z_n) = \lambda^d \hat{f}(z_0, \ldots, z_n)$. Thus $\hat{f}$ is a homogeneous polynomial.

If $f : \mathbb{C}^n \to \mathbb{C}^n$, $f = (f_1, \ldots, f_n)$, $\deg f_i = d_i$, $i = 1, \ldots, n$, is a polynomial system, then by letting $\hat{f}$ equal $(\hat{f}_1, \ldots, \hat{f}_n)$, we obtain a homogeneous system $\hat{f} : \mathbb{C}^{n+1} \to \mathbb{C}^n$. Any zero of f will also be a zero of $\hat{f}$ and justification can be made for the study of such systems in their own right. Thus now we will consider such systems, say $f : \mathbb{C}^{n+1} \to \mathbb{C}^n$ and denote the space of all such f by $\mathcal{H}_d$, $d = (d_1, \ldots, d_n)$, degree $f_i = d_i$.

Recall that an Hermitian inner product on $\mathbb{C}^{n+1}$ is defined by

$$\langle z, w \rangle = \sum_{i=0}^{n} \bar{z}_i w_i, \qquad z, w \in \mathbb{C}^{n+1}.$$

Now, define for degree d homogeneous polynomials f, $g : \mathbb{C}^{n+1} \to \mathbb{C}$,

$$\langle f, g \rangle = \sum_{\alpha} \binom{d}{\alpha}^{-1} \bar{f}_\alpha g_\alpha,$$

where

$$f(z) = \sum_{|\alpha|=d} f_\alpha z^\alpha, \qquad g(z) = \sum_{|\alpha|=d} g_\alpha z^\alpha.$$

Here $\alpha = (\alpha_1, \ldots, \alpha_{n+1})$ is a multi-index and

$$\binom{d}{\alpha} = \frac{d!}{\alpha_1! \cdots \alpha_{n+1}!}, \qquad |\alpha| = \sum_{i=1}^{n+1} \alpha_i.$$

The weighting by the multinomial coefficient is important, and yields unitary invariance of the inner product, as below.

Proposition 3.1 (Reznick 1992) Let $f, N_x : \mathbb{C}^{n+1} \to \mathbb{C}$ be degree d homogeneous polynomials, where $N_x(z) = \langle x, z \rangle^d$. Then $f(x) = \langle f, N_x \rangle$.

Corollary 3.1

$$|f(x)| \le \|f\| \, \|N_x\| \le \|f\| \, \|x\|^d.$$

For f, $g \in \mathcal{H}_d$, define

$$\langle f, g \rangle = \sum \frac{\langle f_i, g_i \rangle}{d_i}, \qquad \|f\| = \langle f, f \rangle^{1/2}.$$

Dedieu has suggested weighting by $1/d_i$ to make the Condition Number Theorem below more natural.

The unitary group $U(n+1)$ is the group of all linear automorphisms of $\mathbb{C}^{n+1}$ which preserve the Hermitian inner product.

There is an induced action of $U(n+1)$ on $\mathcal{H}_d$ defined by

$$(\sigma f)(z) = f(\sigma^{-1} z), \qquad \sigma \in U(n+1), \qquad z \in \mathbb{C}^{n+1}, \quad f \in \mathcal{H}_d.$$

Then it can be proved (see, for instance, BCSS 1997) that

$$\langle \sigma f, \sigma g \rangle = \langle f, g \rangle, \qquad f, g \in \mathcal{H}_d, \qquad \sigma \in U(n+1).$$

This is unitary invariance.

There is a history of this inner product going back at least to Weyl (1932), with contributions or uses in Kostlan (1993), Brockett (1973), Reznick (1992), Bez I–V, Dégot and Beauzamy (1997), Stein and Weiss (1971), Dedieu (1997*a*).

Now we may define the condition number $\mu(f, \zeta)$ for $f \in \mathcal{H}_d$, $\zeta \in \mathbb{C}^{n+1}$, $f(\zeta) = 0$ using the previously defined implicit function context. To be technically correct, one must extend this context to Riemannian manifolds to deal with the implicitly defined projective spaces. See Bez IV for details.

The following is proved in Bez I (but see also Bez III, Bez IV).

Condition Number Theorem 1 Let $f \in \mathcal{H}_d$, $\zeta \in \mathbb{C}^{n+1}$, $f(\zeta) = 0$. Then

$$\mu(f, \zeta) = \frac{1}{d((f, \zeta), \Sigma_\zeta)}.$$

Here the distance d is the projective distance in the space $\{g \in \mathcal{H}_d : g(\zeta) = 0\}$ to the subset where ζ is a multiple root of g.

The proof uses unitary invariance of all the objects. Thus one can reduce to the point $\zeta = (1, 0, \cdots, 0)$, and then to the linear terms, and then to the Eckart–Young theorem.

Dedieu (1997*a*) has generalized this result quite substantially, and has considered sparse polynomial systems (Dedieu 1997*b*). Thus a formula for the eigenvalue problem becomes a special case.

4. Newton's method and point estimates

Say that $z \in \mathbb{C}^n$ is an *approximate zero* of $f : \mathbb{C}^n \to \mathbb{C}^n$ (or $\mathbb{R}^n \to \mathbb{R}^n$, or even for Banach spaces) if there is an actual zero ζ of f (the 'associated zero') and

$$\|z_i - \zeta\| \leq \left(\frac{1}{2}\right)^{2^i - 1} \|z - \zeta\|, \tag{4.1}$$

where z_i is given by Newton's method

$$z_i = N_f(z_{i-1}), \quad z_0 = z, \quad N_f(z) = z - Df(z)^{-1} f(z).$$

Here $Df(z) : \mathbb{C}^n \to \mathbb{C}^n$ is the (Fréchet) derivative of f at z.

An approximate zero z gives an effective termination for an algorithm provided one can determine whether z has the property (4.1).

Towards that end, the following invariant is decisive.

$$\gamma = \gamma(f, z) = \sup_{k \geq 2} \left\| \frac{Df(z)^{-1} D^{(k)} f(z)}{k!} \right\|^{\frac{1}{k-1}}.$$

Here $D^{(k)} f(z)$ is the kth derivative of f considered as a k-linear map and we have taken the operator norm of its composition with $Df(z)^{-1}$; if the expression is not defined, then use $\gamma = \infty$. See Smale (1986), Smale (1987*a*) and Bez I for details of this development.

The invariant γ turns out to be a key element in the complexity theory of non-linear systems. Although it is defined in terms of all the higher derivatives, in many contexts it can be estimated in terms of the first derivative, or even the condition number.

Theorem 4.1 (Smale 1986; see also Traub and Wozniakowski 1979) Let $f : \mathbb{C}^n \to \mathbb{C}^n$, $\zeta \in \mathbb{C}^n$ with $f(\zeta) = 0$. If

$$\gamma(f, \zeta) \|z - \zeta\| \leq \frac{3 - \sqrt{7}}{2},$$

then z is an approximate zero of f with associated zero ζ.

Now let

$$\alpha = \alpha(f, z) = \beta(f, z)\gamma(f, z), \quad \beta(f, z) = \|Df(z)^{-1} f(z)\|.$$

Theorem 4.2 (Smale 1986) There exists a universal constant $\alpha_0 > 0$ such that: if $\alpha(f, z) < \alpha_0$ for $f : \mathbb{C}^n \to \mathbb{C}^n$, $z \in \mathbb{C}^n$, then z is an approximate zero of f (for some associated actual zero ζ of f).

Remark 4.1 This is the result that motivates 'point estimates'. One uses it to conclude that z is an approximate zero f by checking an estimate at the point z only. Nothing is assumed about f in a region or f at ζ.

Remark 4.2 For this definition of approximate zero, the best value of α_0 is probably no smaller than 1/10. See developments, details and discussions in Smale (1987*a*), Wang (1993), Bez I, and BCSS (1997).

Now how might one estimate γ? In Smale (1986, 1987*a*), there is an estimate in terms of the first derivative of f, but an estimate in Bez I seems much more useful. In the context of Section 3, let $f \in \mathcal{H}_d$, $\zeta \in \mathbb{C}^{n+1}$, $f(\zeta) = 0$, and $\gamma_0(f, \zeta) = \|\zeta\|\gamma(f, \zeta)$. The last is to make γ projectively invariant. Recall that $D = \max(d_i)$, $d = (d_1, \ldots, d_n)$, $d_i = \deg f_i$.

Theorem 4.3 (Bez I)

$$\gamma_0(f, \zeta) \leq \frac{D^2}{2} \mu(f, \zeta).$$

Recall that $\mu(f, \zeta)$ is the condition number.

Remark 4.3 One has a similar estimate without assuming $f(\zeta) = 0$.

As a corollary of Theorem 4.3 and a projective version of Theorem 4.1, one obtains the following.

Theorem 4.4 (Separation of zeros, Malajovich-Munoz 1993, BCSS 1997, Dedieu 1997*b*, 1997*d*) Let $f \in \mathcal{H}_d$, and ζ, ζ' be two distinct zeros of f. Then

$$\begin{aligned} d(\zeta, \zeta') &\geq \frac{3-\sqrt{7}}{D^2 \mu(f)}, \\ D &= \max(\deg f_i), \qquad f = (f_1, \ldots, f_n), \\ \mu(f) &= \max_{\zeta,\ f(\zeta)=0} \mu(f, \zeta) \text{ is the condition number of } f, \end{aligned}$$

and d is the distance in projective space.

Remark 4.4 One has also the stronger result

$$d(\zeta, \zeta') \geq \frac{3-\sqrt{7}}{D^2 \mu(f, \zeta)}.$$

Remark 4.5 The strength of Theorem 4.4 lies in its global aspect. It is not asymptotic even though μ is defined just by a derivative.

We end this section by stating a global perturbation theorem (Dedieu (1997*b*)).

Theorem 4.5 Let $f, g : \mathbb{C}^n \to \mathbb{C}^n$, $\zeta \in \mathbb{C}^n$ with $f(\zeta) = 0$. Then, if

$$\alpha(g, \zeta) \leq \frac{|3 - 3\sqrt{17}|}{4} \qquad \text{and}$$

$$\|I - Df(\zeta)^{-1} Dg(\zeta)\| \leq \frac{9-\sqrt{17}}{16},$$

there is a zero ζ' of g such that

$$\|\zeta - \zeta'\| \leq 2\mu(f, \zeta)\|f - g\|.$$

Here everything is affine including $\mu(f, \zeta)$. This uses Theorem 4.2.

5. Linear algebra

Complexity theory is quite implicit in the numerical linear algebra literature. Indeed, numerical analysts have studied the execution time and memory requirements for many linear algebra algorithms. This is particularly true for direct algorithms that solve a problem (such as a linear system of equations) in a finite number of steps. On the other hand, for more difficult linear algebra problems (such as the matrix eigenvalue problem) where iterative methods are needed, the complexity theory is not fully developed. It is our belief that a more detailed complexity analysis is desirable and such a study could help lead to better algorithms in the future.

5.1. Linear systems

Consider the classical problem of a system of linear equations $Ax = b$, where A is a $n \times n$ non-singular matrix, b is a column vector of length n. The standard method for solving this problem is Gaussian elimination (say, with partial pivoting). The number of arithmetic operations required for this method can be found in most numerical analysis textbooks: it is $2n^3/3 + O(n^2)$. Most of these operations come from the LU factorization of the matrix A, with suitable row exchanges. Namely, $PA = LU$, where L is a unit lower triangular matrix (whose entries satisfy $|l_{ij}| \leq 1$), U is an upper triangular matrix, and P is the permutation matrix representing the row exchanges. When this factorization is completed, the solution of $Ax = b$ can be found in $O(n^2)$ operations. Similar operation counts are also available for other direct methods for linear systems, for example, the Cholesky decomposition for symmetric positive definite matrices. Another method for solving $Ax = b$, and, more importantly, for least squares problems, is to use the QR factorization of A. The number of required operations is $4n^3/3 + O(n^2)$. All these direct methods for linear systems involve only a finite number of steps to find the solution. The complexity of these methods can be found by counting the total number of arithmetic operations involved.

A related problem is to investigate the average loss of precision for solving linear systems. It is well known that the condition number κ of the matrix A bounds the relative errors introduced in the solution by small perturbations in b and A. Therefore, $\log \kappa$ is a measure of the loss of numerical precision. To find its average, a statistical analysis is needed. The following result for the expected value of $\log \kappa$ is obtained by Edelman.

Theorem 5.1 (Edelman 1988) Let A be a random $n \times n$ matrix whose entries (real and imaginary parts of the entries, for the complex case) are independent random variables with the standard normal distribution, and let $\kappa = \|A\| \, \|A^{-1}\|$ be its condition number in the 2-norm; then

$$E(\log \kappa) = \log n + c + o(1), \quad \text{for} \quad n \to \infty,$$

where $c \approx 1.537$ for real random matrices and $c \approx 0.982$ for complex random matrices.

The above result on the average loss of precision is a general result valid for any method, as a lower bound. If one uses the singular value decomposition to solve $Ax = b$, the average loss of precision should be close to $E(\log \kappa)$ above. For a more practical method like Gaussian elimination with partial pivoting, the same average could be larger. In fact, Wilkinson's backward error analysis reveals that the numerical solution $\hat{x}$ obtained from a finite precision calculation is the exact solution of a perturbed system $(A+E)\hat{x} = b$. The magnitude of E could be larger than the round-off of A by an extra growth factor $\rho(A)$. This gives rise to the extra loss of precision caused by the particular method used, namely, Gaussian elimination with partial pivoting. Well-known examples indicate that the growth factor can be as large as 2^{n-1}. But the following result suggests that large growth factors only rarely appear exponentially.

Conjecture 5.1 (Trefethen) For any fixed constant $p > 0$, let A be a random $n \times n$ matrix, whose entries (real and complex parts of the entries for the complex case, scaled by $\sqrt{2}$) are independent samples of the standard normal distribution. Then, for all sufficiently large n,

$$\text{Prob}\,(\rho(A) > n^{\alpha}) < n^{-p},$$

where $\alpha > 1/2$.

For iterative methods, we mention that a complexity result is available for the conjugate gradient method (Hestenes and Stiefel 1952). Let A be a real symmetric positive definite matrix, x_0 be an initial guess for the exact solution x_* of $Ax = b$, and x_j be the jth iterate of the conjugate gradient method. Then the following result is well known (Axelsson (1994), Appendix B):

$$\|x_j - x_*\|_A \le 2\left(\frac{\sqrt{\kappa}-1}{\sqrt{\kappa}+1}\right)^j \|x_0 - x_*\|_A,$$

where the A-norm of a vector v is defined as $\|v\|_A = (v^T A v)^{1/2}$. From this, one easily concludes that if

$$j \ge \frac{\log\frac{2}{\epsilon}}{\log\left(\frac{\sqrt{\kappa}-1}{\sqrt{\kappa}+1}\right)} = O\left(\frac{\sqrt{\kappa}}{2}\log\frac{2}{\epsilon}\right),$$

then $\|x_j - x_*\|_A \le \epsilon \|x_0 - x_*\|_A$.

5.2. Eigenvalue problems

In this subsection, we consider a number of basic algorithms for eigenvalue problems. Complexity results for these methods are more difficult to obtain.

For a matrix A, the power method approximates the eigenvector corresponding to the dominant eigenvalue (largest in absolute value). If there is one dominant eigenvalue, for almost all initial guesses x_0, the sequence generated by the power method $x_j = A^j x_0 / \|A^j x_0\|$ converges to the dominant eigenvector. A statistical complexity analysis for the power method tries to determine the average number of iterations required to produce an approximation to the exact eigenvector, such that the angle between the approximate and exact eigenvectors is less than a given small number ϵ (ϵ-dominant eigenvector). These questions have been studied by Kostlan. The average is first taken for all initial guesses x_0 and a fixed matrix A, then extended to all matrices for some distribution.

Theorem 5.2 (Kostlan 1988) For any real symmetric $n \times n$ matrix A with eigenvalues $|\lambda_1| > |\lambda_2| \geq \ldots \geq |\lambda_n|$, the number of iterations $\tau_\epsilon(A)$ required for the power method to produce an ϵ-dominant eigenvector, averaged over all initial vectors, satisfies

$$\frac{\log \cot \epsilon}{\log|\lambda_1| - \log|\lambda_2|} < \tau_\epsilon(A) < \frac{\frac{1}{2}[\psi(n/2) - \psi(1/2)] + \log \cot \epsilon}{\log|\lambda_1| - \log|\lambda_2|} + 1,$$

where $\psi(x) = \Gamma'(x)/\Gamma(x)$.

When an average is taken for the set of $n \times n$ random real symmetric matrices (the entries are independent random variables with Gaussian distributions of zero mean, the variance of any diagonal entry is twice the variance of any off-diagonal entry), the required number of iterations is infinite. However, a finite bound can be obtained if a set of 'bad' initial guesses and 'bad' matrices of normalized measure η are excluded.

Theorem 5.3 (Kostlan 1988) For the above $n \times n$ random real symmetric matrix, with the probability $1 - \eta$, the average required number of iterations to produce an ϵ-dominant eigenvector satisfies

$$\tau_{\epsilon,\eta} < \frac{3n(n+1)}{4\sqrt{2}\eta}\left(\psi(n/2) - \psi(1/2) + 2\log\cot\epsilon\right).$$

Similar results hold for complex Hermitian matrices. Furthermore, a finite bound on random symmetric positive definite matrices is also available. Statistical complexity analysis for a different method of dominant eigenvector calculation can be found in Kostlan (1991).

In practice, the Rayleigh quotient iteration method is much more efficient. Starting from an initial guess x_0, a sequence of vectors $\{x_j\}$ is generated from

$$\mu = \frac{x_{j-1}^T A x_{j-1}}{x_{j-1}^T x_{j-1}}, \quad (A - \mu I)y = x_{j-1}, \quad x_j = \frac{y}{\|y\|}.$$

For symmetric matrices, the following global convergence result has been established.

Theorem 5.4 (Ostrowski 1958, Parlett and Kahan 1969, Batterson and Smillie 1989) Let A be a symmetric $n \times n$ matrix. For almost any choice of x_0, the Rayleigh quotient iteration sequence $\{x_j\}$ converges to an eigenvector and $\lim_{j\to\infty} \theta_{j+1}/\theta_j^3 \leq 1$, where θ_j is the angle between x_j and the closest eigenvector.

A statistical complexity analysis for this method is still not available. In fact, even for a fixed symmetric matrix A, there is no upper bound on the number of iterations required to produce a small angle, say, $\theta_j < \epsilon$ for a small constant ϵ. In general, for a given initial vector x_0, one can not predict which eigenvector it converges to (if the sequence does converge). On the other hand, for nonsymmetric matrices, we have the following result on non-convergence.

Theorem 5.5 (Batterson and Smillie 1990) For each $n \geq 3$, there is a nonempty open set of matrices, each of which possesses an open set of initial vectors for which the Rayleigh quotient iteration sequence does not converge to an invariant subspace.

Practical numerical methods for matrix eigenvalue problems are often based on reductions to the condensed forms by orthogonal similarity transformations. For an $n \times n$ symmetric matrix A, one typically uses Householder reflections to obtain a symmetric tridiagonal matrix T. The reduction step is a finite calculation that requires $O(n^3)$ arithmetic operations. While many numerical methods are available for calculating the eigenvalues and eigenvectors of symmetric tridiagonal matrices, we see the lack of a complexity analysis for these methods.

The QR method with Wilkinson's shift always converges; see Wilkinson (1968). In this method, the tridiagonal matrix T is replaced by $sI + RQ$ (still symmetric tridiagonal), where s is the eigenvalue of the last 2×2 block of T that is closer to the (n, n) entry of T, and $QR = T - sI$ is the QR factorization of $T - sI$. Wilkinson proved that the $(n, n-1)$ entries of this sequence of T always converge to zero. Hoffman and Parlett (1978) gave a simpler proof for the global linear convergence. The following is an easy corollary of their result.

Theorem 5.6 Let T be a real symmetric $n \times n$ tridiagonal matrix. For any $\epsilon > 0$, let m be a positive integer satisfying

$$m > 6 \log_2 \frac{1}{\epsilon} + \log_2(T_{n,n-1}^4 T_{n-1,n-2}^2) + 1.$$

Then, after m QR iterations with Wilkinson's shift, the last subdiagonal entry of T satisfies

$$|T_{n,n-1}| < \epsilon.$$

It would be interesting to develop better complexity results based on the higher asymptotic convergence rate. Alternative definitions for the last subdiagonal entry to be sufficiently small are desirable, because the usual decoupling criterion is based on a comparison with the two adjacent diagonal entries.

The divide and conquer method suggested by Cuppen (1981) calculates the eigensystem of an unreduced symmetric tridiagonal matrix based on the eigensystems of two tridiagonal matrices of half size and a rank-one updating scheme. The computation of the eigenvalues is reduced to solving the following nonlinear equation

$$1+\rho\sum_{j=1}^{n}\frac{c_j^2}{d_j-\lambda}=0,$$

where $\{d_j\}$ are the eigenvalues of the two smaller matrices and $\{c_j\}$ are related to their eigenvectors. This method is complicated by the possibilities that the elements in $\{d_j\}$ may be not distinct and the set $\{c_j\}$ may contain zeros. Dongarra and Sorensen (1987) developed an iterative method for solving the nonlinear equation based on simple rational function approximations. See Bini and Pan (1994) for a complexity analysis of a related algorithm.

A related method for computing just the eigenvalues uses the set $\{d_j\}$ to separate the eigenvalues and a nonlinear equation solver for the characteristic polynomial. In Du, Jin, Li and Zeng (1997*b*), the quasi-Laguerre method is used. An asymptotic convergence result has been established in Du, Jin, Li and Zeng (1997*a*), but a complexity analysis is still not available. The method is complicated by the switch to other methods (the bisection or Newton's method) to obtain good starting points for the quasi-Laguerre iterations.

For a general real nonsymmetric matrix A, the QR iteration with Francis's double shift is widely used to triangularize the Hessenberg matrix H obtained from the reduction by orthogonal similarity transformations from A. In this case, there are simple examples for which the QR iteration does not lead to a decoupling. In Batterson and Day (1992), matrices where the asymptotic rate of decoupling is only linear are identified. For normal Hessenberg matrices, Batterson discovered the precise conditions for decoupling under the QR iteration. See Batterson (1994) for details. To develop a statistical complexity analysis for this method is a great challenge.

6. Complexity in many variables

Consider the problem of following a path, implicitly defined, by a computationally effective algorithm. Let $\mathcal{H}_d$ be as in Section 3.

Let $F:[0,1]\to\mathcal{H}_d\times\mathbb{C}^{n+1}$, $F(t)=(f_t,\zeta_t)$, satisfy $f_t(\zeta_t)=0$, $0\le t\le 1$, with the derivative $Df_t(\zeta_t)$ having maximum rank. For example, ζ_t could be given by the implicit function theorem from f_t and the initial ζ_0 with $f_0(\zeta_0)=0$.

Next, suppose $[0,1]$ is partitioned into k parts by $t_0 = 0$, $t_i = t_{i-1} + \Delta t$, $\Delta t = 1/k$; thus $t_k = 1$.

Define via Newton's method $\hat{N}_{f_{t_i}}$

$$z_i = \hat{N}_{f_{t_i}}(z_{i-1}), \quad i = 1, \ldots, k, \quad z_0 = \zeta_0. \tag{6.1}$$

For sufficiently small Δt, the z_i are well defined and are good approximations of ζ_i. But $k = 1/\Delta t$ represents the complexity, so the problem is to avoid taking Δt much smaller than necessary. What is a sufficient number of Newton steps?

Theorem 6.1 (The main theorem of Bez I) The biggest integer k in

$$cLD^2\mu^2$$

is sufficient to yield z_i by (6.1) which is an approximate zero of f_{t_i} with associated actual zero ζ_{t_i}, each $i = 1, \ldots, k$.

In this estimate c is a rather small universal constant, L is the length of the curve f_t in the projective space, $P(\mathcal{H}_d)$, $0 \le t \le 1$, D is the max of the d_i, $i = 1, \ldots, n$ and $\mu = \max_{0 \le t \le 1} \mu(f_t, \zeta_t)$, where $\mu(f_t, \zeta_t)$ is the condition number as defined in Section 3.

Newton's method and approximate zero have been adapted to projective space. Thus $\hat{N}_f$ for $f \in \mathcal{H}_d$ at $z \in \mathbb{C}^{n+1}$ is the ordinary Newton method applied to the restriction of f to

$$z + \left\{ y \in \mathbb{C}^{n+1} : \langle y, z \rangle = 0 \right\}.$$

As a consequence of the Condition Number Theorem and Theorem 6.1, the complexity depends mainly on how close the path (f_t, ζ_t) comes to the set of ill-conditioned problems. An improved proof of Theorem 6.1 may be found in BCSS (1997).

For earlier work on complexity theory for Newton's method in several variables, see Renegar (1987*a*). Malajovich (1994) has implemented the algorithm and developed some of the ideas of Bez I.

The main theorem of the final paper of the series Bez I–Bez V is as follows.

Theorem 6.2 The average number of arithmetic operations sufficient to find an approximate zero of a system $f : \mathbb{C}^n \to \mathbb{C}^n$ of polynomials is polynomially bounded in the input size (the number of coefficients of f).

On one hand, this result is surprising, because it gives a polynomial time bound for a problem that is almost intractable. On the other hand, the 'algorithm' is not uniform: it depends on the degrees of the (f_i) and even the desired probability of success. Moreover, the algorithm isn't known! It is only proved to exist. Thus Theorem 6.2 cries out for understanding and development. In fact, Mike Shub and I were unable to find a sufficiently good exposition to include in BCSS (1997).

Since deciding if there is a solution to $f : \mathbb{C}^n \to \mathbb{C}^n$ is unlikely to be accomplished in polynomial time, even using exact arithmetic (see Section 8), an astute analysis of Theorem 6.2 can give insight into the basic problem 'What are the limits of computation?' For example, is it 'on the average' that gives the possibility of polynomial time?

A real (rather than complex) analogue of Theorem 6.2 also remains to be found.

Let us give some mathematical detail about the statement of Theorem 6.2. An 'approximate zero' has been defined in Section 4, as, of course, exact zeros cannot be found (Abel, Galois, et al.). Averaging is performed relative to a measure induced by the unitarily invariant inner product on homogenized polynomials of degree $d = (d_1, \ldots, d_n)$, where $d_i = \deg f_i$, $f = (f_1, \ldots, f_n)$ (see Section 3). If $N = N(d)$ is the number of coefficients of such a system f, then unless $n \leq 4$ or some $d_i = 1$, the number of arithmetic operations is bounded by cN^4. If $n \leq 4$ or some $d_i = 1$, then we get cN^5.

An important special case is that of quadratic systems, when $d_i = 2$ and so $N \leq n^3$. Then the average arithmetic complexity is bounded by a polynomial function of n.

'On the average' in the main result is needed because certain polynomial systems, even affine ones of the type $f : \mathbb{C}^2 \to \mathbb{C}^2$, have one-dimensional sets of zeros, extremely sensitive to any (practical) real number algorithm; one would say such f are ill posed.

The algorithm (non-uniform) of the theorem is similar to those of Section 2. It is a continuation method where each step is given by Newton's method (the step size Δt is no longer a constant). The continuation starts from a given 'known' pair $g : \mathbb{C}^{n+1} \to \mathbb{C}^n$ and $\zeta \in \mathbb{C}^{n+1}$, $g(\zeta) = 0$. It is conjectured in Bez V that one could take for g, the system defined by $g_i(z) = z_0^{d_i - 1} z_i$, $i = 1, \ldots, n$ and $\zeta = (1, 0, \ldots, 0)$. A proof of this conjecture would yield a uniform algorithm.

Finally, we remark that in Bez V, Theorem 6.2 is generalized to the problem of finding ℓ zeros, when ℓ is any number between one and the Bézout number $\prod_{i=1}^n d_i$ and the number of arithmetic operations is augmented by the factor ℓ^2.

The proof of Theorem 6.2 uses Theorem 6.1 and the geometric probability methods of the next section.

7. Probabilistic estimates

As described in the Introduction, our complexity perspective has two parts, and the second deals with probability estimates of the condition number. We have already seen some aspects of this in Sections 2 and 5. Here are some further results.

Section 3 describes a condition number for studying zeros of polynomial systems of equations. We have dealt especially with the homogeneous setting and defined projective condition number $\mu(f,\zeta)$ for $f \in \mathcal{H}_d$, $d = (d_1, \ldots, d_n)$, degree $f_i = d_i$, and $\zeta \in \mathbb{C}^{n+1}$ with $f(\zeta) = 0$. Then

$$\mu(f) = \max_{\zeta,\ f(\zeta)=0} \mu(f,\zeta).$$

The unitarily invariant inner product (Section 3) on $\mathcal{H}_d$ induces a probability measure on $\mathcal{H}_d$ (or equivalently on the projective space $P(\mathcal{H}_d)$). With this measure the following is proved in Bez II.

Theorem 7.1

$$\text{Probability}\left\{f \in \mathcal{H}_d\ :\ \mu(f) > \frac{1}{\varepsilon}\right\} \le C_d \varepsilon^4$$

$$C_d = n^3(n+1)(N-1)(N-2)\mathcal{D}, \quad N = \dim \mathcal{H}_d, \quad \mathcal{D} = \prod_{i=1}^{n} d_i$$

In the background of this and a number of related results is a geometric picture (from geometric probability theory), briefly described as follows. It is convenient to use the projective spaces $P(\mathcal{H}_d)$, $P(\mathbb{C}^{n+1})$ and their product for the environment of this analysis. Define V to be the subset of ordered pairs (system, solution):

$$V = \left\{(f,\zeta) \in P(\mathcal{H}_d) \times P(\mathbb{C}^{n+1}) : f(\zeta) = 0\right\}.$$

Let $\pi_1 : V \to P(\mathcal{H}_d)$, $\pi_2 : V \to P(\mathbb{C}^{n+1})$ be the restrictions of the corresponding projections, as shown below.

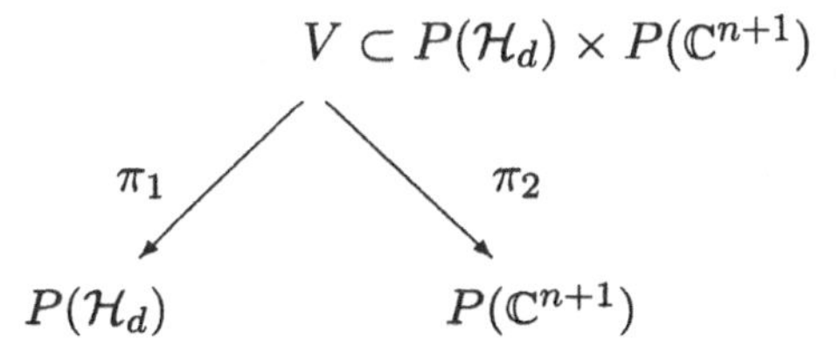

Theorem 7.2 (Bez II) Let U be an open set in V, then

$$\int_{x \in P(\mathcal{H}_d)} \#\left(\pi_1^{-1}(x) \cap U\right) = \int_{z \in P(\mathbb{C}^{n+1})} \int_{(a,z) \in \pi_2^{-1}(z) \cap U} \det\left(DG(a)DG(a)^*\right)^{-1/2}$$

Here $DG(a)$ is the condition matrix, $DG(a)^*$ its adjoint and # means cardinality.

This result and the underlying theory is valid in great generality (see Bez II, IV, V, BCSS (1997)).

There is one aspect of these results and arguments that is quite unsettling and pervades Bez II–V: the implicit existence theory is not very constructive.

Consider the simplest case (Bez III). For the moment, let $d > 1$ be an integer and $\mathcal{H}_d$ the space of homogeneous polynomials in two variables of degree d. It follows from the above geometric probability arguments that there is a subset S_d of $P(\mathcal{H}_d)$ of probability measure larger than one-half such that, for $f \in S_d$, $\mu(f) \leq d$.

Problem 7.1 (Bez III) Construct a family $\{f_d \in \mathcal{H}_d : d = 2, 3, \ldots\}$ so that

$$\mu(f_d) \leq d, \qquad \text{or even} \quad \mu(f_d) \leq d^c,$$

for c any constant.

By 'construct', we mean to provide a polynomial time algorithm (*e.g.* in the sense of the machine of Section 8) which, given input d, outputs f_d satisfying the above condition. (This amounts to constructing elliptic Fekete polynomials.) See also Rakhmanov, Saff and Zhou (1994, 1995).

Another example of an application of the above setting of geometric probability is the following result. For $d = (d_1, \ldots, d_n)$, let $\mathcal{H}_d^{\mathbb{R}}$ denote the space of real homogeneous systems $(f_1, \ldots, f_n)$ in $n+1$ variables with degree $f_i = d_i$. One can average just as before and obtain the following.

Theorem 7.3 (Bez II) The average number of real zeros of a real homogeneous polynomial system is exactly the square root of the Bézout number $\mathcal{D} = \prod_{i=1}^{n} d_i$ ($\mathcal{D}$ being the number of complex solutions).

See Kostlan (1993) for earlier special cases. See also Edelman and Kostlan (1995).

For the complexity results of Bez IV, V, Theorem 7.1 is inadequate. There one has similar theorems where the maximum of the condition number along an interval is estimated.

8. Real machines

Up to now, our discussion might be called the complexity analysis of algorithms, or upper bounds for the time required to solve problems. To complement this theory one needs lower bound estimates for problem solving.

For this endeavour, one must consider all possible algorithms that solve a given problem. In turn this needs a formal definition and the development of algorithms and machines. The traditional Turing machine is ill-suited for this purpose, as is argued in the Manifesto. A 'real number machine' is the most natural vehicle to deal with problem-solving schemes based on Newton's method, for example.

There is a recent development of such a machine in BSS (1989) and BCSS (1997), which we will review very briefly.

Each input is a string y of real numbers of the form

$$\cdots 000 y_1 \cdots y_n 000 \cdots;$$

the size $S(y)$ of y is n. These inputs may be restricted to code an instance of a problem. An 'input node' transforms an input into a state string.

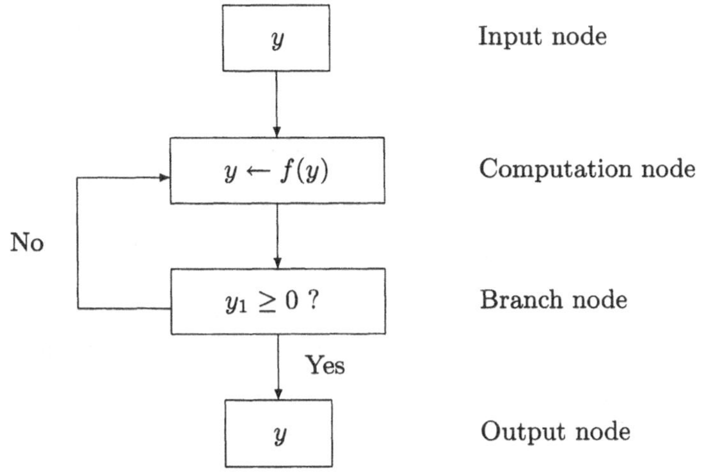

Fig. 1. Example of a real number machine

The computation node replaces the state string by a shifted one, right or left shifted, or does an arithmetic operation on the first elements of the string. The branch nodes and output nodes are self-explanatory.

The definition of a real machine (or a 'machine over $\mathbb{R}$') is suggested by the example and consists of an input node and a finite number of computation, branch, and output nodes organized into a directed graph. It is the flow chart of a computer program seen as a mathematical object. One might say that this real number machine is a 'real Turing machine' or an idealized Fortran program.

The *halting set* of a real machine is the set of all inputs such that, acting on the nodal instructions, we eventually land on an output node. An *input–output map* ϕ is defined on the halting set by 'following the flow' of the flow chart. For precise definitions and developments see BSS (1989) and BCSS (1997).

A machine has *polynomial time complexity* (sometimes with a restricted class of inputs) if it enjoys the property

$$T(y) \leq S(y)^c, \qquad \text{for all inputs } y, \tag{8.1}$$

where c is independent of y. In this estimate, $T(y)$ is the time to the output for the input y measured by the number of nodes encountered in the computation of $\phi(y)$. Recall that the size $S(y)$ of y is the length of the input string y.

If the size of the inputs is bounded, and there are no loops, *i.e.*, the machine is a tree of nodes, then one has a *tame machine*, or an algebraic computation tree. These objects have been used to obtain lower bounds for real number problems. One such development is that of Steele and Yau (1982) and Ben-Or (1983), based on a real algebraic geometry estimate of Oleinik and Petrovski (1949), Oleinik (1951), Milnor (1964) and Thom (1965). Another is that of Smale (1987*b*) and Vassiliev (1992), and based on the cohomology of the braid group.

Lower bounds tend to be modest and difficult to obtain, but are necessary for the understanding of the fundamental problem: 'What are the limits of computation?'

Note that the definition of a real machine is valid with strings of numbers lying in any field if one replaces the branch node with the question, '$y_1 = 0$?' If this field is the field of two elements, one has a Turing machine, and the size becomes the number of bits. If one uses complex numbers, then one has a 'complex machine'.

Side remarks 8.1 The study of zeros of polynomial systems plays a central role in both mathematics and computation theory. Deciding whether a set of polynomial equations has a zero over $\mathbb{R}$ is even universal in a formal sense in the theory of real computation. This problem is called 'NP-complete over $\mathbb{R}$' and hence its solution in polynomial time is equivalent to 'P = NP over $\mathbb{R}$.' For machines over $\mathbb{C}$, this problem is that of the Hilbert Nullstellensatz, and Brownawell's (1987) work was critical in getting the fastest-known algorithm (but not polynomial time!) The relation to NP-complete over $\mathbb{C}$ and 'P = NP over $\mathbb{C}$' is as in the real case. The same applies to the field $\mathbb{Z}_2$ of two elements and 'P = NP over $\mathbb{Z}_2$?' is the same as the classical Cook–Karp problem 'P = NP?' of computer science. See BCSS (1997).

My own belief is that this problem is one of the three great unsolved problems of mathematics (together with the Riemann hypothesis and Poincaré's conjecture in three dimensions).

The rest of Section 8 is more tentative, as we present suggestions in the direction of a 'second generation' real machine.

For an input y of a problem, an extended notion of size still denoted by

$S(y)$ could be convenient. The extended notion would be the maximum of the length of the string (*i.e.* the previously defined size) and other ingredients, as follows:

(i) the condition number $\mu(y)$, or its log, or similar invariants of y
(ii) the precision $\log \varepsilon^{-1}$, where ε is the required accuracy (or perhaps, depending on the problem, ε, or even $\log \log \varepsilon^{-1}$) of the output
(iii) for integer machines, the number of bits.

It is convenient to consider the traditional size of the input as part of the input (BSS 1989, BCSS 1997). Should the same hold for the extended size? We won't try to give a definitive answer here. Part of this answer is a question of convenience, part interpretation. Should the algorithm assume that the condition number is known explicitly? Probably not, at least very generally. On the other hand, if one has a good theoretical result on the distribution, one can make some guess about the condition number. This can to some extent justify taking the condition number of the particular problem as input. It is analogous, for example, to running a path-following program inputing an initial step size as a guess.

Let me give an example of an open problem that fits into this framework. Let $d = (d_1, \ldots, d_m)$ and $\mathcal{P}_{n,d}$ be the space of m-tuples of real polynomials $f = (f_1, \ldots, f_m)$ in n variables with $\deg f_i \leq d_i$. Put some distance D on $\mathcal{P}_{n,d}$. Say that f is *feasible* if the system of inequalities $f_i(x) \geq 0$, all $i = 1, \ldots, m$ has a solution $x \in \mathbb{R}^n$. Let the 'condition number' of f be defined by:

$$\mu(f) = \Big(\inf_{g \text{ not feasible}} D(f,g) \Big)^{-1} \quad \text{if } f \text{ is feasible,}$$

$$\mu(f) = \Big(\inf_{g \text{ feasible}} D(f,g) \Big)^{-1} \quad \text{if } f \text{ is not feasible.}$$

Let the extended size $S(f)$ of $f \in \mathcal{P}_{n,d}$ be the maximum (perhaps ∞) of $\dim \mathcal{P}_{n,d}$ and $\mu(f)$.

Problem 8.1 Is there a polynomial time algorithm deciding the above feasibility problem using the extended size?

The problem is formalized in terms of the real machines described above, using exact arithmetic in particular.

We now propose an extension of the earlier notion of real machine to allow round-off error in the computation.

A *round-off machine over* $\mathbb{R}$ is a real machine, together with a function of inputs that, at each input, computation and output node, adds a state vector of magnitude less than some positive constant δ. One has no *a priori* knowledge of the added state vector (it's an adversary). This idealization has the virtue of simplicity; we hope this compensates for its ignorance of important detail.

A problem will be called *robustly solvable* if it can be solved for inputs of finite extended size by a round-off machine, no matter what the round-off error.

More important is the concept of *robustly solvable* in *polynomial time*. In addition to the estimate (8.1) with extended size, $S(y)$, one adds a requirement such as

$$\frac{1}{\delta(y)} \leq S(y)^c. \tag{8.2}$$

One can now sharpen Problem 8.1 to ask for a decision which is robustly solvable in polynomial time.

The above gives some sense of the notion of a robust or numerically stable algorithm, perhaps improving on the attempts in Isaacson and Keller (1966), Wozniakowski (1977), Smale (1990) and Shub (1993).

9. Some other directions

Many aspects of complexity theory in numerical analysis have not been dealt with in this brief report. We now refer to some of these omissions.

A general reference is Renegar, Shub and Smale (1997), which expands on the previous topics and those below.

There is the important, well-developed field of algebraic complexity theory, which relates very much to some of our account. I have the greatest admiration for this work, but will only mention here Bini and Pan (1994), Grigoriev (1987), and Giusti et al. (1997).

Also well-developed is the area of information-based complexity. In spite of its relevance and importance to our review, I will only mention Traub, Wasilkowski and Wozniakowski (1988), where one will find a good introduction and survey.

Another area in which the mathematical foundation and development are strong is the science of mathematical programming, or optimization. I believe that numerical analysts interested in complexity considerations can learn much from what has happened and is happening in that field. I especially like the perspective and work of Renegar (1996).

REFERENCES

E. Allgower and K. Georg (1990), *Numerical Continuous Methods*, Springer.

E. Allgower and K. Georg (1993), Continuation and path following, in *Acta Numerica*, Vol. 2, Cambridge University Press, pp. 1–64.

O. Axelsson (1994), *Iterative Solution Methods*, Cambridge University Press.

S. Batterson (1994), 'Convergence of the Francis shifted QR algorithm on normal matrices', *Linear Algebra Appl.* **207**, 181–195.

S. Batterson and D. Day (1992), 'Linear convergence in the shifted QR algorithm', *Math. Comp.* **59**, 141–151.

S. Batterson and J. Smillie (1989), 'The dynamics of Rayleigh quotient iteration', *SIAM J. Numer. Anal.* **26**, 624–636.

S. Batterson and J. Smillie (1990), 'Rayleigh quotient iteration for nonsymmetric matrices', *Math. Comp.* **55**, 169–178.

M. Ben-Or (1983), Lower bounds for algebraic computation trees, in *15th Annual ACM Symposium on the Theory of Computing*, pp. 80–86.

D. Bini and V. Pan (1987), 'Sequential and parallel complexity of approximating polynomial zeros', *Computers and Mathematics (with applications)* **14**, 591–622.

D. Bini and V. Pan (1994), *Polynomial and Matrix Computations*, Birkhäuser, Basel.

L. Blum, F. Cucker, M. Shub and S. Smale (1996), 'Complexity and real computation: a manifesto', *Int. J. Bifurcation and Chaos* **6**, 3–26. **Referred to as the Manifesto.**

L. Blum, F. Cucker, M. Shub and S. Smale (1997), *Complexity and Real Computation*, Springer. To appear. **Referred to as BCSS (1997).**

L. Blum, M. Shub and S. Smale (1989), 'On a theory of computation and complexity over the real numbers: NP-completeness, recursive functions and universal machines', *Bull. Amer. Math. Soc.* **21**, 1–46. **Referred to as BSS (1989).**

R. Brockett (1973), in *Geometric Methods in Systems Theory, Proceedings of the NATO Advanced Study Institute* (D. Mayne and R. Brockett, eds), D. Reidel, Dordrecht.

W. Brownawell (1987), 'Bounds for the degrees in the Nullstellensatz', *Annals of Math.* **126**, 577–591.

G. Collins (1975), *Quantifier Elimination for Real Closed Fields by Cylindrical Algebraic Decomposition*, Vol. 33 of *Lect. Notes in Comp. Sci.*, Springer, pp. 134–183.

J. J. M. Cuppen (1981), 'A divide and conquer method for the symmetric tridiagonal eigenproblem', *Numer. Math.* **36**, 177–195.

J.-P. Dedieu (1997*a*), Approximate solutions of numerical problems, condition number analysis and condition number theorems, in *Proceedings of the Summer Seminar on 'Mathematics of Numerical Analysis: Real Number Algorithms', AMS Lectures in Applied Mathematics* (J. Renegar, M. Shub and S. Smale, eds), AMS, Providence, RI. To appear.

J.-P. Dedieu (1997*b*), Condition number analysis for sparse polynomial systems. Preprint.

J.-P. Dedieu (1997*c*), 'Condition operators, condition numbers and condition number theorem for the generalized eigenvalue problem', *Linear Algebra Appl.* To appear.

J.-P. Dedieu (1997*d*), 'Estimations for separation number of a polynomial system', *J. Symbolic Computation*. To appear.

J. Dégot and B. Beauzamy (1997), 'Differential identities', *Trans. Amer. Math. Soc.* To appear.

B. Dejon and P. Henrici (1969), *Constructive Aspects of the Fundamental Theorem of Algebra*, Wiley.

J. Demmel (1987), 'On condition numbers and the distance to the nearest ill-posed problem', *Numer. Math.* **51**, 251–289.

J. J. Dongarra and D. C. Sorensen (1987), 'A fully parallel algorithm for the symmetric eigenvalue problem', *SIAM J. Sci. Statist. Comput.* **8**, 139–154.

Q. Du, M. Jin, T. Y. Li and Z. Zeng (1997*a*), 'The quasi-Laguerre iteration', *Math. Comp.* To appear.

Q. Du, M. Jin, T. Y. Li and Z. Zeng (1997*b*), 'Quasi-Laguerre iteration in solving symmetric tridiagonal eigenvalue problems', *SIAM J. Sci. Comput.* To appear.

C. Eckart and G. Young (1936), 'The approximation of one matrix by another of lower rank', *Psychometrika* **1**, 211–218.

A. Edelman (1988), 'Eigenvalues and condition numbers of random matrices', *SIAM J. Matrix Anal. Appl.* **9**, 543–556.

A. Edelman and E. Kostlan (1995), 'How many zeros of a random polynomial are real?', *Bull. Amer. Math. Soc.* **32**, 1–38.

C. F. Gauss (1973), *Werke*, Band X, Georg Olms, New York.

M. Giusti, J. Heintz, J. E. Morais, J. Morgenstern and L. M. Pardo (1997), 'Straight-line program in geometric elimination theory', *Journal of Pure and Applied Algebra*. To appear.

G. Golub and C. van Loan (1989), *Matrix Computations*, Johns Hopkins University Press.

D. Grigoriev (1987), in *Computational complexity in polynomial algebra, Proceedings of the International Congress Math. (Berkeley, 1986)*, Vol. 1, 2, AMS, Providence, RI, pp. 1452–1460.

P. Henrici (1977), *Applied and Computational Complex Analysis*, Wiley.

M. R. Hestenes and E. Stiefel (1952), 'Method of conjugate gradients for solving linear systems', *J. Res. Nat. Bur. Standards* **49**, 409–436.

M. Hirsch and S. Smale (1979), 'On algorithms for solving $f(x) = 0$', *Comm. Pure Appl. Math.* **32**, 281–312.

W. Hoffman and B. N. Parlett (1978), 'A new proof of global convergence for the tridiagonal QL algorithm', *SIAM J. Numer. Anal.* **15**, 929–937.

E. Isaacson and H. Keller (1966), *Analysis of Numerical Methods*, Wiley, New York.

H. Keller (1978), Global homotopic and Newton methods, in *Recent Advances in Numerical Analysis*, Academic Press, pp. 73–94.

R. Kellog, T. Li and J. Yorke (1976), 'A constructive proof of Brouwer fixed-point theorem and computational results', *SIAM J. Numer. Anal.* **13**, 473–483.

M. Kim (1988), 'On approximate zeros and rootfinding algorithms for a complex polynomial', *Math. Comp.* **51**, 707–719.

E. Kostlan (1988), 'Complexity theory of numerical linear algebra', *J. Comput. Appl. Math.* **22**, 219–230.

E. Kostlan (1991), 'Statistical complexity of dominant eigenvector calculation', *J. Complexity* **7**, 371–379.

E. Kostlan (1993), On the distribution of the roots of random polynomials, in *From Topology to Computation: Proceedings of the Smalefest* (M. Hirsch, J. Marsden and M. Shub, eds), Springer, pp. 419–431.

G. Malajovich (1994), 'On generalized Newton algorithms: quadratic convergence, path-following and error analysis', *Theoret. Comput. Sci.* **133**, 65–84.

G. Malajovich-Munoz (1993), On the complexity of path-following Newton algorithms for solving polynomial equations with integer coefficients, PhD thesis, University of California at Berkeley.

J. M. McNamee (1993), 'A bibliography on roots of polynomials', *J. Comput. Appl. Math.* **47**(3), 391–394.

J. Milnor (1964), On the Betti numbers of real varieties, in *Proceedings of the Amer. Math. Soc.*, Vol. 15, pp. 275–280.

C. Neff (1994), 'Specified precision root isolation is in NC', *J. Comput. System Sci.* **48**, 429–463.

C. Neff and J. Reif (1996), 'An efficient algorithm for the complex roots problem', *J. Complexity* **12**, 81–115.

O. Oleinik (1951), 'Estimates of the Betti numbers of real algebraic hypersurfaces', *Mat. Sbornik (N.S.)* **28**, 635–640. In Russian.

O. Oleinik and I. Petrovski (1949), 'On the topology of real algebraic surfaces', *Izv. Akad. Nauk SSSR* **13**, 389–402. In Russian; English translation in *Transl. Amer. Math. Soc.* **1**, 399–417 (1962).

A. Ostrowski (1958), 'On the convergence of Rayleigh quotient iteration for the computation of the characteristic roots and vectors, I', *Arch. Rational Mech. Anal.* **1**, 233–241.

V. Pan (1997), 'Solving a polynomial equation: some history and recent progress', *SIAM Review*. To appear.

B. N. Parlett and W. Kahan (1969), 'On the convergence of a practical QR algorithm', *Inform. Process. Lett.* **68**, 114–118.

E. A. Rakhmanov, E. B. Saff and Y. M. Zhou (1994), 'Minimal discrete energy on the sphere', *Mathematical Research Letters* **1**, 647–662.

E. A. Rakhmanov, E. B. Saff and Y. M. Zhou (1995), Electrons on the sphere, in *Computational Methods and Function Theory* (R. M. Ali, S. Ruscheweyh and E. B. Saff, eds), World Scientific, pp. 111–127.

J. Renegar (1987*a*), 'On the efficiency of Newton's method in approximating all zeros of systems of complex polynomials', *Math. of Oper. Research* **12**, 121–148.

J. Renegar (1987*b*), 'On the worst case arithmetic complexity of approximating zeros of polynomials', *J. Complexity* **3**, 90–113.

J. Renegar (1996), 'Condition numbers, the Barrier method, and the conjugate gradient method', *SIAM J. Optim.* To appear.

J. Renegar, M. Shub and S. Smale, eds (1997), *Proceedings of the Summer Seminar on 'Mathematics of Numerical Analysis: Real Number Algorithm', AMS Lectures in Applied Mathematics*, AMS, Providence, RI.

B. Reznick (1992), *Sums of Even Powers of Real Linear Forms*, Vol. 463 of *Memoirs of the American Mathematical Society*, AMS, Providence, RI.

J. R. Rice (1966), 'A theory of condition', *SIAM J. Numer. Anal.* **3**, 287–310.

L. Santaló (1976), *Integral Geometry and Geometric Probability*, Addison-Wesley, Reading, MA.

A. Schönhage (1982), The fundamental theorem of algebra in terms of computational complexity, Technical report, Math. Institut der Universität Tübingen.

A. Schönhage (1987), Equation solving in terms of computational complexity, in *Proceedings of the International Congress of Mathematicans*, AMS, Providence, RI.

M. Shub (1993), On the work of Steve Smale on the theory of computation, in *From Topology to Computation: Proceedings of the Smalefest* (M. Hirsch, J. Marsden and M. Shub, eds), Springer, pp. 443–455.

M. Shub and S. Smale (1985), 'Computational complexity: on the geometry of polynomials and a theory of cost I', *Ann. Sci. École Norm. Sup.* **18**, 107–142.

M. Shub and S. Smale (1986), 'Computational complexity: on the geometry of polynomials and a theory of cost II', *SIAM J. Comput.* **15**, 145–161.

M. Shub and S. Smale (1993*a*), 'Complexity of Bézout's theorem I: geometric aspect', *J. Amer. Math. Soc.* **6**, 459–501. **Referred to as Bez I.**

M. Shub and S. Smale (1993*b*), Complexity of Bézout's theorem II: volumes and probabilities, in *Computational Algebraic Geometry* (F. Eyssette and A. Galligo, eds), Vol. 109 of *Progress in Mathematics*, pp. 267–285. **Referred to as Bez II.**

M. Shub and S. Smale (1993*c*), 'Complexity of Bézout's theorem III: condition number and packing', *J. Complexity* **9**, 4–14. **Referred to as Bez III.**

M. Shub and S. Smale (1994), 'Complexity of Bézout's theorem V: polynomial time', *Theoret. Comput. Sci.* **133**, 141–164. **Referred to as Bez V.**

M. Shub and S. Smale (1996), 'Complexity of Bézout's theorem IV: probability of success; extensions', *SIAM J. Numer. Anal.* **33**, 128–148. **Referred to as Bez IV.**

S. Smale (1976), 'A convergent process of price adjustment and global Newton method', *J. Math. Economy* **3**, 107–120.

S. Smale (1981), 'The fundamental theorem of algebra and complexity theory', *Bull. Amer. Math. Soc.* **4**, 1–36.

S. Smale (1985), 'On the efficiency of algorithms of analysis', *Bull. Amer. Math. Soc.* **13**, 87–121.

S. Smale (1986), Newton's method estimates from data at one point, in *The Merging of Disciplines: New Directions in Pure, Applied, and Computational Mathematics* (R. Ewing, K. Gross and C. Martin, eds), Springer, pp. 185–196.

S. Smale (1987*a*), Algorithms for solving equations, in *Proceedings of the International Congress of Mathematicians*, AMS, Providence, RI, pp. 172–195.

S. Smale (1987*b*), 'On the topology of algorithms I', *J. Complexity* **3**, 81–89.

S. Smale (1990), 'Some remarks on the foundations of numerical analysis', *SIAM Review* **32**, 211–220.

J. Steele and A. Yao (1982), 'Lower bounds for algebraic decision trees', *Journal of Algorithms* **3**, 1–8.

E. Stein and G. Weiss (1971), *Introduction to Fourier Analysis on Euclidean Spaces*, Princeton University Press.

R. Thom (1965), Sur l'homologie des variétés algébriques réelles, in *Differential and Combinatorial Topology* (S. Cairns, ed.), Princeton University Press.

J. Traub and H. Wozniakowski (1979), 'Convergence and complexity of Newton iteration for operator equations', *J. Assoc. Comput. Mach.* **29**, 250–258.

J. Traub, G. Wasilkowski and H. Wozniakowski (1988), *Information-Based Complexity*, Academic Press.

L. N. Trefethen (preprint), Why Gaussian elimination is stable for almost all matrices.

V. A. Vassiliev (1992), *Complements of Discriminants of Smooth Maps: Topology and Applications*, Vol. 98 of *Transl. of Math. Monographs*, AMS, Providence, RI. Revised 1994.

X. Wang (1993), Some results relevant to Smale's reports, in *From Topology to Computation: Proceedings of the Smalefest* (M. Hirsch, J. Marsden and M. Shub, eds), Springer, pp. 456–465.

H. Weyl (1932), *The Theory of Groups and Quantum Mechanics*, Dover.

J. Wilkinson (1963), *Rounding Errors in Algebraic Processes*, Prentice-Hall.

J. Wilkinson (1968), 'Global convergence of tridiagonal QR algorithm with origin shifts', *Linear Algebra Appl.* **I**, 409–420.

H. Wozniakowski (1977), 'Numerical stability for solving non-linear equations', *Numer. Math.* **27**, 373–390.